Structure and Function in Excitable Cells

Structure and Function in Excitable Cells

Edited by

Donald C. Chang

Baylor College of Medicine and Rice University
Houston, Texas

Ichiji Tasaki

National Institute of Mental Health
National Institutes of Health
Bethesda, Maryland, and
The Marine Biological Laboratory
Woods Hole, Massachusetts

William J. Adelman, Jr.

National Institute of Neurological and
Communication Disorders and Stroke
National Institutes of Health at the
Marine Biological Laboratory
Woods Hole, Massachusetts

and

H. Richard Leuchtag

Texas Southern University
Houston, Texas, and
University of Texas Medical Branch
Galveston, Texas

PLENUM PRESS • NEW YORK AND LONDON

Library of Congress Cataloging in Publication Data

Main entry under title:

Structure and function in excitable cells.

Bibliography: p.
Includes index.
1. Excitation (Physiology). 2. Cell membranes. 3. Electrophysiology. I. Chang, Donald C. II. Title: Excitable cells.
QP363.S8 1983 599'.087 83-13475

ISBN-13: 978-1-4615-9339-3 e-ISBN-13: 978-1-4615-9337-9
DOI: 10.1007/978-1-4615-9337-9

Softcover reprint of the hardcover 1st edition 1983

A Division of Plenum Publishing Corporation
233 Spring Street, New York, N.Y. 10013

Contributors

William J. Adelman, Jr.

Laboratory of Biophysics
Intramural Research Program
NINCDS
National Institutes of Health at
the Marine Biological Laboratory
Woods Hole, Massachusetts 02543

Gilbert Baumann

Department of Physiology
Duke University Medical Center
Durham, North Carolina 27710

Martin Blank

Department of Physiology
College of Physicians and Surgeons
Columbia University
New York, New York 10032

Donald C. Chang

Department of Physiology
Baylor College of Medicine
Houston, Texas 77030
and Department of Physics
Rice University
Houston, Texas 77251

D.F. Clapin

Electron Microscopy Unit
Department of Anatomy
Faculty of Health Sciences
University of Ottawa
Ottawa, Ontario Canada K1N 9A9
and Marine Biological Laboratory
Woods Hole, Massachusetts 02543

Louis DeFelice

Department of Anatomy
Emory University School of Medicine
Atlanta, Georgia 30322

George Eisenman

Department of Physiology
UCLA Medical School
Los Angeles, California 90024

Mark H. Ellisman

Department of Neurosciences
University of California, San Diego
School of Medicine
La Jolla, California 92093

Sachiko Endo

Department of Biophysics and Biochemistry
Faculty of Science
University of Tokyo
Bunkyo-ku, Tokyo 113, Japan

Harvey M. Fishman

Department of Physiology and Biophysics
University of Texas Medical Branch
Galveston, Texas 77550

Jarl Hagglund

Department of Neurology
University Hospital
S-751 23 Uppsala, Sweden

Nobutaka Hirokawa

Department of Anatomy and Neurobiology
Washington University School of Medicine
St. Louis, Missouri 63110

Alan Hodge

Laboratory of Biophysics
NINCDS
National Institutes of Health at
the Marine Biological Laboratory
Woods Hole, Massachusetts 02543

Kunihiko Iwasa

Laboratory of Neurobiology
National Institutes of Mental Health
Bethesda, Maryland 20205
and Marine Biological Laboratory
Woods Hole, Massachusetts 02543

Michael J. Kell

Department of Anatomy
Emory University School of Medicine
Atlanta, Georgia 30322

B.I. Khodorov

Vishnevsky Surgery Institute
Moscow 113093, U.S.S.R.

Takaaki Kobayashi

Department of Biochemistry
Jikei University
Minato-ku, Tokyo 105, Japan

Harold Lecar

Laboratory of Biophysics
IRP
National Institute of Neurological and
Communicative Disorders and Stroke
National Institutes of Health
Bethesda, Maryland 20205

Graham V. Lees

Département de Biophysique
Laboratoire de Neurobiologie Cellulaire du CNRS
F 91190 Gif sur Yvette, France

H. Richard Leuchtag

Department of Biology
Texas Southern University
Houston, Texas 77004
and Department of Physiology and Biophysics
University of Texas Medical Branch
Galveston, Texas 77550

S. Rock Levinson

Department of Physiology
University of Colorado Health Sciences Center
School of Medicine
Denver, Colorado 80262

James D. Lindsey

Department of Neurosciences
University of California, San Diego
School of Medicine
La Jolla, California 92093

Gilbert N. Ling

Department of Molecular Biology
Pennsylvania Hospital
Philadelphia, Pennsylvania 19107

Gen Matsumoto

Electrotechnical Laboratory
Tsukuba Science City, Ibaraki 305, Japan

Patrick Meares

Chemistry Department
University of Aberdeen
Old Aberdeen AB9 2UE, Scotland

J. Metuzals

Electron Microscopy Unit
Department of Anatomy
Faculty of Health Sciences
University of Ottawa
Ottawa, Ontario, Canada
and Marine Biological Laboratory
Woods Hole, Massachusetts 02543

Hans Meves

I. Physiologisches Institut der Universität des Saarlandes
6650 Homburg, Saar
West Germany

L. E. Moore

Department of Physiology and Biophysics
University of Texas Medical Branch
Galveston, Texas 77550

Catherine Morris

Department of Biology
University of Ottawa
Ottawa, Ontario, Canada K1N 9A9

Hiromu Murofushi

Department of Biophysics and Biochemistry
Faculty of Science
University of Tokyo
Bunkyo-ku, Tokyo 113, Japan

Yves Pichon

Département de Biophysique
Laboratoire de Neurobiologie Cellulaire du CNRS
F 91190 Gif sur Yvette, France

Denis Poussart

Département de Génie électrique
Université Laval
Québec G1K 7P4, Canada

Francisco Rodríguez

Centro de Biofísica y Bioquímica
Instituto Venezolano de Investigaciones Científicas
Caracas 1010A, Venezuela
and Instituto International de Estudios Avanzados
Caracas 1015A, Venezuela

Jack Rosenbluth

Department of Physiology
New York University School of Medicine
New York, New York 10016

Hikoichi Sakai

Department of Biophysics and Biochemistry
Faculty of Science
University of Tokyo
Bunkyo-ku, Tokyo 113, Japan

John Sandblom

Department of Physiology and Medical Biophysics
University of Uppsala
S-751 23 Uppsala, Sweden

Charles L. Schauf

Department of Physiology
Rush University
Chicago, Illinois 60612

Zadila Suárez-Mata

Centro de Biofísica y Bioquímica
Instituto Venezolano de Investigaciones Científicas
Caracas 1010A, Venezuela
and Instituto International de Estudios Avanzados
Caracas 1015A, Venezuela

Ichiji Tasaki

Laboratory of Neurobiology
National Institutes of Health
Bethesda, Maryland 20205
and Marine Biological Laboratory
Woods Hole, Massachusetts 02543

Torsten Teorell

Department of Physiology and Medical Biophysics
Biomedical Center
Uppsala University
S-751 23 Uppsala, Sweden

Gloria Villegas

Centro de Biofísica y Bioquímica
Instituto Venezolano de Investigaciones Científicas
Caracas 1010A, Venezuela
and Instituto International de Estudios Avanzados
Caracas 1015A, Venezuela

Raimundo Villegas

Centro de Biofísica y Bioquímica
Instituto Venezolano de Investigaciones Científicas
Caracas 1010A, Venezuela
and Instituto International de Estudios Avanzados
Caracas 1015A, Venezuela

Clayton Wiley-Livingston

Department of Pathology
University of California, San Diego
School of Medicine
La Jolla, California 92093

Brendan S. Wong

Department of Physiology
Baylor College of Dentistry
Dallas, Texas 75246

Preface

This book is a collection of up-to-date research reviews dealing with various aspects of the structure and function of excitable cells. Its overall objective is to further the search for a better understanding of the mechanism of excitation on a structural and physicochemical basis. The chapters are written by active investigators from a variety of disciplines, representing many different points of view. Their complementary fields of expertise give this book the rare feature of extraordinary breadth.

Excitability is a fundamental property of many biological systems. The mechanisms by which nerve impulses are initiated and propagated, and by which rhythmical activities are produced in nerve, muscle, and cardiac cells, can be fully elucidated only when the process of excitation is derived from fundamental principles applied to known structural forms, at both the macroscopic and the molecular level. The problems of excitation are complex, requiring knowledge of many aspects of cells, including their morphology, electrobiology, chemical physics, and biochemistry.

From recent studies of excitable cells a number of new facts about their structure and function have emerged, so that it became imperative for physiologists interested in studies of cellular excitation processes to take these findings into consideration. Therefore, several investigators working at the Marine Biological Laboratory in Woods Hole decided to bring together for a discussion of these problems a group of electron microscopists, electrophysiologists, physical chemists and biochemists who have made significant contributions to our present-day knowledge of biological membranes. Thanks to financial aid from three U.S. government agencies—the Office of Naval Research, National Institute of Mental Health, and National Institute of Neurological and Communicative Disorders and Stroke—and the administrative support of the Marine Biological Laboratory, this hope became a reality: At the International Conference on Structure and Function in Excitable Cells, from August 31 to September 3, 1981, some 130 researchers from 12 countries gathered in Woods Hole and exchanged their views on various problems related directly or indirectly to the cellular excitation mechanism. The present volume is based on the papers given by the invited lecturers at this Conference, expanded and brought up to date to keep pace with current developments.

This book consists of four more or less distinct but interrelated parts. Part I deals with the results of recent studies of the ultrastructure of excitable cells. Among the many new tools that have become available to provide a clearer view of the structures of these cells is the increasing application of electron microscopy, in conjunction with rapid freezing, freeze–fracture and new fixation techniques. Ellisman, Lindsey, Wiley-Livingston and Levinson discuss the differentiation and maintenance of the membrane cytoskeletal structure in nerve. Rosenbluth studies the structure of the node of Ranvier of myelinated nerve fibers. Metuzals, Clapin and Tasaki present evidence for the existence of a highly organized structure consisting of different types of filamentous elements intimately connected to the plasmalemma and for the importance of this structure for the maintenance of excitability. The chapter by Hodge and Adelman shows that neurofilaments and neurotubules, together with cross bridges, form a three-dimensional lattice. Hirokawa's application of the method of rapid freezing and deep etching provides new views of synaptic and axonal structures.

Part II, introduced by Adelman, deals mainly with recent electrophysiological findings. Lecar, Morris and Wong review the work on the minute current pulses, known as "single-channel currents," that have recently been observed in many membrane structures, and DeFelice and Kell discuss the relation of these to noise and impedance studies. Moore studies the complex admittance of muscle to observe the linear manifestations of its ion conductances, while the results of analyses of axon membrane noise are discussed by Pichon, Poussart and Lees. Chang presents a view of the ion selectivity of the conduction pathway based on a membrane cortex model. Baumann applies an aggregation model to stochastic processes. The use of specific neurotoxins to probe membrane behavior is discussed by Meves, who studies scorpion toxin, and by Khodorov, whose findings relate to batrachotoxin.

In Part III, a physicochemical approach to studies of the cellular excitation mechanisms is adopted. Tasaki and Iwasa relate their findings on the mechanical responses of the axon membrane to structural studies of the axolemma–ectoplasm complex. Torsten Teorell, the founder of the Teorell–Meyer–Sievers theory for ionic membranes, gives an authoritative historical review of the development of physical chemistry of the membrane. Meares analyzes nonlinear membrane phenomena that arise from flux coupling. The universal biological solvent, water, has been replaced with heavy water in the studies of Schauf. Ling studies the effects of intracellular potassium-ion sorption on the production of cell potentials, and Eisenman, Sandblom and Hagglund analyze the behavior of narrow ion-conducting channels. A newly found connection between electrodiffusion and nonlinear waves is explored by Leuchtag and Fishman, and charged layers at the membrane surface are modeled by Blank.

The fourth part describes various observations emphasizing the importance of protein molecules in the nerve membrane. Villegas, Villegas, Suárez-Mata and Rodríguez examine the behavior of reconstituted vesicles containing specific proteins from excitable cells. The last chapter, by Matsumoto, Murofushi, Endo, Kobayashi and Sakai, points to the necessity for the presence of microtubules for the maintenance of excitability in squid axon.

This book is founded on the premise that to achieve a clear understanding of the mechanism of the excitation process, experts from many fields must join forces to attack the problem. Only by developing improved morphological techniques can we hope to learn about the structure of the "functional membrane"; only after we gain enough knowledge about the way ions and solvent molecules interact with protein molecules can we hope to know the mechanisms that control the flow of ions. We will need a better understanding of the chemical physics and physical chemistry of polyelectrolyte systems and of the physics of permeable membranes before we can explain how excitable cells work. Certainly we need further biochemical analyses to work out the chemical structure of the protein and phospholipid molecules in the membrane, so that we will know the kind of molecular apparatus with which we are dealing. The reader will see that the authors emphasize different aspects of the problem as they see them to be the most important ones. As a result, this book explores a vast territory. There is very little doubt that some of these explorations will ultimately prove to be fruitful. We have organized this book in the belief and hope that such a multidisciplinary approach will help unlock the mystery of the excitation process.

The help of Dr. Arthur B. Callahan of the Office of Naval Research, which has been particularly valuable both in organizing the conference and preparing this book, is gratefully acknowledged.

Donald C. Chang
Ichiji Tasaki
William J. Adelman, Jr.
H. Richard Leuchtag

Contents

I. Fine Structure in Excitable Cells

II. Cellular Excitation: Recent Findings and Models

I

Fine Structure in Excitable Cells

1

Differentiation of Axonal Membrane Systems, the Axolemma, and the Axoplasmic Matrix

Mark H. Ellisman, James D. Lindsey, Clayton Wiley-Livingston, and S. Rock Levinson

In this chapter we will discuss the interactions between axonal membrane systems and the "microtrabecular lattice" (MTL). In order to do this a brief review of some relatively new but fundamental observations on the membrane systems within myelinated axons is needed.

I. MEMBRANE SYSTEMS OF THE AXON

The coexistance of several different intra-axonal membrane systems is a concept which has only recently gained general acceptance. Most likely this is due to the complexity and intimacy of these various systems. The first system to be described in depth was the axoplasmic reticulum (Droz *et al.*, 1975; Tsukita and Ishikawa, 1976). This reticulum consists of an anastomotic network of fine tubules which course down the axon (diagramed in Fig. 1). Although well known individually for some time, multivesicular bodies have only recently been grouped with dense lumened and clear lumened cisternae to make up the second major intra-axonal membrane system (Weldon, 1975; Bunge, 1977; Broadwell and Brightman, 1979). This second system characteristically appears able to sequester tracer molecules such as thorium dioxide or horseradish peroxidase from the extracellular environment. A third intra-axonal membrane system has been described but has not yet received the level of general acceptance of the first two. This last group is best seen as distinct from the axoplasmic

Mark H. Ellisman and James D. Lindsey • Department of Neurosciences, University of California, San Diego, School of Medicine, La Jolla, California 92093. *Clayton Wiley-Livingston* • Department of Pathology, University of California, San Diego, School of Medicine, La Jolla, California 92093. *S. Rock Levinson* • Department of Physiology, University of Colorado Health Sciences Center School of Medicine, Denver, Colorado 80262.

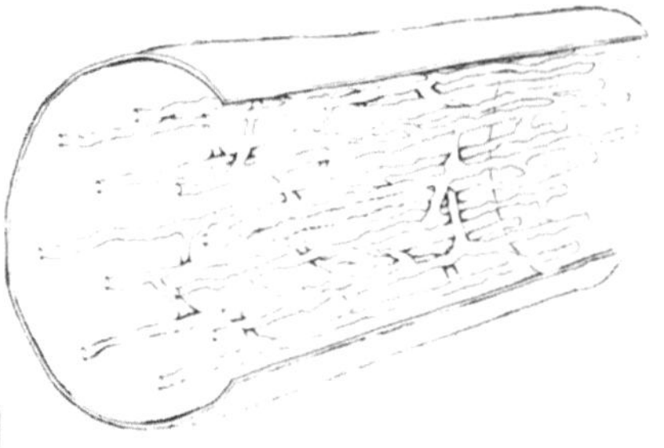

1

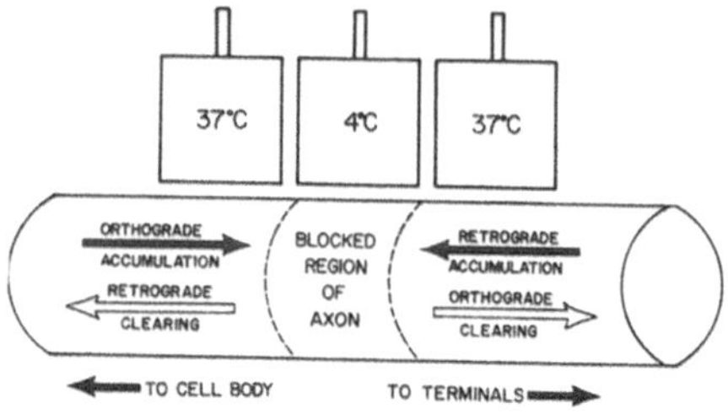

2

3

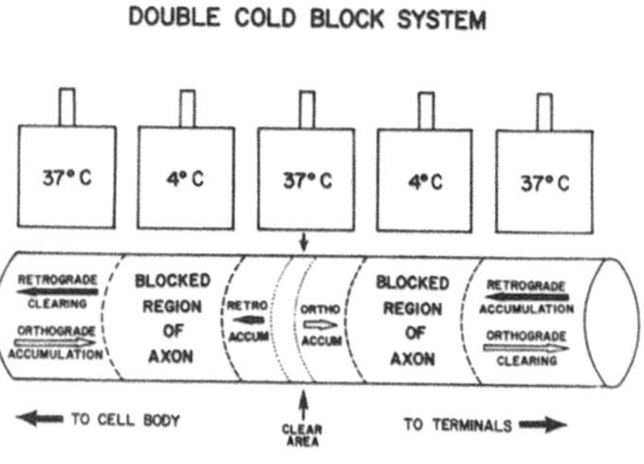

4

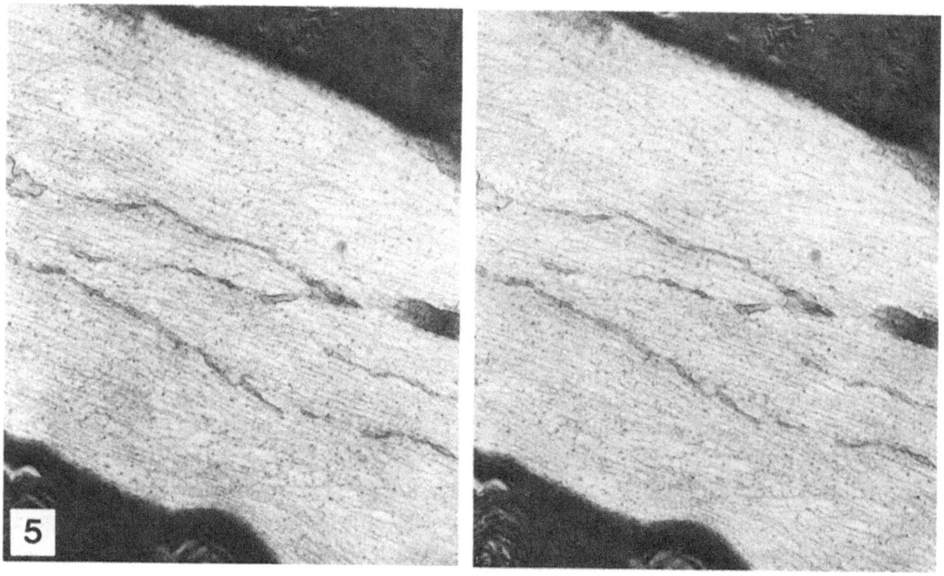

5

Figures 1 and 3. An artist's representation of two of the membrane systems of the axon drawn approximately to scale in the same axoplasmic space. Fig. 1. The axoplasmic reticulum, an interconnected membrane system. Fig. 3. Orthograde transport compartments or vectors, discrete vesicular and vesiculotubular entities of the axon.

Figure 2. Diagram illustrating the design of the cold block system. The cylinder represents axons of the saphenous nerve while the squares represent cold and warm aluminum blocks.

Figure 4. Diagram illustrating the design of double cold block experiments. The cylinder represents the axons within the saphenous nerve while the squares represent the selectively warmed and chilled blocks of aluminum applied to the skin just above the nerve. Beneath the two 4°C blocks, fast axonal transport is arrested whereas under the 37°C blocks it continues in a normal fashion. After 3–4 hr, transported material moving faster than 25 mm/day in either direction will have cleared from the region under the central warm block and no new material will have entered.

reticulum in thick sections and primarily contains clear lumened elongated cisternae and vesicles.

The functional physiology of these various membrane systems remains obscure. It is now generally accepted that rapidly transported macromolecules move in association with membranous structures of some sort. This concept has recently received further support by the finding that all rapidly transported macromolecules pass through the Golgi apparatus in the cell body enroute to destinations along the axolemma or beyond (Hammerschlag *et al.*, 1982). At least one of these macromolecules, the $Na^{+} + K^{+}$ ATPase is known to be rapidly transported (Baitinger and Willard, 1981) and is interesting in the context of this volume since it is known to be involved in contributing to properties of excitable membranes. The axoplasmic reticulum first became suspect as a possible vector for the fast transport of such macromolecules because of its shape and extent (for reviews see Schwartz, 1979; Grafstein and Forman, 1980). Newly obtained evidence, however, is decreasing the attractiveness of this notion.

This new evidence comes from experiments designed to block fast transport focally in a minimally disruptive manner. In a recent electron microscopic study, Tsukita and Ishikawa (1980) showed that blocking axoplasmic transport by gluing a cooled aluminum block over mouse saphenous nerve (Fig. 2), results in an accumulation of membranous material moving in the orthograde direction on the proximal side of the cold zone, while retrogradely moving components on this same proximal side continue to move centrally away from this site. Conversely, retrogradely moving materials distal to the chilled zone accumulate on the distal side of the cold block. With this single system it is possible to examine the morphology of orthograde and retrograde compartments following a suitable period of time for accumulation. A study using similar logic was recently conducted by Smith (1980). Smith's paradigm makes use of a mechanical block on single teased fibers that are observed at both light and electron microscopic levels. We have been conducting similar *in vivo* experiments in our own laboratory, to examine the mobility of the axoplasmic reticulum and also to determine the relationship of motile elements to the cytoskeleton (Ellisman and Lindsey, 1981, 1983).

Micrographs taken of tissue from the proximal side of the blockade, in all studies, demonstrate an accumulation of vesicular or vesiculotubular structures (Tsukita and Ishikawa, 1980; Smith, 1980; Ellisman and Lindsey, 1981, 1983). These types of discrete membrane bounded compartments accumulate where the logic of these experiments would predict the orthogradely moving components to selectively accumulate. Figure 3 depicts the form of these orthograde transport compartments or "vectors" (as we will refer to them) summarized from electron micrographs. The form of these compartments does not resemble that of the axoplasmic reticulum (depicted in Fig. 1), and we do not find axoplasmic reticulum accumulating against either distal or proximal sides of the cold block. Due to these two inconsistencies, we became curious about the relationship of the axoplasmic reticulum to the motile vectors and the actual mobility of the former.

A double cold block apparatus was designed (Fig. 4) and used to examine the mobility of the axoplasmic reticulum. This apparatus maintains a small length of nerve at 37°C between two areas that are cold blocked. After 3–4 hr, any axoplasmic elements moving faster than 25 mm/day should have moved to either the proximal or distal edge of the warm

Figure 5. In the middle of the warm region between the two cold blocks, fast transported elements have presumably cleared from the axoplasm. Here in these stereopair micrographs of a 1 μm thick section, viewed with the high voltage electron microscope (HVEM), the only type membranous elements remaining in the axoplasm are those morphologically and histochemically resembling the axoplasmic reticulum (the 3D effect is best when viewed with the aid of a 2× stereoviewer). ×19,000, 10° tilt.

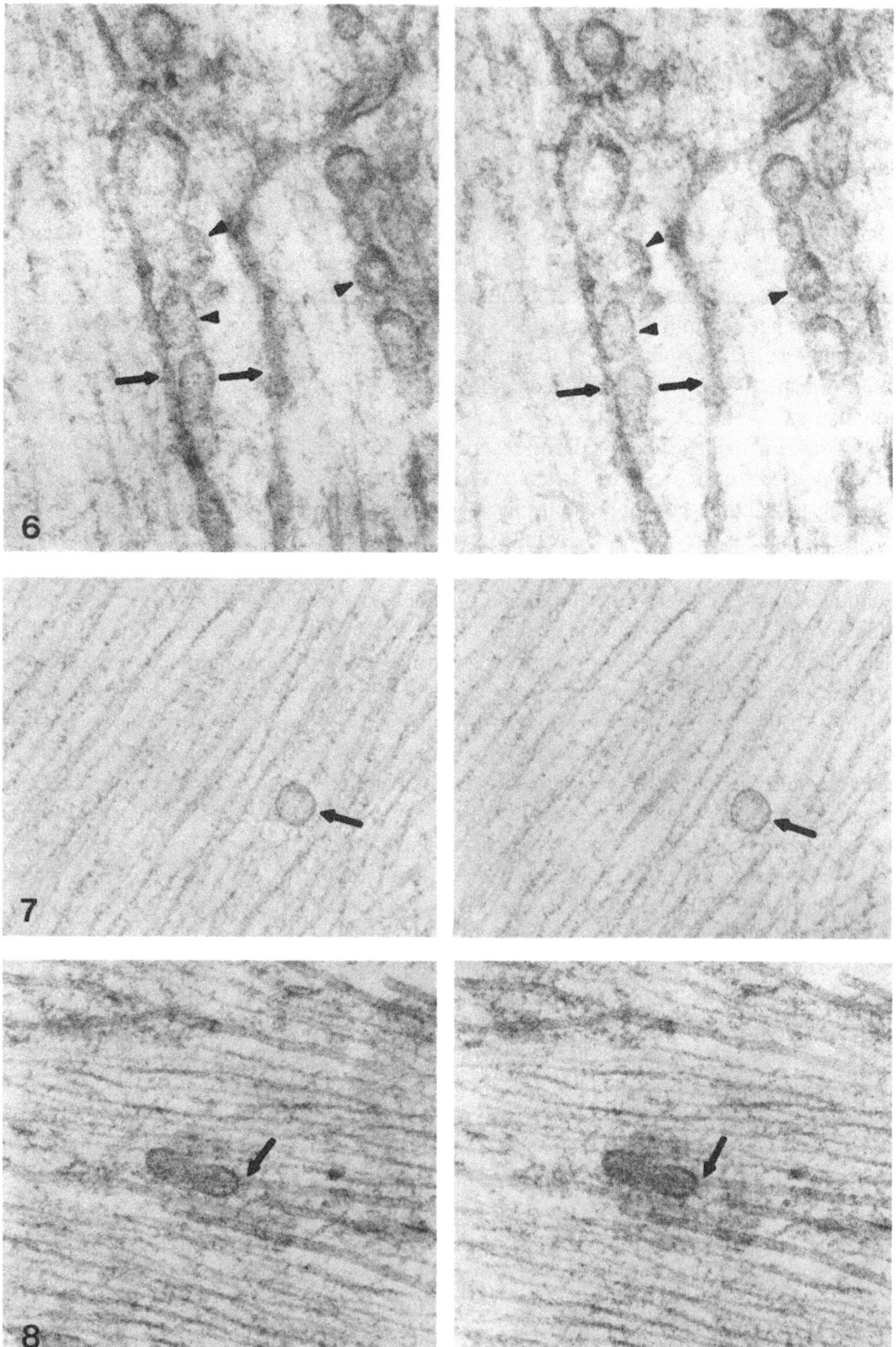

Figure 6. Axoplasm adjacent to proximal side of a cold blocked region illustrating the axoplasmic reticulum elements (arrows) that are clearly separate from discrete orthograde vector type elements (arrowheads). Continuities between these membrane systems were not observed in the large number of micrographs examined. ×73,000.

zone. In the center of the warm zone all rapidly moving organelles should be absent leaving components which would not have been subject to rapid transport (Fig. 5). Using this procedure, we found that numerous clearly defined elements of the axoplasmic reticulum remain in such central regions while discrete elements virtually disappear without apparent fusions (Fig. 6). These observations support the view that the axoplasmic reticulum is a distinct membrane system, separate from the discrete motile membranous compartments (vectors) which carry rapidly transported macromolecules.

The functional significance of this axoplasmic reticulum is not established. Whether a complicated system of membranous tubules within the axon, such as this, would contribute significantly to the electrical properties of the axon is unknown. Serial section analysis of guinea pig ear nerves (Reiter, 1966) and tracer experiments using cultured neurites have fueled speculation that this system has occasional continuity with the axolemma, not unlike the transverse tubule system of muscle (Bunge, 1977; Weldon, 1975). The axoplasmic reticulum may have some properties not unlike the sarcoplasmic reticulum of muscle in that it appears capable of sequestering calcium (Duce and Keen, 1978; Henkart *et al.*, 1978). Thus, this membrane system may be important in regulating the ionic milieu of the axon as well as contribute capacitance to its static electrical properties.

II. THE MICROTRABECULAR LATTICE OF THE AXON AND ITS INTERACTIONS WITH MEMBRANOUS COMPONENTS OF THE AXON

Now that the three different membrane systems in the axon have been illustrated we will describe how the entire axonal cytoskeleton including the MTL specializes upon interaction with the discrete membranous vectors of the axoplasm.

Neurofilaments and microtubules are crosslinked in a periodic manner by MTL crosslinking components in both chemically fixed and rapidly frozen and etched preparations (Ellisman and Porter, 1980; Ishikawa and Tsukita, 1982). In our high voltage electron microscopic (HVEM) studies we found that discrete cisternae often appeared asymmetrically connected by these crosslinkages to microtubules or neurofilaments as may be seen in the example of a small vesicle in Fig. 7. Such asymmetry was commonly observed, with one end and the sides of the vector connected to fibrous components and the other not. The nonconnected end exhibiting an absence of crosslinkages was often associated with a void in the axoplasmic space without neurofilaments, microtubules, or crossbridges. The immediately obvious question here is: does this asymmetry systematically reflect the direction of motion of a vector? More specifically, one might ask, are the voids systematically on the leading or trailing ends of vectors?

By using the single cold block method (illustrated in Fig. 2) to focally stop transport, we were able to look at vectors fixed while moving in a predictable direction. This was accomplished by looking at sections, 4–6 mm proximal to the area of vector accumulation. Here orthograde vectors would be moving toward the dam while retrograde vectors will

---

Figure 7. High magnification HVEM stereopair micrographs of peripheral nerve axoplasm. Many small vesicular cisternae (arrow) are larger in diameter than the average distance between adjacent neurofilaments and microtubules. The crossbridging connections are often disconnected on one end of such cisternae. ×68,000.

Figure 8. An orthograde vector predicted to be moving from left to right. Crossbridges are seen on the presumptive leading end (arrows) and a clear area behind. ×75,000.

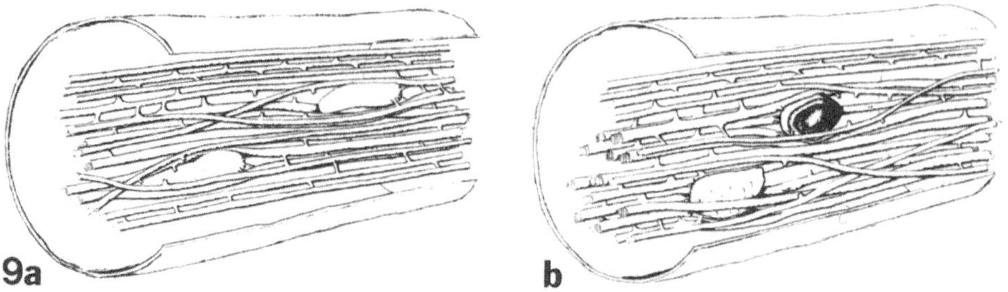

9a b

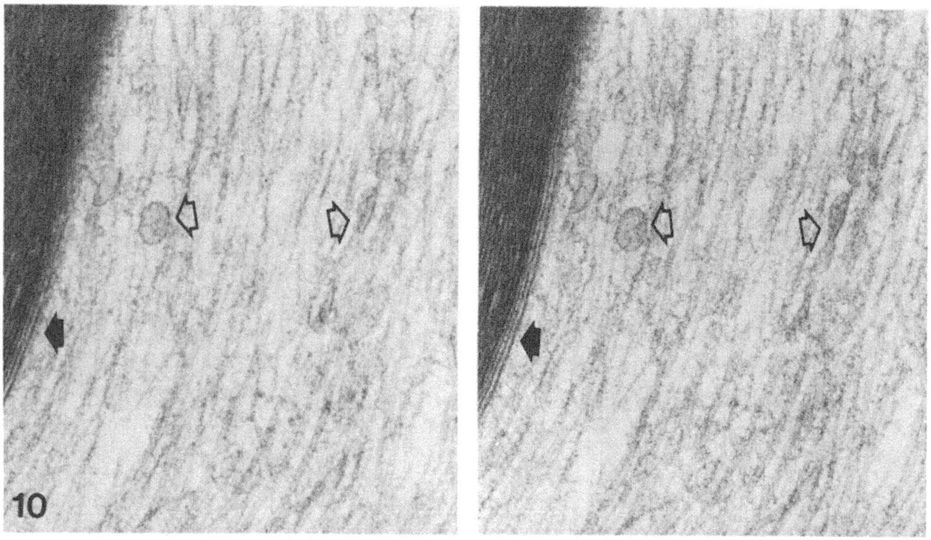

10

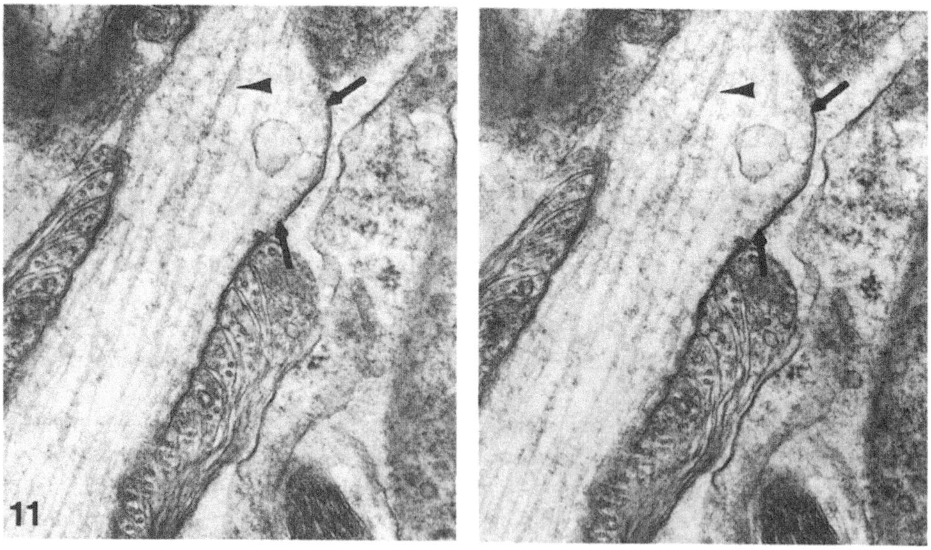

11

Figure 9. An artist's representation of the asymmetry found for orthograde and retrograde vectors. The cell body lies towards the left in each diagram. (a) Orthograde vectors are crosslinked at leading ends and disconnected on trailing ends. (b) Retrograde vectors are also crosslinked to the microtubules and neurofilaments on their leading ends and disconnected on their trailing ends.

Figure 10. HVEM stereopair micrographs of a longitudinal thick section through a myelinated peripheral nerve axon. Linkages to the plasma membrane (solid arrows) and discrete membranous cisternae of axonal smooth endoplasmic reticulum (SER) (open arrows) are visible. ×45,000.

have had sufficient opportunity to clear the region between the analyzed zone and the cold block as well as the analyzed region itself. Likewise a converse argument can be made for the presence of moving retrograde vectors and absence of orthograde vectors in a zone 4–6 mm distal to the distal edge of the cold block.

At ligature lesions, evidence for a delayed turnaround of orthogradely moving material has been found (Bisby, 1977). Whether such a turnaround occurs at a cold block is still unknown. Suggestive evidence against this possibility arises from the dramatic depletion of multivesicular and multilamellar bodies which are known retrograde vectors (Tsukita and Ishikawa, 1980; LaVail *et al.*, 1980; Ellisman and Lindsey, 1983), from the proximal analyzed region. Although further work is still needed, it is thus likely that the discrete compartments within the proximal zone analyzed are predominantly orthograde vectors.

By maintaining the proximodistal orientation of the axon throughout the processing and to the electron micrographs, we were able to determine that the asymmetry of crossbridges, the voids, and the organization of microtubules and neurofilaments, all related to both orthograde and retrograde vector direction. We found that there is a predominance of crossbridges on the leading ends of vectors and an absence of crossbridges (with significant voids creating enlarged spaces between microtubules or neurofilaments) on the trailing ends of both types of vectors. We suspect that this asymmetry has something to do with the actual movement of vectors during rapid axonal transport. Figure 8 is an orthograde vector presumably moving from left to right down this axon. The drawings presented in Figs. 9a and b summarize our observations on the asymmetry of crossbridges on both the orthograde and retrograde vectors.

III. CYTOSKELETAL SPECIALIZATION OF THE AXOLEMMA

A. Internodal Zone

In addition to MTL crossbridges attaching to transport vectors, one also finds crossbridges between the microtubules and neurofilaments (Ellisman and Porter, 1980) and, most importantly for what we will consider next, from microtubules and neurofilaments to the axolemma. Figure 10 is an example of an axolemmal–cytoskeletal connection in an internodal region of a myelinated axon. Here, the subaxolemmal (cortical) specialization of the plasma membrane is relatively simple. A relatively thin layer of cortical material from which wispy elements project deeper into the axoplasm, connect to "core cytoskeletal components," the so-called fibrous proteins, microtubules, and neurofilaments.

B. Nodal Zone

A more complex cortical specialization or axoplasmic matrix specialization is found where the axolemma is highly specialized at the node of Ranvier or in initial segment regions (Peters, 1966). Figure 11 is an HVEM stereopair of a thick section illustrating what is generally referred to as the subaxolemmal densification at the node of Ranvier. Upon close examination of this micrograph, however, one notices that this densification is composed

---

Figure 11. HVEM stereopair micrographs of a central nervous system node of Ranvier. Note that the specialization of the axolemma at the nodal membrane contains a band of fine filaments seen here in cross section (arrows). Also visible in this micrograph is an example of the axoplasmic reticulum (arrowheads). × 13,000.

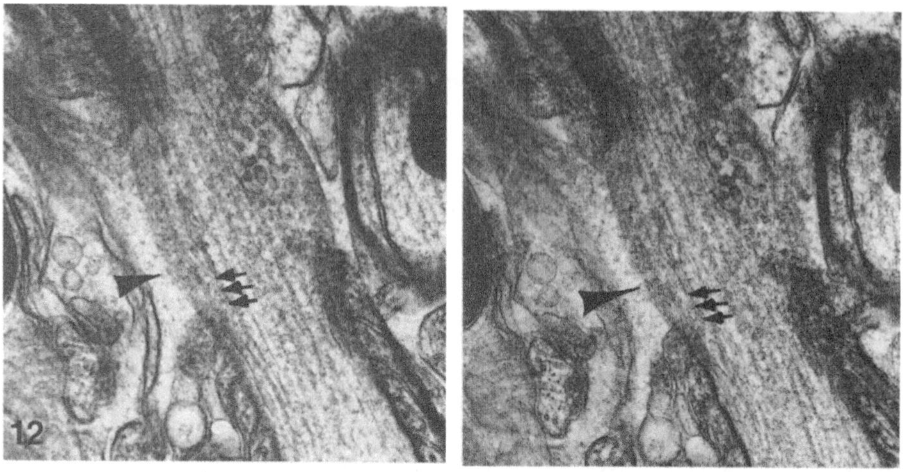

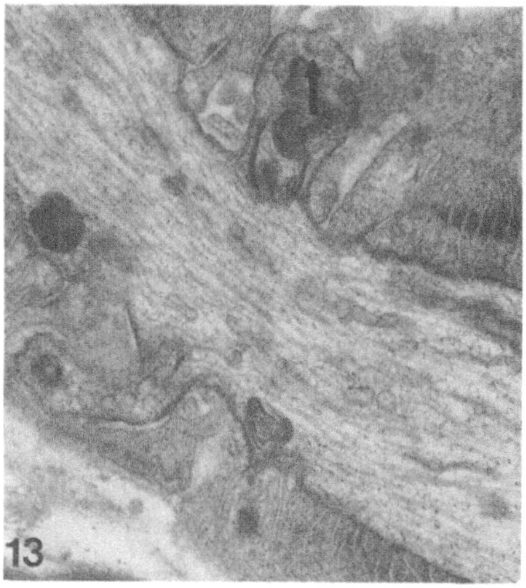

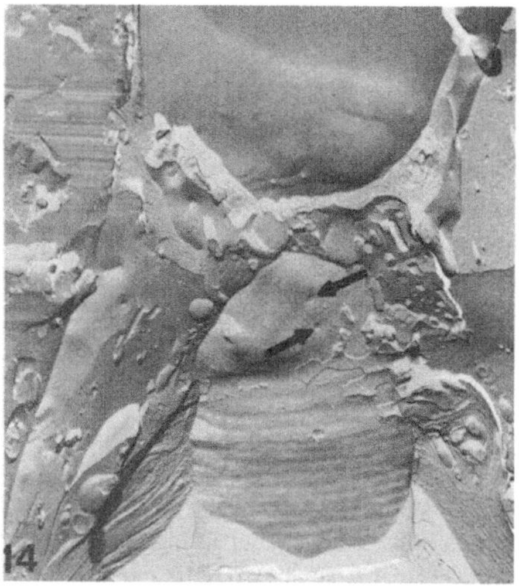

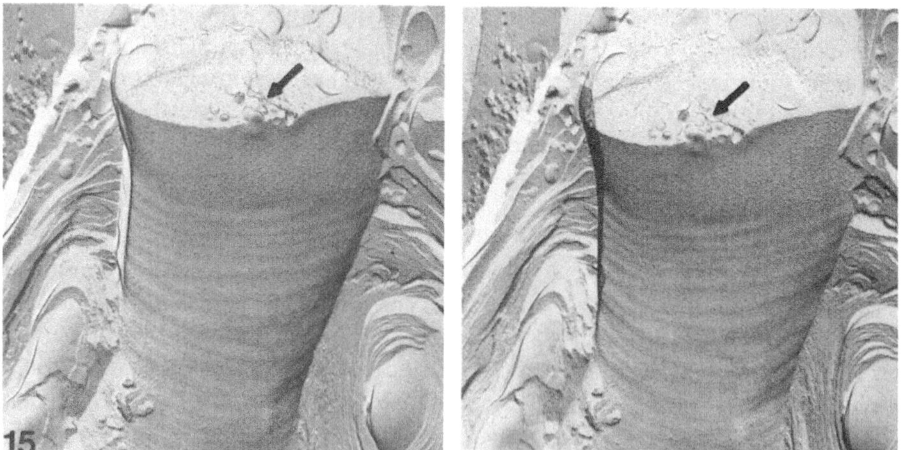

Figure 12. An HVEM stereopair capturing virtually an entire node within the thick section. The nodal membrane is seen *en face* (arrowhead) and here an array of fine filaments like those seen in Fig. 11 are visible (small arrows). ×30,000.

of a series of cross-sectioned filaments, here appearing as a series of 40 Å diameter dots. In fortunately oriented ultrathin sections the periodicity of the "dense layer" beneath the nodal membrane is also noticeable.

The organization of the cortical filaments of the subnodal zone is best appreciated in high voltage electron micrographs of thick sections in which a portion of the nodal membrane is tangentially included within the section (Fig. 12). The nodal membrane thus viewed *en face* appears as a translucency with the filamentous components of the cortical densification just beneath the axolemma. From this perspective the fine filaments are seen to be arranged in parallel, forming a wide band beneath the nodal membrane. This precise arrangement of very fine filaments appears to be restricted to the nodal zone of the axolemma.

Noteworthy and very evident in Fig. 11 is a system of subsurface cisternae characteristic of the paranodal zone of axoplasm (Ellisman, 1977). Several other types of membrane organelles are commonly subjacent to the nodal and paranodal regions of the axolemma. For instance, notice the multivesicular body in Fig. 12 just below the membrane of the nodal zone. Multivesicular bodies are observed frequently at nodes of Ranvier. These organelles are part of the neuronal lysosomal system and we suspect they are involved in the turnover of nodal membrane proteins by endocytocis followed by their retrograde transport to the cell body.

Evidence for endo- and/or exocytotic activity at the node of Ranvier is especially evident during development (Wiley-Livingston and Ellisman, 1980). For example, Fig. 13 is a micrograph of a node of Ranvier from a 9-day-old rat exhibiting a coated vesicle and many smooth vesicles intimately associated with the nodal axolemma. Evidence for continuity between axoplasmic vesicular components and the axolemma in the nodal zone has been obtained in freeze–fracture replicas (Fig. 14). Here, dimples may represent either endo- or exocytosis. Occasionally, in such replicas we find a fracture through a region of the nodal zone, where such depressions occur, that also fracture into the underlying axoplasm (Fig. 15). From these micrographs it is evident that dimples in the axolemma are directly connected to vesicular elements within the axoplasm. Thus, in the nodal area there are several types of cisternae and there is evidence for membrane exchange between the axolemma and membrane bounded components of the axoplasm.

The filaments of the subnodal densification are illustrated (Fig. 16). Also illustrated in this diagram are some of the specializations characteristic of the paranodal zone including the subaxolemmal cisternae and variations in particle size and density exposed in freeze–fracture replicas (Ellisman, 1976, 1979; Rosenbluth, 1976; Wiley and Ellisman, 1980).

C. Paranodal Zone

In the paranodal zone where the glial loops of the myelinating cells attach, connections between the core cytoskeletal structure of the axoplasm and small tufts of wispy material

---

Figure 13. A spinal root from a 7-day-old rat. During the early stages of myelination, evidence of endocytotic and exocytotic events at the nodal membrane includes invaginations and evaginations of the axolemma, the appearance of dense core and coated vesicles (arrow), as well as the presence of multilammelar and multivesicular bodies. ×24,000.

Figure 14. An electron micrograph of a freeze–fracture replica from rat peripheral nerve. Evidence of endo- or exocytotic events are seen here as depressions in the nodal membrane (arrows). ×11,000.

Figure 15. A stereopair of freeze–fractured node of Ranvier, nodal membrane, partially "crossfractured" into axoplasm. Many vesicles (arrows) in the axoplasm show apparent continuity with the nodal membrane axonal P face (AX PF), suggesting that active membrane, and/or membrane protein turnover occurs here. ×21,000.

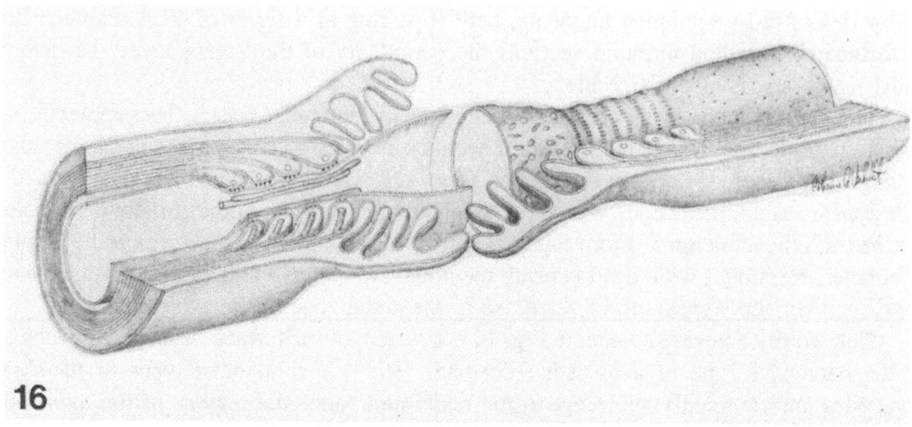

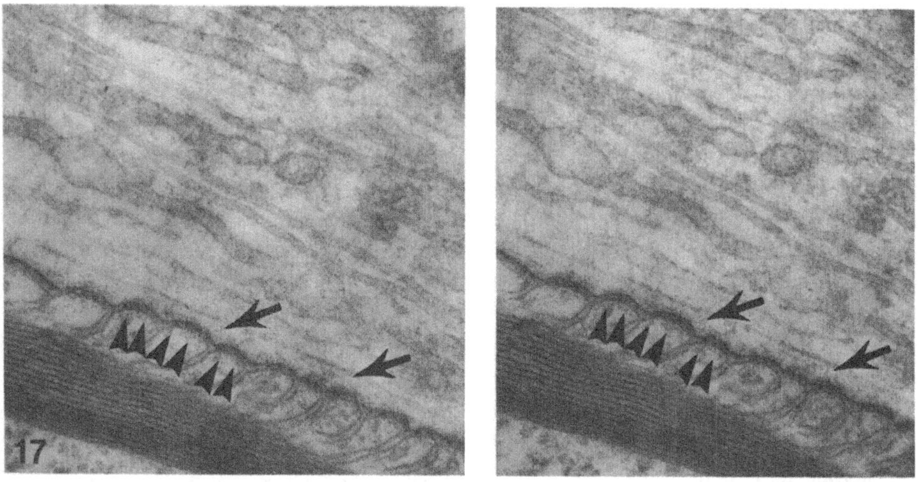

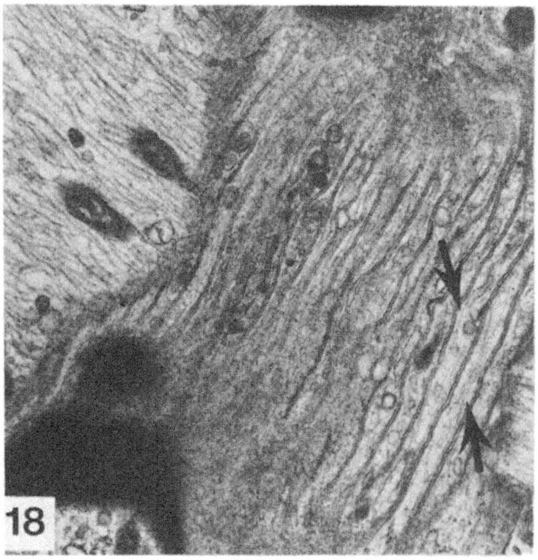

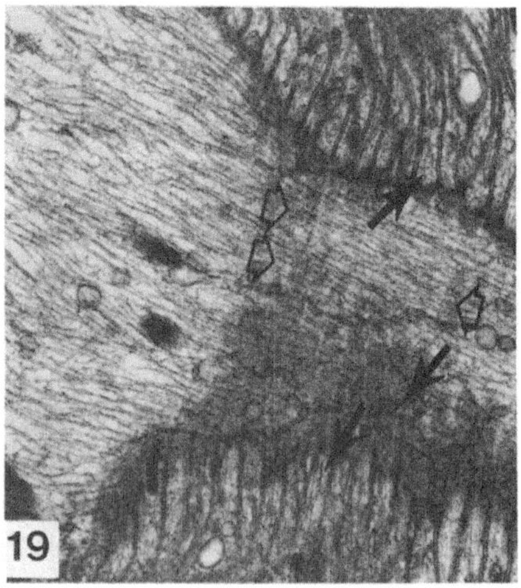

Figure 16. This diagram illustrates some of the structural features of the node of Ranvier observed either in sections (transmission electron microscopy (TEM) and HVEM), on left HVEM or in freeze–fracture replicas, on right (see text and Figs. 11, 12, 21–25).

Figure 17. Cytoskeletal connections at the paranodal zone viewed here in stereo. Filamentous connections from the neurofilaments to the glial–axonal junction (G–A J) are seen in the axon (arrows) while cross-sectioned 70 Å microfilaments are visible within each glial loop (small arrows). ×50,000.

of the axolemma are also seen. Figure 17 is a stereopair of a thin section from the paranodal region illustrating crossbridges between axoplasmic neurofilaments and the wispy material of the axolemma immediately subjacent to a paranodal glial loop. Neurofilaments are found here in parallel arrays, very closely applied to the axolemma, and with a regularity of spacing. These features are detailed in Figs. 17–19, which represent sections through this region of different orientation. Perhaps most notable in Fig. 17 are the cross-sectioned arrays of 70 Å microfilaments distributed just above the glial membrane in the paranodal glial loop. Several of these microfilaments appear to be arranged in the cortex of each glial loop adjacent to the glial–axonal junction (GAJ) as it encircles the axon. The 70 Å glial filaments can be seen best when this area of glial and axonal membrane apposition is captured tangentially or *en face* in thin sections (Figs. 18 and 19). Figures 18 and 19 are serial sections oriented such that they capture the cytoplasm of the series of glial terminal loops just before (Fig. 18) and just after (Fig. 19) they are applied to the axon at the paranodal junction. The glial filaments seen in cross section in Fig. 17 are longitudinally exposed in Figs. 18 and 19, as part of a helical circle within the glial loop around the axon. Deeper into the axon, Fig. 19 includes the paranodal junction and some of the axoplasm. Neurofilaments just below the axolemma in the axoplasm are clearly orthogonally arranged with respect to the 70 Å filaments of the glial loops.

Also notable in this series of figures (Figs. 17–19) are more subaxolemmal cisternae of the paranode, probably a specialization of the axoplasmic reticulum (mentioned above, see also Figs. 11 and 12). The entire area and the superpositioning of the glial and axonal filaments may be appreciated in stereopairs of thick sections viewed with the aid of HVEM (Fig. 20). In this thick section one views the axon from a point outside the paranodal zone. Here, one can pick out filaments just above the glial membrane before this membrane is involved in forming the glial axonal junction of the paranode. Slender tubular elements (subsurface cisternae of the glia) appear in several of these paranodal loops. This is most often seen in young nodes and to best advantage with freeze–fracture as illustrated in Fig. 21. The viewing perspective of Fig. 21 is from the axoplasm looking outward toward parts of the helical glial loops enshrouding the paranodal zone of the axolemma. The tubular cisternae of the glial loop seen in cross section in Fig. 17 are clearly visible in this replica. Thus, the cytoskeleton of the axolemma and the cytoskeleton of the glial loop as well as subsurface cisternal systems of both are highly specialized in the paranodal zone.

IV. SPECIALIZATION OF THE MAMMALIAN PARANODE AND NODE EXPOSED BY FREEZE–FRACTURE

A. Paranodal Zone

The preceding two sections consider the cortical specialization of the nodal and paranodal zones including the characteristic linkages of these regions with the core cytoskeleton. While stereo HVEM techniques reveal the three-dimensional aspects of the cortical zone

Figure 18. The 70 Å microfilaments of the glial loops seen in cross section in Fig. 17 are visible in longitudinal orientation in Fig. 18 which is an *en face* section through the G–A J. The filaments contained in the glial loop are notable in the cortical cytoplasm where the plasma membrane is involved in forming the G–A J (arrows). ×22,000.

Figure 19. This is the next serial section after that shown in Fig. 18, here cutting deeper into the axon, the axoplasmic contribution to the junctional complex is included. Microfilaments of the glial loop are still visible (arrows) and the parallel arrays of neurofilaments in the cortical axoplasm (open arrows) are orthogonally oriented to the glial microfilaments. Also note the fragment of the axoplasmic reticulum (arrowhead) very often observed in this cortical axoplasm of the paranodal zone involved in the junctional complex. ×25,000.

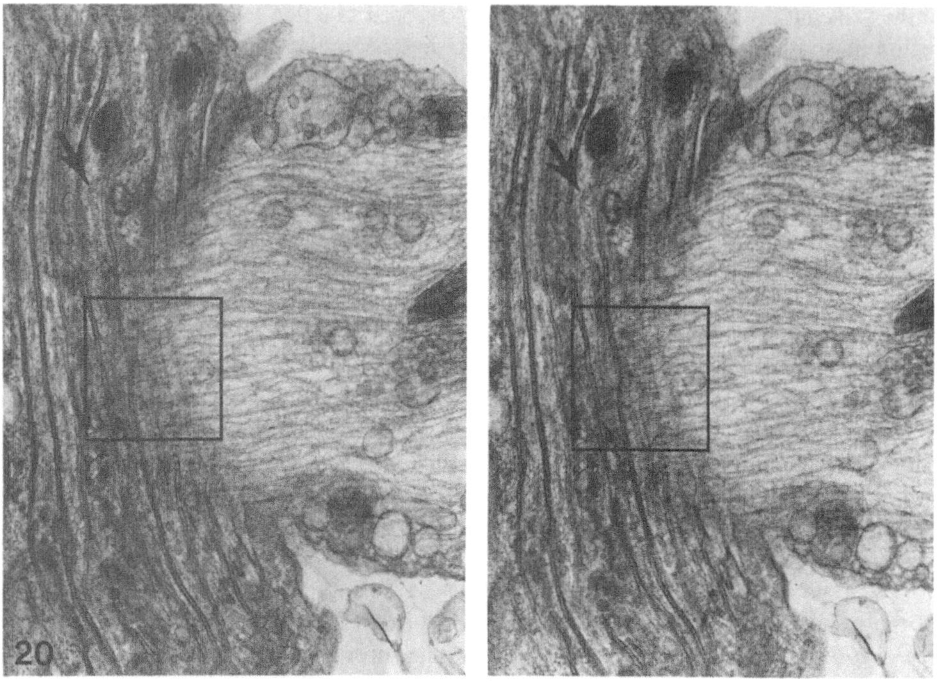

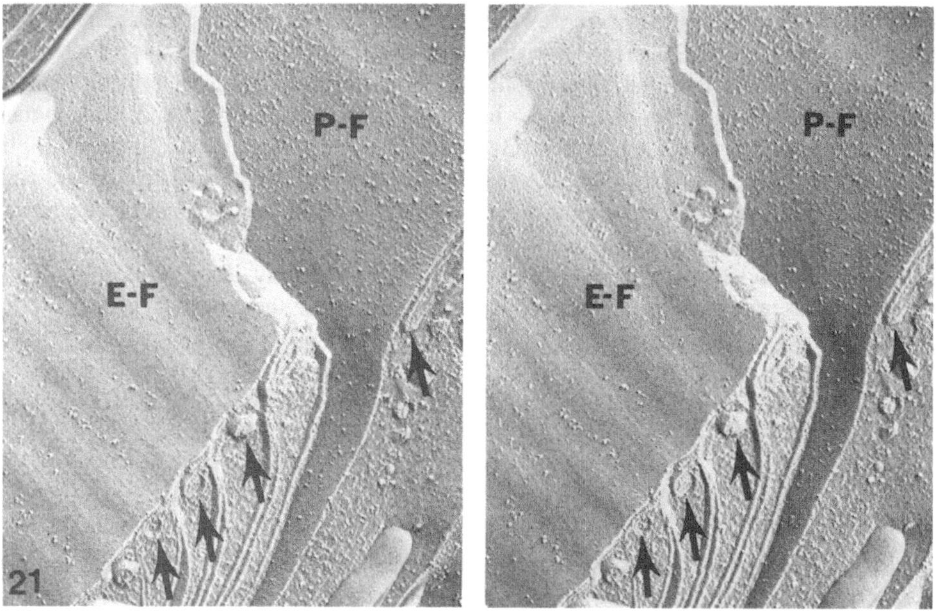

Figure 20. Many of the structural features detailed in the serial sections (Figs. 18 and 19) are visible when thick sections of this paranodal region are viewed with the HVEM. Here, view is from the glial side of the junction into the axoplasm. Orthogonal superpositioning of glial microfilaments and axonal neurofilaments may be observed in the area inscribed by the box. Also notable in such images are elements of a tubular cisternal system contained within the glial loops (arrows). ×26,000.

Figure 21. Freeze-fracture of paranode exposing axonal E face (EF) and glial P face (PF). In this micrograph the tubular cisternae of the glial loops noted above in the stereo pair (Fig. 20), are cross-fractured (arrows). ×41,000.

and subsurface cisternae of this region to best advantage, freeze–fracture replicas reveal the membrane molecular architecture of these zones most effectively. Where the paranodal junction is formed, the membranes of both glia and axon exposed in freeze–fracture replicas contain many particles and associated pits in highly ordered arrays (Wiley and Ellisman, 1980). The protoplasmic fracture face (PF face) of the paranodal axolemma contains rows of slightly elongated particles forming "dimers." We have examined the structural relationship between intramembranous specializations contributed by the myelinating cell (glial PF face and EF face) and elements contributed by the axon (axonal PF face and EF face) in forming the glia–axonal junctional complex with high resolution freeze–fracture techniques. Some of the results of our analysis are shown in Figs. 22–25 and are summarized in a composite diagram (Fig. 26). The junction appears as a highly ordered aggregate of both axonal and glial membrane components. We have speculated that some of these components may be involved in exchanges of ions and/or metabolites between axon and myelinating cell.

B. Nodal Zone

The PF face and EF face of the nodal zone exposed in freeze–fracture replicas contains particles of a much more heterogenous size distribution (Ellisman, 1976, 1979). There are only about 2000 particles/μm^2 exposed in both leaflets of the nodal zone in aldehyde fixed preparations. The types of particles and associated pits exposed in these membranes are illustrated in Figs. 22–25 and detailed in the figure legends. They range in size from 75–200 Å. The most frequent sized particle is 90–100 Å on PF faces while both the PF face and EF face reveal a high density of elongated "rod-shaped" particles. The relationship between Na^+ + K^+ ATPase of the node or the K^+ or Na^+ channels and specific particles is presently unknown. It is known, however, that the Na^+ + K^+ ATPase is a 90–100 Å freeze–fracture particle (Deguchi *et al.*, 1977). In order to define the particle candidates one would like to know whether the distribution of the excitable membrane pumps and channels is limited to the nodal zone which includes the paranode, and/or extends over the entire axolemma (under the myelin sheath as well).

V. IMMUNOCYTOCHEMICAL EVIDENCE FOR THE NODAL MOSAIC

A. Electric Organ

The distribution of at least two of these important components of the excitable membranes within the axolemma, Na^+ channels and the Na^+ + K^+ ATPase, have been examined with immunoelectron microscopic techniques (Wood *et al.*, 1977; Ellisman and Levinson, 1982; Ellisman *et al.*, 1982b). To determine the actual distribution of the Na^+ channel we have raised antibodies to the tetrodotoxin binding component (TTXR), purified from the electric organ of the South American eel *Electrophorus electricus* (Miller *et al.*, 1982; Ellisman and Levinson, 1982). These antibodies were first used to localize the Na^+ channel in light and electron micrographs of eel electric organ and then along myelinated axons of the eel spinal cord (Ellisman and Levinson, 1982).

The distribution of anti-TTXR antibodies is demonstrated by peroxidase reaction product in the light micrograph of the thick epoxy section shown in Fig. 27. The resultant dark staining is limited to the innervated and less convoluted faces of the electroplax (solid arrows) while the highly invaginated noninnervated faces are comparatively unstained (open arrows).

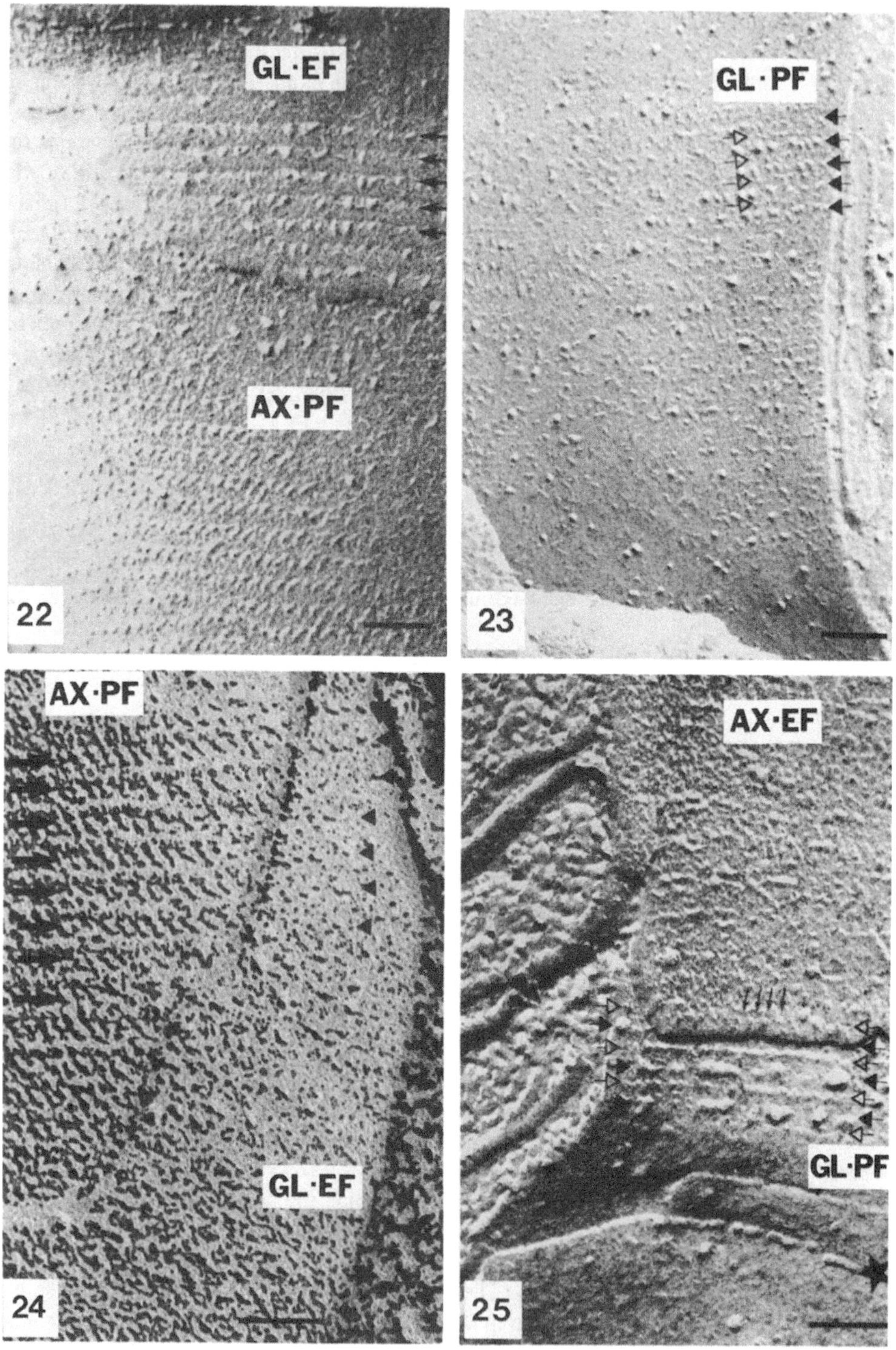

Figure 22. A freeze–fractured E face of a glial loop (GL EF) displaying rows of 160 Å particles. The rows (arrows) are spaced 360 Å apart while the particles within the rows are separated by 200 Å. The underlying axonal P face (AX PF) exhibits rows of dimeric particles. A star indicates the location of a tight junction between adjacent glial loops. 10-day-old rat. ×93,000.

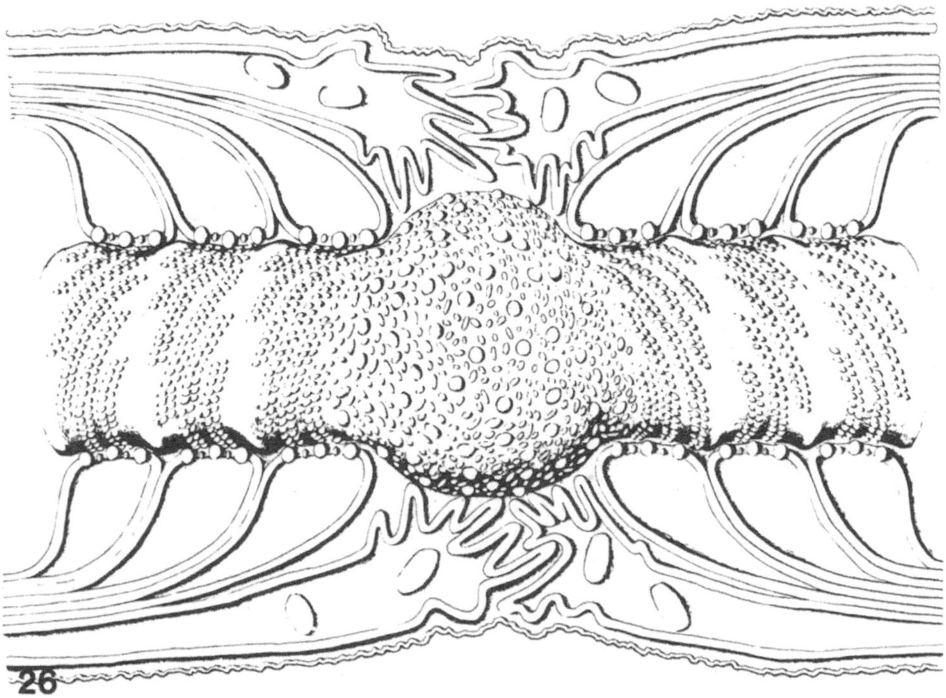

26

Figure 26. In this illustration both P face (dimeric particle distribution) and E face morphology are demonstrated. The paranodal region shows glial loops abutting the scalloped axolemma. Circumferential rows of dimeric particles are shown in the axonal P face. Particles within the glial membranes are shown in cross section as they are positioned above the axolemmal particles. The outermost glial loops interdigitate above the nodal axolemma. The nodal region bulges between the paranodal regions; this region is shown as it would appear in the axonal E face. Large and small particles are closely packed in a uniform annulus around the axon. The basement membrane surrounding the fiber is smooth and closely applied to the Schwann cell membrane.

Figure 23. A P face of a terminal glial loop (GL PF) displaying rows of 160 Å particles (filled arrows). The rows are spaced 360 Å apart, while the particles within the rows are separated by 200 Å. Rows of 75 Å particles (open arrows) are centered between the rows of 160 Å particles. 10-day-old rat. ×93,000.

Figure 24. In a series of transparent overlays, micrographs such as that shown here have been used to project the position of axonal specializations with respect to glial specializations. The rows (arrows) of dimeric particles in the axonal P face (AX PF) are seen to be positioned between the rows of 160 Å particles in the glial E face (GL EF) (arrowheads). The 160 Å glial particles fall on a line of colinearity between dimeric particles of adjacent rows (see text and Fig. 26). 12-day-old rat. ×117,000.

Figure 25. Using transparent overlays as in Fig. 24, it is possible to project the position of specializations on the axonal E face (AX EF) with respect to specializations on the glial P face (GL PF). Rows of large glial particles (closed arrows) alternate with rows of small glial particles (open arrows). The rows of small glial particles are superimposed over the obscure pits of the axonal E face (indicated by small arrows), while the row of large glial particles are superimposed over the ridges of the axonal E face. A star indicates the position of a tight junction between adjacent glial loops. 14-day-old rat. ×117,000.

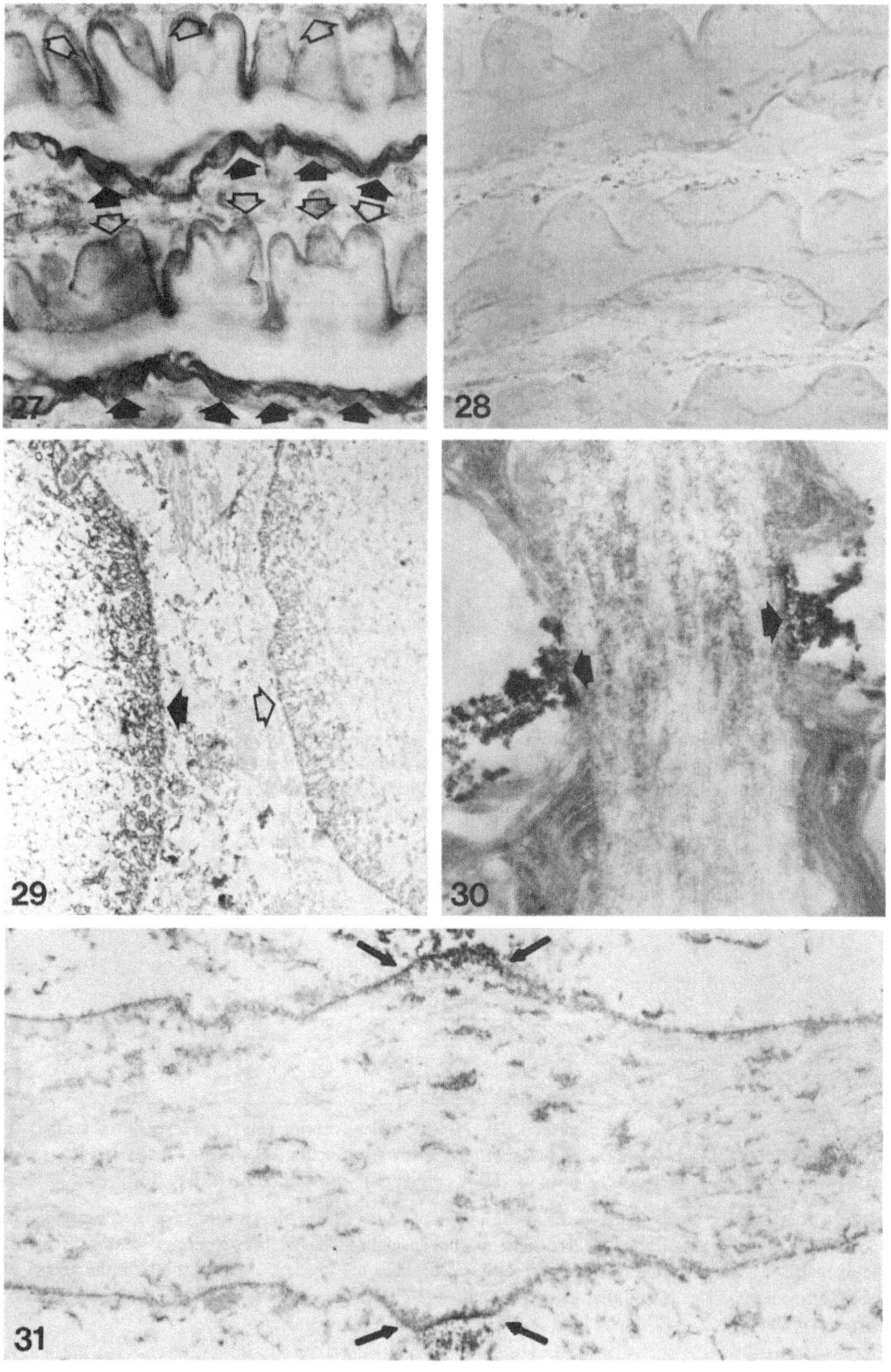

Figures 27 and 28. Light micrographs of thick epoxy sections from electrocytes in the main electric organ. Fig. 27. The peroxidase reaction product localizing the tetrodotoxin binding protein (Na^+ channel) stains the innervated face (solid arrow) while the noninnervated face is unstained (open arrows). Fig. 28. A preadsorbed control is presented for comparison. ×200.

An example of the control employing antibodies against TTXR which have preabsorbed with TTXR is presented in Fig. 28. No significant enhancement of staining in innervated over noninnervated faces was evident in either the pre-adsorbed control or in preparations first incubated in normal rabbit serum.

When viewed at higher magnification with the electron microscope, the microinvaginations (caveole) of both the innervated and noninnervated surfaces may be observed (Fig. 29). Although caveole of both surfaces are approximately the same dimensions, those of the innervated face (in continuity with the plasma membrane of the innervated face) exhibited positive staining for TTXR, while those of the noninnervated face did not. Thus, only the innervated surface, including caveole, appears to contain the antigenic determinant for the anti-TTXR antibodies.

B. Myelinated Nerve

Having established the distribution of sodium channels (TTXR) on the electroplax, and finding this to be in agreement with physiological evidence for their distribution, we applied similar procedures to examine the distribution along myelinated axons. Vibratome sections of dorsal spinal column of electrophorus, exposed to the immunochemical reagents revealed focal sites of very dense staining when examined in thin sections. The focal staining was restricted to the nodal zone of the axolemma in axons retaining paranodal junctions and an intact myelin sheath (Fig. 30). In order to determine whether this focal staining was a result of limited diffusion of immunochemical reagents into the paranodal zone some preparations were exposed to much more vigorous vibratome sectioning. This often resulted in desheathing of axons after primary fixation but prior to immunocytochemical reactions. The staining of axons with the axolemma made accessible in this manner is nonetheless restricted to nodal foci (Fig. 31). No staining was observed in either the preabsorbed IgG control preparation or tissues incubated first in normal rabbit serum instead of anti-TTXR antibodies. Thus, the distribution of antigenic determinants for the anti-TTXR antibodies (sodium channels) appears to be restricted to the nodal zone of the node of Ranvier.

We have also used antibodies raised against rat $Na^+ + K^+$ ATPase to examine the distribution of this protein on myelinated mammalian axons. The distribution at the node is the same as that shown in detail above for the Na^+ channel. Wood and co-workers (1977) used antisera against eel electric organ $Na^+ + K^+$ ATPase to examine its distribution in knife fish. Our experiments have required examining the distribution of the $Na^+ + K^+$ ATPase in rat and mouse nerves. The antibodies raised against the eel protein do not cross react significantly with rat $Na^+ + K^+$ ATPase, so we have purified the rat $Na^+ + K^+$ ATPase and raised antibodies to it (Ariyasu *et al.*, 1982; Schenk *et al.*, 1982). These antibodies raised against the purified rat kidney $Na^+ + K^+$ ATPase cross react with rat

---

Figure 29. Electron micrograph of the immunoreacted electrocytes innervated (solid arrows) and noninnervated surfaces (open arrows). Two apposed surfaces are presented, note that the caveolae of the innervated face stain while those of the noninnervated do not. ×3,800.

Figure 30. Node of Ranvier from the dorsal columns of the eel spinal cord. An experimental node, partially demyelinated by the action of the vibratome is presented here. The arrows indicate the nodal zone reaction product localizing the Na^+ channels. ×13,000.

Figure 31. This node was completely demyelinated prior to the immunocytochemical reaction and only the nodal zone (arrows) is focally stained for Na^+ channels. ×16,000.

neuronal and glial $Na^+ + K^+$ ATPases and also with the enzyme in the mouse nervous system.

Thus, both $Na^+ + K^+$ ATPase and sodium channels appear to be located in the nodal zone of the node of Ranvier. Correlation of the Na^+ channel with a specific group of particles exposed by freeze–fracture electron microscopy at the node of Ranvier would enable direct measurement of the number of channels per μm^2. This may soon be feasible as the TTXR has an apparent molecular weight of 250,000 dalton (Levinson and Ellory, 1973; Agnew *et al.*, 1978; Miller *et al.*, 1982) and on the basis of its solubility and other physical characteristics is likely to be an integral membrane protein (Miller *et al.*, 1982; Ellisman *et al.*, 1982a,b). Thus, by analogy with the acetylcholine receptor complex or rhodopsin (Darszon *et al.*, 1980) one would expect to visualize a correlated freeze–fracture particle of approximately 100 Å. Quantitative freeze–fracture examination of nodal, paranodal, and internodal zones of myelinated axons has revealed approximately 2000 appropriately sized particles per square micron on both protoplasmic (PF face) and external (EF face) fracture faces of the nodal membrane (Ellisman, 1979; Wiley and Ellisman, 1980; Kristol *et al.*, 1977, 1978) where the TTXRs and $Na^+ + K^+$ ATPases are located. Some of the particles found on the nodal PF face undoubtedly represent the $Na^+ + K^+$ ATPase since this protein is known to be exposed by freeze–fracturing (Deguchi *et al.*, 1977).

A correlation of sodium channels with freeze–fracture particles is further complicated by differences between numbers of particles found at nodes and the current predicted values for the number of channels per square micron of nodal membrane. Estimates of the number of sodium channels/μm^2 have been made using physiological (Nonner *et al.*, 1975) and pharmacological (Ritchie and Rogart, 1977) techniques. These estimated density values are approximately 5000/μm^2 and 10,000/μm^2 respectively.

Direct determination of whether the sodium channel is exposed by freeze–fracture and if so, what size and shape it is, including knowledge of which fracture face(s) it partitions to, will undoubtedly help reconcile the differences between density estimates obtained using different techniques. It is noteworthy in this regard that the data of Kristol and co-workers (1977) on excitable and inexcitable nodes of *Stenarchus* implicate EF face particles as correlates for Na^+ channels since far fewer of these are found in the membranes of inexcitable nodes.

Restriction of Na^+ channels to the nodal zone of the axolemma is one aspect of the regional specialization of the axolemma. We do not know what mechanisms are used by the axon and/or myelinating cell to promote or retain such localized membrane specialization. It appears however to be independent of myelination or the "trapping" of nodal components by the paranodal junctions (Ellisman, 1976, 1979; Wiley-Livingston and Ellisman, 1981). Another possible way of restricting the lateral mobility of Na^+ channels at nodes might be through linkages with the extracellular matrix. Perhaps a more attractive proposal would be to suggest that the lateral mobility of membrane components at the node is restricted by specific elements of the axoplasmic cytoskeleton. Dense granular material just beneath the nodal membrane which appears to be composed of fine filaments in high voltage electron micrographs (as detailed above) could be the morphological correlate of such a cytoskeletal anchoring system. Understanding the relation of this cortical material to the Na^+ channel or $Na^+ + K^+$ ATPase of the node will depend upon further study.

The localization of TTX binding protein and $Na^+ + K^+$ ATPase to the nodal zone eliminates the two activities of these proteins as functional correlates for the paranodal zone of the axon. Recent voltage clamping work on demyelinated mammalian preparations (Ritchie and Chui, 1981) has suggested that the paranodal zone may be involved in delayed rectification and thereby one may speculate that the paranodal junction is involved in this activity.

Further work in progress is aimed to define the locations of specific macromolecules involved in nodal excitability and discover how they are anchored or aggregate in their specific locations so as to form a functional mosaic. Axonal transport may play an important role in modulating the turnover of proteins at the node such as the $Na^+ + K^+$ ATPase or the sodium channel. Thus, the cytoskeletal membrane interactions that occur both at the level of the node of Ranvier and at sites where transport vectors interact with microtubules or neurofilaments may be functionally linked by molecular sorting codes which specify the location where axonally transported membrane proteins are to be added to the axolemma (Ellisman, 1982).

ACKNOWLEDGMENTS. The authors wish to acknowledge the assistance of Thomas Deerinck, Derek Leong, Judith Nichol, and Dolores Taitano. This was supported by grants from the NIH to M.H.E. (#NS14718) and S.R.L. (#NS15879), as well as grants from the Muscular Dystrophy Association and the National Multiple Sclerosis Society to M.H.E. M.H.E. is an Alfred P. Sloan Research Fellow. S.R.L. is a Research Career Development awardee. C.W.-L. was supported by the National Institute of General Medical Sciences Training Grant (#GM07198) and the National Multiple Sclerosis Society. J.D.L. was a NSF Predoctoral Fellow (#SP179-22285).

REFERENCES

Agnew, W. S., Levinson, S. R., Brabson, J. S., and Raftery, M. A., 1978, Purification of the tetrodotoxin-binding component associated with the voltage-sensitive sodium channel from *Electrophorus electricus* electroplax membranes, *Proc. Natl. Acad. Sci. USA* **75**:2606–2610.

Ariyasu, R. G., Ellisman, M. H., Nichol, J. A., and Deerinck, T. D., 1982, Immunocytochemical localization of sodium-potassium adenosine triphosphatase in the rat central nervous system, *Soc. Neurosci. Abst.* **8**:415.

Baitinger, C., and Willard, M., 1981, Axonal transport of Na^+K^+ ATPase and protein I in retinal ganglion cells, *Soc. Neurosci. Abst.* **7**:742.

Bisby, M. A., 1977, Retrograde axonal transport of endogenous protein: Differences between motor and sensory axons, *J. Neurochem.* **28**:303–314.

Broadwell, R. D., and Brightman, M. W., 1979, Cytochemistry of undamaged neurons transporting exogenous protein in vivo, *J. Comp. Neurol.* **185**:31–74.

Bunge, M. B., 1977, Initial endocytosis of peroxidase or ferritin by growth cones of cultured nerve cells, *J. Neurocytol.* **6**:407–439.

Darszon, A., Vandenberg, C. A., Schonfeld, M., Ellisman, M. H., Spitzer, N., and Montal, M., 1980, Reassembly of protein-lipid complexes into large bilayer vesicles: Perspectives for membrane reconstitution, *Proc. Natl. Acad. Sci. USA* **77**:239–243.

Deguchi, N., Jorgensen, P. L., and Maunsbach, A. B., 1977, Ultrastructure of the sodium pump, comparison of thin sectioning, negative staining, and freeze-fracture of purified, membrane-bound (Na^+, K^+)-ATPase, *J. Cell Biol.* **75**:619–634.

Droz, B., Rambourg, A., and Koenig, H. W., 1975, The smooth endoplasmic reticulum: Structure and role in the renewal of axonal membrane and synaptic vesicles by fast axonal transport, *Brain Res.* **93**:1–13.

Duce, I. R., and Keen, P., 1978, Can neuronal smooth endoplasmic reticulum function as a calcium reservoir? *Neuroscience* **3**:837–848.

Ellisman, M. H., 1976, The distribution of membrane molecular specializations characteristic of the node of Ranvier is not dependent upon myelination, *Soc. Neurosci. Abst.* **2**:410.

Ellisman, M. H., 1977, High voltage electron microscopy of cortical specializations associated with membranes at nodes of Ranvier, *J. Cell Biol.* **75**:108a.

Ellisman, M. H., 1979, Molecular specializations of the axon membrane at nodes of Ranvier are not dependent upon myelination, *J. Neurocytol.* **8**:719–735.

Ellisman, M. H., 1982, A hypothesis on the interaction between membrane systems within the axon and the microtrabecular cross-linkages of the axoplasmic matrix, in: *Axoplasmic Transport* (D. Weiss, ed.), Springer-Verlag, New York, pp. 55–63.

Ellisman, M. H., and Levinson, S. R., 1982, Immunocytochemical localization of sodium channel (TTX binding protein) distribution in excitable membranes of *Electrophorus electricus*, *Proc. Natl. Acad. Sci. USA*, **79**:6707–6711.

Ellisman, M. H., and Lindsey, J. D., 1981, The axonal reticulum within myelinated axons is not rapidly transported, *J. Cell Biol. Abst.* **91**:91a.

Ellisman, M. H., and Lindsey, J. D., 1983, The axoplasmic reticulum within myelinated axons is not transported rapidly. *J. Neurocytol.* **12**:393–411.

Ellisman, M. H., and Porter, K. R., 1980, The microtrabecular structure of the axoplasmic matrix visualization of cross-linking structures and their distribution, *J. Cell Biol.* **87**:464–479.

Ellisman, M. H., Agnew, W. S., Miller, J. A., and Levinson, S. R., 1982a, Electron microscopic visualization of the tetrodotoxin binding protein from *Electrophorus electricus*, *Proc. Natl. Acad. Sci. USA*, **79**:4461–4465.

Ellisman, M. H., Levinson, S. R., Agnew, W. S., Miller, J., and Deerinck, T. J., 1982b, Visualization of the sodium channel and its immunocytochemical localization, *Soc. Neurosci. Abst.*, **8**:727.

Grafstein, B., and Forman, D. S., 1980, Intracellular transport in neurons, *Physiol. Rev.* **60**:1167–1283.

Hammerschlag, R., Stone, G. C., Lindsey, J. D., and Ellisman, M. H., 1982, Evidence that all newly synthesized proteins destined for fast axonal transport pass through the Golgi apparatus, *J. Cell Biol.* **93**:568–575.

Henkart, M. P., Reese, T. S., and Brinley, F. J., Jr., 1978, Endoplasmic reticulum sequesters calcium in the squid giant axon, *Science* **202**:1300–1303.

Ishikawa, H., and Tsukita, S., 1982, Morphological and functional correlates of axoplasmic transport, *Axoplasmic Transport* (D. Weiss, ed.), Springer-Verlag, New York, pp. 251–259.

Kristol, C., Akert, K., Sandri, C., Wyss, U. R., Bennett, M. V. L., and Moor, H., 1977, The Ranvier nodes in the neurogenic electric organ of the knife fish Sternarchus: A freeze-etching study on the distribution of the membrane-associated particles, *Brain Res.* **125**:17–212.

Kristol, C., Sandri, C., and Akert, K., 1978, Intramembranous particles at the noes of Ranvier of the cat spinal cord: A morphometric study, *Brain Res.* **142**:391–400.

LaVail, J. H., Rapisardi, S., and Sugino, I. K., 1980, Evidence against the smooth endoplasmic reticulum as a continuous channel for the retrograde axonal transport of horseradish peroxidase, *Brain Res.* **191**:3–20.

Levinson, S. R., and Ellory, J. C., 1973, Molecular size of the tetrodotoxin-binding sites estimated by irradiation inactivation, *Nature New Biol.* **245**:122–123.

Miller, J. A., Agnew, W. S., and Levinson, S. R., 1982, The tetrodotoxin binding protein from the electroplax of *E. electricus*: Isolation of milligram quantities of the principle polypeptide and its physicochemical characterization, *J. Biol. Chem.*, in press.

Nonner, W., Rojas, E., and Stampfli, R., 1975, Gating currents in the node of Ranvier: Voltage and time dependence, *Phil. Trans. R. Soc. (London) B* **270**:483–492.

Peters, A., 1966, The node of Ranvier in the central nervous system, *Quart. J. Exp. Physiol.* **51**:229–239.

Reiter, W., 1966, Uber das Raumsystem des endoplasmischen Retikulums von Hautnervenfasern. Untersuchungen an Serienschnitten, *Z. Zellforsch. Mikrosk. Anat.* **72**:446–461.

Ritchie, J. M., and Chui, S. Y., 1981, Distribution of sodium and potassium channels in mammalian myelinated nerve, in: *Demyelinating Disease: Basic and Clinical Electrophysiology* (S. G. Waxman and J. M. Ritchie, eds.), pp. 329–342, Raven Press, New York.

Ritchie, J. M., and Rogart, R. B., 1977, Density of sodium channels in mammalian myelinated nerve fibers and nature of the axonal membrane under the myelin sheath, *Proc. Natl. Acad. Sci. USA* **74**:211–215.

Rosenbluth, J., 1976, Intramembranous particle distribution at the node of Ranvier and adjacent axolemma in myelinated axons of the frog brain, *J. Neurocytol.* **5**:731–745.

Schenk, D., Grosse, R., Ellisman, M. H., Bradshaw, V. B., and Leffert, H., 1982, Na^+, K^+-ATPase: A new assay of Na^+-ATPase reveals covert antipump antibodies, *Anal. Biochem.*, **125**:189–196.

Schwartz, J. H., 1979, Axonal transport: Components, mechanisms, and specificity, *Annu. Rev. Neurosci.* **2**:467–504.

Smith, R. S., 1980, The short term accumulation of axonally transported organelles in the region of localized lesions of single myelinated axons, *J. Neurocytol.* **9**:39–65.

Tsukita, S., and Ishikawa, H., 1976, Three-dimensional distribution of smooth endoplasmic reticulum in myelinated axons, *J. Elec. Microscopy* **25**:141–149.

Tsukita, S., and Ishikawa, H., 1980, Movement of membranous organelles in axons. Electronic-microscopic identification of anterogradely and retrogradely transported organelles, *J. Cell Biol.* **84**:513–530.

Weldon, P. R., 1975, Pinocytotic uptake and intracellular distribution of colloidal thorium dioxide by cultured sensory neurites, *J. Neurocytol.* **4**:341–356.

Wiley, C. A., and Ellisman, M. H., 1980, Rows of dimeric-particles within the axolemma and juxtaposed particles within glia, incorporated into a new model for the glial-axonal junction at the node of Ranvier, *J. Cell Biol.* **84**:261–280.

Wiley-Livingston, C. A., and Ellisman, M. H., 1980, Development of axonal membrane specializations defines nodes of Ranvier and precedes Schwann cell myelin formation, *Dev. Biol.* **79**:334–355.
Wiley-Livingston, C. A., and Ellisman, M. H., 1981, Myelination dependent axonal membrane specializations demonstrated in insufficiently myelinated nerves of the dystrophic mouse, *Brain Res.* **224**:55–67.
Wood, J. G., Jean, D. H., Whitaker, J. N., McLaughlin, B. J., and Albers, R. W., 1977, Immunocytochemical localization of the sodium, potassium activated ATPase in knifefish brain, *J. Neurocytol.* **5**:571-581.

2

Structure of the Node of Ranvier

Jack Rosenbluth

I. INTRODUCTION

The axolemma of myelinated nerve fibers has been shown to be markedly nonuniform with respect to the distribution of ion channels. Specifically, sodium channels are concentrated at the nodes of Ranvier (Nonner *et al.*, 1975; Conti *et al.*, 1976; Ritchie and Rogart, 1977; Chiu, 1980; Sigworth, 1980) but sparse elsewhere (Brismar, 1981; Chiu and Ritchie, 1982), while potassium channels, especially in mammals, appear to have the converse distribution (Kocsis and Waxman, 1980; Sherratt *et al.*, 1980; Smith and Schauf, 1981; Chiu and Ritchie, 1982). Until recently, ultrastructural studies have been rather disappointing in demonstrating counterpart structural specializations. Although the node has selective staining properties (Waxman and Quick, 1977), the nodal axolemma itself is not distinctive in electron micrographs of sectioned fibers and is characterized only by an ill-defined cytoplasmic undercoating associated with its inner surface (Peters, 1966). The freeze–fracture technique, however, reveals striking structural specializations in both nodal (Rosenbluth, 1976; Kristol *et al.*, 1978; Tao-Cheng and Rosenbluth, 1980b) and paranodal (Livingston *et al.*, 1973; Dermietzel, 1974; Schnapp and Mugnaini, 1975) regions of the axolemma, thus reintroducing the possibility of correlations that might permit structural identification of specific ionophores. Such correlations initially appeared not to be straightforward, largely because of inconsistencies in the physiologic data, but recent studies suggest that the sodium channels, at least, may be identifiable (Rosenbluth, 1981b). The purpose of this chapter will be to review the relevant structural and physiological studies and to discuss the correlations that have been proposed. In addition, studies of developing, degenerating, and myelin-deficient nerve fibers will be discussed in relation to the ways in which interactions between axons and glia can influence the structures of the respective cells.

Jack Rosenbluth • Department of Physiology, New York University School of Medicine, New York, New York 10016. Research reported in this chapter was supported in part by NIH grant #NS07495 and a grant from the Muscular Dystrophy Association.

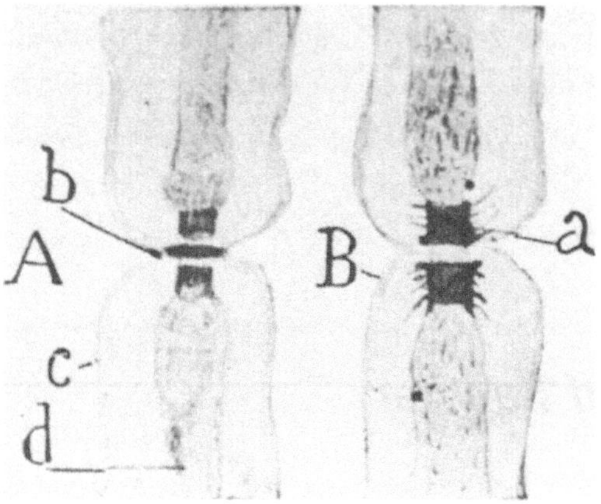

Figure 1. Drawing of silver-stained nodes showing impregnation of paranodal region and nodal "gap" substance (left) and "spinous bracelet" in paranodal region (right). (From Cajal, 1926).

II. HISTORICAL STUDIES

Early studies of nerve fibers stained with osmic acid showed nodes of Ranvier merely as periodic unstained gaps in the myelin sheath, while silver stains, in contrast, impregnate the nodal region selectively, producing the appearance of a "cementing ring" surrounding the axis cylinder. In some preparations (Fig. 1), the adjacent paranodal regions are stained as well and may exhibit transverse striations or a barbed pattern referred to as a "spinous bracelet" (Ranvier, 1875; Nageotte, 1922; Cajal, 1926).

These observations were quite mysterious until electron microscopic studies were carried

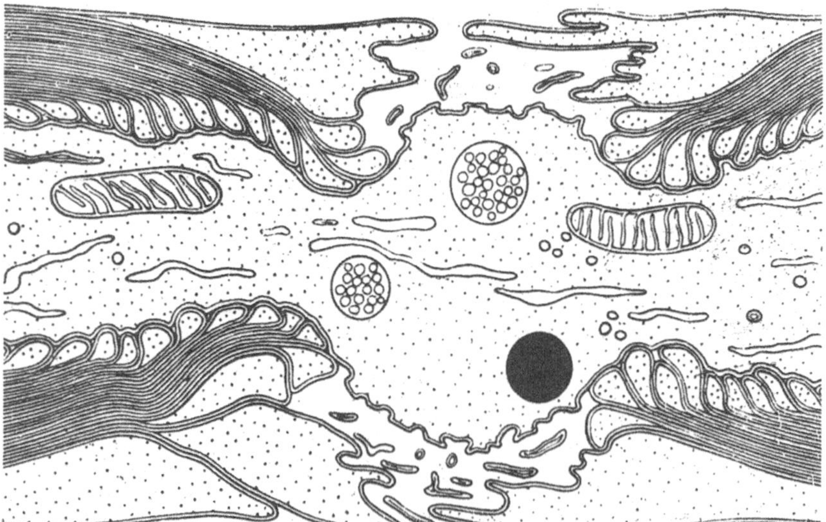

Figure 2. Diagram of nodal and paranodal regions as seen in thin sections. Myelin lamellae terminate in "loops," each of which indents the axolemma producing a scalloped paranodal region. The nodal axolemma is covered by Schwann cell microvilli. No specialization of either the nodal or paranodal axolemma is indicated. (Modified from Robertson, 1959).

out in the 1950s. The spiral wrapping hypothesis of myelinogenesis, formulated by Geren (1954) and subsequently applied to the paranodal region of the sheath by Robertson (1959), provides a remarkably simple explanation for the form and means of generation of the very intricate structure of the paranodal myelin sheath in peripheral nerves and represents one of the first major scientific accomplishments that grew out of the use of electron microscopy in anatomical studies of thin tissue sections. Later studies showed comparable findings in the central nervous system as well (Bunge, 1968).

The transverse paranodal striations and spinous bracelet in particular undoubtedly correspond to the edge of the Schwann cell or oligodendrocyte as it winds helically around the axon at either end of each myelin segment. Indeed, electron micrographs of colloidal lanthanum distribution in the paranodal region, used to demonstrate extracellular pathways between the cell processes (Hirano and Dembitzer, 1969), yield images strikingly similar to the spinous bracelet (Fig. 6).

Robertson's summary diagram of the node (Fig.2) illustrates the major ultrastructural features of both nodal and paranodal regions. Each myelin lamella terminates in a small cytoplasm-containing expansion ("terminal loop"), which is very closely applied to the axolemma and indents it. These loops represent multiple sections through the edge of the myelinating cell as it winds around the axon and are thus continuous with each other and with the outermost layer of the sheath. The latter, in the case of peripheral myelin, forms innumerable microvillous processes that project into the nodal gap yet allow communication between the nodal axolemma and the extracellular space.

III. RECENT ULTRASTRUCTURAL STUDIES

A. Perinodal Region

Since Robertson's description, several new structural details have been added. The perinodal microvilli may form closely spaced paracrystalline arrays (Landon and Hall, 1976). A basal lamina surrounds the Schwann cells and bridges from one internodal segment to the next across the node. The extracellular space surrounding the nodal axolemma is filled with a fine filamentous matrix (Fig. 3) which is quite unlike the collagenous connective tissue external to the basal lamina and probably corresponds to the argentophilic "cementing ring." This matrix contains a polyanionic proteoglycan (Landon and Langley, 1971; Langley, 1979) which may be of considerable functional significance in view of its cation binding properties. The presence of this matrix should be kept in mind also because its binding characteristics could cause misleading results to be obtained in histochemical studies since precipitates from various sources tend to collect nonspecifically in this region. Some controversy about the presence of acetyl cholinesterase at the node of Ranvier already exists and probably arises from nonspecific binding of reaction product. Thus, the interpretation of histochemical and immunohistochemical studies of the node must be interpreted very cautiously (see Landon and Hall, 1976, for discussion).

In the central nervous system, oligodendrocytes do not form perinodal microvilli, and nodes here are often surrounded merely by a widened extracellular space which, however, also contains a fine matrix material whose function is presumably equivalent to that of the peripheral perinodal matrix. In tracts containing large fibers, astrocytes may form perinodal microvilli (Hildebrand, 1971), but these are not a constant feature of central nodes as they are in the case of peripheral fibers.

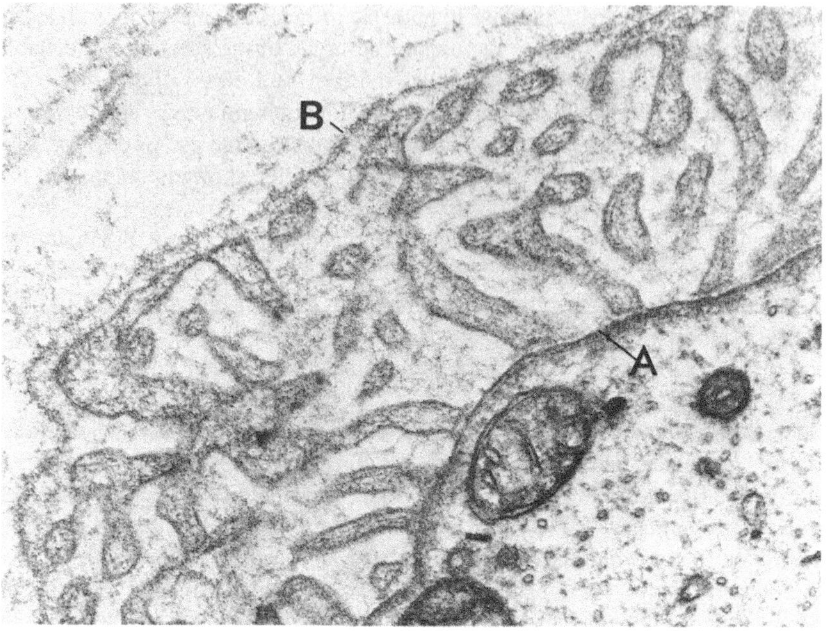

Figure 3. Transverse section through node of Ranvier. The perinodal space surrounding the axolemma (A) contains Schwann cell microvilli in a filamentous (presumably proteoglycan) matrix and is separated from the endoneurial space by a basal lamina (B).

B. Paranodal Junction

Probably the most conspicuous structure that has emerged from examination of the paranodal region by electron microscopy is the axoglial or axon–Schwann cell junction. Studies by Andres (1965) and Bargmann and Lindner (1964) originally demonstrated the distinctive nature of this junction between terminal Schwann cell loops and the axolemma of peripheral fibers, and an equivalent junction was later described in the central nervous system by Peters (1966). In either location the junction is characterized by a gap of approximately 20–40 Å bridged by periodically disposed densities (Fig. 4).

Approximately 3–6 of these densities are associated with each terminal loop. In tangential sections it is clear that these intercellular densities represent ridges of considerable extent which run approximately transversely, at least in mammals. Thus, they are referred to as "transverse bands" although they are, in fact, slightly oblique in most cases.

The degree of obliquity depends upon species. In mammals, the bands form a small angle and thus are virtually transverse, but in frogs they form an angle of ~30° with the edge of each junctional strip. Although the magnitude of the angle may vary slightly from place to place, the sign of the angle is always the same (Tao-Cheng and Rosenbluth, 1980b). As described above, the Schwann cell or oligodendroglial processes that form this junction represent the cytoplasm-containing edge of the myelin sheath continuous ultimately with the perinuclear portion of the cell. Since each myelin lamella is slightly longer than the one underneath, it overhangs the latter and is therefore able to reach and make contact with the axolemma. Thus, it is not only the innermost layer of the myelin sheath that contacts the axon; rather, each layer forms an intimate junction with the axon in the paranodal region, except in the case of thick myelin sheaths where the winding pattern in the paranodal region is not quite so regular and some of the paranodal loops do not reach all the way to the axolemma (Hildebrand, 1971).

Terminal loops often, but not always, form tight junctions with each other and are often interconnected by stacks of desmosomes as well which presumably prevent circumferential slippage of successive lamellae and thereby maintain the diameter of the paranodal collar. The latter, in turn, probably accounts for the marked restriction of axonal diameter in the paranodal region so conspicuous in large caliber fibers where the diameter of the axon in the paranodal region may be less than half that in the internode.

The cytoplasm of the loops tends to be rather dense, and it often contains conspicuous microtubules, which are circumferential in orientation. Membrane-limited vesicles and small dense inclusions are also occasionally seen. Frequently, especially in peripheral nerves. the membrane facing the axolemma has associated with it a flattened membranous sac (Rosenbluth, 1979a) which is also circumferential in orientation (Fig. 5). In appearance, this structure is reminiscent of the "subsurface cisterns" that occur in nerve cells (Rosenbluth, 1962) and of the flattened sacs of sarcoplasmic reticulum at "dyads" and "triads" in striated muscle cells in that it is closely applied to the junctional plasma membrane of the terminal loop and sometimes appears to be connected to it by faint periodic densities. Occasionally, flattened membranous sacs adjoin the paranodal axolemma as well, but these are not seen consistently.

Ribosomes are present on the nonjunctional surface of the terminal loop sacs (Fig. 5), thus identifying them as components of the granular endoplasmic reticulum presumably involved in protein synthesis. These flattened sacs are particularly conspicuous in developing and regenerating peripheral myelin, and they occur most consistently in the loops derived from the outer layers of the myelin sheath, thus corresponding in distribution to that of the transverse bands. In circumstances where axon–Schwann cell junctions are formed in aberrant locations the sacs occur precisely in the region where transverse bands can be detected (Fig. 16).

The function of these sacs is unknown but their precise association with the transverse bands, especially during development, suggests that they may be involved in the synthesis of the latter. An alternative possibility is that they play a role in some form of signaling between the Schwann cell surface and its interior comparable to that believed to occur at skeletal muscle triads and at neuronal subsurface cisterns (Rosenbluth, 1962; Henkart *et al.*, 1976), both of which are thought to be involved in calcium exchange (Henkart *et al.*, 1976). Axons are well known to exert a continuing trophic influence on the cells that myelinate them, and although the site and nature of the influence are not yet known, the intimacy and extent of the association between the respective cell membranes in the paranodal region make this an appropriate candidate for such an interactive site.

In addition, there has long been speculation about the role of glia in regulating the ionic milieu of neurons. In the case of the node of Ranvier, the perinodal microvilli and associated mitochondria, together with the adjacent paranodal junction, could be part of a complex apparatus for maintaining a suitable ionic milieu at the node by, for example, replenishing sodium (Landon and Hall, 1976) or recapturing potassium released during repetitive axonal activity. Indeed, the basic architecture of myelin-forming Schwann cells corresponds to that of epithelial cells involved in ion transport, which have an apical surface bearing microvilli separated from the lateral and basal surfaces by the junctional complex and which maintain an ionic gradient between these respective regions.

The remarkably narrow space between the junctional Schwann cell membrane and the axolemma has led to the assumption that the junction acts as a seal isolating the perinodal extracellular space from the periaxonal extracellular space in the internodal region. However,

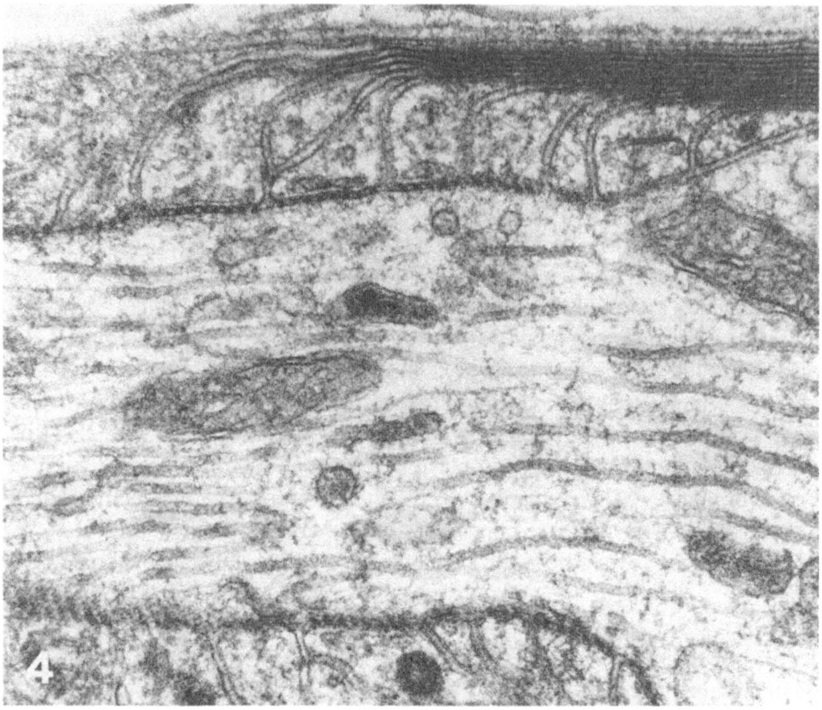

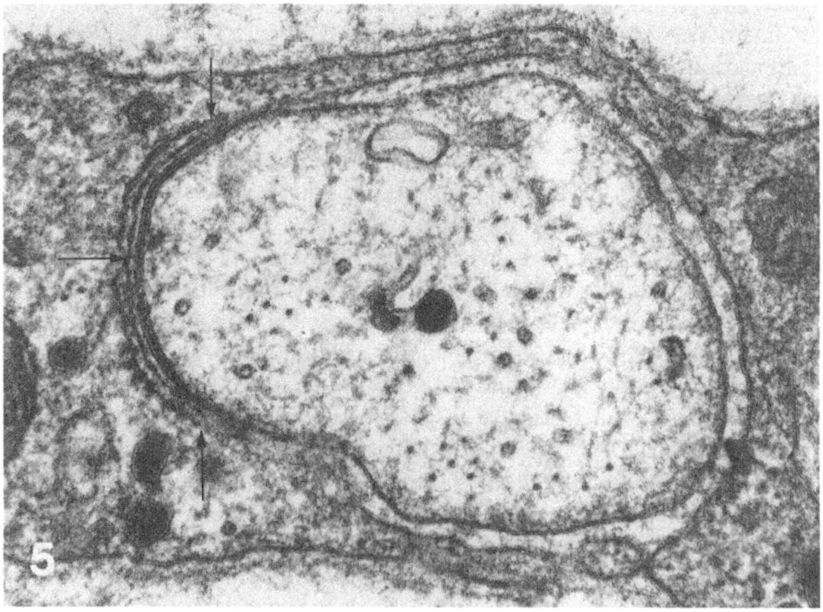

Figure 4. Longitudinal section through the paranodal axon–Schwann cell junction showing terminal loops of myelin adjoining the axolemma closely. The gap between the membranes is approximately 20–40 Å wide and is interrupted by periodic densities cut normally (top) and tangentially (lower left). (From Rosenbluth, 1981b).

Figure 5. Transition between nodal and paranodal regions. The right half of this cross-sectioned axon is separated from the surrounding Schwann cell by a wide (nodal) interspace. At the left the Schwann cell is very closely applied to the axon (paranodal junction) and it contains an elongated flattened sac (arrows) adjacent to the junctional Schwann cell membrane.

tracer studies have been carried out demonstrating that materials of molecular weight as high as ~2000 are able to penetrate into this space. Feder *et al.* (1969) injected microperoxidase into the central nervous system and were able to show that this material could be demonstrated between the glial and axonal membranes in the paranodal regions. Equivalent studies have also been carried out with colloidal lanthanum (Fig. 6) both centrally (Hirano and Dembitzer, 1969) and peripherally (Revel and Hamilton, 1969) showing penetration of this material as well. Presumably, the transverse bands that cross the ~30 Å space between the respective membranes do not prevent the movement of these tracers perhaps because the bands have an oblique course across the junctional strips, thus leaving a pathway from one strip to the next that gradually permits penetration into the entire paranodal region. Although the junction is permeable over the rather long time course of the tracer experiments, the impedance of this narrow pathway may be sufficiently high so that over the very short time course of an action potential the junction is virtually sealed to current flow.

C. *Role in Conduction*

The architecture of this unique junction is such as to suggest that it is indeed designed to play a dual role. Presumably, if the only function of this junction were to seal off the perinodal space from the periaxonal space, then a tight junction would be a more appropriate structure in this location. An elongated, partially open pathway, on the other hand, may serve as a low-pass filter that effectively blocks passage of high frequency currents associated with the action potential to the internode but permit passage of lower frequency activity or direct currents in either direction between node and internode.

A possible physiological role for a paranodal junction of this type is suggested by Barrett and Barrett (1982), who observed depolarizing after-potentials in myelinated fibers, ascribed to discharge of the internodal capacity through the node by way of a resistive pathway. Partial depolarization of the nodal membrane by this means could enhance conduction velocity and reliability of conduction of high-frequency repetitive activity through regions of low safety factor such as branch points or unmyelinated nerve terminals.

Whether fully formed paranodal junctions are essential for saltatory conduction is uncertain. Presumably, before their formation during development or remyelination some shunting of current occurs through the paranodal region. However, since in both cases internodal length is relatively short, even a relatively thin, incompact sheath, lacking mature paranodal junctions, may still be able to support saltatory conduction, albeit with reduced safety factor and conduction velocity. In demyelination, computer simulations suggest that

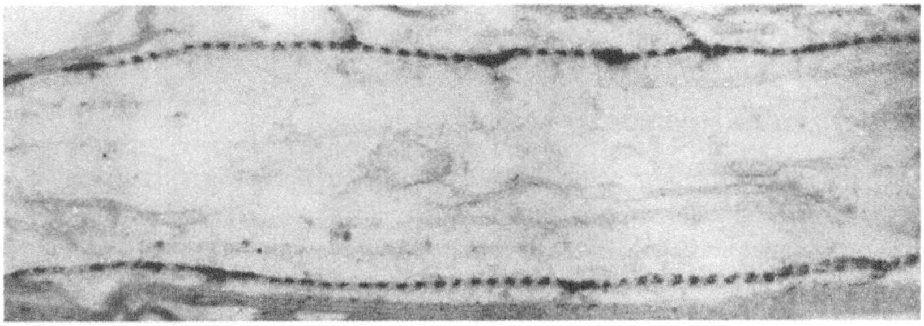

Figure 6. Colloidal lanthanum used as an extracellular tracer fills the paranodal junctional cleft between the "transverse bands" and also fills the triangular extracellular space between adjoining terminal loops and the axolemma. (From Hirano and Dembitzer, 1969).

saltatory conduction may persist in the face of considerable thinning of the myelin sheath (Rasminsky, 1978), but the effect of opening paranodal junctions is not known.

IV. FREEZE–FRACTURE STUDIES

A. Background

The introduction of the freeze–fracture method opened up new possibilities for examination of extensive areas of cell membranes inasmuch as the fracture plane tends to follow the lipid bilayer of biological membranes selectively, splitting it into inner and outer leaflets which can be viewed *en face* (Pinto da Silva and Branton, 1970). The interior of the membrane artificially exposed to view in this way usually appears as a continuous surface containing variable numbers of projections referred to as "intramembranous (or intramembrane) particles," "freeze–fracture particles," or simply "particles." These particles presumably represent integral membrane proteins which, rather than fracturing in the same plane as the lipid bilayer, tend to survive intact and be pulled out of one leaflet of the membrane remaining attached to the other (Branton, 1971; Pinto da Silva *et al.*, 1981). Although entities other than proteins may also appear as particles (Robertson, 1981), the majority ordinarily seen probably represent proteins of substantial molecular weight which traverse the lipid bilayer. Whether the particles correspond to the proteins alone or to some lipoprotein complex is uncertain.

In most membranes, the large majority of particles are seen on the P (protoplasmic) fracture face, i.e., in the membrane leaflet adjacent to the cytoplasm, while relatively few appear in the external leaflet (E fracture face) of the membrane. This unequal distribution may reflect attachment of many integral membrane proteins to intracellular cytoskeletal elements. Particles that do appear in the E face may represent entities whose attachment to structures outside the cell, or associated with its external surface, is relatively stronger. Some structures attached externally, e.g., gap junction particles, which are presumably bound to their counterparts across the gap, nevertheless remain with the P face of the membrane during freeze–fracturing (Peracchia, 1977), perhaps indicating even stronger binding to cytoplasmic components or, alternatively, fracturing through "weak links" within the protein molecule itself.

Study of membrane fracture faces provides a view of the interior of membranes that could not be seen by any other means and from which significant information about regional differentiation and integral membrane protein distribution may be obtained. The recent development of specific labeling methods for identifying particles (Pinto da Silva *et al.*, 1981) promises to increase the usefulness of freeze–fracture studies further.

B. Axolemmal Structure

When the freeze–fracture method was first applied to nerve fibers, attention was focused primarily on the paranodal axoglial junction (Livingston *et al.*, 1973; Dermietzel, 1974; Schnapp and Mugnaini, 1975). Extensive analysis showed the junctional axolemma to have unique structural characteristics that did not permit it to be classified simply as a gap junction, tight junction, or adherent junction. Moreover, its appearance has been found to vary depending on preparative method, and its intrinsic structure in the native state still continues to be uncertain (Wiley and Ellisman, 1980; Rosenbluth, 1981b). Perhaps because of the intriguing nature of this junction, little attention was paid to the nodal axolemma initially.

A study directed at the node itself, however, demonstrated that the axolemma here also has a distinctive structure (Rosenbluth, 1976). The internodal axolemma, like most membranes, has a high concentration of P face particles, and very few E face particles, but the nodal axolemma has a consistently high concentration of E face particles as well (Figs. 8 and 9). Calculation of their concentration yielded a mean value of ~1200/μm^2.

Subsequent studies have shown that such particle concentrations occur in the E face of central (Fig. 8) and peripheral (Fig. 9) nodes, in large as well as small fibers, and in mammals as well as cold-blooded vertebrates (Kristol *et al.*, 1978; Rosenbluth, 1979a; Tao-Cheng and Rosenbluth, 1980b). The concentration of particles is remarkably consistent with the originally published figure of ~1200/μm^2, averaging approximately 1000 to 1500/μm^2 in different locations. An interesting exception occurs in the electric organ of the knifefish, which consists of modified nerve fibers containing both excitable and inexcitable nodes. Although both have high P face particle densities, only the excitable nodes exhibit high concentrations of E face particles (Kristol *et al.*, 1977).

Nodal particles vary in size but on average are larger than most intramembranous particles and some appear irregular in shape as well. Use of the rotary shadowing technique (Margaritis *et al.*, 1977), however, produces smaller particles of more uniform shape (Fig. 7). Attempts have been made to identify subgroups of large (Kristol *et al.*, 1978) or large and tall (Tao-Cheng and Rosenbluth, 1980b) particles among the general population of nodal particles, but since the spectrum of particle sizes in both P and E fracture faces is unimodal, any such subdivision is necessarily arbitrary. Moreover, freeze–fracture replicas prepared from artificial membranes composed of a single purified protein in a purified phospholipid may also show marked variation in particle dimensions (Hong and Hubbel, 1972). Thus, particles of different sizes at the node need not represent different entities.

The simplest view of the nodal membrane specialization based on present data is that it consists of a population of intramembranous particles numbering ~1000–1500/μm^2 that appear in the E fracture face of that membrane compared with ~100 such particles/μm^2 in the internodal axolemma (Kristol *et al.*, 1978; Tao-Cheng and Rosenbluth, 1980b). It is the unusual concentration of E face particles that most clearly characterizes the nodal axolemma and distinguishes it from the internodal membrane and, presumably, underlies some of the physiological properties of the nodal membrane as well. The possibility that the significant nodal particles are the larger ones, which appear primarily in the E face but to some extent in the P face as well, has been considered (Tao-Cheng and Rosenbluth, 1980b), although in other instances virtually all particles of a single species appear in the same fracture face (Peracchia, 1977). Additional possibilities are also discussed in Tao-Cheng and Rosenbluth (1980b).

C. *Correlation with Physiological Data*

High concentrations of sodium channels in the nodal axolemma have been inferred from current fluctuation analyses, gating current measurements, and saxitoxin binding studies. Of these, the first method yields the lowest values. Conti *et al.* (1976) originally reported a nodal sodium channel density of ~2000/μm^2. More recently, Sigworth's (1980) data indicate a concentration of only ~1000/μm^2, although this could be an underestimate if the number of available channels were reduced by ultraslow inactivation (Neumcke *et al.*, 1979) or if the nodal area were smaller than that assumed.

Gating current studies originally led to an estimate of ~5000 channels/μm^2 in the frog (Nonner *et al.*, 1975), but this figure may include dipoles other than sodium channels (Kimura and Meves, 1979) and may thus more properly represent an upper limit rather than an

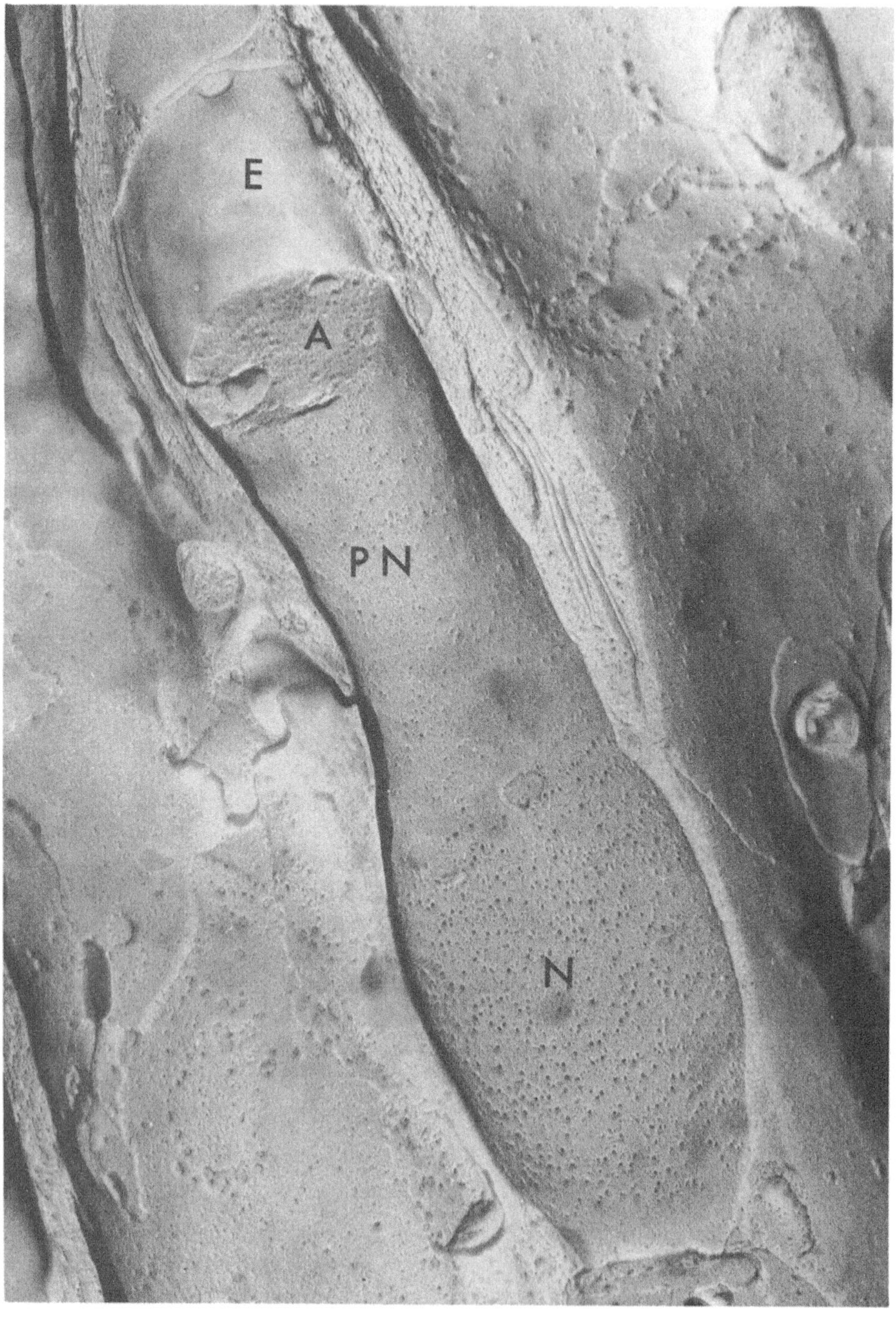

Figure 7. Nodal and paranodal axolemma of a small fiber from rat spinal cord as seen in a rotary shadowed preparation (P face). The nodal membrane (N) contains numerous tall (heavily-shadowed) particles. The particles in the paranodal membrane (PN) are fewer in number and not as tall. They are randomly distributed and exhibit no apparent substructure or characteristic shape. A, cross-fractured axoplasm. E, axolemmal E face. Myelin lamellae are visible on both sides of the paranodal axon.

Figure 8. Nodal and paranodal axolemma from frog spinal cord (E face). The paranodal membrane, characterized by broad stripes, exhibits a diagonal pattern, but contains very few particles. The nodal membrane (N) has no such pattern but contains numerous large particles. Inset: "Lakes" of particles (L) in the paranodal axolemma. (From Rosenbluth, 1976).

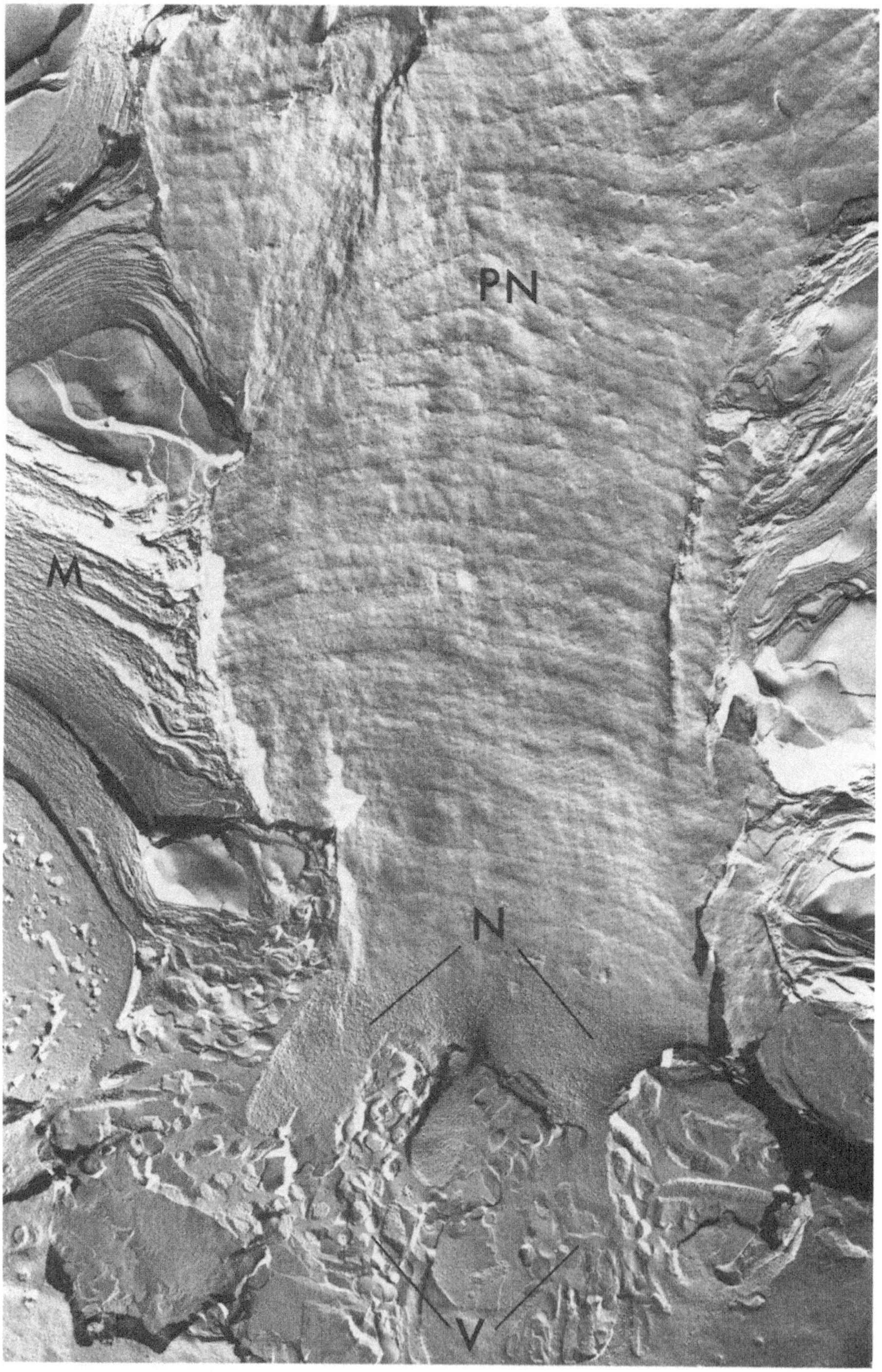

Figure 9. Nodal (N) and paranodal (PN) axolemma of a large fiber (E face). Approximately 40 paranodal indentations are visible. The myelin lamellae (M) approach the axolemma nearly perpendicularly. The nodal membrane containing innumerable particles undulates in and out of the fracture plane and is surrounded by closely spaced microvilli (V).

accurate estimate of channel density. A more recent study by Chiu (1980) reports only ~1600 channels/μm^2 by this method in the rabbit.

The highest estimate of sodium channel density is that based on saxitoxin binding studies from which a nodal concentration of ~12,000/μm^2 was derived (Ritchie and Rogart, 1977), but this figure probably also represents an overestimate growing out of the assumption that only the nodal axolemma is exposed to the toxin. As discussed above, the paranodal junction is not tight, and it undoubtedly admits small molecules such as saxitoxin into paranodal and even internodal regions where extranodal sodium channels presumably also occur. Indeed, the finding that virtually no additional channels were labeled by saxitoxin in fibers stripped of myelin suggests that paranodal and internodal channels are labeled in the intact fibers as well.

For these reasons, the higher estimates of nodal sodium channel density are likely to be overestimates and the lower estimates to represent more accurate reflections of channel concentration. The latter correlate well with nodal E face particle concentration in freeze–fracture replicas and suggest that each such particle might represent one sodium channel (Rosenbluth, 1976, 1981b; Tao-Cheng and Rosenbluth, 1980b). This view is consistent also with biochemical studies on tetrodotoxin-binding component isolated from excitable tissues, which have demonstrated a Stokes radius of ~95Å (Agnew *et al.*, 1980) and a molecular weight of ~2–3 × 10^5 (Barchi *et al.*, 1981). Biophysical studies indicate a molecular weight in the same range (Levinson and Ellory, 1973). Thus, far from being a mere pore in the membrane, the sodium channel is apparently more than adequate in size to be seen in freeze–fracture replicas. It appears to be a protein molecule of some complexity which presumably includes not only the pore for sodium flux, but also the voltage sensing and gating mechanisms.

In short, with respect to location, concentration, and size, there is a good correlation between nodal E face particles and voltage sensitive sodium channels. The freeze–fracture method therefore provides a possible means for following the concentration and distribution of these channels at individual nodes during development, in pathological conditions, or under experimental circumstances. No other method is capable of providing data of such a detailed nature on individual nodes.

D. *Extranodal Particle Aggregates*

Some large particles of the type found in the E face of the nodal axolemma are always present in the paranodal axolemma lined up in the thin, nonjunctional "grooves" between strips of junctional membrane. These are relatively few in number (Fig. 8). Further from the node, however, irregularities often appear in the paranodal windings (Fig. 10), and here large aggregates of these particles may occur in "lakes" (Figs. 8 and 10). In addition, large numbers of such particles may accumulate at the junction between the paranode and internode (Fig. 11). The total number of these paranodal and juxtaparanodal particles may considerably exceed the number at the node itself, but in contrast to the node, which *invariably* has a high density of E face particles, the paranodal and juxtaparanodal membranes do not exhibit large aggregates of particles consistently, and fibers have been seen both with them and without. The frequency of occurrence of large extranodal aggregates has not yet been established.

These extranodal E face particles resemble nodal particles in all respects and thus could also represent sodium channels. Since the axon lacks the organelles for protein synthesis, these could merely be en route to or from the nodes, as part of a normal turnover process (Rosenbluth, 1976). Presumably, such "concealed" channels would not contribute to the

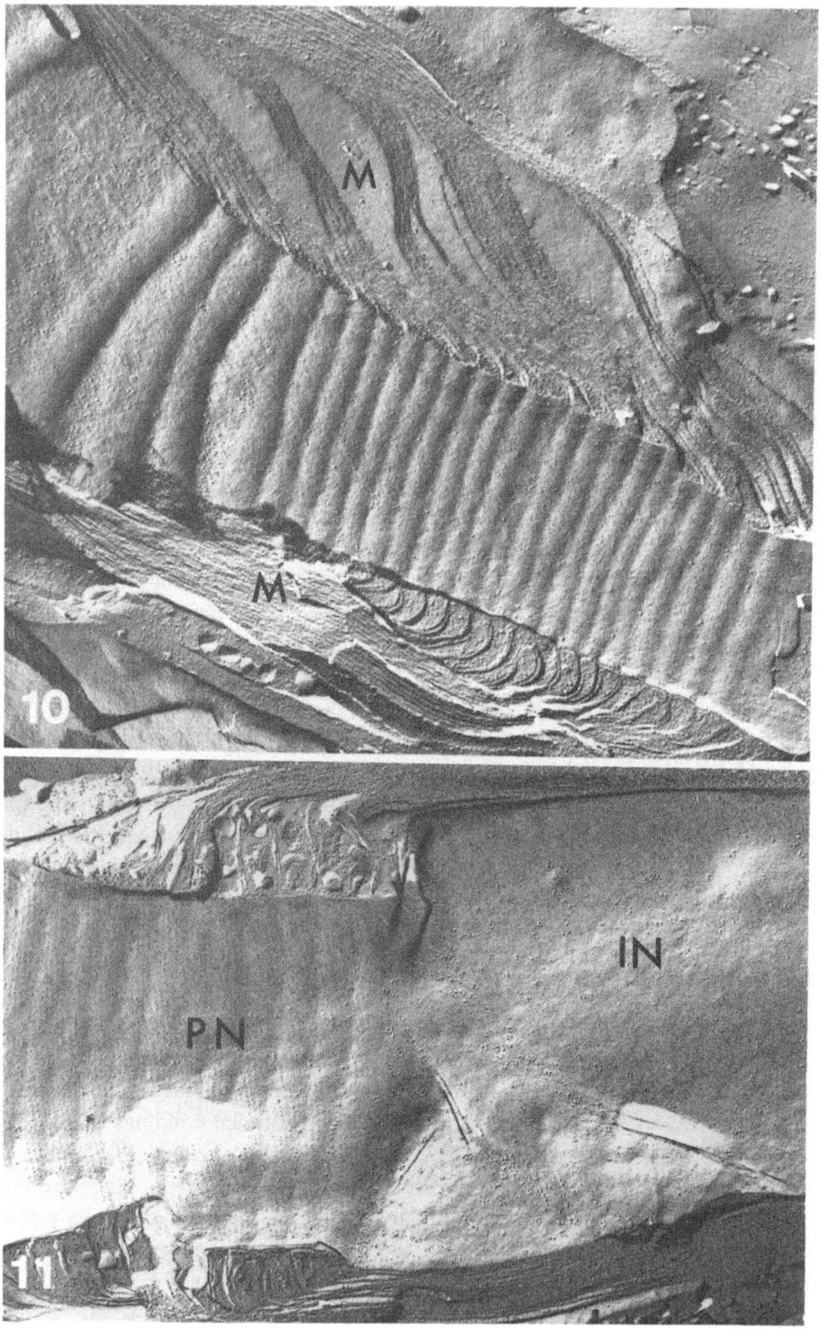

Figure 10. Paranodal axolemma showing irregularities in the winding pattern of the innermost myelin lamellae (those furthest from the node) creating "lakes" containing variable numbers of particles (left). M, myelin.

Figure 11. Juxtaparanodal axolemma (E face). Nodal-type particles are visible in lines between paranodal (PN) indentations and in the internodal axolemma (IN) immediately adjacent to the paranodal region.

nodal action potential because of the intervening paranodal junction but might be "seen" if the junction opened significantly as in paranodal demyelination. The internodal axolemma, exclusive of the juxtaparanodal region, contains a consistently low concentration (~100/μm^2) of such E face particles (Kristol *et al.*, 1978; Tao-Cheng and Rosenbluth, 1980b) even below that in unmyelinated axons (Black *et al.*, 1981), which are known to have a low sodium channel density (Ritchie *et al.*, 1976).

Recent physiological studies indicate that voltage sensitive potassium channels occur in the paranodal and internodal axolemma but are absent at the node of Ranvier in mammalian myelinated fibers (Kocsis and Waxman, 1980; Sherratt *et al.*, 1980; Chiu and Ritchie, 1982) and some amphibian fibers as well (Smith and Schauf, 1981). This localization is clearly inconsistent with the distribution of the axolemmal E face particles and suggests that if potassium channels are visible in the membrane fracture faces, they are probably included among the P face particles. Attempts to identify them will have to await further information about details of their concentration, distribution, size, and shape.

E. Conclusion I

The principal conclusion based on the studies reviewed thus far is that in myelinated fibers the axolemma at the node of Ranvier is characterized by a conspicuous population of relatively large E face intramembranous particles whose concentration corresponds to the density of nodal sodium channels estimated by physiological methods. In the internodal axolemma such particles are extremely sparse, although large aggregates sometimes occur in paranodal "lakes" and in the juxtaparanodal axolemma. These latter regions are separated from the node by the paranodal junction, however, and presumably do not contribute to the ionic events associated with the action potential under normal conditions.

V. FACTORS THAT INFLUENCE AXOLEMMAL PARTICLE DISTRIBUTION

In view of the potential importance of the axolemmal E face intramembranous particles, it is of interest to consider what influences, either intrinsic or extrinsic, might affect their disposition along the axolemma and their concentration in specific regions. Since such aggregates occur only in myelinated axons, one obvious possibility is that the myelin sheath itself is in some way responsible for particle localization. This is consistent also with the fact that such particle aggregates appear not to occur in the axons of certain mutant animals that are virtually devoid of myelin (Rosenbluth, 1979b). Studies of myelinogenesis support this view as well.

A. Developmental Studies

Axons examined at stages before myelin formation has begun, when the axons travel in fascicles without intervening Schwann or glial cell processes, demonstrate no evidence for any of the nodal or paranodal membrane specializations characteristic of adult myelinated fibers (Bray and Oldfield, 1980; Tao-Cheng and Rosenbluth, 1980a, 1982; Waxman *et al.*, 1982). Rare examples of particle aggregates begin to appear about the time axons are enveloped individually by Schwann cells and occur more frequently, after wrapping has begun (Fig. 12), adjacent to the indentations marking the edges of the developing sheaths

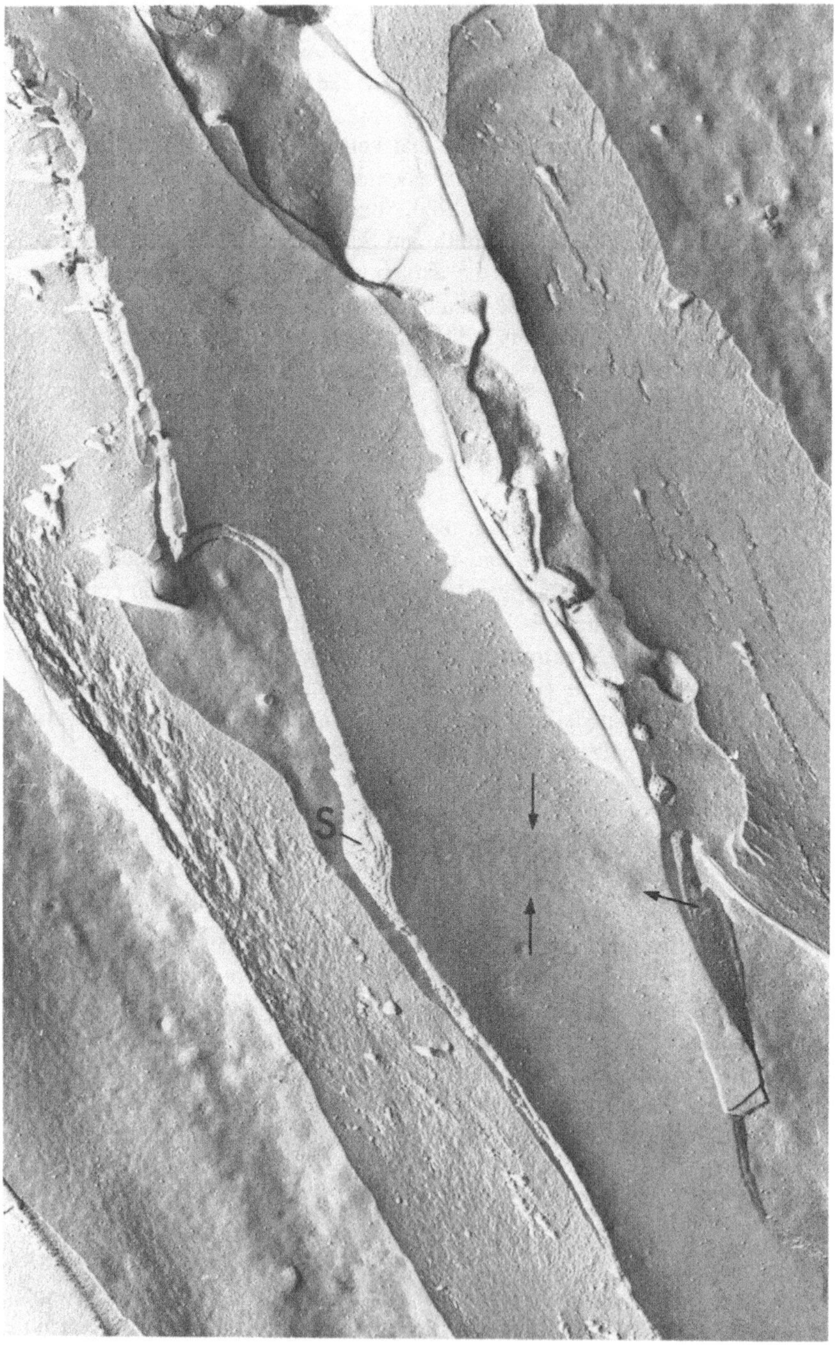

Figure 12. Early stage of myelin formation. An axonal E face is shown with some particle aggregation. The axon is ensheathed by a Schwann cell, the cross-fractured edge of which is shown at S. This edge indents the axon at arrows. No particles are present over the indentation. (From Tao-Cheng and Rosenbluth, 1981a).

(Tao-Cheng and Rosenbluth, 1980a, 1982; Wiley-Livingston and Ellisman, 1980). At these early stages, the aggregates have a lower particle density than that found at adult nodes, and they are spread over a greater length of the axon. As myelin formation proceeds, the paranodal indentations, which are strikingly free of E face particles (Fig. 13), appear to restrict the distribution of the particles to a progressively smaller area.

Adult paranodal indentations, which also contain few large E face particles, often exhibit a "paracrystalline" appearance, especially after etching, that probably represents an embossed pattern imposed by the adjacent transverse bands (Rosenbluth, 1981b). During myelin formation this paracrystalline appearance is usually not seen, presumably because of immaturity of the transverse bands. Nevertheless, the corresponding indentations of the axolemma have equally few particles within them and thus also appear to exclude the characteristic nodal particles. In Fig. 13, for example, the paranodal region on one side of the forming node exhibits several indentations that extend across the full width of the axon and are virtually particle-free. Only occasional particles occur on the internodal side of this "barrier." On the opposite side of the node the paranodal region is less mature, and the indentations do not completely traverse the axolemma. Although the indented region of the axolemma is particle-free, particles extend through the "defect" in this forming paranodal junction. At very early stages the indentations by the ensheathing cell are shallow and easily overlooked but nevertheless also appear to exclude axolemmal E face particles (Fig. 12).

On the basis of such observations it appears that the nodal membrane specialization, characterized by E face particle aggregates, occurs initially in an immature and rather diffuse form, and over the period of myelin formation is shaped by the edges of the myelinating cells gradually until ultimately the particles are confined to a narrow annulus where they exist at high concentration in the fully myelinated fiber (Tao-Cheng and Rosenbluth, 1982).

The critical points in the developmental sequence seem to be the times at which axons become ensheathed individually and then become indented by the ensheathing cells rather than the times at which compaction of myelin lamallae begins or transverse bands form. These latter events occur at various times in different locations and species.

B. Demyelination

Experimental conditions that damage either the myelin sheath or the neuron or both may result in demyelination. In order to examine the significance of the myelin sheath in maintaining axolemmal differentiation, it is essential that an agent be used that selectively affects the sheath and spares the neuron. Otherwise changes seen could represent direct damage to the axolemma rather than secondary changes. Thus, Wallerian degeneration and heavy metal intoxications would be unsuitable.

An immune model based on the use of an antiserum to galactocerebroside (Brown *et al.*, 1980) appears to have the requisite selectivity, and examination of peripheral nerve fibers exposed to it (Rosenbluth *et al.*, 1981; Rosenbluth, 1981b) shows that paranodal axolemmal specializations are altered or lost early in demyelination (within 6 hr), but nodal E face particle accumulations are still present at this time (Fig. 14). After six days, however, characteristic nodal and paranodal specializations could not be found in the demyelinated areas. The rapidity of the changes in the paranodal region is consistent with the view that the paracrystalline pattern there is not intrinsic to that membrane, but rather represents an embossed pattern created by closely applied adjacent structures. Disappearance of the nodal particle aggregates subsequently suggests that these are normally confined by the adjacent paranodal membrane, whose characteristics in turn appear to depend on close association with myelinating cells. Thus, maintenance of the nodal specialization too appears to depend,

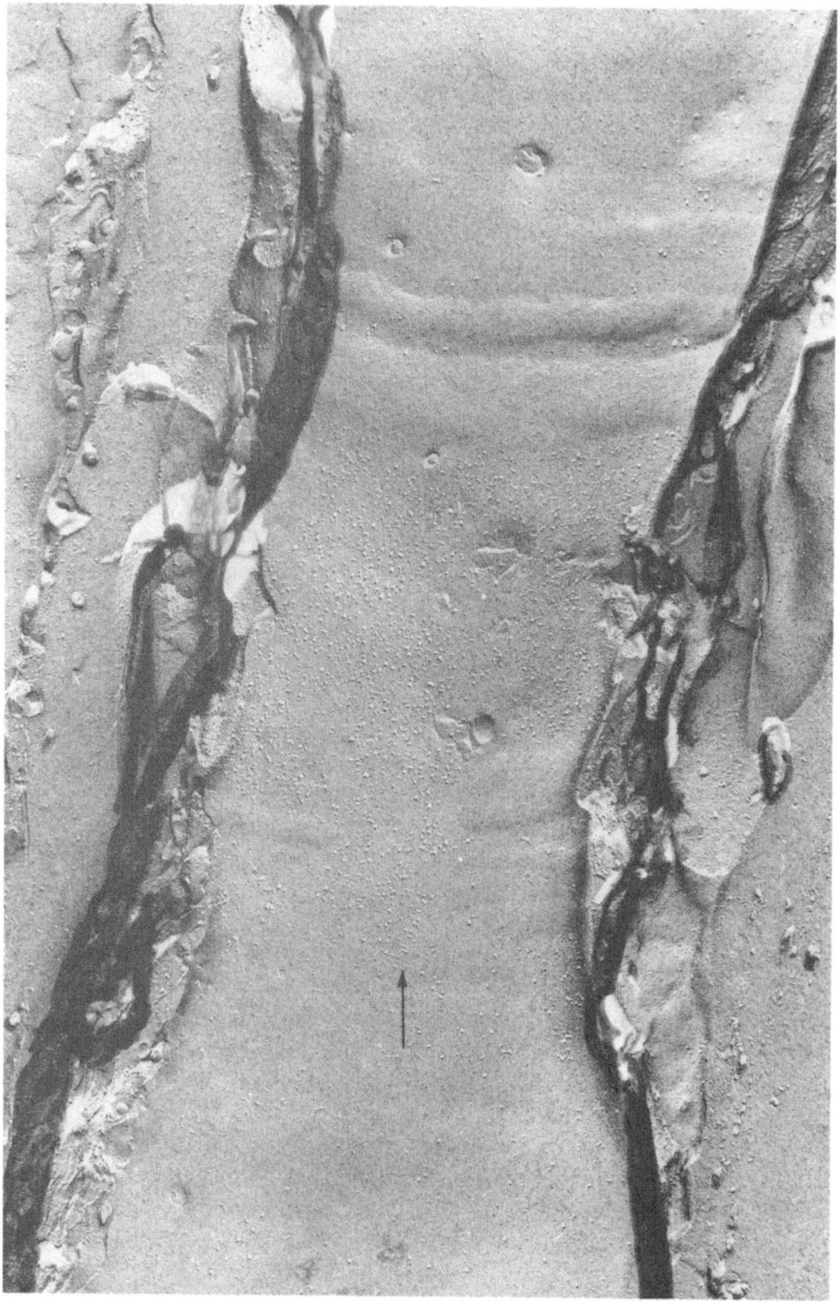

Figure 13. A later stage in myelin formation. The E face particle aggregate is now "confined" on one side by two prominent indentations of the ensheathing Schwann cell. The opposite paranodal area is less developed. Here the indentations do not extend all the way across the axon, and nodal particles extend through the resulting "defect" (arrow) into the paranodal region. (From Tao-Cheng and Rosenbluth, 1981a).

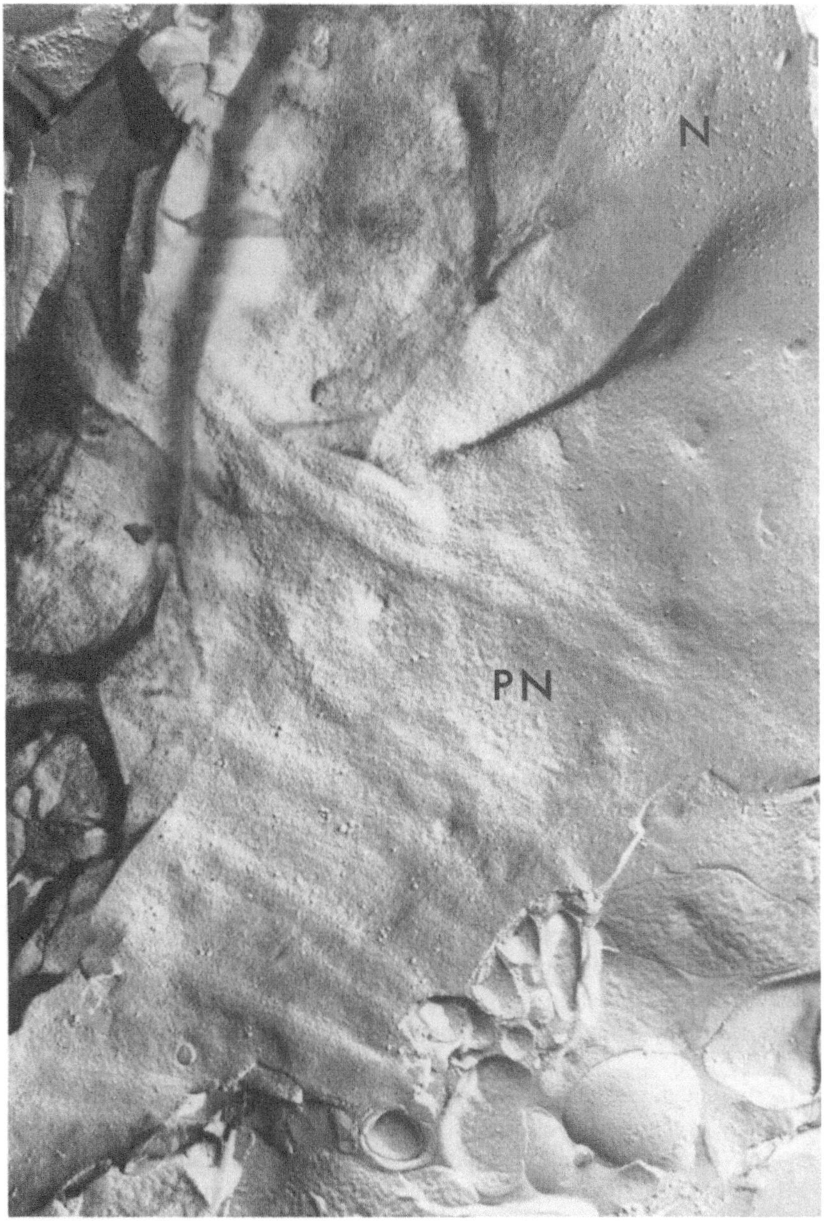

Figure 14. Paranodal region from a rat root, which had been exposed to antigalactocerebroside serum for approximately six hours (E face). The abnormal paranodal region (PN) is separated from the nodal region (N) by a dedifferentiated region of paranodal demyelination. The margin of the node is still well defined. (From Rosenbluth *et al.*, 1981).

albeit indirectly, on continuing close axoglial association, and dedifferentiation occurs when that association is disrupted (Rosenbluth *et al.*, 1981; Rosenbluth, 1981b). In effect, the developmental sequence is reversed upon demyelination.

Previous studies indicate that acute demyelination is accompanied by conduction block but continuous conduction may be seen in some demyelinated fibers after several days

(Bostock and Sears, 1978; Rasminsky, 1978; Sears, 1979). This finding could reflect dedifferentiation with redistribution of sodium channels from the highly concentrated regions in and around the nodes to the more barren internodal regions. Particle counts suggest that the redistribution of focally concentrated sodium channels could play a role in the development of continuous conduction but probably only in fibers with short internodes (Rosenbluth, 1981b). Calculation of the total number of particles present in and around the node in relation to the area of the internode indicates that fibers with an internodal length of about a millimeter would probably gain only a few particles per μm^2 if all particles from the nodal, paranodal, and juxtaparanodal regions on either side of this internode were to be uniformly redistributed. This is hardly a significant change. But if the internodal length were only 100 microns, the increase upon redistribution would be tenfold greater.

C. Myelin-Deficient Mutants

Studies of a variety of inbred mice that have genetically determined defects in myelin confirms the impression that the distribution of axolemmal particle aggregates depends upon the location of closely associated Schwann and glial cell processes (Rosenbluth, 1979b). In the case of the Jimpy mouse peripheral myelin appears normal, but central myelin is virtually absent. Correspondingly, both paranodal and nodal axolemmal specializations occur with normal frequency in peripheral nerves but have not yet been seen in replicas of the brain or spinal cord (Rosenbluth, 1979b). Thus, in the case of a dorsal root ganglion cell, for example, the peripheral process of the neuron apparently differentiates normally even though the central process within the spinal cord remains undifferentiated. Since these processes are derived from the same cell, and indeed are branches of a common axon, it is apparent that the neuron cannot independently bring about differentiation of its entire axon and that local environmental factors must play an important role in determining which sites are to become regionally specialized.

In the Dystrophic mutant as well, nodal and paranodal membrane specializations occur normally in peripheral nerve fibers but not in the amyelinated regions of the spinal roots characteristic of this mutant (Rosenbluth, 1979a). The occasional aberrant membrane specializations seen there (Rosenbluth, 1979a; Bray *et al.*, 1980) probably represent examples of amyelinated axons interacting with Schwann cells belonging to adjacent normally myelinated fibers, although they could also represent recently demyelinated axons (Rosenbluth, 1981b).

The Dystrophic mutant also offers an opportunity to examine "heminodes," which occur at the junction between the amyelinated segments of axons and the adjacent normally myelinated regions. Here too particle distribution is abnormal. Although some degree of axolemmal particle concentration may occur at these sites, typical annular aggregates of E face particles have not been seen, and in the example shown in Fig. 15 the axolemma adjacent to the paranode exhibits no discernable specialization at all. Thus, the presence of myelin at one side of a node only is insufficient to assure normal differentiation of the nodal axolemma.

In addition to the foregoing mutants, in which myelin in particular locations is virtually absent, a number of mutations are now known that result in bizarre myelination either in the peripheral or central nervous system or both. The "dysmyelination" in these animals affords an opportunity to examine the differentiation of the axolemma under conditions in which myelin, although present, is clearly abnormal in form.

One such mutant, referred to as Shiverer, is characterized by a gross diminution and abnormality in central nervous system myelin. Ultrastructural study of central white matter

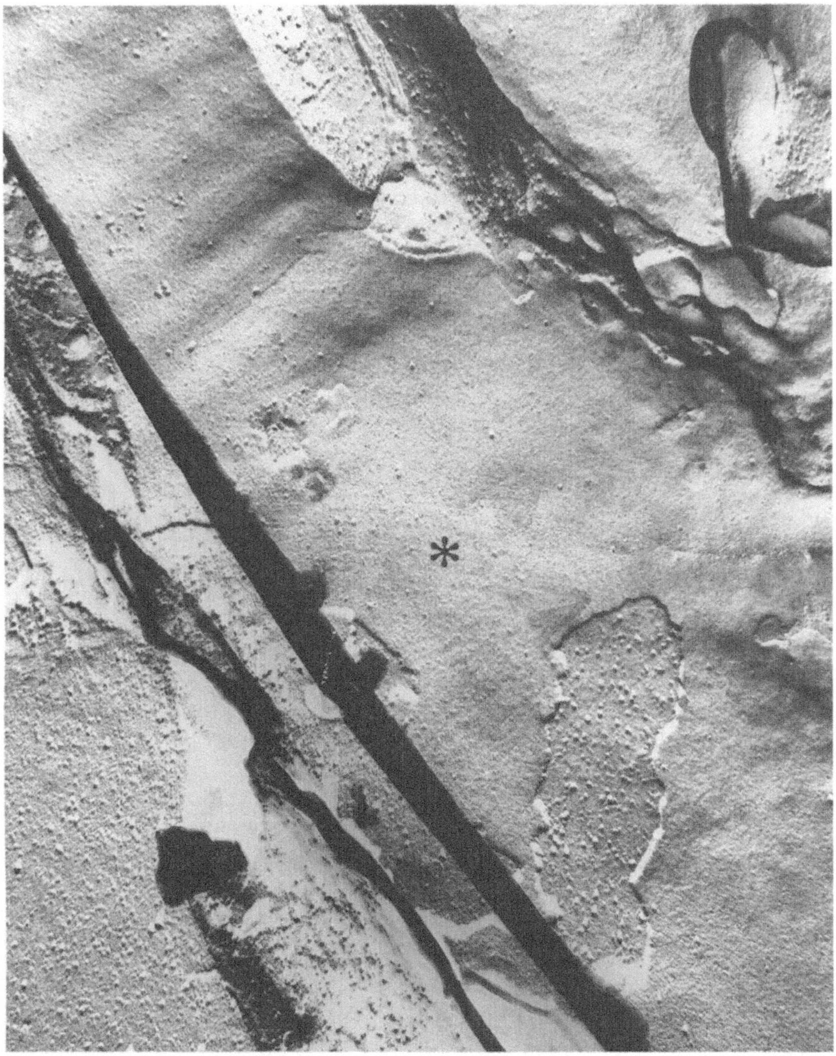

Figure 15. Heminode from Dystrophic mouse root. The paranodal axolemma exhibits a diagonal pattern which extends into the obliquely oriented indentation produced by the outermost terminal loop(*). The adjacent nodal axolemma contains no particle accumulation. (From Rosenbluth, 1981b).

in these animals (Rosenbluth, 1980) reveals inordinate numbers of axoglial junctions containing transverse bands (Fig. 17). These junctions are encountered not only in the vicinity of the paranodal region but at many sites of fortuitous apposition between oligodendrocytes and axons. Analysis of freeze–fracture replicas of Shiverer white matter (Rosenbluth, 1981a) shows a corresponding increase in the number of axolemmal E face patches having the paracrystalline pattern characteristic of such junctions (Fig. 18). These aberrant junctions are, however, bizarre in arrangement and location and tend not to be associated with E face particle aggregates except where they appear either to completely encircle the axolemma or to completely surround and isolate a patch of axolemma (Fig. 19). Only in these latter instances are substantial numbers of particles found as if "trapped" or "blocked" by the

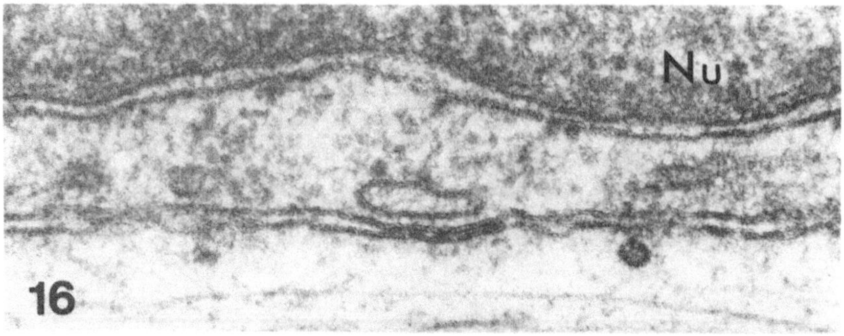

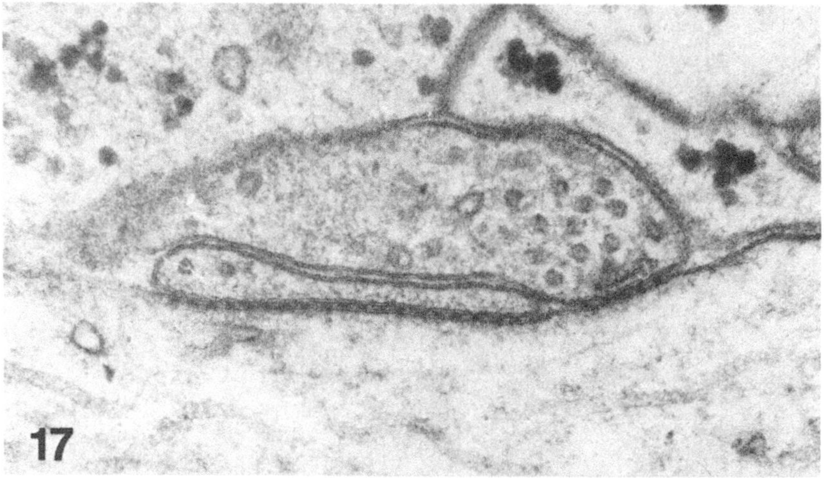

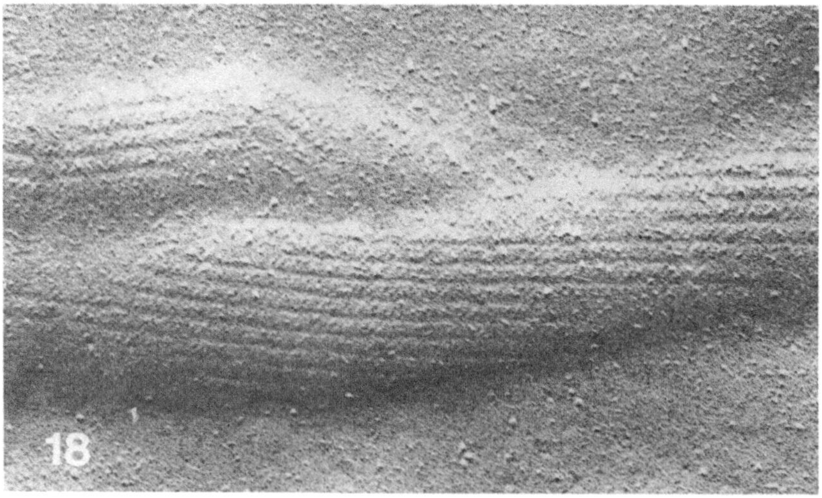

Figure 16. Aberrant axon–Schwann cell junction in Trembler mouse peripheral nerve showing typical characteristics including a narrow intercellular gap containing periodic densities and a flattened sac in the immediately adjacent Schwann cell cytoplasm. Nu, Schwann cell nucleus.

Figure 17. Aberrant axoglial junction in Shiverer mouse brain. Transverse bands are visible, but no myelin. (From Rosenbluth, 1980).

Figure 18. Glial indentations in the E face of a Shiverer axolemma. The paracrystalline structure is typical, but the indentations are oriented longitudinally with respect to the axis of the axon and have few particles associated with them. (From Rosenbluth, 1981a).

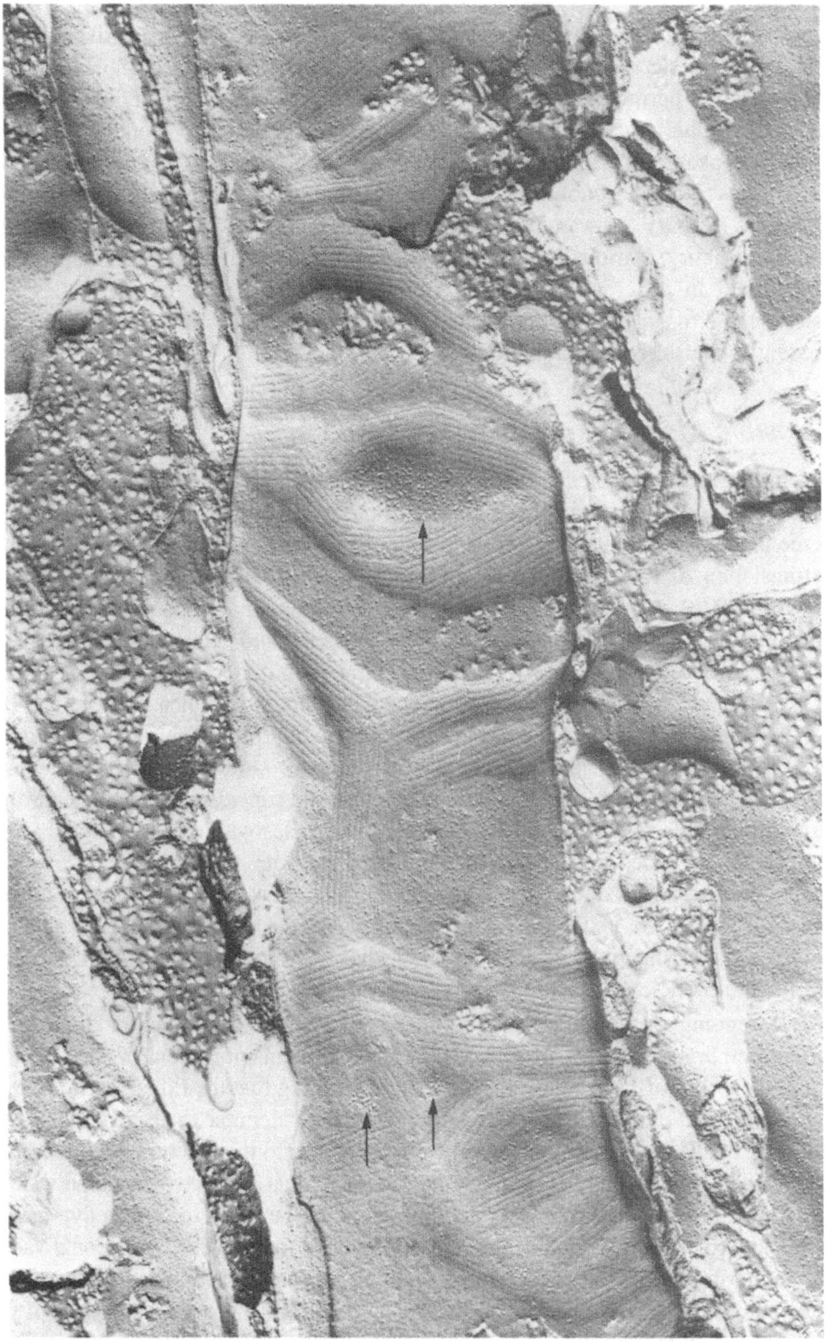

Figure 19. Axolemmal E face from Shiverer mouse central nervous system. The pattern of the paranodal indentations is bizarre and the paracrystalline pattern highly irregular. Areas of membrane that are completely encircled by paracrystalline membrane show aggregates of "trapped" particles (arrows). (From Rosenbluth, 1981a).

junctional membrane. Thus, the distribution of particle aggregates is bizarre in the central fiber tracts of these animals, corresponding to the abnormal distribution of the axoglial junctions there.

Comparable, but somewhat less striking, bizarre axolemmal specializations are also found in the central nervous system of Quaking animals (Rosenbluth, 1979b). In this mutant, as in Shiverer, the peripheral nervous system is much less affected than the central nervous system, and peripheral paranodal and nodal membrane specializations are accordingly close to normal in appearance. The converse is seen in Trembler animals whose peripheral axons (Fig. 16) show frequent abnormalities in the distribution of axolemmal specializations resulting from a primary Schwann cell defect (Aguayo *et al.*, 1979) but whose central axons show no abnormality. Various circumstances in which node-related membrane specializations are inconsistent in different regions are summarized in Fig. 20. In all of these, factors extrinsic to the axon clearly affect axolemmal differentiation, which, therefore, cannot be independently controlled by the neuron alone.

D. Possible Mechanisms of Axoglial Interaction

As discussed above, close association between the glial and axonal membranes at the site of the paranodal junction may affect the mobility of some intramembranous particles within the plane of the axolemma. Thus, a ring of junctional membrane may act as a barrier to the translation of these particles leading to a "pile-up" of particles against that barrier (Rosenbluth, 1976). At normal nodes there is such a junctional region on either side of the nodal axolemma, which could serve to "trap" particles and retain them at the node. At "heminodes," where some degree of particle accumulation may be seen, there is a junctional barrier on one side only; on the side lacking myelin nodal particles can be found spread along a considerable length of the axolemma with no sharply defined border. Particle aggregates that appear in the juxtaparanodal axolemma are similar in distribution, having a sharp boundary at the innermost paranodal indentation and considerable spread in the opposite direction into the internodal axolemma. "Lakes" of particles within the paranodal axolemma are comparable to nodes in that they are bordered on both sides by paranodal junctions, which may thus serve to "trap" particles in this location as well. It is noteworthy that in the latter two locations, the juxtaparanodal axolemma and the paranodal lakes, neither a specialized axolemmal undercoating internally nor glial microvilli externally can be invoked to account for the particle aggregations seen.

The foregoing observations suggest that at the sites of the paranodal junction, the respective membranes interact in such a way as to prevent movement of the characteristic nodal particles, thus affecting their distribution. Assuming that the E face particles represent integral membrane proteins that extend through the axolemma to its external surface or beyond, they could be sterically impeded from moving into the junctional axolemma by the closely apposed glial membrane in much the same way that a boat would be prevented from entering a channel too shallow for its draught. Bonding between the respective membranes across the intercellular gap could also impede particle movement. Additional factors involving anchoring by the cytoskeleton internally or the perinodal microvilli externally (Rosenbluth, 1976, 1981b) have been proposed on morphological grounds but have not yet been supported experimentally.

E. Conclusion II

The distribution of E face particles in the axolemma of myelinated fibers is dependent upon the investing myelin sheath. During the course of myelinogenesis particle aggregates

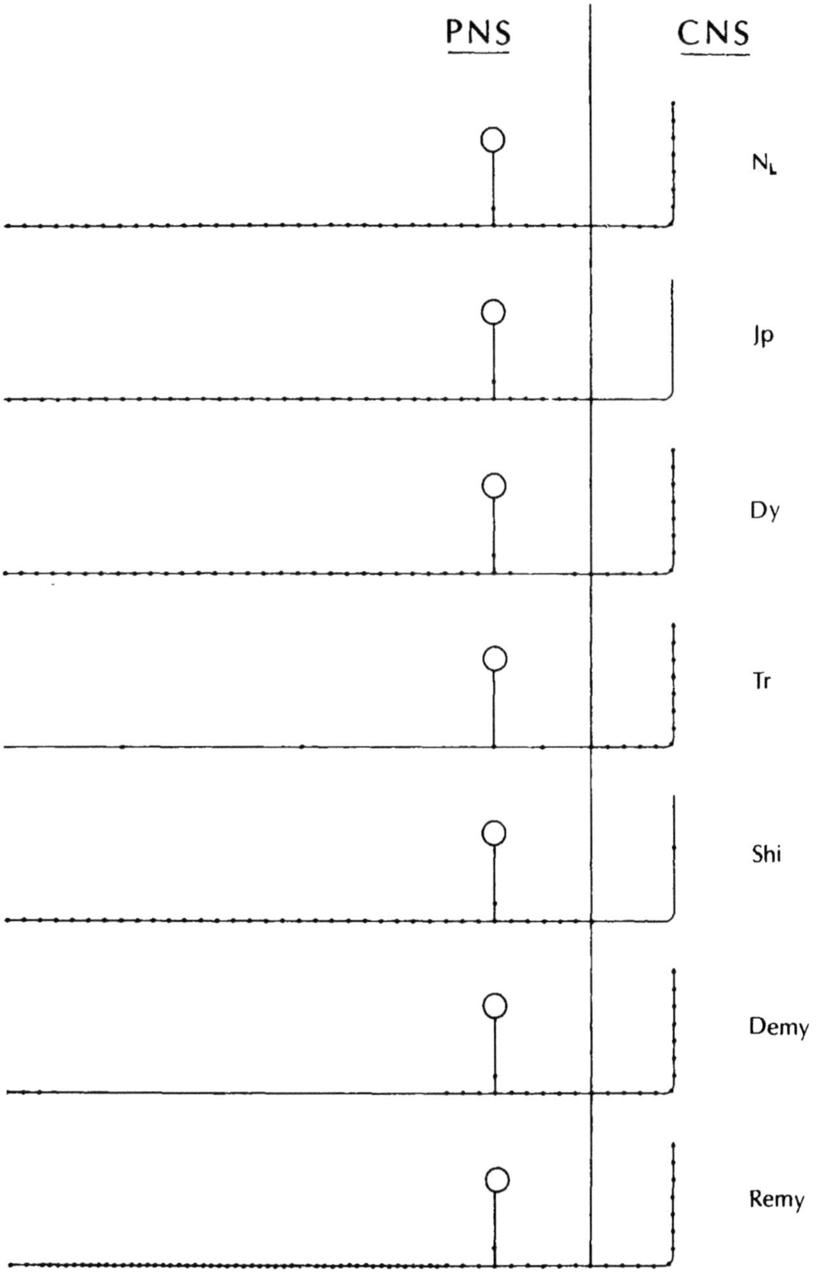

Figure 20. Summary diagram of regional axolemmal abnormalities. In each case, a dorsal root ganglion cell is shown with a peripheral process extending towards the left and a central process extending towards the right into the central nervous system. The normal case (N_L) shows periodic nodal and paranodal specializations (represented by black dots) regularly along central and peripheral segments. The Jimpy (Jp) and Shiverer (Shi) mutants have normal peripheral segments but grossly abnormal central segments. The Dystrophic (Dy) and Trembler (Tr) mutants have normal central segments but abnormalities peripherally. The last two examples show demyelination (Demy), characterized by dedifferentation of the affected portion of the axon, and remyelination (Remy) in which myelin has re-formed but with a shorter internodal spacing than normal (See Rosenbluth, 1981b, for discussion).

appear at first in an immature form distributed over a considerable length of the axolemma. Only gradually do these particles become confined to the typical adult nodal annulus concurrent with the development of the myelin sheath, which appears to restrict them to progressively narrower regions of the membrane. Where myelin does not develop, these annular aggregates fail to form; where myelin develops in a bizarre pattern, the distribution of the particles is correspondingly bizarre; and where myelin breaks down, the axolemma dedifferentiates. Since the same axon may be normally differentiated in one region, but undifferentiated or abnormally differentiated in another, it cannot be the neuron alone that dictates the shape and distribution of its membrane specializations. Rather, these must be determined by interaction with local environmental elements including, presumably, the investing Schwann or glial cells. Thus, myelin-forming cells appear to have functions that go beyond their role in insulating the axolemma. It is clear from the studies presented here as well as from a variety of physiological studies that the axolemma of myelinated nerve fibers is regionally differentiated and in this respect distinctly different from the axolemma of unmyelinated nerve fibers. The interaction between the axon and the investing Schwann or glial cells appears to govern this regional differentiation and, presumably, the periodic distribution of sodium channels in myelinated axons.

REFERENCES

Agnew, W. S., Moore, A. C., Levinson, S. R., and Raftery, M. A., 1980, Identification of a large molecular weight peptide associated with a tetrodotoxin-binding protein from the electroplax of *Electrophorus electricus*, *Biochem. Biophys. Res. Commun.* **92:**860–866.

Aguayo. A. J., Bray, G. M., and Perkins, C. S., 1979, Axon-Schwann cell relationships in neuropathies of mutant mice, *Ann. N.Y. Acad. Sci.* **317:**512–531.

Andres. K. H., 1965, Über die Feinstruktur besonderer Einrichtungen in markhaltigen Nervenfasern des Kleinhirn der Ratte, *Z. Zellforsch.* **65:**701–712.

Barchi, R. L., Weigele, J. B., and Cohen, S. A., 1981, Isolation of sodium channels from excitable tissues, in: *Demyelinating Diseases: Basic and Clinical Electrophysiology* (S. G. Waxman and J. M. Ritchie, eds.), *Adv. Neurol.* **31:**377–390, Raven Press, New York.

Bargmann, W., and Lindner, E., 1964, Über den Feinbau des Nebennierenmarkes des Igels (*Erinaceus europaeus L), Z. Zellforsch.* **64:**868–912.

Barrett, E. F., and Barrett, J. N., 1982, Intracellular recording from vertebrate myelinated axons: Mechanism of the depolarizing afterpotential, *J. Physiol.* **323:**117–144.

Black, J. A., Foster, R. E., and Waxman, S. G., 1981, Freeze-fracture ultrastructure of rat C.N.S. and P.N.S. nonmyelinated axolemma, *J. Neurocytol.* **10:**981–993.

Bostock, H., and Sears, T. A., 1978, The internodal axon membrane: Electrical excitability and continuous conduction in segmental demyelination, *J. Physiol.* **280:**273–301.

Branton, D., 1971, Freeze-etch studies of membrane structure, *Phil. Trans. R. Soc. B* **261:**133–138.

Bray, G. M., and Oldfield, B. J., 1980, A freeze-fracture study of the developing axolemma in rat optic nerves, *Neurosci. Abst.***6:**289.

Bray, G. M., Cullen, M. J., Aguayo, A. J., and Rasminsky, M., 1980, Node-like areas of intramembranous particles in the unsheathed axons of dystrophic mice, *Neurosci. Lett.* **13:**203–208.

Brismar, T., 1981, Specific permeability properties of demyelinated rat nerve fibres, *Acta Physiol. Scand.* **113:**167–176.

Brown, M. J., Sumner, A. J., Saida, T., and Asbury, A. K., 1980, The evolution of early demyelination following topical application of anti-galactocerebroside serum in vivo, *Neurology* (Minneapolis) **30:**371.

Bunge, R. P., 1968, Glial cells and the central myelin sheath, *Physiol. Rev.* **48:**197–251.

Cajal, S. R., 1926, *Elementos de histologia normal y de tecnica micrografica*, p. 521, Tipografia artistica, Madrid.

Chiu, S. Y., 1980, Asymmetry currents in the mammalian myelinated nerve, *J. Physiol.* **309:**499–519.

Chiu, S. Y., and Ritchie, J. M., 1982, Evidence for the presence of potassium channels in the internode of frog myelinated nerve fibres, *J. Physiol.* **322:**485–500.

Conti, F., Hille, B., Neumcke, B., Nonner, W., and Stämpfli, R., 1976, Conductance of the sodium channels in myelinated nerve fibres with moderate sodium inactivation, *J. Physiol. (London)* **262:**729–742.

Dermietzel, R., 1974, Junctions in the central nervous system, II. A contribution to the tertiary structure of the axonal-glial junctions in the paranodal region of the node of Ranvier, *Cell Tissue Res.* **148:**577–586.

Feder, N., Reese, T. S., and Brightman, M. W., 1969, Microperoxidase, a new tracer of low molecular weight. A study of the interstitial compartments of the mouse brain, *J. Cell Biol.* **43:**35A–36A.

Geren, B. B., 1954, The formation from the Schwann cell surface of myelin, *Exp. Cell Res.* **7:**558–562.

Henkart, M., Landis, D., and Reese, T. S., 1976, Similarity of junctions between plasma membranes and endoplasmic reticulum in muscle and neurons, *J. Cell Biol.* **70:**338–347.

Hildebrand, C., 1971, Ultrastructural and light-microscopic studies of the nodal region in large myelinated fibres of adult feline spinal cord white matter, *Acta Physiol. Scand. Suppl.* **364:**43–80.

Hirano, A., and Dembitzer, H. M., 1969, The transverse bands as means of access to the periaxonal space of the central myelinated nerve fiber, *J. Ultrastruct. Res.* **28:**141–149.

Hong, D., and Hubbel, W. L., 1972, Preparation and properties of phospholipid bilayers, *Proc. Natl. Acad. Sci. USA* **69:**2617–2621.

Kimura, J. E., and Meves, H., 1979, The effect of temperature on the asymmetry charge movements in squid giant axons, *J. Physiol.* **289:**479–500.

Kocsis, J. D., and Waxman, S. G., 1980, Absence of potassium conductance in central myelinated axons, *Nature* **287:**348–349.

Kristol, C., Akert, K., Sandri, C., Wyss, U. R., Bennett, M. V. L., and Moor, H., 1977, The Ranvier nodes in the neurogenic electric organ of the knifefish *Sternarchus:* A freeze-etching study on the distribution of membrane-associated particles, *Brain Res.* **125:**197–212.

Kristol, C., Sandri, C., and Akert, K., 1978, Intramembranous particles at the nodes of Ranvier of the cat spinal cord: A morphometric study, *Brain Res.* **142:**391–400.

Landon, D. N., and Hall, S., 1976, The myelinated nerve fibre, in: *The Peripheral Nerve* (D. N. Landon, ed.), pp. 1–105, Chapman & Hall, London.

Landon, D. N., and Langley, O. K., 1971, The local chemical environment of the node of Ranvier: A study of cation binding, *J. Anat.* **108:**419–432.

Langley, O. K., 1979, Histochemistry of polyanions in peripheral nerve, in: *Complex Carbohydrates of Nervous Tissue* (R. V. Margolis and R. K. Margolis, eds.), pp. 197–207, Plenum Press, New York.

Levinson, S. R., and Ellory, J. C., 1973, Molecular size of the tetrodotoxin-binding side estimated by irradiation inactivation, *Nature (New Biol.)* **245:**122–123.

Livingston, R. B., Pfenninger, K., Moor, H., and Akert, K., 1973, Specialized paranodal and inter-paranodal glial-axonal junctions in the peripheral central nervous system: A freeze-etching study, *Brain Res.* **58:**1–24.

Margaritis, L., Elgsaeter, A., and Branton, D., 1977, Rotary replication for freeze-etch, *J. Cell Biol.* **72:**47–56.

Nageotte, J., 1922, *L'organisation de la matiere*, pp. 235–238, Felix Alcan, Paris.

Neumcke, B., Schwarz, W., and Stämpfli, R., 1979, Slow actions of hyperpolarization on sodium channels in the membrane of myelinated nerve, *Biochim. Biophys. Acta* **558:**113–118.

Nonner, W., Rojas, E., and Stämpfli, R., 1975, Gating currents in the node of Ranvier: Voltage and time dependence, *Phil. Trans. R. Soc. London B* **270:**483–492.

Peracchia, C., 1977, Gap junctions. Structural changes after uncoupling procedures, *J. Cell Biol.* **72:**629–641.

Peters, A., 1966, The node of Ranvier in the central nervous system, *Quart. J. Exp. Physiol.* **51:**229–236.

Pinto da Silva, P., and Branton, D., 1970, Membrane splitting in freeze-etching. Covalently bound ferritin as a membrane marker, *J. Cell Biol.* **45:**598–605.

Pinto da Silva, P., Parkison, C., and Dwyer, N., 1981, Fracture-label: Cytochemistry of freeze-fracture faces in the erythrocyte membrane, *Proc. Natl. Acad. Sci. USA* **78:**343–347.

Ranvier, L., 1875, *Traite Technique d'Histologie*, F. Savy, Paris.

Rasminsky, M., 1978, Physiology of conduction in demyelinated axons, in: *Physiology and Pathobiology of Axons* (S. G. Waxman, ed.), pp. 361–376, Raven Press, New York.

Revel, J. P., and Hamilton, D. W., 1969, The double nature of the intermediate dense line in peripheral nerve myelin, *Anat. Rec.* **163:**7–16.

Ritchie, J. M., and Rogart, R. B., 1977, Density of sodium channels in mammalian myelinated nerve fibers and nature of the axonal membrane under the myelin sheath, *Proc. Natl. Acad. Sci. USA* **74:**211–215.

Ritchie, J. M., Rogart, R. B., and Strichartz, G., 1976, A new method for labeling saxitoxin and its binding to nonmyelinated fibers of the rabbit vagus, lobster walking leg and garfish olfactory nerves, *J. Physiol.* **261:** 477–494.

Robertson, J. D., 1959, Preliminary observations on the ultrastructure of nodes of Ranvier, *Z. Zellforsch.* **50:**553–560.

Robertson, J. D., 1981, A review of membrane structure with perspectives on certain transmembrane channels, in: *Demyelinating Disease: Basic and Clinical Electrophysiology* (S. G. Waxman and J. M. Ritchie, eds.), pp. 419–477, Raven Press, New York.

Rosenbluth, J., 1962, Subsurface cisterns in neuronal plasma membranes, *J. Cell Biol.* **13:**405–421.

Rosenbluth, J., 1976, Intramembranous particle distribution at the node of Ranvier and adjacent axolemma in myelinated axons of the frog brain, *J. Neurocytol.* **5**:731–745.
Rosenbluth, J., 1979a, Aberrant axon-Schwann cell junctions in dystrophic mouse nerves, *J. Neurocytol.* **8**:655–672.
Rosenbluth, J., 1979b, Freeze-fracture studies of nerve fibers: Evidence that regional differentiation of the axolemma depends upon glial contact, in: *Current Topics in Nerve and Muscle Research*, ICS No. 455 (A. J. Aguayo and G. Karpati, eds.), pp. 200–209, Excerpta Medica, Amsterdam.
Rosenbluth, J., 1980, Central myelin in the mouse mutant Shiverer, *J. Comp. Neurol.* **194**:639–648.
Rosenbluth, J., 1981a, Axoglial junctions in the mouse mutant Shiverer, *Brain Res.* **208**:283–297.
Rosenbluth, J., 1981b, Freeze-fracture approaches to ionophore localization in normal and myelin-deficient nerves, in: *Demyelinating Disease: Basic and Clinical Electrophysiology* (S. G. Waxman and J. M. Ritchie, eds.), pp. 391–418, Raven Press, New York.
Rosenbluth, J., Sumner, A., and Saida, T., 1981, Dedifferentiation of the axolemma associated with demyelination, in: *Proc. 39th EMSA Mtg.* (G. W. Bailey, ed.), pp. 496–497, Claitor's, Baton Rouge.
Schnapp, B., and Mugnaini, E., 1975, The myelin sheath: Electron microscopic studies with thin section and freeze fracture, in: *Golgi Centennial Symposium Proceedings* (M. Santini, ed.), pp. 209–233, Raven Press, New York.
Sears, T. A., 1979, Nerve conduction in demyelination, amyelination and early regeneration, in: *Current Topics in Nerve and Muscle Research (Int. Congr. Ser. 455)* (A. J. Aguayo and G. Karpati, eds.), pp. 181–188, Excerpta Medica, Amsterdam.
Sherratt, R. M., Bostock, H., and Sears, T. A., 1980, Effects of 4-aminopyridine on normal and demyelinated mammalian nerve fibers, *Nature* **283**:570–572.
Sigworth, F. J., 1980, The variance of sodium current flunctuations at the node of Ranvier, *J. Physiol.* **307**:97–129.
Smith, K. J., and Schauf, C. L., 1981, Size-dependent variation of nodal properties in myelinated nerve, *Nature* **293**:297–299.
Tao-Cheng, J.-H., and Rosenbluth, J., 1980a, Membrane specializations in developing nodes of Ranvier, in: *Proc. 38th EMSA Mtg.* (G. W. Bailey, ed.), pp. 626–627, Claitor's, Baton Rouge.
Tao-Cheng, J.-H., and Rosenbluth, J., 1980b, Nodal and paranodal membrane structure in complementary freeze-fracture replicas of amphibian peripheral nerves, *Brain Res.* **199**:249–265.
Tao-Cheng, J.-H., and Rosenbluth, J., 1982, Development of nodal and paranodal membrane specializations in amphibian peripheral nerves. *Dev. Brain Res.* **3**:577–594.
Waxman, S. G., and Quick, D. C., 1977, Cytochemical differentiation of the axon membrane in A- and C-fibers, *J. Neurol. Neurosurg. Psychiat.* **40**:379–386.
Waxman, S. G., Black, J. A., and Foster, R. E., 1982, Freeze-fracture heterogeneity of the axolemma of premyelinated fibers in the C. N. S., *Neurology* **32**:418–421.
Wiley, C. A., and Ellisman, M. H., 1980, Rows of dimeric particles within the axolemma and juxtaposed particles within glia incorporated into a new model for the paranodal glial-axonal junction at the node of Ranvier, *J. Cell Biol.* **84**:261–280.
Wiley-Livingston, C. A., and Ellisman, M. H., 1980, Development of axonal membrane specializations defines nodes of Ranvier and precedes Schwann cell myelin elaboration, *Dev. Biol.* **79**:334–355.

3

The Axolemma–Ectoplasm Complex of Squid Giant Axon

J. Metuzals, D. F. Clapin, and I. Tasaki

I. INTRODUCTION

The giant nerve fiber of the squid provides a unique preparation for the study of the structure and function of the plasma membrane, the cytoplasmic ground substance, and of their interrelationships in excitability. The axoplasm is organized into a three-dimensional network of filamentous and globular proteins assembled in a helical fashion. The globular subunit proteins, actin and α- and β-tubulins, are assembled into microfilaments and microtubules. The filamentous subunit proteins of neurofilaments make up the four-stranded 10-nm-wide intermediate filaments and their filamentous crosslinks (Metuzals, 1969; Krishnan *et al.*, 1979). The three-dimensionality and network character of the axoplasm is expressed in the terminology used by various authors to refer to its ultrastructure: filamentous or neurofilamentous network (Metuzals, 1969; Metuzals and Mushynski, 1974), neuroplasmic network (Hodge and Adelman, 1980), microtrabecular lattice (Ellisman and Porter, 1980), and axoplasmic matrix filaments (Ochs and Burton, 1980).

The cylindrical geometry of the axon greatly facilitates the study of the relative contributions of the different filament systems to the shape and functions of the cell (Metuzals *et al.*, 1983). Neurofilaments and microtubules tend to be oriented roughly parallel with the long axis of the axon. They may be referred to as "axial" components of the network. Associating these axial filaments are crosslinking proteins: microtubule associated proteins (Kim *et al.*, 1979), and higher molecular weight components of the neurofilament complex isolable from squid giant axon (Metuzals and Clapin, 1981; Metuzals *et al.*, 1982) and vertebrate axons (Geisler and Weber, 1981). Proteins forming filamentous projections from the axial filaments are in a nominal sense "radial" filaments in the reference frame of the squid giant axon. It is these radially directed components that we propose differentiate the

J. Metuzals and D. F. Clapin • Electron Microscopy Unit, Department of Anatomy, Faculty of Health Sciences, University of Ottawa, Ottawa, Ontario, Canada K1N 9A9 and Marine Biological Laboratory, Woods Hole, Massachusetts 02543. *I. Tasaki* • Laboratory of Neurobiology, National Institutes of Health, Bethesda, Maryland 20205; and Marine Biological Laboratory, Woods Hole, Massachusetts 02543.

network into a loose endoplasmic zone on the interior of the axon and a denser, ectoplasmic or cortical zone immediately apposed to the axolemma (Metuzals *et al.*, 1983). In this chapter we summarize some of the morphological data on the axoplasmic network of proteins, especially concerning its differentiation into ectoplasm and endoplasm.

Classical cytological experiments have demonstrated that most types of cells have a peripheral region of cytoplasm which displays the properties of a viscoelastic gel and which functions in motility and shape determination (Brown and Danielli, 1964). The existence of a peripheral, dense ectoplasmic region in squid giant axon and a central core of less dense endoplasm was inferred from microinjection experiments (Chambers, 1947; Chambers and Kao, 1952). In perfused squid giant axons the axolemma appears to be protected by a thin, robust film of axoplasm (Baker *et al.*, 1962). Metuzals and Izzard (1969) have investigated the squid giant axon *in situ,* and in fresh and fixed preparations by differential interference microscopy and electron microscopy. A peripheral ectoplasmic region can be distinguished by differences in the orientation, arrangement, and packing density of filamentous structures in the axoplasm. In the ectoplasm, a cross lattice of filaments appears more regular and more densely packed than in the central core where a looser, reticular array predominates. The filaments are twisted into a right-hand helix and can be resolved into a hierarchy of decreasing order of size.

The ectoplasmic network is firmly attached to the inner face of the axolemma and is composed of the neurofilamentous network, actin filaments, and microtubules (Metuzals, 1969; Metuzals and Tasaki, 1978; Hodge and Adelman, 1980). Structural, chemical, and electrophysiological studies of intracellularly perfused giant axons have demonstrated that the ectoplasm has an important role in the maintenance of the normal morphological characteristics of the axolemma and of excitability (Metuzals and Tasaki, 1978; Tasaki, 1982). Based on results of voltage-clamp studies on squid axons, Chang (1979, 1980) has also suggested that the cytoplasmic protein network attached to the axolemma is functionally important to the excitation process of the nerve cell. It is justified, therefore, to introduce the term *axolemma–ectoplasm complex* in order to define this entity (Metuzals *et al.*, 1980a, 1981b). In the lens fibers, Ramaekers *et al.* (1982) have described the intimate association between the intermediate-sized filaments and the lipid bilayer as a plasma membrane–cytoskeleton complex. These authors suggested, furthermore, that the protein of the intermediate filaments, vimentin, may be contained within the lipid membrane core as an integral membrane component.

We have used the following experimental approaches in the study of the axolemma–ectoplasm complex of the squid giant axon: (1) removal of the Schwann cell sheath to expose the outer surface of the axolemma after 15 minutes of fixation in glutaraldehyde, (2) double grid mounting of isolated portions of the complex obtained by microdissection from desheathed axons, (3) an examination of extruded axoplasm by various electron microscopic techniques including double grid mounting, and (4) determination of the protein composition and structural roles of the neurofilamentous network.

II. DISTRIBUTION OF MICROTUBULES IN THE SQUID GIANT AXON

Microtubules can be easily identified in electron micrographs taken at medium magnification of cross-sectioned squid giant fibers (Fig. 1). A quantitative estimate of microtubule distribution was made by tracing their locations on a transparent overlay of cross-sectioned material such as that illustrated in Fig. 1. An uneven distribution of the microtubules in the squid giant axon is clearly demonstrated in Fig. 2. Maximum density is observed in the

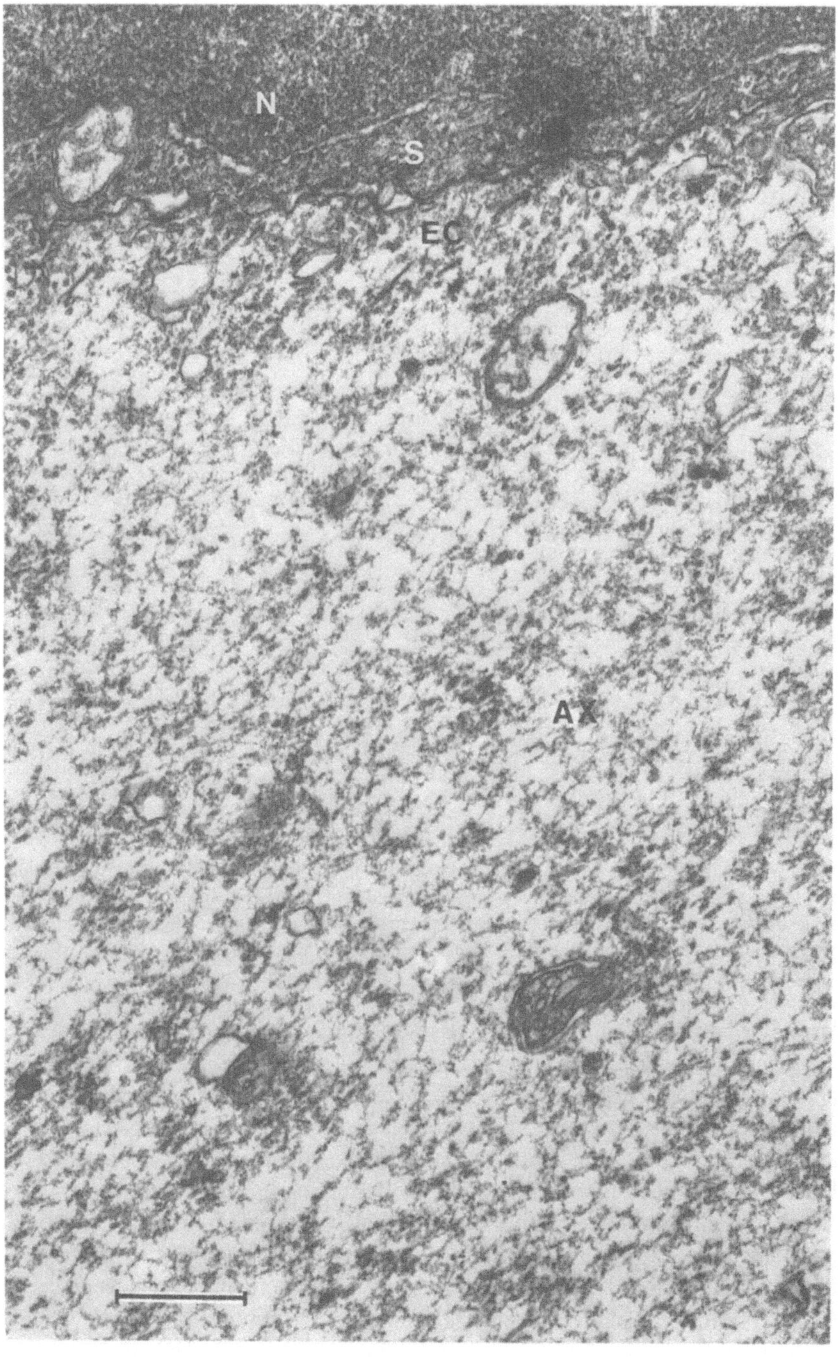

Figure 1. Electron micrograph of a cross section through the squid giant nerve fiber illustrating the distribution pattern of the microtubules. Axoplasm (AX), ectoplasm (EC), cytoplasm (S), and nucleus (N) of the Schwann cell. The fiber was treated for 1 hr in 1 mg/ml trypsin dissolved in natural sea water and fixed afterwards in 1% paraformaldehyde and 1% glutaraldehyde dissolved in 0.5 *M* sodium cacodylate/HCl (pH 7.3). Bar: 0.5 μm.

Figure 2. Tracing of the electron micrograph illustrated in Fig. 1. Profiles of cross-sectioned microtubules have been marked on an overlay. Note the nonuniform distribution of the microtubules arranged in groupings. The ectoplasmic zone adjoining the axolemma (AL) displays greater density of microtubule distribution than the regions deeper inside the axon (AX). Cytoplasm (S) and nucleus (N) of the Schwann cell. Bar: 1 μm.

ectoplasmic or cortical region adjoining the axolemma. The microtubules are distributed in irregular groupings or patches. The greatest density of microtubule distribution exists in a zone approximately 0.5 μm thick adjoining the axolemma. About 110 microtubules can be counted in a square micrometer of this part of the axon. Figure 3 illustrates graphically the decline of microtubule density in the direction of the endoplasmic core of the axon.

The same general pattern of microtubule distribution has been observed in giant nerve fibers fixed according to different protocols. Whole fibers fixed by immersion in aldehyde fixatives containing Co (II) ions yielded essentially the same result as described above. Cobalt ions were included in the fixative in an attempt to improve the preservation of microtubules. While it might be argued that fixation by immersion may be subject to selectively improved preservation of the microtubules in the periphery of the axon due to differential rates of diffusion of the aldehydes and other components of the fixative, particularly Ca (II) ions, it was noted that a similar distribution is obtained when the fixative is injected directly into the axon (Hodge and Adelman, 1980; Metuzals *et al.*, 1983). The distinctive pattern of distribution of microtubules as illustrated in Fig. 2 is very similar to that of cross-sectioned filament profiles illustrated previously in Fig. 20 by Metuzals and

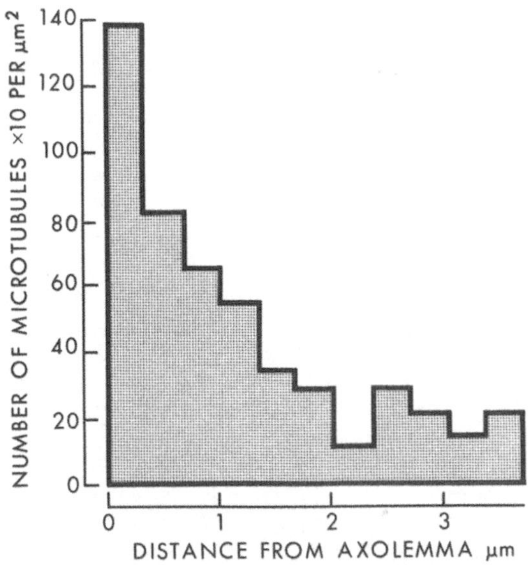

Figure 3. Quantitative representation of microtubule distribution in a cross section of squid giant axon. The tracing in Fig. 2 was divided into 0.28 μm squares. Numbers of microtubules in each square were scored and added together with all squares of equal distance from the axolemma (±0.14 μm). Maximum microtubule density expressed as number of microtubules per $(\mu m)^2$ is observed within the ectoplasmic zone, approximately 0.3–0.5 μm from the axolemma.

Izzard (1969; see also Sakai *et al.*, 1981). The conclusion is justified that the distribution of microtubules as illustrated in Figs. 1–3 is closely related to their arrangement in living axons.

Microtubules apposing the axolemma have been observed in squid giant axon (Hodge and Adelman, 1980), in nematode axons (Chalfie and Thomson, 1979), and in rat sensory axons (Bray and Bunge, 1981).

The increased density of microtubule distribution in the axolemma–ectoplasm complex represents both a structural and functional specialization of the cytoskeleton, thought to be related directly to membrane excitability (see Sakai *et al.*, 1981; Fukuda *et al.*, 1981).

III. OUTER SURFACE OF THE AXOLEMMA OF DESHEATHED AXONS

Investigations of the axolemma–ectoplasm complex are facilitated by the use of preparations in which the Schwann-cell layer has been removed from fixed fibers and the outer surface of the membrane of the axon exposed (Metuzals *et al.*, 1981b). In these specimens the outer surface of the axolemma can be investigated in the scanning electron microscope. It is also possible to remove a portion of the axolemma together with its adjoining axoplasm for preparation by double grid mounting for transmission electron microscopy. Unfortunately, since this procedure involves the use of fixatives, the protein composition of the complex cannot be obtained from these microdissected pieces.

A portion of the desheathed axon and the transected and everted Schwann sheath is illustrated in Fig. 4. The surface of the desheathed axon appears to be free of any remnants of the Schwann sheath (Fig. 5). It displays rope-like striations formed by a ridge-and-groove

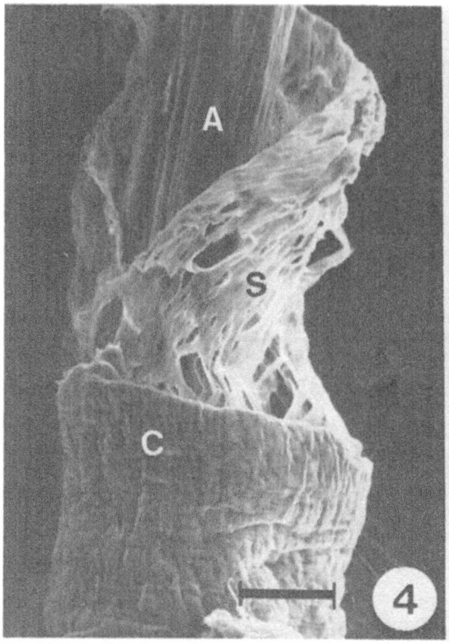

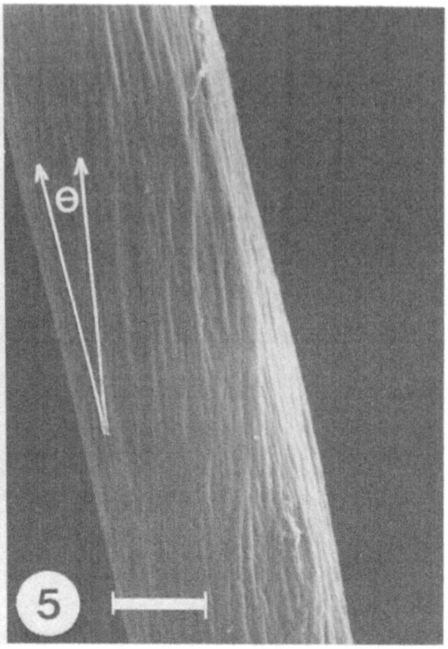

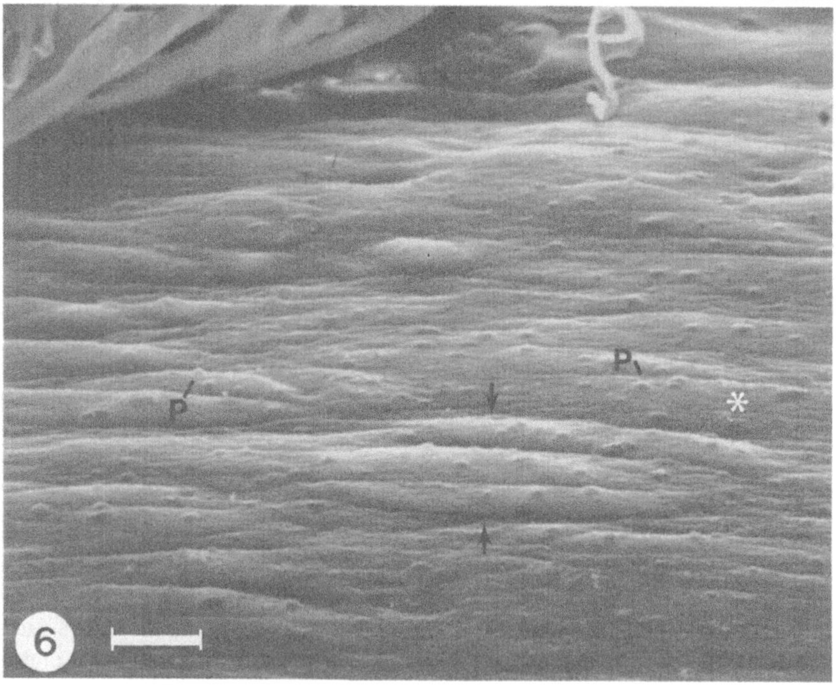

Figure 4. Scanning electron micrograph of desheathed giant nerve fiber of the squid: exposed surface of the axon (A), Schwann sheath (S), and everted layer of Schwann sheath (C). Bar: 0.1 mm (Figs. 4–9, 11 and 12 are published with permission from Springer-Verlag, Heidelberg).

Figure 5. Scanning electron micrograph of axonal surface at survey magnification: a ridge and groove pattern on the surface of the desheathed axon can be seen. The ridges are oriented in a right-handed helix with a tilt angle (θ) of about 10°. Bar: 50 μm.

Figure 6. Scanning electron micrograph of surface of the desheathed axon at high magnification: groupings of ridges (opposed arrows) oriented approximately parallel to the long axis of the axon, protuberances (P), and finer ridge and groove pattern (asterisk). Bar: 10 μm.

pattern. The ridges are oriented in a right-handed helix with a tilt angle of about 10°. The axons of the right and left giant fibers of the squid have the same right-handed helicity, indicating the intrinsic nature of this type of helical symmetry of the neuron. A finer structure of the surface of the desheathed axons can be revealed at higher magnification. Scanning electron micrographs show characteristic groupings of ridges about 5 μm wide (opposed arrows, Fig. 6). Protuberances (P, Fig. 6), approximately 1.5 μm wide at their base, project from the surface of the ridges. A much finer pattern of ridges about 50 nm across can be discerned everywhere on the surface of the axon (asterisk, Fig. 6).

The surface texture of the axon as observed in scanning electron micrographs of desheathed axons may be related to the bundling and unbundling pattern of the helical filaments of the ectoplasm as described by Metuzals and Izzard (1969; see also Fig. 13). The observation that both the right and left axons have the same right-handed helicity is in agreement with the general concept of inherent helicity of the cytoplasm (Metuzals *et al.*, 1981a) which has to be the same in cells from both sides of the body. This structural characteristic is probably imposed on the axoplasm by the helical symmetry of individual cytoskeletal filaments, particularly neurofilaments.

The nature of the protuberances on the surface of the desheathed axon remains an open question. Investigations of the electric properties of the giant axon using very fine glass electrodes have revealed domains of membrane in an excitable state coexisting side by side with zones in a resting state. On the outer surface of the membrane such excitable zones were estimated to be about 2 μm in diameter (Inoue *et al.*, 1974). Here we raise a question. Can the surface protuberances revealed in this study be the structural correlate of this electrophysiological phenomenon?

These views of the surface topography of the axon imply that the underlying ectoplasm is composed of filamentous structures that have bundled and unbundled forms. Surface protuberances of the axolemma are probably maintained through some localized aggregation of the associated ectoplasmic cytoskeleton (see also Fig. 9).

The presence of surface protuberances, in itself, suggests a local contractile phenomenon possibly initiated by calcium activation of a submembranous filamentous system. The micropapillae of *Beroe ovata* giant smooth muscle fibers (maximal height 40 nm and maximal diameter 55 nm) display intracellular concentrations of calcium ions and a differentiation of the cell coat above these calcium deposits (Nicaise *et al.*, 1982). It is interesting to note that these excitable cells display a surface "helicoidal" ridge pattern and that successive lengths of the same fiber may coil both clockwise and counterclockwise, implying the existence of two sets of helical attachment points for actin having opposite handedness (Hernandez-Nicaise *et al.*, 1982).

IV. ULTRASTRUCTURE OF THE AXOLEMMA–ECTOPLASM COMPLEX IN CROSS SECTIONS OF DESHEATHED AXONS AND WHOLE FIBERS

No traces of Schwann sheath are found in thin sections of desheathed axons prepared by standard techniques of thin sectioning and transmission electron microscopy, as in Fig. 7. In a relatively thick cross section of the desheathed axon, the axoplasm appears to be differentiated into dense and less dense regions (Fig. 7). The boundary between the dense and less dense regions is not sharply defined; rather these areas grade into one another. A dense region is present circumferentially in association with the axolemma and corresponds to the cortex or ectoplasm of the axon (EC, Fig. 7). This ectoplasmic region is about 0.3–0.5

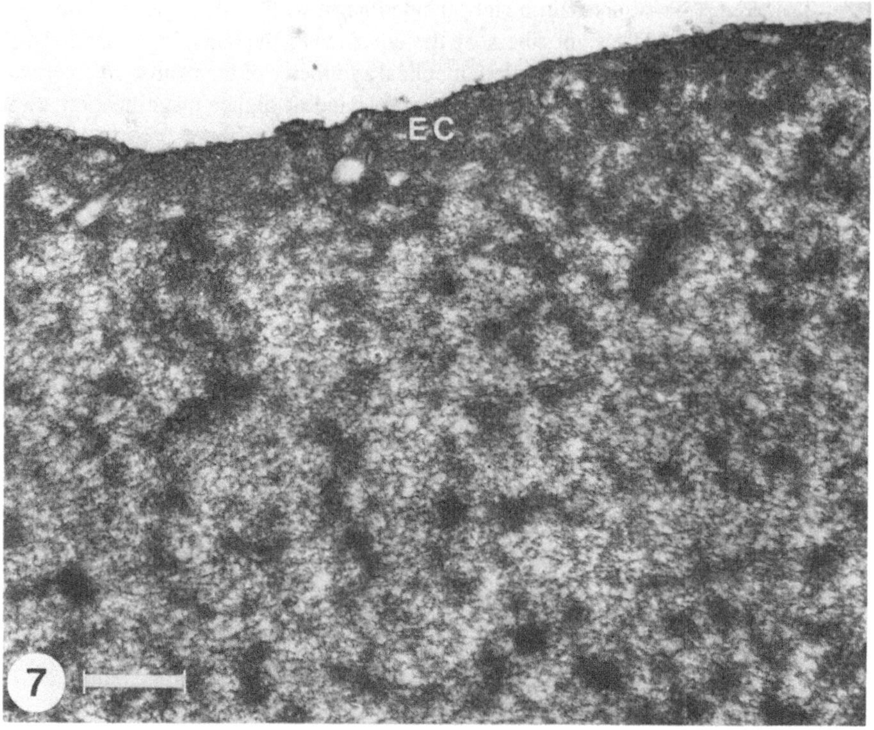

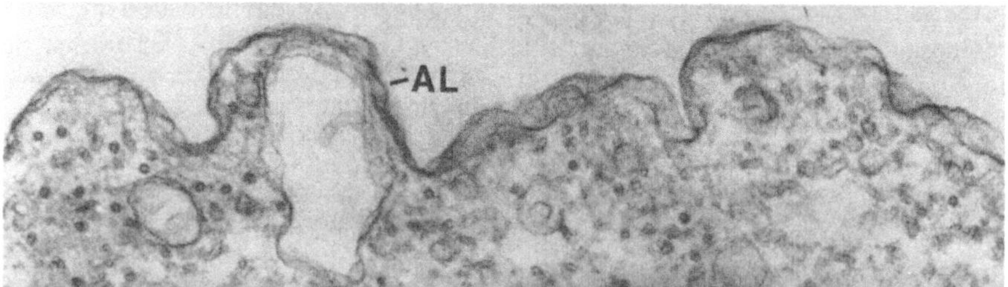

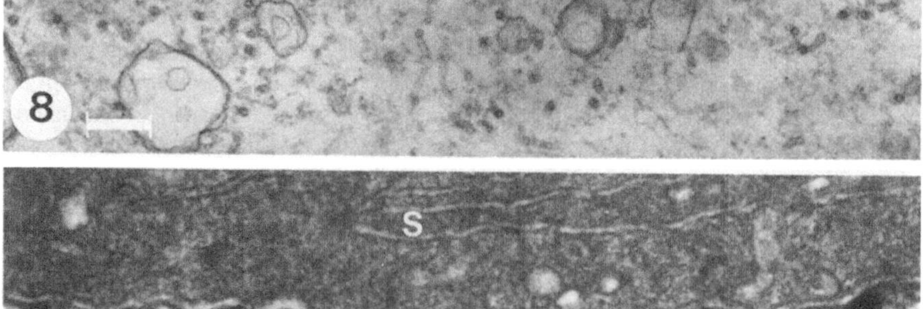

Figure 7. Electron micrograph of a thin section through desheathed and embedded axon; ectoplasm (EC). Fixed in 1% paraformaldehyde and 1% glutaraldehyde dissolved in artifical seawater. Bar: 0.5 μm.

μm thick and extends into the axon in the form of interconnected columns. The principal components of the dense as well as the less dense regions is the neurofilamentous network. Because of the thickness of the section, the existence of microtubules cannot be clearly identified in this figure. In thinner sections (Fig. 8), it is apparent that microtubule distribution is the same as in intact fibers. They are concentrated predominantly in the ectoplasm and more sparsely distributed within the rest of the axon. Thin sections unequivocally demonstrate the continuity of the axolemma in the desheathed axons.

When $CoCl_2$ is included in the fixative instead of Ca (II) ions, dense bodies associated with the axolemma are clearly visualized in intact nerve fibers (arrowhead, Fig. 9). A dense web of fine filament matrix associated with the longitudinally oriented components of the neuroplasmic network, is very well preserved (Metuzals *et al.*, 1983). Filaments are associated with the axolemma (arrow, Fig. 9) and with the dense bodies (arrowhead, Fig. 9). Similar structures can also be observed in desheathed axons (Metuzals *et al.*, 1981b).

These electron dense structures correspond to the calcium-containing electron-dense globules and plaques identified by Hillman and Llinas (1974) and Oschman *et al.* (1974; see also Villegas et al., 1972). Where these are observed in desheathed axons it may be that the desheathing process exposes calcium binding sites of the axolemma to Ca (II) ions present in the fixative buffer. The resulting saturation of these sites with calcium causes them to become electron opaque. Dense bodies having the same size and distribution can be identified in whole fibers fixed in the presence of cobaltous ions. The substitution of Co (II) ions has no detrimental effect on the action potential of lobster axons and central neurons (Blaustein and Goldman, 1968; Pitman *et al.*, 1972). Cobalt (II) ions do not activate the calcium-activated protease which is capable of rapidly degrading neurofilaments (Pant and Gainer, 1980). Since elastic scattering of the electron beam increases with atomic number of the atoms of the specimen, Co (II) ions are a sensitive marker of calcium binding sites (Williams, 1970). The functional significance of interactions between the neurofilamentous network and membrane-associated calcium-binding sites remains to be elucidated. However, calcium densities at the plasma membrane of *Dictyostelium discoideum* have been implicated in the functioning of contractile proteins (De Chastellier and Ryter, 1981). These electron dense calcium binding sites may correspond to the location of a calcium–magnesium-dependent ATPase and to areas of increased concentration of actin microfilaments. It should be noted that the structural complexes associated with the membrane of squid giant axon observed by Villegas and Villegas (1976), have a similar morphology to the electron dense bodies described above but may be distinguished by certain features. These ouabain-sensitive sites probably correspond to the sodium–potassium-dependent ATPase locations in the plasma membrane (Sabatini *et al.*, 1968).

The association of filaments with dense bodies (Fig. 9) may be morphological evidence for the kinds of interactions between F-actin filaments and sodium channels as reported by Fukuda *et al.* (1981). The giant axon of *Mercierella enigmatica* Fauvel possesses hemidesmosomelike structures, associated with the axolemma, which are connected to the neurofilamentous network (Skaer *et al.*, 1978).

Figure 8. Electron micrograph of a thin section of desheathed axon. Axolemma (AL), note numerous profiles of microtubules in the ectoplasm. Fixed in 2.5% glutaraldehyde dissolved in seawater. Bar: 0.2 μm.

Figure 9. Transmission electron micrograph of a thin section of an intact nerve fiber fixed in a solution containing 0.5% glutaraldehyde, 10 mM $CoCl_2$, 0.74 M sucrose and 0.1 M s-collidine buffer (pH 7.3). Schwann cell (S), axolemma (AL), axoplasm (AX), filaments associated with the axolemma (arrow) and with the dense bodies (arrowhead). Bar: 0.1 μm.

V. FREEZE–FRACTURING OF THE AXOLEMMA AND THE ADJOINING AXOPLASM

The squid giant nerve fiber was investigated by the standard technique of freeze–fracturing (or freeze–etching) in order to compare the structures observed in replicas with those disclosed by other methods of electron microscopy (Peracchia, 1974; Stolinski *et al.*, 1981). The fractured axoplasm displays ridge-like elevations which may correspond to elements of the neurofilamentous network (arrowhead, Fig. 10). A layer of axoplasm about 0.3 μm wide adjoining the boundary of the fractured P face (Branton *et al.*, 1975) of the axolemma,

Figure 10. Freeze–fracture replica showing the P face (inner leaflet) of the axolemma (P) and the fractured axoplasm (AX). Ridges (arrowhead) corresponding to neurofilaments, ectoplasm (EC), striations (arrow) in the ectoplasm, strands (S) and Y-shaped (Y) intramembrane particle aggregates. Fixed for 15 min in 1% paraformaldehyde and 1% glutaraldehyde dissolved in seawater and treated afterwards for 1 hr with 5% glycerol in seawater before freezing. Bar: 0.2 μm.

differs from the rest of the axoplasm by a more even and denser texture. It may correspond to the ectoplasmic region of the axoplasm as noted in cross-sectioned specimens. A texture of fine striation slanting towards the boundary of the fracture P face (arrow, Fig. 10) can be observed. Intramembrane particles are distributed in the P face in a characteristic pattern: individual intramembrane particles are distributed between particle aggregates which are of varying shape and size. Strands and Y-shaped intramembrane particle aggregates (S, Y, Fig. 10) can be identified among aggregates of irregular appearance. The largest aggregates measure up to 0.1 μm in diameter. These intramembrane particles and particle aggregates are in the same size range and approximate distribution as the electron-dense calcium-binding sites noted on the axolemma of desheathed axons and cobalt-treated whole nerve fibers. This identification is not conclusive. The formation of aggregates is probably controlled by the cytoskeletal elements of the underlying ectoplasm. Fine filament attachments were noted in thin sections showing electron dense bodies associated with the axolemma (Fig. 9). Rodlike structures associated with the protoplasmic fracture face of the plasma membrane have been observed in freeze–fracture replicas of isolated lens fiber plasma membrane–cytoskeleton complex (Ramaekers *et al.*, 1982).

VI. A COMPARISON OF ISOLATED AXOLEMMA–ECTOPLASM COMPLEX AND EXTRUDED AXOPLASM

A portion of the axolemma, together with adhering ectoplasm, was removed from the fixed and desheathed axons by microdissection, spread on a grid, stained with uranyl acetate, and air dried. These specimens were found to be difficult to handle because of a tendency to float free of grids. However, in successful preparations a positively stained, fine filament network is observed (Fig. 11).

Isolated axolemma–ectoplasm complex was prepared for electron microscopy by a double grid mounting procedure. This technique was developed in order to investigate three-dimensional structural relationships in relatively thick samples and to identify the chemical nature of the components. The double grid method eliminated problems encountered in earlier studies in the handling and staining of these specimens (Metuzals *et al.*, 1981b, 1982, 1983). The complex, isolated by microdissection, was mounted between folding double grids, fixed, stained with 1% uranyl acetate solution, and critical point dried. In such preparations the ectoplasm is revealed as a network of 10-nm-wide filaments having a predominantly longitudinal orientation (Fig. 12). The 10 nm filaments have a center to center spacing of approximately 40 nm. Cross-associating filaments 5–7 nm in diameter can be seen in these preparations. Survey stereo electron micrographs of freshly extruded axoplasm, treated by double grid mounting technique, is illustrated in Fig. 13. Helical bundles of filaments extend in the direction of the longitudinal axis of the extruded axoplasmic rod.

Extruded axoplasm extracted in a physiological buffer (Morris and Lasek, 1981) for 140 min and double grid mounted is shown in Fig. 14. The polypeptide composition of the preparation was determined by sodium dodecyl sulfide–polyacrylomide gel electrophoresis (SDS–PAGE) analysis. It consists mainly of two peptides, one in the range of 200,000 to 220,000 daltons, the other 60,000 daltons, which comigrate with the peptides of isolated neurofilament protein (Morris and Lasek, 1981). Extraction for 140 min results in an intact, partially purified preparation of the neurofilamentous network (Metuzals *et al.*, 1980b, 1982, 1983).

In thick areas of the neurofilamentous network it becomes more difficult to detect individual filaments because of superposition. However, the continuous and dense nature of the neurofilamentous network is made much more obvious than in similar preparations viewed

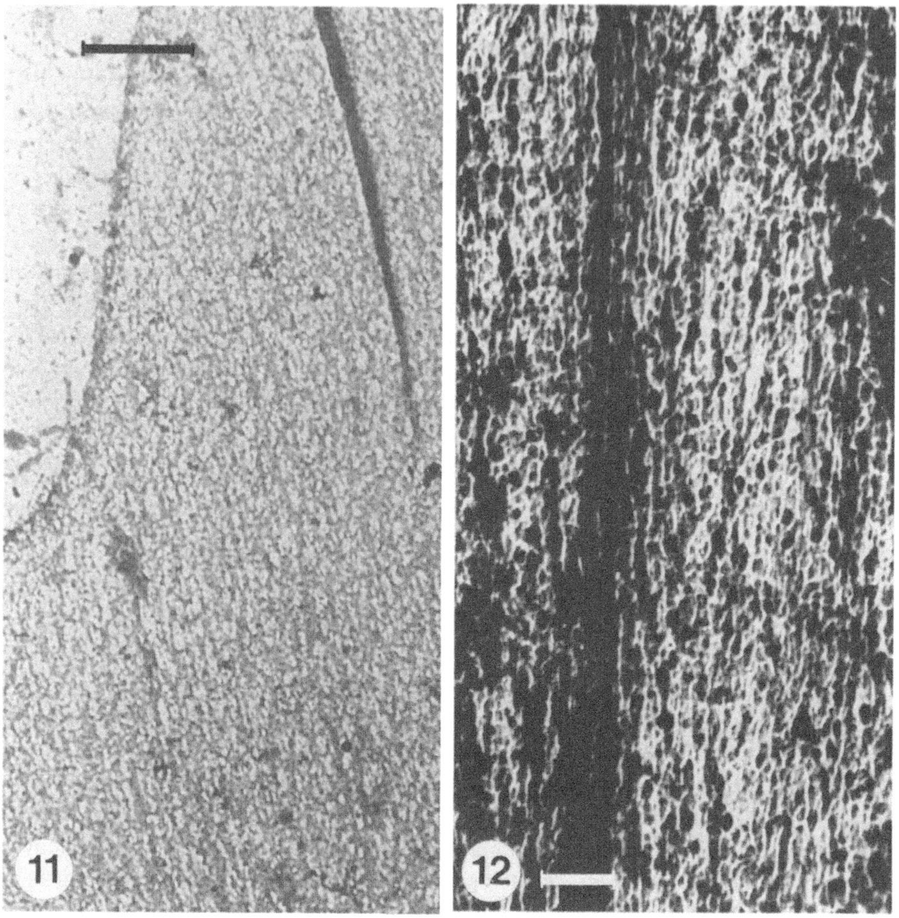

Figure 11. Isolated axolemma–ectoplasm complex, spread on the grid, stained with 1% uranyl acetate at pH 4.4 for 1 min, air dried. Bar: 0.5 μm.

Figure 12. Double grid mounted axolemma–ectoplasm complex, fixed for 30 min in 1% glutaraldehyde and 1% paraformaldehyde solution dissolved in 0.5 M sodium cacodylate/HCl (pH 7.3), stained with 1% uranyl acetate (pH 7.3), and critical point dried. Bar: 0.1 μm.

in conventional thin sections (Fig. 14). Mitochondria (large, densely stained, round bodies in Fig. 14) are present at different levels of the approximately 1 μm thick specimen. An obvious striation slanting upward to the left in Fig. 14 is approximately parallel to the long axis of the extruded axoplasmic rod. On close inspection fine filaments are apparent extending at various angles from the axially deposed filaments.

VII. CROSSLINKING COMPONENTS OF THE NEUROFILAMENTOUS NETWORK

The character of the neurofilamentous network and thus, to a great extent of the ectoplasm, is determined by the nature of its crosslinking elements (Metuzals *et al.*, 1983).

We have developed a simple procedure for the isolation of a high molecular weight component of the neurofilament complex from homogenates of squid axoplasm (Figs. 15,16). Electron microscopy of this component indicates that it may be a part of the crosslinking filamentous system of the neurofilamentous network (Metuzals and Clapin, 1981) and also of the matrix of fine filaments (Metuzals *et al.*, 1983).

It has been demonstrated that the two major neurofilament proteins of squid axoplasm (60 K and 200 K) copurify with a high molecular weight protein NF-1 (Pant *et al.*, 1978; Lasek *et al.*, 1979; Roslansky *et al.*, 1980). This protein has the appearance of sheets of filaments 2–5 nm in diameter forming network arrays, which in some areas are polygonal figures about 20 nm long at each edge (Fig. 17). These fine filaments are connected with 10 nm neurofilaments giving the latter a lampbrush-like appearance (Fig. 18). A few of the intact surviving 10 nm filaments, present as a contaminant of the preparation, display a distinct helical substructure (arrow, Fig. 19) of four intercoiled 2-nm-wide unit filaments

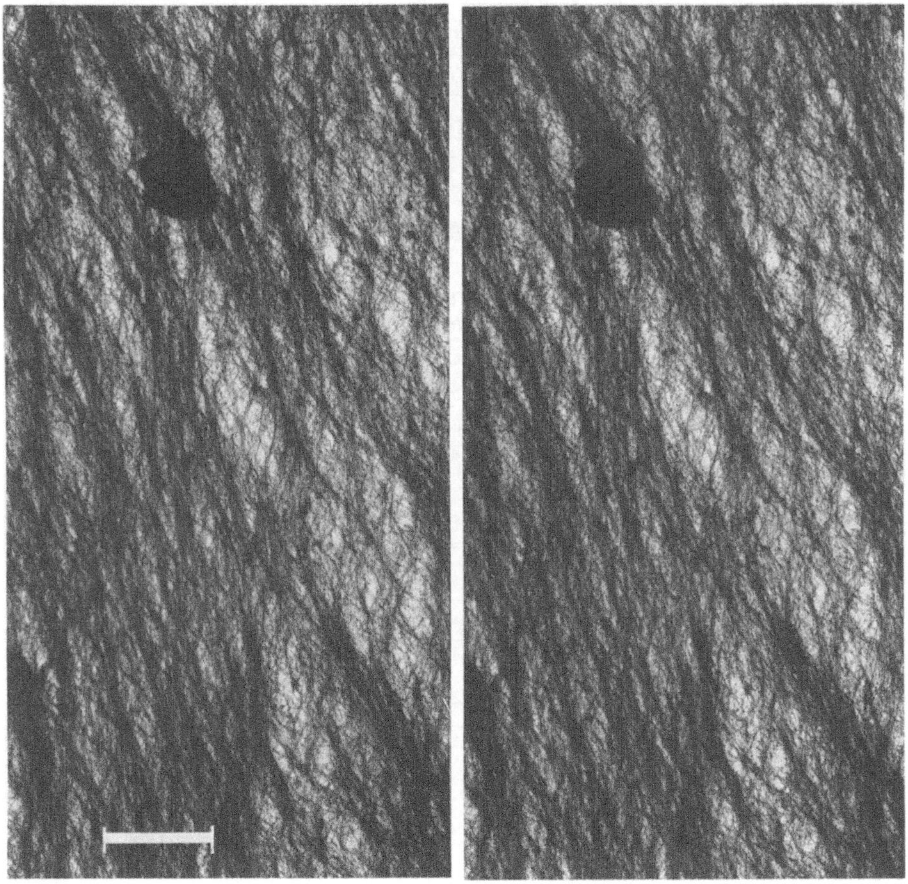

Figure 13. Stereopair micrographs of axoplasm extruded directly into the fixative and mounted in a double grid. The thickness of the preparation varies from 0.8–1.2 μm. Note the bundling and unbundling of the filaments and their helical course approximately parallel to the long axis of the extruded axoplasmic rod. Fixed for 30 min in 1% glutaraldehyde and 1% paraformaldehyde solution dissolved in 0.5 M sodium cacodylate/HCl (pH 7.3), stained with 1% uranyl acetate (pH 7.3) and critical point dried. ±6° tilt. Bar: 2 μm.

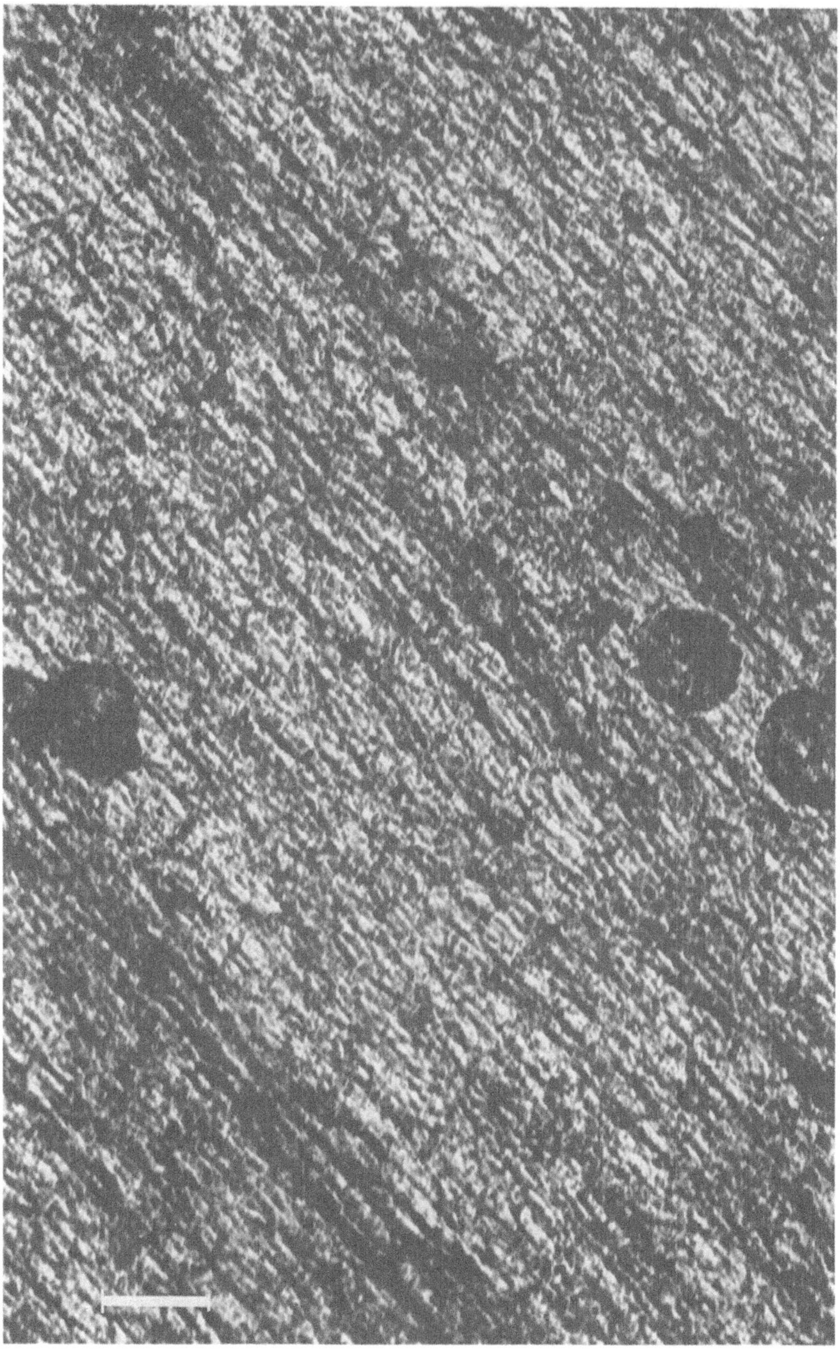

Figure 14. Electron micrograph of a relatively thick (≈1 μm) region of extruded and extracted axoplasm. The extruded axoplasmic rod was extracted for 140 min in a physiological buffer (Morris and Lasek, 1981) and fixed afterwards for 30 min in 1% glutaraldehyde and 1% paraformaldehyde solution dissolved in 0.5 M sodium cacodylate/HCl (pH 7.3), stained with 1% uranyl acetate (pH 7.3) and critical point dried. The extracted axoplasm consists of highly oriented longitudinal elements linked into a network by thin transverse bridges. The longitudinal elements, slanting upwards to the left, are oriented approximately parallel to the long axis of the extruded axoplasmic rod. The round, dense bodies are mitochondria. Micrograph taken in collaboration with Dr. Alan Hodge. Bar: 0.1 μm.

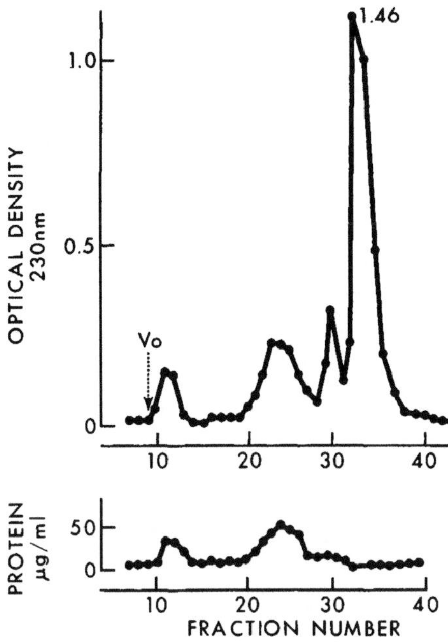

Figure 15. Fractionation of axoplasm proteins from squid giant axon: 14 axons extruded into 0.15 M KCl, 5 mM EGTA, 1 mM 2-mercaptoethanol, 10 mM Tris, pH 7.4, homogenized at 0°C, centrifuged 15,000 × g for 20 min at 4°C, supernatant applied to column of Bio Gel A 5 m equilibrated with homogenizing buffer. NF-1 appears as the principal component of the V_0 fraction. Upper curve, absorbance at 230 nm; lower curve, protein concentration as determined by Lowry method.

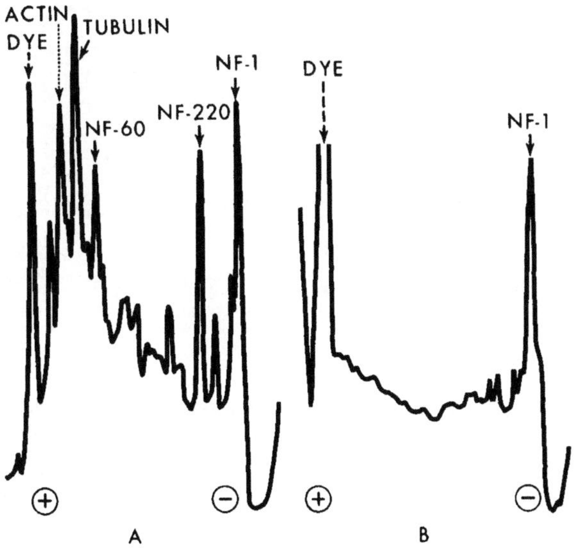

Figure 16. Densitometric scans at 575 nm of 5% SDS–PAGE gels stained with Coomassie brilliant blue R. A: whole axoplasm, B: V_0 fraction obtained by preparation scheme described above except using Bio Gel A 15 m.

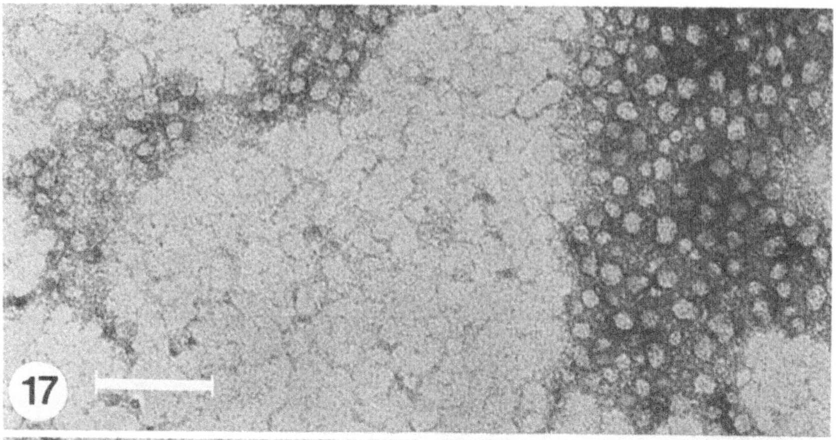

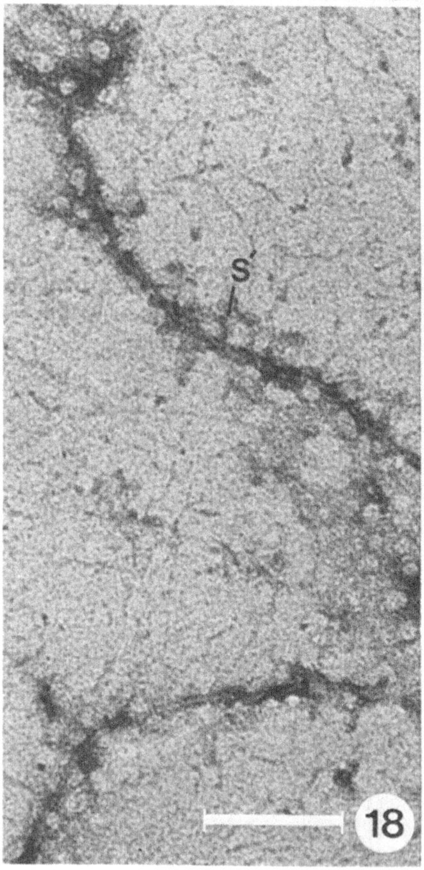

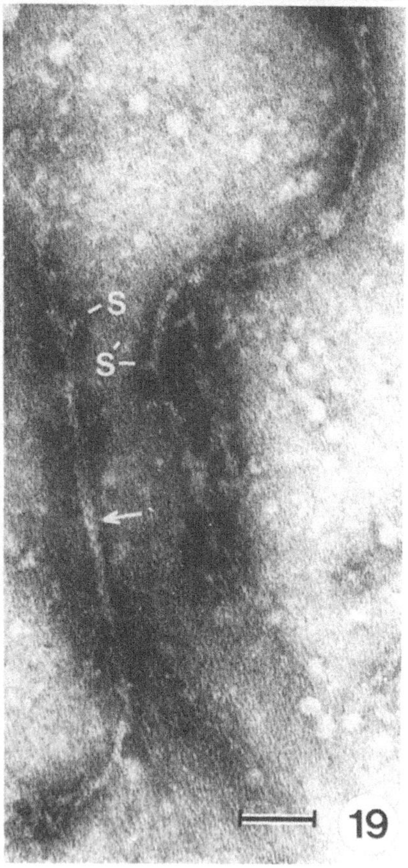

Figure 17. Electron micrograph of 380 K eluate from axoplasm homogenate fractionated on Bio Gel A 15 m column; filaments 2–5 nm in diameter aggregate in network arrays and polygons. Stained with 1% uranyl acetate (pH 4.4) for 3 min. Bar: 0.2 μm.

Figure 18. Electron micrograph of 380 K eluate from axoplasm homogenate, 2–5 nm wide filament network connected with the 10 nm neurofilaments giving the latter a lampbrush-like appearance, sidearms (S). Stained with 1% uranyl acetate (pH 4.4) for 3 min. Bar: 0.2 μm.

Figure 19. High resolution electron micrograph of negatively stained neurofilaments in the 380 K eluate, oblique sidearms (S), orthogonal sidearms (S′), helical intercoiling of the 2-n-wide unit filaments (arrow). Bar: 50 nm.

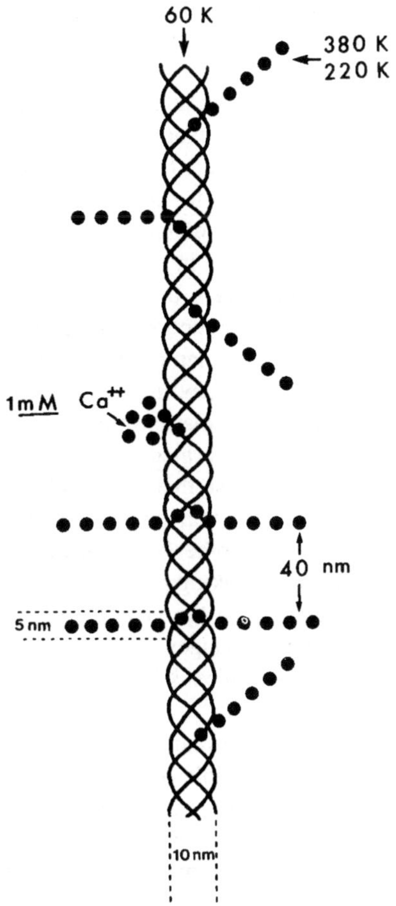

Figure 20. Model of the neurofilament. The 10-nm-wide core is composed of four intercoiled unit filaments (60,000 daltons). The 5-nm-wide sidearms (380 K and 220 K) are associated with the core filament at different angles and are spaced 35–45 nm apart. A number of the sidearms are paired. Sidearms become globular in the presence of 1 mM Ca^{++} ions.

and periodically distributed pairs of sidearms oriented at various angles (S, S′, Fig. 19). We assume that the four-stranded, 10 nm filaments are composed principally of the 60 K protein.

A model of neurofilament structure based on electron microscopic and biochemical data is presented in Fig. 20. The high molecular weight protein is at least a portion of the crosslinking system of the neurofilamentous network and may be implicated in its association with the axolemma and other membranous structures of the cell. The conformation of crosslinks is altered in the presence of 1 mM calcium solutions to a globular form (J. Metuzals and J. R. Morris, unpublished observations).

VIII. CONCLUSION

The ectoplasm of the squid giant axon is analogous to structures in many other cell types and represents a plasma membrane-apposed differentiation of the cytoskeleton. Such a differentiation has been termed "membrane cytoskeleton" (Kirkpatrick, 1979), "membrane matrix" (Koppel *et al.*, 1981) or "plasma membrane–cytoskeleton complex" (Ramaekers *et al.*, 1982). The principal components of the ectoplasm are the neurofilaments, the microtubules, the actin microfilaments and the dense bodies (Figs. 2,7, and 9). In the ectoplasm the subunit stoichiometry of the three principal neurofilamentous network proteins is probably shifted to more crosslinking protein, i.e., to a higher crosslinking density. In the endoplasm there is less of the crosslinking structures, an arrangement resulting in a more open network structure.

The membrane cytoskeleton or matrix is a dominant factor in determining the diffusion and distribution characteristics of integral membrane proteins represented by intramembrane particles in Fig. 10. Such a viewpoint removes the focus of membrane events from the bilayer *per se* to the nature of the high molecular weight matrix associated with the bilayer. Thus, the lipid bilayer provides a solvent phase for diffusion of the integral membrane proteins to move in two dimensions laterally through the matrix (Koppel *et al.*, 1981). The membrane matrix or membrane cytoskeleton is modeled by these authors as a dynamic network with junction points, or crosslinks, continually breaking down and reforming. Changes in the nature and extent of matrix crosslinking could be introduced by phosphorylation or the binding of divalent cations to any of the matrix components. This general concept of the relationship between the plasma membrane and the adjoining portion of the cytoplasm, elaborated in many types of cells, is consistent with our data and conclusions.

We propose that the axoplasm projects its organization through the ectoplasm into the axolemma and that this organization is reflected in the surface pattern of the axon (Figs. 5 and 6). On the other hand, the activities of the axolemma exert their influence on the axoplasm via the crosslinking properties of the neurofilamentous network through changes of the ionic surroundings of the neurofilamentous network proteins during excitation. Besides the conformational changes of the neurofilamentous network, assembly–disassembly of actin and tubulin molecules is involved in the reciprocal relationships between the axolemma and the axoplasm (Sakai *et al.*, 1981; Fukuda *et al.*, 1981).

The molecular model of the neurofilaments shown in Fig. 20 represents the basic assembly principle of the axoplasm organization and is repeated at all the hierarchic levels of the axon structure. The pattern of filamentous organization of the axoplasm is clearly revealed in the stereo electron micrographs of extruded axoplasm (Fig. 13) and in the surface texture of axons observed in scanning electron micrographs of desheathed axons (Figs. 4, 5, and 6). The order of the filamentous axon components should somehow be reflected in the distribution pattern of the intramembrane particles in the axolemma as revealed by freeze–etching (Fig. 10). This pattern must be related to (1) the excitability of the axon, and (2) the mechanisms of interactions between the axon and the surrounding Schwann cell (Peracchia, 1974; Stolinski *et al.*, 1981).

Our data, together with a review of the pertinent literature, suggest that in squid giant axon the ectoplasm plays a vital role in excitability and, thus, the term "axolemma–ectoplasm complex" may be coined to denote a possible meaningful structural and functional entity of the axon.

Measurements of ion conduction noise (Fishman, 1981) indicate that the internal surface of the axolemma may contain structures that extend into the axoplasm of the squid giant fiber to form a matrix that is in series with membrane ion channels.

Future investigations of the nature of the axolemma–ectoplasm complex must be directed toward the chemical characterization of its protein and lipid components and determination of the three-dimensional arrangement of these components and their dynamic changes in living axons during excitation. Investigations of the swelling of nerve fibers associated with action potentials (Iwasa *et al.*, 1980) and the examination of fast axonal transport by Allen video-enhanced contrast–differential interference contrast (AVEC–DIC) microscopy (Allen *et al.*, 1981, 1982) provide two important methods to observe changes in the axolemma–ectoplasm complex in living fibers.

ACKNOWLEDGMENTS. The authors would like to acknowledge the collaboration in this investigation of Drs. A. J. Hodge, R. J. Lasek, J. R. Morris, J. L. Oschman, and S. Terakawa. This investigation was supported by grant MA-1247 from the Medical Research Council of Canada.

REFERENCES

Allen, R. D., Metuzals, J., and Tasaki, I., 1981, Fast axonal transport in the squid giant axon, *Biol. Bull.* **161:**303.

Allen, R. D., Metuzals, J., Tasaki, I., Brady, S. T., and Gilbert, S. P., 1982, Fast axonal transport in squid giant axon, *Science* **218:**1127–1129.

Baker, P. F., Hodgkin, A. L., and Shaw, T. I., 1962, Replacement of the axoplasm of giant nerve fibers with artificial solutions, *J. Physiol. (London)* **164:**330–354.

Blaustein, M. P., and Goldman, D. E., 1968, The action of certain polyvalent cations on the voltage-clamped lobster axon, *J. Gen. Physiol.* **51:**279–291.

Branton, D., Bullivant, S., Gilula, N. B., Karnovsky, M. J., Moor, H., Mühlethaler, K., Northcote, D. H., Packer, L., Satir, B., Satir, P., Speth, V., Staehlin, L. A., Steere, R. L., and Weinstein, R. S., 1975, Freeze-etching nomenclature, *Science* **190:**54–56.

Bray, D., and Bunge, M. B., 1981, Serial analysis of microtubules in cultured rat sensory axons, *J. Neurocytol.* **10:**589–605.

Brown, F., and Danielli, J. F., 1964, The cell surface and cell physiology, in: *Cytology and Cell Physiology* (G. H. Bourne, ed.), pp. 239–310, Academic Press, New York.

Chalfie, M., and Thomson, J. N., 1979, Organization of neuronal microtubules in the nematode Caenorhabditis elegans, *J. Cell Biol.* **82:**278–289.

Chambers, R., 1947, The shape of oil drops injected into the axoplasm of the giant nerve of the squid, *Biol. Bull.* **93:**191.

Chambers, R., and Kao, C. Y., 1952, The effect of electrolytes on the physical state of the nerve axon of the squid and of stentor, a protozoon, *Exp. Cell Res.* **3:**564–573.

Chang, D. C., 1979, A physical model of nerve axon. II: Action potential and excitation currents. Voltage-clamp studies of chemical driving forces of Na^+ and K^+ in squid giant axon, *Physiol. Chem. Phys.* **11**(3)**:**263–288.

Chang, D. C., 1980, Possible role of cytoplasmic microtubule structure in the excitation properties of nerve axon, *Eur. J. Cell Biol.* **22:**304.

De Chastellier, C., and Ryter, A., 1981, Calcium-dependent deposits at the plasma membrane of *Dictyostelium discoideum* and their possible relation with contractile proteins, *Biol. Cell.* **40:**109–118.

Ellisman, M. H., and Porter, K. R., 1980, Microtrabecular structure of the axoplasmic matrix: Visualization of cross-linking structures and their distribution, *J. Cell Biol.* **87:**464–479.

Fishman, H. M., 1981, Material from the internal surface of squid axon exhibits excess noise. Implications in modeling membrane noise, *Biophys. J.* **35:**249–255.

Fukuda, J., Kameyama, M., and Yamaguchi, K., 1981, Breakdown of cytoskeletal filaments selectively reduces Na and Ca spikes in cultured mammal neurones, *Nature* **294:**82–85.

Geisler, N., and Weber, K., 1981, Self-assembly *in vitro* of the 68,000 molecular weight component of the mammalian neurofilament triplet proteins into intermediate filaments, *J. Mol. Biol.* **151:**565–571.

Hernandez-Nicaise, M-L., Bilbaut, A., Malaval, L., and Nicaise, G., 1982, Isolation of functional giant smooth muscle cells from an invertebrate: Structural features of relaxed and contracted fibers, *Proc. Natl. Acad. Sci. USA* **79:**1884–1888.

Hillman, D. E., and Llinas, R., 1974, Calcium-containing electron-dense structures in the axons of the squid giant synapses, *J. Cell Biol.* **61:**146–155.

Hodge, A. J., and Adelman, W. J., Jr., 1980, The neuroplasmic network in *Loligo* and *Hermissenda* neurons, *J. Ultrastruct. Res.* **70:**220–241.

Inoue, I., Tasaki, I., and Kobatake, Y., 1974, A study of the effects of externally applied sodium-ions and detection of spatial non-uniformity of the squid axon membrane under internal perfusion, *Biophys. Chem.* **2:**116–126.

Iwasa, K., Tasaki, I., and Gibbons, R. C., 1980, Swelling of nerve fibers associated with action potentials, *Science* **210:**338–339.

Kim, H., Binder, L. I., and Rosenbaum, J. L., 1979, The periodic association of MAP_2 with brain microtubules *in vitro, J. Cell Biol.* **80:**266–276.

Kirkpatrick, F. H., 1979, New models of cellular control: Membrane cytoskeletons, membrane curvature potential, and possible interactions, *Biosystems* **11**:93–109.

Koppel, D., Sheetz, M. P., and Schindler, M., 1981, Matrix control of protein diffusion in biological membranes, *Proc. Natl. Acad. Sci. USA* **78**(6):3576–3580.

Krishnan, N., Kaiserman-Abramoff, I. R., and Lasek, R. J., 1979, Helical substructure of neurofilaments isolated from *Myxicola* and squid giant axons, *J. Cell Biol.* **82**:323–335.

Lasek, R. J., Krishnan, N., and Kaiserman-Abramoff, I. R., 1979, Identification of the subunit proteins of 10-nm neurofilaments isolated from axoplasm of squid and *Myxicola* giant axons, *J. Cell Biol.* **82**:336–346.

Metuzals, J., 1969, Configuration of a filamentous network in the axoplasm of the squid (*Loligo pealii* L.) giant nerve fiber, *J. Cell Biol.* **43**:480–505.

Metuzals, J., and Clapin, D. F., 1981, Modes of crosslinking in neurofilament protein isolated from squid giant axon: Electron microscopic evidence for paracrystalline arrays, *Biol. Bull.* **161**(2):308.

Metuzals, J., and Izzard, C. S., 1969, Spatial patterns of thread-like elements in the axoplasm of the giant nerve fiber of the squid (*Loligo pealii* L.) as disclosed by differential interference microscopy and by electron microscopy, *J. Cell Biol.* **43**:456–479.

Metuzals, J., and Mushynski, W. E., 1974, Electron microscope and experimental investigations of the neurofilamentous network in Deiter's neurons, *J. Cell Biol.* **61**:701–722.

Metuzals, J., and Tasaki, I., 1978, Subaxolemmal filamentous network in the giant nerve fiber of the squid (*Loligo peali* L.) and its possible role in excitability, *J. Cell Biol.* **78**:597–621.

Metuzals, J., Terakawa, S., and Tasaki, I., 1980a, The axolemmina–ectoplasm complex investigated in desheathed squid giant axons, in: *Electron Microscopy 1980, Vol. 2 Biology*, Publ. by the Seventh European Congress on Electron Microscopy Foundation, Leiden, The Netherlands.

Metuzals, J., Lasek, R. J., and Hodge, A. J., 1980b, Stereo electron microscopy of the neurofilamentous network prepared by a sandwich technique, *Eur. J. Cell Biol.* **22**(1):380.

Metuzals, J., Montpetit, V., and Clapin, D. F., 1981a, Organization of the neurofilamentous network, *Cell Tissue Res.* **214**:455–482.

Metuzals, J., Tasaki, I., Terakawa, S., and Clapin, D. F., 1981b, Removal of the Schwann sheath from the giant axon of the squid: An electron microscopic study of the desheathed axolemma and of associated axoplasmic structures, *Cell Tissue Res.* **221**:1–15.

Metuzals, J., Clapin, D. F., and Chapman, G. D., 1982, Axial and radial filamentous components of the neurofilamentous network, *Cell Tissue Res.* **223**:507–518.

Metuzals, J., Hodge, A. J., Lasek, R. J., and Kaiserman-Abramof, I. R., 1983, Neurofilamentous network and filamentous matrix preserved and isolated by different techniques from squid giant axon, *Cell Tissue Res.* **228**:415–432.

Morris, J. R., and Lasek, R. J., 1981, Stable polymers of the axonal cytoskeleton: The axoplasmic ghost, *J. Cell Biol.* **92**:192–198.

Nicaise, G., Hernandez-Nicaise, M-L., and Malaval, L., 1982, Electron microscopy and X-ray microanalysis of calcium-binding sites on the plasma membrane of *Beroe* giant smooth muscle fibre, *J. Cell Sci.* **55**:353–364.

Ochs, R. L., and Burton, P. R., 1980, Distribution and selective extraction of filamentous components associated with axonal microtubules of crayfish nerve cord, *J. Ultrastruct. Res.* **73**:169–182.

Oschman, J. L., Hall, T. A., Peters, P. D., and Wall, B. J., 1974, Association of calcium with membranes of squid giant axon: Ultrastructure and microprobe analysis, *J. Cell Biol.* **61**:156–165.

Pant, H. C., and Gainer, H., 1980, Properties of a calcium-activated protease in squid axoplasm which selectively degrades neurofilament proteins, *J. Neurobiol.* **11**:1–12.

Pant, H. C., Shecket, G., Gainer, H., and Lasek, R. J., 1978, Neurofilament protein is phosphorylated in the squid giant axon, *J. Cell Biol.* **78**(2):R23–R27.

Peracchia, C., 1974, Excitable membrane ultrastructure, *J. Cell Biol.* **61**:107–122.

Pitman, R. M., Tweedle, C. D., and Cohen, M. J., 1972, Branching of central neurons: Intracellular cobalt injection for light and electron microscopy, *Science* **176**:412–414.

Ramaekers, F. C. S., Dunia, I., Dodemont, H. J., Benedetti, E. L., and Bloemendal, H., 1982, Lenticular intermediate-sized filaments: Biosynthesis and interaction with plasma membrane, *Proc. Natl. Acad. Sci. USA* **79**:3208–3212.

Roslansky, P. F., Cornell-Bell, A., Rice, R. V., and Adelman, W. J., Jr., 1980, Polypeptide composition of squid neurofilaments, *Proc. Natl. Acad. Sci. USA* **77**:404–408.

Sabatini, M. T., Dipolo, R., and Villegas, R., 1968, Adenosine triphosphatase activity in the membrane of the squid nerve fiber, *J. Cell Biol.* **38**:176–183.

Sakai, H., Matsumoto, G., Endo, S., and Kobayashi, T., 1981, Microtubules in squid giant axon: *In vitro* assembly, distribution in the axon and maintenance of the membrane excitability, in: *Nerve Membrane: Biochemistry and Function of Channel Proteins* (G. Matsumoto and M. Kotani, eds.), pp. 185–202, University of Tokyo Press, Tokyo.

Skaer, H. le B., Treherne, J. E., Benson, J. A., and Moreton, R. B., 1978, Axonal adaptations to osmotic and ionic stress in an invertebrate osmoconformer (*Mercierella enigmatica* Fauvel). I. Ultrastructural and electrophysiological observations on axonal accessibility, *J. Exp. Biol.* **76:**191–204.

Stolinski, C., Breathnach, A. S., Martin, B., Thomas, P. K., King, R. H., and Gabriel, G., 1981, Associated particle aggregates in juxtaparanodal axolemma and adaxonal Schwann cell membrane of rat peripheral nerve, *J. Neurocytol.* **10:**679–691.

Tasaki, I., 1982, *Physiology and Electrochemistry of Nerve Fibers, Biophysics and Bioengineering Series* (A. Noordergraff, ed.), Academic Press, New York, London.

Villegas, G. M., and Villegas, J., 1976, Structural complexes in the squid giant axon membrane sensitive to ionic concentrations and cardiac glycosides, *J. Cell Biol.* **69:**19–28.

Villegas, G. M., Villegas, J., and De Weer, P., 1972, Proteolytic enzyme and strophanthoside effects on the ultrastructure and membrane electrical potentials of the squid giant nerve fiber, in: *Fourth International Congress IUPAB,* Volume 3 (Sections IX–XV), p. 31, Biophysics, Moscow.

Williams, R. J. P., 1970, The biochemistry of sodium, potassium, magnesium and calcium, *Quart. Rev. Chem. Soc. (London)* **24:**:331–365.

4

The Neuroplasmic Lattice

Structural Characteristics in Vertebrate and Invertebrate Axons

Alan J. Hodge and William J. Adelman, Jr.

I. ORDERED STRUCTURE IN AXONS

Observations made in this laboratory over the past several years utilizing a variety of electron microscopic methods have demonstrated that the longitudinally oriented filamentous elements in axons from a wide variety of organisms, namely, neurofilaments and neurotubules (microtubules) are, in general, linked together by periodically distributed and transversely oriented bridges to form a well-defined three-dimensional grid structure, at least under the rigor-inducing fixation conditions used. This we have termed the *neuroplasmic lattice* (Hodge and Adelman, 1978a,b, 1980a,b, 1981, 1983). The term "lattice" has been used deliberately in a quasicrystalline sense to indicate the presence of order, not of the very high degree found in atomic or small molecular crystalline arrays, but rather approaching the somewhat lower degree found in macromolecular arrays such as that present in fibrils of collagen, fibrin, and in the thick and thin filament lattices in various types of muscle. In these examples, *inter alia,* the presence of order was convincingly established by the pioneering x-ray diffraction work of Astbury and colleagues (Astbury, 1947), in establishing the various classes of fibrous proteins, particularly the keratin–myosin–epidermin–fibrin (kmef) group, and further elucidated by electron microscopy (see reviews by Hodge, 1959, 1960, 1967; Squire, 1981) and low angle x-ray diffraction (Bear, 1942, 1945; Huxley, 1953; Huxley and Brown, 1967; Traub and Piez, 1971).

Unfortunately, no such demonstration of order in whole axoplasm *per se* has as yet been achieved by low angle x-ray diffraction, probably because of the great difficulty associated with attaining a reasonable signal-to-noise ratio in such a highly hydrated and attenuated structure. However, Day and Gilbert (1972) showed that dried and stretched strands of *Myxicola* axoplasm gave a typical α-pattern. Gilbert (unpublished), moreover,

Alan J. Hodge and William J. Adelman, Jr. • Laboratory of Biophysics, NINCDS, National Institutes of Health at the Marine Biological Laboratory, Woods Hole, Massachusetts 02543.

before his recent untimely death, was able to obtain good high and low angle x-ray diffraction patterns from artificial fibers drawn from centrifuged pellets of *Myxicola* neurofilaments subjected to the mild action of endogenous calcium-activated proteases. These clearly showed that the neurofilament "core" protein was predominantly of the α-type, and that the chains were packed together with sufficient order to yield a well-developed meridional pattern as yet unindexed but corresponding to an axial period of several tens of nanometers. This suggests that the filaments are assembled from very long fibrous α-helical subunits. Recently, Milam and Erickson (1982) have observed a left-handed helix of 21 nm pitch in keratin filaments and neurofilaments. Earlier, Kallman and Wessells (1967) had observed a 22 nm axial periodicity of keratin filament bundles found in thin sections of epithelial cells. These values are of great interest because they are compatible with being the second order spacing of the 43 nm longitudinal bridge spacing in axoplasm observed independently during electron microscopic studies of both vertebrate and invertebrate nerve tissue (Hodge and Adelman, 1978a,b, 1980a,b, 1981, 1983). Ellisman and Porter (1980) also observed a similar period in mammalian neurons.

The presence of lateral projections extending from, and/or bridges between, elements of the axoplasm as seen in single projection micrographs of thin sections from a variety of species is not novel. For example, crossbridges have been observed between neighboring neurofilaments (Metuzals, 1966, 1969; Berthold, 1978), between neurofilaments and adjacent neurotubules (Yamada *et al.*, 1971; Roslansky *et al.*, 1980; Rice *et al.*, 1980), between neighboring neurotubules (Tani and Ametani, 1970; Burton and Fernandez, 1973; Ochs and Burton, 1980), and between the filamentous elements and nearly all membrane surfaces exposed to the axoplasm. The concept of a network of bridges, not necessarily periodically ordered, appears to have been tacitly assumed by many from the appearance of thin sections, which at best give a very limited sampling of the lattice. Direct visualization of the periodic three-dimensional characteristics of the neuroplasmic lattice has only proved feasible by examination of relatively thick sections (0.2–0.5 μm) using stereoscopic methods.

The presence of an ordered lattice comprising neurofilaments and some neurotubules, when present, linked together by a periodically distributed system of bridges was shown stereoscopically by Hodge and Adelman (1978a, 1980b) for *Loligo* (squid) and *Hermissenda* (a nudibranch) axons, and extended to include *Bufo* (toad) and *Homarus* (lobster) axons (Hodge and Adelman, 1980a,b, 1981, 1983). In the case of *Homarus,* as already noted by Burton *et al.*, (1975), there appeared to be very few, if any, neurofilaments present so that the neuroplasmic lattice comprised only neurotubules and crossbridges. Ochs and Burton (1980) have also described the presence of a network in crayfish axons, using the term "axoplasmic matrix filaments" to describe the bridges. Ellisman and Porter (1980) have described the array of crossbridges in relatively thick sections of rat axons examined by high voltage electron microscopy (HVEM). Their term, "microtrabecular lattice," which was extended to the axon from observations on other cells (Porter, 1976; Wolosewick and Porter, 1979), appears to be the equivalent of the "neuroplasmic lattice" of Hodge and Adelman (1978a,b, 1980a,b, 1981, 1983). Ellisman and Porter (1980) also noted that "the wispy strands of trabeculae appeared to radiate into the ground substance at right angles to the microtubules and neurofilaments with a periodicity of ~42.5 nm." Thus, as we have long believed, and as mounting evidence concerning their specificity suggests, the crossbridges appear to be formed elements in their own right and should not be considered as part of the ground substance or matrix. In other words, there does not seem to be a justifiable basis for making an arbitrary distinction between formed elements and those structures found in the "ground substance" or "axoplasmic matrix," the term used by some authors.

As we shall see, most elements of the axoplasm appear to exist in states of moderately high order or quasicrystallinity, at least in the "rigor state," and even the so-called soluble proteins, commonly assumed to be randomly distributed in the aqueous phase, are more likely to exist in some kind of loosely bound condition relative to the so-called cytoskeletal elements. While some of the filamentous elements of axoplasm must subserve a skeletal function, especially in relation to maintenance of overall neuronal form, perhaps mediated by the crossbridges seemingly linking the membrane and especially the axolemmal surface proteins with the neuroplasmic lattice, nevertheless, even their skeletal role must be a dynamic one, as must be their likely participation in such functions of the axoplasm as transport of metabolites and transmitter substances (Schwartz, 1979).

II. ELECTRON STEREOMICROSCOPY OF AXONS

A. General Aspects

The presentation of three-dimensional information derived from electron micrographs is still in its infancy. The most practical and widely used method is the viewing of stereo-pairs of micrographs produced by appropriate tilting of the specimen in the electron microscope. Its most severe limitation is the limited in-depth information content allowable in viewing only two projections of the specimen before "confusion" (due to obscuration of structural detail) results in the fused binocular image. Stereoscopic viewing lends itself well to the study of surfaces since they contain much less visual information than do "in-depth" specimens such as thick sections of tissue. The risk of stereoscopic confusion arising during observation of specific specimen details is much less, and much of this can be circumvented by such techniques as "rotary shadowing." It is not surprising, therefore, that the technique has found much favor in the study of thin metal shadowed replicas of surfaces produced by such methods as freeze–fracture and freeze–etching (see the chapters in this volume by Rosenbluth and Hirokawa).

In the case of solid specimens such as thick sections of embedded tissue, the problem of visual information density is further compounded by the unfavorable electron scattering properties of the specimen. In particular, the low atomic number atoms making up the embedding medium cause a large component of forward inelastic electron scattering, i.e., of electrons suffering appreciable energy loss and hence increasing their wavelength. At conventional accelerating voltages (50–100 kV), the inherent chromatic aberration characteristics of transmission electron microscopes cause diffuse focusing of these electrons, resulting in a severe loss of image contrast, most of which is normally achieved by a combination of accurate focusing of paraxial elastically scattered electrons together with the exclusion of those outside the acceptance aperture of the electron optical system.

These problems, which are quite severe at conventional accelerating voltages (40–120kV), can be overcome by utilizing a TEM such as the Philips EM400, which is programmed to minimize the effects of chromatic aberration or, by recourse to HVEM (~1 MeV), where the radial distribution functions for electron scattering are not as severe. It should be noted that in both cases there remains in stereoscopy the problem of confusion arising from excessive information densities in the specimen. Indeed, it is the information density of the specimen itself which dictates the maximum useful thickness of a section for adequate three-dimensional sampling in stereo regardless of accelerating voltage. It seems clear that further progress in three-dimensional electron microscopy of cellular structure must await the development of techniques for both reconstituting the structure from multiple projections in TEM or scanning transmission electron microscopy (STEM) and visualizing it conveniently and unambiguously. Clearly, digital computers will be involved and it is to be hoped that their application will lead to improved methods for three-dimensional visualization of fine structure, including the production of holographic displays.

B. *Loligo* Axons

The smaller axons of *Loligo* were fixed by simple immersion in an appropriate fixative, balanced with respect to pH, tonicity, and ionic strength and preferably containing a small amount of ethylene glycol tetraacetic acid (EGTA) to minimize the disruptive effects of free Ca^{++} (Hodge and Adelman, 1980a). Small concentrations of Mg^{++} were also added in order to promote formation and preservation of the "rigor state." These small axons contained the usual assortment of organelles (mitochondria, elements of the smooth endoplasmic reticulum, various kinds of vesicles, etc.) found in other axons (Peters *et al.*, 1976), and the bulk of the axoplasms consisted of a crossbridged oriented lattice of neurofilaments, with the incorporation of variable numbers of neurotubules. This ordered neuroplasmic lattice is shown stereoscopically in Fig. 1, where the crossbridges confer a finely striated appearance on the axoplasm.

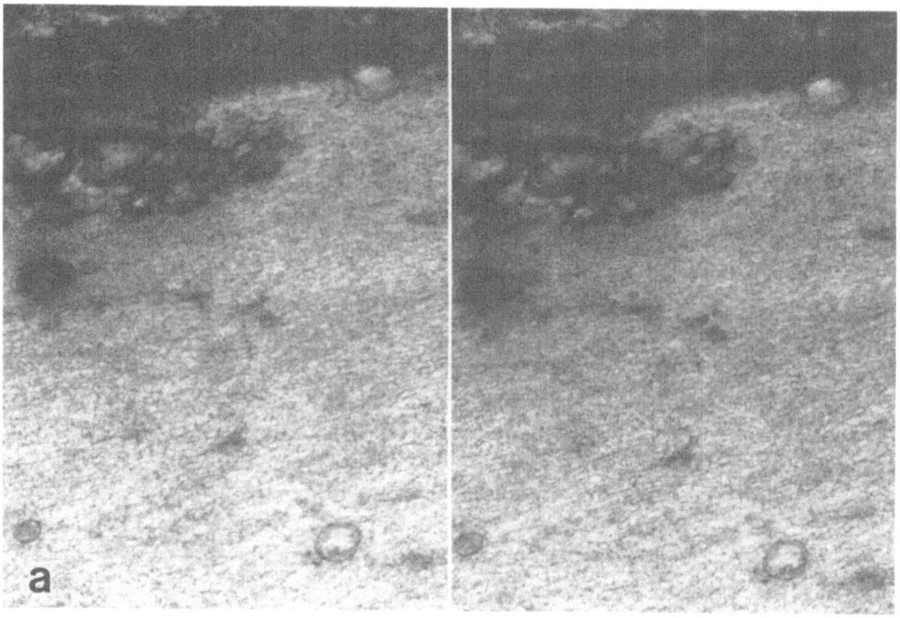

Figure 1. (a) Stereomicrographs of an ~0.3 mm thick longitudinal section through a small fiber accompanying a third-order giant axon in *Loligo* mantle showing the well-ordered neuroplasmic lattice comprising mostly neurofilaments and thin transverse bridge elements. Section stained with uranyl acetate only. ×15,000. (b) Stereomicrographs of a relatively thick (~0.3 μm) longitudinal section through a *Loligo* giant axon following cannulation, irrigation with an isotonic solution of rabbit myosin fraction S_1, then fixation by irrigation as described by Hodge and Adelman (1980a). Note that the neuroplasmic lattice consists primarily of an essentially longitudinal array of neurofilaments and neurotubules, the latter tending to occur in bundles. Note also that all structures appear to be linked by crossbridges and that there is a higher concentration of interstitial material associated primarily but not exclusively with the neurotubule bundles. Axolemma and sheath complex at top. Uranyl–lead staining. ×17,000. (c) The same *Loligo* specimen as shown in (b) but shown in a thinner uranyl–lead-stained section with better resolution. The small arrows indicate adventitious lead deposits, while the larger arrow indicates the higher density of interstitial material in a typical neurotubule bundle. Crossbridging is highly evident throughout. ×23,000.

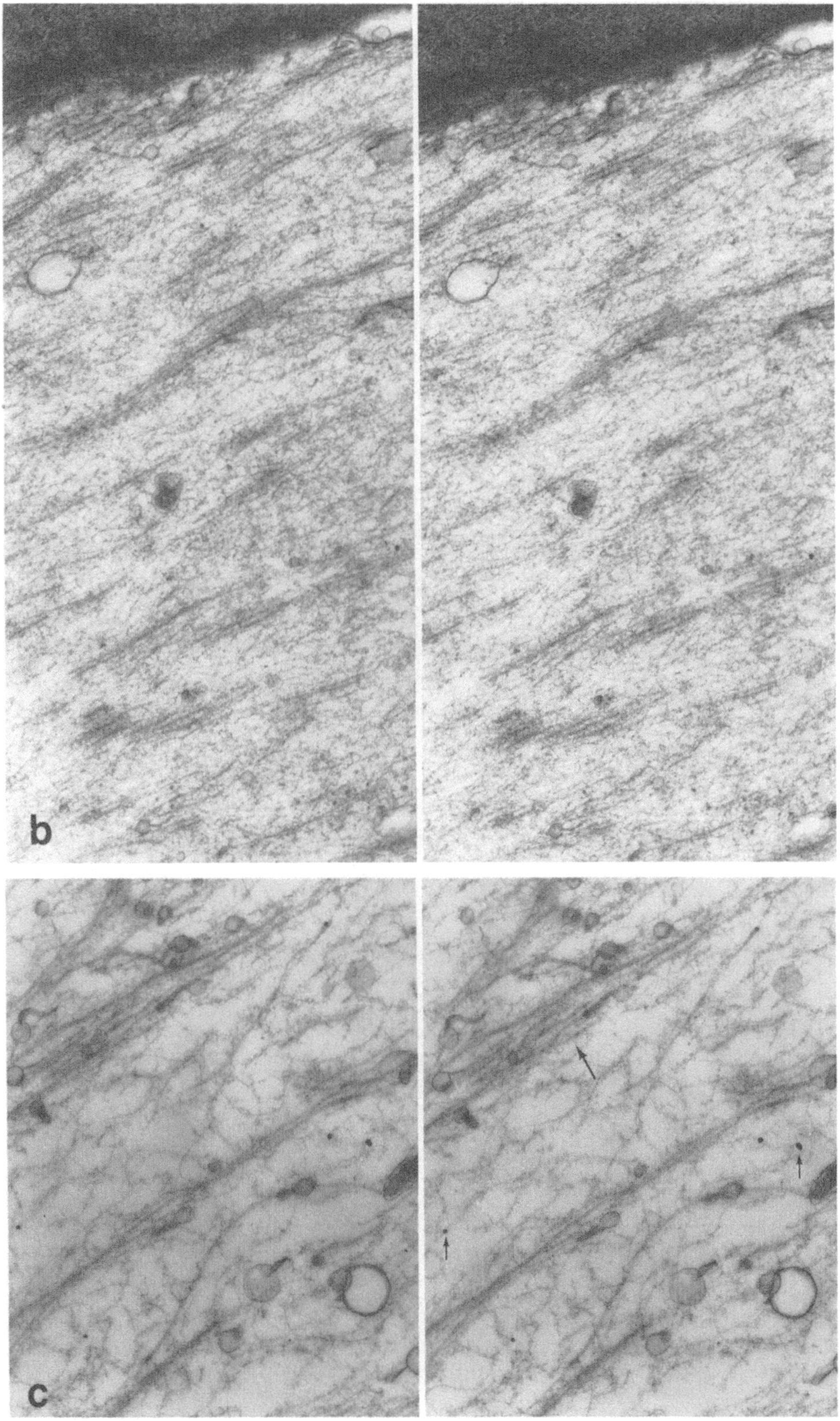

Figure 1. Continued

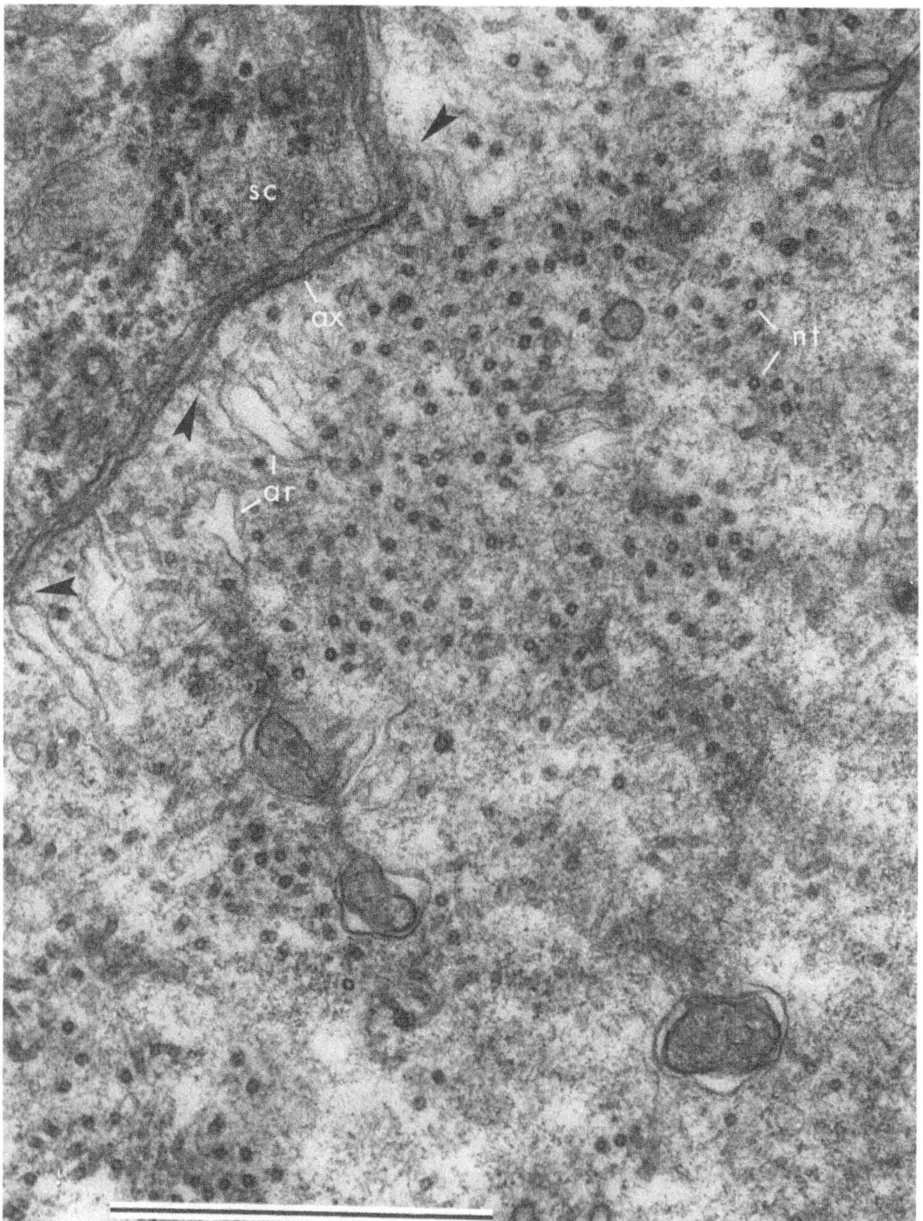

Figure 2. Transverse section through a *Loligo* giant axon, after fixation by internal irrigation following cannulation, showing the high concentration of neurotubules (nt) in the subaxolemmal cortical zone of the axoplasm and the intricate relationship (arrowheads) of the agranular reticulum (ar) with the cytoplasmic face of the axolemma (ax) bordered by a Schwann cell (sc). The bar represents 1 μm. ×53,000. (From Hodge and Adelman, 1980a).

In the case of the giant axons (~50–500 μm in diameter) radiating from the stellate ganglion of the mantle, immersion fixation gave poor and slow preservation as evidenced by the length of time (minutes) required for the yellowing reaction with glutaraldehyde, and the paucity of neurotubules in thin sections. Rapid fixation with good preservation, as judged by immediate yellowing of the axoplasm and the presence of numerous neurotubules (microtubules) as well as neurofilaments, was achieved by internal irrigation with fixative following cannulation (Hodge and Adelman, 1978a,b, 1980a). In transverse sections of giant axons (Figs. 2 and 3), the general appearance of the axoplasm was similar to that already described for small fibers of *Loligo* in that a well-developed neuroplasmic lattice of crossbridged neurofilaments and neurotubules was found, in which were embedded mitochondria, smooth endoplasmic reticulum (SER), vesicles and other cytoplasmic elements. Crossbridging by thin transverse bridges appeared to be virtually universal. In particular, the fibrous elements were commonly linked to all of the various membrane surfaces, particularly the axolemma. The neurotubules occurred both singly and in small clusters of various sizes which, although in general distributed relatively homogeneously throughout the axoplasm, were found in large numbers in a subaxolemmal cortical zone also containing an abundance

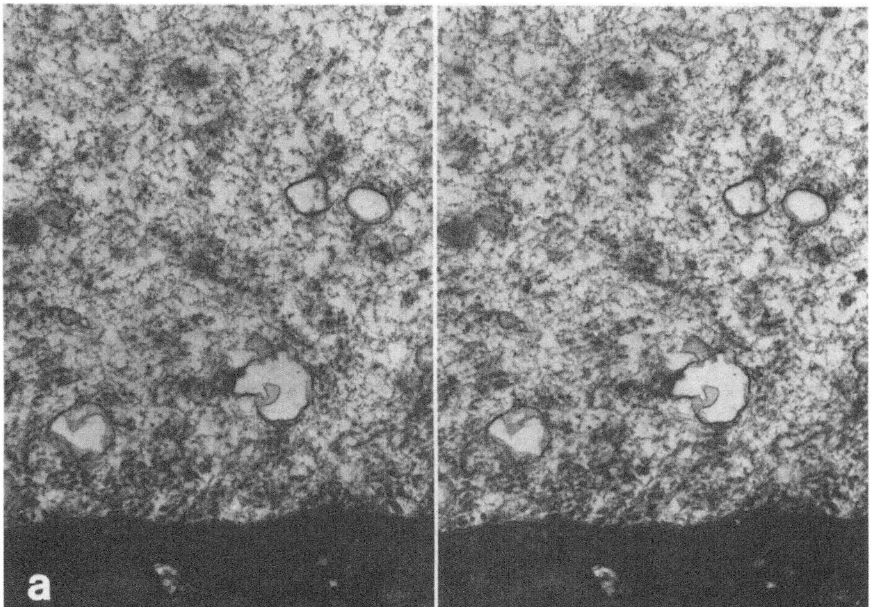

Figure 3. (a) Stereopair of *Loligo* giant axon. Relatively thick (~0.3 μm) transverse section of an axon fixed by internal irrigation with the standard fixative but containing 0.2% tannic acid. Section stained with uranyl–lead. Note the lattice extending throughout, and the concentration of agranular reticulum elements and neurotubules in the subaxolemmal zone (axolemma and Schwann sheath complex are at bottom of picture). ×17,100. (From Hodge and Adelman, 1980a). (b) Uranyl–lead-stained thick transverse section (~0.5 μm) of a *Loligo* giant axon fixed as in (a) but after irrigation with an isotonic solution containing fraction S_1 from rabbit myosin. Note that a finely granular appearing fibrous material appears preferentially associated (arrows) with the subaxolemmal region (arrowhead) and clumps of neurotubules. Bridges between fibrous elements are universal and extend to membrane surfaces such as the SER elements (small arrow). 100kV accelerating voltage; ×23,000. (c) Schematic representation of the *neuroplasmic lattice* illustrating a possible arrangement of the crossbridge array between longitudinal elements. Although not shown, similar bridges link these elements to cytoplasmic membrane surfaces and are also observed running between neurotubules. In the arrangement shown, the fundamental period (d) is 6 times the more readily apparent ~43 nm spacing between bridges.

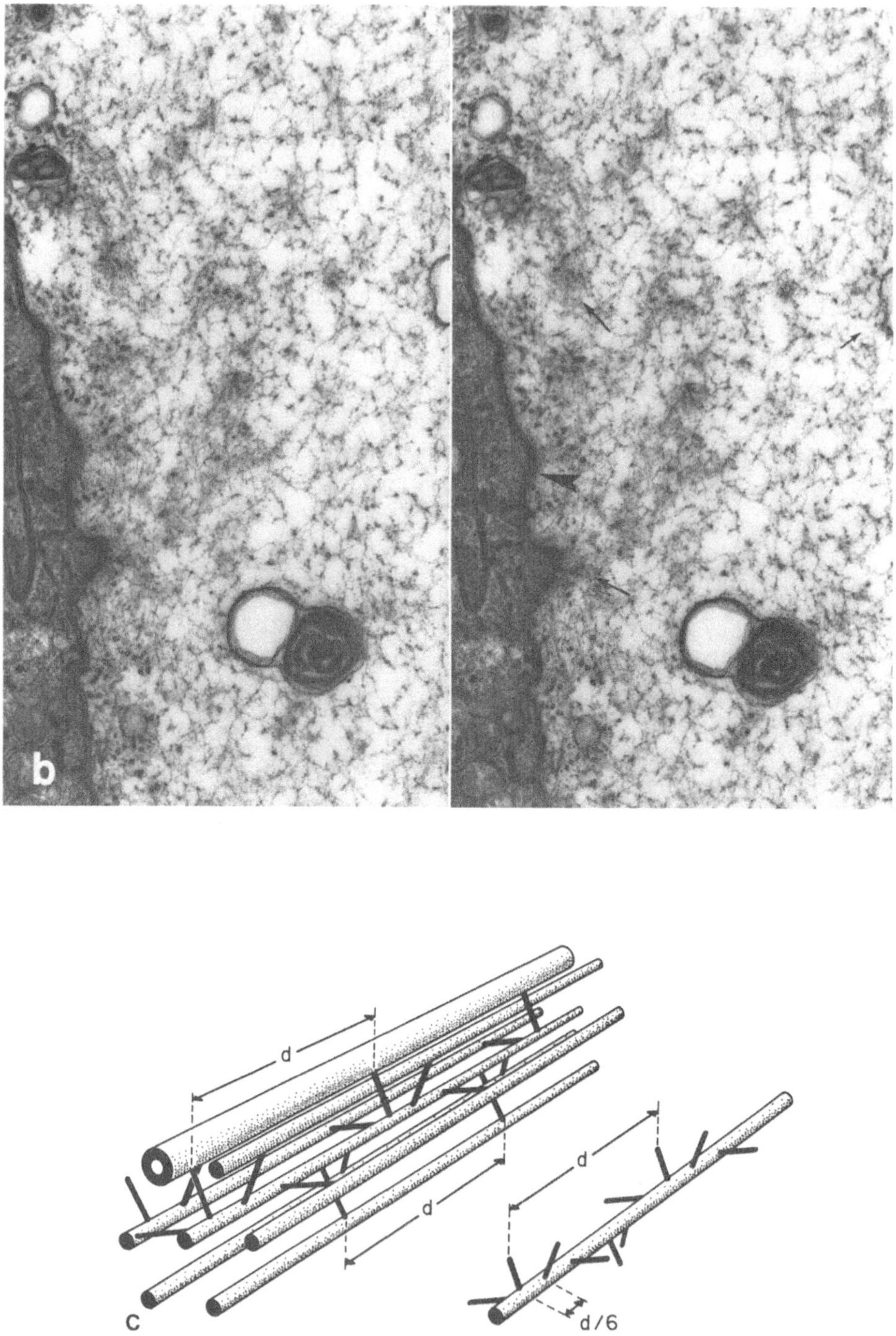

Figure 3. Continued

of tubular elements of the SER (Figs. 2 and 3). As pointed out by Hodge and Adelman (1978a, 1980a), the subaxolemmal zone appears to be specialized in that there appears to be close juxtaposition and possibly direct structural continuity of some neurotubules and elements of the SER (Fig. 2). Recent investigations by gel electrophoresis and electron microscopy on taxol-treated extruded squid axoplasm (Morris *et al.*, 1981) indicated that the neurotubules retained their mutual crossbridges as long as a putative trypsin-sensitive 300 K microtubule-associated protein (MAP) remained, even when the neurofilament lattice was removed by activation of a calcium-activated protease specific for neurofilaments. The neurotubule-rich regions also contain an unknown interstitial finely granular component (Fig. 3), which was also observed in axons of the other species to be described.

In longitudinal sections of both small and giant *Loligo* axons, the thin and sometimes "wispy" or "fluffy" crossbridges were spaced about 40–45 nm along the filamentous elements (Hodge and Adelman, 1978a,b, 1980a,b). However, the type of order present has proved difficult to determine precisely. In particular, although stereo examination of thick sections

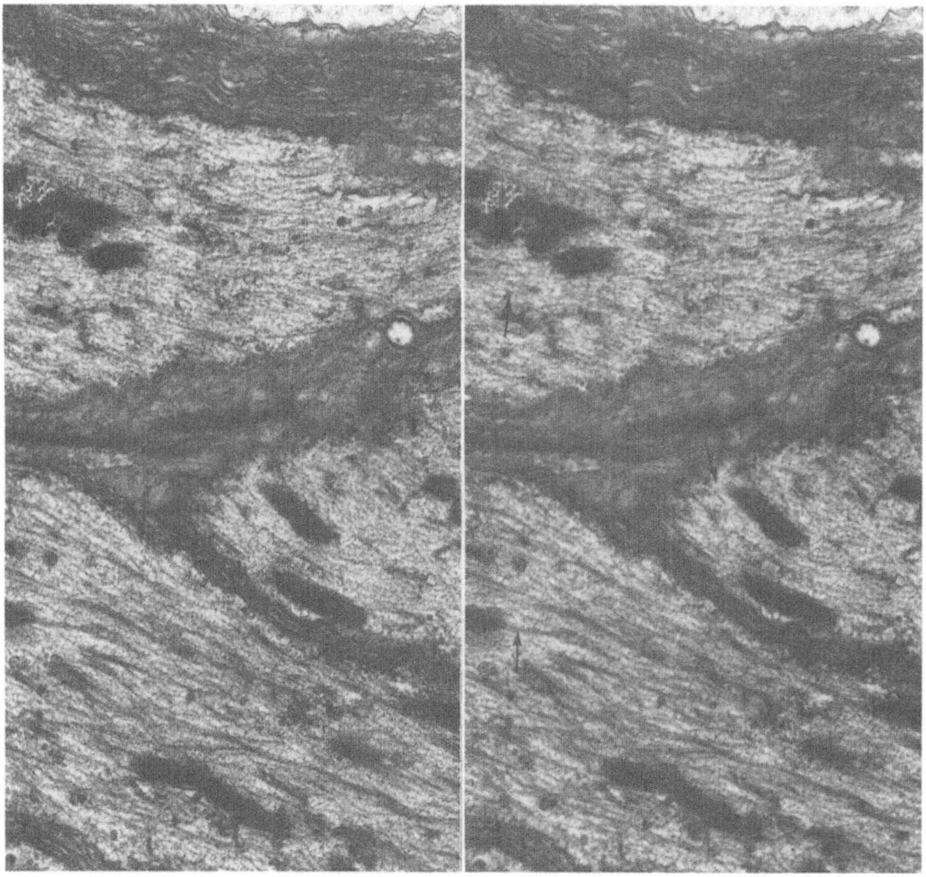

Figure 4. Stereomicrographs of an ~0.2 μm thick section through a *Hermissenda* brain preparation passing nearly longitudinally through three adjacent axons. Note the neuroplasmic lattice filling all free axonal space and that the crossbridges are oriented essentially normal (arrows) to the neurofilaments and/or neurotubules, which tend to occur in bundles (especially in lower axon). Section stained with ethanolic uranyl acetate, aqueous lead citrate. ×16,500. (From Hodge and Adelman, 1980a).

suggests that they are located with some type of screw axis symmetry, most likely three- or six-fold (see Fig. 3), it has not yet proved possible to resolve this question either by direct observation or by analytical methods such as optical autocorrelation of images and Fourier analysis of suitably oriented STEM traces. However, there is no doubt that the bridges are arranged with a relatively high degree of order consistent with the spatial limitations imposed by the presence of structures such as mitochondria and other cytoplasmic inclusions. While these can clearly be regarded as causing "defects" in an otherwise ordered neuroplasmic lattice, they are perforce included by virtue of being linked to the filamentous elements by crossbridges.

C. *Hermissenda* Axons

The neuroplasmic lattice of *Hermissenda* axons in small connectives and in the cerebropleural ganglia has been described in some detail (Hodge and Adelman, 1980a) as seen in relatively thick sections (~0.2 μm). As in the case of squid axons, those in *Hermissenda* were filled with a neuroplasmic lattice consisting primarily of periodically crossbridged longitudinally-oriented neurofilaments and neurotubules, with the numerical ratio between these elements varying considerably from axon to axon, presumably for functional reasons. The axial spacing between bridges as in the squid was found to be ~40–45 nm.

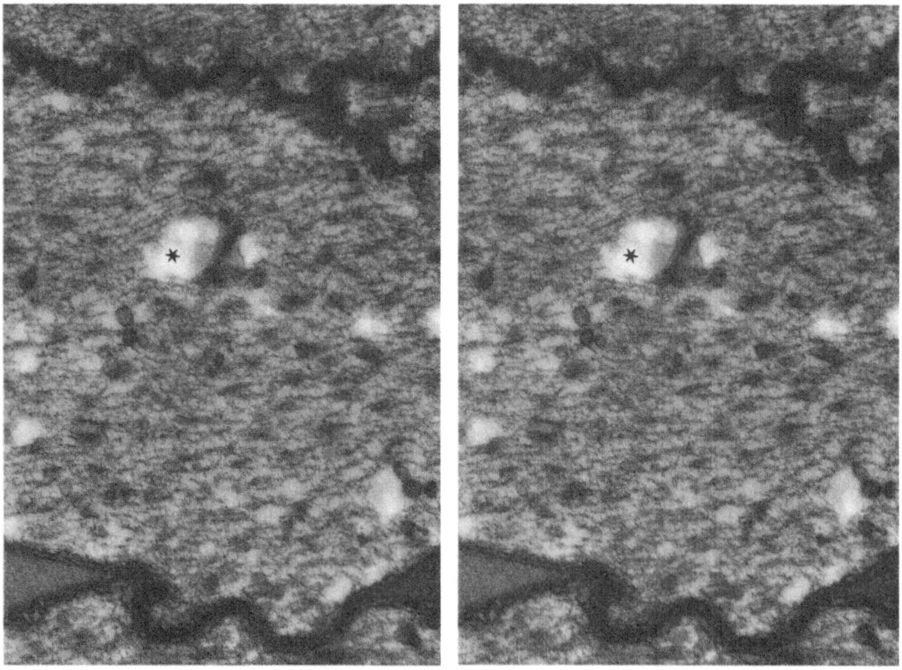

Figure 5. Stereopair of an ~0.2 μm thick section passing obliquely through axons in a *Hermissenda* cerebropleural ganglion. Note the thin transverse bridges (2–3 nm diameter) joining the longitudinal filamentous elements of the neuroplasmic lattice. Fixative included 0.2% tannic acid. Section stained with aqueous uranyl–lead. Starred regions are local artifacts. ×33,000. (From Hodge and Adelman, 1980a).

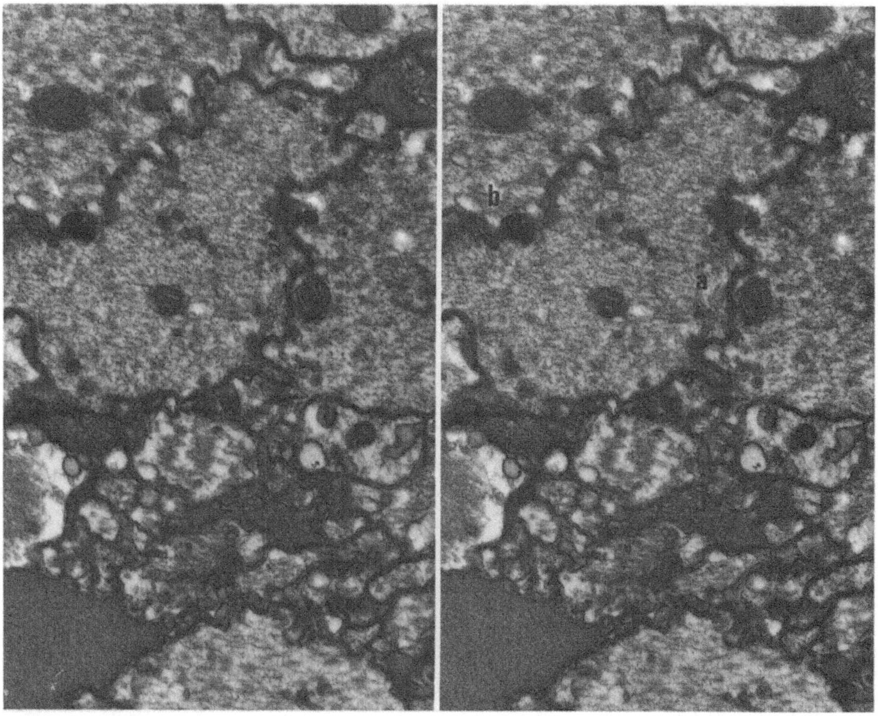

Figure 6. Stereomicrographs of a region close to the axon shown in Fig. 5. Note the order present in the neuroplasmic lattices of the axons (also dendrites and/or small axonal branches), and the variations in apparent type of order between different axons (a and b), although all show numerous thin crossbridges. Specimen preparation as in Fig. 5. ×23,500. (From Hodge and Adelman, 1980a).

The overall characteristics of the neuroplasmic lattice were best observed in oblique sections which were closer to longitudinal than transverse. In Fig. 4, the crossbridged lattice can be seen throughout the axoplasm of the three axons in the field of view, with only slight perturbation of the order resulting from the presence of mitochondria and other organelles. In the lowermost axon in the figure, the longitudinal dark streaks are due to the presence of groups of neurotubules. The tendency of the neurotubules to occur in clusters has been alluded to in the case of squid axons, and it is also a feature of vertebrate axons (see *Bufo* axons). Almost any angle of sectioning will give rise to superposition effects in single projection images, but these ambiguities can usually be resolved by stereoscopic examination. The orderly disposition of the crossbridges is readily visible in Fig. 5, where the obliquity was calculated to be ~45°.

The characteristics of the lattice often appeared to differ from axon to axon, even when they were adjacent (Fig. 6), and the ratio of neurotubules to neurofilaments varied widely (Fig. 7) as did their relative abundance and that of other organelles and axoplasmic components. While much of this variation was undoubtedly due to different angular sampling of the lattice, some must be attributed to unknown functional differences. It seems unlikely that this variation was due to differences in fixation of individual axons, which was rapid for these small specimens as indicated by the immediate yellowing reaction on exposure to glutaraldehyde.

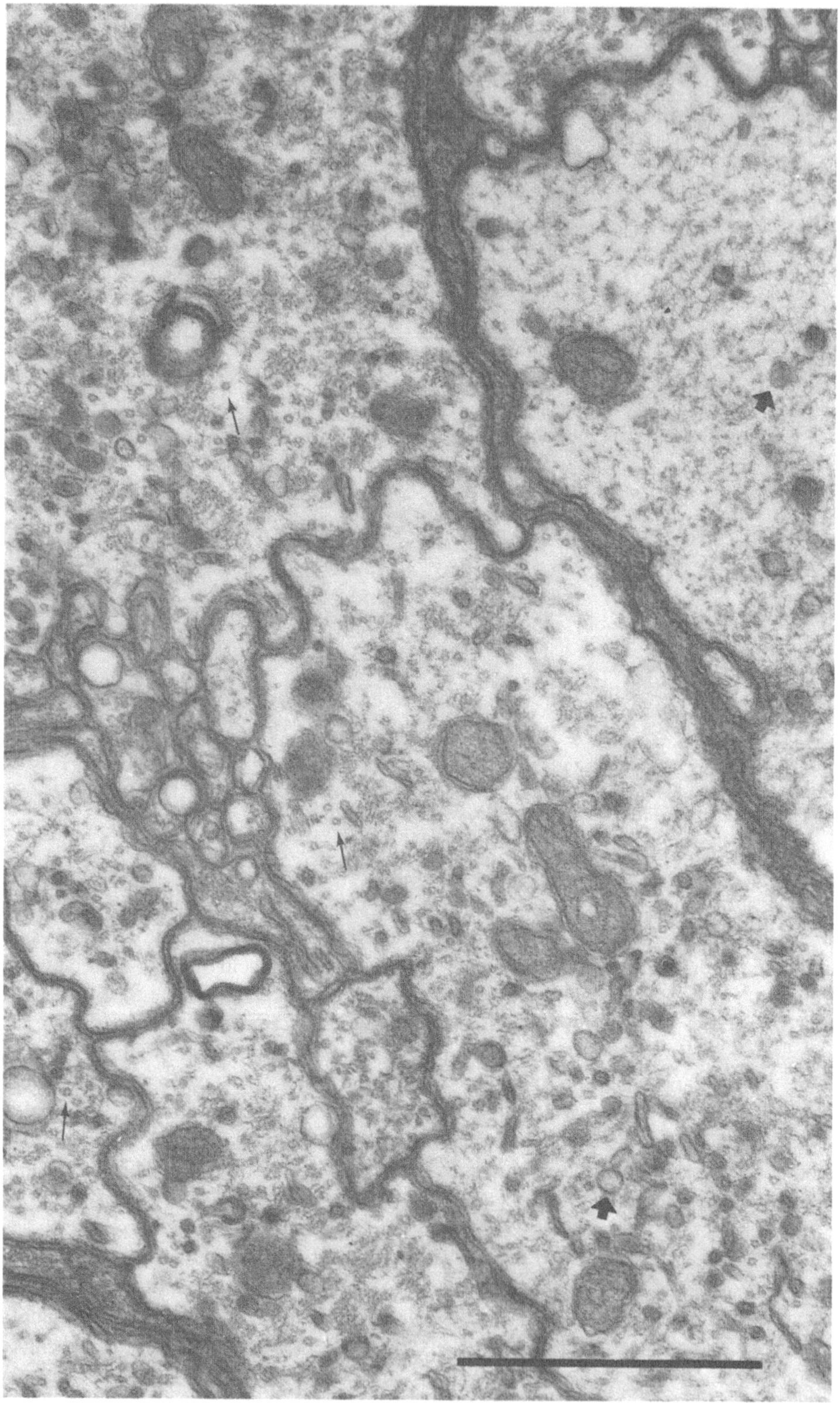

Figure 7. Micrograph of an area close to that shown in Fig. 6 showing the considerable differences between neighboring axons, particularly in the relative numbers of neurotubules (thin arrows) and neurofilaments present. Vesicles (thick arrows) are present in all axons. Bar represents 1 μm. ×46,300.

D. *Bufo Axons*

Various peripheral nerves of the toad *Bufo woodhousii fowleri*, when fixed by immersion using an appropriately lower (~290 milliosmolar) tonicity of the glutaraldehyde fixative used for marine mollusks (Hodge and Adelman, 1980a), showed the presence of an ordered crossbridged neuroplasmic lattice (Hodge and Adelman, 1980b) in both myelinated and unmyelinated axons which was essentially indistinguishable from that already described for the axons of *Loligo* and *Hermissenda*.

Figure 8 shows stereoscopically the disposition of the axoplasmic structures in a node of Ranvier (RN). The axonal space is filled with the crossbridged neuroplasmic lattice in which are embedded the usual organelles. It is evident that, while some of the neurotubules present appear to be an integral part of the lattice as evidenced by their orientation and bridging, others seem to be relatively independent of this order. Where directly measurable,

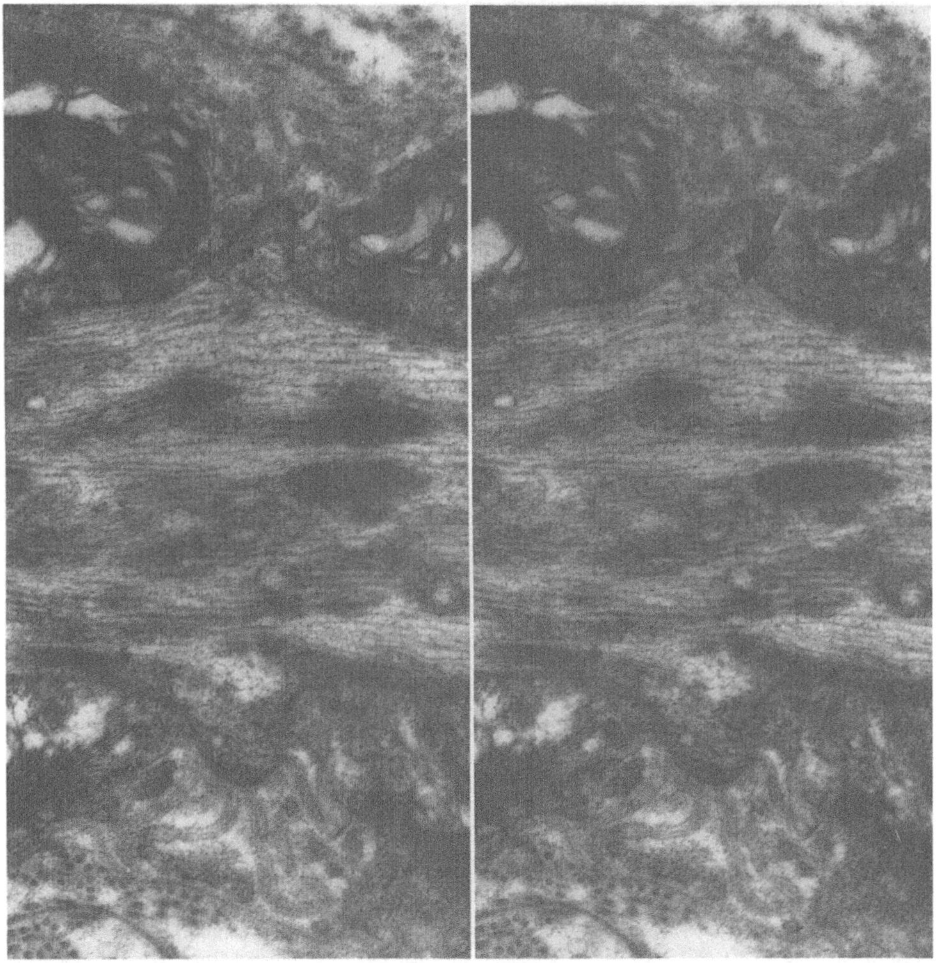

Figure 8. Stereomicrographs of an ~0.2 μm thick longitudinal section through a node of Ranvier in *Bufo* peripheral nerve. Note the numerous periodically disposed transverse bridges of the neuroplasmic lattice linking adjacent longitudinal elements (especially in the area indicated by the large arrow), and extending also to the axolemmal surface. Specimen fixed in the presence of 0.2% tannic acid, section stained with ethanolic uranyl acetate, aqueous lead citrate. ×28,500.

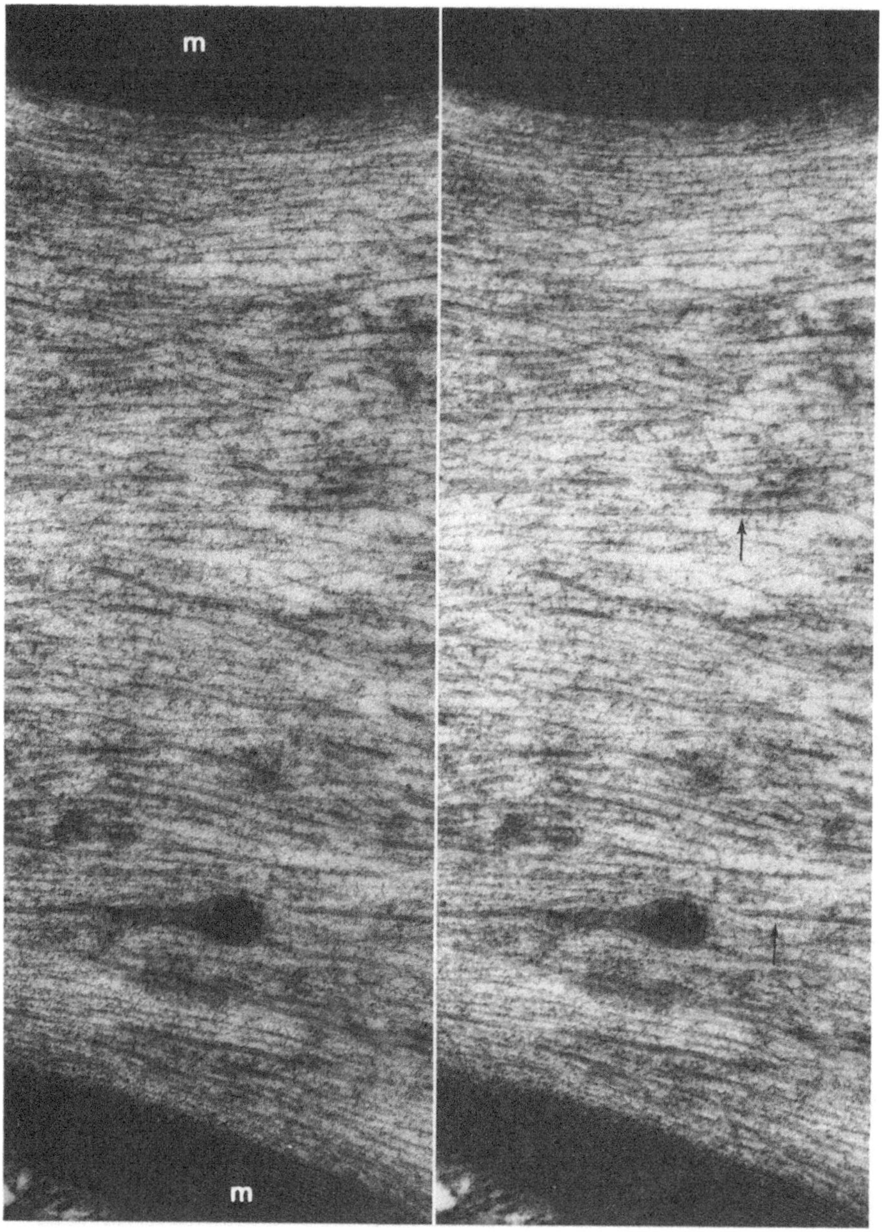

Figure 9. Stereomicrographs of an ~0.2 μm thick longitudinal section of *Bufo* peripheral nerve passing through a constricted zone (SLN) associated with a Schmidt–Lanterman cleft. The ordered bridge system of the lattice is comparable with that shown in Fig. 8. Note that neurotubules (arrows), as well as neurofilaments, are linked by bridges. The myelin sheath is indicated with m. Section stained with ethanolic uranyl acetate and aqueous lead citrate. ×39,000.

the axial spacing of the crossbridges was found to be ~40–45 nm, a value indistinguishable from that found for *Loligo* and *Hermissenda* axons.

Early in this study, it was observed that the regions of the Schmidt–Lanterman clefts were almost universally associated with a more or less pronounced axonal constriction. In these nodal regions, which will be termed Schmidt–Lanterman nodes (SLN), the neurofilaments and neurotubules, while not differing appreciably in total numbers from nonconstricted regions, were packed together much more closely. This seemed to be a major factor in the better preservation of the lattice order observed in these SLN regions, at least as seen in sections of Epon-embedded material. In longitudinal sections of *Bufo* axons in the SLN regions, the orderly arrangement of thin transverse bridges can be clearly seen (Figs. 9 and 10). Unfortunately, in many cases, the individual bridges appear to have suffered breakage and retraction, probably because of the imperfections of specimen preparation techniques. Experience suggests that the most damaging parts of the protocol are the dehydration and polymerization stages, but definitive evidence is scarce, primarily because there are currently no objective and practical methods for continued observation at high resolution of cellular structures during fixation, dehydration, embedding, and sectioning. X-ray diffraction methods are only applicable to very highly ordered structures such as myelin sheath and muscle, and require vastly greater sample volumes than does electron microscopy. The effects of fixation, dehydration, and polymerization of embedding medium on the structure of the myelin sheath have been studied by x-ray diffraction (Kirschner and Hollingshead, 1980). In any event, it seems likely that the neuroplasmic lattice order originally present in the axons must have been a good deal better than that finally seen in electron micrographs of sections obtained from fixed, dehydrated, and embedded material.

In Figs. 8–10, it is evident that the crossbridges of the lattice also extend to the various membrane surfaces facing the axoplasm proper. In particular, the bridges interact with the inner surface of the axolemma, where there appears to be some specialization of structural elements, sometimes visible as thin microfibrillar structures. This specialized subaxolemmal dense zone (~20 nm thick) between the axolemma and the neuroplasmic lattice, which can be seen more clearly in transverse sections (Figs. 11 and 12), presumably acts as an interface

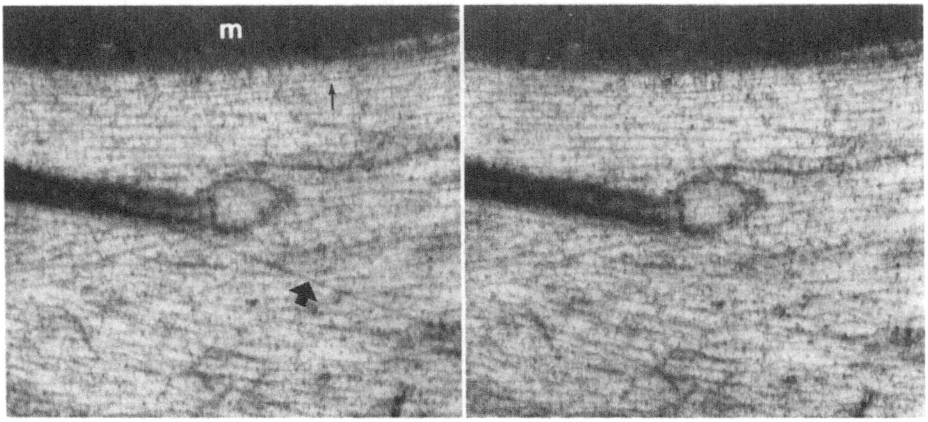

Figure 10. Stereomicrographs of an ~0.2 μm thick longitudinal section through a SLN region similar to that of Fig. 9 to illustrate the periodic order of the crossbridge system and its interaction with the axolemma (thin arrow). The diagonal line (thick arrow) is an artifact produced by STEM scanning. The myelin sheath is indicated by m. Fixation and staining as in Fig. 8. ×31,400.

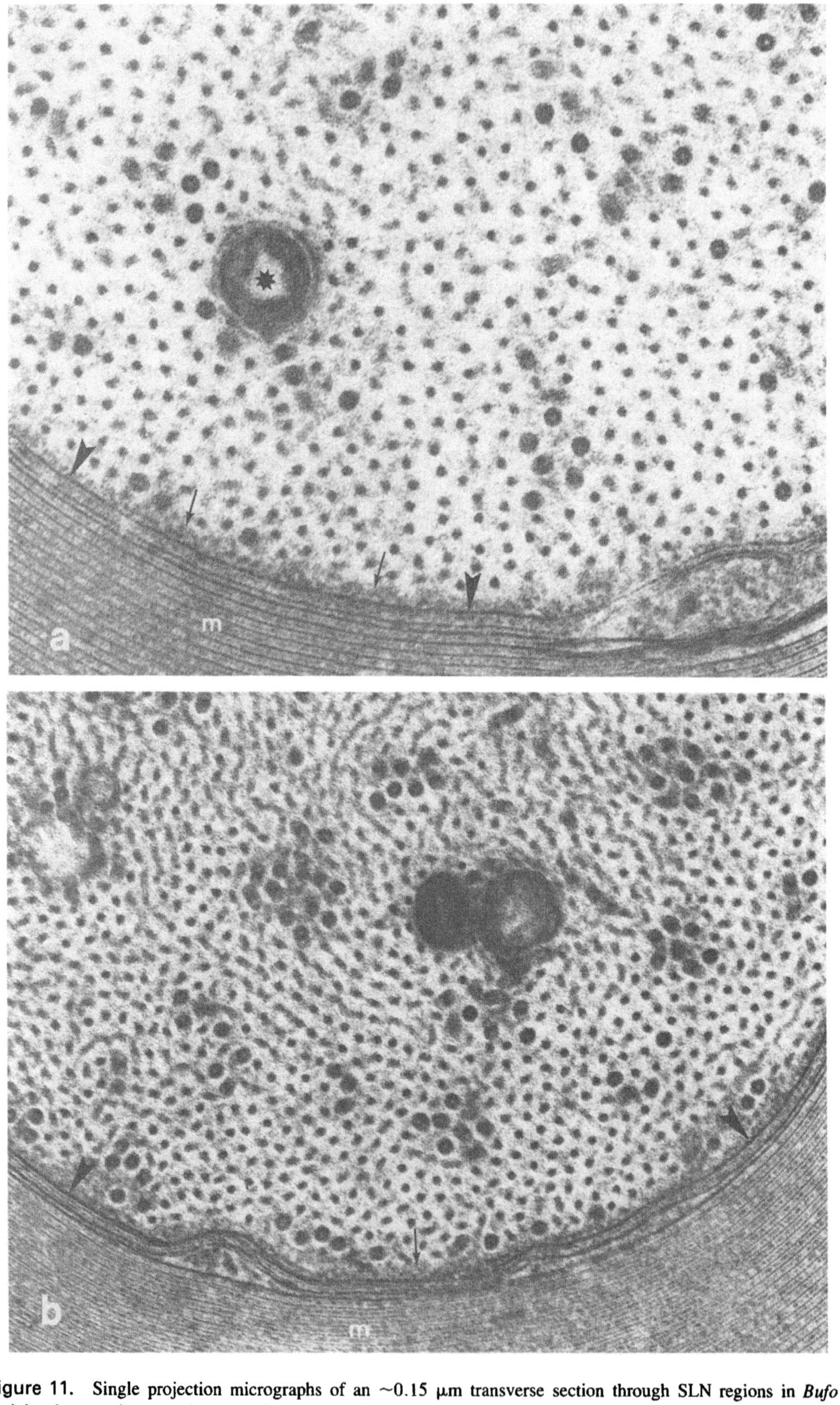

Figure 11. Single projection micrographs of an ~0.15 μm transverse section through SLN regions in *Bufo* peripheral axons in a specimen similar to that of Figs. 9 and 10. Fixation and staining as in Fig. 8. Note the tendency of the neurotubules to occur in clusters, the close juxtaposition of both neurofilaments and neurotubules

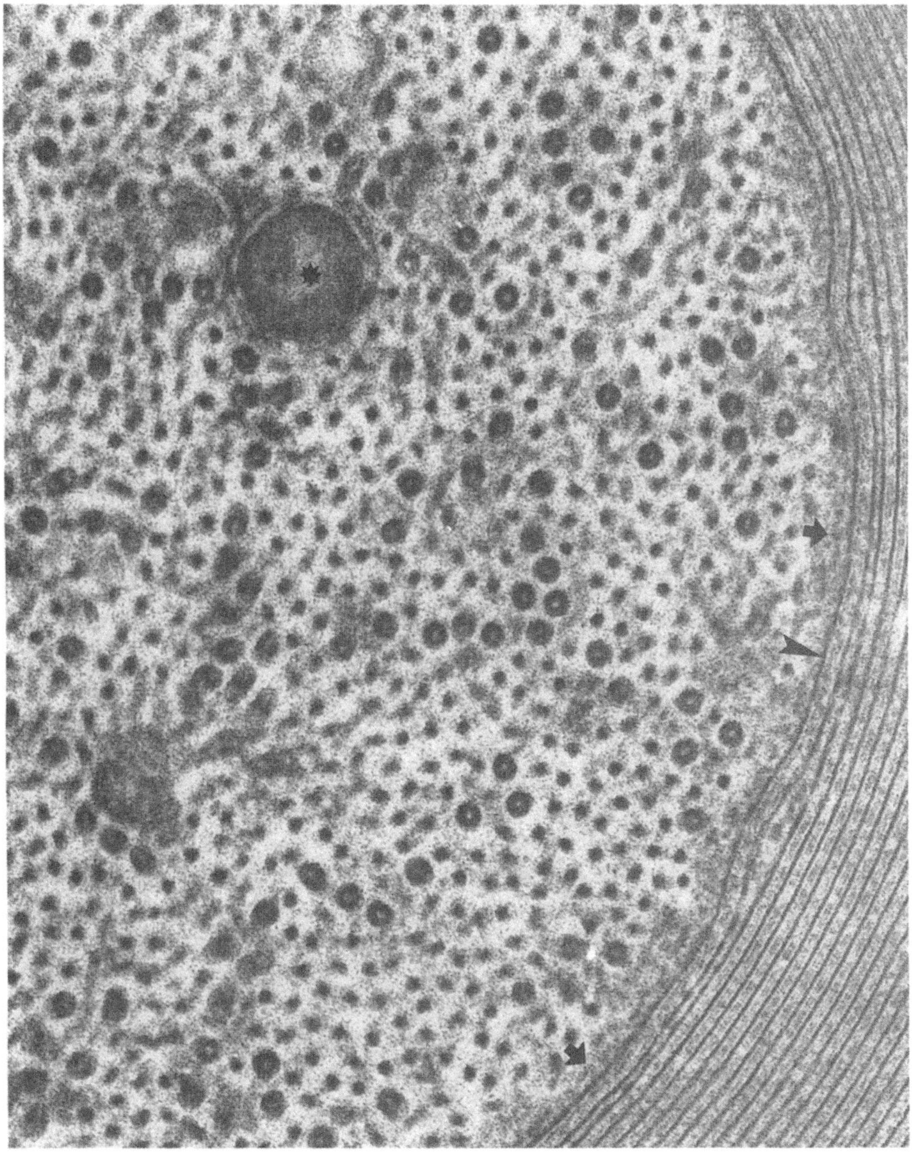

Figure 12. Single projection micrograph of an ~0.2 μm transverse section through another SLN region in a *Bufo* peripheral axon showing a higher degree of constriction and hence closer packing of the longitudinal elements than that present in Fig. 11. In this case, many of these are at closest approach, i.e., spaced ~20 nm apart (center-to-center). The subaxolemmal "ectoplasmic" zone, from which neurofilaments and neurotubules are excluded, shows a characteristically granular dense appearance (arrows) suggesting the presence of longitudinally oriented microfibrillar (actin?) structures. Note the high density of the cytoplasmic axolemmal face (arrowheads). A presumed mitochondrion is labeled with an asterisk. Fixation and staining as in Fig. 8. ×142,400.

---

with mitochondrion (asterisk) present in (a), and the granular appearing "ectoplasmic" zone (arrows) immediately subjacent to the axolemma (arrowheads). In (a), most neurofilaments and neurotubules are spaced 25–50 nm apart (center-to-center), with the closest being ~20 nm apart. Compare with (b) and Fig. 12, where the longitudinal elements are much more closely packed. (a) ×114,000, (b) ×104,000.

subserving the cytoskeletal function of the lattice in maintaining axonal form. It is clear from these figures that the components of this subaxolemmal cortical zone are closely contiguous to, or structurally continuous with, the densely stained and presumably proteinaceous inner face of the axolemma itself. Neurofilaments and microtubules are apparently excluded from it, but bridges emanating from them appear to penetrate it. These results appear to be substantially in agreement with the findings of Rosenbluth (this volume) and of Hirokawa (1982 and this volume), using freeze–etch techniques. It is of interest to note that, while the often rather "wispy" and attenuated nature of the neuroplasmic lattice crossbridge array seems to require rather thick sections for adequate sampling and analysis, the analysis of bridge distribution in striated muscle with its well defined and dense arrays of crossbridges appear to be better carried out using quite thin transverse (Luther *et al.*, 1981; Squire, 1981) or longitudinal sections (Reedy, 1968).

Several other features visible in the transverse sections through the constricted SLN regions (Figs. 11 and 12) are worthy of note. Even in the most constricted cases observed, the neurofilaments and neurotubules, while crowded together, remained separated from one another by a minimum distance of ~20 nm (center to center). Presumably, this was because the presence of transverse arms or bridges precluded any closer approach. A clear parallel to this observation can be seen in the findings of Dentler *et al.* (1975) and Kim *et al.* (1979), who showed that microtubules, equipped with projections as a result of *in vitro* polymerization of tubulin in the presence of certain MAPs, did not pack as tightly together on ultracentrifugation as did those polymerized in the absence of MAPs. Furthermore, they were able to show that the MAP-associated microtubules formed arrays apparently joined together by periodically spaced bridges under these conditions. In the most closely packed SLN regions of *Bufo* axons, we found (Hodge and Adelman, 1983) that the total number of longitudinal filamentous elements (both neurofilaments and neurotubules) per unit area was about 1000 μm^{-2}, a value not far removed from that (1280 μm^{-2}) calculated for hexagonally packed elements 30 nm apart. A nearest neighbor computer analysis of the arrays of neurofilaments and neurotubules in cross sections of myelinated axons clearly indicated that the packing of these elements is hexagonal (Adelman *et al.*, 1982; Hodge *et al.*, 1983). The highest packing density we observed corresponded to an 8–10-fold increase in the number of filamentous elements per unit area over that present in the average axon beyond the limits of the constriction with the total number remaining constant.

In all cross sections of *Bufo* axons, regardless of longitudinal location, the neurotubules were found in clusters of various sizes (Figs. 11 and 12), but were also commonly observed as single entities with no discernible distribution pattern. While neurotubules frequently appeared in close juxtaposition, and were often bridged to mitochondria and other membraneous elements, so too were the neurofilaments. Thus, the overall picture of the neuroplasmic lattice as seen in sections did not allow any specific correlation with the movements of mitochondria and vesicular structures so vividly evident in live axons as seen by darkfield time-lapse cinephotomicrography (Forman *et al.*, 1977; Grafstein and Forman, 1980), and more recently by enhanced video contrast observation techniques in real time using differential interference optics (Allen *et al.*, 1981, 1982a,b; Brady *et al.*, 1981, 1982; Breuer *et al.*, 1982; Fahim *et al.*, 1982; Hodge *et al.*, 1982b).

E. *Homarus* Axons

The axons of crayfish ventral nerve cord appear to contain no neurofilaments but, instead, are endowed with a crossbridged lattice of neurotubules (Burton and Fernandez, 1973; Ochs and Burton, 1980). Burton *et al.* (1975) observed the same kind of axonal structure in *Homarus*, and also noted that the microtubules in axons of contiguous supporting cells contained characteristically different numbers of subunits as seen in cross section. This observation, together with much primarily immunological and chemical evidence showing that there exist many distinct varieties within the class of microtubules and also in the class of 10 nm filaments (Yen and Fields, 1981), has prompted us to retain the term neurotubule rather than use microtubule for these elements when they occur within the cytoplasm of neurons.

Our findings in *Homarus* axons (Hodge and Adelman, 1981, 1983) are similar to those of Ochs and Burton (1980). The axons in the ventral nerve cord appeared to be filled by a neuroplasmic lattice consisting of longitudinal neurotubules linked together transversely by an orderly array of structurally complex bridges. The usual organelles were present, with the very long mitochondria being confined mostly to the cortical region. The neurotubule packing density varied considerably from axon to axon (Fig. 13), and was often so high

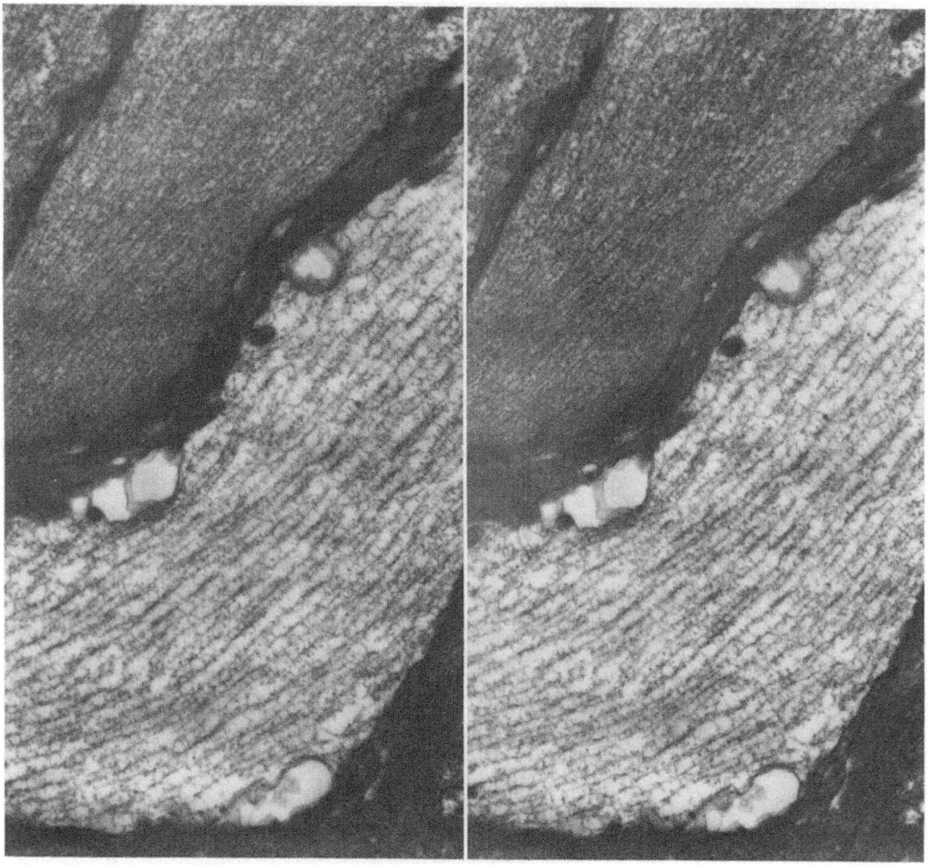

Figure 13. Low magnification electron stereomicrographs of an ~0.2 μm section passing longitudinally through several axons in *Homarus* ventral nerve cord, using similar fixation as for *Loligo* and *Hermissenda*. Note the array of neurotubules linked by regularly spaced transverse bridges to form a neuroplasmic lattice. The axons at the upper left have concentrations of neurotubules and bridges too high for adequate visualization at this section thickness, i.e., they possess too much information density. The mitochondria and the presumed elements of the smooth endoplasmic reticulum appear to be confined to a cortical zone of the axoplasm. En bloc staining with ethanolic uranyl acetate–phosphotungstic acid and section stained with aqueous lead citrate. ×8,600.

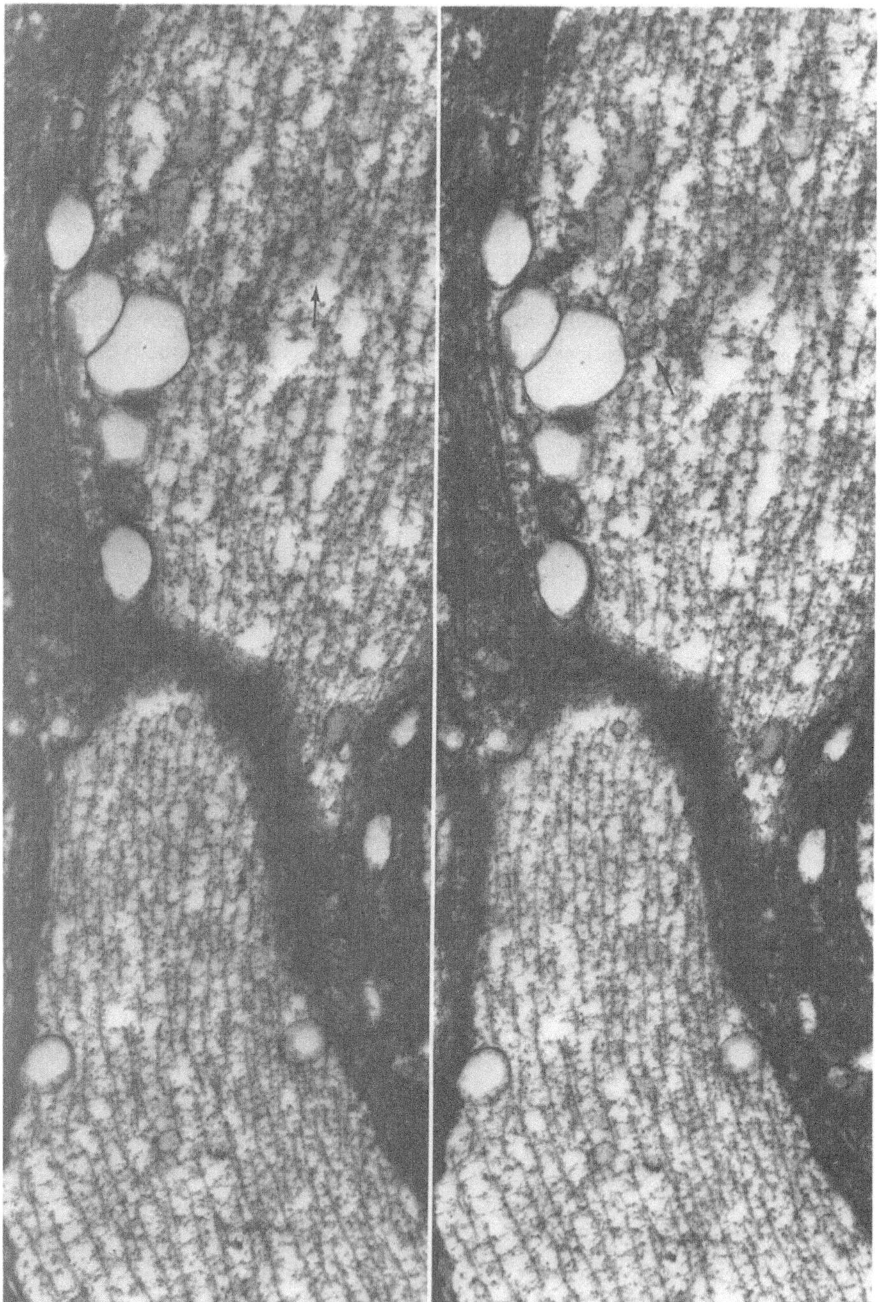

Figure 14. Stereomicrographs of an oblique longitudinal section through axons in *Homarus* ventral cord showing the neuroplasmic lattice in more detail. Note the regularity of the crossbridge distribution and the rows of small vesicles (arrows) occupying interstitial positions between neurotubules in the axon at top. Specimen and section treatment as in Fig. 13. ×17,000.

that optical confusion resulted in viewing stereo images of thick sections, e.g., axon at upper left in Fig. 13. However, a highly ordered lattice was easily visible in less densely packed axons (Figs. 13 and 14). Transverse sections (Fig. 15) have so far been of little stereoscopic value in determining the spatial, and particularly the angular, distribution of the bridges in the lattice, although it evidently closely approximates hexagonal.

The distribution of the bridges, especially as seen in Figs. 14 and 16, is suggestive of screw-axis symmetry, but we cannot be definite at this stage other than to state that the strong tendency for hexagonal packing evident in transverse sections (Fig. 15) suggests the presence of a three- or six-fold screw axis. Computer analysis of these arrays gives evidence that the packing is hexagonal (Hodge *et al.*, 1983). Higher magnification micrographs (Fig. 16) reveal that the bridges, although usually "fuzzy" in appearance, sometimes exhibit a zig-zag, perhaps helical, character. It is clear that structural evaluation of these MAP structures must await further chemical separation and characterization. A start in this direction has been made by Ochs and Burton (1980) using selective extraction procedures. As was the case for the molluscan and vertebrate axons, it has proved difficult to determine the axial bridge spacing directly, even though considerable order was apparent. Fortunately, the application of autocorrelative methods allowed evaluation of the periodicity (Fig. 19) or, at least, an integrally related value, in this case of ~88 nm. The bridge spacing in these axons will be further explored using Fourier analytical techniques in STEM as we have done for *Bufo* axons (Figs. 20–22). Evaluation of the micrographs in terms of lattice parameters was complicated further by the presence of small uniform oblate spheroidal particles, 20–25 nm in diameter (Figs. 15 and 16). These rather smooth appearing particles were often seen in

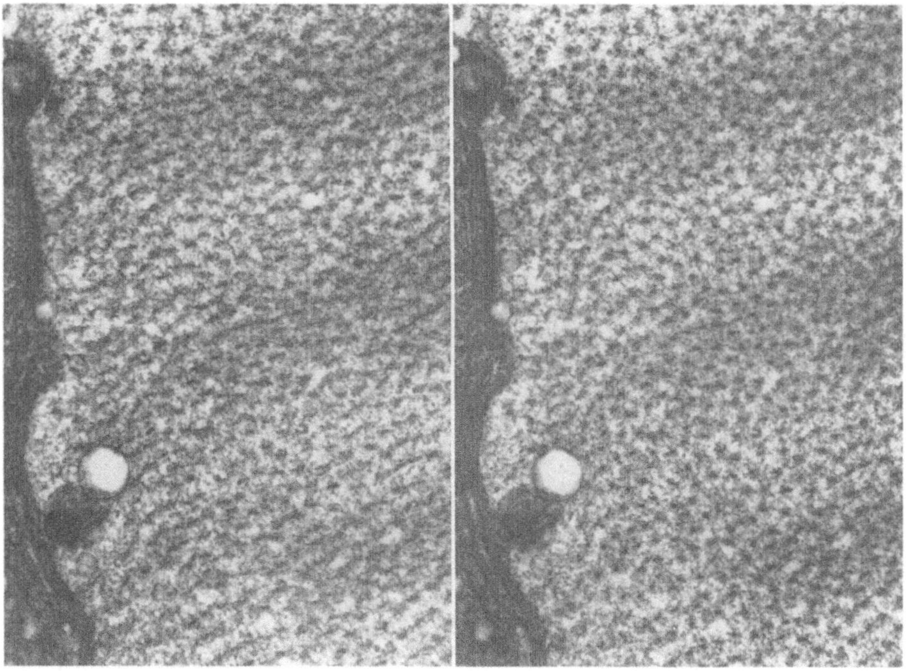

Figure 15. Stereomicrographs of an ~0.2 μm transverse section through a *Homarus* ventral cord axon. Note the strong tendency toward hexagonal packing, and the difficulty in interpreting the bridge distribution in this relatively thick section. Same specimen preparation and section staining as in Fig. 13. ×19,400.

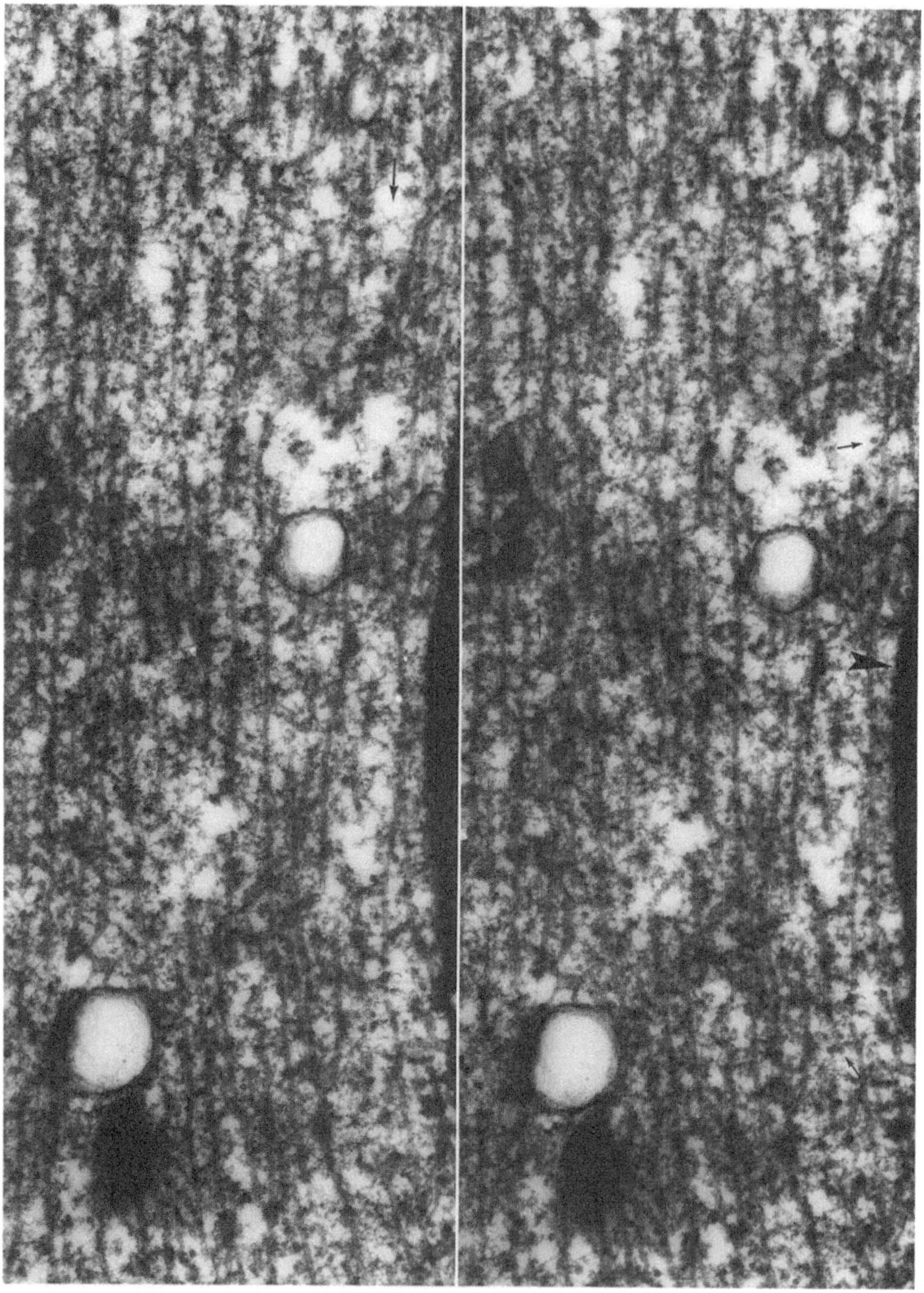

Figure 16. Higher magnification stereomicrographs of a longitudinal section through a *Homarus* ventral cord axon to illustrate details of the crossbridges. Note their regular distribution, linkage with the axolemma (arrowhead), and complex and seemingly helical type of structure (arrows), also the presence of small oblate spheroidal particles, probably in course of being transported (small arrows). Fixation and section staining same as in Fig. 13. ×38,000.

large numbers and appeared to be distributed in a fairly orderly manner with respect to the neuroplasmic lattice. Their composition and function remains unknown, but since they were not always present, and hence, presumably not a structural component of the lattice itself, they probably represent particles undergoing transport.

III. OPTICAL AUTOCORRELATION OF NEUROPLASMIC LATTICE STRUCTURE

An obvious way of determining the parameters of the neuroplasmic lattice would be to apply optical diffraction techniques to the TEM or STEM images, but our initial efforts along these lines (Hodge and Adelman, 1983) proved unsuccessful. The lattice, because of its small contribution to the overall Fourier transform, appeared to be lost in the large background components. However, two-dimensional image enhancement methods, either directly during acquisition of STEM images, or later applied to recorded TEM or STEM images using a digital computer, are currently under way.

In the meantime, attention was turned toward enhancement of periodicity in the neuroplasmic lattice using optical autocorrelation methods (Hodge and Adelman, 1978b, 1980b). Of the many methods available, including more recently, that of Gilev (1979), we have so far investigated the use of several multiple displacement techniques as well as the application of Ronchi gratings, a method originated by S. Inoué and described by Erickson (1973). However, for the most part, we have chosen to use the inherently more reliable single displacement technique described by Inoué and Ritter (1975), which has been widely used for emphasizing periodicities in a wide variety of biological subsystems (Kim *et al.*, 1979).

Figure 17 is an original TEM image showing the crossbridge arrangement in a longitudinal section of a *Bufo* axon together with its corresponding single displacement image in which the displacement was determined by direct observation of maximal reinforcement prior to making a double exposure print (Hodge and Adelman, 1983). Strong reinforcement is clearly visible at a displacement corresponding to a longitudinal spacing of ~40 nm in the axon. We have used this technique on micrographs of longitudinal sections from axons in nerves of *Loligo, Hermissenda,* and *Bufo.* Despite the vicissitudes of the method, in which reinforcement could be observed at values corresponding to integral mutliples or submultiples of the bridge spacing, results have been consistent with our direct observations that the bridge spacing, but not necessarily the fundamental period, was in the range of 40–45 nm. The following example illustrates the above point. Figure 18 shows an original STEM micrograph of a *Bufo* peripheral myelinated axon in a Schmidt–Lanterman nodal region (a) together with its displacement image, and (b) corresponding to a spacing of about 40 nm. The reinforcement is quite satisfactory at this displacement value.

Hodge and Adelman (1983) have found that the single displacement procedure can be implemented directly in the electron microscope prior to recording an image by making use of the wobbler device often used as a focusing aid. In the Philips EM400 at least, linear image "wobbling" in either of two selectable directions 90° apart is accomplished by introducing a 30 Hz square wave signal into the imaging lens system. Depending on the amplitude of this signal, the observer simultaneously sees two images which are separated by a distance dependent on the degree of over- or underfocus. Since most electron micrographs are taken in a somewhat underfocused condition, the wobbler amplitude can be used as a continuous image displacement device to observe reinforcement of periodic structure in suitably oriented specimens. A rotary stage considerably facilitates such orientation. In practice, with the image at optimal underfocus, the wobbler amplitude is varied until maximal reinforcement is achieved and films then exposed both with and without the wobbler. Figure 19 is an example of this procedure, the specimen in this case being a longitudinal section of a *Homarus* ventral cord axon. Reinforcement of the quite highly ordered transverse bridge structure is apparent, in this case with a displacement corresponding to ~88 nm in object space. Another axon in the same section gave good reinforcement with a displacement equivalent to ~40 nm. In general, our observations indicate that the particular values obtained are closely related, not only to the angle between the section plane and the longitudinal axis of the

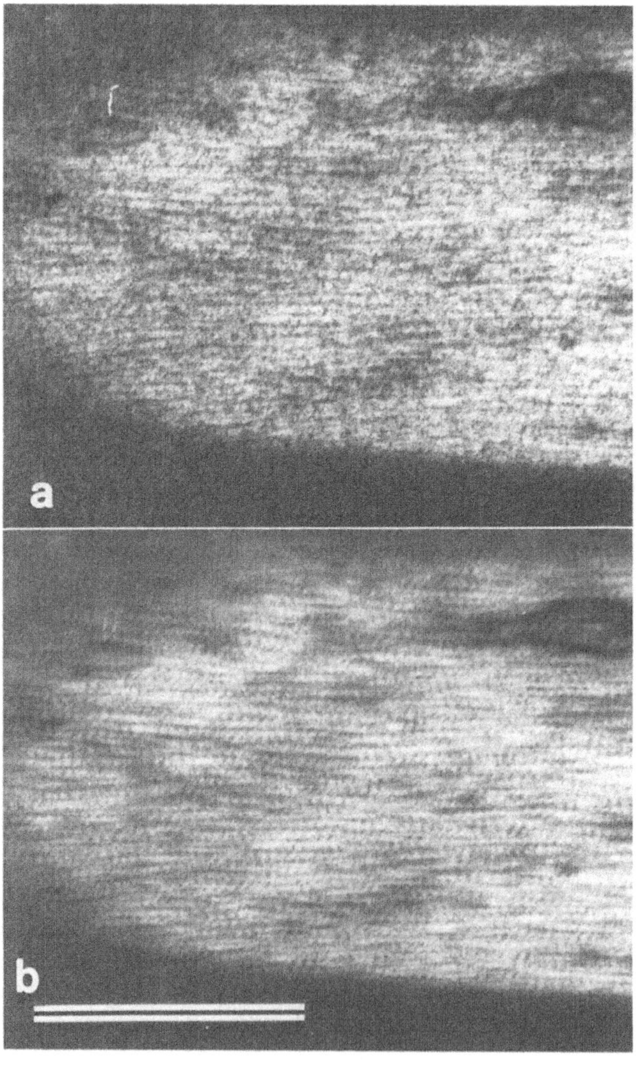

Figure 17. (a) Single projection micrograph of an ~0.2 μm longitudinal, almost tangential, section through a SLN region in *Bufo* peripheral nerve. The section passed through a peripheral region of the axoplasm where a high degree of order was preserved in the neuroplasmic lattice. In (b), the same micrograph was doubly exposed using a technique (Hodge and Adelman, 1981) in which the actual displacement was first determined by obtaining minimal contrast while projecting the negative onto a movable aligned positive print. Note that displacement direction is horizontal in the picture. The value corresponded to a spacing of ~40 nm in the specimen. Bar represents 1 μm. Specimen and section treatment as in Fig. 8. ×36,800.

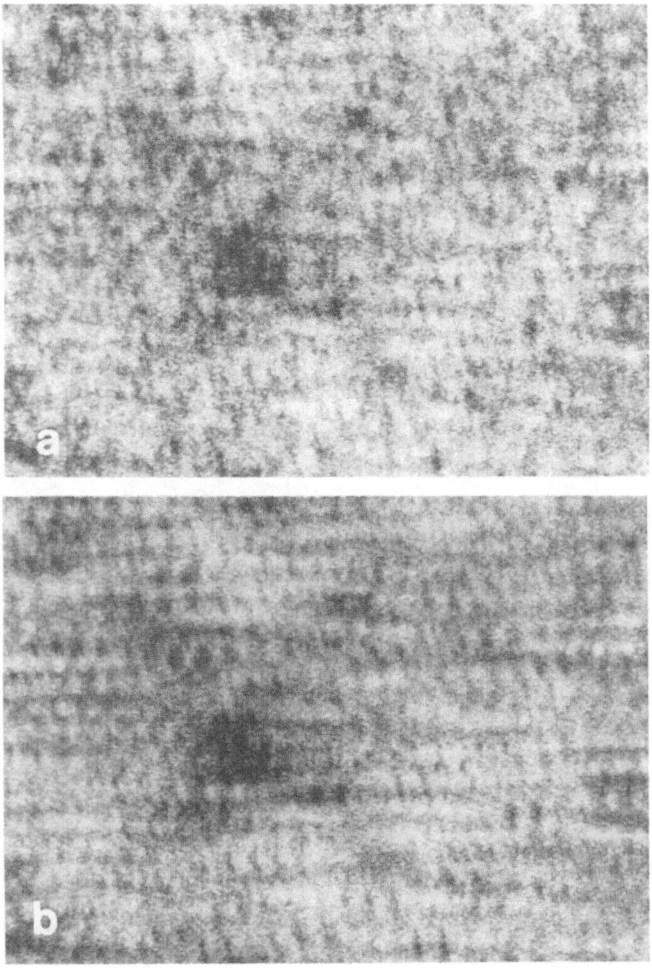

Figure 18. (a) STEM micrograph of an ~0.2 μm section through the axoplasm in a SLN region of a *Bufo* peripheral axon. The neurofilaments running horizontally are clearly discernible, as are the vertical linear densities which represent the crossbridges and/or superposition images of them. Displacement (horizontal in the picture) determined by the method described in Fig. 17 gave satisfactory reinforcement (b), corresponding to a spacing of ~40 nm. Specimen and section treatment as in Fig. 8. ×87,000.

axon, but also to the section thickness relative to the lattice parameters. Analogous results have been observed in computer Fourier analysis of STEM video line signals, as indicated in the next section.

IV. STEM VIDEO LINE ANALYSIS

There are many advantages inherent in using STEM for the acquisition of images of sections. There is much less specimen damage from electron irradiation since the area of interest is not under constant bombardment as it is in TEM. The transmitted electrons are also less subject to lens aberrations and, as a result, image contrast is better. Furthermore, since the image is formed by raster line scanning using a fine electron probe, the electron scattering properties of the specimen directly determine the intensity of the serial electronic

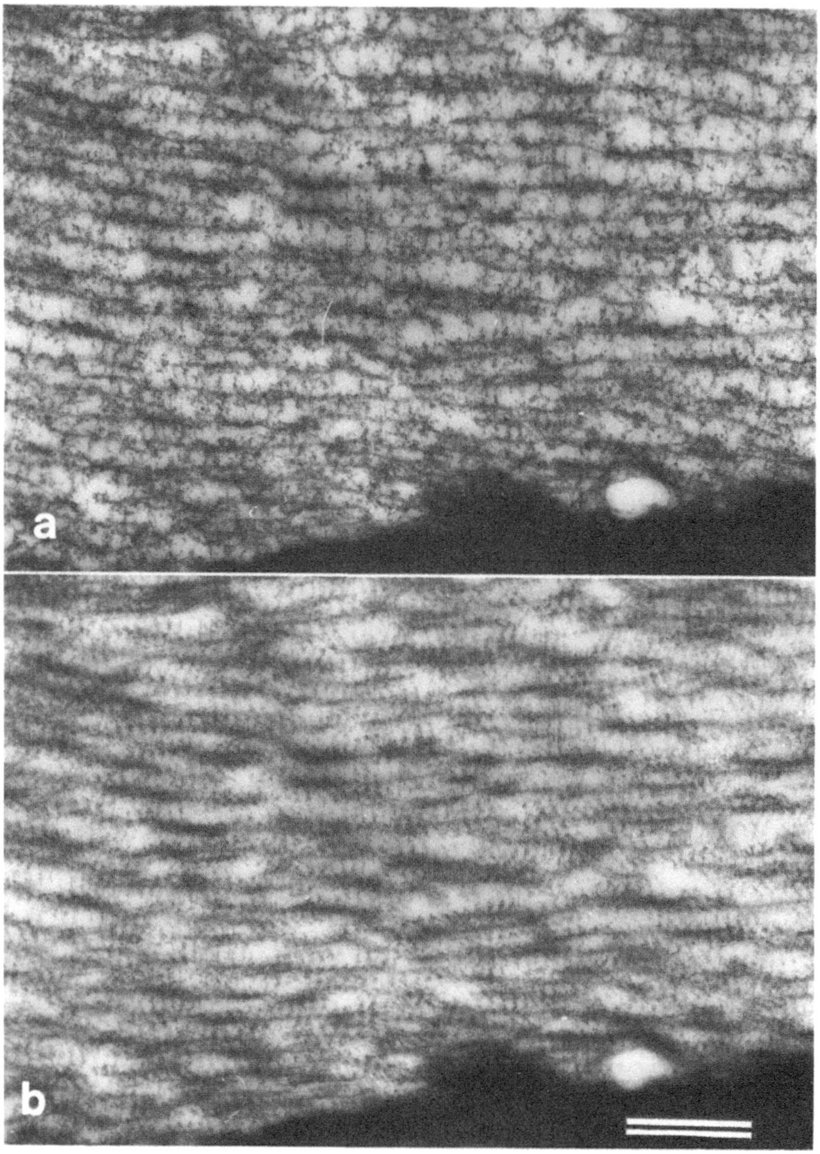

Figure 19. (a) shows a region of the major axon in Fig. 13 at higher magnification. (b) is an image of the same area obtained after adjustment of the image "wobbler" to give maximal reinforcement with image displacement along the fiber axis, i.e., horizontal in the micrograph (see text). Bar indicates 1 μm.

signal obtained. These video line signals can be recorded electrically, and are then available for various processing procedures. In particular, video line information from a suitably oriented structure can be analyzed using Fourier methods in a digital computer. Examination of the power spectrum allows evaluation of the specimen in terms of periodicities present. We have recently presented the results of applying this technique to a number of model systems including myelin sheath, *Loligo* mantle muscle, and rabbit tropomyosin crystals (Adelman *et al.*, 1981; Hodge *et al.*, 1982a), and its use in the evaluation of neuroplasmic lattice parameters (Hodge and Adelman, 1980a,b, 1981, 1983).

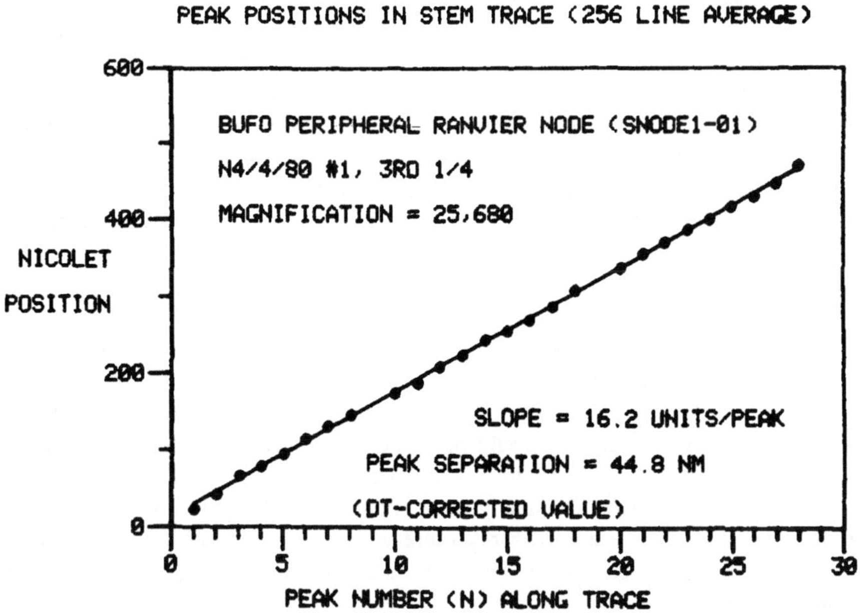

Figure 20. Plot of the peak positions obtained by direct observation of a STEM line trace as scanned through the axoplasm in a node of Ranvier parallel with the axonal longitudinal axis. The STEM trace was recorded on a Nicolet digital CRT (see reconstituted trace in Fig. 22) and the peak positions determined visually after horizontal and vertical expansion. The vertical axis represents the position in memory of peak locations, the horizontal the assigned peak number starting from an arbitrary position. The small deviation of peak positions from the least-squares fit line is strong evidence for the presence in the trace of a well-defined periodicity corresponding to the line slope.

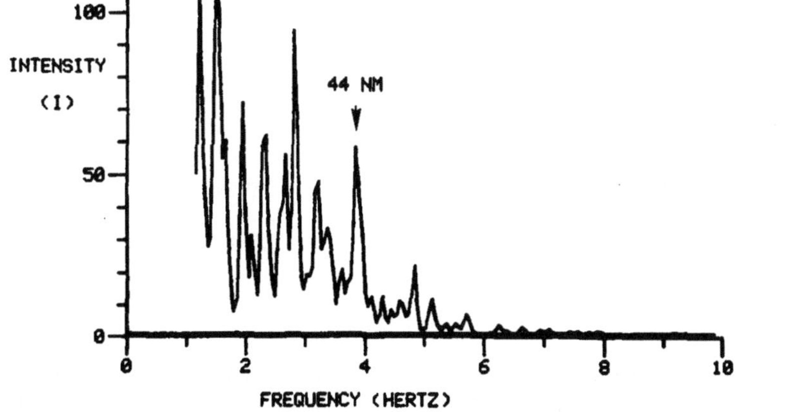

Figure 21. Power spectrum of the same STEM line trace as shown in reconstituted form in Fig. 22, and analyzed for peak positions in Fig. 20. The peak at 3.93 Hz corresponds to a periodicity in the specimen of 44.8 nm, in good agreement with the value derived from the line plot of Fig. 20 and with direct measurements of apparent bridge spacings in electron micrographs.

As already mentioned, it is often difficult to obtain direct measurements of periodicity in a micrograph which clearly shows evidence of it. Fortunately, the main period can often be evaluated by the simple expedient of recording a single video line signal (or a number of repeats to reduce noise) on a digital recording oscilloscope (the instrument we used had 2047 memory locations for signal intensities), and then manually determining and plotting the memory location of peaks in the line signal. Figure 20 shows such a plot obtained while scanning axially along a longitudinal section of axoplasm in a node of Ranvier. The straight line derived from a least-square fit to at least 90% of the points obtained in the memory range shown (1/4 of the line) gave a value (*d*) of the period equal to 44.8 nm, in good agreement with other determinations on this and other specimens. The power spectrum obtained from this run is shown in Fig. 21. It shows a peak at 3.93 Hz corresponding to a periodicity in the specimen of 44 nm, in good agreement with the value of 44.8 nm obtained by video signal peak plotting. It should be noted that this value of 44 nm may represent a major subperiod of the lattice structure rather than the true fundamental since at least one other integrally related lower frequency peak is present in the power spectrum. Other non-integrally related peaks present probably represent moiré superposition effects due to misalignments of the scan relative to the lattice axes or, more likely, arise from localized distortions in the lattice.

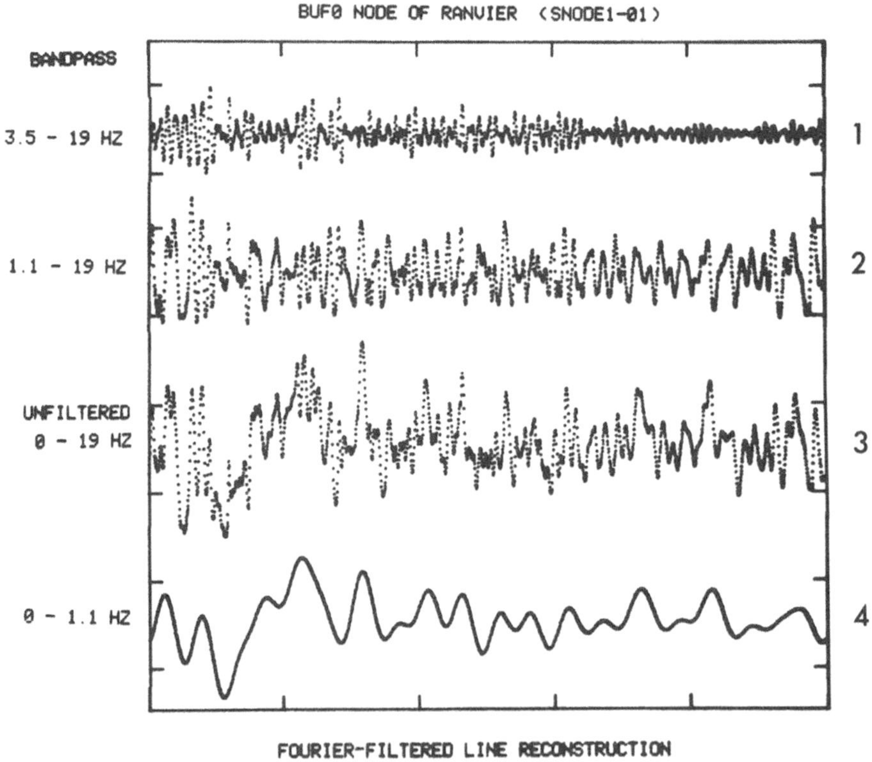

Figure 22. Reconstituted line traces generated by carrying out reverse Fourier transformations on the forward FFT data used to obtain the power spectrum of Fig. 21 with notch filtering as indicated at left. The unfiltered trace (3) was indistinguishable from the original trace as recorded on the Nicolet digital CRT and used to obtain the line plot shown in Fig. 20. It can be seen that removal of the lower frequency components (trace 4) results in a smoother trace (2) allowing better visualization of the periodic structure plotted in Fig. 20. Further narrowing of the window shows the period still more clearly (trace 1). Vertical scale factor was identical for all four traces.

Once evaluation of the power spectrum had been completed, unfiltered reverse Fourier transformation of the forward Fourier transform already carried out on the video line data resulted in reconstitution of the original video line trace (Fig. 22). Appropriate windowing (notch filtering) allowed the reconstitution of traces in which large scale variations, represented mostly by low frequency components, were virtually eliminated, thus allowing clearer visualization of the ~40 nm periodicity than could be seen in the unfiltered trace. The single line method is, of course, also applicable to TEM images, requiring use of a high resolution microdensitometer to obtain optical density traces. We are currently extending these techniques to two-dimensional analysis of STEM images.

V. DISCUSSION

A. Specimen Preparation in Relation to Structural Findings

The massive assault currently under way with the aid of the electron microscope to determine the fine structure of cells has, as always in the past, run headlong into the deeply fundamental and practical problems of first, attempting to stabilize the instantaneous status quo of the membranes and other cytoplasmic subsystems and, then, by suitable manipulation, preparing a specimen of high contrast (electron scattering power) and small thickness appropriate to the formation of reasonably high resolution images without disruption of fine structure. With the exception of methods aimed at dehydration of the specimen while maintaining its three-dimensional integrity, e.g., freeze drying (Williams, 1953) and critical point drying (Anderson, 1951), most cell structure studies since about 1948 have involved one or more fixation steps, followed by dehydration in an organic solvent miscible with water, infiltration with a resin monomer, followed by polymerization. The blocks thus obtained have been sectioned at various thicknesses and the sections examined with or without further staining with heavy metal atoms. A common variation has been *en bloc* staining in either the aqueous or dehydration stages of the protocol. It is reasonable to ask, therefore, how well has this general method preserved cytoplasmic structural detail? Probably, answers would range from the ecstatic to the unprintable! Nevertheless, it seems reasonable to assume that the bulk of current knowledge of cellular ultrastructure has been garnered using these basic techniques.

Since the early days of thin sectioning, osmium tetroxide has been used as the major component of the fixative, oftentimes the only active one. While this has sufficed to stabilize the various membrane systems (however, see Kirschner and Hollingshead, 1980), and thus allowed overall mapping of compartmentation, etc., it eventually became apparent that OsO_4 itself was insufficient for stabilization of protein components. This led to renewed emphasis on the use of aldehydes as crosslinking agents, notably the dialdehyde, glutaraldehyde, which with its two independently reactive groups, could form reasonably long linkages between available reactive amino acid side-chains. However, it was soon found that treatment with OsO_4 was still necessary after glutaraldehyde fixation for good overall preservation of ultrastructure. Similarly, the early use of methacrylates as embedding media gave way to epoxy type resins in order to minimize shrinkage and other artifacts arising during the polymerization stage.

That glutaraldehyde is a good fixative for protein structures in general is indicated by x-ray diffraction studies on protein crystals (Wishner *et al.*, 1975) and on myelin sheath (Kirschner and Hollingshead, 1980). In the former instance, the x-ray diffraction patterns showed (1) almost no change in the lattice parameters of protein crystals after fixation with glutaraldehyde, and (2) changes in intensities of diffraction spots interpretable as being due to an introduction into the lattice of an ordered array of density consistent with the formation of intermolecular bridges. In the case of myelin, the results also clearly show a stabilization of proteins at the surface of the lipid bilayers. It seems likely, therefore, that glutaraldehyde will rapidly stabilize protein structures in the cell provided (1) that sufficient side-chains are available for reaction and (2) that reaction can take place before degradation of the structure occurs. The latter condition would be especially important for polymeric structures maintained in an equilibrium situation, either by concentration effects or by active biochemical processes. An example of this occurred in the fixation of *Loligo* giant axons, where it was found (Hodge and Adelman, 1978a,b, 1980a) that, when fixation of the axoplasm was slow as occurred in immersion fixation, very few neurotubules were found, whereas internal irrigation with fixative following cannulation resulted in very rapid fixation and an abundance of neurotubules.

Thus, while there can be little doubt that glutaraldehyde stabilizes existing fibrous protein structures such as the neuroplasmic lattice, the situation with respect to the so-called soluble proteins of the axoplasm is less clear. In the first place, many of these are bound more or less loosely, but possibly in an orderly array, to existing "cytoskeletal" structures, rather than being in free solution. In that case, the reaction with glutaraldehyde could reasonably be expected to secure them *in situ*. Protein molecules in solution, on the other hand, would first be "activated" by reaction of certain side-chains with one of the aldehyde groups of the glutaraldehyde, leaving the other reactive group of the dialdehyde free to react with any available ε-amino group encountered during diffusion. Clearly, this would have the effect of condensing free proteins onto existing stable structures or, at the very least, of causing some polymerization of them. With the exception of this condensation of free soluble proteins, it seems fair to conclude that glutaraldehyde fixation, under controlled conditions of ionic strength, pH, and the presence of "stabilizers" such as EGTA to counter the undesirable effects of Ca^{++} ions, results in fairly good preservation of cell structure.

However, there remains the possibility of serious distortions of structure occurring during dehydration and embedding, as well as during the sectioning itself. The effects of swelling and dehydration in the myelin system has been studied by x-ray diffraction, but there appear to have been no comparable studies on proteinaceous cytoplasmic components. The process of dehydration, because of the drastic environmental change caused by it, would seem to be the most likely step for artifact generation while the infiltration with resin does not appear likely to cause problems. Polymerization, on the other hand, was often a severe problem with methacrylate embedding, and there is some evidence of small-scale structural disruption even with epoxy-type resins. However, these small defects do not appear to seriously hamper reasonable interpretation of micrographs.

It has been obvious for a long time that the best probable solution to the first part of the problem posed at the beginning of this section would be freezing of the specimen, the faster the better, but what then? Clearly, the best solution might be to somehow reduce the specimen to a thin layer appropriate for electron microscopy, that is, to section it. However, the mechanical properties of frozen cytoplasm at low temperatures do not readily allow this and, in any case, the specimen, consisting almost entirely of low atomic number elements, first, would require examination at cryogenic temperatures and second, would give very poor contrast. In principle, this problem might yield to computer contrast enhancement techniques or to the development of electron microscopes based on fine discrimination of electron energy loss characteristics. However, such techniques have yet to be developed to the practical stage.

Several practical solutions have been developed, particularly for thinner specimens such as the peripheral regions of tissue culture cells. Osmium tetroxide vapor fixation has been widely used for many years and was the method used by K. R. Porter and his colleagues in their pioneering demonstration of the endoplasmic reticulum as a distinct component of the cytoplasm (ca. 1950). Another promising development, now widely and effectively used, is rapid freezing of the tissue followed by substitution of water and fixation of the tissue at relatively low temperature, i.e., freeze substitution. For example, van Harreveld and Crowell (1964) froze the tissue on a polished silver block and used 2% OsO_4 in acetone at $-85°C$ as the substitution medium. In all cases, the very poor thermal conductivity of amorphous ice limits the thickness of good preservation to a few micrometers. The actual freezing process has been considerably advanced by the development of a machine using liquid helium and ensuring rapid contact of the specimen with the cooling block. Kopstad and Elgsaeter (1982a,b) recently analyzed cooling rates during impact and liquid-jet specimen freezing. Many researchers have followed the freezing procedure by fracturing the specimen to generate surfaces and then forming a metal replica of these while cold and under vacuum, either immediately or after a period of etching of the fracture faces (see structural chapters, this volume). These methods are basically extensions of a "shadow casting" technique used much earlier for the visualization of crystal surfaces (Hall, 1950) and other structures by forming a metal surface replica, usually by thermal evaporation of a heavy metal in vacuum (Williams and Wyckoff, 1946), i.e., the method of shadow casting, followed by deposition of a supporting film.

With the exception of freeze-substitution, the above freezing and fracturing methods have the aim of producing a thin replica of a surface. This heavy metal replica can then be examined at fairly high resolution since it constitutes an almost ideal specimen. However, the problem is basically the interpretation of images derived from these artificially created surfaces. This is particularly difficult where deep etching has been employed in highly filamentous regions of the cytoplasm. Many published micrographs show strong indications of at least partial structural collapse and "condensation" of thinner or more fragile elements onto the more structurally robust. Another drawback in applying this method to cytoplasmic structure is the strong possibility that sampling is far from random, with the crack most probably proceeding along limited planes of least structural rigidity. Thus, longitudinal fractures along a structure such as an axon could easily give an erroneous representation of the three-dimensional distribution of longitudinally oriented filamentous elements such as neurofilaments and neurotubules, unless supported by independent evidence from transverse sections of the same region, clearly a difficult requirement.

The critical point method of dehydration followed by direct TEM observation has found favor in the investigation of certain types of specimen, particularly in conjunction with high voltage electron microscopy (HVEM; Porter, 1976; Byers and Porter, 1977; Porter *et al.*, 1979; Wolosewick and Porter, 1979). Their general finding was that the various cytoplasmic organelles and inclusions were attached to and suspended in a network of microtrabeculae which were also observed to terminate on the plasma membrane. For this seemingly random network, they chose to use the term "microtrabecular lattice." As pointed out earlier, it is our feeling (Hodge and Adelman, 1978a,b, 1980a,b) that the term "lattice" should be reserved for a periodically ordered network, a connotation consonant with common crystallographic usage. In this sense, we agree, and so do our results, with the use by Ellisman and Porter (1980) of the term microtrabecular lattice to describe the periodic distribution of crossbridges between the longitudinal filamentous elements of axoplasm, but disagree that the crossbridge array should be represented as part of the "axoplasmic matrix." We feel that the crossbridges should be considered as formed structural axoplasmic elements, which are possibly structurally intrinsic in the case of neurofilaments, and not as a component of a residual "matrix" or ground substance. It should be noted also that the findings of Ellisman and Porter relating to periodicity of the crossbridges were apparently based primarily on their appearance in sections of epoxy embedded material, rather than on polyethylene glycol (PEG)-embedded sections from which the PEG had been removed prior to being critical point dried, or on regularity seen in freeze–etched replicas.

B. Reality of the Neuroplasmic Lattice

It seems clear from the above discussion that examination of thicker sections of epoxy-embedded material, after glutaraldehyde/OsO_4 or freeze–substitution fixation, is still the most trustworthy technique for determining the three-dimensional ordered characteristics of cytoplasmic filamentous networks such as the neuroplasmic lattice, with valuable supplementary information being provided by "surface-producing methods" such as freeze–fracturing and freeze–etching (Hirokawa, 1982). The evidence presented here and in other chapters of this volume leave little doubt that, in the neuroplasmic lattice, we are dealing with a dynamic ordered array of crossbridged elements, not unlike that present in muscle, but still of largely unknown composition and function other than an obvious cytoskeletal role, although here too, we can only guess at its necessarily dynamic implementation in such active cell processes as axonal transport.

The mechanical and optical properties of axoplasm, which can be obtained readily in pure form from *Myxicola* and *Loligo* giant axons, definitely support the presence of a three-dimensionally crossbridged lattice, rather than a simple longitudinal array of neurofilaments and/or neurotubules. Gilbert (1975a,b) has described in considerable detail the structure of *Myxicola* axoplasm at the light and electron microscopical levels, its mechanical properties and composition, together with some pertinent observations on squid axoplasm, which has many characteristics in common. Perhaps the most notable difference is the near absence of tubulin and of neurotubules in *Myxicola* axoplasm. Crossbridges were also found to be present (see also Gilbert *et al.*, 1975). The results of birefringence studies on squid giant axons and extruded axoplasm (Bear *et al.*, 1937), *Myxicola* axoplasm (Gilbert, 1975a), and frog sciatic nerve axoplasm (Thornburg and deRobertis, 1956) provide independent support for the presence of an ordered neuroplasmic lattice. For squid and *Myxicola,* the measured birefringence was essentially identical, $\sim 1.5 \times 10^{-4}$; for frog sciatic, 2.4×10^{-4}. These values are only about 1.5% that of wool (Woods, 1955) and 3% that of the A-band of striated muscle (Bear *et al.*, 1937), and since both these values are too low to be due entirely to differences in protein concentration, they were taken by Gilbert (1975a) to indicate a lower degree of orientation of the protein filaments in axoplasm. However, as pointed out by Hodge and Adelman (1980b), it seems likely that the strong positive axial birefringence to be expected of the relatively large amount (~3.5% protein) of oriented filamentous elements in the squid giant axon would be compensated to a considerable extent by the presence of the ordered system of radially oriented crossbridges. It would also clearly be of interest to explore other anisotropies of the axoplasm as they may relate to the presence and orientation of the crossbridges.

X-ray diffraction patterns of dried humidified axoplasm (Day and Gilbert, 1972; Gilbert, 1975a) showed clearly that the neurofilament protein is predominantly of the α-helical coil type with indications (1.8 nm and 2.5 nm meridional arcs) of a longer spacing. Gilbert (unpublished) subsequently obtained striking low angle x-ray diffraction patterns using fibers made from centrifugal pellets of neurofilaments mildly treated with endogenous calcium-activated protease. These showed many meridional reflections possibly indexing on a period of ~43 nm.* This result is of interest, not only in relation to the bridge spacings found in the neuroplasmic lattice, but also because it corresponds very closely with the myosin reflections in x-ray diffraction patterns of vertebrate striated muscle (Huxley, 1953; Huxley and Brown, 1967), as well as the crossbridge spacing as seen by electron microscopy (Draper and Hodge, 1949; Hodge *et al.*, 1954). Another striking homology is represented by the amino acid composition of neurofilaments (Gilbert, 1975b). There is a high proportion of polar residues, comparable to that found in the α-class of muscle proteins: myosin, tropomyosin, and paramyosin. Thus, there can be little doubt that the major component of the neurofilament protein, i.e, the 68 K polypeptide observed by gel electrophoresis (Hoffman and Lasek, 1975; Lasek *et al.*, 1979), is composed primarily of coiled coil α-helical structures, probably packed together by ordered charge interactions and nonpolar forces into a highly structured filament in a manner analagous to that by which the thick filaments of muscle appear to be constructed from myosin and tropomyosin or paramyosin molecules in various species.

This still leaves the composition and function of the crossbridges to be considered. In the case of the bridges between neurotubules, the work of Dentler *et al.* (1975) and of Kim *et al.* (1979) has shown that relatively high molecular weight microtubule-associated proteins (MAPs) are capable of orderly interaction with neurotubules and of forming an ordered array of crossbridges between them. It has also been shown that the gel character of squid axoplasm is at least partially dependent on bridges between neurotubules because liquefaction results on trypsin treatment of the neurofilament free gel, coincident with release of cleavage products of a high molecular weight putative MAP (Morris *et al.*, 1981). In this work, the neurofilament-free axoplasmic gel was obtained by induction of the resident neurofilament-specific calcium-activated protease. The situation is less clear with respect to neurofilament arms or bridges between them or to neurotubules. The work of Lasek and colleagues (Lasek *et al.*, 1979) and others has shown by transport rate and other criteria that neurofilaments comprise a triplet of peptides (68 K, 120 K, and 180 K), and it seems likely that the two larger peptides are involved in bridge formation and function.

Clearly, much more work is required in this area to resolve the situation. However, even at this stage, the evidence supports the concept that the neurofilament *in vivo* possesses a native periodicity of ca. 43 nm, which is expressed in the form of an orderly array of interactive side-arms. These may be either intrinsic structures, i.e, they represent a covalently linked moiety of one of the neurofilament protein subunits, or they may be extrinsic, and comprise one or more neurofilament-associated proteins (NAPs) interacting with periodically distributed surface sites, as seems to be the case for MAPs and neurotubules.

It seems appropriate to comment briefly on the question of order in cytoplasmic systems. The history of cytoarchitectural studies has been one of finding higher degrees of order in virtually all subsystems than was hitherto thought to be the case. This understanding has been brought about largely through the application of increasingly sophisticated electron

* This value is based on our interpretation of a print of an x-ray diffraction pattern obtained from Dr. Gilbert in 1979 in which at least six of the meridional reflections (those at ~7.2, 3.6, 2.54, 1.8, 1.4, and 1.2 nm) could be indexed on an axial repeat of 43.2 nm. He had indicated in a note included with the pattern that the many meridionals were not all Bragg-related, but could be indexed on repeats of ⩾35 nm, with the coherence length of the brightest meridional reflection (~7.1 nm) indicating axial ordering over distances greater than 60 nm. The presence of such long spacing order in the neurofilament "core" is entirely compatible with our view that the true axial repeat generated by a six-fold screw axis distribution of cross-bridges would be 6 × 43 nm ≃ 260 nm. Structural similarities and differences amongst various neurofilaments have been noted recently by Wais-Steider *et al.* (1983).

microscopic techniques and substantiated in those limited instances where it could be applied usefully, by x-ray diffraction methods. Since function in all active systems clearly implies changes in state of structural elements, usually in some cyclic fashion involving energy-coupled systems, it is clear that function must involve some temporary perturbation of structural order. This could be viewed as a short-term shift from a ground or resting state in which there is maximal interaction between structural elements consistent with viability. The highest degree of order would be expected in the "rigor" state where all specific sites are occupied and there is no bridge activity. Perhaps the clearest example of this can be seen in the myofibrils of striated muscle, for it is well known that the order present in this highly ordered crossbridge lattice system is lessened during bridge activity, as evidenced by characteristic intensity changes in the x-ray diffraction pattern. However, these are of necessity, time averaged patterns, and it seems likely that with adequate temporal resolution, the attainment of which requires enormously intense x-ray sources, crossbridge activity will be seen to be a temporally ordered cyclic phenomenon. Efforts to define these changes are currently under way in several laboratories. It should therefore come as no great surprise that the concensus of opinion presented at this conference on structural aspects of neurons seems to be that axonal cytoplasm is highly ordered. Clearly, much remains to be done in order to understand how this structural order subserves the various functions of the axon, in particular, its electrical activity and transport properties.

ACKNOWLEDGMENTS. The authors wish to thank Richard B. Waltz, Ruthanne Mueller, and Clyde L. Tyndale for their very able assistance during this work, Dr. Kenneth Edds for kindly supplying us with rabbit fraction S_1, and Dorothy Leonard and Betty Hodge for their invaluable aid in creating order out of the primordial manuscript matrix.

REFERENCES

Adelman, W. J., Jr., Hodge, A. J., and Waltz, R. B., 1981, Characterization of periodic structure in subcellular macromolecular arrays by Fourier processing of stem video signals, *Biophys. J.* **33**(2, pt. 2):92a.

Adelman, W. J., Jr., Hodge, A. J., and Waltz, R. B., 1982, Trigonometric nearest neighbor analysis of the neuroplasmic lattice arrays in axons, *Biol. Bull.* **163**:379.

Allen, R. D., Metuzals, J., and Tasaki, I., 1981, Fast axonal transport in the squid giant axon, *Biol. Bull.* **161**:303.

Allen, R. D., Lasek, R. J., Gilbert, S. P., Hodge, A. J., and Govind, C. K., 1982a, Fast axonal transport in lobster axons, *Biol. Bull.* **163**:379–380.

Allen, R. D., Metuzals, J., Tasaki, I., Brady, S. T., and Gilbert, S. P., 1982b, Fast axonal transport in squid giant axon, *Science* **218**:1127–1129.

Anderson, T. F., 1951, Techniques for the preservation of three-dimensional structure in preparing specimens for the electron microscope, *Trans. N.Y. Acad. Sci.* **13**:130–134.

Astbury, W. T., 1947, On the structure of biological fibres and the problem of muscle, *Proc. R. Soc. (London)* **B134**:303–328.

Bear, R. S., 1942, Long x-ray diffraction spacings of collagen, *J. Am. Chem. Soc.* **64**:727.

Bear, R. S., 1945, Small-angle x-ray diffraction studies on muscle, *J. Am. Chem. Soc.* **67**:1625–1626.

Bear, R. S., Schmitt, F. O., and Young, J. Z., 1937, The ultrastructure of nerve axoplasm, *Proc. R. Soc. (London)* **B123**:505–519.

Berthold, C-H., 1978, Morphology of normal peripheral axons, in: *Physiology and Pathobiology of Axons* (S. G. Waxman, ed.), pp. 3–63, Raven Press, New York.

Brady, S. T., Lasek, R. J., and Allen, R. D., 1981, Fast axonal transport in extruded axoplasm from squid giant axon, *Biol. Bull.* **161**:304.

Brady, S. T., Lasek, R. J., and Allen, R. D., 1982, Fast axonal transport in extruded axoplasm from squid giant axon, *Science* **218**:1129–1131.

Breuer, A. C., Eagles, P. A. M., Gilbert, S. P., Allen, R. D., Metuzals, J., Clapin, D. E., and Sloboda, R. D., 1982, Fast transport in isolated axoplasm of *Myxicola infundibulum*, *Biol. Bull.* **163**:381.

Burton, P. R., and Fernandez, H. L., 1973, Delineation by lanthanum staining of filamentous elements associated with the surfaces of axonal microtubules, *J. Cell Sci.* **12**:567–583.

Burton, P. R., Hinckely, R. E., and Pierson, G. B., 1975, Tannic acid-stained microtubules with 12, 13 and 15 protofilaments, *J. Cell Biol.* **65**:227–233.

Byers, H. R., and Porter, K. R., 1977, Transformations in the structure of the cytoplasmic ground substance in erythrophores during pigment aggregation and dispersion. I. A study using whole-cell preparations in stereo high voltage electron microscopy, *J. Cell Biol.* **75**:541–558.

Day, W. A., and Gilbert, D. S., 1972, X-ray diffraction pattern of axoplasm, *Biochim. Biophys. Acta* **285**:503–506.

Dentler, W. L., Granett, S., and Rosenbaum, J. L. 1975, Ultrastructural localization of the high molecular weight proteins associated with *in vitro* assembled brain microtubules, *J. Cell Biol.* **65**:237–241.

Draper, M. H., and Hodge, A. J., 1949, Studies on muscle with the electron microscope. I. The ultrastructure of toad striated muscle, *Austral. J. Exp. Biol. Med. Sci.* **27**:465–484.

Ellisman, M. H., and Porter, K. R., 1980, Microtrabecular structure of the axoplasmic matrix: Visualization of cross-linking structures and their distribution, *J. Cell Biol.* **87**:464–479.

Erickson, R. O., 1973, Tubular packing of spheres in biological fine structure, *Science* **181**:705–716.

Fahim, M. A., Brady, S. T., Hodge, A. J., and Lasek, R. J., 1982, EM and AVEC-DIC analyses of membranous organelle transport in squid giant axons and isolated axoplasm, *Biol. Bull.* **163**:382–383.

Forman, D. S., Padjen, A. L., and Siggins, G. R., 1977, Axonal transport of organelles visualized by light microscopy: Cinemicrographic and computer analysis, *Brain Res.* **136**:197–213.

Gilbert, D. S., 1975a, Axoplasmic architecture and physical properties as seen in the *Myxicola* giant axon, *J. Physiol.* **253**:257–301.

Gilbert, D. S., 1975b, Axoplasmic chemical composition in *Myxicola* and solubility properties of its structural protein, *J. Physiol.* **253**:303–319.

Gilbert, D. S., Newby, B., and Anderton, B., 1975, Neurofilament disguise, destruction and discipline, *Nature (London)* **256**:586–589.

Gilev, V. P., 1979, A simple method of optical filtration, *Ultramicroscopy* **4**:323–336.

Grafstein, B., and Forman, D. S., 1980, Intracellular transport in neurons, *Physiol. Rev.* **60**:1167–1283.

Hall, C. E., 1950, Electron microscopy of crystalline catalase, *J. Biol. Chem.* **185**:749–754.

Hirokawa, N., 1982, Cross-linker system between neurofilaments, microtubules, and membranous organelles in frog axons revealed by the quick-freeze, deep-etching method, *J. Cell Biol.* **94**:129–142.

Hodge, A. J., 1959, The fibrous proteins of muscle, *Rev. Mod. Phys.* **31**:409–425.

Hodge, A. J., 1960, Principles of ordering in fibrous systems, in: *Proc. Fourth International Conference on Electron Microscopy*, Berlin, Sept. 10–17, 1958, pp. 119–139, Springer-Verlag, Berlin.

Hodge, A. J., 1967, Structure (of collagen) at the electron microscopic level, in: *Treatise on Collagen, Vol. 1. Chemistry of Collagen* (G. N. Ramachandran, ed.), pp. 185–205, Academic Press, London and New York.

Hodge, A. J., and Adelman, W. J., Jr., 1978a, The ordered transverse bridge lattice system associated with neurofilaments and microtubules in axonal cytoplasm of *Loligo* and *Hermissenda*, *Soc. Neurosci. 8th Annual Meeting*, St. Louis, Nov. 5–9, 1978, p. 331.

Hodge, A. J., and Adelman, W. J., Jr., 1978b, The ordered fine structure of neuronal cytoplasm in *Loligo* and *Hermissenda*, *Biol. Bull.* **155**:443–444.

Hodge, A. J., and Adelman, W. J., Jr., 1980a, The neuroplasmic network in *Loligo* and *Hermissenda* neurons, *J. Ultrastruct. Res.* **70**:220–241.

Hodge, A. J., and Adelman, W. J., Jr., 1980b, Characterization of periodic order in the neuroplasmic lattice of *Bufo, Loligo* and *Hermissenda* axons by a Fourier analytical technique in scanning transmission electron microscopy (STEM), *Biol. Bull.* **159**:471.

Hodge, A. J., and Adelman, W. J., Jr., 1981, Neuroplasmic lattice order in vertebrate and invertebrate axons: Demonstration by stereoscopic and autocorrelation electron microscopy and Fourier analytical techniques in STEM, *Biophys. J.* **33**(2, pt. 2):93a.

Hodge, A. J., and Adelman, W. J., Jr., 1983, The structure of the neuroplasmic lattice in vertebrate and invertebrate neurons, in preparation.

Hodge, A. J., Huxley, H. E., and Spiro, D., 1954, Electron microscope studies on ultrathin sections of muscle, *J. Exp. Med.* **99**:201–206.

Hodge, A. J., Adelman, W. J., Jr., Waltz, R. B., and Tyndale, C. L., 1982a, Analysis of periodic structure in model subcellular macromolecular arrays by Fourier processing of single line video signals in scanning transmission electron microscopy, *IEEE Trans. Biomed. Eng.* **BME-29**:439–447.

Hodge, A. J., Govind, C. K., Lasek, R. J., and Allen, R. D., 1982b, Correlation of electron microscopic fine structure with videomicroscopic observations in identified lobster axons during glutaraldehyde fixation, *Biol. Bull.* **163**:384–385.

Hodge, A. J., Adelman, W. J., Jr., and Waltz, R. B., 1983, Nearest neighbor analysis of neurofilaments and neurotubules in lobster cord and toad peripheral axons is consistent with an hexagonally ordered neuroplasmic lattice, *Biophys. J.* **41:**23a.

Hoffman, P. N., and Lasek, R. J., 1975, The slow component of axonal transport. Identification of major structural peptides of the axon and their generality among mammalian neurons, *J. Cell Biol.* **66:**351–366.

Huxley, H. E., 1953, X-ray analysis and the problem of muscle, *Proc. R. Soc. (London)* **B141:**59–62.

Huxley, H. E., and Brown, W., 1967, The low-angle x-ray diagram of vertebrate striated muscle and its behaviour during contraction and rigor, *J. Mol. Biol.* **30:**383–434.

Inoué, S., and Ritter, H., 1975, Dynamics of mitotic spindle organization and function, in: *Molecules and Cell Movement* (S. Inoué and R. E. Stephens, eds.), pp. 3–30, Raven Press, New York.

Kallman, F., and Wessels, N. K., 1967, Periodic repeat units of epithelial cell tonofilaments, *J. Cell Biol.* **32:**227–231.

Kim, H., Binder, L. I., and Rosenbaum, J. L., 1979, The periodic association of MAP2 with brain microtubules *in vitro, J. Cell Biol.* **80:**266–276.

Kirschner, D. A., and Hollingshead, C. J., 1980, Processing for electron microscopy alters membrane structure and packing in myelin, *J. Ultrastruct. Res.* **73:**211–232.

Kopstad, G., and Elgsaeter, A., 1982a, Theoretical analysis of the ice crystal size distribution in frozen aqueous specimens, *Biophys. J.* **40:**155–162.

Kopstad, G., and Elgsaeter, A., 1982b, Theoretical analysis of specimen cooling rate during impact freezing and liquid-jet freezing of freeze-etch specimens, *Biophys. J.* **40:**163–170.

Lasek, R. J., Krishnan, N., and Kaiserman-Abramof, I. R., 1979, Identification of the subunit proteins of 10 nm neurofilaments isolated from axoplasm of squid and *Myxicola* giant axons, *J. Cell Biol.* **82:**336–346.

Luther, P. K., Munro, P. M. G., and Squire, J. M., 1981, Three-dimensional structure of the vertebrate muscle A-band. III. M-region structure and myosin filament symmetry, *J. Mol. Biol.* **151:**703–730.

Metuzals, J., 1966, Electron microscopy of neurofilaments, *Proc. VI Int. Congress for Electron Microscopy,* Kyoto, Japan, pp. 459–460.

Metuzals, J., 1969, Configuration of a filamentous network in the axoplasm of the squid *(Loligo pealeii)* giant nerve fiber, *J. Cell Biol.* **43:**480–505.

Milam, L., and Erickson, H. P., 1982, Visualization of a 21-nm axial periodicity in shadowed keratin filaments and neurofilaments, *J. Cell Biol.* **94:**592–596.

Morris, J. R., Hodge, A. J., and Lasek, R. J., 1981, The microtubule network in the squid giant axon, *Biol. Bull.* **161:**308.

Ochs, R. L., and Burton, P. R., 1980, Distribution and selective extraction of filamentous components associated with axonal microtubules of crayfish nerve cord, *J. Ultrastruct. Res.* **73:**169–182.

Peters, A., Palay, S. L., and Webster, H., de F., 1976, *The Fine Structure of the Nervous System,* Saunders, Philadelphia.

Porter, K. R., 1976, Introduction: Motility in cells, *Cold Spring Harbor Conf. Cell Proliferation* **3**(Book A):1–28.

Porter, K. R., Byers, H. R., and Ellisman, M. H., 1979, The cytoskeleton, in: *The Neurosciences, Fourth Study Program* (F. O. Schmitt and F. G. Worden, eds.), pp. 703–722, MIT Press, Cambridge, Mass.

Reedy, M. K., 1968, Ultrastructure of insect flight muscle. I. Screw senses and structural grouping in the rigor cross-bridge lattice, *J. Mol. Biol.* **31:**155–176.

Rice, R. V., Roslansky, P. F., Pascoe, N., and Houghton, S. M., 1980, Bridges between microtubules and neurofilaments visualized by stereo-electron microscopy, *J. Ultrastruct. Res.* **71:**303–310.

Roslansky, P. F., Cornell-Bell, A., Rice, R. V., and Adelman, W. J., Jr., 1980, Polypeptide composition of squid neurofilaments, *Proc. Natl. Acad. Sci. USA* **77:**404–408.

Schwartz, J. H., 1979, Axonal transport: Components, mechanisms, and specificity, *Annu. Rev. Neurosci.* **2:**467–504.

Squire, J. M., 1981, *Structural Basis of Muscular Contraction,* Plenum Press, New York.

Tani, E., and Ametani, T., 1970, Structure of microtubules in brain nerve cells as revealed by ruthenium red, *J. Cell Biol.* **46:**159–165.

Thornburg, W., and deRobertis, E., 1956, Polarization and electron microscope study of frog nerve axoplasm, *J. Biophys, Biochem. Cytol.* **2:**475–482.

Traub, W., and Piez, K. A., 1971, The chemistry and structure of collagen, *Adv. Protein Chem.* **25:**243–352.

van Herreveld, A., and Crowell, J., 1964, Electron microscopy after rapid freezing on a metal surface and substitution fixation, *Anat. Rec.* **149:**381–386.

Williams, R. C., 1953, A method of freeze-drying for electron microscopy, *Exp. Cell Res.* **4:**188–201.

Williams, R. C., and Wyckoff, R. W. G., 1946, Applications of metallic shadow-casting to microscopy, *J. Appl. Phys.* **17:**23–33.

Wishner, B. C., Ward, K. B., Lattman, E. E., and Love, W. E., 1975, Crystal structure of sickle-cell deoxy-hemoglobin at 5A resolution, *J. Mol. Biol.* **98:**179–194.

Wolosewick, J. J., and Porter, K. R., 1979, Microtrabecular lattice of the cytoplasmic ground substance. Artifact or reality, *J. Cell Biol.* **82:**114–139.

Woods, H. J., 1955, *Physics of Fibres*, Institute of Physics, London.

Yamada, K. M., Spooner, B. S., and Wessells, N. K., 1971, Ultrastructure and function of growth cones and axons of cultured nerve cells, *J. Cell Biol.* **49:**614–635.

Yen, S-H, and Fields, K. L., 1981, Antibodies to neurofilament, glial filament, and fibroblast intermediate filament proteins bind to different cell types of the nervous system, *J. Cell Biol.* **88:**115–126.

Wais-Steider, C., Eagles, P. A. M., Gilbert, D. S., and Hopkins, J., 1983, Structural similarities and differences amongst neurofilaments, *Nature (London),* in press.

5

Membrane Specialization and Cytoskeletal Structures in the Synapse and Axon Revealed by the Quick-Freeze, Deep-Etch Method

Nobutaka Hirokawa

According to present concepts of synaptic transmission, nerves discharge the chemical transmitter in response to the action potential propagated in the axon, by a mechanism that is activated by the entry of calcium into the presynaptic terminal. This calcium entry triggers the exocytosis of transmitter-containing vesicles at specialized release sites named the "active zone." The chemical transmitter diffuses through the synaptic cleft to be accepted by receptor molecules in the postsynaptic membranes. In the case of synapses such as neuromuscular junctions (NMJ), the receptor molecules form receptor–ion-channel complexes. The presence of the transmitter induces conformational change of the channel molecules to open the channels and cause an endplate potential to appear at the muscle membrane, which triggers muscle contraction.

In this chapter, current information on the membrane specialization and cytoskeletal elements at the NMJ and axons is presented. We focus particularly on the structural arrangement of the acetylcholine (ACh) receptor–ion-channel complexes on the postsynaptic membranes and on the cytoskeletal structure of the axon and its relationship with the axolemma revealed by the new quick-freeze, deep-etch technique. More detailed information about the quick-freeze and deep-etching method is obtainable in Heuser and Salpeter (1979), Heuser and Kirschner (1980), Hirokawa and Heuser (1981b, 1982a,b), Hirokawa (1982) and Hirokawa *et al.* (1982).

I. PRESYNAPTIC STRUCTURE

Since DeRobertis and Bennett (1955) and Palay (1956) clearly demonstrated the presence of synaptic vesicles in the presynaptic terminals and proposed that they were secretory

Nobutaka Hirokawa • Department of Anatomy and Neurobiology, Washington University School of Medicine, St. Louis, Missouri 63110.

droplets of neurotransmitter, these structures have been generally observed in various kinds of synapses by the thin-section method. Palay further observed that synaptic vesicles clustered over regions where synaptic membrane appeared dense and proposed that such region could contain specialized sites for vesicle discharge (Fig. 1). Later, Couteaux and Pecot-Dechavassine (1970) named the specialized region "active zones" and reported invaginations of the plasmalemma at the active zones that looked like synaptic vesicles in the process of discharging their contents. This specialization of presynaptic membrane was further visualized as a number of dense projections arranged in triagonal array by specific staining methods using phosphotungstic acid (PTA) or bismuth iodide, which are supposed to react with positively charged proteins by Gray (1963), Bloom and Aghajanian (1966, 1968) and Pfenninger (1971). The intimate relationship between the active zone and synaptic vesicle discharge was most clearly demonstrated by freeze–fracture studies. Freeze–fracture studies revealed accumulation of large intramembranous particles at the active zone. These particles are characteristically arranged in double rows in the NMJ (Figs. 2,3, and 4; Dreyer *et al.*, 1973; Heuser *et al.*, 1974).

When the nerves were stimulated during fixation, plasmalemmal deformations of the same size as the synaptic vesicles were observed to line up adjacent to the membrane specializations (Heuser *et al.*, 1974; Figs. 3 and 4). Recently, Heuser *et al.* (1979) beautifully confirmed this without the use of chemical fixation using the quick-freeze, freeze–fracture method. These data strongly suggest that the transmitter is released by exocytosis of vesicles. Because the characteristic double rows of large intramembrane particles are so close to the sites of exocytosis, the idea has arisen that these large particles may be the sites of calcium channels (Heuser, 1976; Hirokawa and Heuser, 1981a).

Thus, freeze–fracture offers us useful information about intramembranous structures. However, because of its inherent technical limitation, freeze–fracture as it is conventionally carried out cannot be used to study the true outside and inside of the membrane, extracellular structures, or cytoskeletal structures that are buried in ice crystals (Fig. 5A and B). This is because the freezing step, traditionally carried out in Freon quenched with liquid nitrogen, is very slow. In order to prevent the formation of ice crystals of damagingly large sizes, cryoprotectants must be used. Conventional cryoprotectants, such as glycerol and DMSO, are unfortunately nonetchable at freeze–fracture temperatures, so that etching (sublimation) of ice to reveal true surfaces is impossible. This problem has been recently overcome through the development of quick-freezing, i.e., freezing of samples by appression against a copper block cooled to the temperature of liquid nitrogen (Hirokawa and Kirino, 1980) or liquid helium (Heuser *et al.*, 1979). The rapidity of freezing makes it possible to freeze specimens without cryoprotectants. Very deep "etching" is therefore possible and one is able to observe, for the first time, true biological surfaces at very high resolution.

Figure 5C provides a quick-freeze, deep-etch image of frog NMJ fractured parallel to the long axis of a nerve terminal, which is an equivalent view with Fig. 1. Numerous synaptic vesicles are clustered in the terminal and some nonetchable substances are found at the active zone. This picture also shows clearly how the basal lamina looks. The basal lamina covers the postsynaptic membrane as a veil from which many spines project to connect pre-and postsynaptic membranes. Thus, in a cross section it looks like a barbed wire. Recently, McMahan *et al.* (1978) have proved that ACh esterase is localized in the basal lamina and it has been clearly shown that the basal lamina at the junctional area is antigenically different from that at the extrajunctional area and works as recognition site for reinnervation (Sanes *et al.*, 1978; Sanes and Hall, 1979). Concerning the cytoskeletal structure in the presynaptic terminal, however, clusters of synaptic vesicles unfortunately interfere with clear visualization. Figures 6 and 7 depict the inside of the presynaptic terminal of the electric organ of

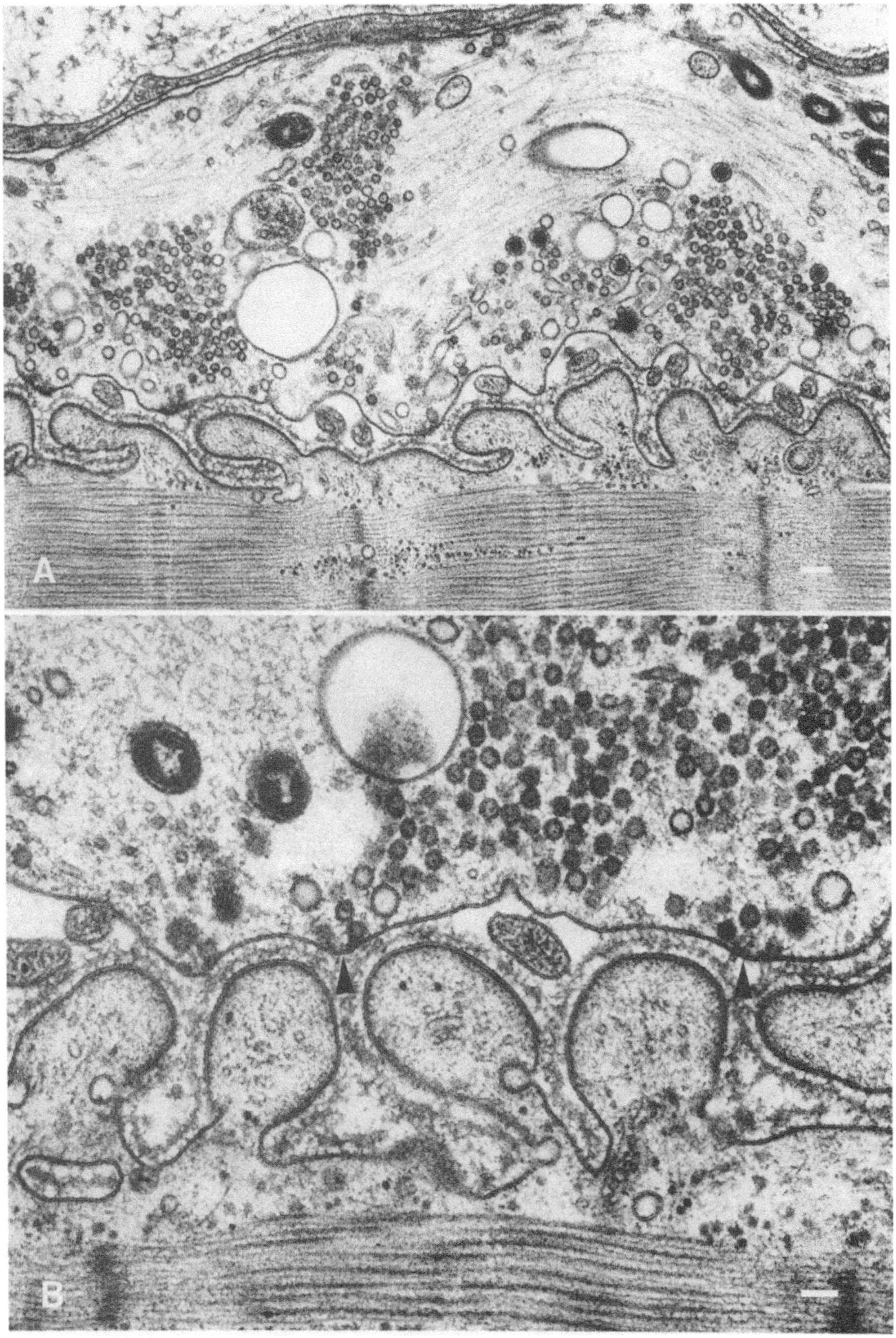

Figure 1. (A) A low magnification view of a frog neuromuscular junction longitudinally sectioned. This sample was quick-frozen directly from life and then freeze-substituted in acetone and osmium tetroxide. The presynaptic terminal is filled with synaptic vesicles and cisternae. Postsynaptic folds, which are parallel with each other and perpendicular to the long axis of the nerve terminal, appear as finger-like protrusions in this section. Bar 0.2 μm. (B) A high magnification view of a frog NMJ prepared by freeze-substitution and sectioned longitudinally. In the presynaptic terminal, synaptic vesicles are clustered at the active zone (arrowheads) which locate opposite to the infoldings. Fuzzy materials are undercoating the postsynaptic membrane. Extracellular spaces are filled with the basal lamina which bisect the pre- and postsynaptic parts. Bar 0.1 μm.

Figure 2. A very low magnification view of a freeze–fractured frog NMJ. The nerve was stimulated in the weak fixative paraformaldehyde. The protoplasmic face (P face) of the nerve terminal is exposed as a long tube. The active zone (arrows) are identified as double rows parallel with each other on the P face. Bar 1 μm.

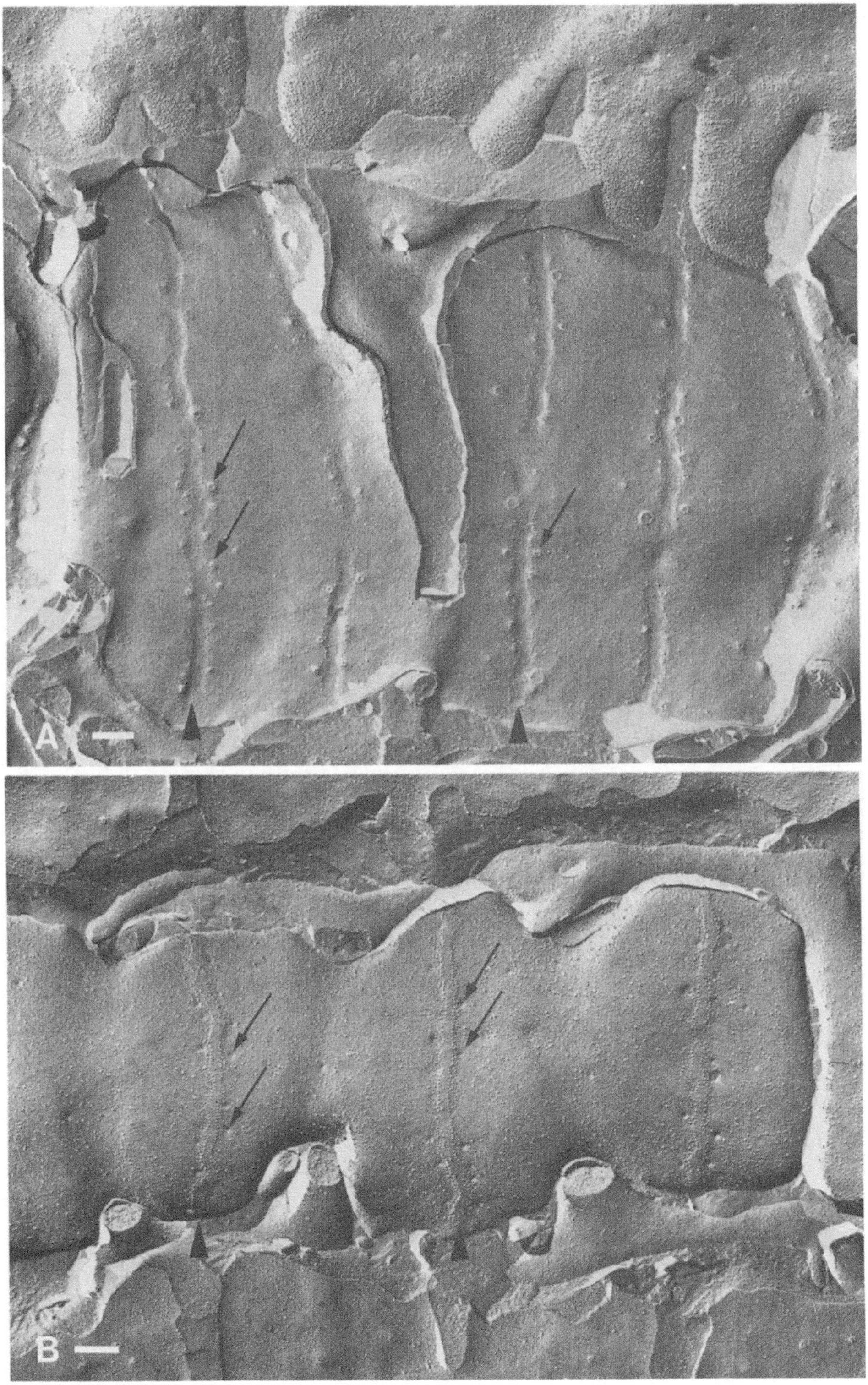

Figure 3. Higher magnification views of presynaptic terminals prepared as in Fig. 2. The active zones are characterized as furrows parallel with each other (arrowheads) on the exoplasmic face (E face) (A), and ridges (arrowheads) which are delineated by double rows of large particles on the P face (B). Possible fusion sites of the synaptic vesicles are recognized as protuberances (arrows) next to the furrow on the E face (A) and as dimples (arrows) next to the ridges on the P face (B). At the upper part of Fig. 3A the P face of a postsynaptic membrane is observed. It is studded with numerous P face particles. Bar 0.1 μm.

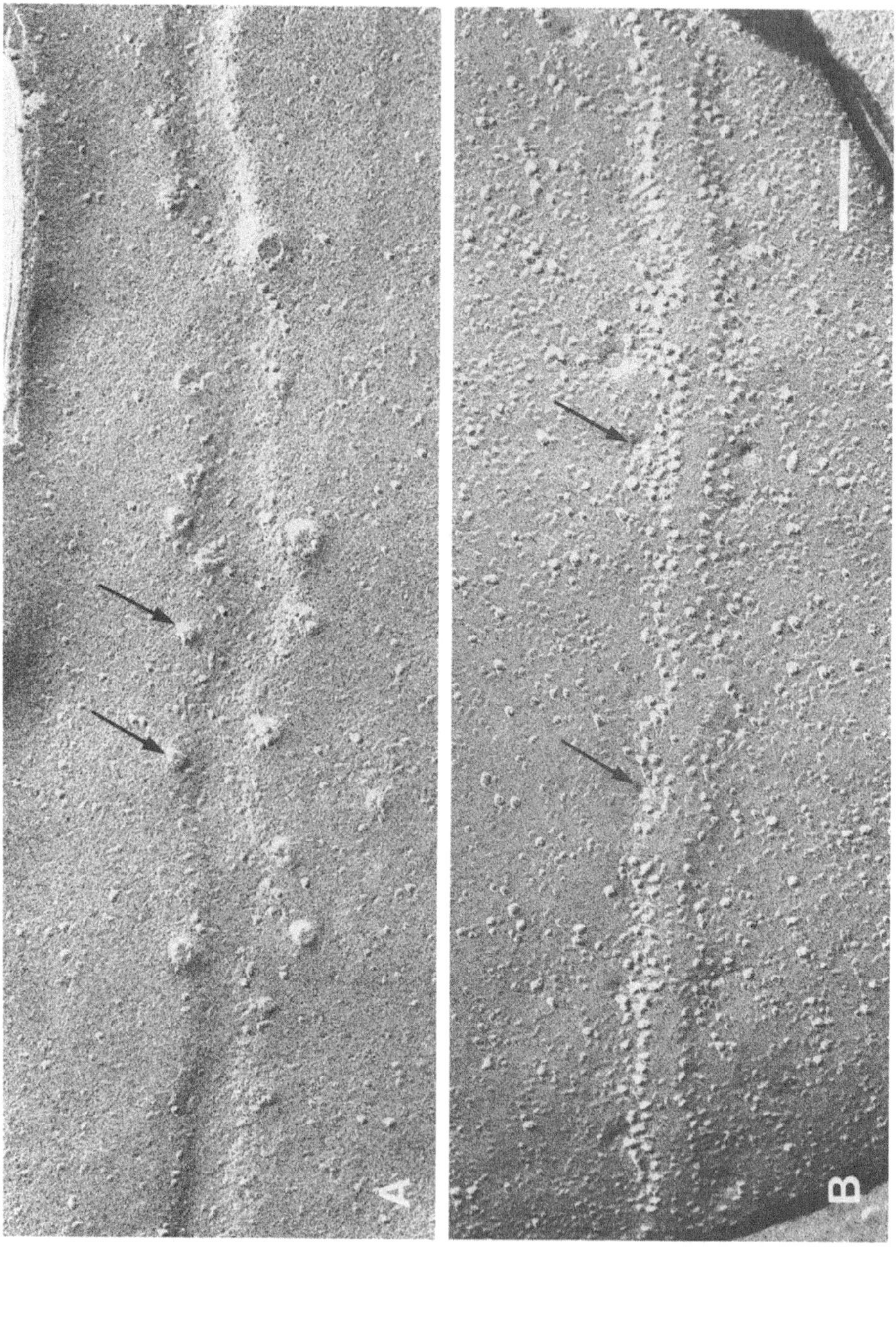

Figure 4. Very-high-magnification views to show a single active zone. The nerve is stimulated in the weak fixative paraformaldehyde. Protuberances (arrows) next to the furrow on the E face (A), and dimples (arrows) next to the ridges on the P face (B) are observed after nerve stimulation. Bar 0.1 μm.

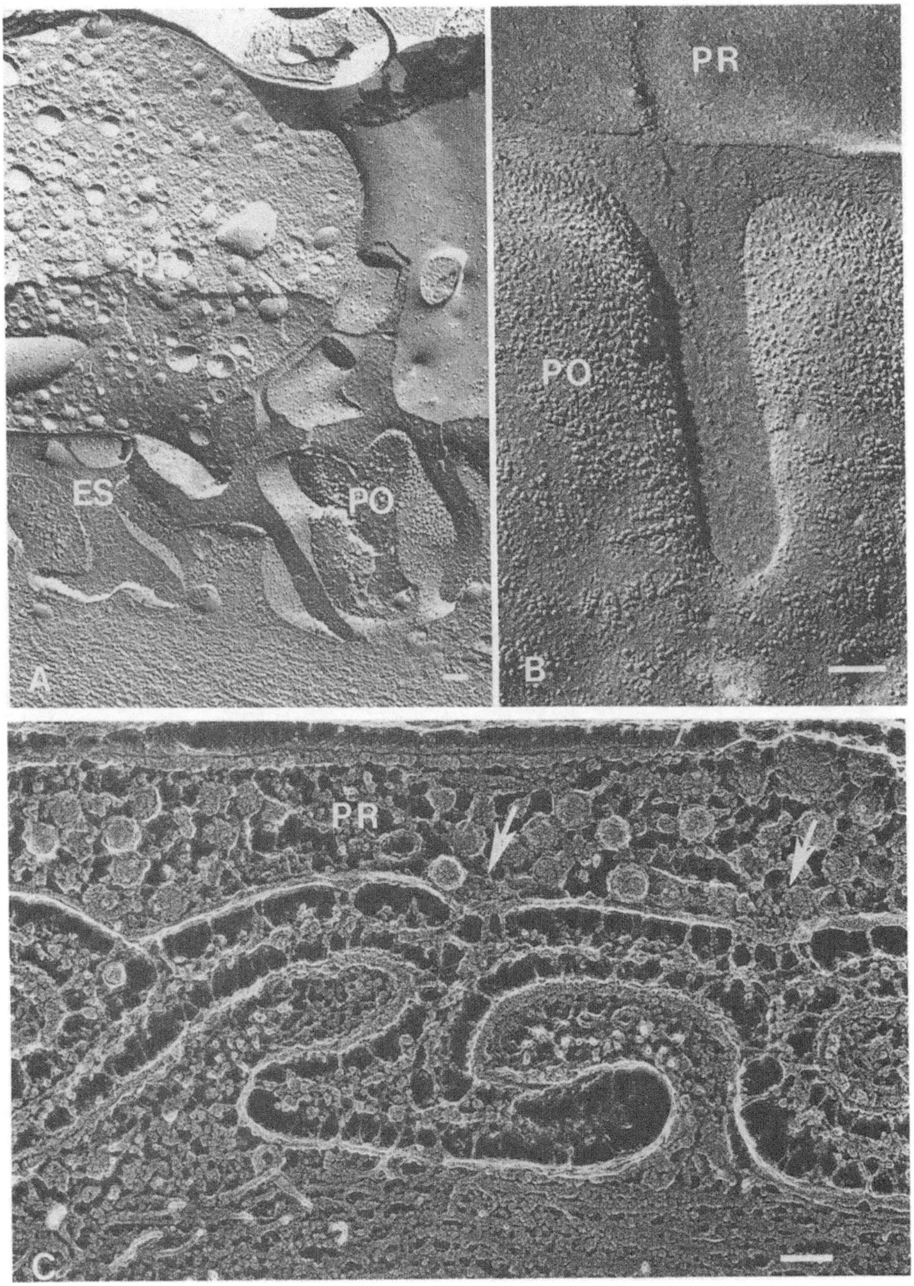

Figure 5. Comparison of the conventional freeze–fracture (A,B), and the quick-freeze, freeze–fracture, deep-etch, rotary replication (C). Structures inside the presynaptic cytoplasm (PR), extracellular space (ES), and the postsynaptic foldings (PO) cannot be learned much from by conventional freeze-fracture, because the cryoprotectant, which is necessary to avoid ice crystal damage for the conventional method, prevent etching of the ice in those spaces (A). In Fig. 5B presynaptic E face (PR), infoldings and P face of the postsynaptic membrane (PO) are observed. Although conventional freeze–fracture is sufficient to observe the intramembranous structure, nothing can be learned about the extracellular structure and true surface structures of the membrane. In contrast, because the quick-freeze method can freeze samples fast enough so that the cryoprotectants are no longer necessary, deep-etching can be performed to reveal three-dimensional structures in the presynaptic terminal (PR) and the basal lamina (Fig. 5C). Arrowheads point to the active zones. However, simple quick-freeze, deep-etching is still not enough to display the cytoplasmic structure in some cases, because the soluble proteins interfere with clear visualization of cytoskeletal elements. The soluble proteins should be removed as mentioned in the text.

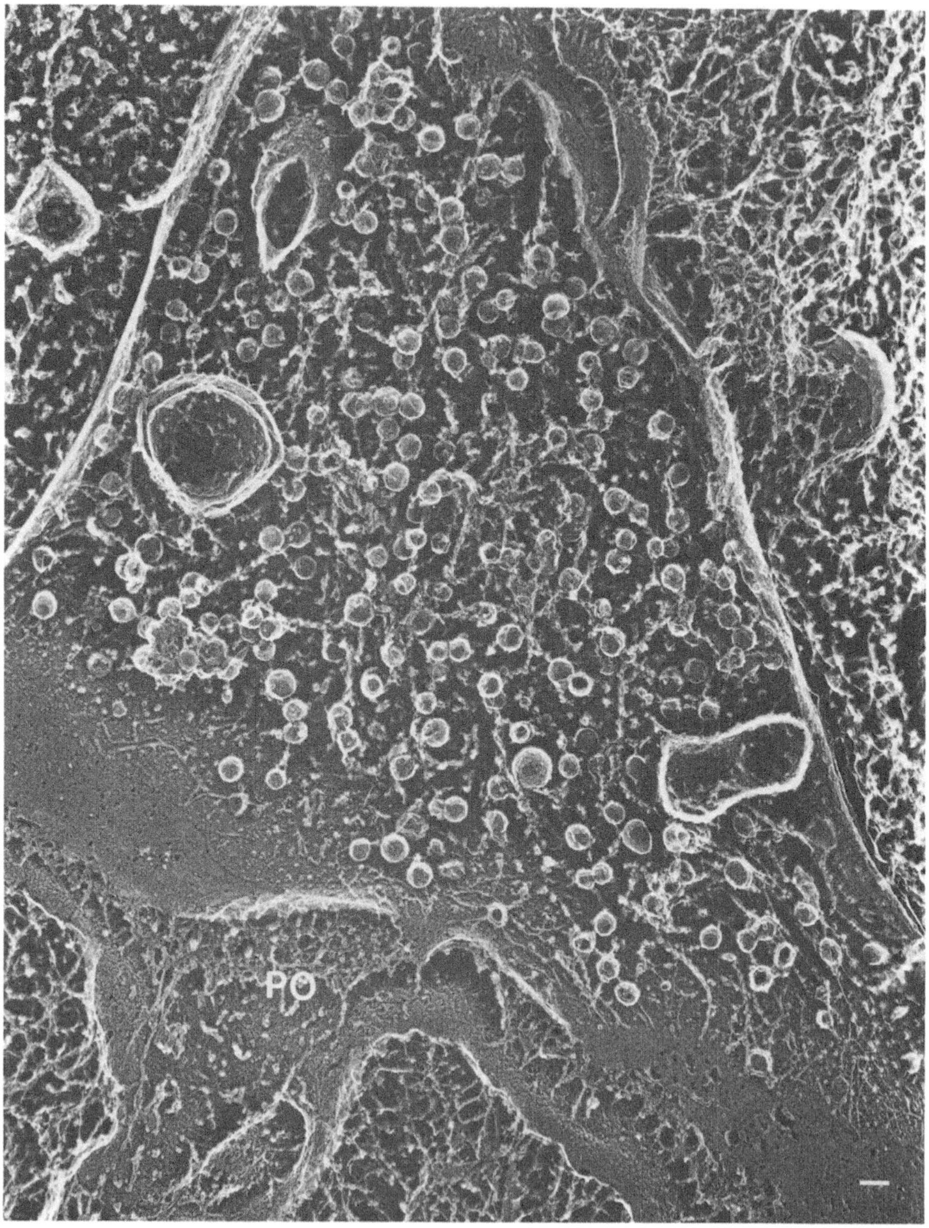

Figure 6. Inside of the presynaptic terminal of the *Torpedo* electric organ. The tissues were homogenized gently and incubated with subfragment 1 (S_1) of myosin. There are networks of actin filaments in the presynaptic terminal. PO: true outside surface of the postsynaptic membrane. Bar 0.1 μm.

Torpedo. This preparation has been homogenized to expose the cytoplasm, and then incubated with subfragment 1 (S_1) of myosin to localize actin. Actin filaments are decorated by S_1 to show a characteristic double helical rope-like structure. These micrographs demonstrate that actin is present in the presynaptic terminal, possibly forming a framework. Some of the actin filaments are closely associated with the presynaptic membrane, but nondecorated,

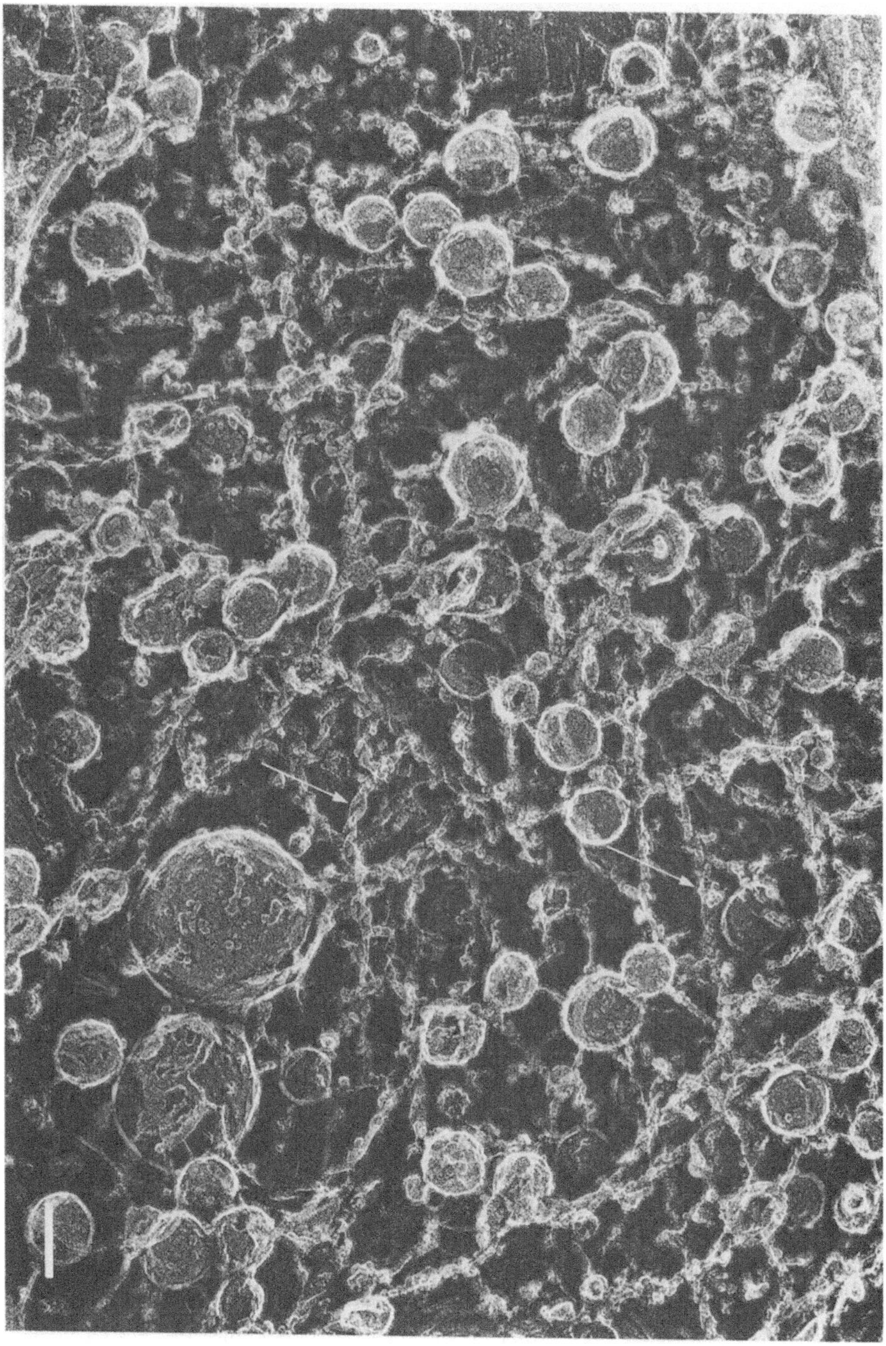

Figure 7. Inside of a presynaptic terminal of the *Torpedo* electric organ after decoration with S_1. The actin filaments are decorated with S_1 to show double helical structure (arrows).

filamentous structures also exist in this region. Whether these actin filaments play an active role in synaptic vesicle exocytosis is still an open question.

II. POSTSYNAPTIC STRUCTURE

There is now abundant physiological and biochemical evidence that the postsynaptic membranes of chemical synapses possess clusters of receptor molecules which are somehow immoblized at the immediate synaptic sites (Peper and McMahan, 1972; Anderson and Cohen, 1974; Kuffler and Yoshikami, 1975; Fertuck and Salpeter, 1976). These receptor-molecule–ion-channel complexes change their conformation by acceptance of chemical transmitters and somehow modulate their permeability to inorganic ions.

A number of thin section studies have observed localized thickening of the postsynaptic membrane. At that region the postsynaptic membrane is coated by a fuzzy layer of dense, granular material. This material appeared more filamentous than granular, so it has been called a "subsynaptic web." The material is stained selectively by PTA. Birks *et al.* (1960) and Couteaux and Pecot-Dechavassine (1968) have noticed that nearly all the muscle membrane at the junctional region is coated with fuzzy dense material except for the deepest recesses of the postsynaptic folds at the neuromuscular junction. Electron microscopic autoradiography using α-bungarotoxin (which binds specifically with ACh receptors) clearly demonstrated that α-bungarotoxin binding sites are located predominantly in regions where the postsynaptic membrane is underlaid by dense material and restricted to the top of the sarcoplasmic protrusions (Fertuck and Salpeter, 1976).

On the other hand, freeze–fracture studies have shown an abundance of intramembranous particle clusters on the postsynaptic muscle membrane at top of the sarcoplasmic protrusions (Heuser *et al.*, 1974; Rash and Ellisman, 1974; Fig. 5B). Since it is generally accepted that intramembranous particles revealed by freeze–fracture represent protein molecules, it is quite reasonable to speculate most of these particles are ACh receptors. However, there is little information available on how these clusters of ACh receptors first gather or how they are stabilized during the life of the organism. An intriguing aspect of this problem is that receptors remain clustered even though individual molecules within the clusters are constantly being exchanged during normal turnover. In this regard it is natural to assume that some cytoplasmic structure could anchor receptors and thus immobilize them against the fluidity of the membrane. Heuser and Salpeter (1979) have demonstrated that using the quick-freeze, deep-etch method one can visualize ACh receptors in *Torpedo* electric organ with unusual clarity. Taking advantage of this finding we have studied the organization of the receptors from outside and inside of the postsynaptic membrane in NMJ preparations of frogs and other animals. If one applies the quick-freeze, freeze–fracture, deep-etch method to intact muscles, it soon becomes apparent that only very limited areas of the true outside surface of postsynaptic membrane can be studied. This is because the basal lamina is closely applied to the postsynaptic membrane and extensively covers it (Fig. 5C). Hence, it is necessary to remove the basal lamina by collagenase treatment before muscles are frozen. Figure 8 displays a low magnification view of several postsynaptic folds at a collagenase-treated frog neuromuscular junction. On the right side of this field the fracture passed through the synaptic cleft, while at the lower left it broke into the cytoplasm of the muscle endplate. Deep-etching removed the remaining ice from the synaptic cleft and exposed the true surface of the postsynaptic folds at the right. The top surfaces of the postsynaptic fold are covered with numerous protrusions aligned parallel with each other. In contrast, only a few protrusions can be observed at the bottom of the infoldings (Fig. 8). Higher magnification further reveals

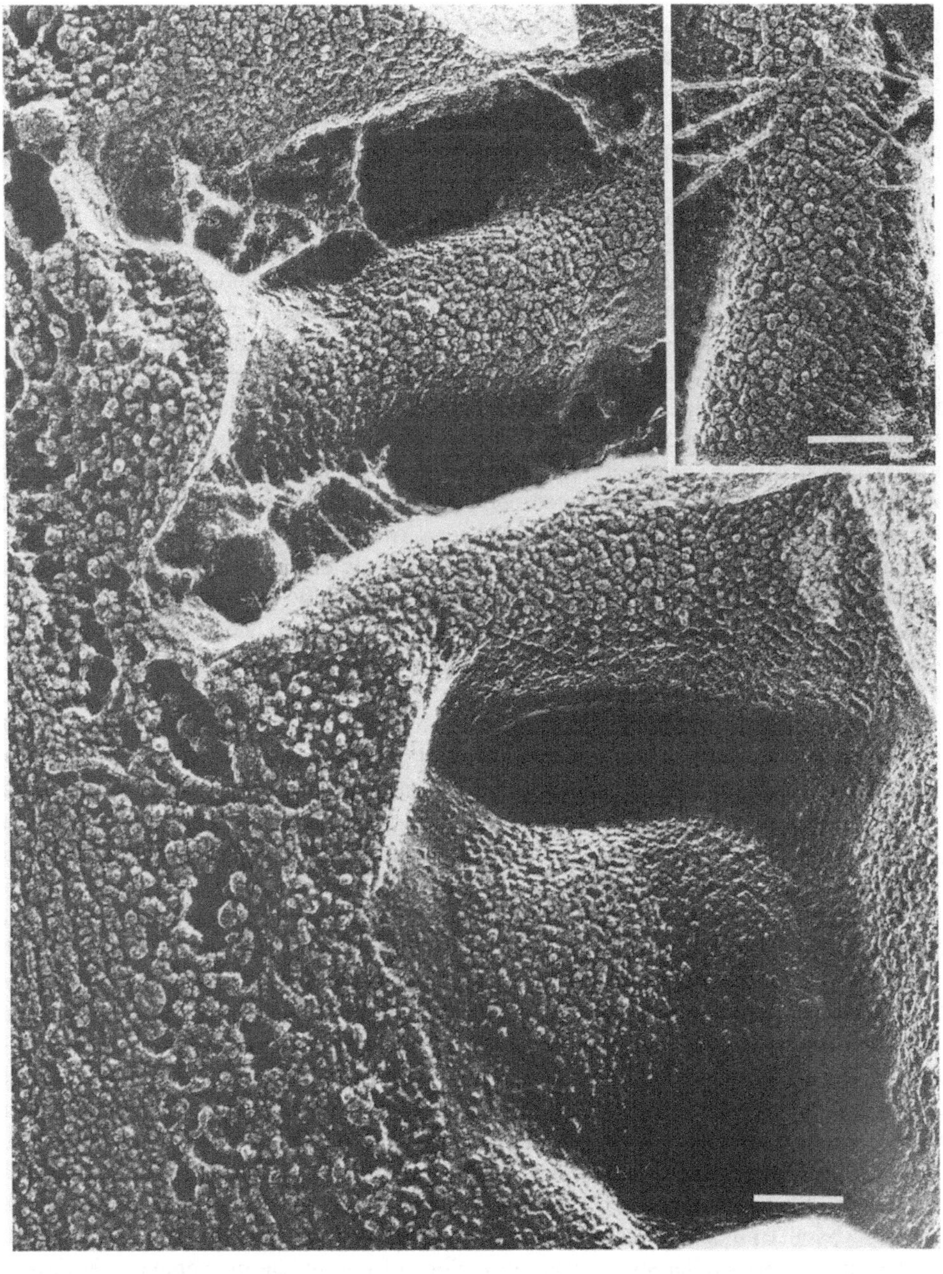

Figure 8. Quick-freeze, deep-etch, *en face* view of the true outside surface of the postsynaptic folds in a frog NMJ. The folds are aligned parallel with each other and the true outer surface of the postsynaptic membrane is clustered by numerous 8–9 nm protrusions. However, there are only a few bumps on the bottom of the infoldings. Inset shows higher magnification view of the true surface protrusions which tend to be aligned as parallel rows two abreast. The basal lamina is removed by collagenase before freezing. Bar 0.1 μm.

that the protrusions, 8–9 nm in diameter, are organized into irregular plaques separated by relatively particle-free furrows. Each plaque is composed of hundreds of surface protrusions which display in many areas a quasicrystalline order, consisting of parallel rows of protrusions two abreast (Figs. 8 and 9). To compare structural differences between junctional and extrajunctional muscle membranes, Fig. 10 provides views of the true surface and fractured P face of the extrajunctional (Fig. 10A) and junctional muscle membranes (Fig. 10B). The surface protrusions on the extrajunctional membrane are more diffusely distributed and they are much fewer in number than those observed on the junctional muscle membrane. The surface protrusions at the junctional area number $\sim$10,000/μm^2 (Hirokawa and Heuser, 1982a, b), which is about twice more than the number of P face particles per unit area. We assume most of these surface protrusions could be heads of the ACh receptors, because it is well known that ACh receptors are accumulated at the junctional muscle membrane, especially at the top of the postsynaptic folds and because biochemically (Fertuck and Salpeter, 1976) and physiologically (Katz and Miledi, 1972) estimated density of the receptors corresponds very well with this value. Furthermore, the morphology and arrangement of these surface protrusions resemble well-characterized ACh receptors on the postsynaptic membranes of *Torpedo* electric organ (Fig. 11). There are two characteristic aspects of the arrangement of ACh receptors. First, the receptors are only localized at the junctional area; more specifically they are accumulated only on the top of the postsynaptic folds. Second, the receptors tend to be arranged as parallel rows two abreast. As mentioned above, such regional differentiation of postsynaptic membrane raises the intriguing question of how a muscle cell maintains localized clusters of receptors in specific loci beneath the nerve, even while it is turning over these receptors. The receptors have little lateral mobility, according to fluorescence photobleach-recovery studies (Axelrod *et al.*, 1976). What prevents their diffusion away from the endplate?

Three mechanisms could be considered: (1) lateral association of the receptors with each other in the plane of the membrane would contribute to a parallel arrangement of receptors, (2) direct or indirect attachment of the receptors to extracellular structures would hold them in place, if these extracellular structures were present only in the endplate, and (3) attachment of the receptors to localized intracellular structures would also hold them in place.

In support of the first mechanism we found that treatment with β-mercaptoethanol disturbs the regular arrangement of the receptors (Fig. 12). Thus, the tendency for the receptors to associate with each other may be in part due to the disulfide bonds that they form with each other, which keep them together as dimers even when they are extracted from the membrane with detergents (Hamilton *et al.*, 1979). In part, this association could also be due to the involvement of other molecules. Cartand *et al.*, (1981) have recently shown that the characteristic crystals seen in *Torpedo* postsynaptic membrane disappear when a receptor-associated 43 K protein is extracted from the underside of the membrane.

In support of the second mechanism for immobilizing receptors, that of anchoring them to external structures, we observed that the postsynaptic membrane is connected to the overlying basal lamina in the synaptic cleft by many thin spines (Fig. 5C). Although the basal lamina forms similar spines with all regions of the muscle membrane, it is possible that the chemical specificity of these connections is different at different locations. In fact, it has been shown that the junctional basal lamina bears antigens distinct from those at the extrajunctional basal lamina (Sanes and Hall, 1979). Moreover, in regeneration experiments the synaptic basal lamina, even in the absence of the nerve, has been shown to possess sufficient chemical information to induce redifferentiation of the muscle and presumably recollection of the receptor molecules in its immediate locale (Burden *et al.*, 1979). However,

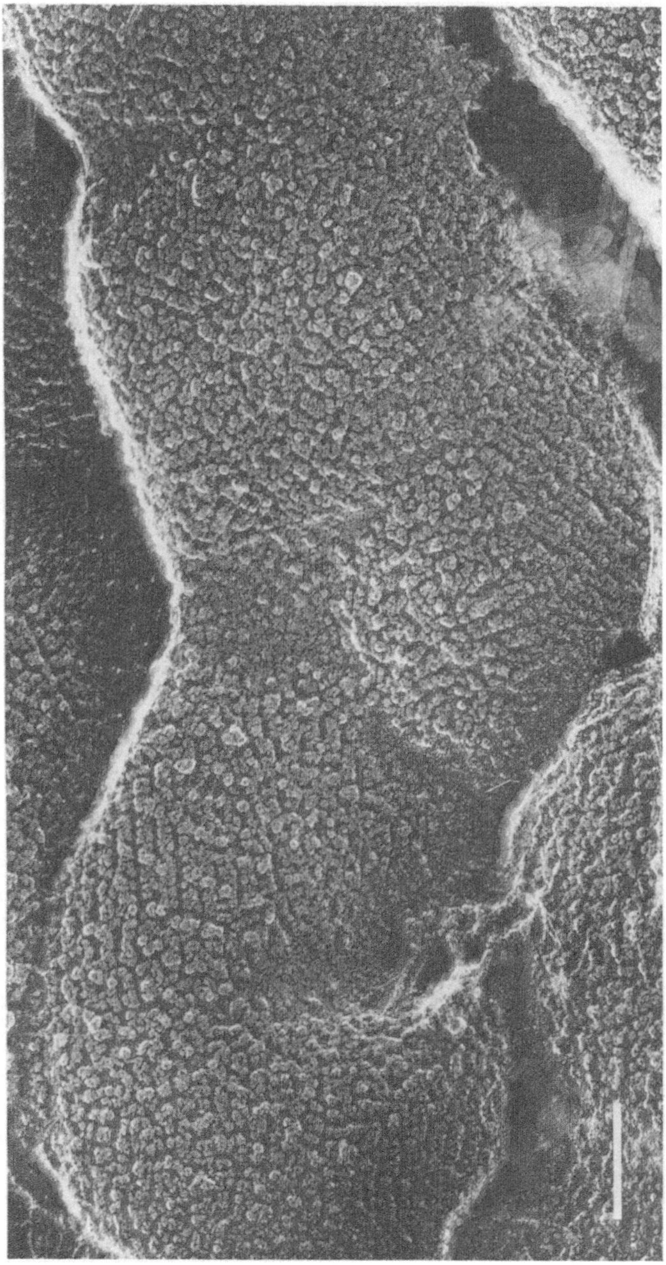

Figure 9. The true outside view of the convoluted postsynaptic fold of a frog NMJ which shows confluent plaques composed of masses of tiny surface protrusions in quasicrystalline patterns. Bar 0.1 μm.

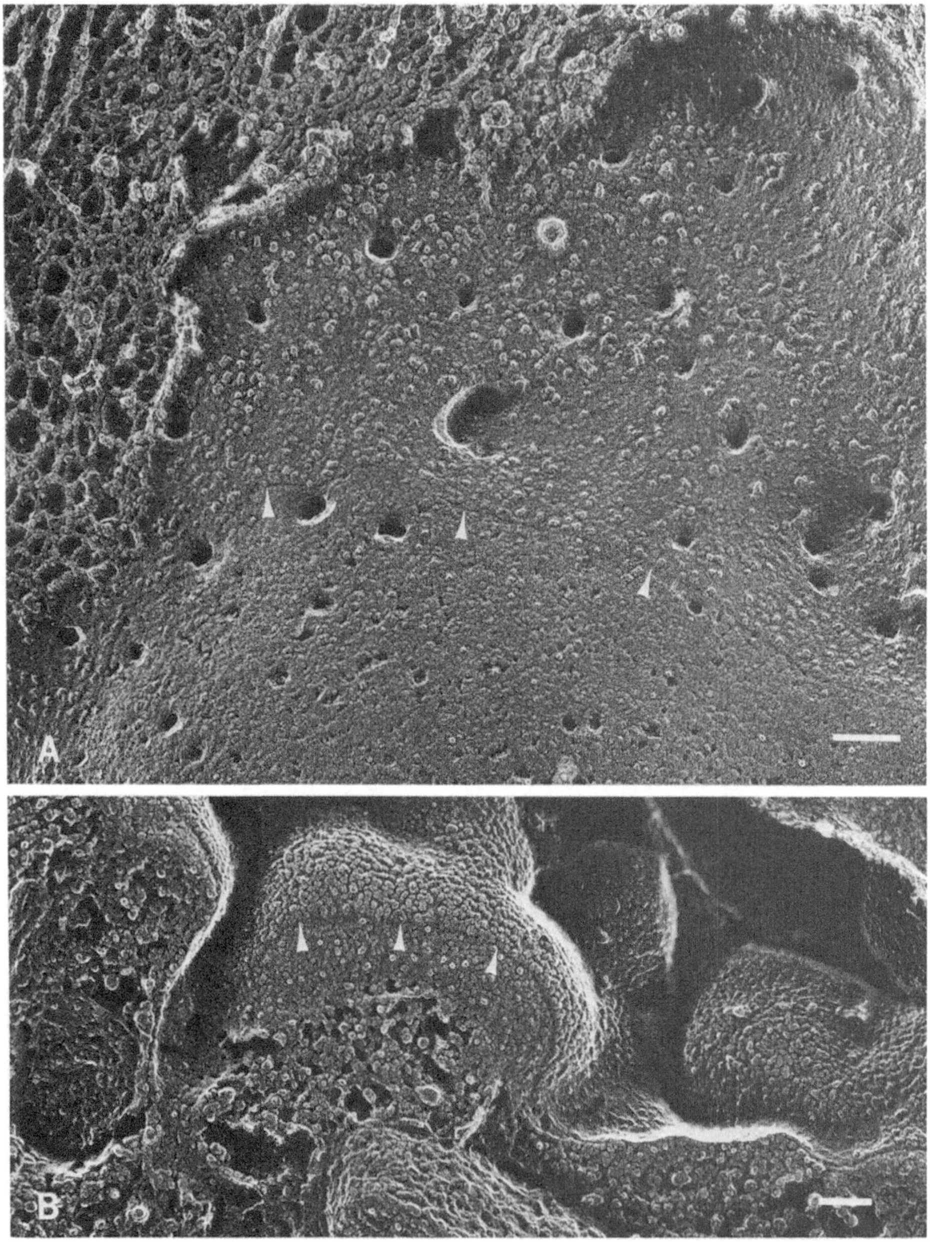

Figure 10. Comparison between the true outer surface of the muscle membranes at the extrajunctional area (A) and at the junctional area (B). The basal lamina is removed by incubation with collagenase. Arrowheads show boundary between fractured P face (lower) and true outer surface (upper) in Fig. 10A and B. Density of the surface protrusions at the extrajunctional area is much less than it is at the junctional area. Bar 0.1 μm.

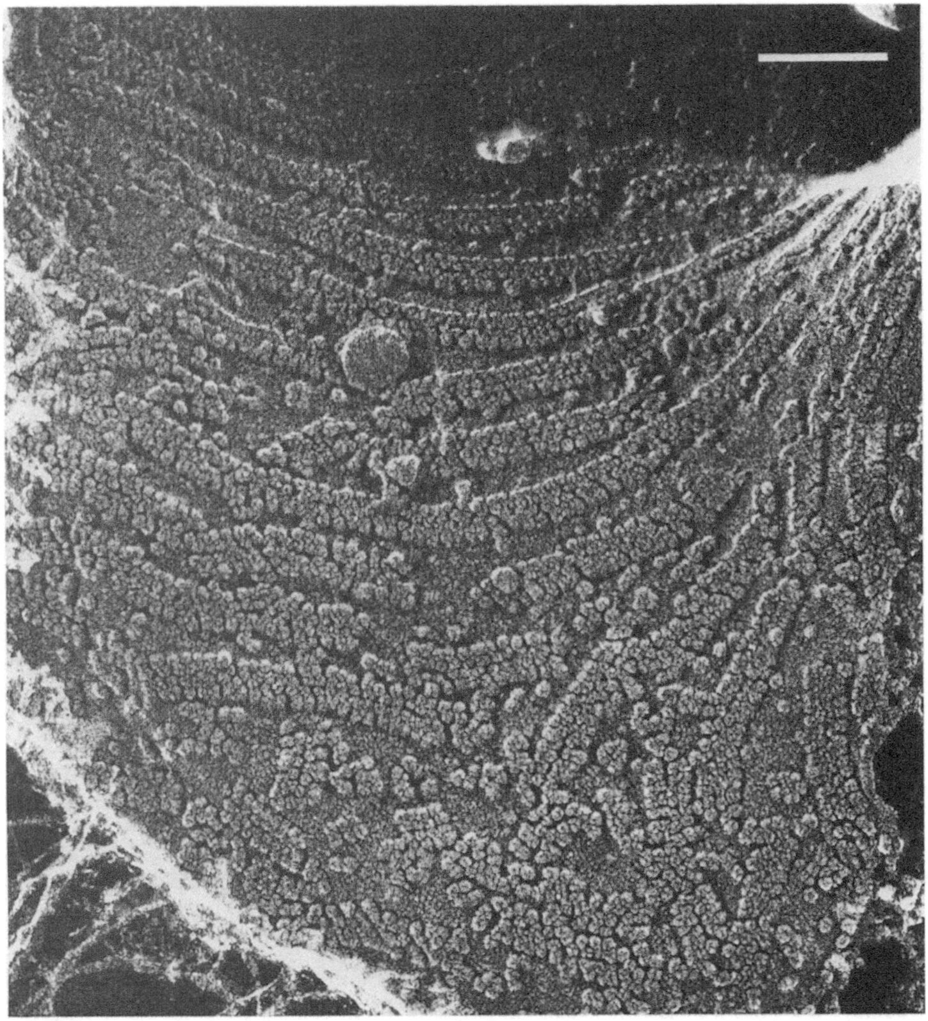

Figure 11. The true outer surface of the postsynaptic membrane of the *Torpedo* electric organ. Heads of the ACh receptors are aligned as parallel rows, two or four abreast. An individual ACh receptor looks like a doughnut. Bar 0.1 μm.

this mechanism alone cannot fully explain the immobility of receptors, because the receptors were still concentrated at the apical part of the postsynaptic folds even after 12 h of collagenase treatments (junctional basal lamina begin to detach after 2–3 hr). This mechanism could be responsible more in the developmental process than in the adult tissue. A third mechanism, possibly the most important for restricting receptor mobility, might be their attachment to structures inside the cell.

Although freeze-etching of unfixed or fixed whole muscle fibers provides views of extracellular basal lamina or external surface of postsynaptic membranes, it does not give us much useful information about the intracellular cytoskeletal structures at the NMJs. This is because most of the cytoplasm (except some specialized regions) is filled with soluble proteins that appear as granular substances and interfere with clear visualization of cytoskeletal elements in deep-etched replicas. To overcome this problem plasma membranes

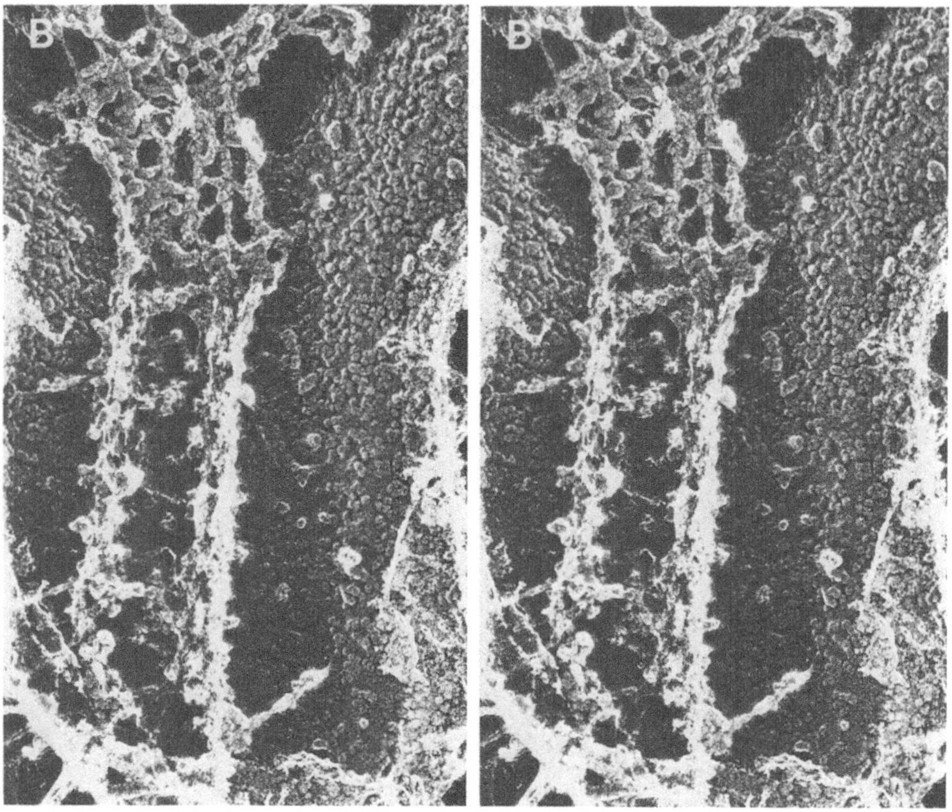

Figure 12. The regular arrangement of ACh receptors in the postsynaptic membrane of *Torpedo* electric organ (A) is perturbed by treatment with β-mercaptoethanol (B). Bar 0.1 μm.

should be permeabilized chemically or physically to allow egress of the soluble proteins. To achieve this goal we used 0.1% saponin or gentle homogenization on the NMJ preparation before quick-freezing. Figure 13 demonstrates the internal structure of several postsynaptic protrusions in a muscle treated briefly with saponin before freezing. In the center of each sarcoplasmic protrusions are found 12 nm filaments that form loose bundles which run parallel to the adjacent infoldings. Immediately underneath the postsynaptic membrane is a characteristic meshwork of short thin strands. This submembranous web is not found underneath any other part of the muscle membrane; thus, it presumably corresponds to the postsynaptic density seen in thin sections. Figure 14 provides *en face* view of the submembranous web which is viewed from the center of the folding. Similar networks of delicate strands are also observed in rat NMJ (Fig. 15). This sample was obtained by isolating single myofibrils and collecting them after brief fixation with a weak fixative (1% paraformaldehyde). The process for isolation of single fibers broke the muscle membrane and induced leakage of soluble proteins. To aid the reader, a corresponding figure obtained by conventional thin sectioning is also shown. In thin sections, the finger-like postsynaptic sarcoplasm appears to contain a vaguely dense material, particularly under the regions of postsynaptic membrane that approach the nerve most closely, which is exactly where the postsynaptic receptors are known to cluster (Fertuck and Salpeter,1976). In deep-etched replicas, this submembranous material can be resolved into a distinct meshwork of branched and anastomotic stands which attach to the inside of the postsynaptic membrane. Other cytoskeletal elements in these sarcoplasmic protrusions including actin, intermediate filaments, and microtubules (stabilized by taxol) are also attached at many placed along their length to the submembranous meshwork and hence indirectly to the postsynaptic membrane. This meshwork of thin strands was also observed in high-voltage electron micrographs of thick-sectioned NMJs (Ellisman *et al.*, 1976).

We have noted that the meshwork persists in denervated muscle, as do the clusters of receptor molecules. Burrage and Lentz (1979) have reported that during synaptic develop-

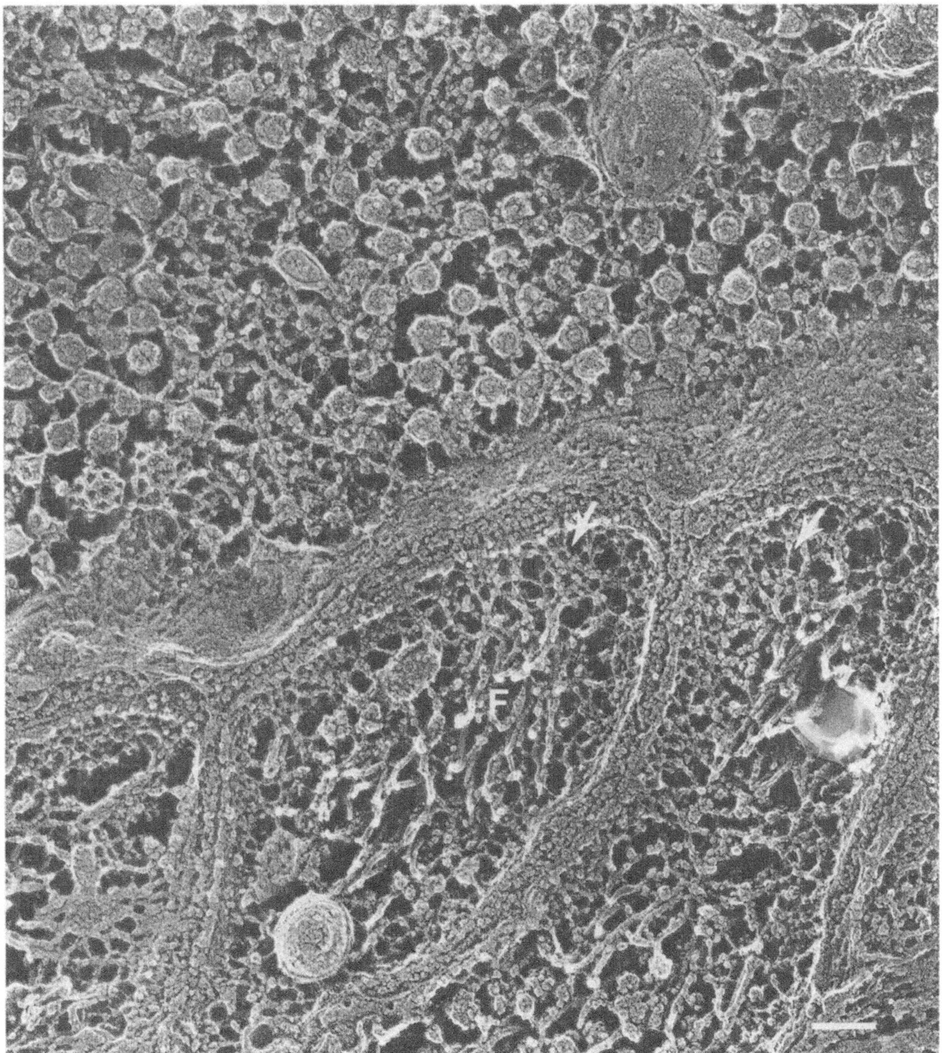

Figure 13. Inside of the presynaptic terminal and the postsynaptic folds of a frog NMJ quick-frozen, deep-etched after brief treatment with 0.1% saponin which allows egress of soluble protein in the cytoplasm. Just underneath the postsynaptic membrane observed are networks of fine strands (arrows) which connect postsynaptic membrane with bundles of 12 nm filaments (F) running parrallel to the infoldings as a core of postsynaptic folds. Bar 0.1 μm.

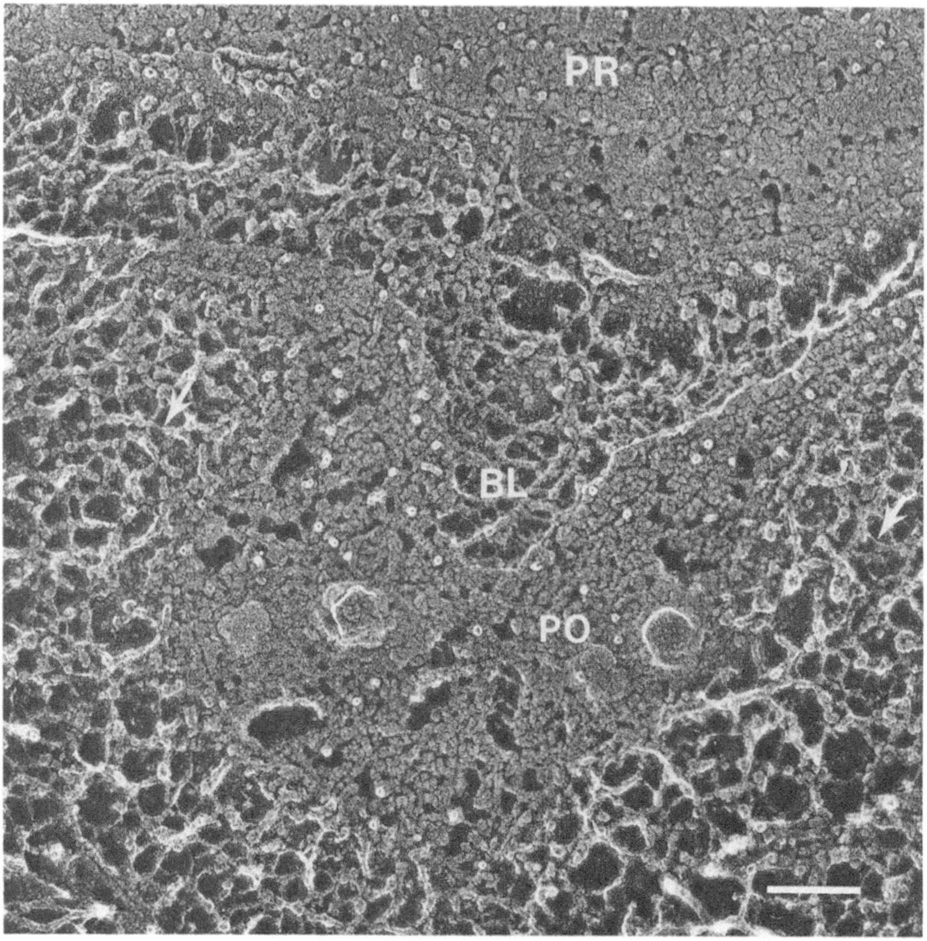

Figure 14. Inside of the postsynaptic membrane of a frog NMJ. This micrograph shows, from the top to the bottom, P face of a presynaptic membrane (PR), basal lamina (BL), E face of the postsynaptic membrane (PO), and networks of fine strands inside the postsynaptic membrane (arrows). Bar 0.1 μm.

ment, postsynaptic densities appear in thin sections as soon as receptor clusters begin to be stabilized. Detergent extraction of developing muscle fibers does not remove such receptor clusters (Prives *et al.*, 1979). Thus, from these structural and experimental data it is suggested that the subsynaptic meshwork could play an important role in immobilizing receptors. However, because the meshwork does not appear to contact every receptor molecule directly, most receptors would have to be tethered indirectly, possibly by an undercoating of other proteins such as 43 K protein or possibly by a tendency to associate with each other (mechanisms 1).

Thus, current structural evidence would support the following view. A number of relatively discrete receptor aggregates are each attached in a few places to the submembranous meshwork. They are further held in place via attachments to cytoskeletal elements, including actin, intermediate filaments, and microtubules (Hirokawa and Heuser, 1980, 1982a, b; Hall *et al.*, 1981; Gulley and Reese, 1981). An immediate challenge in this area would be to characterize the biochemical nature of the submembranous meshwork.

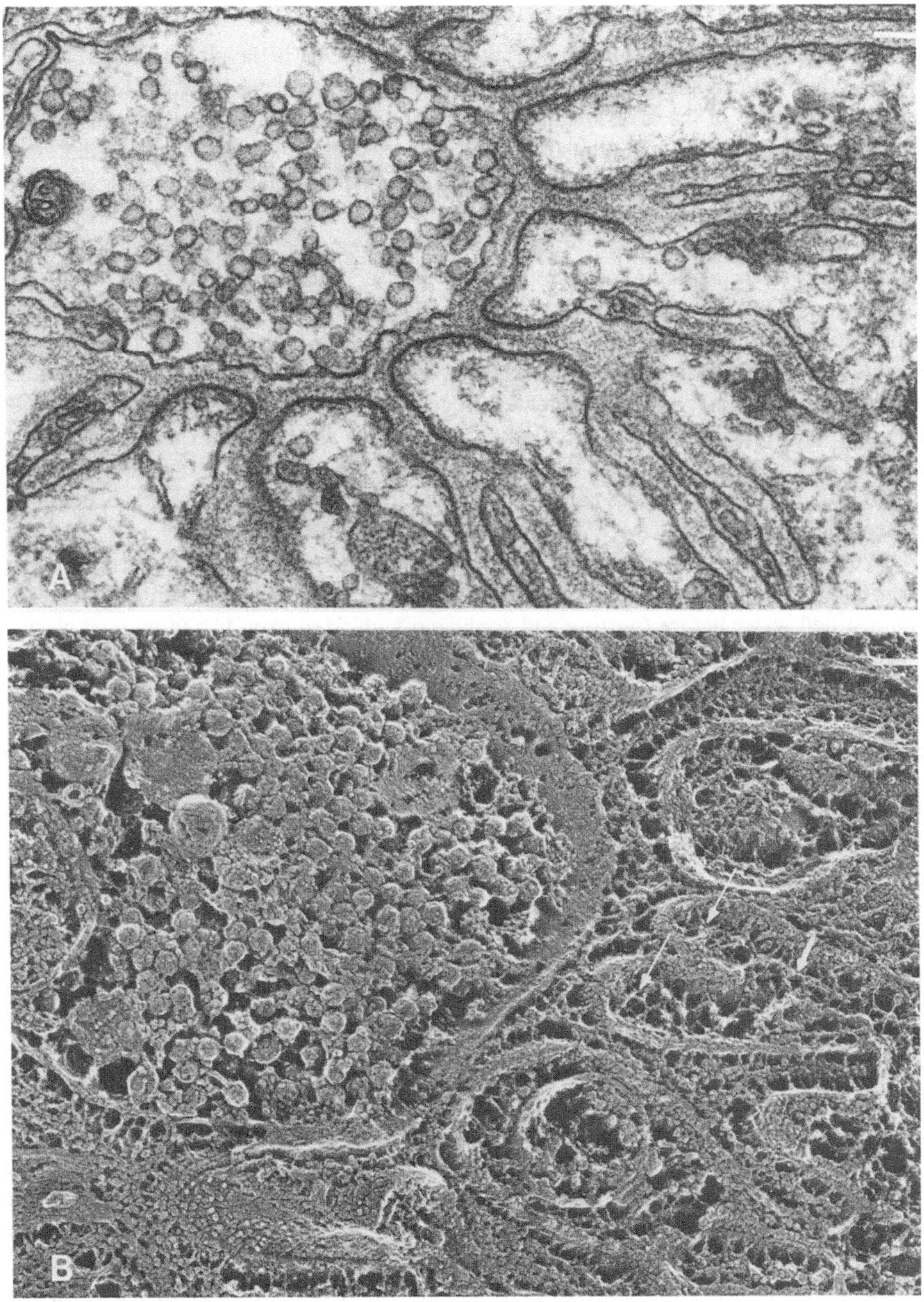

Figure 15. Conventional thin-section micrograph (A) and quick-freeze, deep-etch view (B) of rat NMJs. The long arrows point to networks of fine strands and the short arrow indicates cytoskeletal filaments in the sarcoplasmic protrusion. Bar 0.1 μm.

III. CYTOSKELETAL STRUCTURE OF AXONS AND ITS RELATIONSHIP WITH THE AXOLEMMA

Axons are differentiated mainly for two functions. One is the propagation of impulses and the other is the transport of substances. Recently, there has been considerable interest in the cytoskeleton of the axon because it has been thought to participate in both functions. A historical account of axonal ultrastructure is given elsewhere in this volume. In this chapter we will limit ourselves to a discussion of new information which has been obtained using the quick-freeze, deep-etch procedure.

Quick-freeze, deep-etching of unfixed axons revealed an elaborate system of cross-connections among axolemma, microtubules (MT), neurofilaments (NF) and membranous organelles (MO). As shown in Fig. 16 axons contain many longitudinally oriented NF connected to each other by numerous fine crosslinkers (4–6 nm in diameter, 20–50 nm in length). Two specialized regions are apparent within the axons. One is composed of fascicles of MT linked with each other by fine crossbridges, and the other in the subaxolemmal space, consisting of a network of long crosslinks (50–150 nm) which connect axolemma with NF and MT. MO are found mainly in these two specialized regions and are intimately associated with MT via fine short (10–20 nm in length) crossbridges. In unfixed axons granular materials are seen on the surfaces of MO, MT, NF, and cross-connections.

We tried several different procedures to test whether these cross-connections are real or artifactual. To check whether salt could contribute to them, fixed axons were washed with distilled water before freezing and compared with fixed axon frozen without washing. Because axons look identical under both sets of conditions, we conclude the cross-connections are not salt-induced artifacts. The only difference between fixed and unfixed samples is that the fine surface substructrue of MT is obscured in fixed samples. To examine whether soluble components in the axoplasm precipitated and condensed to form the cross-connections after deep-etching, the axolemma was permeabilized to allow egress of soluble components physically (by gentle homogenization) or chemically (by incubation with 0.1% saponin). Figure 17 compares neurofilaments in a fresh axon with those in a saponin-treated axon. The cross-connections are found clearly after removal of soluble components, but the granular materials attached on the MT, NF, MO, and cross-connections are no longer present so that cross-connections look smooth in contour. Thus, the cross-connections are real structures and the granular materials attached on the cytoskeletal elements in intact samples could be soluble substances, probably protein in nature. In this permeabilizing process, it is necessary to add taxol in the incubation medium to preserve MT. This is probably because tubulin pools in the axoplasm become drastically diluted by the surrounding medium when the axon is permeabilized, MT would thus depolymerize if a stabilizing agent is not present.

The cross-connections in the axoplasm are categorized into three groups: NF associated crosslinkers, MT-associated crossbridges, and long crosslinks in the subaxolemmal space. NF-associated crosslinkers (4–6 nm in diameter, 20–50 nm in length) connect NF with each other, thus giving them the appearance of ladders. The distance between adjacent crosslinkers vary from 25–100 nm but is mostly from 30–50 nm. Similar crosslinkers are found between NF and MO and between NF and MT (Figs. 18 and 19). The chemical nature of these crosslinkers is unknown. NF is composed of a triplet of proteins (MW 200 K, 145 K, 73 K) (Hoffman and Lasek, 1975). Recently, direct decoration studies of isolated NF or NF in cultured neurons with antibodies against the triplet proteins indicated that 73 K protein seems to form the backbone of the filaments, but the 200 K protein is periodically arranged in a more peripheral position (Willard and Simon, 1981; Sharp *et al.*, 1981). These studies raise the possibility that the 200 K protein could form crosslinker-like structures. Although

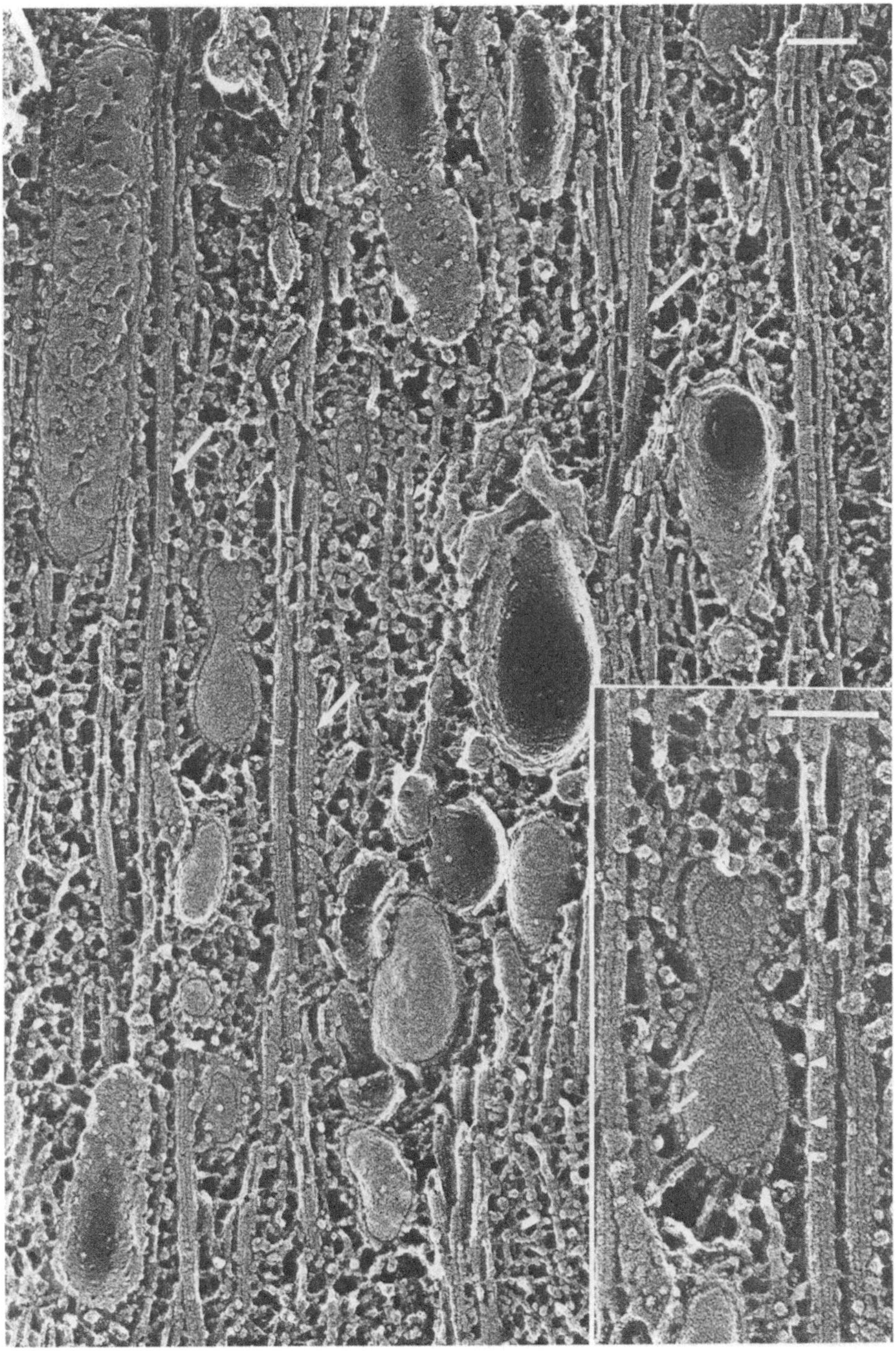

Figure 16. Overall view of unfixed axon directly frozen from life. The axoplasm is filled with longitudinally oriented neurofilaments (NF) (thin arrows) and microtubules (MT) (thick arrows). Membranous organelles (MO) are localized in the columns delineated by MT. Elaborate cross-connections are observed among MT, NF, and MO. The crossbridges between MT and MO are indicated by arrowheads and crosslinkers between NF and MO are pointed by short arrows in the inset. Bar 0.1 μm.

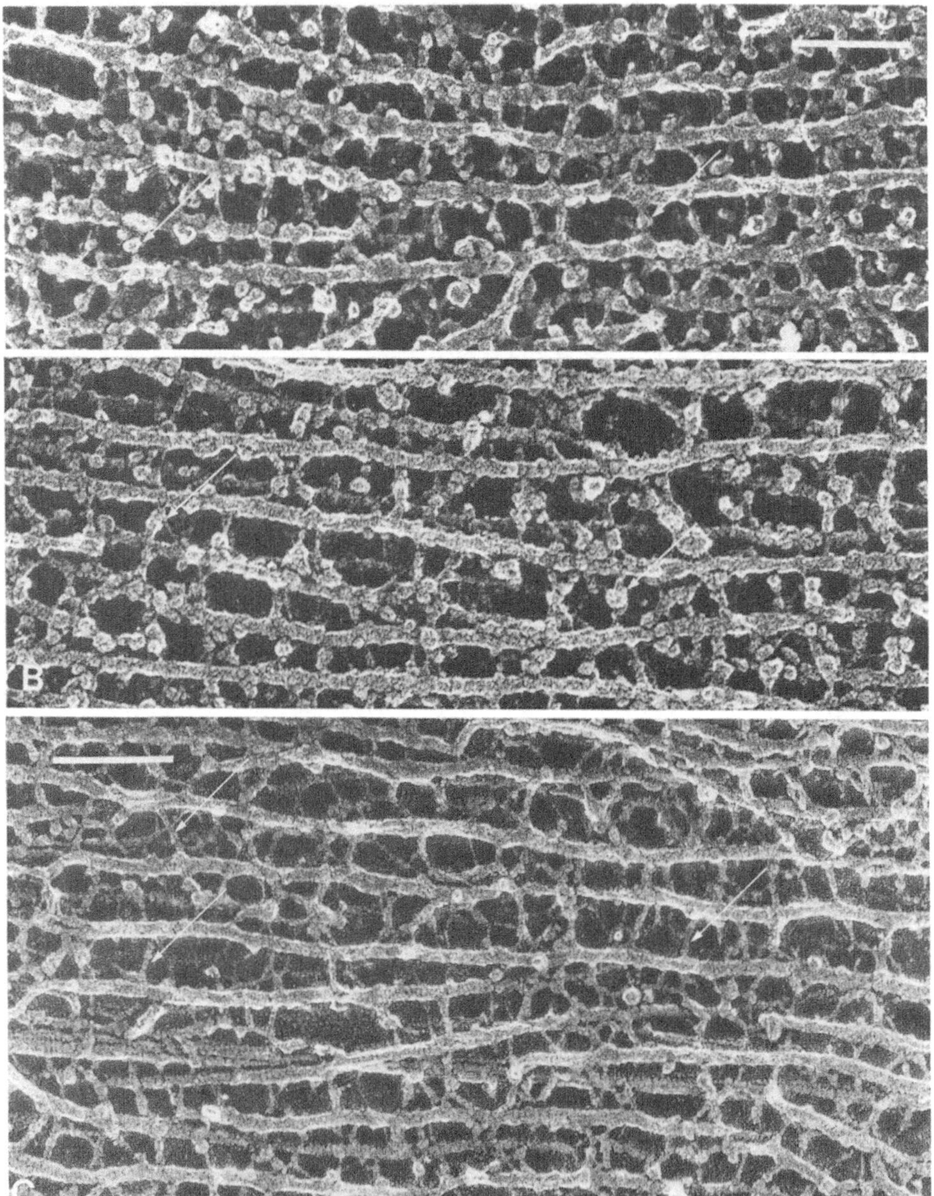

Figure 17. Comparison of quick-frozen, deeply-etched neurofilaments after various treatments. (A) Fresh frog axon, (B) frog axon fixed with glutaraldehyde and paraformaldehyde and washed with distilled water, (C) frog axon incubated with 0.1% saponin for 20 min before freezing. NF are aligned parallel with each other. Numerous crosslinkers (4–6 nm in width, 20–50 nm in length) (arrows) connect adjacent NF. The fresh axon and the fixed axon washed with distilled water look the same. The crosslinkers certainly exist after saponin treatment, but granular materials attaching on NF and crosslinkers are gone after this treatment. Thus, the crosslinker is a real structure, not an artifact produced by condensation or precipitation of soluble proteins or salts after fixation or deep-etching; however, the granular materials could be soluble proteins. Bar 0.1 μm.

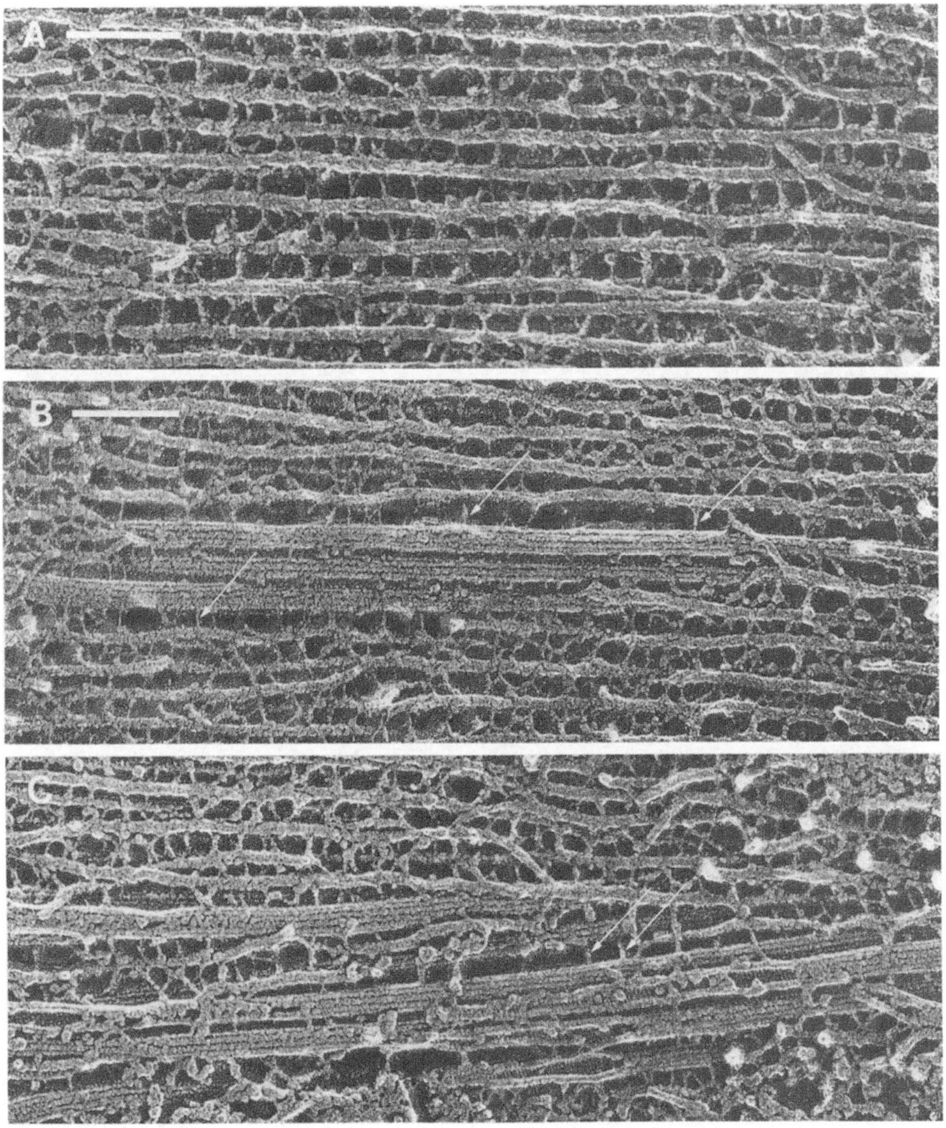

Figure 18. Cytoskeletal structure of frog axons treated with 0.1% saponin. (A) Numerous crosslinkers connect adjacent NF, mostly perpendicularly, like ladders. Granular material is almost removed. (B,C) MT were stabilized by taxol in the incubation medium. MT tend to run as a group. The outer surface of the MT is characterized by longitudinal substructures showing an arrangement of protofilaments and at the inner surface regular 4 nm oblique stripes representing a staggered arrangement of protofilaments are found (C). Sometimes crossbridges (arrows in C) connect adjacent microtubules and very often crosslinkers (arrows in B) connect the MT almost perpendicularly to the NF (B). Bar 0.1 μm.

the crosslinkers between NF were not preserved in these studies, the speculation offered by the authors fits well with the ladder-like structural characteristics of the crosslinkers. Immunocytochemistry combined with the quick-freeze, deep-etch method should give us the answer. MT associated crossbridges are usually short (<20 nm) and less numerous than crosslinkers between NF (Fig. 18). They crosslink MT with each other and MT with MO,

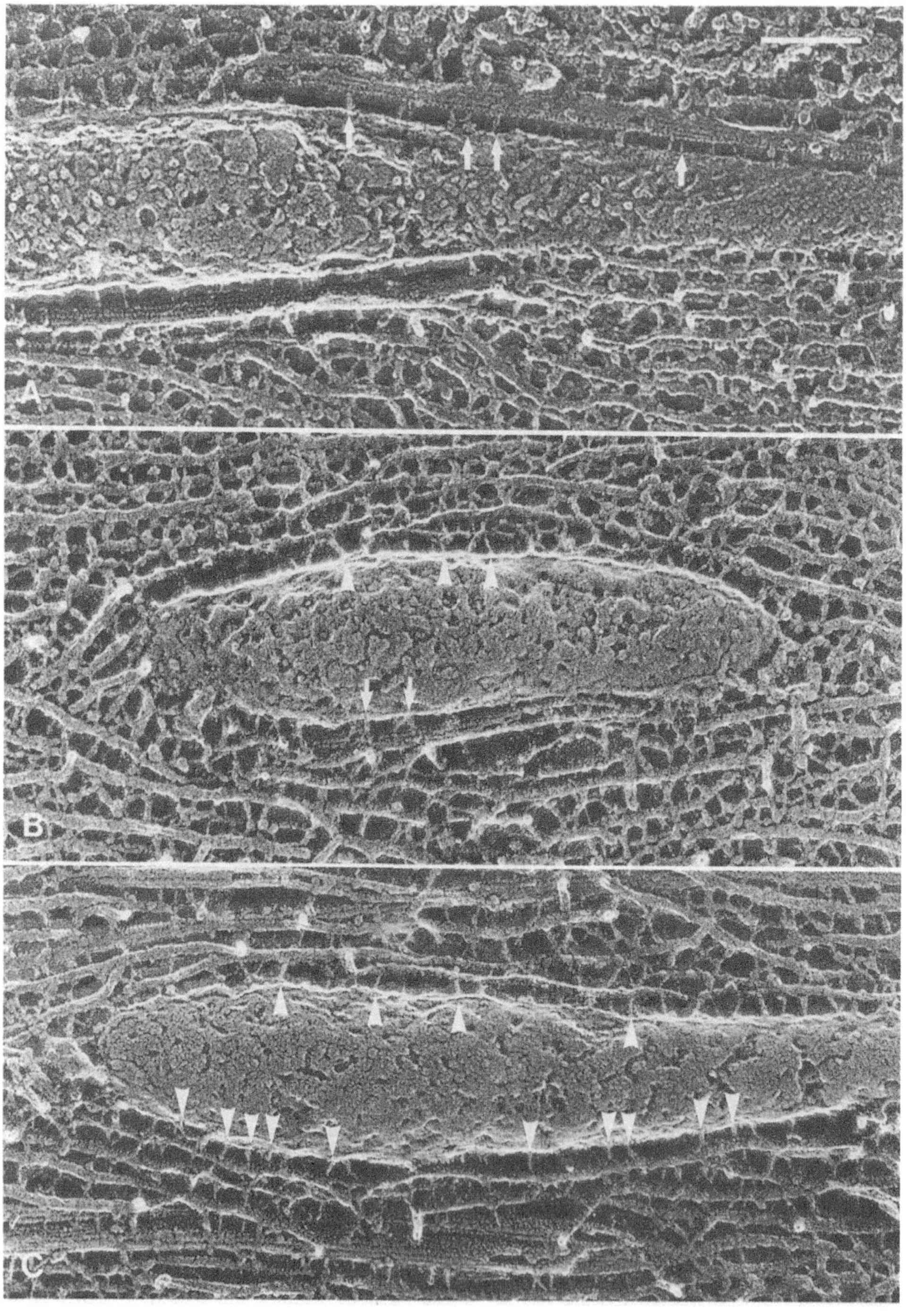

Figure 19. These three micrographs show crossbridges (arrows) between MT and mitochondria and crosslinkers (arrowheads) between NF and mitochondria in saponin-treated frog axons. Bar 0.1 μm.

although occasionally, longer crossbridges are seen between MT and MO (Fig. 20). Figure 19 displays crossbridges between a mitochondrion and MT. The crossbridges associated with MT could be composed of MT-associated proteins (MAPs). In fact, it has been suggested that MAPs could form crossbridge-like structures (Heuser and Kirschner, 1980).

These cross-connections are extensively developed throughout the axoplasm. They could constitute a labile and elastic framework. In addition, the fact that NF are connected by crosslinkers with each other and with MT into a unit may have interesting implications regarding the slow transport of the NF and MT. It is also possible that the crossbridges between MO and MT may have some ATPase activity to generate a force for MO sliding on the MT. However, biochemical analyses are necessary to elucidate the function of the crossbridges observed here.

Long crosslinks in the subaxolemmal space could be more related to the excitability of the axolemma. Metuzals and Tasaki (1978) observed a subaxolemmal network of filaments including actin by scanning electron microscopy and thin-sectioning and suggested a possible relationship between axolemmal excitability and cytoskeletal elements. Matsumoto and Sakai (1979) also suggested that associated with the internal surface of the plasma membrane there are microtubules which regulate in part both resting and action potentials.

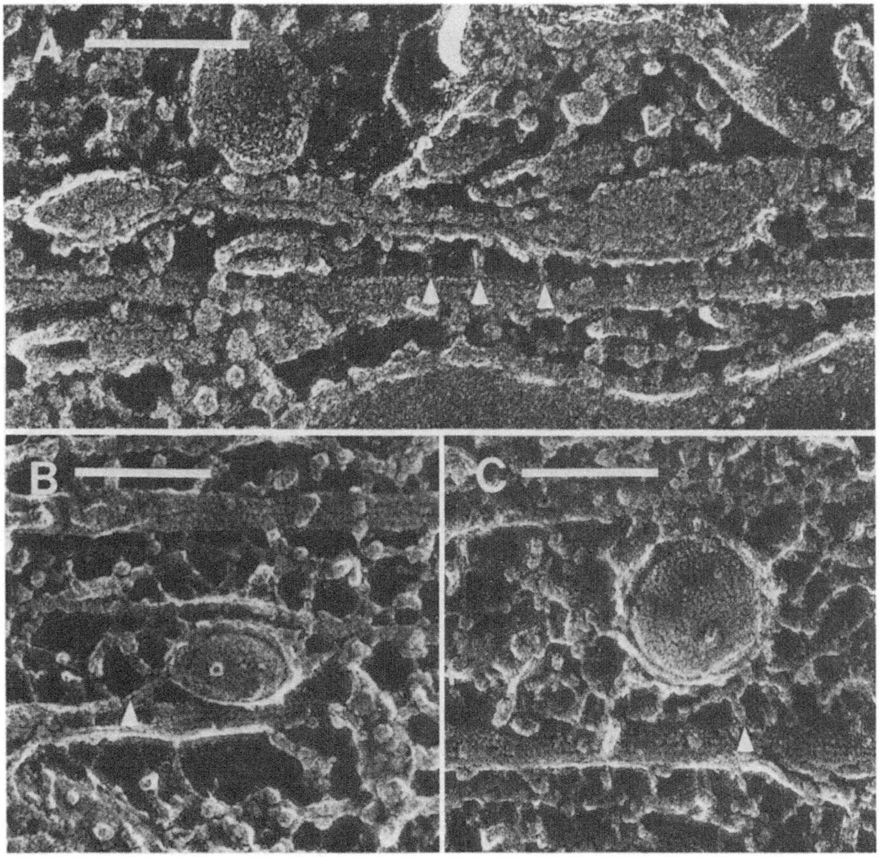

Figure 20. These micrographs display crossbridges (arrowheads) between MT and membrane organelles possibly conveyed by rapid transport in fresh frog axons. Bar 0.1 μm.

Quick-freeze, deep-etching reveals the subaxolemmal space (~100 nm from the axolemma) as a differentiated region composed mainly of a dense network of long (50–150 nm) crosslinks connecting MT and NF with the axolemma and subaxolemmal filamentous networks (Fig. 21). Occasionally, MT are enmeshed in the long crosslink network. The cytoplasmic surface of the axolemma is decorated with numerous 8–9 nm bumps, but the granularity of the axoplasm due to the presence of soluble proteins makes it difficult to analyze clearly the structure just inside the axolemma in the fresh axons quick-frozen from the living state (Fig. 21A). Physical rupture of a part of the axolemma by homogenization clearly demonstrates the existence of a network of filaments 8 nm or thinner just inside the axolemma (Fig. 21B and C). Some filaments are in the appropriate size range (8–9 nm) to be actin. One candidate for the long crosslinks is MAPs. The other candidate could be fodrin (a 240 K, 235 K dimer), a high-molecular-weight actin-binding protein which is very similar in nature to spectrin (a 240 K, 220 K dimer) in erythrocytes and the TW 260–240 protein

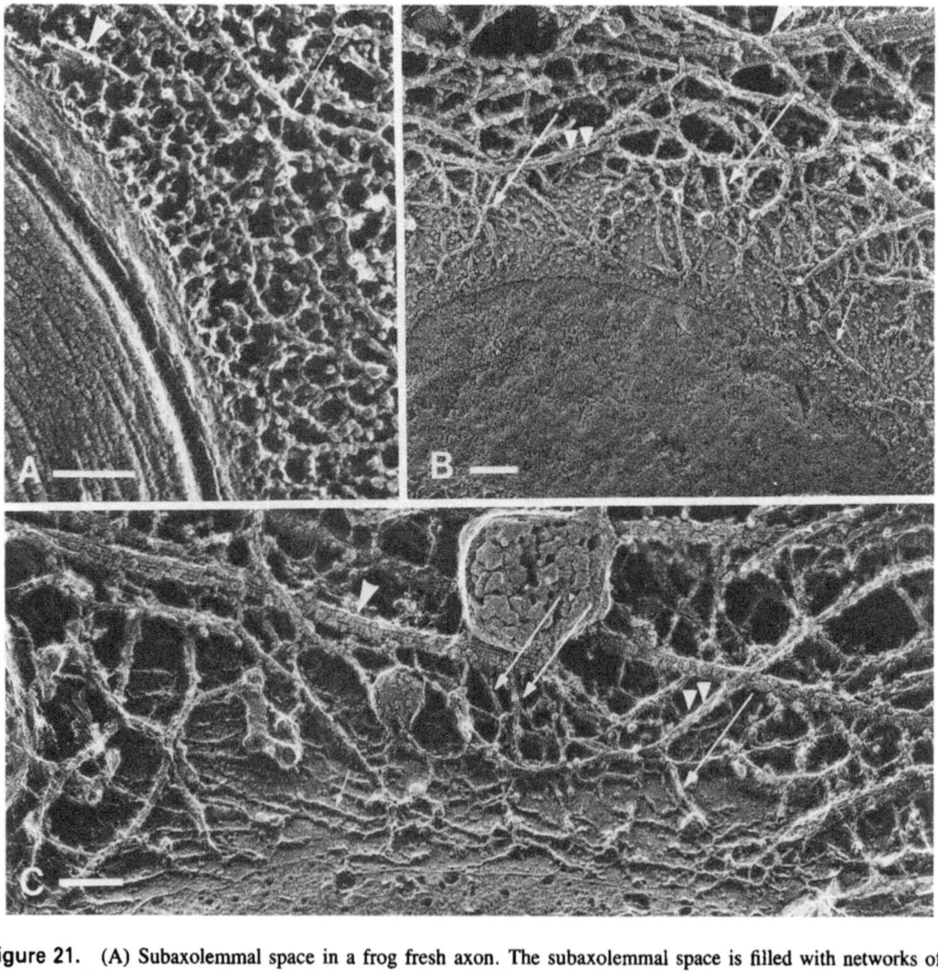

Figure 21. (A) Subaxolemmal space in a frog fresh axon. The subaxolemmal space is filled with networks of longer crosslinks among axolemma, MT (arrowheads) and NF (arrows). (B,C) Gently homogenized *Torpedo* axon. The inside of the axolemma is studded with numerous 8–9 nm bumps. Networks of 8 nm actin-like filaments and thinner filaments (short arrows) exist and long crosslinks (long arrows) connect NF (double arrowheads) and MT (single arrowheads) with the axolemma and subaxolemmal filamentous networks. Bar 0.1 μm.

(a 260 K, 240 K dimer) in the brush border of intestinal epithelial cells (Glenney *et al.*, 1982). Levine and Willard (1981) have demonstrated the presence of fodrin at the axolemma using immunofluorescence microscopy. It crosslinks actin filaments with each other and may be able to crosslink actin with plasma membrane, other membrane organelles, MT or NF. Furthermore, on the analogy from spectrin, it could be anchored to some intrinsic membrane proteins through a specialized molecule such as "ankyrin" (Branton *et al.*, 1981). Thus, it is possible that subaxolemmal filamentous networks containing actin and fodrin could play a role in the anchoring of ion channel proteins and the generation of excitability of the axolemma.

Future studies by the quick-freeze, deep-etching method combined with immunocytochemistry are in progress to elucidate the chemical composition of these filamentous networks.

ACKNOWLEDGMENTS. The author wishes to thank J. E. Heuser for his kind support throughout this work and W. Ip for his great help in preparing the manuscript. This work was supported by grants to the author and J. E. Heuser from the U.S. Public Health Service.

REFERENCES

Anderson, M. J., and Cohen, M. W., 1974, Fluorescent staining of acetylcholine receptors in vertebrate skeletal muscle, *J. Physiol. (London)* **237**:385–400.

Axelrod, D., Ravdin, P., Koppel, D. E., Schlessinger, J., Webb, W. W., Elson, E. L., and Podleski, T. R., 1976, Lateral motion of fluorescently labelled acetylcholine receptors in membranes of developing muscle fibres, *Proc. Natl. Acad. Sci. USA* **73**:4594–4598.

Birks, R., Huxley, H. E., and Katz, B., 1960, The fine structure of the neuromuscular junction of the frog, *J. Physiol. (London)* **150**:134–144.

Bloom, F. E., and Aghajanian, G. K., 1966, Cytochemistry of synapses: Selective staining for electron microscopy, *Science* **154**:1575–1577.

Bloom, F. E., and Aghajanian, G. K., 1968, Fine structural and cytochemical analysis of the staining of synaptic junctions with phosphotungstic acid, *J. Ultrastruct. Res.* **22**:361–367.

Branton, D., Cohen, C. M., and Tyler, J., 1981, Interaction of cytoskeletal proteins on the human erythrocyte membrane, *Cell* **24**:24–32.

Burden, S. J., Sargent, P. B., and McMahan, U. J., 1979, Acetylcholine receptors in regenerating muscle accumulate at original synaptic sites in the absence of the nerve, *J. Cell Biol.* **82**:412–425.

Burrage, T. G., and Lentz, T. L., 1979, Surface specialization associated with high density accumulations of acetylcholine receptors in embryonic chick muscle, *Abstracts of the Society for Neuroscience, USA* **5**:877.

Cartand, J., Sobel, A., Rousselet, A., Devaux, P. F., and Changeux, J.-P., 1981, Consequences of alkaline treatment for the ultrastructure of the acetylcholine receptor-rich membranes from Torpedo marmorata electric organ, *J. Cell Biol.* **90**:418–426.

Couteaux, R., and Pecot-Dechavassine, M., 1968, Particularites structurales du sarcoplasme sous-neural, *Compt. Rend.* **266**:8–10.

Couteaux, R., and Pecot-Dechavassine, M., 1970, Vasicles synaptiques et poches au niveau des zones actives de la jonction neuromusculaire, *Compt. Rend.* **271**:2346–2349.

DeRobertis, E. D. P., and Bennett, H. S., 1955, Some features of the submiscroscopic morphology of synapses in frog and earthworm, *J. Biophys. Biochem. Cytol.* **1**:47–58.

Dreyer, F., Peper, K., Akert, K., Sandri, C., and Moor, H., 1973, Ultrastructure of the active zone in the frog neuromuscular junction, *Brain Res.* **62**:373–380.

Ellisman, M. H., Rash, J. E., Staehelin, L. A., and Porter, K. R., 1976, Studies of excitable membranes. II. A comparison of specializations at neuromuscular junctions and nonjunctional sarcolemmas of mammalian fast and slow twitch muscle fibres, *J. Cell Biol.* **68**:752–774.

Fertuck, H. C., and Salpeter, M. M., 1976, Quantitation of junctional and extrajunctional acetylcholine receptors by electron microscopic autoradiography after ^{125}I-α-bungarotoxin binding at mouse neuromuscular junctions, *J. Cell Biol.* **69**:144–158.

Glenney, J. R., Glenney, P. Osborn, M., and Weber, K., 1982, An F actin and calmodulin-binding protein from isolated intestinal brush borders has a morphology related to spectrin, *Cell* **28:**843–854.

Gray, E. G., 1963, Electron microscopy of presynaptic organelles of the spinal cord, *J. Anat.* **97:**101–106.

Gulley, R. L., and Reese, T. S., 1981, Cytoskeletal organization at the postsynaptic complex, *J. Cell Biol.* **91:**298–302.

Hall, Z. W., Lubit, B. W., and Schwartz, J. H., 1981, Cytoplasmic actin in postsynaptic structures at the neuromuscular junction, *J. Cell Biol.* **90:**789–792.

Hamilton, S. L., McLaughlin, M., and Karlin, A., 1979, Formation of disulfide-linked dimers of acetylcholine receptor in membrane from Torpedo electric tissue, *Biochemistry* **18:**155–163.

Heuser, J. E., 1976, Morphology of synaptic vesicle discharge and reformation at the frog neuromuscular junction, in: *Motor Innervation of Muscles* (S. Thesleff, ed.), pp. 51–115, Academic Press, London.

Heuser, J. E., and Kirschner, M., 1980, Filament organization revealed in platinum replicas of freeze-dried cytoskeletons, *J. Cell Biol.* **86:**212–234.

Heuser, J. E., and Salpeter, S. R., 1979, Organization of acetylcholine receptors in quick-frozen deep-etched, and rotary-replicated Torpedo postsynaptic membrane, *J. Cell Biol.* **82:**150–173.

Heuser, J. E., Reese, T. S., and Landis, D. M. D., 1974, Functional changes in frog neuromuscular junctions studied with freeze-fracture, *J. Neurocytol.* **3:**109–131.

Heuser, J. E., Reese, T. S., Dennis, M. J., Jan, Y., Jan, L., and Evans, L., 1979, Synaptic vesicle exocytosis captured by quick freezing and correlated with quantal transmitter release, *J. Cell Biol.* **81:**275–300.

Hirokawa, N., 1982, Crosslinker system between neurofilaments, microtubules, and membranous organelles in frog axons revealed by the quick-freeze, deep etching method, *J. Cell Biol.* **94:**129–142.

Hirokawa, N., and Heuser, J. E., 1980, Possible anchoring structures for the organized arrays of receptors seen at neuromuscular junctions, *J. Cell Biol.* **87:**75a.

Hirokawa, N., and Heuser, J. E., 1981a, Structural evidence that botulinum toxin blocks neuromuscular transmission by impairing the calcium influx that normally accompanies nerve depolarization, *J. Cell Biol.* **88:**160–171.

Hirokawa, N., and Heuser, J. E., 1981b, Quick-freeze, deep-etch visualization of the cytoskeleton beneath surface differentiations of intestinal epithelial cells, *J. Cell Biol.* **91:**399–409.

Hirokawa, N., and Heuser, J. E., 1982a, Internal and external differentiations of the postsynaptic membrane at the neuromuscular junction, *J. Neurocytol.* **11:**487–510.

Hirokawa, N., and Heuser, J. E., 1982b, The inside and outside of gap junction membranes visualized by deep etching, *Cell* **30:**395–406.

Hirokawa, N., and Kirino, T., 1980, An ultrastructural study of nerve and glial cells by freeze-substitution, *J. Neurocytol.* **9:**243–254.

Hirokawa, N., Tilney, L. G., Fujiwara, K., and Heuser, J. E., 1982, The organization of actin, myosin, and intermediate filaments in the brush border of intestinal epithelial cells, *J. Cell Biol.* **94:**425–443.

Hoffman, P. N., and Lasek, R. J., 1975, The slow component of axonal transport. Identification of major structural polypeptides of the axon and their generality among mammalian neurons, *J. Cell Biol.* **66:**351–356.

Katz, B., and Miledi, R., 1972, The statistical nature of the acetylcholine potential and its molecular components, *J. Physiol. (London)* **224:**665–699.

Kuffler, S. W., and Yoshikami, D., 1975, The distribution of acetylcholine sensitivity at the postsynaptic membrane of vertebrate skeletal twitch muscles: Iontophoretic mapping in the micron range, *J. Physiol. (London)* **244:**703–730.

Levine, J., and Willard, M., 1981, Fodrin: Axonally transported polypeptides associated with the internal periphery of many cells, *J. Cell Biol.* **90:**631–643.

Matsumoto, G., and Sakai, H., 1979, Microtubules inside the plasma membrane of squid giant axons and their possible physiological function, *J. Membr. Biol.* **50:**1–14.

McMahan, U. J., Sanes, J. R., and Marshall, L. M., 1978, Cholinesterase is associated with the basal lamina at the neuromuscular junction, *Nature (London)* **271:**172–174.

Metuzals, J., and Tasaki, I., 1978, Subaxolemmal filamentous network in the giant nerve fiber of the squid (*Loligo pealei L.*) and its possible role in excitability, *J. Cell Biol.* **78:**597–621.

Palay, S. L., 1956, Synapses in the central nervous system, *J. Biochim. Biophys. Acta* **2:**193–201.

Peper, K., and McMahan, U. J., 1972, Distribution of acetylcholine receptors in the vicinity of nerve terminals on skeletal muscle of the frog. *Proc. R. Soc. London B* **181:**431–440.

Pfenninger, K. H., 1971, The cytochemistry of synaptic densities. II. Proteinaceous components and mechanism of synaptic connectivity, *J. Ultrastruct. Res.* **35:**451–475.

Prives, J., Daniels, M. P., Neal, F. M., Bauer, H.-C., Penman, S., and Christian, C. N., 1979, Two classes of acetylcholine receptors on cultured muscle cells distinguished by detergent extraction, *Abstracts of the Society for Neuroscience USA* **6:**487.

Rash, J. E., and Ellisman, M. H., 1974, Studies of excitable membranes. Macromolecular specializations of the neuromuscular junction and the nonjunctional sarcolemma, *J. Cell Biol.* **63**:467–586.
Sanes, J. R., and Hall, Z. W., 1979, Antibodies that bind specifically to synaptic sites on muscle fiber basal lamina, *J. Cell Biol.* **83**:357–370.
Sanes, J. R., Marshall, L. M., and McMahan, U. J., 1978, Reinnervation of muscle fiber basal lamina after removal of myofibers. Differentiation of regenerating axons at original synaptic sites, *J. Cell Biol.* **78**:176–198.
Sharp, G. A., Shaw, G., and Weber, K., 1981, Immunoelectron microscopical localization of the three neurofilament triplet proteins along neurofilaments of cultured dorsal root ganglion neurons, *Exp. Cell Res.* **137**:403–413.
Willard, M., and Simon, C., 1981, Antibody decoration of neurofilaments, *J. Cell Biol.* **89**:198–205.

II

Cellular Excitation: Recent Findings and Models

6

An Introduction to Membrane Conductances

William J. Adelman, Jr.

I. EXCITABILITY AND CONDUCTANCE

Acceptance of the idea that electrical excitability in neural membranes results from the voltage-dependent opening and closing of ion-specific membrane channels is widespread among electrophysiologists. Much of the evidence that has led to this acceptance has been based on measurements of current flow through excitable membranes (Conti and Neher, 1980; Sigworth and Neher, 1980). These channels have been described in terms of their individual unit conductances or their conductances in ensemble. This chapter is written from the viewpoint of a "channel" advocate. For other views of bioelectrical phenomena, see the chapters in this book by Tasaki and Iwasa, Ling, Leuchtag and Fishman, and Chang.

In 1939, Cole and Curtis published a paper which focused attention on nerve membrane conductance. This paper showed that there was a large increase in membrane conductance associated with the appearance of an action potential. Using an electrical bridge which was balanced across the resting axon, Cole and Curtis determined that a bridge unbalance occurs during the action potential and follows a time course similar to the voltage changes. Cole and Curtis (1938) had previously suggested that there was a conductance increase associated with the appearance of an action potential in the large cylindrical single cell of the alga *Nitella*, but it was the squid giant axon experiment which proved the most convincing.

Throughout World War II, the question as to the exact nature of the relation between the voltage change during the nerve impulse and the membrane conductance change remained unanswered. At the end of the war, Cole established a Department of Biophysics at the University of Chicago where he and George Marmont devised methods and equipment to answer this question. Making use of a long internal wire electrode to greatly lower the internal axon resistance, thereby making the axon interior isopotential, they were able to "space clamp" the squid giant axon so that an action potential once initiated would appear simultaneously over a large membrane area without propagation. Marmont (1949) devised a method whereby the internal wire was made a current-carrying electrode, and current flow

---

William J. Adelman, Jr. • Laboratory of Biophysics, Intramural Research Program, NINCDS, National Institutes of Health at the Marine Biological Laboratory, Woods Hole, Massachusetts 02543.

across a central length of axon could be controlled by electronic feedback. Measuring the transmembrane potential, Marmont showed that following a brief effective current stimulus an action potential would occur even when the net membrane current was maintained at zero throughout the course of the membrane action potential.

Cole has claimed (personal communication) that once the zero net current action potential was obtained it was immediately apparent to him that the membrane conductance was some function of membrane potential and that the current clamp device should be reconfigured so as to measure membrane current as a function of a set of constant amplitude voltage steps. This approach was suggested by the method employed by Bartlett (1945) to measure the electrical characteristics of the iron wire (Heathcote, 1907) nerve model (see Lillie, 1936, for a general description).

The zero net current action potential in the space clamped giant axon implied that the membrane capacity current (derivable from the derivative of the action potential) was balanced by a membrane ionic current flowing through a variable membrane conductance (Cole, 1968). One could determine the characteristics of this ionic current if one could rapidly charge the membrane capacity, C_m, by stepping the membrane potential, V, to a constant value making $C_m(dV/dt) = 0$. It was this reasoning that lead to the development of the voltage clamp method (Cole, 1949; Hodgkin *et al.*, 1949, 1952). Most measurements of membrane, multiple channel, or single channel conductances utilize this method applied in a variety of technical forms. By using square pulses of voltage under membrane potential control, the membrane conductance of the squid giant axon was shown to be time variant and voltage-dependent. Before considering the implications of these "voltage clamp" experiments, a brief simplified description of conductances, in general, seems necessary.

II. ELECTRICAL CONDUCTANCE

If we consider an isolated metallic conductor, we can think of the free electrons in the material as being in constant random motion. This motion is analogous to the motion of the molecules of a gas confined to a closed container. The flux of electrons across any arbitrary plane within the material will be the same in one direction as it is in the opposite direction. If a potential difference is set up across the conductor an electric field will exist at every point within the material. The field, E, will act on the free electrons imparting a resultant motion in the direction opposite to the field. The flow of current I through any imaginary plane perpendicular to the field is given by

$$I = dq/dt \tag{1}$$

and for a steady field $I = q/t$, where, in mks units, I is given in amperes, the charge transferred q in coulombs and the time t in seconds. The free electron motion is constant because of a sequence of accelerations by the field and decelerations by collisions of the free electrons with atoms in the material lattice.

In 1827, Georg S. Ohm stated that the current flow through a wire was a direct function of the potential difference between its ends if there were no internal sources in the wire itself. In the simplest form of Ohm's Law, the constant of proportionality between the current, I, and the voltage difference, V, is called the conductance, g, and the law is expressed as (assuming constant temperature)

$$I = gV \tag{2}$$

In a wire of uniform material and constant cross section, the value of the conductance is directly proportional to the cross-sectional area and inversely proportional to the length. The conductance changes when the temperature or the material is changed. The constant of this proportionality is called the conductivity, σ, which is characteristic of the temperature and the material and not of a particular specimen of the material. The conductivity is given by

$$\sigma = i/E \tag{3}$$

and is defined only for isotropic materials whose properties do not vary with direction in the material. For a wire of constant cross-sectional area, A, and given length, ℓ, carrying a steady current, I

$$\sigma = i/E = (I/A)/(|V|/\ell) \tag{4}$$

Thus, i and E have the dimensions of current density and potential gradient, respectively.

Using Eq. (2), we can rewrite Eq. (4) by letting $I/V = g$ so that $\sigma = g(\ell/A)$. Therefore, conductivity is in units of conductance per unit length. In most applications considered in this chapter, g will be used rather than σ.

III. IONIC CONDUCTANCE

The conductivity of aqueous electrolyte solutions depends on the concentration of the solute or electrolyte (Robinson and Stokes, 1959). In the same notation as was used for metallic conductors, the conductivity of an electrolyte solution is simply the specific conductivity, σ_s, divided by the concentration of the electrolyte, c. Thus, the equivalent conductivity, Λ, is given by:

$$\Lambda = \sigma_s/c \tag{5}$$

If an electrical potential is applied across a volume of an electrolyte solution, the resultant current flow through the solution will result from the motion in opposite directions of oppositely charged ions. The equivalent conductivity is the sum of two equivalent ionic conductivities, λ_1 and λ_2, and

$$\Lambda = \lambda_1 + \lambda_2 \tag{6}$$

At infinite dilution the values of the equivalent conductivities have limiting values and

$$\Lambda^\circ = \lambda_1^\circ + \lambda_2^\circ \tag{7}$$

At infinite dilution one can consider that the ions are so far apart that they are without influence on each other. Ionic motion then depends only on the nature of the ionic species, the properties of the medium, and the temperature. The values of λ_1° and λ_2° are independent of each other.

Ions in solution move under the influence of an external field in a manner analogous to the motion of free electrons in a metal. If a constant potential difference is applied across a volume of an electrolyte solution, the ions are almost instantaneously accelerated to a constant velocity at which their motion is limited by frictional or viscous forces in the solvent. For ordinary fields this velocity is directly proportional to the impressed field which implies that Ohm's Law can be applied to electrolytes for most ordinary fields. This limiting velocity per unit force is defined as the mobility, u.

Considering a unidimensional treatment for systems not far from equilibrium (Schwartz, 1971), this velocity can be expressed as

$$v_i = -u_i \frac{\partial \hat{\mu}_i}{\partial x} \tag{8}$$

where v_i is the average diffusion velocity of species i, and $\hat{\mu}_i$ is the electrochemical potential with independence among ionic species assumed. If we consider that the flux of the diffusing species is j_i, then

$$j_i = c_i v_i \tag{9}$$

where c_i is the concentration of the diffusing species in moles per unit volume. Thus

$$j_i = -u_i c_i \frac{\partial \hat{\mu}_i}{\partial x} \tag{10}$$

Consider an ionic species in equilibrium across a planar membrane. For an isothermal and isobaric regime

$$\frac{\partial \hat{\mu}_i}{\partial x} = \left[RT \frac{\partial}{\partial x} \ln c_i + z_i F \frac{\partial V}{dx} \right] \tag{11}$$

and

$$j_i = -u_i c_i \left[RT \frac{\partial}{\partial x} \ln c_i + z_i F \frac{\partial V}{dx} \right] \tag{12}$$

where R is the gas constant, T is the absolute temperature, F is Faraday's constant, z_i is the valence of the species i, and V is the electrical potential. Equation (12) is the Nernst–Planck relation (Nernst, 1889; Planck, 1890).

Integrating Eq. (12) from one membrane boundary to the other and considering that at equilibrium $j_i = 0$ gives

$$V_2 - V_1 = \frac{-RT}{z_i F} \ln \frac{c_i(2)}{c_i(1)} \tag{13}$$

which is the Nernst equilibrium potential relation.

Equation (13) can be formulated to calculate the equilibrium potential for sodium ions, E_{Na}, (for example, across a cell membrane) as follows:

$$E_{na} = \frac{RT}{z_i F} \ln \frac{[NA]_o}{[Na]_i} \tag{14}$$

where $[Na_o]$ is the external sodium concentration and $[Na_i]$ is the internal sodium concentration. Properly, $[Na_o]$ and $[Na_i]$ should be expressed as activities (Lewis and Randall, 1923). However, if the external and internal activity coefficients are similar, concentration may be used without significant error. In Eq. (14), E_{Na} is measured internally with respect to the external solution and, therefore, E_{Na} is positive for $[Na_o] > [Na_i]$.

Equation (12) can be rearranged to give for univalent ions

$$\frac{-j_i}{Fu_ic_i} = \frac{RT}{F}\frac{\partial}{\partial x}\ln c_i \pm \frac{\partial V}{\partial x} \tag{15}$$

In the steady state (Schwartz, 1971)

$$\frac{\partial j_i}{\partial x} = -\frac{\partial c_i}{\partial t} = 0 \tag{16}$$

and Eq. (15) may be integrated between boundary 2 assumed to be the inside of the membrane, i, and boundary 1 assumed to be the outside, o, to give

$$-(j_iF)\int_{o}^{i}\frac{dx}{F^2uc_i} = \frac{RT}{F}\ln\frac{c_i}{c_o} \pm V_m \tag{17}$$

where $V_m = V_i - V_o$. The sign of V_m is positive if i is a cation and negative if i is an anion. If i is a univalent cation then, for unit area,

$$I_i = -j_iF \tag{18}$$

Modern convention takes outward current being positive with an external voltage reference. If the ionic conductance is denoted as

$$g_i = 1\Bigg/\left(\int_{o}^{i}\frac{dx}{F^2u_ic_i}\right) \tag{19}$$

and

$$E_i = -\frac{RT}{F}\ln\frac{c_i}{c_o} \tag{20}$$

then Eq. (17) gives

$$I_i = g_i\,[V_m - E_i] \tag{21}$$

Equation (21) will be used later to describe voltage dependent ionic conductance in excitable membranes.

IV. EXCITABLE CONDUCTANCES IN MEMBRANES

As this book is concerned with excitable systems, it seems appropriate to consider the nerve impulse. Investigators in the past have focused on the generation of the nerve impulse and its propagation along the axon as a general model system for electrical excitability. A description of this system is complicated and involves many parameters, each of which is a function of one or more variables. The voltage clamp method described earlier in this chapter is basically a means by which some variables are held constant and others are allowed to change in a controlled manner. The conventional technique involves maintaining the membrane potential constant in time by means of electronic feedback and maintaining the potential difference across the membrane constant over a given membrane area (or length of axon) by means of axial external and internal metal electrodes (space clamp). Thus, in a voltage clamped giant axon the voltage difference between the inside and the outside of a space clamped axon is stepped to a new value chosen by the investigator and held constant while membrane currents are recorded. Because $C(dv/dt) = 0$ for a step clamp, any changes in currents, I_m, across the axon surface are attributed to changes in the conductance, g_m, of the axon membrane (Hodgkin and Huxley, 1952a), the voltage across the membrane, V_m, being held constant. Therefore, assuming

$$I_m = g_m V_m \tag{22}$$

any changes in measured current must be reflecting changes in membrane conductance. Conductance was chosen by Hodgkin and Huxley (1952b) because it is directly proportional to the ease with which ions are expected to cross the membrane. This expectation came from previous radioactive isotope experiments which indicated an exchange of sodium and potassium ions between the inside and outside of axons during the generation of nerve impulses or action potentials (Rothenberg, 1950; Keynes and Lewis, 1951).

As the radioactive tracer experiments were interpreted in terms of fluxes occurring across the axon membrane, currents flowing during a voltage clamp experiment were immediately conceived of as ionic membrane currents flowing through membrane ionic conductances (Hodgkin and Huxley, 1952a). The early experiments on squid giant axon using the voltage clamp method (Cole, 1949; Hodgkin *et al.*, 1949, 1952) suggested that there were two major components of membrane current flowing across the axon membrane in response to step changes in membrane potential. These were an early transient current and a delayed steady state current.

Hodgkin and Huxley (1952a) suggested that the early transient current was carried predominantly by sodium ions. They also presented a set of reasons for suggesting that the delayed outward current was carried by potassium ions. Hodgkin (1951) had reviewed the evidence for an ionic basis for the generation of the action potential. On the basis of external ion substitutions (Hodgkin and Katz, 1949; Huxley and Stämpfli, 1951) and the radioisotope flux measurements (Rothenberg, 1950; Keynes and Lewis, 1951), it had been suggested that the impedance changes seen during the action potential were related to transient sequential changes in the axon membrane permeability to sodium and potassium ions (Hodgkin, 1951). Hodgkin and Huxley (1952a), following a suggestion of Teorell (1949), stated that any "definition of permeability which takes no account of electrical forces is meaningless in connection with the movements of ions, though it may well be appropriate for uncharged solutes." In this regard they considered that permeability was related to conductance and current flow to ionic flux.

With regard to the early transient current, Hodgkin and Huxley (1952a) showed that this current was inward over a given voltage (V_m) range, zero at a given voltage and outward at voltages greater than the zero current (reversal) potential. They were able to correlate the values of the reversal potential, V_{Na}, for the initial transient current with the value of the sodium equilibrium potential, E_{Na}, as calculated by the Nernst relation (see Eq. (14)) from estimated values of the internal and external sodium ion concentrations (activities). In their own words,

The driving force for a particular ion species is clearly zero at the equilibrium potential for that ion. The driving force may therefore be measured as the difference between the membrane potential and the equilibrium potential. . . . the driving force for sodium ions will be $(E - E_{Na})$, which is also equal to $(V - V_{Na})$. The permeability of the membrane to sodium ions may therefore be measured by $I_{Na}/(E - E_{Na})$. This quotient, which we denote by g_{Na}, has the dimensions of a conductance (current divided by potential difference), and will therefore be referred to as the sodium conductance of the membrane. Similarly, the permeability of the membrane to potassium ions is measured by the potassium conductance g_K, which is defined as $I_K/(E - E_K)$. Conductances defined in this way may be called chord conductances and must be distinguished from slope conductance (G) defined as $\partial I/\partial E$.

V. MEMBRANE CONDUCTANCE GATING

Hodgkin and Huxley (1952a) were able to separate the early transient membrane current from the delayed outward current by substituting choline chloride (assumed to be impermeant) for sodium chloride in the external medium bathing the squid axon. Varying the external sodium concentration changed V_{Na} and altered the amplitude of the initial transient current. The changes in V_{Na} brought about by changes in $[Na_o]$ correlated well with predictions made from Nernst potential calculations, and the alterations in current density correlated well with predictions made from calculations based on the assumption that the major charge carriers through the early transient conductance were sodium ions. Therefore, Hodgkin and Huxley (1952a) called the early transient conductance the sodium conductance, g_{Na}. They also called the delayed conductance the potassium conductance, g_K. A small time invariant and voltage independent conductance was called the leakage conductance, g_L, as they assumed it was related to nonspecific leaky holes (damaged regions) in the axolemma.

Both g_{Na} and g_K are time variant and voltage-dependent (Hodgkin and Huxley, 1952b). For a depolarizing step in membrane potential the separated values of g_{Na} increased to a maximum and then declined towards a constant low value. On the other hand, g_K increased from a low value to a steady maximum for the same depolarizing step. While the rising phase of g_{Na} was faster than g_K, both conductances increased in a sigmoidal manner having an initial slope of zero. Different pulse potentials had different maxima. Therefore, the maximum values of g_{Na} and g_K were found to vary with the membrane potential. The rates of rise and fall of g_{Na} and the rate of rise of g_K were also shown to be voltage dependent.

Hodgkin and Huxley (1952c) made an assumption that there were a finite number of conductance units in a given membrane area and when all of these units are in the conducting state there is a maximal conductance, $\bar{g}_i$. The maximal value for the sodium conductance was taken as $\bar{g}_{Na}$ and that for the potassium conductance was taken as $\bar{g}_K$ (Hodgkin and Huxley, 1952d).

In order to describe the variaton in g_{Na} and g_K with membrane potential, $\bar{g}_{Na}$ and $\bar{g}_K$ were multiplied by dimensionless parameters having values which varied between zero and unity. One parameter, m, described the rising phase of g_{Na}, another parameter, h, described the falling phase of g_{Na}, and a single parameter, n, described the rising phase of g_K. These parameters depended on voltage and time. In order to fit the sigmoidal shape of the rise of g_{Na} and g_K, m and n were raised to the third and fourth powers, respectively.

The membrane current was described by the following relation (Hodgkin and Huxley, 1952d):

$$I_M = C\frac{dV}{dt} + \bar{g}_{Na}\, m^3h(V - V_{Na}) + \bar{g}_K\, n^4(V - V_K) + \bar{g}_L(V - V_L). \tag{23}$$

The variaton of m was described by the following differential equation

$$\frac{dm}{dt} = \alpha_m(1 - m) - \beta_m m \tag{24}$$

and similarly h and n by the differential equations:

$$\frac{dh}{dt} = \alpha_h(1 - h) - \beta_h h \tag{25}$$

and

$$\frac{dn}{dt} = \alpha_n(1 - n) - \beta_n n \tag{26}$$

The rate constants, α's and β's, were shown to be voltage-dependent and Hodgkin and Huxley (1952b,c) were able to obtain sets of data for these voltage-dependencies from voltage clamp data.

The solution of Eq. (24) is

$$m = m_\infty - (m_\infty - m_o)\exp(-t/\tau_m) \tag{27}$$

where

$$m_o = \alpha_m/(\alpha_m + \beta_m) \tag{28}$$

$$m_\infty = \alpha_m'/(\alpha_m' + \beta_m') \tag{29}$$

and

$$\tau_m = 1/(\alpha_m + \beta_m) \tag{30}$$

where the primes refer to values obtained at long times. The solutions to Eqs. (25) and (26) are identical except that the subscripts h and n substitute for m.

The variations of m_∞, h_∞, and n_∞ with V are shown in Fig. 1 and the variations τ_m, τ_h,

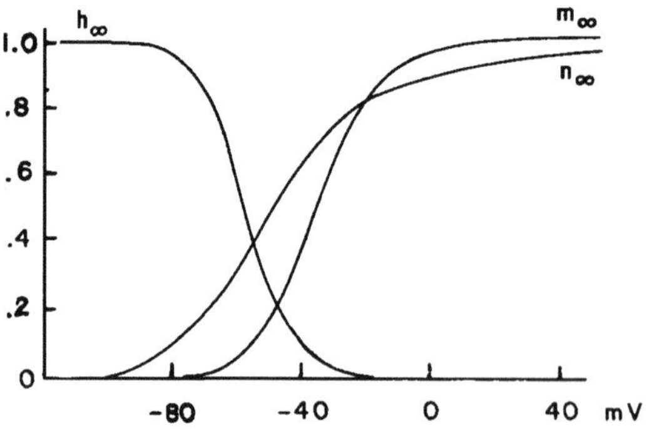

Figure 1. Steady state values of m, n, and h as a function of the membrane voltage, V, in mV, where V is defined in absolute terms. Hodgkin and Huxley defined the resting potential as zero, and took voltages from this zero such that depolarizations had a negative sign. V equals $-(V + 60)$ in the Hodgkin–Huxley convention.

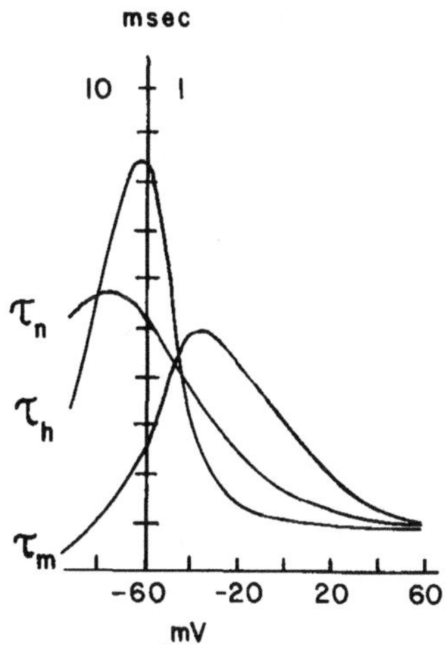

Figure 2. Conductance parameter time constants, τ_m, τ_n, and τ_h in msec as a function of membrane voltage, V, in mV, as defined in the legend for Fig. 1. Values of τ_m are plotted on the 0–1.0 msec scale and values of τ_n and τ_h are plotted on the 0–10 msec scale.

and τ_n with voltage are shown in Fig. 2. For a correlation between computed values of m, h, and n and I_M, I_{Na}, and I_K for a sample step change in membrane potential (see Fig. 6-3 in Palti, 1971). The sets of Eqs. (23–26) describe the voltage-dependent "gating" of the membrane conductances. Hodgkin and Huxley (1952d) using data obtained from squid axon voltage clamp experiments were able to predict the locus of the action potential by numerically solving Eq. (23) for V for conditions (stimuli) which produce an action potential experimentally.

Several years ago, a modification of the τ_h (and hence α_h and β_h) function on V was presented (Adelman and Palti, 1969). They showed that the rate of sodium inactivation (Hodgkin and Huxley, 1952c) was a function of the external potassium concentration (Adelman and Palti, 1969). Taking into consideration the properties of the Schwann cell sheath enclosing the giant axon in modifying the periaxonal potassium concentration as a function of outward potassium ion flow through potassium channels (Frankenhaeuser and Hodgkin, 1956), Adelman and FitzHugh (1975) modified the original Hodgkin and Huxley (1952d) equations so as to incorporate both the K^+ accumulation and the $\tau_h = f(K_o)$ effects. These modified equations were good predictors (Adelman and FitzHugh, 1975) of a number of measurable excitation phenomena not predicted by the original Hodgkin and Huxley equations.

VI. UNIT CONDUCTANCES

In the opening paragraph of this chapter it was stated that there is now wide acceptance of the idea that there are microscopic units of conductance called channels in excitable membranes. On the basis of many experiments (see the chapter in this book by Lecar *et al.* for references), these channels are conceived to be in two general states, an open conducting state and a closed nonconducting state. In the open state each channel has a unit conductance, γ. If one of these channels is examined over a lengthy period of time, it is found that the channel is likely to fluctuate in a somewhat random manner between its nonconducting and conducting states, the transitions between states occurring virtually instantaneously. At any one instant in time the channel conductance is either zero or some value. Over the time chosen the average conductance is less than the unit channel conductance except for the case where the channel remains in its open state continually.

It is also conceived that for a channel permeable in its open state to a specific ion species one can express the unit conductance, γ, as

$$\gamma_{open} = I_{open}/(V - E_i) \tag{31}$$

I_{open} is the current flowing through the open channel impelled by a driving force given by $(V - E_i)$ where E_i is the equilibrium potential of the specific charge carrier contributing to the current flow. This relationship implies that there exists a linear open channel current voltage relation. Such a relation seems improbable, particularly if there is ion interaction and competition for sites within the channel (French and Adelman, 1976). Equation 31 should be generalized to

$$\gamma_{open} = I_{open}/\mathrm{f}(V - E_i) \tag{31a}$$

Lecar (1981) has considered this problem in detail in deriving his equation 17. The average conductance, $\bar{\gamma}$, of the single fluctuating channel may be given by

$$\bar{\gamma} = p\gamma \tag{32}$$

where p is the probability of finding the channel in the open state.

If we assume that the single channel has the same reversal potential and the same driving force over the relatively long time interval during which $\bar{\gamma}$ is measured, then the average current is

$$\bar{i} = \bar{\gamma}\mathrm{f}(V - E_i) \tag{33}$$

or

$$\bar{i} = p\gamma_{open}\mathrm{f}(V - V_{rev}) \tag{34}$$

If we consider that a given area of membrane contains N channels and that for a given set of constant conditions each of the N channels has the same probability of being open, then the conductance, g, of the ensemble of channels is given by

$$g = Np\gamma_{open} \tag{35}$$

In electrically excitable membranes the ionic conductances are usually found to be functions of membrane voltage. For the case where γ_{open} can be shown not to be a function of membrane voltage, the voltage-dependence of g is assumed to follow from the voltage dependence of the probability, p.

In the chapter by DeFelice *et al.* in this book, it is pointed out that given a large ensemble of channels, i.e., N is large, it is very difficult to determine whether $p = f(V)$ or $\gamma = f(V)$ on the basis of g measurements alone. The relation between an ensemble of stochastic unit conductances and macroscopic conductances as measured by means of voltage clamp has been discussed by Lecar *et al.* (1975), Ehrenstein and Lecar (1977), DeFelice (1981), Fohlmeister and Adelman (1981) and in the chapter by Lecar *et al.* in this book.

VII. CONDUCTANCE MEASUREMENTS AT PRESENT

The chapters in this section are concerned with electrical studies of excitable cells. In the first of these, Lecar *et al.* review the general evidence and arguments for the existence of membrane channels and for the study of channels in terms of unit conductances. This chapter should be consulted for pertinent citations to the channel literature. They present an extensive list of channels now being studied in real and artificial membranes. Inasmuch as many workers consider that a channel is a protein or polypeptide (Montal *et al.*, 1981) creating a hydrophilic pathway across a predominantly hydrophobic barrier (a lipid bilayer membrane), one finds that the study of channel phenomena involves voltage clamp methods, gating measurements, variation of electrochemical driving forces, measurement of ion selectivity, the production of current blockade by competitors, agents, drugs and toxins, and the direct variation of conductance behavior by manipulating the chemistry of the channels themselves.

Most of these studies have produced unit conductance behavior that can be used to predict the macroscopic behavior of excitable cells on the basis of the time and spatial summation of the measured unit events. In the chapter by DeFelice and Kell, a simple channel model is presented which can be used to predict such complicated phenomena as the higher order kinetics found in the Hodgkin and Huxley (1952d) description of the potassium conductance, membrane noise spectra (DeFelice, 1981) and impedance characteristics (DeFelice *et al.*, 1981). Earlier in this introductory chapter I considered only the properties of the open and closed states of a single channel. DeFelice and Kell present a complete scheme for transition through a set of intermediate states which cannot be measured directly from the end states (open or closed) but can be obtained from the transition kinetics. They also consider the problem of activation and inactivation at the single channel level. In this connection, their treatment should be compared with the experiment of Sigworth and Neher (1980) who showed that patch electrode measurements of voltage clamped single sodium channels in muscle cells are related to macroscopic measurements of sodium currents from these cells. In the Sigworth and Neher experiment the response of a single sodium channel to a step in membrane potential difference repeated and summed many times has the same shape as the macroscopic membrane response to a single voltage clamp step which presumably arose from the spatial summation of a set of similar unit channels. Sigworth and Neher's (1980) superposition summation experiment is taken by many researchers as the most convincing evidence that excitable phenomena have their basis in single channel stochastic kinetics.

The chapter by L. E. Moore returns to impedance characteristics of excitable cells and presents an elegant treatment of complex admittance in terms of voltage-dependent sodium and potassium conductances. This paper places much of the muscle membrane literature into a framework whereby an understanding of complex behavior can be obtained in terms of the familiar conductance kinetics represented by such model systems as that first proposed by Hodgkin and Huxley (1952d). This work should be compared with such general treatments as that of Poussart *et al.* (1977), FitzHugh (1981), DeFelice (1981), DeFelice *et al.* (1981), and Fishman *et al.* (1981).

Pichon *et al.* in their chapter in this book report on the relation between membrane ionic currents and their expression in terms of noise and admittance in the cockroach axon. Tracing the origin of their work back to that begun by Derksen and Verveen in 1966, they are able to use 4-aminopyridine (4-AP) to change both the admittance and the noise spectra in a manner consistent with expectations that 4-AP selectively blocks the potassium conductance in axons. Again, they compare these expectations with those obtained by others (Fishman *et al.*, 1981). Their conclusion is that there is a continuum between macroscopic conductance measurements obtained by conventional voltage clamp methods and such measurements as noise and admittance of excitable membranes. They conclude that such a connection lies in the ensemble properties of single channel fluctuations and their controlled variation with membrane voltage.

The chapter by Chang takes a different view of conductance pathways in excitable membranes. Focusing on submembrane cytoplasmic protein components, he presents a view that this "cell cortex" material and that of the membrane form a fixed-charge system with ion-exchange properties. According to Chang, a membrane channel operates as a conduit to the cytoplasmic protein underlying the membrane and it is the conformational changes in this later structure that account for time variation in the electrical properties of the cell. For other views on the possible role of cytoplasmic proteins see the chapters in this book by Tasaki *et al.*, Ling, Matsumoto *et al.*, and Tasaki (1982).

Another view of membrane conductance is taken in the chapter by Baumann in this book. For many years Baumann has modeled the behavior of excitability inducing material (EIM) in artificial lipid bilayer membranes. This chapter presents the aggregation model of membrane channels (Baumann and Mueller, 1974) in a form in which the microscopic fluctuation phenomena of unit conductances and the macroscopic electrical behavior of excitable cells can be predicted on the basis of a channel composed of subunits which aggregate when the membrane electric field changes. The model has gating and selectivity both linked to channel assembly. Baumann presents several criteria for judging the adequacy and validity of any model system. Several years ago, Goldman (1971) presented a similar set of criteria and it is interesting to compare these with Baumann's.

The final two chapters in this section report on the effects of toxins on the axon sodium conductance. The chapter by Meves considers scorpion toxins which have interesting affects on the activation and inactivation of the sodium conductance. The chapter by Khodorov reviews the literature describing sodium conductance modification by batrachotoxin and presents evidence for this agent having the ability to bind to a receptor whose characteristics are dependent on the state of the sodium channel. This finding might prove valuable in specifying the nature of the sodium channel. It certainly should be considered as one of the specific tests for identification as sodium channels that isolated channel material must pass when incorporated in lipid bilayers.

In conclusion, it is clear that electrical measurements of excitable systems continue to be a valuable tool in assessing the properties of the elements composing these systems. It is also clear that any ensemble of unit elements must possess the properties detailed for cells and cell membranes. While more and more interest is directed toward discovery and identification of unit conductances, membrane channels, channel forming proteins, and other molecular structures associated with or modifying these unit events, there must be a rational continuum between these miniature electrical events and the macroscopic conductances that have demanded the attention of workers in the excitation field over the last three decades.

REFERENCES

Adelman, W. J., Jr., and FitzHugh, R., 1975, Solutions of the Hodgkin–Huxley equations modified for potassium accumulation in a periaxonal space, *Fed. Proc.* **34:**1322–1329.

Adelman, W. J., Jr., and Palti, Y., 1969, The influence of external potassium on the inactivation of sodium currents in the giant axon of the squid, *Loligo pealei, J. Gen. Physiol.* **53:**685–703.

Bartlett, J. H., 1945, Transient anode phenomena, *Trans. Electrochem. Soc.* **87:**521–545.

Baumann, G., and Mueller, P., 1974, A molecular model of membrane excitability, *J. Supramol. Struct.* **2**:538–557.
Cole, K. S., 1949, Dynamic electrical characteristics of the squid axon membrane, *Arch. Sci. Physiol.* **3**:253–258.
Cole, K. S., 1968, *Membranes, Ions and Impulses,* University of California Press, Berkeley.
Cole, K. S., and Curtis, H. J., 1938, Electric impedance of *Nitella* during activity, *J. Gen. Physiol.* **22**:37–64.
Cole, K. S., and Curtis, H. J., 1939, Electric impedance of the squid giant axon during activity, *J. Gen. Physiol.* **22**:649–670.
Conti, F., and Neher, E., 1980, Single channel recordings of K^+ currents in squid axons, *Nature* **285**:140–143.
DeFelice, L. J., 1981, *Introduction to Membrane Noise,* Plenum Press, New York.
DeFelice, L. J., Adelman, W. J., Jr., Clapham, D. E., and Mauro, A., 1981, Second-order admittance in squid axon, in: *The Biophysical Approach to Excitable Systems* (W. J. Adelman, Jr. and D. E. Goldman, eds.), pp. 37–63, Plenum, New York.
Derksen, H. E., and Verveen, A. A., 1966, Fluctuations of resting membrane potential, *Science* **151**:1388–1389.
Ehrenstein, G., and Lecar, H., 1977, Electrically gated ionic channels in lipid bilayers, *Quart. Rev. Biophys.* **10**:1–34.
Fishman, H. M., Moore, L. E., and Poussart, D., 1981, Squid axon K conduction: Admittance and noise during short- versus long-duration step clamps, in: *The Biophysical Approach to Excitable Systems* (W. J. Adelman, Jr. and D. E. Goldman, eds.), pp. 65–95, Plenum, New York.
FitzHugh, R., 1981, Nonlinear sinusoidal currents in the Hodgkin–Huxley model, in: *The Biophysical Approach to Excitable Systems* (W. J. Adelman, Jr. and D. E. Goldman, eds.), pp. 25–35, Plenum, New York.
Fohlmeister, J. F., and Adelman, W. J., Jr., 1981, Gating kinetics of stochastic single K channels, in: *The Biophysical Approach to Excitable Systems* (W. J. Adelman, Jr. and D. E. Goldman, eds.), pp. 123–132, Plenum, New York.
Frankenhaeuser, B., and Hodgkin, A. L., 1956, The after-effects of impulses in the giant nerve fibers of *Loligo, J. Physiol. (London)* **131**:341–376.
French, R. J., and Adelman, W. J., Jr., 1976, Competition, saturation and inhibition—Ionic interactions shown by membrane ionic currents in nerve muscle and bilayer systems. *Current Topics in Membranes and Transport,* **8**:161–207.
Goldman, D. E., 1971, Excitability models, in: *Biophysics and Physiology of Excitable Membranes* (W. J. Adelman, Jr., ed.), pp. 337–358, Van Nostrand Reinhold, New York.
Heathcote, H. L., 1907, The passivifying, passivity and activifying of iron, *J. Soc. Chem. Ind. (London)* **26**:899–917.
Hodgkin, A. L., 1951, The ionic basis of electrical activity in nerve and muscle, *Biol. Rev.* **26**:339–409.
Hodgkin, A. L., and Huxley, A. F., 1952a, Currents carried by sodium and potassium ions through the membrane of the giant axon of *Loligo, J. Physiol. (London)* **116**:449–472.
Hodgkin, A. L., and Huxley, A. F., 1952b, The components of membrane conductance in the giant axon of *Loligo, J. Physiol. (London)* **116**:473–496.
Hodgkin, A. L., and Huxley, A. F., 1952c, The dual effect of membrane potential on sodium conductance in the giant axon of *Loligo, J. Physiol. (London)* **116**:497–506.
Hodgkin, A. L., and Huxley, A. F., 1952d, A quantitative description of membrane current and its application to conduction and excitation in nerve, *J. Physiol. (London)* **117**:500–544.
Hodgkin, A. L., and Katz, B., 1949, The effect of sodium ions on the electrical activity of the giant axon of the squid, *J. Physiol. (London)* **108**:37–77.
Hodgkin, A. L., Huxley, A. F., and Katz, B., 1949, Ionic currents underlying activity in the giant axon of the squid, *Arch. Sci. Physiol.* **3**:129–150.
Hodgkin, A. L., Huxley, A. F., and Katz, B., 1952, Measurement of current-voltage relations in the membrane of the giant axon of *Loligo, J. Physiol. (London)* **116**:424–448.
Huxley, A. F., and Stämpfli, R., 1951, Effect of potassium and sodium on resting and action potentials of single myelinated fibres, *J. Physiol (London)* **112**:496–508.
Keynes, R. D., and Lewis, P. R., 1951, The sodium and potassium content of cephalopod nerve fibres, *J. Physiol. (London)* **114**:151–182.
Lecar, H, 1981, Single-channel conductances and models of transport, in *The Biophysical Approach to Excitable Systems* (W. J. Adelman, Jr. and D. E. Goldman, eds.) pp. 109–121, Plenum, New York.
Lecar, H., Ehrenstein, G., and Latorre, R., 1975, Mechanism for channel gating in excitable bilayers, *Ann. N.Y. Acad. Sci.* **264**:304–313.
Lewis, G. N., and Randall, M., 1923, *Thermodynamics,* p. 255, McGraw-Hill, New York.
Lillie, R. S., 1936, The passive iron wire model of protoplasmic and nervous transmission and its physiological analogues, *Biol. Rev.* **11**:181–209.
Marmont, G., 1949, Studies on the axon membrane; I. A new method, *J. Cell. Comp. Physiol.* **34**:351–382.
Montal, M., Darszon, A., and Schindler, H., 1981, Functional reassembly of membrane proteins in planar lipid bilayers, *Quart. Rev. Biophys.* **14**:1–79.

Nernst, W., 1889, Die electromotorische Wirksamkeit der Jonen, *Z. Phys. Chem.* **4**:129–181.

Ohm, G. S., 1827, *The Galvanic Circuit Investigated Mathematically*, translated by W. Francis, New York, 1891.

Palti, Y., 1971, Description of axon membrane ionic conductances and currents, in: *Biophysics and Physiology of Excitable Membranes* (W. J. Adelman, Jr., ed.), pp. 168–182, Van Nostrand Reinhold, New York.

Planck, M., 1890, Ueber die Erregnng von Electricitat und Warme in Electrolyten, *Ann. Phys. Chem.* **40**:561–576.

Poussart, D., Moore, L. E., and Fishman, H., 1977, Ion movements and kinetics in squid axon. I. Complex admittance, *Ann. N.Y. Acad. Sci.* **303**:355–379.

Robinson, R. A., and Stokes, R. H., 1959, *Electrolyte Solutions*, Butterworths, London.

Rothenberg, R. A., 1950, Studies on permeability in relation to nerve function. II. Ionic movements across axonal membranes, *Biochim. Biophys. Acta* **4**:96–114.

Schwartz, T. L., 1971, The thermodynamic foundations of membrane physiology, in: *Biophysics and Physiology of Excitable Membranes* (W. J. Adelman, Jr., ed.), pp. 47–95, Van Nostrand Reinhold, New York.

Sigworth, F. J., and Neher, E., 1980, Single Na^+ channel currents observed in cultured rat muscle cells, *Nature* **287**:447–449.

Tasaki, I., 1982, *Physiology and Electrochemistry of Nerve Fibers*, Academic Press, New York.

Teorell, T., 1949, Permeability, *Annu. Rev. Physiol.* **11**:545–564.

7

Single-Channel Currents and the Kinetics of Agonist-Induced Gating

Harold Lecar, Catherine Morris, and Brendan S. Wong

I. GATED IONIC CHANNELS AND EXCITATION

The two-state gated ion-selective channel is the central concept of a paradigm for explaining membrane excitation throughout the nervous system. According to this concept, ionic channels are membrane proteins which undergo conformational transitions between nonconducting and conducting states in response to a stimulus. Electrical excitability, hence, has its origin in ion-selective channels whose opening and closing rates are controlled by the transmembrane electric field; chemical excitability at synapses is mediated by channels that are controlled by the binding of agonist molecules to receptor sites. With the advent of the patch electrode technique (Neher and Sakmann, 1976), the current pulses associated with the activation of the individual ionic channels can be observed directly in excitable cell membranes.

A question to be considered in this chapter is that of relating the stochastic properties of the random single-channel current jumps to the macroscopic kinetics of conductance change during excitation. This question was first taken up a number of years ago when it was discovered that certain peptide additives could induce a kind of excitability in lipid bilayer membranes (Mueller *et al.*, 1962). For the lipid bilayer systems, the excitability induced by the bacterial protein, EIM (excitability-inducing material), was early explained in terms of discrete channels which could undergo conductance transitions with rates modulated by the electric field across the membrane (Ehrenstein *et al.*, 1970; Latorre *et al.*, 1972; Ehrenstein *et al.*, 1974; Lecar *et al.*, 1975). For these channels, the transition rates for opening vary with voltage in such a manner as to produce the macroscopic voltage-dependent conductance and attendant voltage-dependent conductance relaxation times. A number of gated ionic channels have since been studied in lipid bilayer membranes, and

Harold Lecar • Laboratory of Biophysics, IRP, National Institute of Neurological and Communicative Disorders and Stroke, National Institutes of Health, Bethesda, Maryland 20205. *Catherine Morris* • Department of Biology, University of Ottawa, Ottawa, Ontario, Canada K1N 9A9. *Brendan S. Wong* • Department of Physiology, Baylor College of Dentistry, Dallas, Texas 75246.

there appear to be a variety of types of voltage-dependent behavior, some quite simple and some even more complicated than the gating of natural channels (Ehrenstein and Lecar, 1977; Latorre and Alvarez, 1981).

The early EIM studies demonstrated the mechanism for the steep voltage-dependent permeabilities of electrically excitable membranes. This particular bacterial product was chosen for study because its electrical behavior appeared uniquely simple. Two-state gated channels randomly fluctuate open and closed with transition rates that are exponential functions of membrane potential. From this phenomenon alone, one can deduce the kinetics of macroscopic excitation, as will be discussed later.

Originally, it was thought that EIM was a bizarre substance which accidentally formed channels having voltage-sensitive gating. It is now known that EIM is not unique; Diphtheria toxin (Misler *et al.*, 1982; Donovan *et al.*, 1981), Gonnococcus toxin (Greco, 1981), and an extract from *E. coli* (Schindler and Rosenbusch, 1978), all produce similar gated channels. These systems may provide general insights about the structural basis of gating phenomena. In the last few years, it has become possible to reconstitute the functioning ionic channels of natural excitable membranes. At this writing, channel currents are being measured for acetylcholine (ACh) receptor channels extracted from electroplax (Nelson *et al.*, 1980; Schindler and Quast, 1980; Boheim *et al.*, 1981; Montal *et al.*, 1981) and for Ca^{++}-induced K^+ channels from sarcoplasmic reticulum (Latorre *et al.*, 1982a,b). For a review of reconstitution of membrane protein structures, see Montal *et al.* (1981).

II. THE PATCH CLAMP: OBSERVING SINGLE CHANNELS IN VIVO

The development of the patch clamp technique (Neher and Sakmann, 1976) initiated the present period of intense activity in which single-channel currents are being observed for most of the channels thought to operate in excitable membranes. In order to detect the current jump caused by the opening of an individual molecular channel, it is necessary to keep the background current noise passing through the membrane lower than the channel signal. Since the sources of background noise in the membrane, such as thermal noise, scale with membrane area, the primary requirement is to record current from an isolated small area of membrane.

We can estimate the membrane area required by comparing the natural membrane to a synthetic lipid bilayer. Since membrane current noise is proportional to the product of specific conductance times area, and bilayers have specific conductances 10^{-5}–10^{-4} that of natural membrane, an isolated membrane patch would have to be at least 10^{-5} smaller in area than the bilayer to obtain comparable background noise. In practice, an area of 1 square micron is isolated by pressing a blunt micropipette onto the cell surface or by applying gentle suction.

A major consideration for the patch-clamp technique is obtaining a high-resistance seal between the pipette rim and the cell surface. The tight seal not only ensures that all the current generated within the isolated patch passes through the detector, but also reduces the major source of extraneous noise, thermal noise generated in the seal resistance itself. At present, seal resistances of the order of gigaohms (10^9 ohms) can be obtained, and with such tight seals, sensitivity is limited only by noise emanating from the current-detecting amplifiers.

An important innovation in patch-clamp technology is the "tear-off" method (Hamill and Sakmann, 1981; Horn and Patlak, 1980), which allows an isolated patch of membrane to be excised from the cell in functioning condition. Fig. 1 shows an example of single

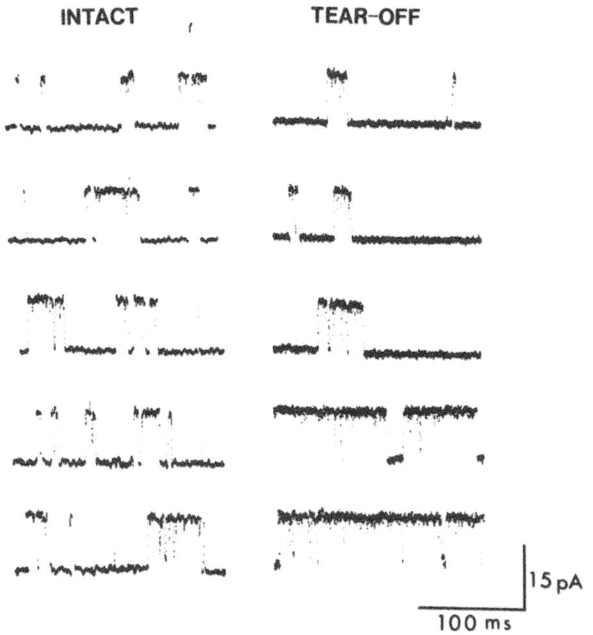

Figure 1. Single Ca^{2+}-dependent K^+ channels in an anterior pituitary tumor cell (AtT-20/D16-16). Channel jumps are seen in an intact cell and an excised inside-out membrane patch. Upward deflections represent outward currents through the channels. Unitary conductance for the excised patch is 130 pS with 5.4 mM K^+ in pipette and 140 mM K^+ at inner membrane surface; Ca^{2+} concentration, 10^{-7} M. (Wong, Lecar and Adler, unpublished data.)

channels as recorded in a membrane patch before and after excision. The torn-off patch is, in effect, a natural planar bilayer with the channels functioning normally. The torn-off membrane patch has its cytoplasmic side accessible to experimental manipulation. The channel illustrated in Fig. 1 is a Ca^{++}-sensitive K^+ channel. Experiments on intact cells (Meech, 1978; Meech and Strumwasser, 1970) suggest that Ca^{++} acts as an agonist of sorts at receptors on the inner membrane surface. Tear-off experiments show that application of Ca^{++} to the cytoplasmic membrane surface directly causes bursts of channel activity (Pallota *et al.*, 1981; Wong *et al.*, 1982,3). A variant of the tearoff technique also allows "outside-out" patches to be made (Hamill *et al.*, 1981).

The patch-clamp techniques, thus, are not only a way of observing single channels but also open up a new general means to isolate and characterize the functioning of a localized region of almost any cell membrane. Readers interested in the methodology are referred to two excellent reviews (Neher *et al.*, 1978; Hamill *et al.*, 1981).

III. RELATION BETWEEN CHANNEL FLUCTUATIONS AND MACROSCOPIC EXCITATION

The main focus of this chapter, is the way single-channel measurements can be used to ferret out kinetic details of the gating process. The kinetic data that emerge from looking at the elementary stochastic process itself are complementary to the facts learned from macroscopic voltage-clamp or membrane noise experiments. As a particular example, we will discuss some recent work on the kinetics of cholinergic channel activation in the presence of different agonists. However, first we backtrack to review the interpretation of single-channel measurements in a simple system for which the correspondence between microscopic and macroscopic kinetics is straightforward.

The agonist-induced current jumps in a muscle-cell membrane (Fig. 2) are discrete

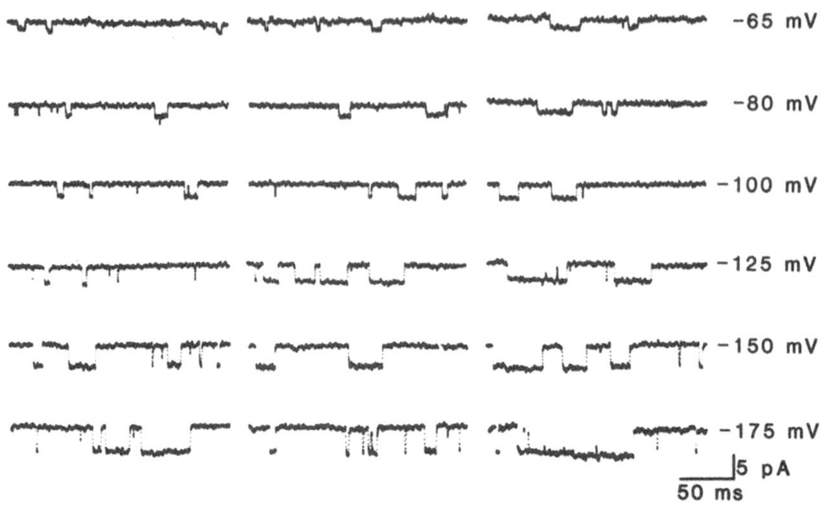

Figure 2. Carbachol-induced channels in cultured rat muscle. Patch pipette is filled with saline and 10^{-6} M carbachol. For membrane potential clamp at hyperpolarizing pulse from −65 mV to −175 mV, as indicated, jump frequency, duration, and "bursting" properties change with membrane potential.

rectangular current jumps of uniform amplitude but random duration and random interjump interval. Let us briefly recall why such current jumps are regarded as conclusive evidence for transmembrane channels and are inconsistent with other permeation mechanisms, such as carriers.

Consider the flux represented by a single random event. The postsynaptic current jumps of Fig. 2 have a current amplitude of 4–5 pA or a flux of $\sim 3 \times 10^7$ ions/sec under a 100 mV driving force. Such a flux certainly is beyond the capacity of a single carrier molecule shuttling back and forth across the membrane (Stein, 1968). To preserve something like a carrier picture, the discrete events would have had to be interpreted as random epochs of the activity of cooperative groups of carrier units. Then there would be no natural explanation for the uniform amplitudes, since cooperative groups of specialized carriers would be expected to have a spread of sizes. Similarly, the constant jump amplitudes appear to us to argue against an interpretation in terms of induced domains, such as extensive regions of ion exchanger, created by some phase transition.

The discrete quanta of current, thus, correspond most simply to the activation and inactivation of discrete, independent macromolecular pores distributed over the membrane. This is certainly not surprising in view of the physical requirements for excitability. During axonal excitation, for example, the membrane conductance increases 2 to 3 orders of magnitude with no concomitant change in membrane capacitance (Cole and Curtis, 1939). This is only possible if a minute proportion of the membrane area is devoted to high-dielectric constant ion-permeable regions. In fact, energy calculations show that only a pore mechanism can come close to lowering the barrier to permeation through a lipid bilayer sufficiently to explain the observed unit currents (Parsegian, 1969; Levitt, 1980; Lecar, 1981).

Single-channel current records, such as those of Fig. 2, provide a view of the molecular stochastic process which underlies the action of the channels. For the EIM channel in lipid bilayers, the relation between these stochastic events and the macroscopic voltage-dependent gating is particularly straightforward. As voltage is varied through the region of steep voltage-dependent conductance, the jump amplitudes merely increase linearly. Excitability does not reside in the current–voltage relation of individual open channels. The gating process is manifest only in the variation of jump frequencies and durations with voltage.

Figure 3 summarizes the simple kinetics of EIM gating. The open and closed state durations can be plotted as histograms. From these histograms, one obtains conditional probabilities, the probability of an open channel closing, β, and the probability of a closed channel opening, α. The figure shows these rates as functions of potential. Since the mean time spent by a channel in the open state, T_o, and the mean time spent in the closed state, T_c, are both exponential functions of voltage, the transition probabilities, which are just the reciprocals of these mean dwell times, can be written simply as

$$\begin{aligned} \alpha &= T_c^{-1} = \lambda \exp[(V - V_0)/2V_1] \\ \beta &= T_o^{-1} = \lambda \exp[-(V - V_0)/2V_1] \end{aligned} \tag{1}$$

The three empirical parameters of Eq. (1) are V_0, the potential at which the opening and closing rates are equal, V_1 (mV), which indicates the steepness of the potential dependence, and λ (sec^{-1}), the absolute rate of opening or closing at V_0.

The two parameters which describe the macroscopic excitation phenomena, the voltage-dependent steady state conductance $\bar{g}(V)$ and the voltage-dependent conductance relaxation time $\tau(V)$, are readily obtained from the transition probabilities. These properties can be deduced from the single-channel observations embodied in Eq. (1). Hence, the probability of being in the open state is given by

$$\bar{p}_{op}(V) = \alpha/(\alpha + \beta) = [1 + \exp(-(V - V_0)/V_1)]^{-1} \tag{2}$$

This expression has the sigmoid form characteristic of the voltage-dependent conductances of excitable membranes, as shown in Fig. 3C. For a membrane with N independent channels having unit conductance, γ, the steady state conductance is

$$\bar{g}(V) = N\gamma\bar{p}_{op}(V) \tag{3}$$

Figure 3C shows that for EIM, the steady state voltage-dependent conductance does, in fact, coincide with the voltage-dependent probability that a single channel is in the open conformation.

Equation (2) can also be given the interpretation of a Boltzmann distribution of open and closed channels in equilibrium, in which case the parameter V_1 is related to the effective charge, Q, which must be moved across the membrane to open the channel (Ehrenstein *et al.*, 1970), $V_1 = kT/Q$, where k is Boltzmann's constant and T is absolute temperature. In this interpretation, V_0 is related to the intrinsic conformational energy difference between the open and closed states in the absence of an applied electric field.

The EIM system has particularly simple relaxation kinetics. Since the "on" and "off" probabilities are exponentially distributed, the opening and closing of channels are Poisson processes. Consequently, the probability of being open, p_{op}, evolves in time according to a first-order rate equation

$$\frac{dp_{op}}{dt} = \alpha p_{cl} - \beta p_{op} = \alpha\,(1 - p_{op}) - \beta p_{op} = (\alpha + \beta)\left(\frac{\alpha}{\alpha + \beta} - p_{op}\right) \tag{4}$$

For an ensemble of N independent channels Eq. (4) can be multiplied by $N\gamma$ to obtain

$$\frac{dg}{dt} = \frac{\bar{g}(V) - g}{\tau(V)} \tag{5}$$

Here we see the identification of $\tau(V)$, the relaxation time with the quantity $1/(\alpha + \beta)$. Substituting from Eq. (1), the voltage dependence of the relaxation time is

$$\tau(V) = (\alpha + \beta)^{-1} = (1/2\lambda)\text{sech}[(V - V_0)/2V_1].$$

the characteristic bell-shaped curve of macroscopic conductance relaxation.

The single-channel data of Fig. 3 gives rise to the essential ingredients of electrical excitability, the steep voltage-dependent conductance accompanied by a voltage-dependent relaxation time. For this simple system the correspondence between the stochastic behavior of a single channel and the macroscopic conductance-change kinetics is perfectly deductive. The analogy between this description and the microscopic interpretation of the Hodgkin–Huxley (HH) equations is obvious (Ehrenstein *et al.*, 1970; Lecar *et al.*, 1975). Axonal excitation is described in terms of hypothetical gating subunits which have the same type of electric field-dependent relaxation kinetics. Of course, since the axonal K conductance, for example, exhibits significant delays in activation, the dynamics of conductance change in response to a voltage step will be more complex than the simple two-parameter conductance of the EIM system, which obeys the Poisson distribution. The delay can be represented by

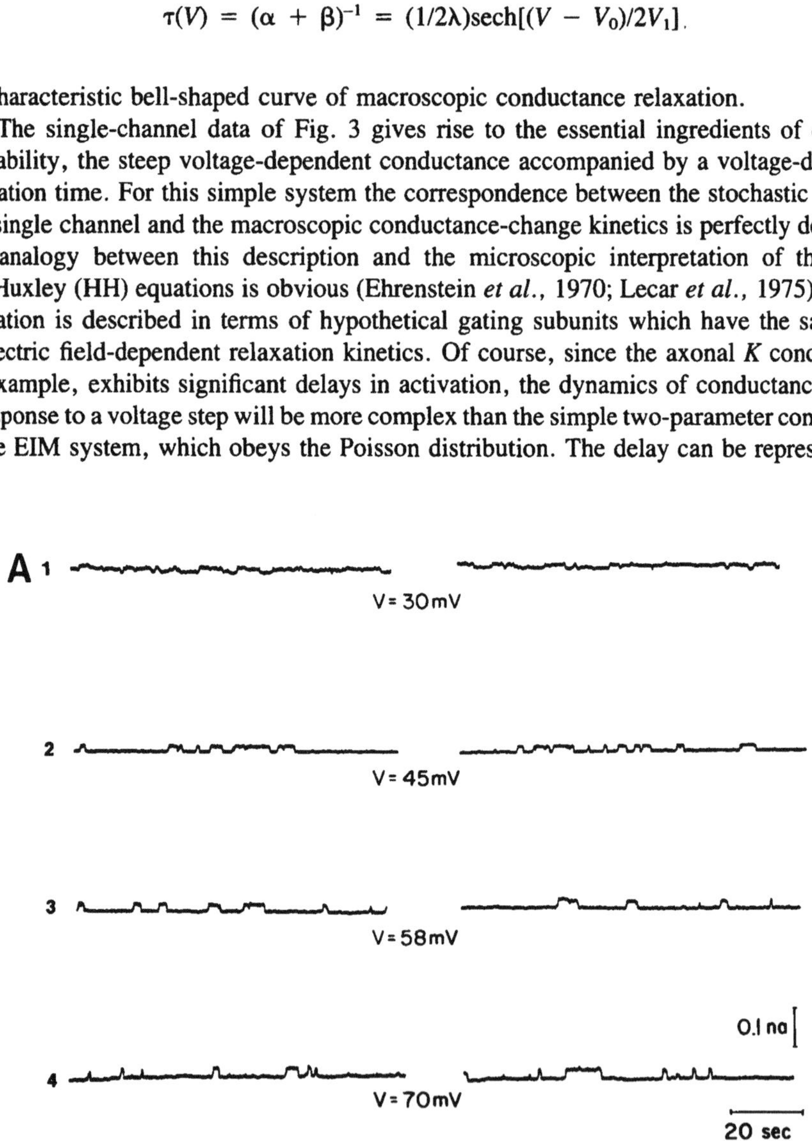

Figure 3. Scheme of EIM-induced electrical excitability. (A) Channel fluctuations show changes in duration and interpulse interval as voltage is varied (Ehrenstein *et al.*, 1970). (B) Transition rates for opening, α, and for closing, β, are exponential functions of voltage (Ehrenstein *et al.*, 1974). (C) Steady-state voltage-dependent conductance (dashed curve) coincides with probability of channel being in open state (numbers next to points indicate number of channels in experiment; Latorre *et al.*, 1972).

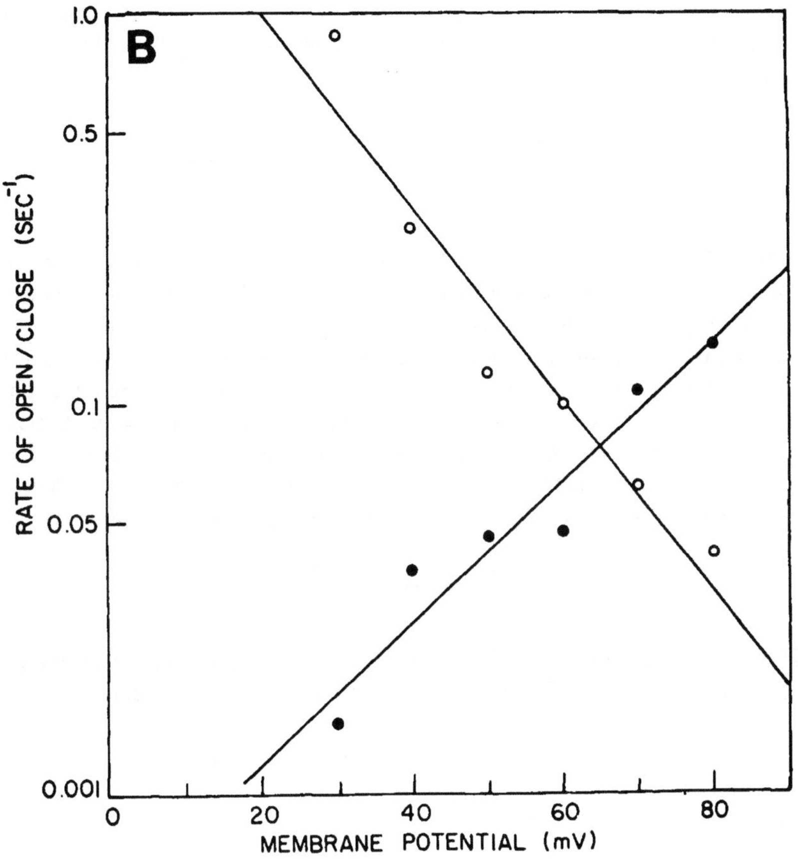
B
1.0
0.5
0.1
0.05
0.001
RATE OF OPEN / CLOSE (SEC⁻¹)
0
20
40
60
80
MEMBRANE POTENTIAL (mV)

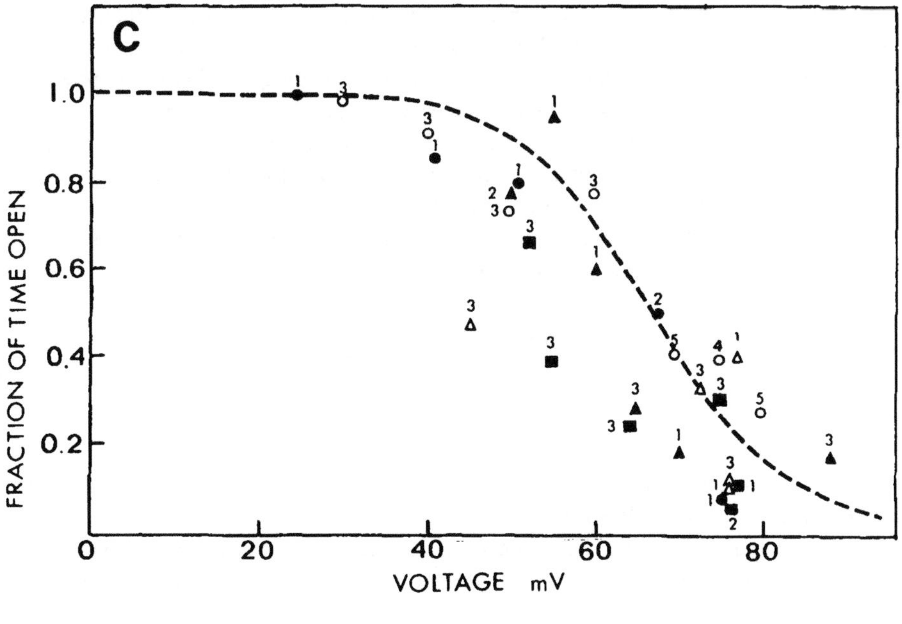
C
1.0
0.8
0.6
0.4
0.2
FRACTION OF TIME OPEN
0
20
40
60
80
VOLTAGE mV

a sequence of closed states (Cl_n) through which the channel proceeds in order to reach the open conformation (Op). The simple two-state scheme is now replaced by an N-state sequential scheme,

$$Cl_1 \underset{k_{-1}}{\overset{k_{+1}}{\rightleftharpoons}} Cl_2 \underset{k_{-2}}{\overset{k_{+2}}{\rightleftharpoons}} Cl_3 \cdots \rightleftharpoons Cl_{N-1} \underset{k_{-(N-1)}}{\overset{k_{+(N-1)}}{\rightleftharpoons}} Op \tag{6}$$

for which there is still only one open state, but $N - 1$ unobservable precursor states.

The most obvious properties that would change when the unobservable states are considered are the statistics of the intervals between openings. The distribution of interjump intervals should reflect the more complex kinetics. In particular, if the rates to and from the next-to-last step are in the right ratio, we might observe a certain number of rapid reopenings, or reversions to the open state from the last closed state. Thus, a system momentarily in state $N - 1$ will branch between the open state and the next more remote closed state with rates $k_{+(N-1)}$ and $k_{-(N-2)}$ respectively. Such behavior will be observed as a tendency to open in "bursts" with each burst having, on the average a number of interruptions, I, given by

$$I = k_{+(N-1)}/k_{-(N-2)} \tag{7}$$

For the Hodgkin–Huxley model for K conductance, Eq. (6) can be rewritten in terms of the parameters α_n and β_n, defined in macroscopic voltage-clamp experiments,

$$N = 5;\ k_{+j} = (N - j)\alpha_n;\ k_{-j} = j\beta_n$$

and the number of interruptions is

$$I = \alpha_n/3\beta_n$$

Thus, a literal interpretation of the Hodgkin–Huxley K kinetics in terms of four independent gating subunits leads to a specific testable hypothesis about the number of interruptions per burst. This prediction has been tested on the squid giant axon (Conti and Neher, 1980). The bursting was observed with a rate of interruption far exceeding the literal HH prediction.

Postsynaptic excitation by agonists can be described by kinetic schemes of channel transitions in which some of the rates represent binding of agonist molecules to receptors. The currently most favored scheme for postsynaptic activation (Adams, 1981), in which two agonist molecules (A) bind to a receptor (R) to reach a state from which the channel opens by some conformation change or isomerization, is a 4-state Markov process with three closed states that are not directly observable,

$$\underset{(Cl_0)}{R} \underset{k_{-1}}{\overset{Ak_{+1}}{\rightleftharpoons}} \underset{(Cl_1)}{AR} \underset{k_{-2}}{\overset{Ak_{+2}}{\rightleftharpoons}} \underset{(Cl_2)}{A_2R} \underset{\alpha'}{\overset{\beta'}{\rightleftharpoons}} \underset{(Op)}{A_2R^*} \tag{8}$$

(Here we have a problem with the usual notation in the literature, which reverses α and β from the electrically excitable case).

In the scheme of Eq. (8), the two association steps have rates proportional to agonist concentration. The isomerization step is generally thought to have voltage-dependence in its activation step. The need for voltage-dependence is illustrated in Fig. 2, where the opening frequencies and open-state durations clearly vary with membrane potential. If we assume, as with the electrically excitable K^+ channel, that bursting might come primarily from reversions to the immediate precursor of the open state (two-agonist bound closed state), the number of interruptions per burst would be

$$I = \beta'/k_{-2} \quad (9)$$

a quantity that does not depend on agonist concentration but might depend on membrane potential.*

The preceding discussion has been somewhat oversimplified. Bursts were defined only for the case in which reopenings are primarily reversions from the final closed state. The agonist system might have rates such that a sojourn in the family of states Cl_1, Cl_2, and Op constituted a burst. In this case, the number of interruptions would obey a more complicated relation in terms of the rates, as can be predicted by following the theory of Colquhoun and Hawkes (1981).

Bursting is one graphic example of a phenomenon which can only be seen with single-channel measurements and can be used to discriminate between contending kinetic models. The types of single-channel records obtainable by the patch-clamp method offer opportunities for sorting out these kinetic schemes in novel ways.

The kinetic schemes of Eqs. (6) and (8) are examples of the type of discrete-state schemes which can be written for channel gating. An immense variety of voltage-clamp, current-noise, gating-current (for Na channels) and dose-response (for agonist-activated channels) data (Adams, 1975; Dreyer *et al.*, 1978) exists. This data gives a detailed account of the dynamics of conformation change, but kinetic models involving numerous unobservable closed states are not likely to be unique. To arrive at definitive kinetic models there is a need for additional constraints based on molecular structure. However, by trying to use stochastic schemes to predict single-channel behavior one can arrive at qualitative features of the fluctuations which can discriminate between candidate schemes.

In gating schemes, such as those of Eqs. (6) and (8), some of the transition rates representing changes in conformation must be strongly voltage-dependent and some rates, representing agonist binding steps must be agonist-concentration dependent. In Fig. 2, the changes in frequency and duration at a fixed concentration of agonist when membrane potential is varied could be consistent with the notion that the conformational rates, α' and β' vary in the transmembrane electric field. However, it is also possible that the agonist binding rates for a charged agonist might be somewhat field-dependent. For the Ca^{++}-sensitive K^+ channel a fit to the probability of channel opening as a function of concentration and potential is obtained by assuming voltage-sensitive Ca^{++} (Wong *et al.*, 1983).

Many statistical properties, such as the distribution of jump durations, the distribution of intervals between jumps, the fraction of time spent in each state, and the flickering or bursting properties that the jumps might have for certain assumptions can be calculated. In a recent theoretical review, Colquhoun and Hawkes (1981) show how many observable consequences of such stochastic models can be predicted by formally solving the appropriate Markov equation for the appropriate statistical quantities. For example, one simple result for the three kinetic models listed is the distribution of channel open times. Each of these

* Nelson and Sachs (1982) have recently suggested that k_{-2} must be voltage-dependent in order for the scheme of Eq. (8) to be consistent with their data on chick muscle.

models has only one open state and consequently the distribution of open times depends only upon the total rate of departure from that state. Thus, the distribution of open times for these three schemes is given by

$$f(t) = (T_{op}^{-1})\exp(-t/T_{op}) \tag{10}$$

where T_{op}, the mean open-state lifetime is given by

$$T_{op} = \beta^{-1} \text{ for EIM}$$
$$T_{op} = (4\beta_n)^{-1} \text{ for HH potassium}$$
or
$$T_{op} = \alpha'^{-1} \text{ for agonist}$$

Although many single-channel experiments on agonist-activated channels report single exponential distributions, there have recently been a number of reports of more complex open-time distributions, requiring two exponentials (Colquhoun and Sakmann, 1981; Jackson *et al.*, 1982; Jackson and Lecar, 1982; Lecar *et al.*, 1982; Cull-Candy and Parker, 1982). Such distributions require either kinetic models with two distinct open states or the coexistence of separate channel populations. In principle, the two possibilities can be distinguished in single-channel experiments designed to see the sequential activation of an individual channel. One kinetic possibility for explaining a two-exponential distribution (Dionne *et al.*, 1978) is to have channel activation by both one and two bound agonists, with the one-agonist state being perhaps less stable than the two, so that it has a shorter lifetime. Such a model would predict a linear combination of two exponentials whose ratio would change with agonist concentration, a feature which should be accurately testable in the near future, using outside-out patches.

Chemical excitation at synapses shows many other complicated kinetic features which can give information about the various drug interactions with the postsynaptic receptor (Rang, 1973; Colquhoun, 1979; Steinbach, 1980; Adams, 1981). Analysis of kinetic models focuses on some of the assumptions that went into Eq. (8). Questions arise such as: Are one, two, or more agonist molecules required to open a channel? Does agonist binding directly cause the opening transition or are there thermal transitions during the drug activated state? What is the nature of drug induced desensitization? Why do some substances behave as partial agonists? Can a simple scheme explain the potentiation of one drug by another? Undoubtedly, single-channel experiments should provide the answers to some of these questions.

IV. APPLICATION TO CHARACTERIZING PARTIAL AND WEAK AGONISTS

Let us illustrate the analysis of kinetic schemes with some single-channel experiments designed to elucidate the differences in behavior of strong, weak, and partial agonists. In particular, we ask whether the kinetic model of Eq. (8) can be tested by single-channel experiments. We wish to find a simple kinetic scheme which shows why some agonists do not produce the full depolarization regardless of concentration applied. Before single-channel experiments, it would have been possible to simply posit that different agonists opened channels to states of different intrinsic conductance, and that weak agonists just produce low-conductance channels. Figure 4 shows that this is not the case. Current jumps are shown

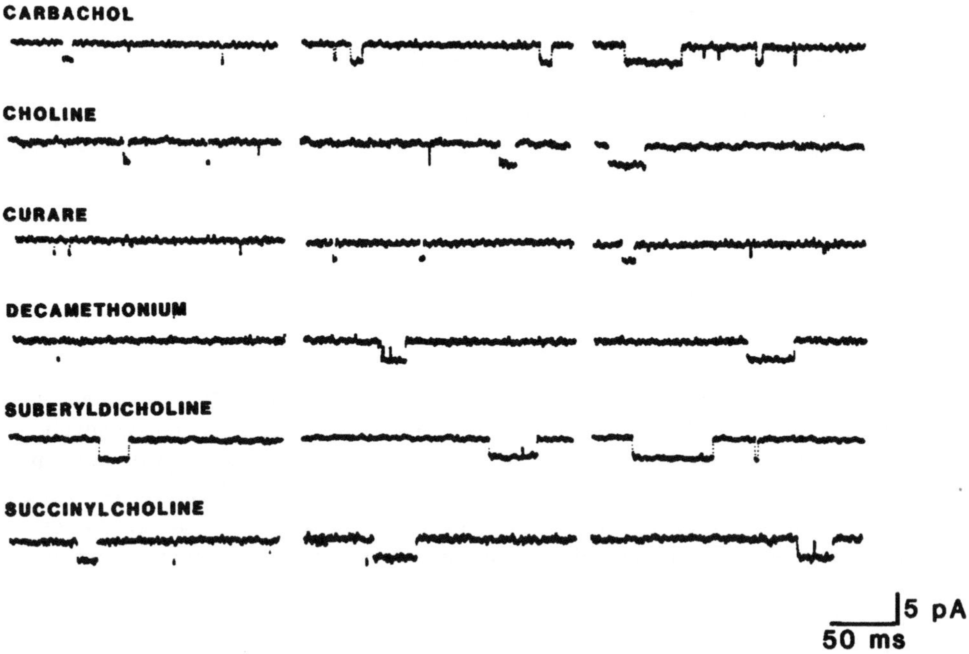

Figure 4. Single-channel currents for strong agonists, weak agonists, and an antagonist. Data recorded a cultured rat muscle, 10^{-6} M concentration of agonist, −100 mV membrane potential. Conductance is the same for all of these agonists, but frequency and durations differ.

for both weak and strong agonists. The jump amplitudes are always the same. The difference resides in the kinetics of channel activation. The reader may be surprised by the inclusion of the classical antagonist curare in the figure, but curare has been shown to act as a weak depolarizing agent in embryonic muscle (Ziskind and Dennis, 1978) and we have shown that curare does indeed activate very short-lived channel openings (Morris *et al.*, 1982).

The scheme of Eq. (8) has a feature which could explain the difference in intrinsic efficacy for the different agonists. Because the last step in the scheme is an isomerization step whose rates are independent of agonist concentration, it represents a rate-limiting step which can determine the maximum probability that a channel can be open. That is, regardless of how many agonist adsorption steps precede this last step, the channel cannot be open for a greater fraction of the time than

$$p_{\text{op,max}} = \beta'/(\alpha' + \beta') \tag{11}$$

the limit of saturating agonist dose. According to this view different agonists produce different activated complexes for which the isomerization rates, and consequently the open-state duty cycle, differ.

Any scheme containing a final isomerization step makes a further prediction. Since β' varies from agonist to agonist, then it is likely that the ratio β'/k_{-2} varies, and there might be agonists for which $\beta' \gg k_{-2}$ so that the consequence of agonist binding is a flickering event wherein a channel reopens several times. Such events had been reported for the agonist suberyldicholine when applied to tissue culture muscle (Nelson and Sachs, 1979; Colquhoun and Sakmann, 1981). A nice way to check for flickering is to work at very low agonist concentration, so that individual "bursts" are well separated in time. This phenomenon is illustrated in Fig. 5 for 5 × 10^{-8} M suberyldicholine.

We should not leave the impression that the isomerization step is the only way to explain the phenomenon of partial agonists. Another kinetic phenomenon, which was discovered in noise experiments on the agonist decamethonium, is self-blocking (Adams and Sakmann, 1978). That is, the agonist can block the lumen of the channel it activates. A scheme for self blocking is

$$Cl_0 \overset{A}{\rightleftharpoons} Cl_1 \overset{A}{\rightleftharpoons} \cdots Cl_{N-1} \underset{\alpha'}{\overset{\beta'}{\rightleftharpoons}} O_p \underset{K_{-b}}{\overset{AK_b}{\rightleftharpoons}} Bl. \tag{12}$$

This scheme too is capable of predicting flickering jumps and partial agonist action. Thus, the two explanations might be indistinguishable qualitatively. However, quantitatively the schemes give different predictions about the agonist concentration-dependence of the rates.

In this discussion of single-channel studies of agonist-induced gating, we can see fragments of a strategy leading to a unified scheme of agonist action. The outline of a general scheme capable of explaining partial agonists, flickering, and concentration-dependent distributions is shown in Fig. 6.

Much more work is needed to convincingly test such schemes. Here, we have attempted to show some ways by which new information about the molecular dynamics of channel proteins is gained from studying the stochastic process directly. A scheme such as that of Fig. 6, while conceptually simple and involving no unlikely molecular transitions, is still kinetically fairly complicated. Yet, one should not be misled by the complexity of the

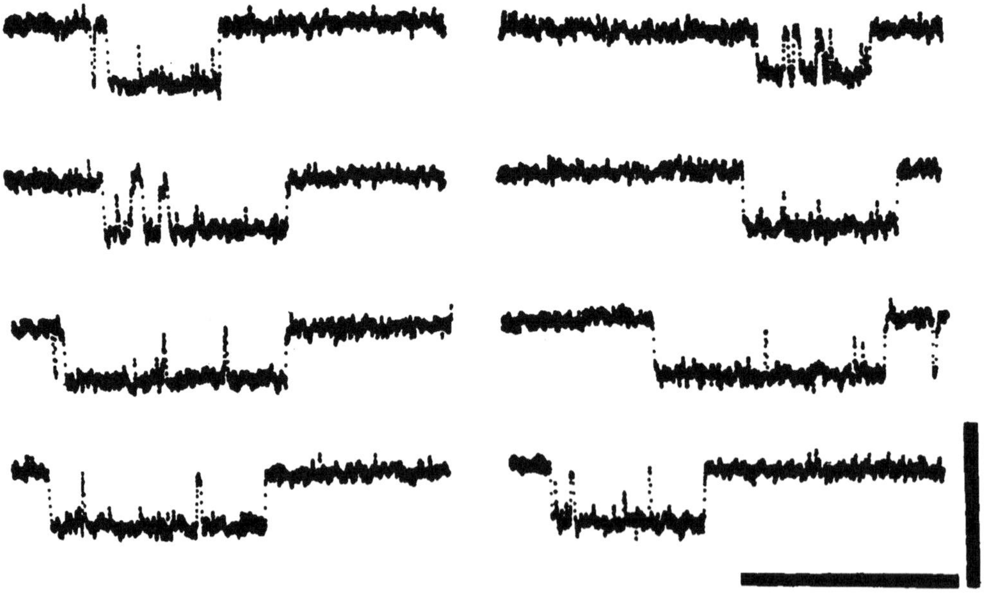

Figure 5. Suberyldicholine-induced single channel recordings made from cultured chick muscle hyperpolarized to −100 mV. Suberyldicholine concentration, 5 × 10^{-8} M. Data filtered at 2 kHz and sample at 1 point/100 μsec. Scale: horizontal, 100 msec; vertical, 10 pA. Although data clearly show fast flickering, all such data are limited by the bandwidth of the recording instrument, and newer data with present state-of-the-art bandwidth (~10 kHz) may reveal additional fast flickering.

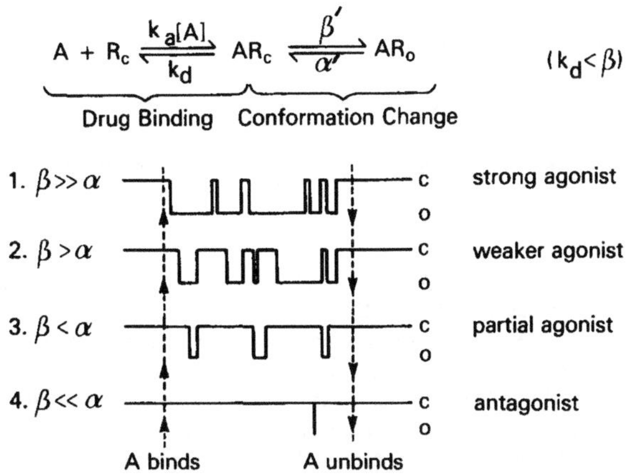

Figure 6. A unified model for partial agonist action. A simplified kinetic scheme is shown as an example of a scheme in which agonist binding leads to stochastic conformation changes. The conclusions would hold true for more complicated schemes involving two binding sites, but the simplified one-agonist scheme illustrates the interplay between association/dissociation and isomerization rates. In this three-state model for drug-induced channel activation and competitive blockade, the receptor–channel complex (*R*) has open (O) and closed (C) conformations. The binding rate (k_a) for the drug (*A*) is not rate-limiting. For simplicity, *A* is small enough to ensure that only one (or none) of the channels under study is bound at any given time, allowing for one to several conformation changes during a single binding episode. The hypothetical traces for Cases 1–4 represent the activity of a single channel, making it useful to think of the rate constants as probabilities. Case 1. Strong agonist: the channel is open during most of the time the drug is bound, but intermittent flickers to the closed state occur. Case 2. Weaker agonist: the channel spends a smaller fraction of its time in the open state, but is open sufficiently long to carry enough current to discharge the membrane voltage. Case 3. Partial agonist: during a single binding episode the channel is closed most of the time, but intermittent flickers to the open state occur, producing a weak agonist action. But because the bound channel spends most of its time in the nonconducting state, the drug has a competitive blocking action against any other agonist present. Case 4. Antagonist: the channel is closed during virtually the entire binding episode. If the channel opens, it closes again so rapidly that no detectable current is conducted.

kinetics and miss the basic simplicity of the phenomenon of channel gating.* The study of single-channel kinetics should lead to the accurate probabilistic equations underlying the various forms of excitability.

ACKNOWLEDGMENT. Dr. Morris was supported in part by a grant from the Scottish Rite Foundation.

REFERENCES

Adams, P. R., 1975, An analysis of the dose-response curve at voltage-clamped frog endplate, *Pflügers Arch.* **360:**145–153.

Adams, P. R., 1981, Acetylcholine receptor kinetics, *J. Membr. Biol.* **58:**161–174.

Adams, P. R., and Sakmann, B., 1978, Decamethonium both opens and blocks endplate channels, *Proc. Natl. Acad. Sci. USA* **75:**2994–2998.

Auerbach, A., and Sachs, F., 1982, Flickering of the nicotinic ion channel to a subconductance state, *Biophys. J.*, submitted.

* New findings keep adding complexity to the pattern of apparent channel transitions. For example, there are reports of open-channel substates having different conductance values (Colquhoun and Sakmann, 1981; Auerbach and Sachs, 1982) and also a report that channel closing may have complex kinetics (Gration *et al.*, 1982).

Boheim, G., Hanke, W., Barrantes, F. J., Eibl, H., Sakmann, B., Fels, G., and Maelicke, A., 1981, Agonist-activated ionic channels in acetylcholine receptor reconstituted into planar lipid bilayers, *Proc. Natl. Acad. Sci. USA* **78**:3586–3590.

Cole, K. S., and Curtis, H. J., 1939, Electric impedance of the squid giant axon during activity, *J. Gen. Physiol.* **22**:649–670.

Colquhoun, D., 1979, The link between drug binding and response: Theories and observations, in: *The Receptors: A Comprehensive Treatise* (R.D. O'Brien, ed.), pp. 93–144, Plenum Press, New York.

Colquhoun, D., and Hawkes, A. G., 1981, On the stochastic properties of single ion channels, *Proc. R. Soc. London B* **211**:205–235.

Colquhoun, D., and Sakmann, B., 1981, Fluctuations in the microsecond time range of the current through single acetylcholine receptor ion channels, *Nature* **294**:464–466.

Conti, F., and Neher, E., 1980, Single channel recording of K^+ currents in squid axons, *Nature* **285**:140–143.

Cull-Candy, S. G., and Parker, I., 1982, Rapid kinetics of single glutamate-receptor channels, *Nature* **295**:410–412.

Dionne, V. E., Steinbach, J. H., and Stevens, C. F., 1978, Analysis of the dose-response relationship at voltage-clamped frog neuromuscular junctions, *J. Physiol.* **281**:421–444.

Donovan, J. J., Simon, M., Draper, R. K., and Montal, M., 1981, Diphtheria toxin forms transmembrane channels in planar lipid bilayers, *Proc. Natl. Acad Sci. USA* **78**:172–176.

Dreyer, F., Peper, K., and Sterz, R., 1978, Determination of dose-response curves by quantitative ionophoresis at the frog neuromuscular junction, *J. Physiol.* **281**:395–419.

Ehrenstein, G., and Lecar, H., 1977, Electrically gated ionic channels in lipid bilayers, *Quart. Rev. Biophys.* **10**:1–34.

Ehrenstein, G., Lecar, H., and Nossal, R., 1970, The nature of the negative resistance in bimolecular lipid membranes containing activity-inducing material, *J. Physiol.* **55**:119 –133.

Ehrenstein, G., Blumenthal, R., Latorre, R., and Lecar, H., 1974, Kinetics of the opening and closing of individual EIM channels in a lipid bilayer, *J. Gen. Physiol.* **63**:707–721.

Gration, K. A. F., Lambert, J. J., Ramsey, R. L., Rand, R. P., and Usherwood, P. N. R., 1982, Closure of membrane channel gated by glutamate receptors may be a two-step process, *Nature* **295**:599–601.

Greco, F. A., 1981, The formation of channels in lipid bilayers by gonococcal major outer membrane protein, *Ph.D. Thesis*, Rockefeller University.

Hamill, O. P., and Sakmann, B., 1981, A cell-free method for recording single-channel currents from biological membranes, *J. Physiol.* **312**:41P–42P.

Hamill, O. P., Marty, A., Neher, E., Sakmann, B., and Sigworth, F. J., 1981, Improved patch-clamp techniques for high-resolution current recording from cells and cell-free membrane patches, *Pflügers Arch.* **391**:85–100.

Horn, R., and Patlak, J., 1980, Single channel currents from excised patches of muscle membrane, *Proc. Natl. Acad. Sci. USA* **77**:6930–6934.

Jackson, M. B., and Lecar, H., 1982, Double-exponential channel current lifetime distributions in cultured nerve and muscle, *Biophys. J.* **37**:3102a.

Jackson, M. B., Lecar, H., Mathers, D. A., and Barker, J. L., 1982, Single channel currents activated by GABA, muscimol and (−) pentobarbital in mouse spinal neurons, *J. Neurosci.*, in press.

Latorre, R., and Alvarez, O., 1981, Voltage-dependent channels in planar lipid bilayers, *Physiol. Rev.* **61**:77–150.

Latorre, R., Ehrenstein, G., and Lecar, H., 1972, Ion transport through excitability-inducing material (EIM) channels in lipid bilayer membranes, *J. Gen. Physiol.* **60**:72–85.

Latorre, R., Vergara, C., and Coronado, R., 1982a, Incorporation of ion channels from biological membranes into planar lipid bilayer, *Biophys. J.*, **37**:170a.

Latorre, R., Vergara, C., and Hidalgo, C., 1982b, Reconstitution in planar lipid bilayers of Ca^{2+}-dependent K^+ channel from transverse tubule membranes isolated from rabbit skeletal muscle, *Proc. Natl. Acad. Sci. USA* **79**:805–809.

Lecar, H., 1981, Single-channel conductances and models of transport, in: *The Biophysical Approach to Excitable Systems* (W.J. Adelman, Jr. and D.E. Goldman, eds.), pp. 109–121, Plenum Press, New York.

Lecar, H., Ehrenstein, G., and Latorre, R., 1975, Mechanism for channel gating in excitable bilayers, *Ann. N. Y. Acad. Sci.* **264**:304–313.

Lecar, H., Morris, C. E., and Wong, B. S., 1982, Single channel recording of weak cholinergic agonists, *Biophys. J.* **37**:313a.

Levitt, D. G., 1980, Electrostatic calculations for an ion channel, *Biophys. J.* **22**:209–220.

Meech, R. W., 1978, Calcium-dependent potassium activation in nervous tissues, *Annu. Rev. Biophys. Bioeng.* **7**:1 –18.

Meech, R. W., and Strumwasser, F., 1970, Intracellular calcium injection activates potassium conductance, *Fed. Proc.* **29**:835.

Misler, S., Kagan, B. L., and Finkelstein, A., 1982, Single channel conductances formed by Diphtheria toxin fragments in planar bilayers, *Biophys. J.* **37**:323a.

Montal, M., Darszon, A., and Schindler, H., 1981, Functional reassembly of membrane proteins in planar lipid bilayers, *Quart. Rev. Biophys.* **14**:1–79.

Morris, C. E., Jackson, M. B., Lecar, H., Wong, B. S., and Christian, C. N., 1982, Activation of individual acetylcholine channels by curare in embryonic rat muscle, *Biophys. J.* **37**:192a.

Mueller, P., Rudin, D. O., Tien, H. T., and Wescott, W. C., 1962, Reconstitution of excitable cell membrane structure in vitro, *Circulation* **26**:1167–1177.

Neher, E., and Sakmann, B., 1976, Single-channel currents recorded from membrane of denervated frog muscle fibers, *Nature* **260**:799–802.

Neher, E., Sakmann, B., and Steinbach, J. H., 1978, The extracellular patch clamp: A method for resolving currents through individual open channels in biological membranes, *Pflügers Arch.* **375**:219–228.

Nelson, D. J., and Sachs, F., 1979, Single ionic channels observed in tissue-cultured muscle, *Nature* **282**:861–863.

Nelson, D. J., and Sachs, F., 1982, Agonist and channel kinetics of the nicotinic acetylcholine receptor, *Biophys. J.* **37**:327a.

Nelson, N., Anholt, R., Lindstrom, J., and Montal, M., 1980, Reconstitution of purified acetylcholine receptors in planar lipid bilayers, *Proc. Natl. Acad. Sci. USA* **77**:3057–3061.

Pallota, B. S., Magleby, K. L., and Barret, J. N., 1981, Single channel recordings of Ca^{2+}-activated K^+ currents in rat muscle cell culture, *Nature* **293**:471–474.

Parsegian, A., 1969, Energy of an ion crossing a low dielectric membrane: Solutions to four relevant electrostatic problems, *Nature* **221**:844–846.

Rang, H. P., 1973, Acetylcholine receptors, *Quart. Rev. Biophys.* **7**:283–399.

Schindler, H., and Rosenbusch, O., 1978, Matrix protein from *Escherichia coli* outer membranes forms voltage-controlled channels in lipid bilayers, *Proc. Natl. Acad. Sci. USA* **75**:3751–3755.

Schindler, H., and Quast, U., 1980, Functional acetylcholine receptor from *Torpedo marmorata* in planar membranes, *Proc. Natl. Acad. Sci. USA* **77**:3052–3056.

Stein, W. D., 1968, Turnover numbers of membrane carriers and the action of the polypeptide antibiotics, *Nature* **218**:570–571.

Steinbach, J. H., 1980, Activation of nicotinic acetylcholine receptors, in: *The Cell Surface and Neuronal Function* (C.W. Cotman, G. Poste, and G.L. Nicolson, eds.), pp. 119–156, Elsevier/North-Holland, Amsterdam.

Wong, B. S., Lecar, H., and Adler, M., 1982, Differentiation of the two potassium channels in clonal anterior pituitary cells by the patch clamp technique, *Biophys. J.* **37**:325a.

Wong, B. S., Lecar, H., and Adler, M., 1983, Single calcium-dependent potassium channels in clonal anterior pituitary cells, *Biophys. J.*, **42**:109–114.

Ziskind, L., and Dennis, M. J., 1978, Depolarising effect of curare on embryonic rat muscles, *Nature* **276**:622–623.

8

Effects of Voltage-Dependent Ion-Conduction Processes on the Complex Admittance of Single Skeletal Muscle Fibers

L. E. Moore

I. CONDUCTANCE MEASUREMENTS IN MUSCLE FIBERS

The analysis of the conductance properties of the skeletal muscle membrane has proven to be considerably more difficult than earlier investigations of axon excitability. The principal reason for this difficulty is the unique internal membrane structure of muscle, consisting of an infolding of the surface membrane to form a transverse tubular membrane structure, as well as the sarcoplasmic reticulum, which appears to be electrically coupled to the tubular system. A voltage-clamp measurement of this complex system is always beset with the difficulty of charging the tubular membrane in a uniform manner so as to allow measurement of the associated membrane currents. The voltage-clamp analyses that have been done probably have had varying amounts of uncontrolled tubular membrane current. As a result, the quantitative analyses of these experiments may not be entirely valid.

Since the muscle membrane structure is complicated and its conductance highly nonlinear, an alternative approach lies in a partial characterization of the ionic processes by means of linear analysis (Moore *et al.*, 1980). The most straightforward approach is perhaps a small-signal step-clamp analysis, which would elicit a relaxation response without significant change in the kinetic parameters from their initial values. A small-step analysis for different membrane potentials would thus provide a linear description of the ionic currents analogous to the linearized Hodgkin and Huxley (1952) equations for the squid axon. The difficulty with this approach is that multiple relaxation times are often difficult to resolve in responses to step functions. Admittance or impedance measurements are far more sensitive for parameter estimation and are therefore the methods of choice (Eisenberg, 1967).

Impedance measurements have been elegantly applied to biological tissues, including muscle, by K. S. Cole and his co-workers (Cole, 1941; Cole, 1968; Cole and Baker, 1941).

L. E. Moore • Department of Physiology and Biophysics, University of Texas Medical Branch, Galveston, Texas 77550.

Until recently these measurements have required relatively long times and therefore are not easily implemented during short membrane depolarizations, which activate the voltage-dependent ionic conductances. Using "white noise" techniques, Poussart *et al.* (1977) have done impedance and admittance measurements on voltage-clamped squid axons within the minimum theoretical time possible for the frequency band investigated (Fishman *et al.*, 1979). These results have shown that the voltage-dependent ionic processes are clearly present in admittance functions measured with voltage perturbations of less than 1 mV. Thus, impedance and admittance functions, determined by a linear approximation to a highly nonlinear excitable membrane system, not only reflect the passive membrane properties, but also include the behavior of the specific sodium and potassium conductance processes. This is possible for two reasons: (1) The sodium and potassium conductance processes are activated during a quasi-steady state and (2) the admittance measurements can be done during a standard voltage-clamp measurement by superimposing a perturbation of less than 1 mV on a large step-clamp command potential. Thus, a linear frequency analysis is possible at different membrane potentials, and it provides linear kinetic parameters, which are themselves voltage dependent. Although this analysis can give information over a wide potential range, it cannot predict the nonlinear response to a large step. In the case of the Hodgkin–Huxley model, it would be possible to determine all of the rate constants but not the multiplicative nature of the *m, h,* and *n* variables.

Previous impedance measurements on muscle have emphasized passive equivalent circuit models which describe the complex membrane system of muscle (Eisenberg, 1967; Falk and Fatt, 1964; Fatt, 1964; Freygang *et al.*, 1967; Schneider, 1970; Valdiosera *et al.*, 1974a). Of particular importance has been the use of models of the transverse tubular system to understand the propagation of action potentials radially into the muscle fiber (Adrian *et al.*, 1969a) and the excitation–contraction coupling process that follows (Constantin, 1975; Peachey and Adrian, 1973). The purpose of this paper is to demonstrate the effect of the voltage-dependent ionic conductances on the muscle-membrane admittance and to describe the admittance behavior over a limited potential range. The effects of the sodium and potassium ion admittances were separated by ion substitution and by partially blocking the potassium conductance with tetraethylammonium ions. The effects of the potassium-ion admittance were observed by replacing all the external sodium ions with tetramethylammonium ions. The measurements reported here appear consistent with some of the earlier results since the ionic conductances are essentially turned off at the large negative resting potential (< − 90 mV) of frog skeletal muscle. However, step-clamp depolarizations that lead to a slight activation of the voltage-dependent ionic conductances bring about significant changes in the magnitude and phase functions.

II. VASELINE-GAP VOLTAGE-CLAMP TECHNIQUE

The Vaseline-gap voltage-clamp for single skeletal muscle fibers (Frankenhaeuser *et al.*, 1966) previously described (Moore, 1972) was used for these experiments, with the addition that the current was recorded by measuring the potential differentially across a 10 kΩ resistor in series with the output of the clamp amplifier.

The single fibers used in these experiments were carefully dissected from the semitendinosus muscle of the frog, *Rana pipiens*, and mounted in a potassium phosphate solution, which contained 67 mM K_2HPO_4 and 33 mM KH_2PO_4 buffered at pH 7.2. The mounting procedure and seal application were complete within 3 min, after which the solution in the recording pool was changed to a normal Ringer's solution. The fibers were maintained at their normal slack length during the entire mounting period, then cut at each end, 300–400 μm from the seal edge. The fibers were rigidly held by the vaseline seals at a tension that maintained a normal sarcomere length of about 2 μm, as measured by laser diffraction. Fibers that were excessively shortened (< 1.8 μm) were specifically avoided in these experiments. A careful dissection and maintenance of sarcomere length appear to assure preservation of the delayed potassium currents in the cut-fiber preparation (Kovacs and Schneider, 1978; Vergara *et al.*, 1978).

Conventional step-voltage-clamp measurements were done on all fibers to assure the presence of transient inward and steady state outward currents. At the temperature at which all of these measurements were done, 12°C, uncontrolled tubular currents have been found (Mandrino, 1977) in addition to the presumably controlled surface-membrane currents. This finding is consistent with the observation of Vergara and Bezanilla (1979), who reported that fluorescent signals from the tubular membranes indicate a significant uncontrolled tubular sodium current when the surface membrane is voltage clamped. Although lowering the temperature does remove the obvious appearance of the tubular current (Hille and Campbell, 1976), it seems unlikely that the tubular membrane can be adequately clamped at any temperature during the sodium transient current. The graded appearance of voltage-clamp currents cannot be taken as a measure of the adequacy of the clamp since differential changes in the tubules, which can occur because of deterioration or partial inactivation of the membrane, or because of low temperature, may allow sufficient control to prevent the marked threshold behavior of the tubular system seen in physiologically normal fibers at about 10°C.

The external solutions were as follows: normal Ringer: 120 mM NaCl, 2.5 mM KCl, 2.0 mM $CaCl_2$, 1 mM Na_2HPO_4, and 0.5 mM NaH_2PO_4; TMA-Ringer: 120 mM tetramethylammonium chloride in place of NaCl; TEA-Ringer: 10 mM tetraethylammonium chloride added to the normal Ringer. The pH of all solutions was adjusted to 7.2.

III. ADMITTANCE-MEASUREMENT METHODS

The basic principles of the admittance measurements used in these experiments have been described elsewhere (Poussart *et al.*, 1977; Fishman *et al.*, 1979). The additional features described here were originally developed in squid-axon experiments with Drs. D. Poussart and H. M. Fishman. In contrast to the squid-axon experiments, the present series of experiments was implemented entirely on an LSI-11 microcomputer system with 12-bit analog–digital (A/D) and digital–analog (D/A) converters. The admittance measurements were made in combination with the step-voltage-clamp method by summing the voltage step with a 1 mV peak-to-peak deterministic "white noise" signal. Voltage-clamp pulses of 1600 msec duration were used. The current responses during the last 800 msec were used for the admittance determination. This protocol established a quasi-steady state for the response both to the step and the small-perturbation inputs. The deterministic "white noise" signal was computer generated by summing 400 sine waves having specified (usually unity) amplitudes, equally spaced frequencies, and random phases. The real-time function contained 1024 discrete values and was repeated once for a total of 2048 points, which were read from a digital–analog converter at 800 μsec intervals. The analog signal was high-pass filtered at 500 Hz with an active Butterworth filter.

The response of the preparation was recorded through a low-pass filter of the same type as the above. The data were recorded with an analog–digital converter in synchrony with the D/A output at a rate of one sampling per 800 μsec. This interval corresponds to a frequency range of 1.25–500 Hz. The output of the current amplifier was high-pass filtered at 1 Hz (6 dB/octave). The reference signal used in computing the transfer functions was always measured after passing through the low- and high-pass filters. A Fast Fourier Analysis routine (Bendat and Piersol, 1971) was used for computation of the real and imaginary components from the time data.

In order to improve the dynamic range of the measurement, the "white-noise" source was modified by changing the amplitude of sine waves to a Lorentzian function with a corner at the 67th frequency (84 Hz for the 500 Hz band). The response of the voltage-clamped fiber to this source was relatively flat over the frequency range measured, thus allowing a much improved dynamic response. This source will be called "Lorentzian noise."

A further improvement in the admittance was made by a procedure termed "coherence elimination" (Fishman *et al.*, 1981). In this procedure two pulses with identical amplitudes on which the noise was superimposed were applied to the voltage clamp approximately 5 sec apart; however, during the second pulse the Lorentzian noise was inverted. The second response was subtracted from the first, thus removing any coherent trends while adding the relaxation responses. The system was calibrated by measuring the transfer function of a model system with a membrane resistance of 6 MΩ. The estimated surface membrane area was 3.4×10^{-4} cm^2.

IV. EXPERIMENTAL RESULTS

The most striking result of these experiments is the observation of ionic admittances in a voltage-clamped single muscle fiber. The three admittance measurements shown in Fig. 1 show the marked sensitivity of the ionic admittance to the membrane potential. Curves labeled "−95" were taken near the normal resting potential of muscle, −95 mV. The magnitude and phase behavior at this membrane potential are similar to that reported by others (Falk and Fatt, 1964; Freygang *et al.*, 1967; Schneider, 1970; Valdiosera *et al.*, 1974a,c). However, a depolarization to −35 mV, seen in the curves labeled "−35," brings about a decrease in the magnitude at low frequencies and a more rapid change in the phase. The observed decrease in the magnitude at low frequencies suggests that the total ionic conductance has decreased with depolarization. This occurs because a depolarizing pulse activates the sodium-conductance system sufficiently to give a net decrease in the total conductance as measured by the magnitude at low frequencies. A further depolarization to −21 mV, curves labeled "−21," shows an antiresonance in the magnitude and a sharp transition region in the phase, characteristic of antiresonance behavior (Fishman *et al.*, 1979). These results clearly demonstrate that the ionic kinetic processes of the muscle membrane are detectable by small-signal linear analysis using perturbations below 1 mV.

The durations of the depolarizations used in the admittance determinations are much longer than typically used in a voltage-clamp analysis of the sodium conductance, 1600 msec vs. a few msec. Since the measurement period is from 800–1600 msec the sodium conductance is presumably inactivated to a constant steady state level. The steady state inactivation level for all depolarizations investigated is clearly not zero, for otherwise no sodium-dependent process would be observed (see below). Admittance measurements at high frequencies and shorter pulse durations show qualitatively the same behavior but significant quantitative differences.

The negative-conductance behavior of the sodium system is dramatically illustrated in Fig. 2, where much of the delayed potassium current has been blocked by tetraethylammonium (TEA) ions (Stanfield, 1970). The three superimposed admittance curves shown in

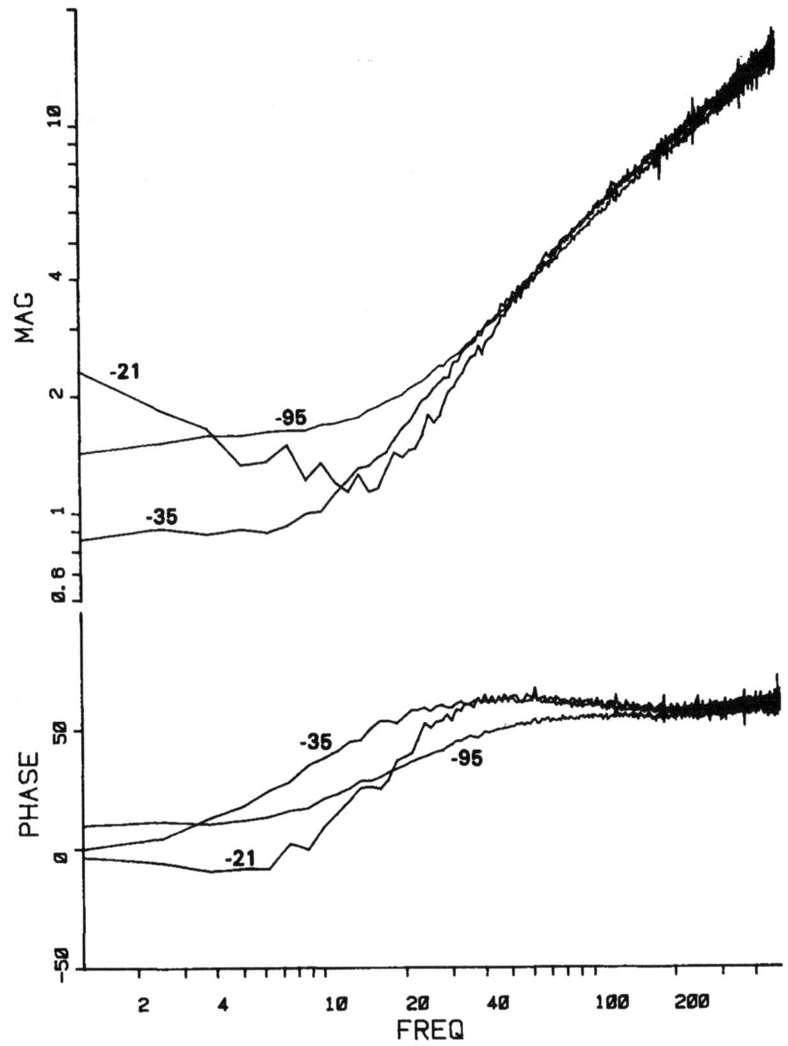

Figure 1. Membrane admittance in normal Ringer's solution. Ordinate, magnitude in mS/cm^2, phase angle in degrees; abscissa, frequency in Hz. Three superimposed measurements are shown, each of which is taken at the indicated membrane potential in mV. The holding potential was -75 mV. In this and all subsequent figures the magnitude calibration is based on an estimated gap surface membrane area in the gap of 3.4×10^{-4} cm^2. The numbers associated with each curve represent the membrane potential in mV during the admittance measurement.

Fig. 2 have behavior generally similar to Fig. 1; however, the depolarized membrane potentials, $+5$ mV and -25 mV, are quite different. The depolarization to 5 mV, compared to the holding level of -75 mV, leads to a large decrease in the magnitude at low frequencies and a slight increase at high frequencies. The pronounced bump visible in the phase around 30–40 Hz is always seen in the phase for these conditions.

The differences between Figs. 1 and 2, caused by removal of most of the potassium conductance, are consistent with similar observations from squid axons as follows: (1) the decrease in the admittance magnitude with moderate depolarization is greater and present

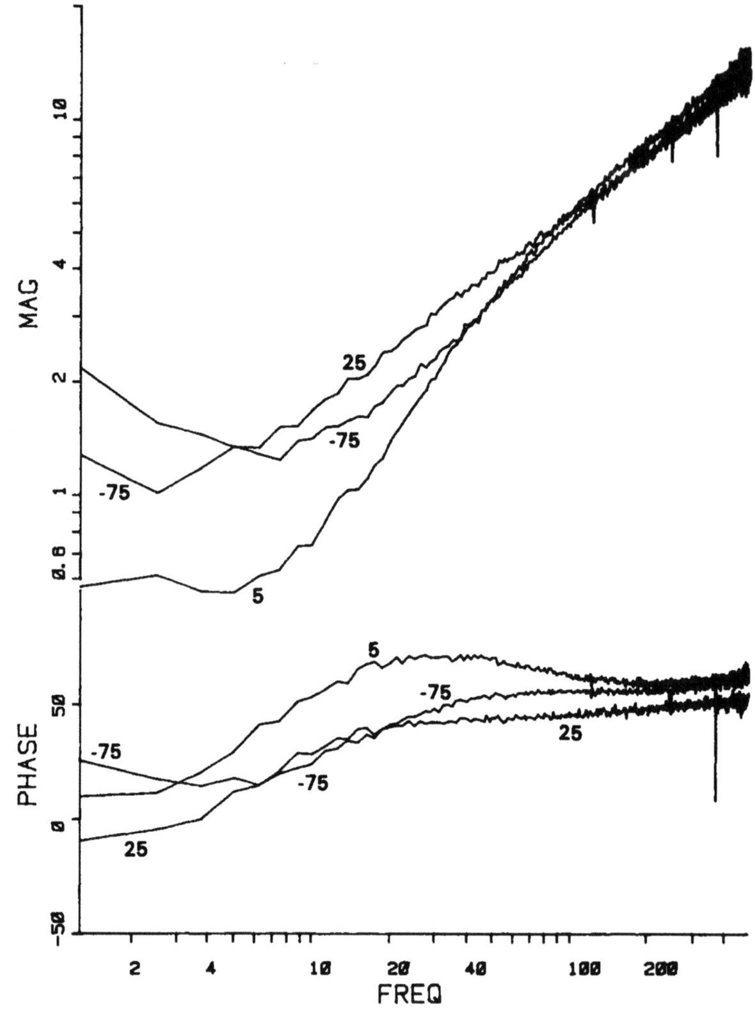

Figure 2. Complex admittance in TEA Ringer's solution. Ordinate, magnitude in mS/cm^2, phase angle in degrees; abscissa, frequency in Hz. The three superimposed admittance measurements were taken at the three indicated membrane potentials in mV.

over a wider potential range if the negative-conductance increase is not offset by a corresponding potassium conductance increase, (2) the level of depolarization at which an antiresonance appears is greater in the presence of TEA because of the increased effect of the negative sodium conductance, and (3) the frequency of the minimum in the antiresonance is lower when K conductance is reduced.

The potential dependence of the ionic admittances for three conditions are shown in the "three-dimensional" plots of Figs. 3, 4, and 5, in which the traces have been displayed diagonally. Note that the voltage scale is linear except at the highest depolarization. Fig. 3 shows the combined admittances in the normal Ringer's solution of the ionic and passive processes for five membrane potentials, −75, −55, −45, −35, and −25 mV. A clear antiresonance appears at −25 mV. Figure 4 shows that in the presence of TEA no pronounced antiresonance is seen for this potential range. Finally, Fig. 5 shows little or no decrease in the magnitude for any depolarization if the external sodium ion concentration is replaced with tetramethylammonium (TMA) ions. This point is somewhat difficult to visualize in the plot shown; however, careful consideration of the third dimension (voltage axis) will confirm the conclusion. In the sodium-free solution the magnitude at low frequencies increases with depolarization and shows relatively little antiresonance behavior for the potentials shown. The phase functions in Fig. 5 show less variation with frequency as the level of depolarization increases than those seen in the solutions containing sodium.

Figures 6 and 7 are similar to Figs. 3 and 4 except that the potential axis is more

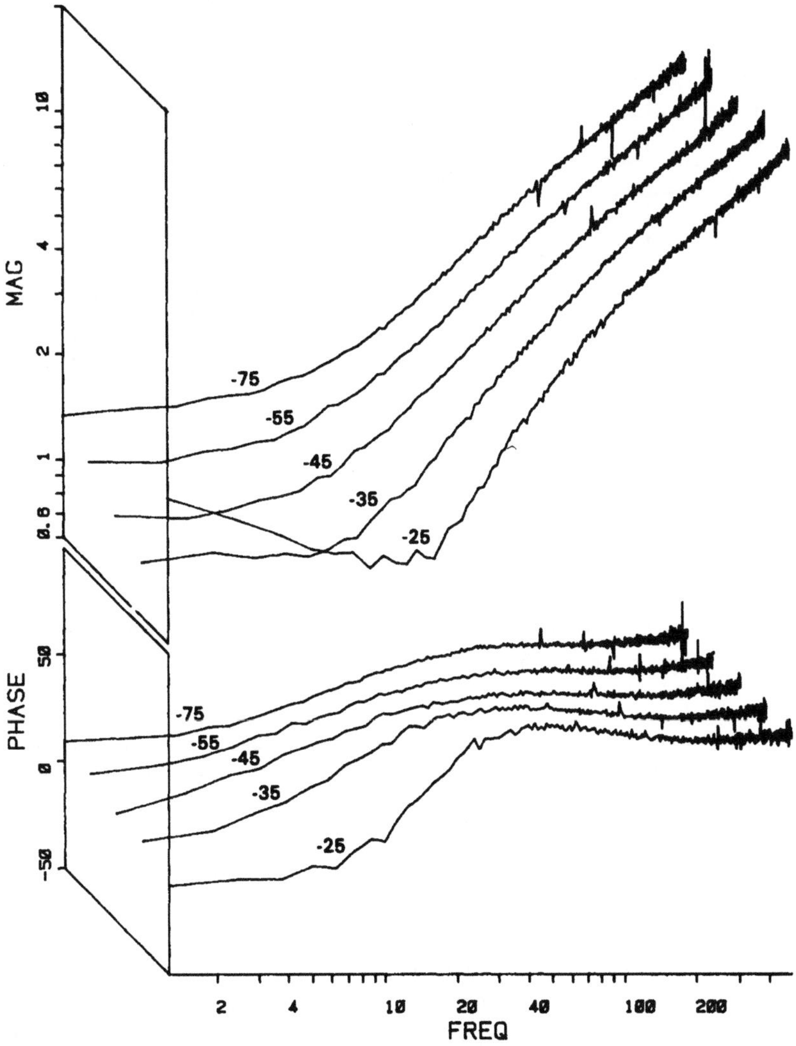

Figure 3. Three-dimensional plots of membrane admittance at different membrane potentials in normal Ringer's solution. Ordinate, magnitude in mS/cm^2, phase angle in degrees; abscissa, offset frequency in Hz. The admittance plots for depolarizations from the −75 mV holding level are offset at 45° in equal increments. The beginning point of each curve indicates the 1.25 Hz values. Note that the magnitude and phase are offset equally at 45°.

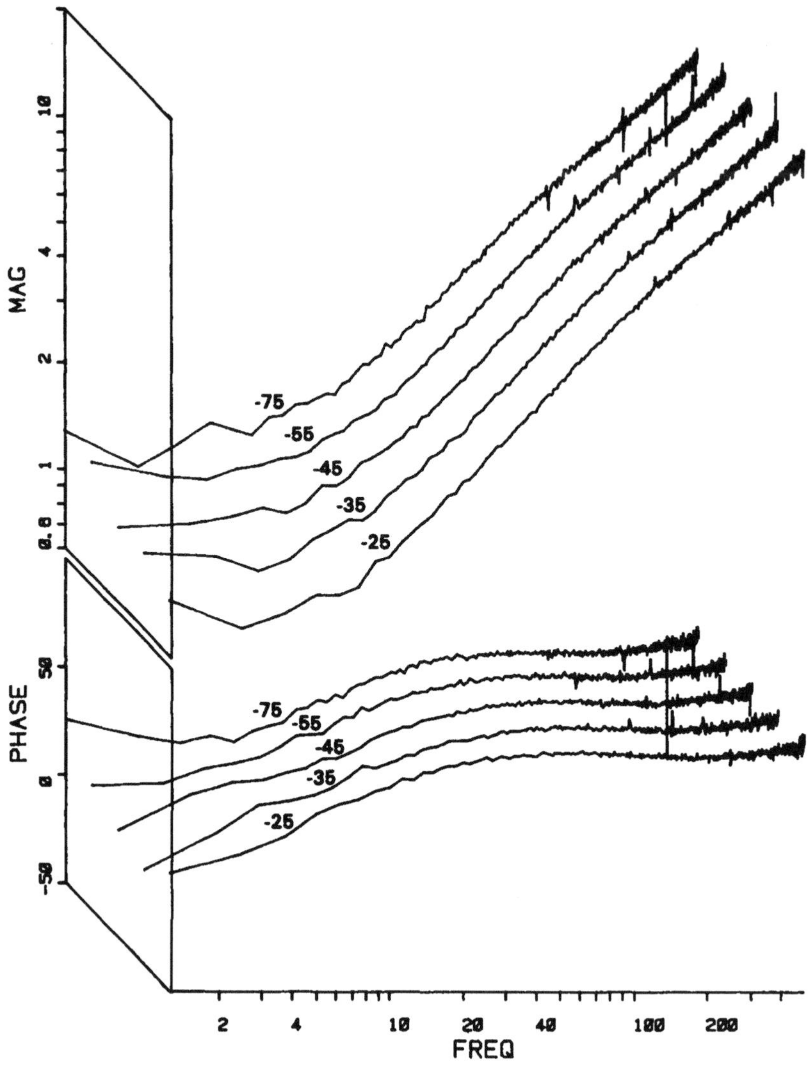

Figure 4. Three-dimensional plots of complex admittance in TEA Ringer's. Ordinate, magnitude in mS/cm^2, phase angle in degrees; abscissa, offset frequency in Hz. The plots in this figure are at the same membrane potentials and offsets as in Fig. 3.

nonlinear and many more admittance plots are presented. Visualization of the magnitude and phase surfaces show the graded changes that occur with potential and frequency. This presentation shows that both the magnitude and phase functions are extremely useful in interpreting the data. At moderate depolarizations the decrease observed in the magnitude indicates a negative sodium conductance, which manifests itself in the phase function by a more pronounced maximum and steeper transition region from low to high frequencies. The phase function alone would have been more difficult to interpret; however, for moderate depolarizations the form of the phase is more sensitive to changes in the membrane potential and therefore relatively more useful for equivalent-circuit parameter estimation than the magnitude.

The effect of ion substitution and partial blockage of the K-conducting system is illustrated in Fig. 8. The curves labeled "K + Na" indicate a pronounced antiresonance at a membrane potential of −25 mV. With the addition of TEA to the Ringer's solution, the curves labeled Na show that the antiresonance of the magnitude is essentially abolished and the low-frequency magnitude decreases. The curves labeled K indicate that, in a sodium-free solution, the magnitude at low frequencies is greatly increased due to removal of the negative sodium conductance. The phase functions for these three solutions show that the sodium system generally increases the angle for most frequencies. Figures 9 and 10 show similar data for membrane potentials of −35 and −45 mV. In Fig. 10 the admittance is nearly unaffected by TEA, whereas blockage of the Na conductance caused marked changes. As the rest potential was approached, ion blockage or substitution had relatively little effect

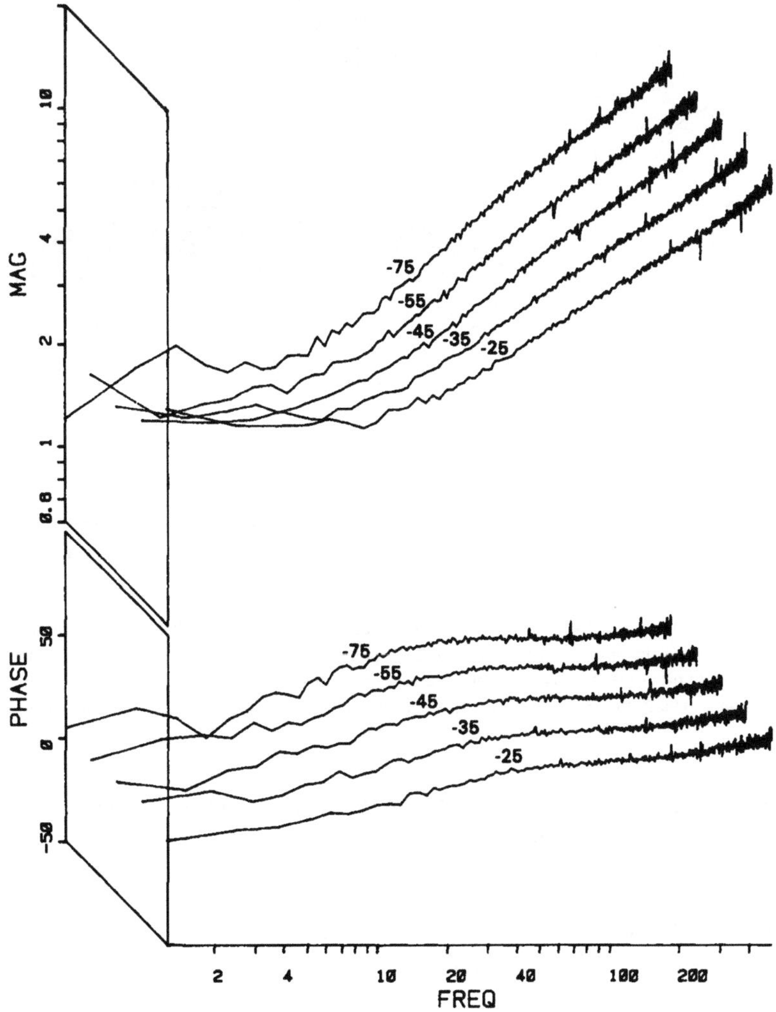

Figure 5. Three-dimensional plots of complex admittance in TMA Ringer's. Ordinate, magnitude in mS/cm^2, phase angle in degrees; abscissa, offset frequency. The plots in this figure are at the same membrane potentials and offsets as in Figs. 3 and 4.

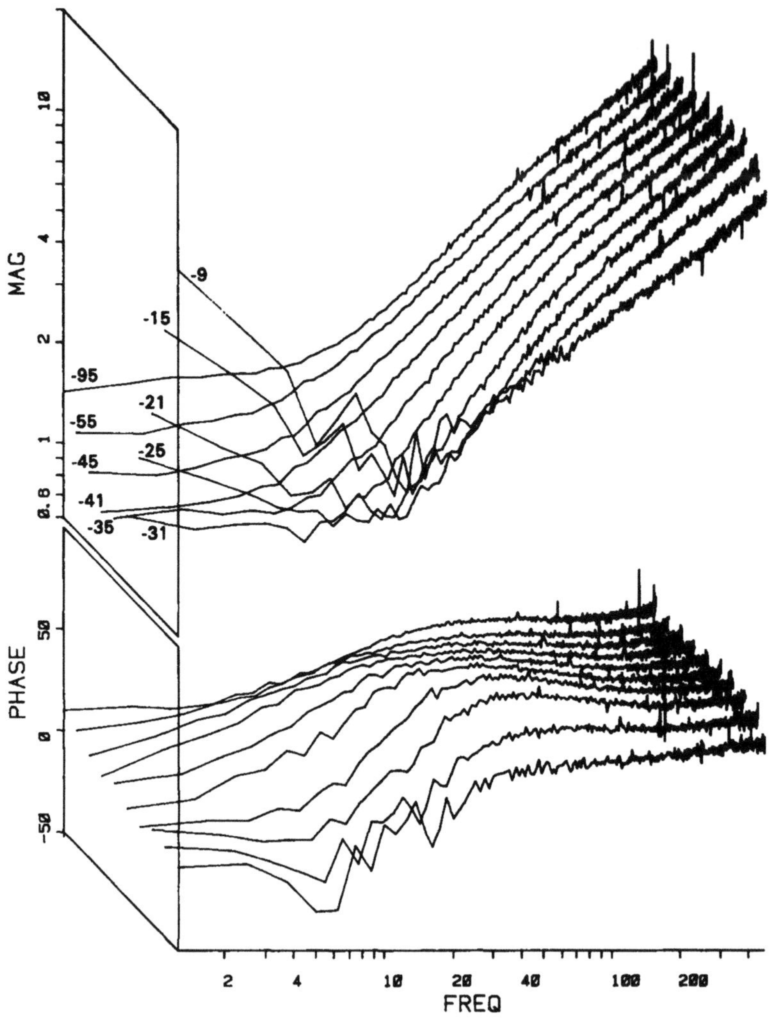

Figure 6. Three-dimensional plots of membrane admittance in normal Ringer's solution. Ordinate, magnitude, in mS/cm^2, phase angle in degrees; abscissa, offset frequency in Hz. The ten complex admittance plots at the indicated membrane potentials are offset in equal increments similar to Fig. 3. The low-frequency surface described by the magnitude shows a drop followed by a rise and the appearance of an antiresonance at about -25 mV.

on the admittance curves. This latter result suggests that the voltage-dependent ion conduction systems are nearly turned off at the level of the resting potential.

V. SIMULATION OF THE CONDUCTANCES BY A LUMPED-CIRCUIT MODEL

The experiments reported in this chapter show that the voltage-dependent specific sodium and potassium conductances are reflected in the complex admittance. The addition of voltage-dependent ionic conductances to the normally passive circuit elements clearly complicates the analysis; however, it also provides a means of characterizing the conductive properties of the tubular and surface membranes.

The effects of the voltage-dependent ion conductance processes on the muscle admittance are here simulated by a lumped-circuit model of the surface and tubular excitable membranes. The parameters in the model are based on experimental data, principally those of Adrian *et al.* (1970) and Adrian and Peachey (1973). The model is composed of an access resistance of 200 Ω cm^2 in series with the lumped tubular system, which has an admittance three times that of the surface membrane. (Capacitance values are generally normalized to external surface area.) In the passive lumped model this would give a capacitance for the

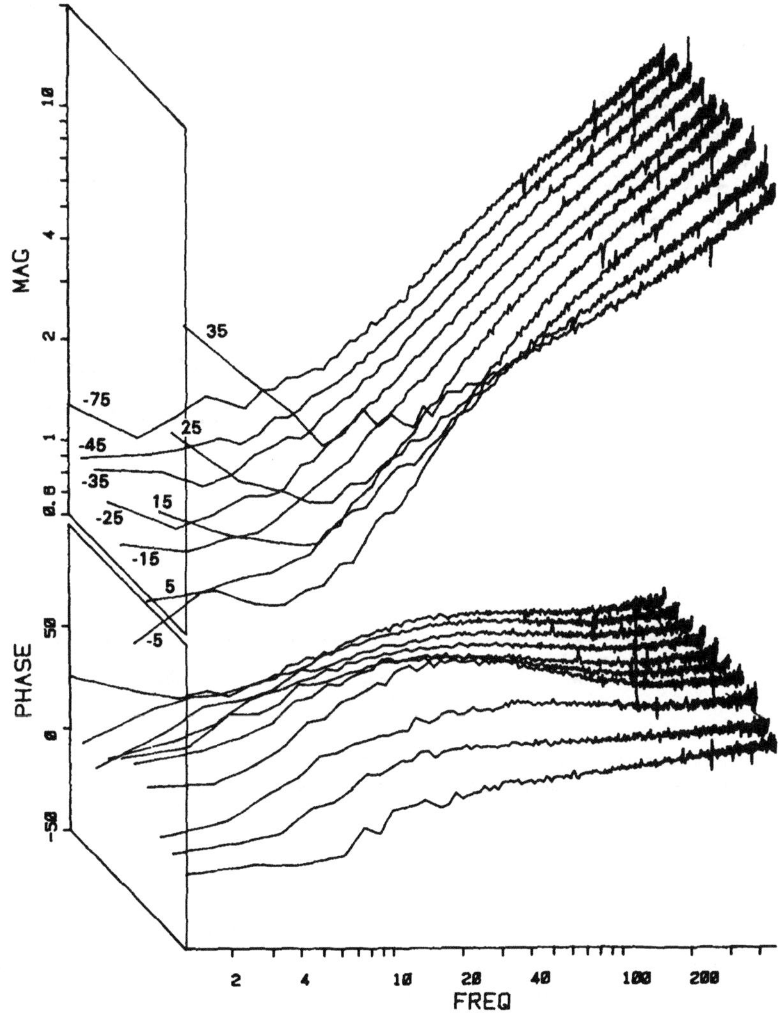

Figure 7. Complex admittance in TEA Ringer's plotted in three dimensions. Ordinate, magnitude in mS/cm^2, phase angle in degrees; abscissa, offset frequency in Hz. The magnitude and phase plots at the indicated membrane potentials are offset as in Fig. 6. Note that the low-frequency magnitude surface decreases followed by an increase and a clear appearance of an antiresonance at the more depolarized potential of about 15 mV than seen in normal Ringer's solution (Fig. 6).

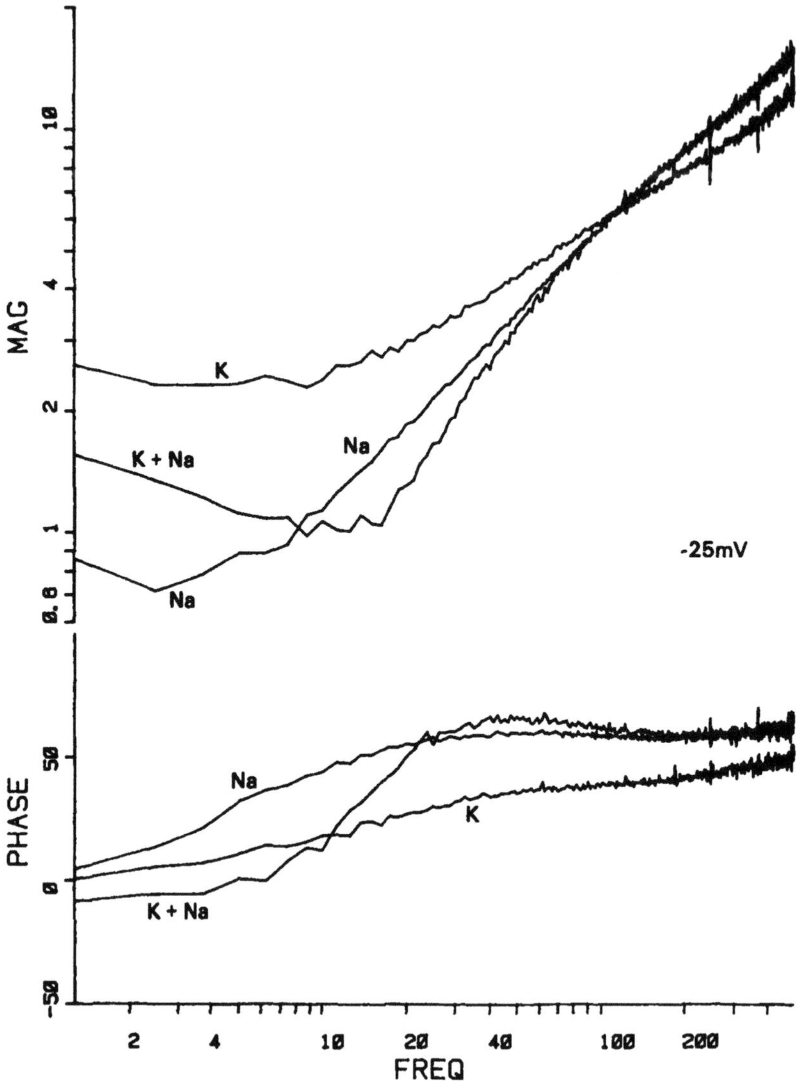

Figure 8. Complex admittance in different solutions at a membrane potential of −25 mV. Ordinate, magnitude in mS/cm^2, phase angle in degrees; abscissa, frequency in Hz. The three superimposed curves are taken at the same membrane potential in normal Ringer's solution labeled K + Na, TEA Ringer's labeled Na, and TMA Ringer's labeled K.

tubules of 6 μF per cm^2 of surface area for a surface capacitance of 2 μF/cm^2 (Gage and Eisenberg, 1969). In addition, the tubular membrane contains circuit elements for the linearized Hodgkin–Huxley equations (Chandler *et al.*, 1962). In the hybrid model of Adrian and Peachey (1973), the limiting ion conductances of the tubular membrane are more than an order of magnitude less than the surface-membrane values. The experiments of Jaimovich *et al.* (1976) suggest that the density of tubular TTX-sensitive sites is about half of the surface density. Therefore, as a first approximation the tubules are simply scaled to the surface membrane by a factor of three to account for the total membrane capacitance.

Variation of this factor affects the quantitative behavior of the admittance, but does not alter the general behavior of the model discussed below.

The inset of Fig. 11 shows the total circuit with branches containing the membrane resistance (R), membrane capacitance (C), a resistance (R_h) and inductance (L_h) in series representing the h or n process (or both), a resistance (R_m) and capacitance (C_m) in series representing the m process, and the tubular system. The values chosen for the voltage-dependent ion-conduction elements are consistent with a moderate depolarization in which the sodium conductance is partially activated. The potassium conductance in this simplified model is arbitrarily set equal to zero. The conductance of this model is composed of the

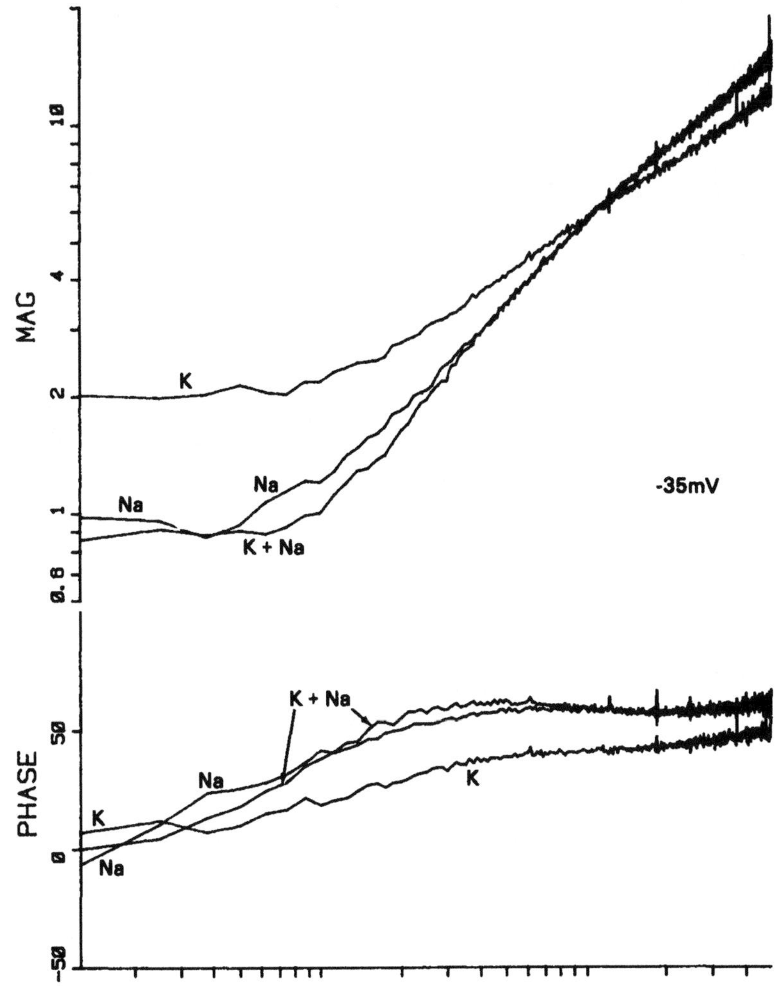

Figure 9. Complex admittance in different solutions at a membrane potential of -35 mV. Ordinate, magnitude in mS/cm^2, phase angle in degrees; abscissa, frequency in Hz. The three superimposed curves are taken at the same membrane potential in normal Ringer's solution (labeled K + Na), TEA Ringer's (labeled Na), and TMA Ringer's (labeled K).

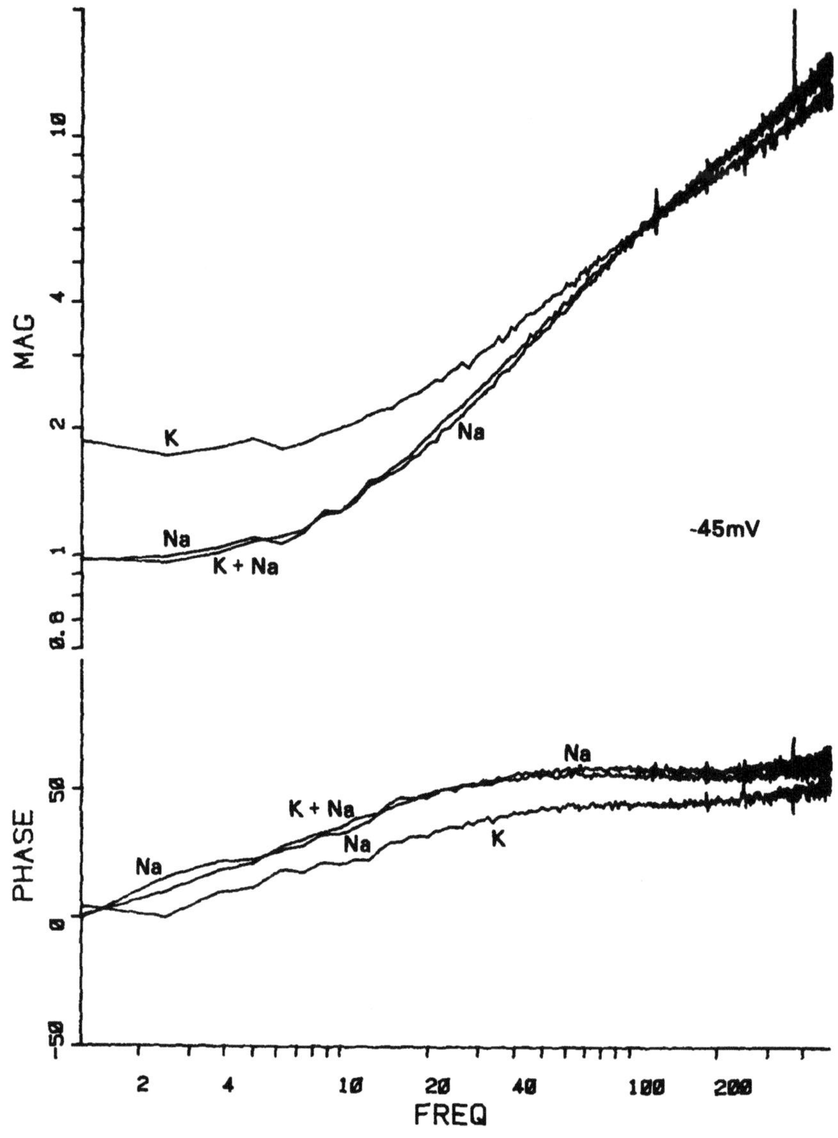

Figure 10. Complex admittance in different solutions at a membrane potential of −45 mV. Ordinate, magnitude in mS/cm^2, negative phase angle in degrees; abscissa, frequency in Hz. The three superimposed curves are taken at the same membrane potential in normal Ringer's solution (labeled K + Na), TEA Ringer's (labeled Na), and TMA Ringer's (labeled K).

leakage conductance, the steady state sodium conductance and a term that reflects the negative conductance of the sodium system. For a complete discussion of these parameters the reader is referred to the papers by Fishman *et al.* (1977), and Mauro *et al.* (1970). The time constant for the *m* process, τ_m, is related to the circuit parameters by the relation, $\tau_m = R_m C_m$, which in this case is 0.8 msec. The τ_h is given by L_h/R_h and is 3 msec. These values for τ_m and τ_h are consistent with the data of Adrian *et al.* (1970) for a moderate depolarization. The value of 8 kΩ cm² for the total resistance is higher than the usual 3–4 kΩ cm² for resting muscle because of the effect of a negative conductance, leading to an increase in R.

If the two branches for m and h are removed, the equivalent circuit is essentially that of the typical lumped-circuit model. Qualitatively, this passive form of the model shows behavior similar to that described by Valdiosera *et al.* (1974b) in their simulation of the lumped model. However, the presence of the specific active ion-conduction elements sig-

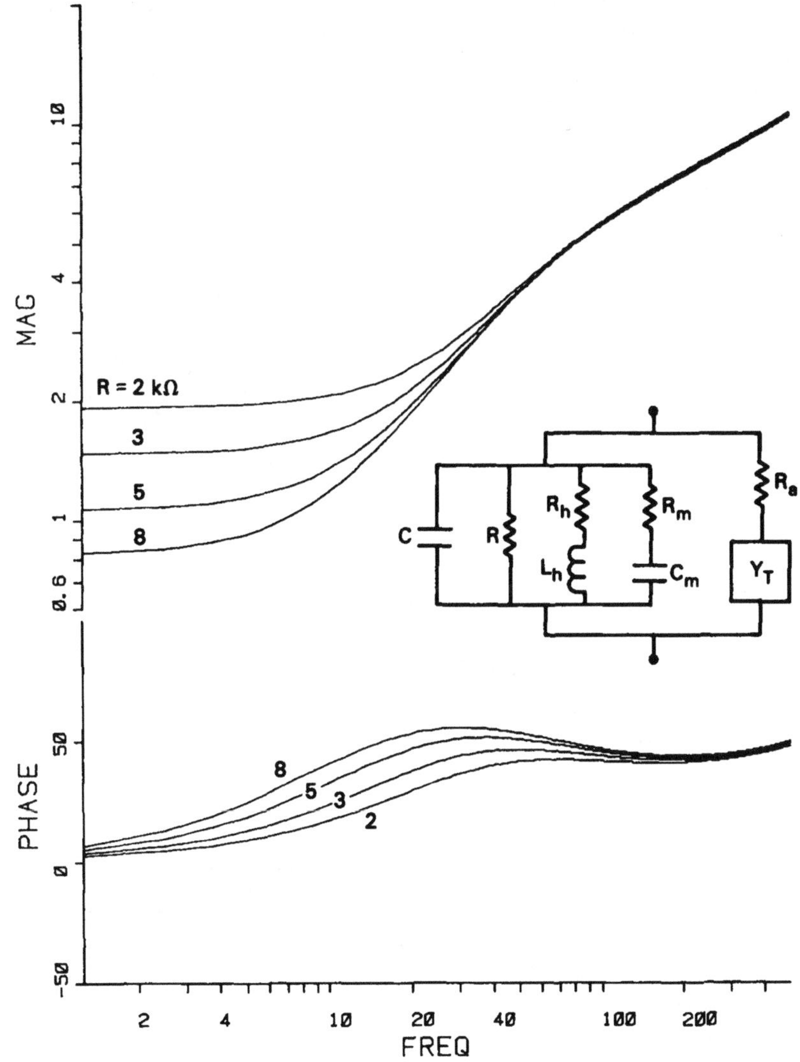

Figure 11. Complex admittance of Hodgkin–Huxley equivalent circuit representation of surface and tubular membrane. The elements of the circuit shown in the inset represent a slightly activated sodium system where the lumped tubular admittance after the 200 Ω access resistance is three times that of the surface membrane. For the purpose of this and the following figures the standard depolarized values of the circuit elements are varied one at a time. These values were R = 8 kΩ, C = 2 μF, R_a = 200 Ω, R_h = 10 kΩ, L_h = 30 H, R_m = 400 Ω, and C_m = 2 μF. In this figure the total resistance (R) is varied from 8 kΩ to 2 kΩ. Ordinate, magnitude in mS/cm², phase angle in degrees; abscissa, frequency in Hz. The scales on all subsequent figures are identical.

nificantly complicates the model and quantitatively alters the phase function. It is noteworthy that for a moderate depolarization the surface membrane alone is rather similar to the lumped passive model. This is apparent by considering the depolarized values of the m circuit elements; namely, 400 Ω in series with 2 μF, values that are remarkably close to those that describe many of the passive impedance data. At normal resting potentials the m and h processes contribute very little to the equivalent circuit.

The curve labeled "8" of Fig. 11, as compared to "3" shows a marked enhancement of the low-frequency phase maximum, which is analogous to a similar phenomenon seen in the data of Fig. 2. The increase in the resistance simulates an increase in the negative conductance of the sodium system. Obviously a more accurate simulation could be made by changing the m and h processes, although at the expense of clarity in understanding of how the circuit elements affect the admittance.

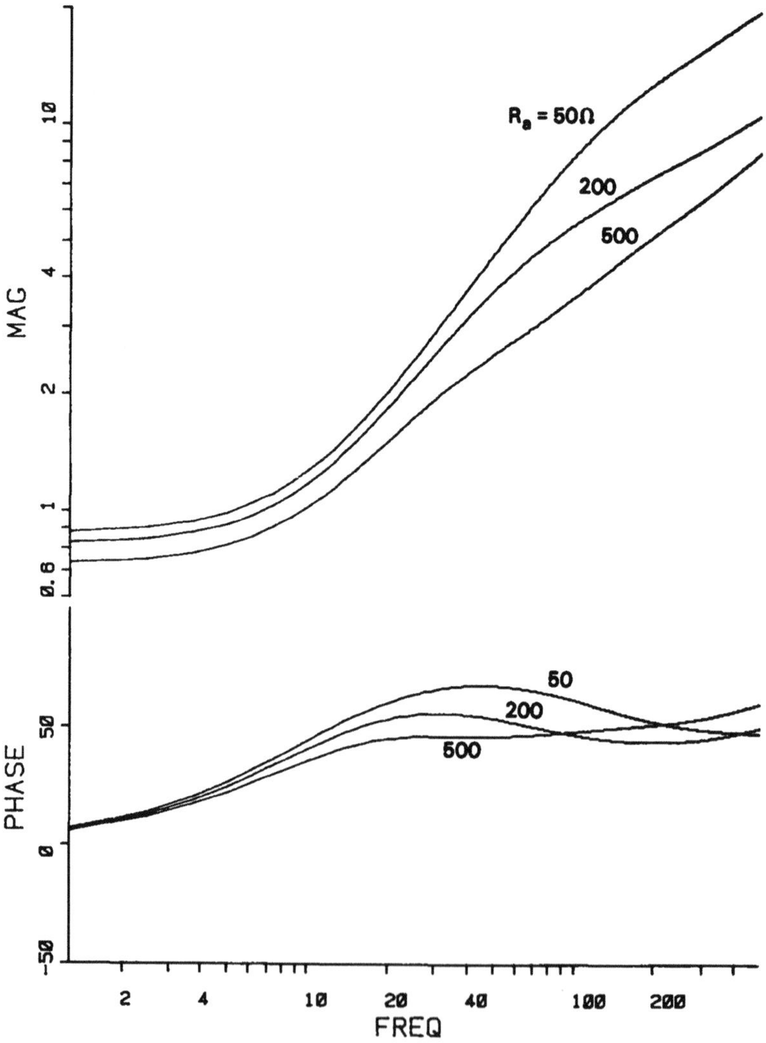

Figure 12. Variation in simulated admittance with tubular access resistance. The values of the access resistance, R_a, are given on the figure. The ordinate and abscissa are the same as in Fig. 11.

Figure 12 illustrates the effect of varying the access resistance on the total admittance of the moderately depolarized surface and tubular membranes. It is clear that the complicated effects seen from these changes require an estimate of the passive circuit elements under conditions in which the active conductances are totally blocked. The effects of the access resistance are further elucidated in Fig. 13, where the curve labeled "$R_a = 0$" shows that despite the complete absence of a tubular membrane, the phase function shows a low-frequency maximum. (In the lumped model, if $R_a = O$, then all the tubular membrane is attributed to the surface since there is no access resistance to the tubules.) This maximum is due to the series *RC* branch of the *m* process and is indistinguishable from a series *RC* that would model the passive tubules. The effects of the tubules alone are shown by the

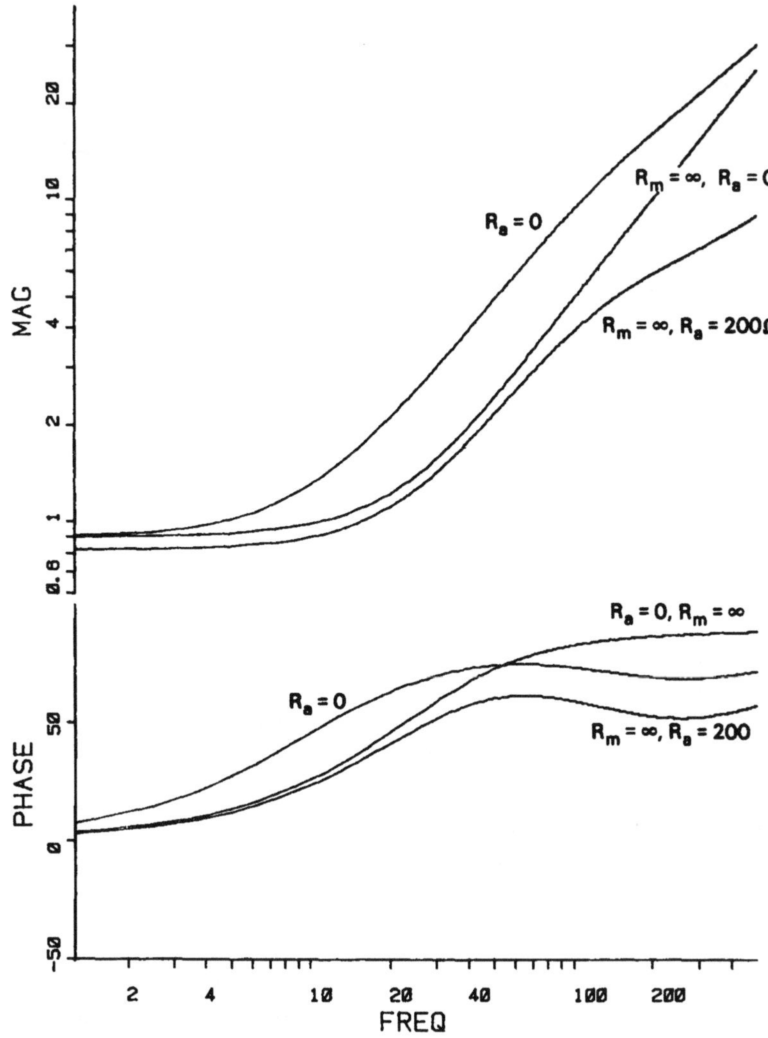

Figure 13. Complex Hodgkin–Huxley admittance with and without access resistance. The curve labeled "$R_a = 0$" shows the effect of the *m* process on the admittance. The curve labeled "$R_m = \infty$, $R_a = 200\Omega$" shows the effect of the tubules without the *m* process. The curve labeled "$R_m = \infty$, $R_a = 0$" shows essentially an *RC* admittance in which the *m* process is removed and there is no access resistance. The scales are identical to Fig. 11.

curves labeled "$R_m = \infty$," and $R_a = 200\ \Omega$" of the Fig. 13. The characteristic low-frequency maximum of the phase is apparent. Because of the large value of R_h, the continued presence of the R_hL_h branch in this simulation is inconsequential. Removal of the tubules as well as the series R_mC_m (*m* process) is shown by the curves labeled "$R_m = \infty$" and "$R_a = 0$." The magnitude and phase functions for this condition are essentially that of a parallel *RC* network.

The appearance of an antiresonance in the magnitude was only observed for large depolarizations of the muscle preparation. This is in contrast to observations on squid axons which show a pronounced antiresonance at rest (Fishman *et al.*, 1977). However, in cesium-perfused axons the antiresonance was not present for a variety of depolarizations (Fishman *et al.*, 1979) because of a substantial steady state negative sodium conductance. In squid,

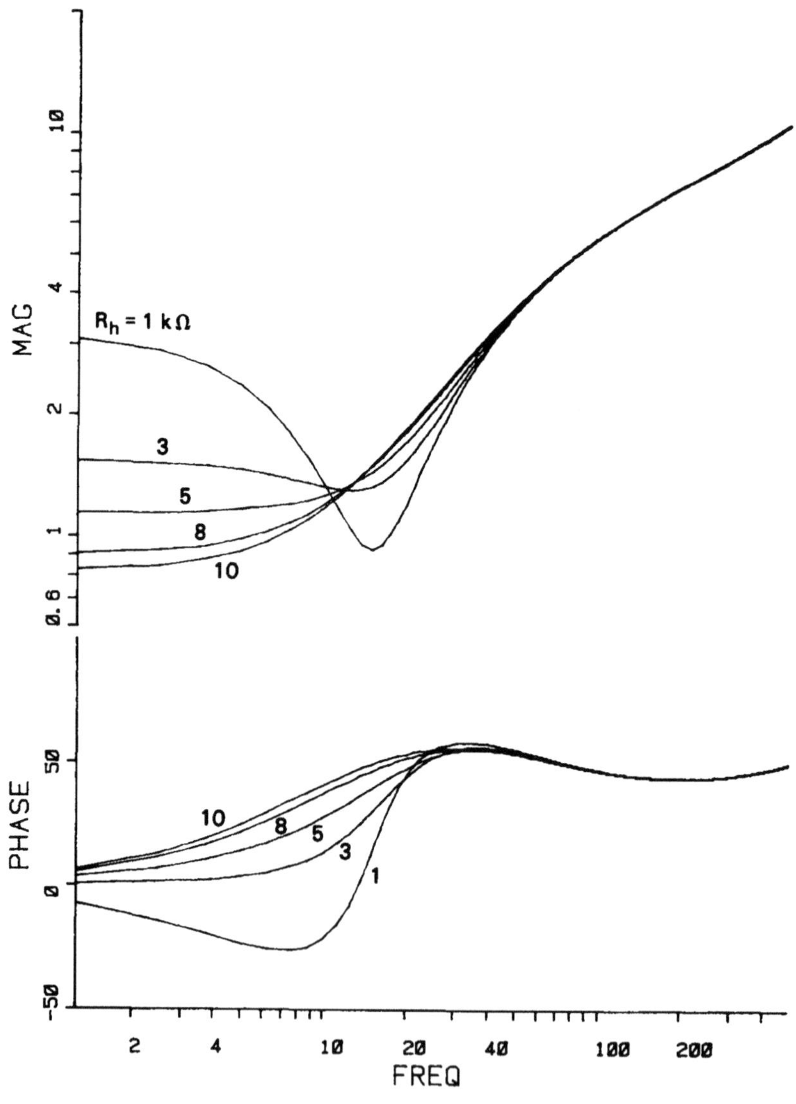

Figure 14. Complex admittance as a function of inactivation. The Hodgkin–Huxley current element, R_h, is varied from 10 kΩ to 1 kΩ to give τ_h values 3 msec to 0.3 msec. Note the appearance of an antiresonance as τ_h decreases. The scales are identical to Fig. 11.

the inclusion of an additional shunt or leakage conductance led to the appearance of an antiresonance with moderate depolarizations. The muscle data with TEA is similar to the latter condition due to the shunting effect of the chloride conductance as well as the remaining potassium conductance.

Figure 14 shows the effect of decreasing R_h from 10 to 1 kΩ. This leads to a range in τ_h from 3 msec to 0.3 msec. Although these values exceed those reported for muscle membrane (Adrian *et al.*, 1970), they show the effect of a decreasing τ_h, which generally occurs with depolarization. These simulations show that the model system can easily accommodate the presence or absence of antiresonance depending on the exact values of the kinetic parameters and structural elements chosen.

Variation of the m process parameters is shown in Fig. 15. The three values of τ_m for

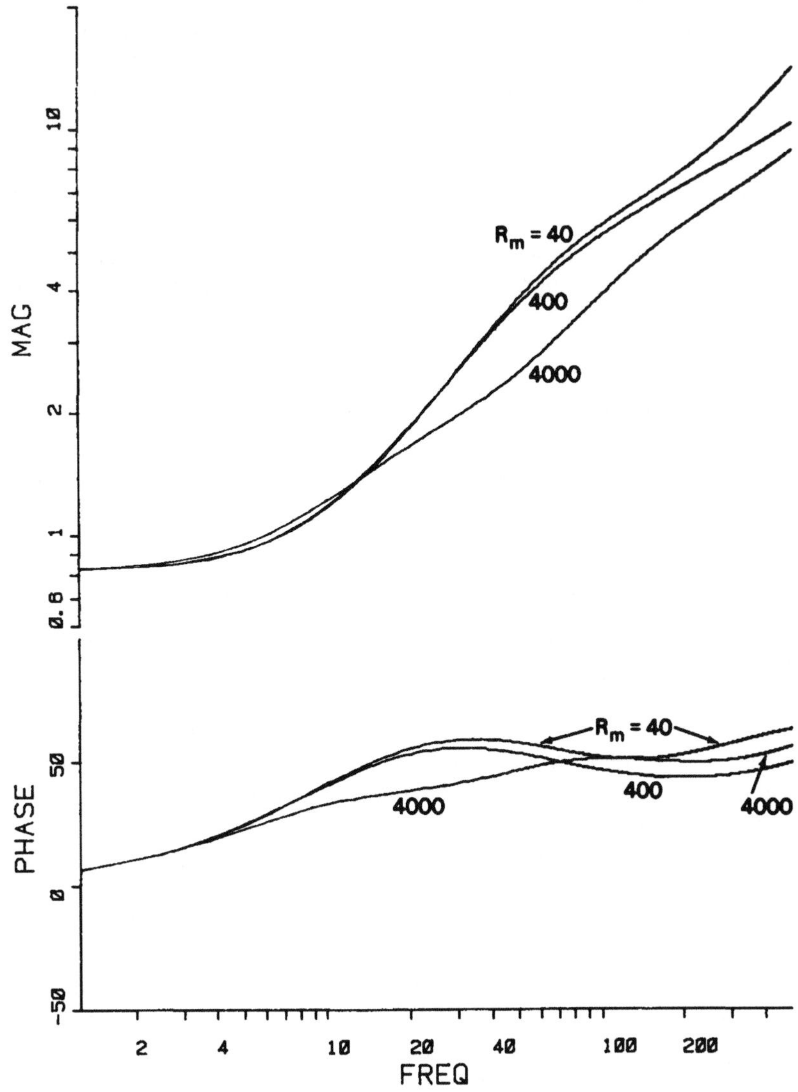

Figure 15. Complex admittance as a function of sodium-conductance activation. Hodgkin–Huxley circuit element, R_m, is varied from 4 kΩ to 40 Ω to give τ_m values 8–80 μsec. The scales are identical to Fig. 11.

these curves are 80 μsec, 0.8 msec and 8 msec corresponding to R_m values of 40 Ω, 400 Ω, and 4000 Ω. These simulations further illustrate the necessity of knowing the passive parameters before accurate estimates of the voltage-dependent terms can be made.

VI. COMPARISON WITH EARLIER WORK

The lumped model treated in these simulations is clearly a first approximation. Compared to the surface membrane, the tubules should be modeled with a significantly lower chord conductance, due to the absence of a chloride component (Eisenberg and Gage, 1969). The distributed cable structure (Adrian *et al.*, 1969b; Falk, 1968; Peachey and Adrian, 1973; Schneider, 1970) should be included since it seems virtually certain that action potentials propagate radially. Finally, the voltage-dependent ion-conductance parameters may be different in the tubules. Nevertheless, the simplified model discussed is useful in considering some of the previous conclusions about tubular effects and the correct equivalent circuit.

The simulation by Adrian and Peachey (1973) of action potentials and voltage-clamp currents were made based on data taken in hypertonic solutions (Adrian *et al.*, 1970). The analysis of Valdiosera *et al.* (1974c) suggests that hypertonic solutions significantly increase the access resistance. This would tend to diminish the effect of the tubules on the membrane capacitance, although the voltage control would be significantly worse. In their simulations, Adrian and Peachey (1973) used limiting sodium and potassium conductance values of 1/20 and 1/33 of the surface membrane values, respectively. Consequently the tubular currents in their model are relatively insignificant. The TTX-binding studies of Jaimovich *et al.* (1976) suggest that the density of sodium channels in the tubules is about half their density in the surface membrane. This result, coupled with numerous reports of notches or late inward currents in voltage clamp records from skeletal muscle, suggest that the tubular membrane currents may be substantial enough to invalidate a complete quantitative analysis of ionic currents for the normal muscle membrane. On the other hand, in experiments that have been designed to nullify the tubular currents, such as by detubulation or possibly by increasing the access resistance, adequate control may have been achieved to measure the ionic currents accurately.

Previous analyses of impedance data have been based on passive circuit elements reflecting only the structural elements of the surface and tubular membranes. This appears to be a reasonable assumption at the normal resting potential of about −90 mV, since the voltage-dependent ionic conductances are relatively inactive. This conclusion is not generally true for excitable membranes and should be experimentally confirmed by making impedance measurements at rest with all the specific ion conductance systems blocked. The importance of this point is well demonstrated in squid-axon measurements, which do show prominent sodium and potassium effects in the admittances at the normal rest potential.

The impedance measurements of Schneider (1970) were done on partially depolarized muscles bathed in 7.5 mM external potassium-ion concentration. Thus, the impedance measurements used for the passive equivalent circuit analysis were done at a membrane potential of −66 mV. The measurements shown in Fig. 1 show that changes in the total admittance occur in this potential range due to activation of the voltage-dependent conductances. The two sets of measurements are not directly comparable since Schneider's (1970) result is a long-term steady state condition, which may have led to complete inactivation of the specific ion-conduction mechanisms. The results of Fig. 1 were taken during a step-voltage-clamp depolarization lasting 1600 msec.

The impedance studies of Valdiosera *et al.* (1974c) in solutions made hypertonic by the addition of sucrose showed changes in the equivalent-circuit values that may be related to specific ion-conduction phenomena. As suggested by the authors, the conclusion that the increase in both the radial resistance and the tubular capacitance may be a consequence of a different equivalent circuit due to the low membrane potential of a −65 mV measured in this solution. These authors showed that measurements in solutions of varying conductivity and decreased sodium ion concentrations fit with hybrid and disk models usually lead to reasonable values for the surface and tubular capacitances. However, the use of the passive lumped model led to large variations in the surface and tubular capacities. In general, it appears that variants of the distributed model fit experimental data from different solutions better than the lumped model (Mathias *et al.*, 1977; Valdiosera *et al.*, 1974c). Although it seems likely that some type of distributed model is correct, some of the variations in the impedance results are probably due to the removal of sodium ions and the complicated interaction of the passive and voltage-dependent ion conductances on the impedance.

VII. CONCLUSIONS

The measurements reported here demonstrate that voltage-dependent ion-conductance processes do contribute to the linearized admittance of muscle. Furthermore, this approach can provide analysis of the highly nonlinear surface and tubular membrane properties by linearization of the response at different membrane potentials.

ACKNOWLEDGMENTS. I am indebted to Professors H. M. Fishman and D. J. M. Poussart for helpful discussion of the technical development of on-line admittance measurements throughout the course of this work and to Dr. T. Iwazumi who designed and constructed the voltage-clamp apparatus used in the experiments. I thank Dr. H. R. Leuchtag for comments on the manuscript.

This work was partially supported by NIH Grant NS-13520.

REFERENCES

Adrian, R. H., and Peachey, L. D., 1973, Reconstruction of the action potential of frog sartorious muscle, *J. Physiol.* **235**:103–131.

Adrian, R. H., Chandler, W. K., and Hodgkin, A. L., 1969a, The kinetics of mechanical activation in frog muscle, *J. Physiol.* **204**:207–230.

Adrian, R. H., Costantin, L. L., and Peachey, L. D., 1969b, Radial spread of contraction in frog muscle fibres, *J. Physiol.* **204**:231–257.

Adrian, R. H., Chandler, W. K., and Hodgkin, A. L., 1970, Voltage clamp experiments in striated muscle fibres, *J. Physiol.* **208**:607–744.

Bendat, J. S., and Piersol, A. G., 1971, *Random Data: Analysis and Measurement Procedures*, Wiley-Interscience, New York.

Chandler, W. K., FitzHugh, R., and Cole, K. S., 1962, Theoretical stability properties of a space-clamped axon, *Biophys. J.* **2**:105–128.

Cole, K. S., 1941, Rectification and inductance in the squid axon, *J. Gen. Physiol.* **25**:29.

Cole, K. S., 1968, *Membranes, Ions and Impulses*, 1972 revised edition, University of California Press, Berkeley.

Cole, K. S., and Baker, R. F., 1941, Transverse impedance of the squid giant axon during current flow, *J. Gen. Physiol.* **24**:535.

Costantin, L. L., 1975, Contractile activation in skeletal muscle, *Prog. Biophys. Mol. Biol.* **29**:197–224.

Eisenberg, R. S., 1967, The equivalent circuit of single crab muscle fibers as determined by impedance measurements with intracellular electrodes, *J. Gen. Physiol.* **53**:279–297.

Eisenberg, R. S., and Gage, P. W., 1969, Ionic conductances of the surface and transverse rubular membranes of frog sartorius fibers, *J. Gen. Physiol.* **53**:279–297.

Falk, G., 1968, Predicted delays in the activation of the contractile system, *Biophys. J.* **8**:608–625.

Falk, G., and Fatt, P., 1964, Linear electrical properties of striated muscle fibers observed with intracellular electrodes, *Proc. R. Soc. B* **160**:69–123.

Fatt, P., 1964, Analysis of the transverse electrical impedance of striated muscle, *Proc. R. Soc. B* **159**:606–651.

Fishman, H. M., Poussart, D. J. M., Moore, L. E., and Siebenga, E., 1977, K^+ conduction description from the low frequency impedance and admittance of squid axon, *J. Membr. Biol.* **32**:255–290.

Fishman, H. M., Poussart, D., and Moore, L. E., 1979, Complex admittance of Na^+ conduction in squid axon, *J. Membr. Biol.* **50**:43–63.

Fishman, H. M., Moore, L. E., and Poussart, D., 1981, Squid axon conduction: Admittance and noise during short versus long-duration step clamps, in: *The Biophysical Approach to Excitable Systems* W. J. Adelman and D. E. Goldman, eds., Plenum, New York.

Frankenhaeuser, B., Lindley, B. D., and Smith, R. S., 1966, Potentiometric measurement of membrane action potentials in frog muscle fibres, *J. Physiol.* **183**:152–166.

Freygang, W. H., Jr., Rapoport, S. I., and Peachey, L. D., 1967, Some relations between changes in the linear electrical properties of striated muscle fibers and changes in ultrastructure, *J. Gen. Physiol.* **50**:2437–2458.

Gage, P. W., and Eisenberg, R. S., 1969, Capacitance of the surface and transverse tubular membrane at frog sartorius muscle fibers, *J. Gen. Physiol.* **53**:265–278.

Hille, B., and Campbell, D. T., 1976, An improved vaseline gap voltage clamp for skeletal muscle fibers, *J. Gen. Physiol.* **67**:265–293.

Hodgkin, A. L., and Huxley, A. F., 1952, A quantitative description of membrane current and its application to conduction and excitation in nerve, *J. Physiol.* **117**:500–544.

Jaimovich, E., Venosa, R. A., Shrager, P., and Horowicz, P., 1976, Density and distribution of tetrodotoxin receptors in normal and detubulated frog sartorius muscle, *J. Gen. Physiol.* **67**:399–416.

Kovacs, L., and Schneider, M. F., 1978, Contractile activation by voltage clamp depolarization of cut skeletal muscle fibres, *J. Physiol.* **277**:483–506.

Mandrino, M., 1977, Voltage-clamp experiments on frog single skeletal muscle fibres; evidence for a tubular sodium current, *J. Physiol.* **269**:605–625.

Mathias, R. T., Eisenberg, R. S., and Valdiosera, R., 1977, Electrical properties of frog skeletal muscle fibers interpreted with a mesh model of the tubular system, *Biophys. J.* **17**:57–94.

Mauro, A., Conti, F., Dodge, F., and Schor, R., 1970, Subthreshold behavior and phenomenological impedance of the squid giant axon, *J. Gen. Physiol.* **55**:497–523.

Moore, L. E., 1972, Voltage clamp experiments on single muscle fibers of *Rana pipiens*, *J. Gen. Physiol.* **60**:1–19.

Moore, L. E., Fishman, H. M., and Poussart, D. J. M., 1980, Small-signal analysis of K^+ conduction in squid axons, *J. Membr. Biol.* **54**:157–164.

Peachey, L. D., and Adrian, R. H., 1973, Electrical properties of the transverse tubular system, in: *The Structure and Function of Muscle,* Vol. III (G. H. Bourne, ed.), Academic Press, New York.

Poussart, D., Moore, L. E., and Fishman, H., 1977, Ion movements and kinetics in squid axon. I. Complex admittance, *Ann. N.Y. Acad. Sci.* **303**:355–379.

Schneider, M., 1970, Linear electrical properties of the transverse tubules and surface membrane of skeletal muscle fibers, *J. Gen. Physiol.* **56**:640–671.

Stanfield, P. R., 1970, The effect of tetraethylammonium ion on the delayed currents of frog skeletal muscle, *J. Physiol.* **209**:209–229.

Valdiosera, R., Clausen, C., and Eisenberg, R. S., 1974a, Measurement of the impedance of frog skeletal muscle fibers, *Biophys. J.* **14**:295–315.

Valdiosera, R., Clausen, C., and Eisenberg, R. S., 1974b, Circuit properties of the passive electrical properties of frog skeletal muscle fibers, *J. Gen. Physiol.* **63**:432–459.

Valdiosera, R., Clausen, C., and Eisenberg, R. S., 1974c, Impedance of frog skeletal muscle fibers in various solutions, *J. Gen. Physiol.* **63**:460–491.

Vergara, J., and Bezanilla, F., 1979, Tubular membrane potentials monitored by a fluorescent dye in cut single muscle fibers, *Biophys. J.* **25**:201a.

Vergara, J., Bezanilla, F., and Salzberg, B., 1978, Nile bile fluorescence signals from cut muscle fibers under voltage or current clamp conditions, *J. Gen. Physiol.* **72**:775–800.

9

Noise, Impedance, and Single-Channel Currents

Louis J. DeFelice and Michael J. Kell

I. THE STUDY OF SINGLE-CHANNEL CURRENTS

Our purpose in studying single-channel currents from excitable membranes is twofold. First, we want to describe the voltage- and time-dependence of the ionic currents that make up the action potential. Second, we want to understand how the channels actually work at the molecular level. When we achieve this we can answer the following questions. What channel properties are responsible for macroscopic properties of membranes (threshold, action potential shape, velocity of propagation, refractory period, spontaneous firing, anesthetic block, and so on)? How do macromolecules flip randomly between conducting and nonconducting states? What causes the kinetics of this random flipping to be voltage-dependent and time-dependent? How do macromolecules select some ions over others? What causes their conductance to be voltage-dependent? How do anesthetics and toxins work?

These are also the goals of macroscopic voltage-clamp studies, and of noise and impedance measurements; however, in experiments that involve many channels, the properties of single channels can only be inferred.

In this paper, we discuss the properties that single channels need for excitability, that is, we show how voltage-dependence and time-dependence are represented in single channels. This model is based on recent direct measurements of single-channel currents in excitable membranes (Conti and Neher, 1980; Sigworth and Neher, 1980).

To illustrate the difference between macroscopic and microscopic properties, consider a transistor based analogue model of an excitable membrane such as that devised by Jerry Lettvin (personal communication). Such a circuit is composed of batteries, transistors, resistors, and capacitors, and is designed to produce an action potential when stimulated. Suppose that the Na, K, and other ionic currents are modeled by separate transistor pathways. These pathways represent voltage-dependent and time-dependent conductances that mimic ionic conductances in nerve. This circuit can be voltage-clamped, Na current can be blocked, leakage can be increased to produce spontaneous firing, the anodal break response can be

---

Louis J. DeFelice and Michael J. Kell • Department of Anatomy, Emory University School of Medicine, Atlanta, Georgia 30322.

demonstrated, and so on. Virtually all macroscopic phenomena observed in membranes can be mimicked by this transistor-based electronic circuit.

What about impedance, noise, and single-channel measurements from such a circuit? The impedance is measured by stimulating the circuit with a current and measuring the voltage reponse, the ratio of voltage to current defines the impedance. The current should be small, so that the voltage response is linear. In this approximation, a nonlinear circuit that mimics action potentials acts like a linear circuit of resistors, capacitors, and inductors. (Admittance is measured by voltage-clamping the circuit, stimulating with a small perturbing voltage, and measuring the current response. The ratio of this response to the voltage perturbation defines the admittance.) The same result is obtained if impedance (or admittance) measurements are made on a real membrane or on the analog circuit. In this sense, impedance is similar to the other macroscopic properties of excitable membranes.

Instead of stimulating the circuit, suppose we measure its intrinsic noise. Analogue circuits like the one mentioned above fail to reproduce the voltage or current noise found in excitable membranes. They also fail to reproduce single-channel currents.

The components of our hypothetical circuit are, of course, noisy. Some even display burst noise similar to single-channel currents (DeFelice, 1980, p. 271). This does not imply, however, that such devices are good models of ionic conductance. Circuits like the one we have been describing are designed only to reproduce the action potential, the total Na current, etc. Their ability to do so is based on the whole circuit, not on the properties of fluctuations that occur in the devices themselves. This paper takes the opposite view, for we require that all electrical phenomena, large and small, be derivable from fluctuations inherent in the channels.

II. PROPERTIES OF AN IDEAL CHANNEL

A. Opening and Closing

The channel postulated in this chapter has either an open or a closed conductance state. At any constant transmembrane voltage across the membrane, the channel flips randomly between these two states. This random flipping occurs spontaneously. The jump from open to closed, or from closed to open, is assumed to occur instantaneously. In the steady state, the channel has an average open time τ_o and an average closed time τ_c. Figure 1 illustrates a series of idealized cycles between openings. The cycle time is $\tau_o + \tau_c$. The probability that a channel is open is the fraction of time spent in the open state,

$$p = \frac{\tau_o}{\tau_o + \tau_c}$$

For example, if $\tau_o = 1$ msec and $\tau_c = 4$ msec, then $p = 1/5$.

The frequency of channel openings is defined as the number of times a channel opens per second,

$$r = \frac{1}{\tau_o + \tau_c}$$

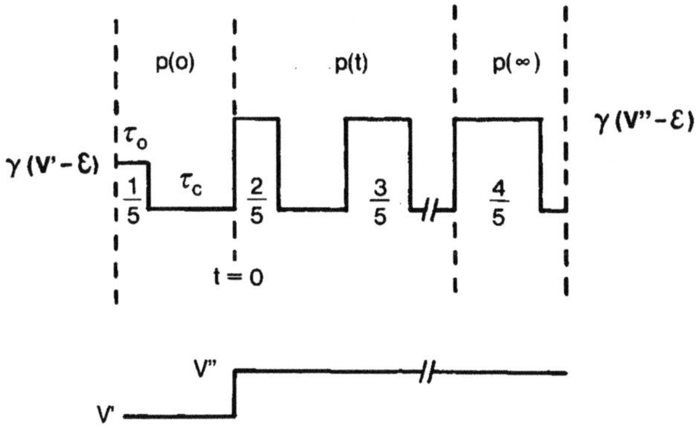

Figure 1. An idealized view of the average current through a single channel before and after a voltage change. Before the voltage change, the average open time is 1 msec and the average closed time is 4 msec. This region is called $p(0)$. Well after the voltage change, the average open time is 4 msec and the average closed time is 1 msec. This region is called $p(\infty)$. $p(t)$ refers to the transition between these two states. The probable fraction of time a channel spends in the open state is shown for each cycle. A single open–close sequence represents an idealized probable cycle, and the length of the cycle was kept constant for simplicity. Although it takes time for open-channel probability to reach the new steady state, the current through the open channel changes instantaneously. See Clay and DeFelice (1983) and DeFelice and Clay (1983) for a complete description.

Keeping to the same example, $r = 200\ \text{sec}^{-1}$. This is not how fast any given opening occurs; we have assumed that to be instantaneous. The probability of finding a channel in the open state can be written as the frequency of channel opening times times the mean open time, or

$$p = r\tau_o$$

If the current through an open channel is i, the current through a channel open a fraction p of the time is pi, and the average current through N such channels is Npi,

$$I = Npi$$

Another way to look at this is as follows. N is the total number of channels (open plus closed) and p is the probability of finding one open; therefore, pN is the number of open channels. The total current is the open-channel current times the number of open channels, or ipN.

B. Voltage-Dependence

The steady state conductance of excitable membranes depends on the voltage V across the membrane. How is this voltage-dependence reflected in individual channels?

Assume that the ideal channel is ohmic and that it is permeable to only one type of ion with equilibrium potential E. The current through the channel is zero when it is closed and i when it is open,

$$i = \gamma(V-E)$$

V is the potential difference across the channel and γ, the single-channel conductance, does not depend on V. The current through N channels is

$$I = Np\gamma(V-E)$$

This is the average macroscopic current through the membrane due to one ionic species. Macroscopic conductance is defined by

$$I = G(V-E)$$

so $G = Np\gamma$. This can be written $G = Nr\tau_o\gamma$. Conductance is proportional to the number of channels, their frequency of opening, how long they stay open, and the conductance of a channel once it does open. If γ and N are constant, G is a function only of r and τ_o. Nonohmic behavior results when the frequency of opening or the average open time depend on voltage. Collecting the constant factors in front yields

$$G(V) = N\gamma r(V)\tau_o(V)$$

The factors r and τ_o are not independent since $r\tau_o$ cannot be greater than one. This follows because τ_o is always less than or equal to $\tau_o + \tau_c$. For example, if $\tau_o = 1$ msec then the largest value r may have is 1000 sec^{-1}. In this case the channel is open all the time. Although $0 \leqslant p \leqslant 1$, it is not required that the limits of 0 and 1 ever be reached.

C. Time-Dependence

What happens to a single channel if the voltage across the membrane changes? To answer this we need a kinetic equation for $p(t)$. In the steady state,

$$\frac{dp}{dt} = 0$$

The value of p in the steady state is

$$p = \frac{\tau_o}{\tau_o + \tau_c}$$

An equation that satisfies these conditions is

$$\frac{dp}{dt} = \frac{1}{\tau_c}(1-p) - \frac{1}{\tau_o}p$$

τ_c and τ_o depend on V. If V changes from V' to V'' at $t = 0$, the solution to the above equation is

$$p(t) = p(0)e^{-t/\tau} + p(\infty)(1 - e^{-t/\tau})$$

where

$$\tau = \frac{\tau_o\tau_c}{\tau_o + \tau_c}$$

τ is one half the harmonic mean of τ_o and τ_c. $p(0)$ is the initial value of p and $p(\infty)$ is the final value.$p(0)$ is evaluated at V' and $p(\infty)$ at V''. $p(t)$ goes from $p(0)$ to $p(\infty)$ with time constant τ, where τ is evaluated at V''.

Figure 1 illustrates how the current through a single channel changes when V changes instantaneously. For simplicity, consider an idealized average cycle. Suppose the channel is initially in the steady state and that $\tau_o = 1/4\tau_c$. Then $p(0) = 1/5$. Suppose that finally $\tau_o = 4\tau_c$. Then $p(\infty) = 4/5$. Two intermediate values of $p(t)$ are shown to illustrate the gradual change from $p = 1/5$ to $p = 4/5$.

III. VOLTAGE-DEPENDENCE OF SINGLE-CHANNEL CONDUCTANCE

We have assumed that γ and E are independent of V, that is, i depends on V only through the relationship $i = \gamma(V-E)$. Single-channel experiments measure the current through a single channel directly, and the value of this current is unrelated to the membrane area.

The $i(V)$ relationship can also be determined from macroscopic experiments. To illustrate this we describe a method used by Noble and Tsien (1968) to measure $i(V)$ curves through K channels in heart cell membranes. The notation has been changed to conform with this chapter.

Consider a voltage-clamp step from the holding potential V_H to the clamp potential V. Assume a population of N channels that conducts a specific ionic current I. From the previous discussion,

$$I = ipN = iN_o$$

where N_o is the number of open channels. We define I_A as the difference between the current at (3) and the current at (2) in Fig. 2, and I_B as the difference between the current at (4) and the current at (1). Then

$$\frac{I_A}{I_B} = \frac{i(V)\,N_o(V) - i(V)\,N_o(V_H)}{i(V_H)\,N_o(V) - i(V_H)\,N_o(V_H)} = \frac{i(V)}{i(V_H)}$$

This is called the rectifier ratio. If V_H is held constant and V is varied, the shape of the current-voltage relationship for open channels is determined:

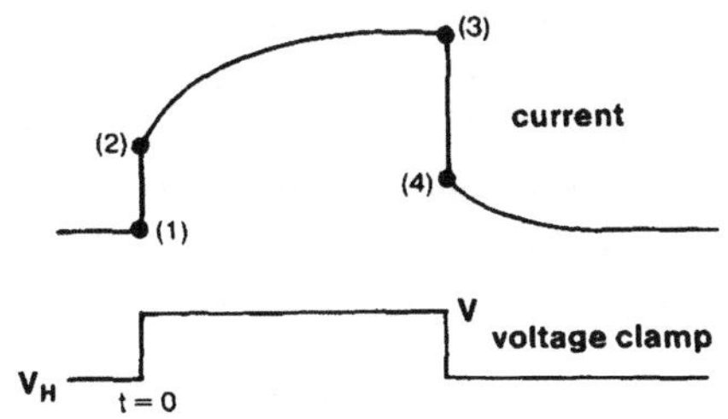

Figure 2. The current at position (2) is equal to the number of open channels at V_H times the single-channel current at the new potential V (the change in potential is instantaneous, so the number of open channels is still the original number). Similar reasoning gives the following currents at each of the four positions indicated. N_o is the number of open channels at V_H or V. (1) $i(V_H)\,N_o(V_H)$, (2) $i(V)\,N_o(V_H)$, (3) $i(V)\,N_o(V)$, and (4) $i(V_H)\,N_o\,(V)$.

$$i(V) = i(V_H)\,\frac{I_A}{I_B}$$

Since $i(V_h)$ is unknown, only the dependence of i on V is found by this method, the absolute value of i is undetermined.

To find $i(V_H)$, consider positions (4) and (1) in Fig. 2. Suppose V_H is chosen so all channels are closed ($p = 0$) and V is chosen so all channels are open ($p = 1$). This assumes that the extreme values of p are available to the channel.

$$\begin{array}{lll} \text{For } p = 1 & (4)\ i(V_H)\,N_o(V) = i(V_H)N \\ \phantom{\text{For }} p = 0 & (1)\ i(V_H)\,N_o(V_H) = 0 = \text{baseline current} \end{array}$$

The difference between (4) and (1) is $i(V_H)N$. Multiplying both sides of the previous equation by N gives

$$i(V)N = \left[i(V_H)N\right]\left[\frac{I_A}{I_B}\right]$$

The two quantities in brackets are measured, and therefore the absolute value of $i(V)N$ is known at the potential V where $p = 1$. This sets the scale for the rest of the curve. This method does not give the value of $i(V)$, but $i(V)$ times the total number of channels.

Instantaneous $i(V)N$ relationships measured from macroscopic experiments such as the one described above are often curved. This is usually interpreted as a voltage-dependent open-channel conductance,

$$\gamma = \gamma(V)$$

Although this is the rule rather than the exception, we ignore voltage-dependent conductances in the rest of this chapter. Few experiments have been done on the same preparation to compare macroscopic $Ni(V)$ relationships with $i(V)$ curves measured directly from single-channel experiments.

IV. ACTIVATION, INACTIVATION, AND RECTIFICATION

In excitable membranes, currents are activated by membrane voltage. If the voltage changes instantaneously, the current changes instantaneously, and then gradually, as shown in Fig. 2. Both effects are explained at the level of single channels by Fig. 1. The instantaneous change in the current, (1) to (2), occurs as a result of the change in the driving force and because a fraction of the channels are already open when the voltage is switched. The gradual change, from (2) to (3), occurs because the probability of finding a channel open changes with time. The gradual change (activation) is described by

$$I(t) = Np(t)\gamma(V-E)$$

Inactivation is a decrease in current that occurs while V is held constant. When the voltage is switched across a channel that inactivates, $p(t)$ first increases and then decreases to a final steady-state value.

Consider only steady-state current-voltage relationships. Rectification occurs if the macroscopic *I(V)* plot is curved, and inward rectifier curves downward (*G* becomes smaller as the membrane is depolarized), and an outward rectifier curves upward (*G* becomes larger as the membrane is depolarized). Rectification can be achieved in two ways in our model. Either the single-channel current can rectify, i.e., $\gamma = \gamma(V)$, or the steady-state variable *p* can rectify. Let us ignore the first possibility. There are two ways that *p* can depend on *V*. The mean channel open time can be voltage-dependent or the frequency of channel openings can be voltage-dependent. These are not independent since

$$0 \leqslant \tau_o(V) r(V) \leqslant 1$$

Another way to look at this is to think of $\tau_o(V)$ and $\tau_c(V)$ as being the voltage-dependent quantities. They too are restricted by

$$0 \leqslant \frac{\tau_o(V)}{\tau_o(V) + \tau_c(V)} \leqslant 1$$

The molecular parameters that actually sense and respond to voltage are unknown. τ_o, τ_c, and *i* are simply the parameters we observe.

The steady-state voltage-dependence of the gating variable can be derived from macroscopic and from microscopic data. *I(V)* and *Ni(V)* are determined from voltage-clamp experiments such as Fig. 2. *r(V)* and $\tau_o(V)$ are measured directly from single-channel experiments. The two expressions for *p(V)* are

$$p(V) = I(V)/Ni(V) \qquad \text{macroscopic}$$

and

$$p(V) = r(V)\tau_o(V) \qquad \text{microscopic}$$

V. CHANNELS WITH HIDDEN STATES

So far we have discussed only the observable parameters *i*, *r*, τ_c, and τ_o. Suppose that a channel has one way of being open but several ways of being closed. Although we only observe one closed state, $i = 0$, there are several configurations of the channel that give $i = 0$. This extension of our model channel is closer to reality. To illustrate this, consider the K channel of the Hodgkin–Huxley model. Our objective is to explain the relationship between things observable from macroscopic experiments and things observable from microscopic experiments.

Assume the K channel is composed of a fixed number of particles or switches. Each switch is described by the following postulated properties: (1) the opening and closing of a single switch can be treated as if it were a reversible, first-order kinetic reaction, (2) any switch has only two states, open or closed, (3) each switch is independent of the state of other switches, (4) a channel composed of *x* switches has a number of configurational states that are equivalent and equally probable in which a certain number of switches are open and a certain number are closed, (5) intrastate configurational changes are unlikely relative to interstate configurational changes since two simultaneous and opposite elementary events are required in the former, (6) a channel can have only one configurational state at a time, and (7) all switches in the channel must be open for current to flow through the channel.

The concept of several configurations existing for a channel is explained in Fig. 3. In this figure, all possible configurations are shown for a four-subunit Hodgkin–Huxley K channel. The number of ways a channel can exist in a particular configuration, in which y switches are open and $(x - y)$ are closed, is

$$\binom{x}{y} = \frac{x!}{y!(x-y)!}$$

The kinetic constants in Fig. 3 come from our assumption that all subunits are identical. The opening and closing of a single switch can be thought of as a conformational change and can be represented by a reversible, first-order kinetic equation. The conformational change for the ith switch in a channel is

$$x_i^{\text{closed}} \underset{\beta}{\overset{\alpha}{\rightleftarrows}} y_i^{\text{open}}$$

In the Hodgkin–Huxley formulation, this relation is written in terms of the probability that a switch is open (n) and the probability that a switch is closed $(1 - n)$,

$$(1-n) \underset{\beta}{\overset{\alpha}{\rightleftarrows}} n$$

The first-order differential rate equation for n is

$$\frac{dn}{dt} = \alpha(1-n) - \beta n$$

In the steady state, $dn/dt = 0$ and $n = \alpha/(\alpha + \beta)$.

The multiples of the basic single switch rate constants that appear in Fig. 3 come about because each state may contain several indistinguishable and equivalent configurations. For

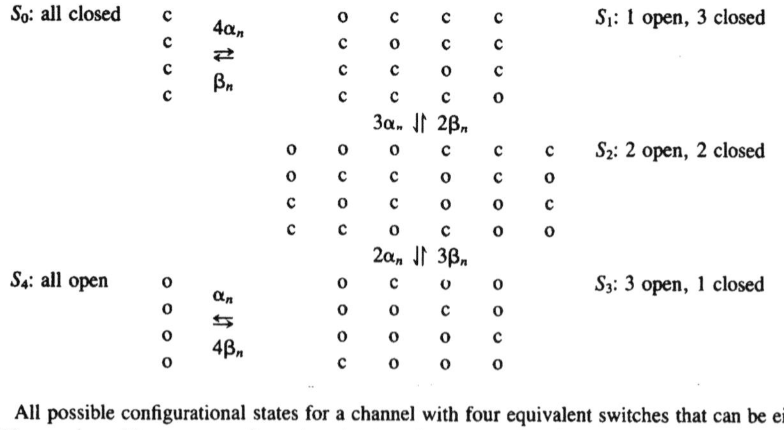

Figure 3. All possible configurational states for a channel with four equivalent switches that can be either open or closed. The number of intrastate configurations is given by permutation theory, e.g., for S_2, $\binom{4}{2}$ = six states. Rate constants are multiples of the basic rate constants for a single switch.

example, there are four ways to go from S_4 to S_3. Kinetically, it would appear that S_4 is disappearing four times more quickly than would be expected by the kinetic scheme $(1-n) \underset{\beta}{\overset{\alpha}{\rightleftarrows}} n$. Consequently, the apparent rate constant is 4β instead of β.

The kinetic scheme for a channel containing x subunits is

$$\begin{matrix} S_0 & \underset{\beta}{\overset{x\alpha}{\rightleftarrows}} & S_1 & \underset{2\beta}{\overset{(x-1)\alpha}{\rightleftarrows}} & \cdots & \underset{(x-1)\beta}{\overset{2\alpha}{\rightleftarrows}} & S_{x-1} & \underset{x\beta}{\overset{\alpha}{\rightleftarrows}} & S_x \\ p_0 & & p_1 & & & & p_{x-1} & & p_x \end{matrix}$$

where p represents the probability of a channel being in a specified configuration.

To analyze these reactions kinetically, we need expressions for the probabilities as a function of time. The basic forms for the probability expressions are generated by assuming a value for the open, conductive-channel state in which x subunits are functioning. Using this value, we express the previous interstate configurational probabilities in terms of the appropriate first-order differential rate expression $\frac{dp_x}{dt}$. An example is given below.

The steady-state probability p_x of a channel with x switches being conductive is

$$p_x = n_\infty^x = \left(\frac{\alpha}{\alpha+\beta}\right)^x$$

where n is the probability that a single switch is open. The probability p_{x-1} is found from the rate expression for the formation of p_x,

$$\frac{dp_x}{dt} = \alpha p_{x-1} - x\beta p_x$$

In the steady state,

$$\frac{dp_x}{dt} = 0$$

Rearranging and solving for p_{x-1} gives

$$p_{x-1} = \frac{x\beta}{\alpha} p_x$$

Since $p_x = \left(\frac{\alpha}{\alpha+\beta}\right)^x$,

$$\begin{aligned} p_{x-1} &= \frac{x\beta p_x}{\alpha} = \frac{x\beta}{\alpha}\left(\frac{\alpha}{\alpha+\beta}\right)\left(\frac{\alpha}{\alpha+\beta}\right)^{x-1} \\ &= x\left(\frac{\alpha+\beta-\alpha}{\alpha+\beta}\right)\left(\frac{\alpha}{\alpha+\beta}\right)^{x-1} \\ &= x\left(1 - \frac{\alpha}{\alpha+\beta}\right)\left(\frac{\alpha}{\alpha+\beta}\right)^{x-1} \end{aligned}$$

or finally,

$$p_{x-1} = x(1-n)n^{x-1}$$

Similar manipulations can be done for p_{x-2}, etc., remembering to account for all gains and losses to each configurational state. A general formula can be written in terms of x, n, and $(1 - n)$ for generating these probabilities. This formula is the Bernoulli distribution

$$p_i = \binom{x}{i} n^i (1-n)^{x-i}$$

where $i = 0, 1, 2, \ldots, x$ and $\binom{x}{i}$ is the binomial coefficient.

These results can be applied to both single-channel experiments and noise analyses in which the observable state (the noisy, conductive state) only occurs for the last conformational change. The states to the left of S_{x-1} only contribute to overall closed times for a channel. If the last state is the only observable event, the kinetics should be first order. The rate expression for the conductive state is

$$\frac{dp_x(t)}{dt} = \alpha p_{x-1}(t) - x\beta p_x(t)$$

But $p_x(t) = [n(t)]^x$; therefore

$$\frac{d[n(t)]^x}{dt} = \alpha x[n(t)]^{x-1}\,[1-n(t)] - x\beta[n(t)]^x$$

By the chain rule for derivatives

$$\frac{d[n(t)]^x}{dt} = x[n(t)]^{x-1}\frac{dn(t)}{dt}$$

and

$$x[n(t)]^{x-1}\frac{dn(t)}{dt} = \alpha x[n(t)]^{x-1}[1-n(t)] - x\beta[n(t)]^x$$

This, on dividing by $x[n(t)]^{x-1}$, reduces to the equivalent rate expression for an isolated switch,

$$\frac{dn(t)}{dt} = \alpha[1-n(t)] - \beta n(t)$$

The results of the above model can be used in two ways in analyzing single-channel records. One method is to consider steady-state values for the probabilities and relate these to the kinetic constants governing the mean channel-open time τ_o and mean channel-closed time τ_c. This is equivalent to adding the separate elementary reactions such that only two states are considered, open and closed. When adding chemical equations, the steady-state rate constant for the overall reaction is found by multiplying the rate constants of each separate reaction. Applying this procedure we obtain

$$S_0 \underset{\beta}{\overset{x\alpha}{\rightleftarrows}} S_1 \qquad K_{01} = \frac{[S_1]}{[S_0]} = \frac{x\alpha}{\beta}$$

$$S_1 \underset{2\beta}{\overset{(x-1)\alpha}{\rightleftarrows}} S_2 \qquad K_{12} = \frac{[S_2]}{[S_1]} = \frac{(x-1)\alpha}{2\beta}$$

$$\vdots \qquad \vdots$$

$$S_{x-1} \underset{x\beta}{\overset{\alpha}{\rightleftarrows}} S_x \qquad K_{(x-1)x} = \frac{[S_x]}{[S_{x-1}]} = \frac{\alpha}{x\beta}$$

which, when the reactions are added and the rate constants are multiplied, will be equivalent to

$$S_0 \underset{\beta^x}{\overset{\alpha^x}{\rightleftarrows}} S_x \qquad K_{0x} = \frac{[S_x]}{[S_0]} = \frac{\alpha^x}{\beta^x}$$

where $[S]$ represents the concentration of each state. This expression can be written in terms of steady state probabilities; p_x is the probability of a channel being open and $(1 - p_x{}^{1/x})^x$ is the probability of a channel being totally closed. Therefore

$$(1 - p_x{}^{1/x})^x \underset{\beta^x}{\overset{\alpha^x}{\rightleftarrows}} p_x$$

$$K_{0x} = \frac{p_x}{(1 - p_x{}^{1/x})^x} = \left(\frac{\alpha}{\beta}\right)^x$$

Consequently, one can determine the ratio of the basic kinetic constants (α and β) by (1) measuring the mean open and closed time constants (τ_o and τ_c) or measuring τ_o and the frequency of opening r, and (2) having a value for x. In terms of τ_o and r,

$$\frac{\alpha}{\beta} = \frac{p_x{}^{1/x}}{1 - p_x{}^{1/x}} = \frac{(r\tau_o)^{1/x}}{1 - (r\tau_o)^{1/x}}$$

One way to determine x from single-channel data is to do an ensemble average. Choose an appropriate length of time, say $10(\tau_o + \tau_c)$, step the voltage many times, and add the currents. x can be found by fitting the average microscopic response the same way one fits the macroscopic response (Hodgkin and Huxley, 1952).

The average open time in this model is

$$\tau_o = \frac{1}{x\beta}$$

The average closed time is

$$\tau_c = \frac{1}{x\beta}\left[\left(1 + \frac{\beta}{\alpha}\right)^x - 1\right]$$

VI. IMPEDANCE

Impedance is a linear property. Excitable membranes are approximately linear for small perturbations. The extent to which the approximation is valid depends strongly on the membrane voltage and the frequency of perturbation (DeFelice *et al.*, 1981). Membrane impedance is generally thought of as being made up of two very different properties. One is the actual resistive and capacitive components of the membrane, such as open-channel resistance and lipid bilayer capacitance. The other is the time-variant conductance of membranes, which appears as an impedance in the frequency domain. Membrane current takes time to change even if the voltage change occurs instantly (Fig. 2). This can be modeled by capacitors and inductors. Such models are only analogies of the behavior and do not represent actual circuit elements present in the membrane.

Membrane impedance is measured traditionally from a large population of channels. We may ask whether impedance can be defined for a single channel. Since part of the impedance is due to a time-variant conductance, the question reduces to this: Does the conductance of a single channel change with time? The question may cause confusion because in our model it always does. The channel flips between zero conductance and γ and is, of course, always changing. But this is not the change to which we refer. Impedance is an average property and a channel in the steady state has an average conductance that is time-invariant. What we are referring to is the time-variance in single-channel probability illustrated in Fig. 1. If the voltage across the channel changes, the new steady-state probability takes time to develop, that is, the conductance is time-variant.

The response shown in Fig. 2 is analogous to the response of an RrL-circuit. Consider a resistance r in series with an inductor L, and let this branch be in parallel with a resistor R. If the voltage is stepped across this circuit, the current response is like the one shown in Fig. 2. The fast part of the response is due to R, the gradual part of the response is due to the entire circuit, and the plateau is due to the resistors rR in parallel. This is explained, at the level of single channels, by Fig. 1. The impedance of a single channel is defined only in this sense.

VII. NOISE

The connection between our single-channel model and noise should now be clear. If each channel in a population of N channels is a random, two-state switch, the total current will fluctuate in the steady state. The spectral density of the current noise can be directly related to the two-state channel model, either the simple model or the more complex model described in Section V. The correspondence between spectral density and channel models has been given many times and will not be repeated here. It is described in detail in DeFelice (1981).

ACKNOWLEDGMENTS. We would like to thank David Clapham and John Clay for their help with part of this work. This work was supported by NIH grant 1-PO1-HL27385.

REFERENCES

Conti, F., and Neher, E., 1980, Single channel recordings of K current in squid axon, *Nature* **285:**140–143.

Clay, J. R. and DeFelice, L. J., 1983, Relationship between membrane excitability and single channel open-close kinetics. *Biophys. J.* **42:**151–157.

DeFelice, L. J., 1981, *Introduction to Membrane Noise,* Plenum Press, New York.

DeFelice, L. J., Adelman, W. J., Jr., Clapham, D. E., and Mauro, A., 1981, Second order admittance in squid axon, in: *The Biophysical Approach to Excitable Systems,* Plenum Press, New York.

DeFelice, L. J. and Clay, J. R., 1983. Membrane current and membrane potential from single-channel kinetics, in: *Single-Channel Recording* (B. Sakmann and E. Nelor, eds.) Plenum Press, New York.

Hodgkin, A. L., and Huxley, A. F., 1952, A quantitative description of membrane current and its application to conduction and excitation in nerve, *J. Physiol.* **117**:500–544.

Noble, D., and Tsien, R. W., 1968, The kinetics and rectifier properties of the slow potassium current in cardiac Purkinje fibers, *J. Physiol.* **195**:185–214.

Sigworth, F. J., and Neher, E., 1980, Single Na channel currents observed in cultured rat muscle cells, *Nature* **287**:447–449.

10

Membrane Ionic Currents, Current Noise, and Admittance in Isolated Cockroach Axons

Yves Pichon, Denis Poussart, and Graham V. Lees

I. ADVANTAGES OF THE COCKROACH AXON

The application of fluctuation analysis to conduction processes in excitable membranes has developed rapidly since the early work of Derksen and Verveen (1966) on the node of Ranvier. First recordings of membrane current noise were made on the giant axon of the lobster (Poussart, 1969, 1971) and showed that the spontaneous noise consisted mainly of a $1/f$ component with an intensity related to the driving force for potassium ions. A second noise component with the apparent form of a relaxation process, $1/[1 + (f/f_c)^2]$, was later observed in both squid axons (Fishman, 1973; Conti *et al.*, 1975; Fishman *et al.*, 1975a) and frog nodes of Ranvier (Siebenga *et al.*, 1973). Since then, numerous experiments have been done on these last two preparations and noise spectra arising from the transitions between "open" and "closed" states of sodium and potassium channels have been characterized. The relationship between the frequency characteristics of this noise and the observed or computed kinetics of the sodium and potassium conductances is, however, still controversial. In large-area noise measurements of squid axon, several technical problems such as low input impedance, potassium accumulation in the periaxonal space, and electrode polarization provide some impediments to a good quantitative analysis of ionic channel noise. In nodes of Ranvier, the situation is better but extrinsic noise is quite high and potassium also accumulates externally during long-lasting depolarizations.

In this chapter, we describe new results obtained with the isolated giant axon of the American cockroach, *Periplaneta americana*, concerning the macroscopic membrane currents, spontaneous noise, and admittance. Compared with the more "classical" preparations, the cockroach axon in the double-oil-gap technique exhibits low access resistance, high input impedance, very little potassium accumulation, greatly reduced electrode polarization due to easier access to the preparation, and very good stability. Another value of this

Yves Pichon and Graham V. Lees • Département de Biophysique, Laboratoire de Neurobiologie Cellulaire du C.N.R.S., F 91190 Gif sur Yvette, France. *Denis Poussart* • Département de Génie électrique, Université Laval, Québec G1K 7P4, Canada.

preparation, which was introduced in 1967 (Pichon, 1967) and is now relatively well known (see Pichon, 1974), is its use in pharmacological experiments.

II. EXPERIMENTAL METHODS

Giant axons (diameter ranging from 30–50 μm) were isolated from the abdominal nerve cord of male cockroaches, *Periplaneta americana,* according to the technique described by Pichon and Boistel (1967). Cross sections of such axons are illustrated in Fig. 1. It can be seen that the Schwann-cell layer which surrounds the axons under normal physiological conditions is almost totally disrupted during the dissection, ensuring direct access of ions or molecules to the axonal membrane.

The single axon and adjacent connectives were then placed in a double oil-gap chamber similar to that described by Pichon (1970). The active patch of membrane was adjusted to 2–3 × 10^{-4} cm^2 and excitability tested with short constant current test pulses and, if the action potential exceeded 70mV, the patch was studied under voltage-clamp conditions. The experimental set-up illustrated in Fig. 2 enabled successive recordings of (1) transmembrane currents corresponding to 5 msec square depolarizations of the axonal membrane (these currents were digitized at a sampling frequency of 125 kHz and stored on tape), (2) spontaneous noise spectra for frequencies between 10 and 2560 Hz at various potential levels, and (3) admittance magnitude and phase for this same bandwidth and the same potential

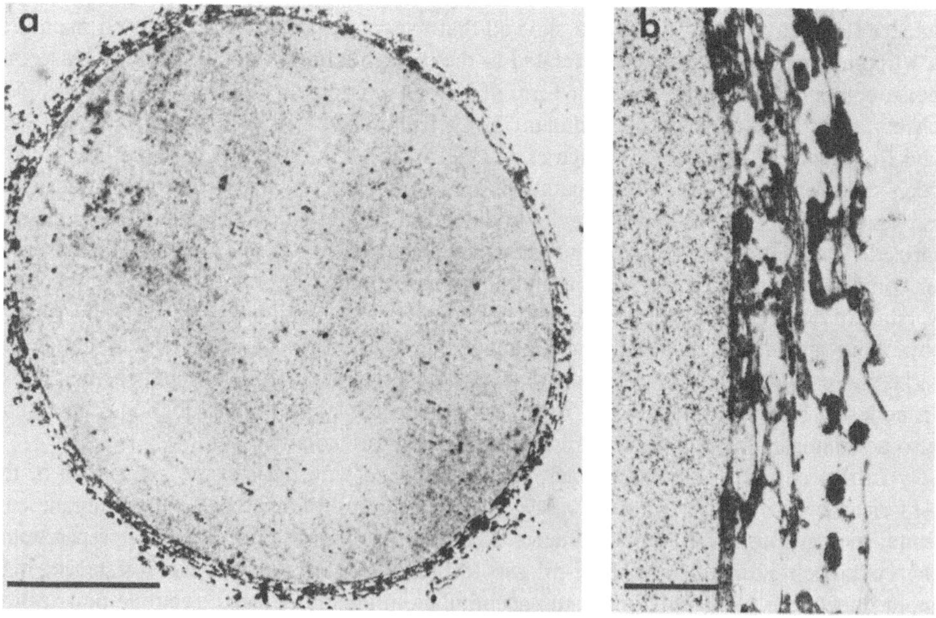

Figure 1. Electron micrographs of cross sections of two giant axons isolated from the abdominal nerve cord of the cockroach, *Periplaneta americana*. The layered glial cells (mesaxon) that normally surround the axons and regulate the ionic content in the vicinity of the excitable membrane (see Treherne and Pichon, 1972) under physiological conditions is almost totally disrupted during the dissection, allowing direct contact between the bathing solution and the nerve membrane. These axons have been fixed in 3% glutaraldehyde in cacodylate with added sucrose followed by postosmication and embedding in araldite. Sections were stained in uranyl acetate and lead citrate. Calibration bars: 10 μm in (a), 1 μm in (b). Courtesy of N. J. Lane (unpublished).

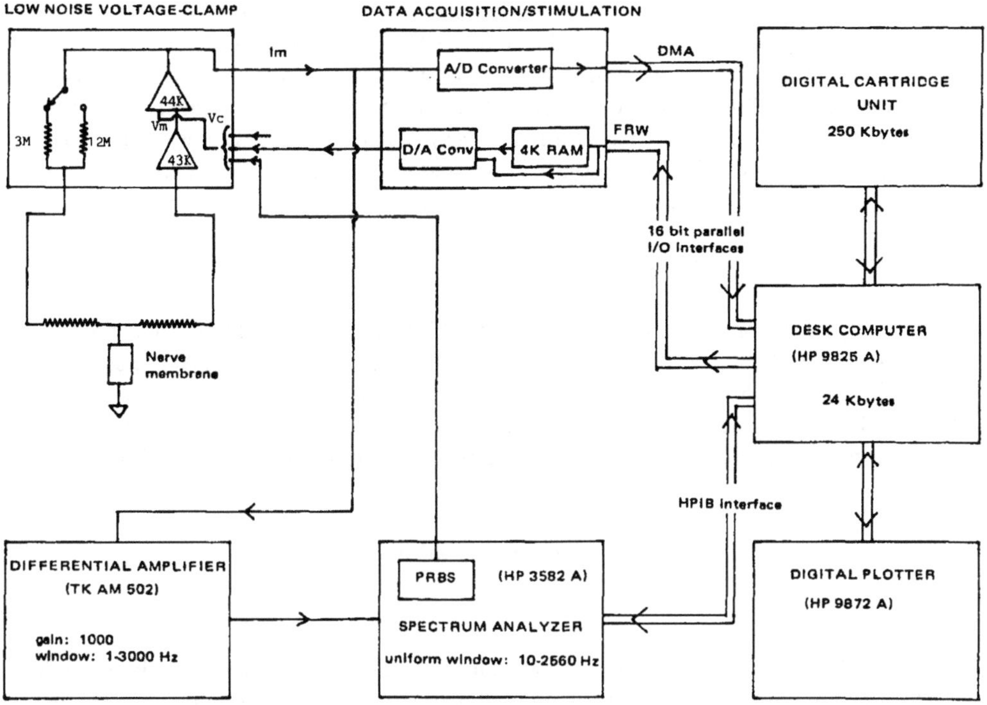

Figure 2. Diagrammatic representation of the set-up used for the study of membrane ionic currents and ionic current noise in cockroach axons. The analog part of the system is shown on the left-hand part of the figure, the digital one on the right. Digital to analog and analog to digital conversions were performed on twelve-bit resolution in the data acquisition and stimulation module (upper middle) and/or the spectrum analyzer (lower middle). All experiments were done under computer control through two 16-bit parallel I/O interfaces and one HPIB (IEEE-488) interface.

levels. Power spectra of spontaneous noise were averaged 30–150 times before storage on tape, whereas admittance magnitude and phase spectra (which were obtained directly with the pseudorandom binary signal, PRBS, of the spectrum analyzer) were obtained in one run (Poussart and Ganguly, 1977; Poussart *et al.*, 1977). All data were processed off-line from tape.

The saline used for these experiments contained 210 mM NaCl, 3.1 mM KCl, 10 mM $CaCl_2$ and was buffered at pH 7.2 by a phosphate-carbonate buffer (Pichon and Treherne, 1974). When needed, 10^{-7} M tetrodotoxin (TTX) and/or 1 mM 4-aminopyridine (4-AP) were added to the solution. Experiments were performed at temperatures ranging from 7 to 12°C.

III. RESULTS

A. Action Potential, Ionic Currents, and Conductances

Since the first experiments of Pichon (1967), it has been clearly established that the ionic currents that underlie excitability in insect axons are very much the same as in squid axons (Pichon, 1974). The action potential is related to a transient increase of the membrane conductance to sodium ions (g_{Na}) bringing the membrane potential towards the equilibrium potential for sodium ions (E_{Na}). This is followed by a decrease in g_{Na} accompanied by a large increase in the potassium conductance (g_K) which return the membrane potential towards its resting value. Changes in conductance have been found to be related to membrane potential, time, and calcium concentration in very much the same way as in squid axons. The Hodgkin–Huxley (1952) equations can be used to describe the ionic currents with only minor modifications. The Hodgkin–Huxley equation for the membrane current I_m is

$$I_m = C_m\,(dV_m/dt) + g_K\,(V_m - E_K) + g_{Na}\,(V_m - E_{Na}) + g_l\,(V_m - E_l) \qquad (1)$$

in which C_m stands for the membrane capacitance (3.3 μF cm^{-2}), V_m the membrane potential, g_l the nonspecific "leakage" conductance (about 0.5 mS cm^{-2}) and E_l the leakage equilibrium potential (around −60 mV). Curve fitting of Eq. (1) to the ionic current traces has shown (Pichon, 1974) that the best results are obtained if the sodium activation parameter (m) is raised to the fifth power and the potassium activation parameter (n) to the third power, so that

$$g_{Na} = g_{Na}\,m^5\,h \qquad (2)$$

$$g_K = g_K\,n^3 \qquad (3)$$

where g_{Na} is the maximum sodium conductance (around 150 mS cm^{-2}), h is the sodium inactivation parameter, and g_K is the maximum potassium conductance (around 25 mS cm^{-2}). Following a step change in the membrane potential, the three parameters m, n, and h were

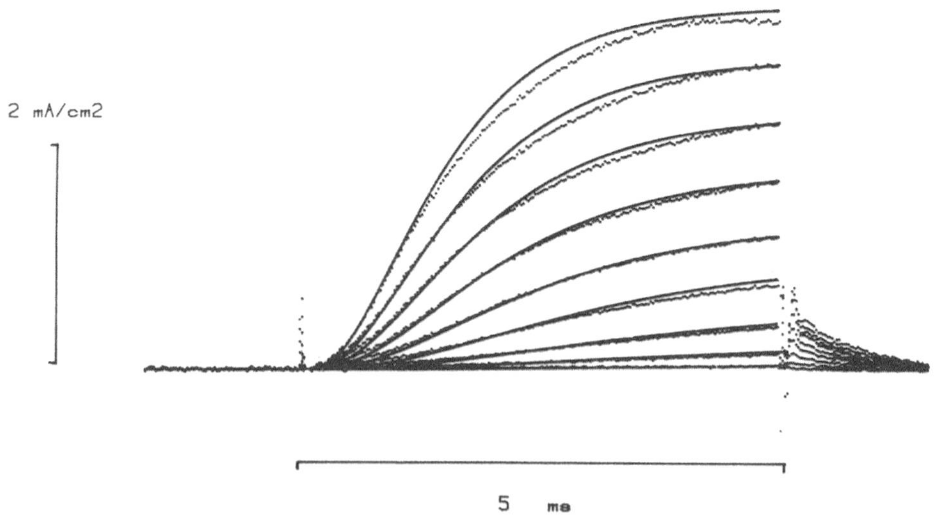

Figure 3. Superimposed tracings of the potassium current corresponding to 5 msec step depolarizations of the nerve membrane from its holding level (−60 mV) to −50, −40, −30, −20, −10, 0, 10, 20, and 30 mV. Continuous curves are computed according to Eq. (3) and (4) after curve fitting of the linearized experimental points with E_K = 74.5 mV and g_K = 35 mS cm^{-2}. Note the outward tails of potassium current at the cessation of the pulse. Calculation of E_K from the values of I_K at the end of the pulse and at the beginning of the tail indicate that for short duration pulses and depolarizations up to 30 mV there is no accumulation (temperature: 11°C). 10^{-7} M TTX was added to the bath to remove sodium current. The leak current was subtracted from the digitized traces.

found to vary according to a single exponential time course from their steady-state values before the pulse to their new steady-state values. Thus, for the potassium activation parameter, n

$$n(t) = n_\infty - (n_\infty - n_0) \exp(-t/\tau_n) \tag{4}$$

where n_0 is the initial value of n and n_∞ its final value at $t = \infty$.

As will be shown later in this chapter, the main noise component that could be observed under our experimental conditions is a potassium channel noise. For this reason, curve fits such as the one illustrated in Fig. 3 have been done only on potassium currents recorded in the presence of 10^{-7} M TTX. As shown in Figure 10, very good agreement was found between experimentally measured τ_n and those calculated from the original Hodgkin–Huxley (1952) equations for squid axons at a similar temperature.

B. Spontaneous Current Noise under Normal Physiological Conditions

Spontaneous membrane current noise was measured at rest and during long duration hyperpolarizing or depolarizing pulses. As shown in Fig. 4, noise was analyzed 1.5 sec

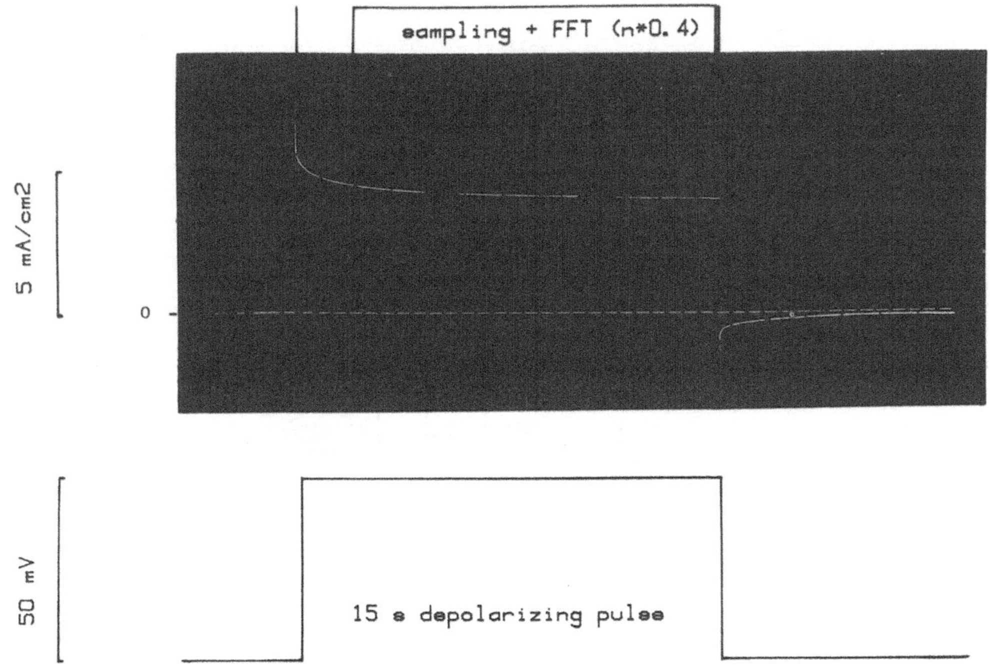

Figure 4. Ionic currents corresponding to a 50 mV, 15 sec depolarizing pulse during spontaneous noise measurement. The transient current was too fast to be recorded. Sampling and processing (which took respectively 200 and 400 msec) were started 1.5 sec after the beginning of the depolarizing step. The duration of the pulses (and thereby the number of averaged spectra) was automatically adjusted to the size of the pulse to avoid secondary effects due to potassium accumulation, electrode polarization or membrane deterioration. In this case, 34 spectra were averaged and the equilibrium potential for potassium ions (E_K) was found to shift from its resting value of about −80 to 52 mV.

after the beginning of the clamp pulses, the duration of which was adjusted to the size of the pulse (see figure caption). This noise was found to be one or two orders of magnitude larger than the background instrumental noise obtained for a similar input impedance of 1 megohm (Fig. 5). It decreased with frequency from 10–500 Hz and increased for higher frequencies. This secondary (probably artifactual) increase in noise was eliminated by subtracting spectra from a reference spectrum (most often the spectrum obtained at rest or for a slightly hyperpolarized membrane). The absence of large antiresonances in admittance records (see below) indicates that this procedure is unlikely to produce large distortions in the spectra. Membrane noise was found to decrease during moderate membrane hyperpo-

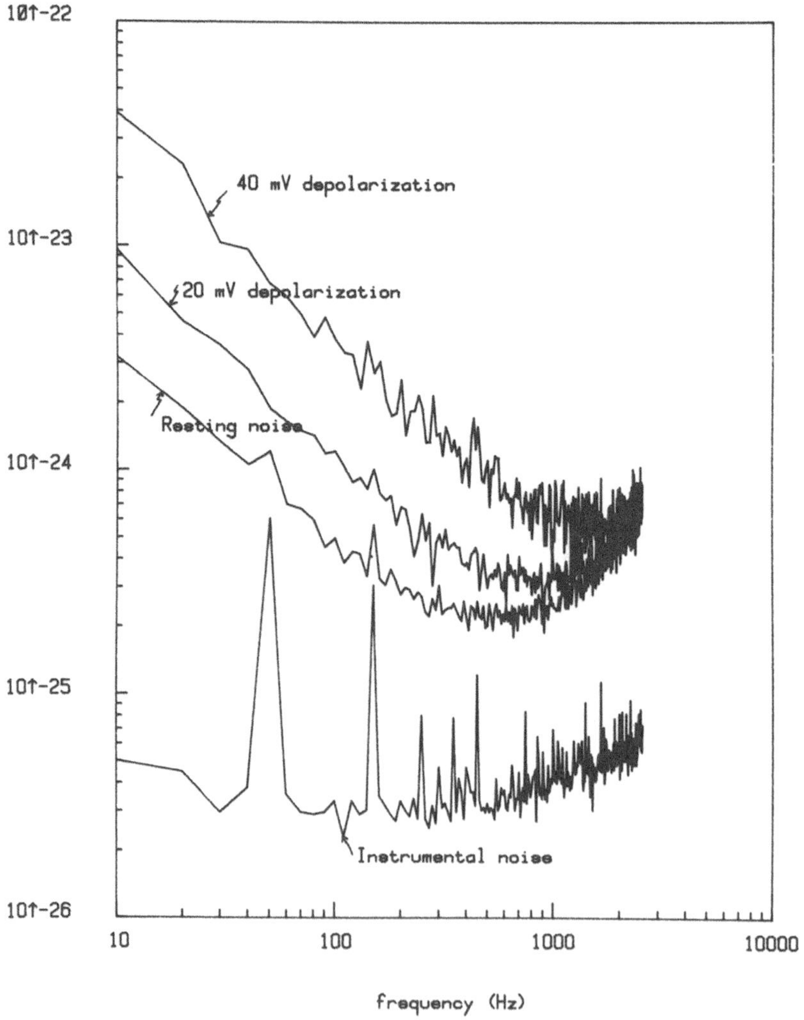

Figure 5. Power-density spectra of membrane current noise from a patch of membrane (about 10^{-4} cm^2) of a giant axon of *Periplaneta* at three potential levels (upper tracings) and instrumentation noise (lower tracing) obtained under similar experimental conditions with the membrane replaced by a 1 megohm resistor model. Note (1) that the instrumentation noise is 1–2 orders of magnitude smaller than the resting noise, (2) that the membrane noise increases with membrane depolarization, and (3) that power spectrum density decreases between 10–500 Hz and increases again. Original unretouched tracings. In this and subsequent figures, ↑ denotes an exponent, for example, 10 ↑ − 22 is equivalent to 10^{-22}.

larizations (up to 30 mV) but to increase for larger hyperpolarizations. However, these changes were small compared to those induced by membrane depolarizations. Depolarizations of the axonal membrane of up to 80 mV were found to increase noise by several orders of magnitude. This increase was always quickly and completely reversible following return to normal holding potential so that several "families" of current noise could be obtained on the same preparation under identical or different experimental conditions.

Noise spectra can be tentatively expressed in terms of three different components: $1/f$ noise, regarded as resulting from the flow of ions across open channels, with $S_K/[1 + (f/f_c^K)^2]$ corresponding to the fluctuation in the number of open and closed potassium channels, and $S_{Na}/[1 + (f/f_c^{Na})^2]$ corresponding to the fluctuation in the number of open and closed sodium channels. The respective contributions of each of these components in a given spectrum has been analyzed with ionic channel blockers.

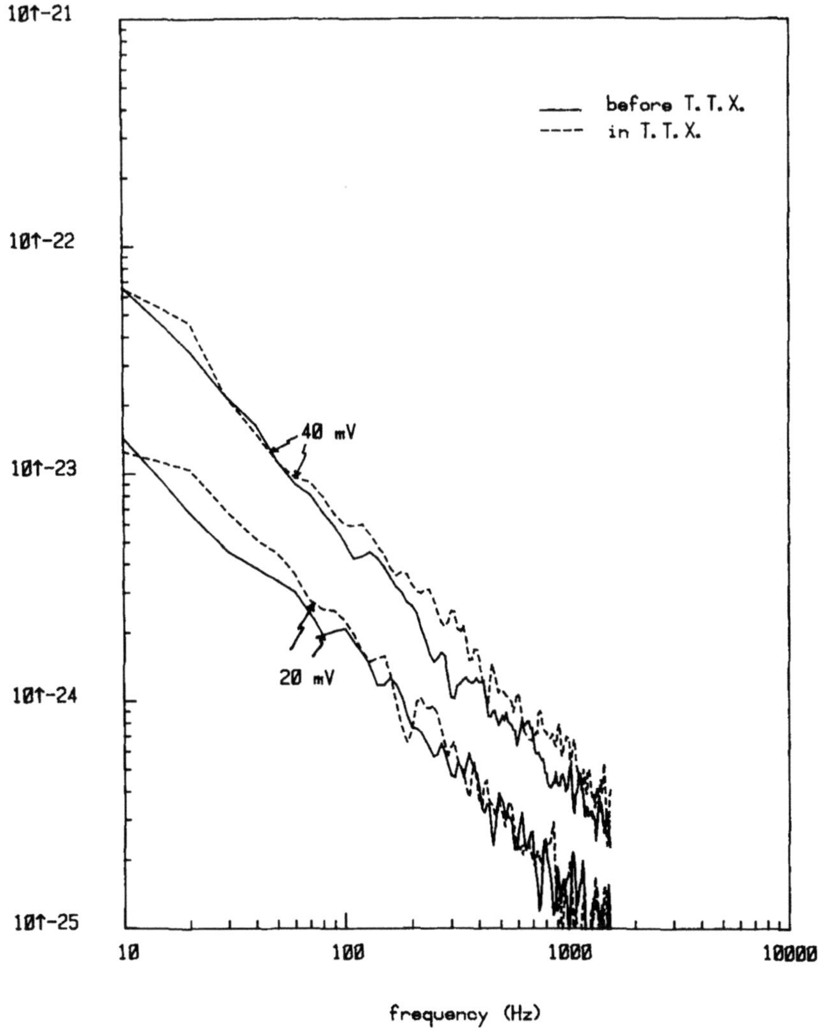

Figure 6. Difference spectra of current noise from a patch of cockroach axon before (continuous lines) and during (interrupted lines) application of 10^{-7} M TTX at two potential levels. Neither the amplitude nor the frequency distribution of the power spectra is modified by the sodium channel blocker, indicating that there is no sodium component in spontaneous current fluctuations under these conditions.

C. Effects of Tetrodotoxin on Membrane Current Noise

Addition of 10^{-7} M TTX, which completely blocks the sodium current (Fig. 3), was found to have no effect on the amplitude of the frequency distribution of spontaneous membrane noise. This absence of effect, which is illustrated in Fig. 6, was to be expected since the resting sodium conductance is very small (about 0.15 ms cm^{-2} according to Pichon, 1974) and sodium current inactivates after a few milliseconds when the membrane is depolarized.

D. Effects of 4-Aminopyridine on Membrane Current Noise

The agent 4-aminopyridine (4-AP) is known to selectively inhibit the potassium conductance in insect axons (Pelhate and Pichon, 1974). Addition of 1 mM 4-AP to the bathing

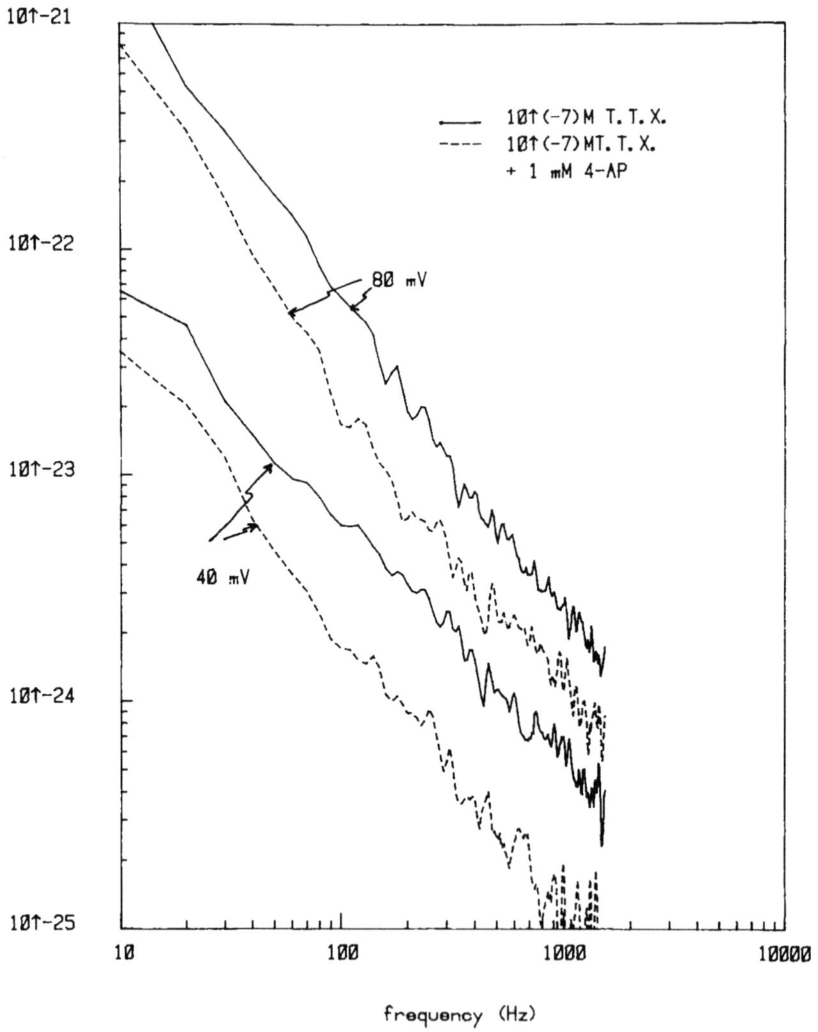

Figure 7. Difference spectra of spontaneous current fluctuations from a patch of cockroach axon depolarized by 40 and 80 mV from the holding potential before (continuous lines) and during (interrupted lines) external application of 1 mM 4-AP. The mean amplitude of the current noise is reduced by a factor of about three but the inhibition is larger for high than for low frequencies, suggesting that 4-AP induces a low-frequency noise.

solution induces a marked decrease in the amplitude of the spontaneous noise at all potential levels. This decrease which is illustrated in Fig. 7 was significantly larger for high frequencies than for low frequencies suggesting that, besides its blocking effect, 4-AP induces a low frequency noise as expected (Fishman *et al.*, 1977) if the molecule can bind to closed channels and be released when the channel opens as suggested by Meves and Pichon (1975, 1977) and Pichon *et al.* (1982).

E. Description of Membrane Current Noise and Potassium Relaxation Kinetics

Difference spectra such as those illustrated in Figs. 6 and 7 have been analyzed in terms of a $1/f$ component and a $1/[1 + (f/f_c^K)^2]$ Lorentzian component. Curve fits of these two components to individual difference spectra were done after elimination of artifactual peaks and smoothing. An approximate value of the corner frequency (f_c) of the Lorentzian

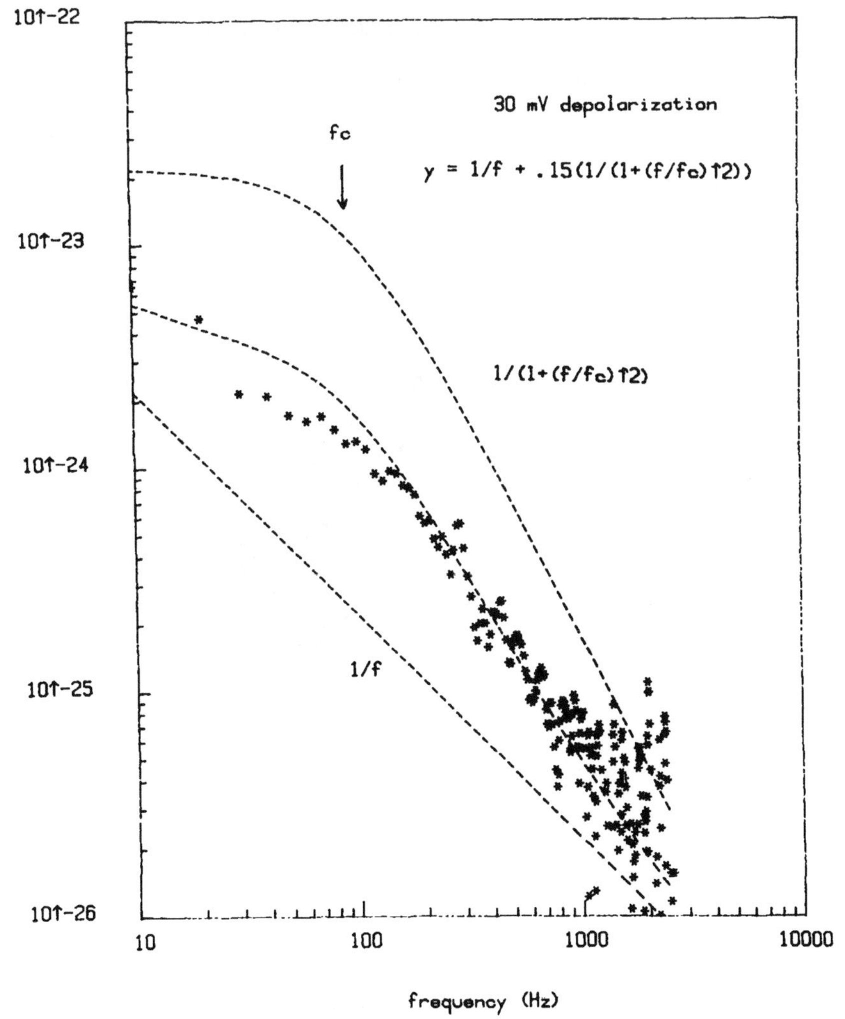

Figure 8. Curve fit of a difference spectrum of spontaneous noise from a patch of cockroach axon with the combination of $1/f$ and Lorentzian components. Original data points.

was then obtained by taking the frequency for which the appropriate $1/f$ straight line was tangent to the smoothed curve. In a third step, this value together with (1) the relative proportion of Lorentzian over $1/f$ components, and (2) a scale factor were entered into the computer. These three values were then adjusted until a reasonably good fit was found between the experimental data and the theoretical curve

$$y = b\left[\frac{1}{f} + \frac{a}{1 + (f/f_c^K)^2}\right]$$

The results of such a fit is illustrated in Fig. 8, which also shows the $1/f$ and the Lorentzian components used for the fit. In all experiments, the $1/f$ component was found to dominate

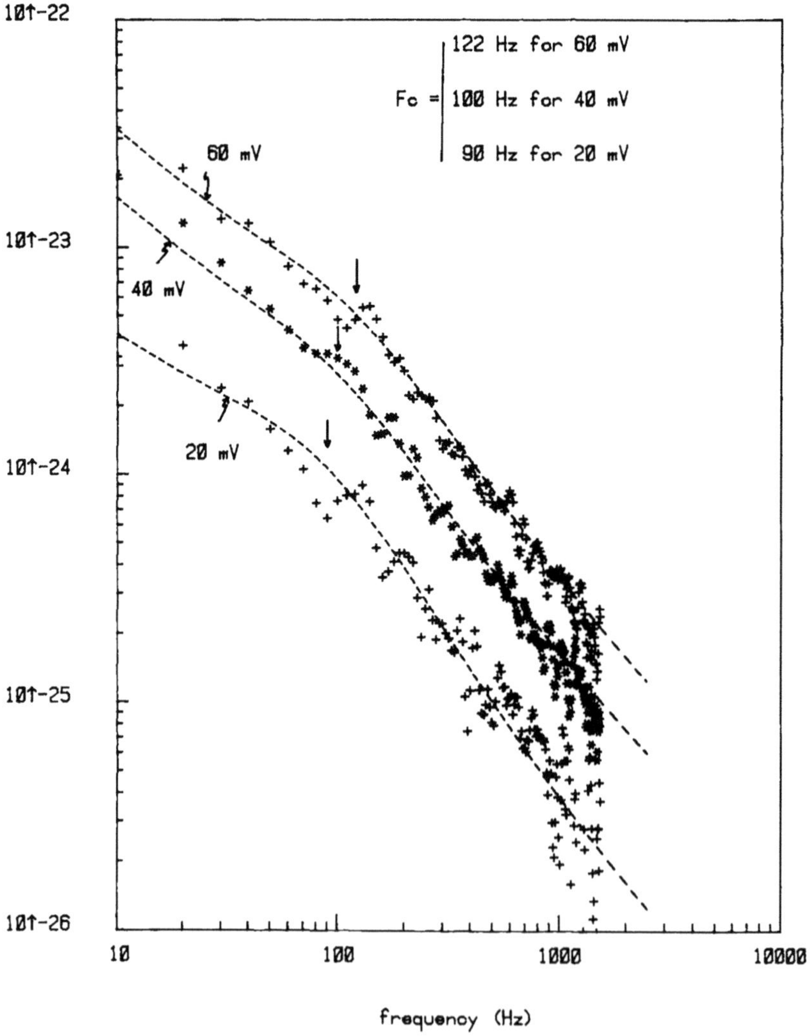

Figure 9. Family of fitted difference spectra of membrane noise from a patch of cockroach axon. Peaks corresponding to 50 Hz (AC line) and its harmonics were eliminated and the corrected data smoothed once. The relative proportion of Lorentzian noise decreases with increased membrane depolarization, whereas the corner frequency increases.

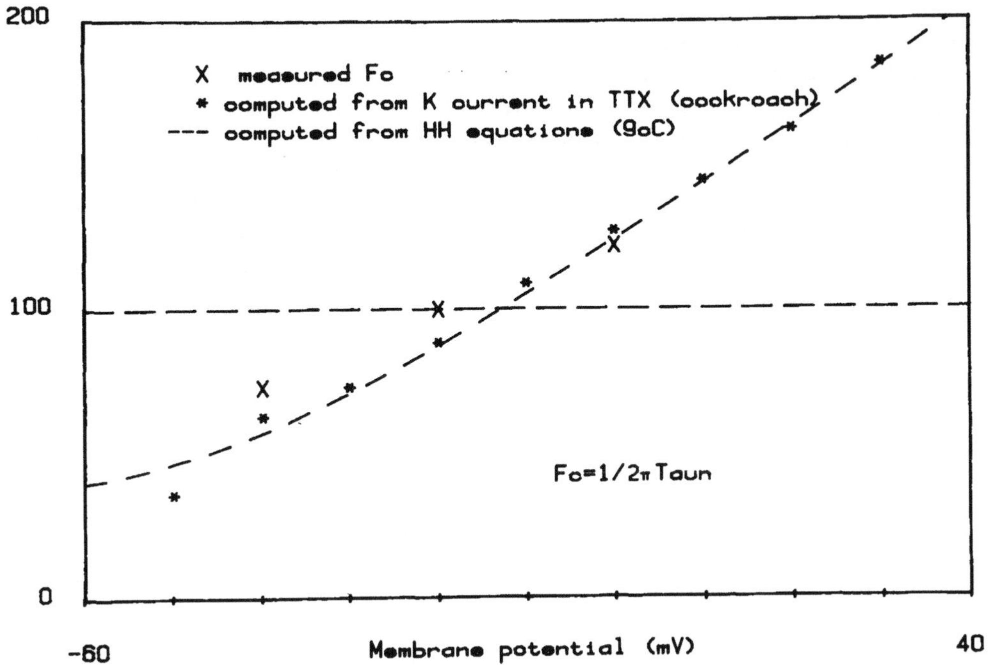

Figure 10. Measured (crosses) and computed (stars and interrupted line) corner frequencies against membrane potential. Data from Figs. 3 and 9. Note the good correspondence between measured and computed f_c suggesting that the Lorentzian component of the noise is related to the relaxation of the potassium channels of the axonal membrane (temperature: 11°C).

the records, especially for large depolarizations in healthy axons, and the *a* parameter never exceeded 0.3. Corner frequencies, f_c, were found to vary between 80 and 150 Hz. For a given axon, the corner frequency was found to increase with the value of the depolarization. An example of a family of three fitted curves corresponding to 20, 40, and 60 mV depolarizations is illustrated in Fig. 9. These values were compared with those predicted from the relaxation kinetics of the potassium system ($f_c = 1/2\pi\tau_n$) and, as illustrated in Fig. 10, very good agreement was found between measured and calculated values. This result, together with the previous observations that spontaneous current noise is insensitive to TTX and strongly reduced by 4-AP, suggest that the Lorentzian component of the noise is indeed due to the opening and closing of single potassium channels of the Hodgkin–Huxley type.

F. Single Channel Conductance and Potassium Channel Density

The previous results suggest that membrane noise recorded in our experiments results from the opening and closing of potassium channels, justifying an attempt to evaluate a single-channel conductance from our noise data. If one makes the assumptions that the *N* potassium channels of the active patch of membrane are independent and have only two states, one open, with a conductance γ, and one nonconducting (closed), and if one knows the probability *p* for one channel to be open, the mean current *I*, the mean variance $\sigma_I^2 = (I - \bar{I})^2$, and the driving force for potassium ions ($V_m - E_K$), it is possible to get an evaluation of γ from the equation

$$\gamma = \sigma_I^2/[\bar{I}\ (1 - p)\ (V_m - E_K)]$$

The variance σ_I^2 was obtained from the area under the current-noise density spectra, E_K from the current records such as those illustrated in Fig. 3, and p from the literature (Pichon, 1974). $\bar{I}$ was obtained either directly from oscilloscope traces such as the one illustrated in Fig. 4, from families such as that illustrated in Fig. 3 (after correction for potassium inactivation), or indirectly from admittance values. The single potassium-channel conductance was found to be of the order of 2.5 pS for small depolarizations, if no correction was made for unreliabilities in the measurement of σ_I^2 due to the limited bandwidth and the contribution of $1/f$ noise. Division of the maximum potassium conductance (g_K) by this value yields a potassium-channel density of $100/\mu m^2$. There are some indications that single-channel conductance may vary with membrane voltage but more experiments are needed to clarify this point.

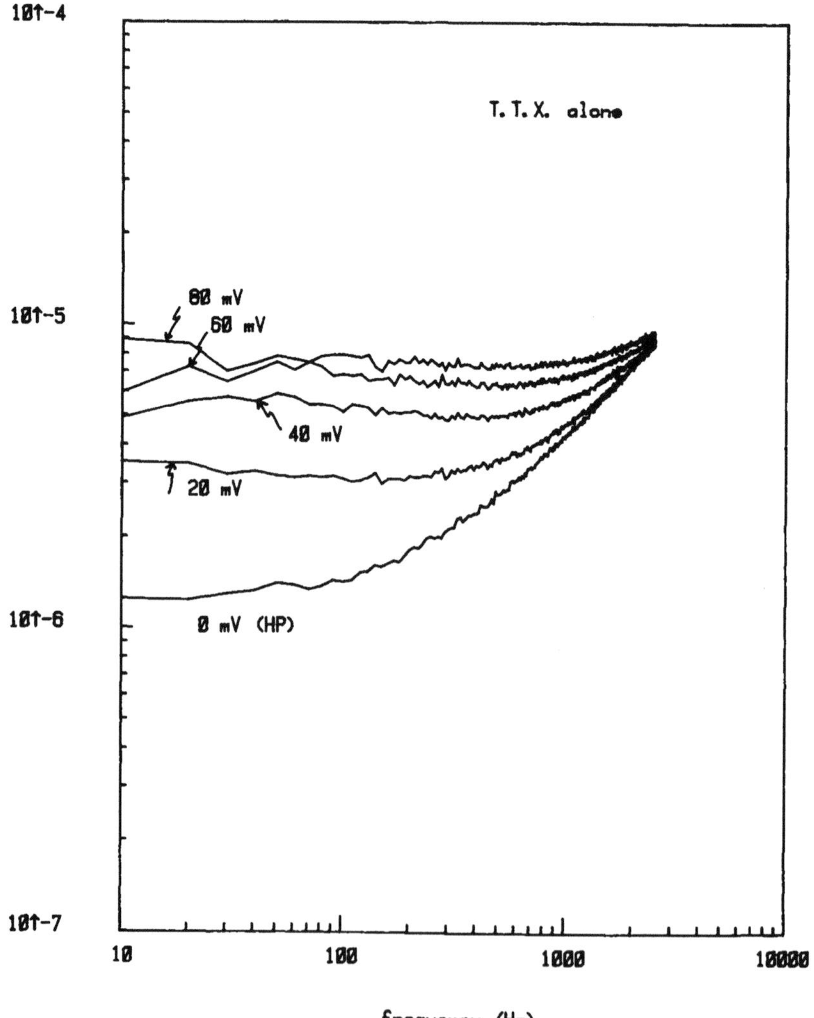

Figure 11. Admittance magnitude against frequency of a patch of cockroach axon at rest (0 mV depolarization) and during various depolarizations in the presence of 10^{-7} M TTX in the external solution. As expected, admittance increases with membrane depolarization but little antiresonance is observed.

G. *Complex Admittance of the Axonal Membrane*

As mentioned in the section on experimental methods, we have measured admittance magnitude and phase of the axonal membrane under various experimental conditions. The pseudorandom source of noise of the spectrum analyzer (PRBS) was adjusted to 1 mV peak-to-peak at the membrane and added to the command voltage signals (holding and steps). The noise sequence (which contained all frequency components necessary for the analysis) was processed at the same time as the corresponding membrane current to obtain both amplitude and phase of the transfer function (256 points each) of the membrane.

The membrane was found to behave as predicted from the linearized Hodgkin–Huxley (1952) potassium conductance, as calculated by Fishman *et al.* (1979). The admittance for

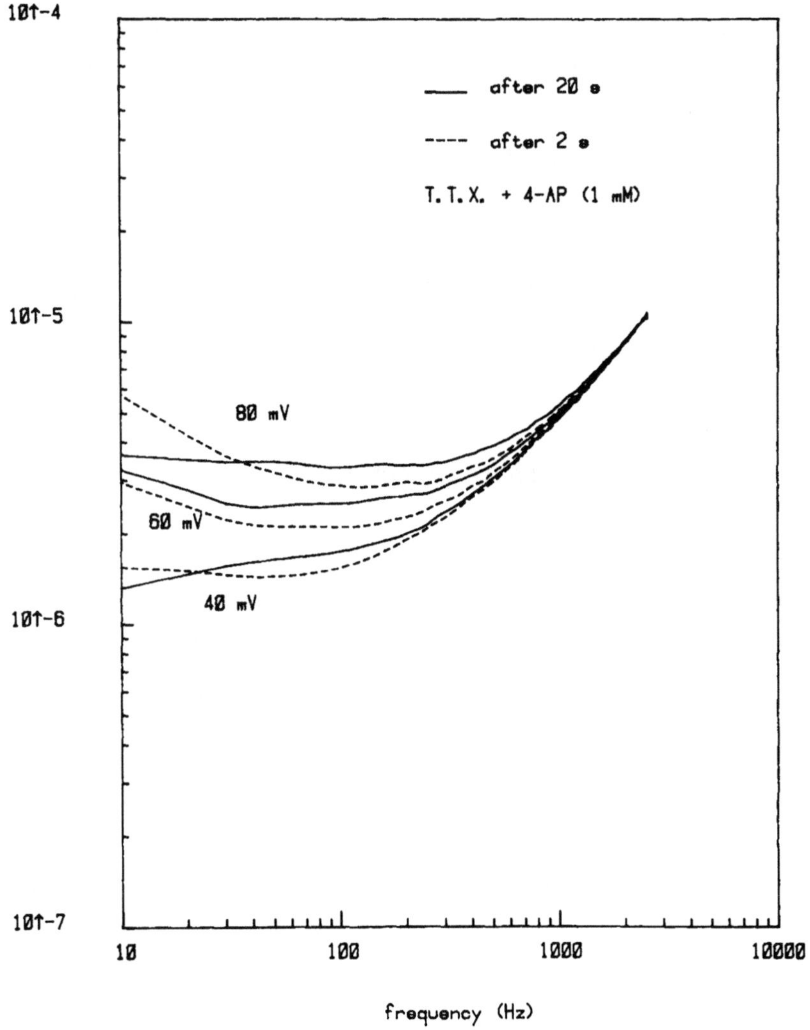

Figure 12. Effects of the duration of the depolarization on admittance of a patch of cockroach axon in TTX (10^{-7}M) and 4-AP (1 mM). Admittance at zero frequency as well as antiresonance peaks are much more important at the end of relatively short (2 sec, interrupted lines) depolarizations than at the end of 20 sec (continuous lines) depolarizations, suggesting time-dependent modifications of the potassium system.

the potassium system can be represented as being made of three components in parallel: (1) potassium chord conductance ($g_K\ n_\infty^3$ in cockroach axons), (2) the membrane capacitance (C), and (3) the frequency-domain manifestation of the voltage- and time-dependent properties of the potassium system, $g_n/(1 + j\omega\tau_n)$. At low-frequency values, the amplitude shows an asymptote that is essentially $g_K n_\infty^3 + g_n$. At frequencies higher than about 50 Hz and up to about 1000 Hz, the curve drops to a minimum, the antiresonance, due to the interactions between the capacitive and the inductive branches of the circuit. The antiresonance is substantially damped by lack of full isolation, which introduces a parallel (resistive) pathway (Fishman *et al.*, 1975b). For higher frequencies, the capacitance dominates and is responsible for the rise in the admittance amplitude. At the resting potential or for slightly hyperpolarized values of potential, the antiresonance is minimum. It becomes clearly visible when the membrane is depolarized and its peak is shifted towards higher frequencies as the membrane is more depolarized, reflecting changes in the inductive component (Fig. 11).

Addition of TTX (10^{-7} M) to the bathing medium had no significant effect on the admittance curves, whereas 1 mM 4-AP shifted the curve towards lower admittance values by a factor of about three.

Since it is possible to record complete admittance curves in only one run, we have decided to see whether or not the dynamic properties of the potassium system change with time during long depolarizations. The result of such an experiment in the presence of TTX and 4-AP is illustrated in Fig. 12. It clearly indicates that, as in single nodes of Ranvier of the frog (van den Berg *et al.*, 1977), the potassium system (which is modified by 4-AP) inactivates during long depolarizations, and this is reflected in admittance magnitude curves as a decrease in both zero-frequency chord conductance and antiresonance amplitude. These data were obtained with 4-AP present and it cannot be excluded that the drug molecule plays some role in this inactivation.

IV. DISCUSSION AND CONCLUSION

The data presented in this chapter illustrate the point that the isolated giant axon of the cockroach in a double-oil-gap arrangement is a useful preparation for the study of membrane ionic currents, spontaneous noise, and admittance.

The results confirm the observations made on squid axon membrane (Conti *et al.*, 1975; Fishman *et al.*, 1975a,b) and the node of Ranvier of the frog (van den Berg *et al.*, 1977; Neumcke *et al.*, 1980). Fast inactivation of the sodium conductance and the relative paucity of sodium channels in the cockroach axonal membrane (as estimated from g_{Na}) explain why, under our experimental conditions, no sodium noise was observed. As in squid axons, the spectral characteristics of the spontaneous membrane current fluctuations can be reasonably well fitted with the combination of a $1/f$ component and one Lorentzian. This kind of fit has been recently questioned on several grounds, the most important one being that the $1/f$ component could be an artifact linked to the poor physiological state of the preparation; an alternative model has been proposed by Neumcke *et al.* (1980). According to this model, noise data for potassium can be fitted with the combination of a diffusion spectrum originating from the opening and closing of the potassium channels and a plateau. No attempt has yet been made to fit this model to our data. Nevertheless, there are reasons to believe that the $1/f$ component is not artifactual but corresponds to restricted diffusion of moving ions at the narrow opening of the potassium channels, although one cannot exclude the possibility that a fraction of the $1/f$ noise is related to the leakage conductance. One argument in favor of the $1/f$ component not being an artifact is that this component dominated in preparations exhibiting good excitability, small leakage, and large ionic currents (the reverse of what one would expect from a preparation in bad physiological condition).

The good fit between the measured corner frequencies of the Lorentzian component of the noise and the values calculated from the kinetics of the potassium current indicates that the Lorentzian component reflects the opening and closing of the potassium channels, but it is too early to conclude that there is complete agreement between the experimental data and a two-state (open–closed) conductance model. More data and reliable curve-fitting procedures are needed to clarify this point. Our estimation of single potassium channel conductance (2.5 pS) compares well with those reported in the literature for squid axons (12 pS according to Conti *et al.*, 1975, 2 pS according to Moore *et al.*, 1979) or nodes of Ranvier of the frog (4 pS according to Begenisich and Stevens, 1975, 2.9 pS according to van den Berg *et al.*, 1977, and 2.7 pS for motor fibers and 4.6 pS for sensory fibers according to Neumcke *et al.*, 1980). This estimation should, however, be taken with some caution since it relies on a small number of experiments and has not been corrected for the contribution of $1/f$ noise.

Our admittance data are in good agreement with those of Fishman *et al.* (1982), the only difference being that antiresonance appears to be smaller in cockroach axons. This difference may be due to the fact that we usually measured admittance at the end of long (several seconds) depolarizing pulses and that, as illustrated in our Fig. 12, antiresonance was reduced in comparison with that measured after only two seconds. In their experiments, Fishman *et al.* (1982) measured admittance even earlier during the depolarizations (20 msec or 1000 msec).

The insect axon is a very suitable preparation for the study of the microscopic events underlying excitability. Since the axonal membrane of the cockroach is directly accessible from the outside, it should be possible, if not easy, to record single-channel activity from very small patches of membrane and to correlate this activity with membrane noise and kinetics of membrane conductance changes. This technique should also be very useful for pharmacological studies.

ACKNOWLEDGMENTS. The authors wish to thank Dr. Nancy J. Lane for the electron micrographs of single axons. This work was supported in part by grants from the Conseil de Recherche en Science et Genie (Canada), grant A5274, the Centre National de Recherche Scientifique (France), and the Délégation Générale à la Recherche Scientifique et Technique (France), grant 79. 7.1068. G. V. Lees was supported by a European Molecular Biology Organization fellowship.

REFERENCES

Begenisich, T., and Stevens, C. F., 1975, How many conductance states do potassium channels have? *Biophys. J.* **15:**843–846.

Conti, F., DeFelice, L. J., and Wanke, E., 1975, Potassium and sodium ion current noise in the membrane of the squid giant axon, *J. Physiol. London* **248:**45–82.

Derksen, H. E., and Verveen, A. A., 1966, Fluctuations of resting neural membrane potential, *Science* **151:**1388–1389.

Fishman, H. M., 1973, Relaxation spectra of potassium channel noise from squid axon membrane, *Proc. Natl. Acad. Sci. USA* **70:**876–879.

Fishman, H. M., Moore, L. E., and Poussart, D., 1975a, Potassium-ion conductance noise in squid axon membrane, *J. Membr. Biol.* **24:**305–328.

Fishman, H. M., Poussart, D., and Moore, L. E., 1975b, Noise measurements in squid axon membrane, *J. Membr. Biol.* **24:**281–304.

Fishman, H. M., Moore, L. E., and Poussart, D., 1977, Ion movements and kinetics in squid axon. II. Spontaneous electrical fluctuations, *Ann. N.Y. Acad. Sci.* **303**:399–423.

Fishman, H. M., Poussart, D., and Moore, L. E., 1979, Complex admittance of Na^+ conduction in squid axon, *J. Membr. Biol.* **50**:43–63.

Fishman, H. M., Moore, L. E., and Poussart, D., 1982, Squid axon K conduction: Admittance and noise in short versus long-duration step clamps, in: *The Biophysical Approach to Excitable Membranes* (W. J. Adelman, Jr. and D. E. Goldman, eds.), pp. 65–95, Plenum Press, New York.

Hodgkin, A. L., and Huxley, A. F., 1952, Quantitative description of membrane current and its application to conduction and excitation in nerve, *J. Physiol., London* **117**:500–544.

Meves, H., and Pichon, Y., 1975, Effects of 4-aminopyridine on the potassium current in internally perfused giant axons of the squid, *J. Physiol., London* **251**:60–62P.

Meves, H., and Pichon, Y., 1977, The effect of internal and external 4-amino-pyridine on the potassium currents in intracellularly perfused squid giant axons, *J. Physiol., London* **268**:511–532.

Moore, L. E., Fishman, H. M., and Poussart, D., 1979, Chemically induced K conduction noise in squid axon, *J. Membr. Biol.* **47**:99–112.

Neumcke, B., Schwarz, W., and Stämpfli, R., 1980, Difference between K channels in motor and sensory nerve fibres of the frog as revealed by fluctuation analysis, *Pflügers Arch.* **387**:9–16.

Pelhate, M., and Pichon, Y., 1974, Selective inhibition of potassium current in the giant axon of the cockroach, *J. Physiol., London* **242**:90–91P.

Pichon, Y., 1967, Application de la technique du voltage imposé à l'étude de la fibre nerveuse isolée d'insecte, *J. Physiol., Paris* **9**:282.

Pichon, Y., 1970, Voltage-clamp study of the cockroach giant axon: A simple method, *J. Physiol., London* **210**:86–88P.

Pichon, Y., 1974, Axonal conduction in insects, in: *Insect Neurobiology* (J. E. Treherne, ed.), pp. 73–117, North Holland, Amsterdam.

Pichon, Y., and Boistel, J., 1967, Current-voltage relations in the isolated giant axon of the cockroach under voltage-clamp conditions, *J. Exp. Biol.* **47**:343–355.

Pichon, Y., and Treherne, J. E., 1974, The effects of sodium-transport inhibitors and cooling on membrane potentials in cockroach central nervous connectives, *J. Exp. Biol.* **61**:203–218.

Pichon, Y., Meves, H., and Pelhate, M., 1982, Effects of aminopyridines on ionic currents and ionic channel noise in unmyelinated axons, in: *Effects of Aminopyridines and Similarly Acting Drugs on Nerve, Muscle and Synapse* (P. Lechat, S. Thesleff and W. C. Bowman, eds.), pp. 53–68, Pergamon Press, Oxford.

Poussart, D., 1969, Nerve membrane current noise: Direct measurement under voltage clamp, *Proc. Natl. Acad. Sci. USA* **64**:95–99.

Poussart, D., 1971, Membrane current noise in lobster axon under voltage clamp, *Biophys. J.* **11**:211–234.

Poussart, D., and Ganguly, U., 1977, Rapid measurement of system kinetics, *Proc. IEEE* **65**:741–747.

Poussart, D., Moore, L. E., and Fishman, H. M., 1977, Ion movement and kinetics in squid axon. I. Complex admittance, *Ann. N.Y. Acad. Sci.* **303**:355–379.

Siebenga, E., Meyer, A., and Verveen, A. A., 1973, Membrane shot noise in electrically depolarized nodes of Ranvier, *Pflügers Arch.* **341**:87–96.

Treherne, J. E., and Pichon, Y., 1972, The insect blood-brain barrier, in: *Advances in Insect Physiology, IX* (J. E. Treherne, M. J. Berridge, and V. B. Wigglesworth, eds.), pp. 257–308, Academic Press, London.

van den Berg, R. J., Siebenga, E., and de Bruin, G., 1977, Potassium ion noise currents and inactivation in voltage-clamped nodes of Ranvier, *Nature* **265**:177–179.

11

Ion-Selectivity and "Gating" Properties of the Conductance Pathways in Squid Axon

The View of a Membrane–Cortex Model

Donald C. Chang

I. INTRODUCTION

The objectives of this chapter are two-fold: (1) to review the experimental findings of the ionic effects on the ion-selectivity and "gating" properties of the conductance pathways of the nerve cell, and (2) to discuss how these findings can be used to differentiate various hypotheses of "channel" structure, based either on the traditional membrane view or an alternative view, the membrane–cortex model.

We will use the word "channel" in a very general sense to represent a conductive pathway for the permeant ions. It does not necessarily have to be a single pore. In fact, the single-pore model is only one of the possible models of channel structure, and we will have an extensive discussion about this particular model in this chapter.

This chapter is divided into six sections. The first section is a summary of the chapter. The second section is an introduction to the membrane–cortex model. This model is the result of an attempt to integrate the features of two historically antagonistic views, the membrane-permeation view and the cytoplasm-adsorption view. This work takes into strong consideration the recent morphological findings that cytoplasm is a highly structured protein complex and the membrane proteins are clearly connected to the cytoplasmic protein network (see Part I of this book). We will give a summary of the basic features of this model and its supporting evidence.

The third section is a review of the ion-selectivity of the channel based on the reversal-potential measurements. The results and method of determining the "permeability ratio" of ions in a conductance pathway are discussed. The effects of internal ions on the ion-selectivity and the implication of the lability of the "permeability ratio" will also be discussed. We

Donald C. Chang • Department of Physiology, Baylor College of Medicine, Houston, Texas 77030; and Department of Physics, Rice University, Houston, Texas 77251.

will point out how consideration of the cortex structure can help to explain the experimental data.

The fourth section concerns the ionic effects of internal K^+ on the kinetics of the early channel. A summary report on the study of correlation between the effects of internal ions (Cs^+ and Na^+) on the early current and the late current is given in the fifth section. The data bear significantly on the structure and properties of the ionic conductance pathways, including a suggestion of possible coupling between the early and late channels.

Finally, the sixth section summarizes the evidence showing that the "excitable membrane" is a polyelectrolyte system. It explains how the experimental findings discussed in the earlier sections can be explained based on such a view. A conceptual model of the channel structure different from the single-pore model is suggested.

II. THE MEMBRANE–CORTEX MODEL

A. Basic Features

Unlike the general concept of the ionic theory, which regards the excitable membrane as a thin (60–80 Å) lipid bilayer decorated with proteins in a manner described by the fluid–mosaic model (Keynes, 1972), we think the functional "physiological membrane" is much more complicated. Specifically, the membrane–cortex model maintains:

1. The "physiological membrane" of the excitable cell consists not only of the cell membrane (plasmalemma), but also a thick layer of sublemmal protein structure called the "cell cortex."
2. This cortical protein structure forms a fixed-charge system and hence possesses some properties of a biological ion-exchanger.
3. The ionic channels are membrane proteins connected to the cortical protein structures. These proteins can undergo conformational changes during excitation, and thus exhibit time-variable electrical properties.

Since the cortex is much thicker than the plasmalemma, the physical properties of the membrane–cortex complex can be significantly different from those of the plasmalemma alone. As will be seen in the latter part of this chapter, properties of the cortical complex can explain some of the electrophysiological findings that are difficult to explain in the traditional membrane (plasmalemma) view.

B. Supporting Evidence

A considerable amount of evidence supports the existence of a functionally important cortex.

1. Morphological Evidence

Electron microscopic studies of many types of biological cells have found that the layer of cytoplasm adjacent to the plasmalemma has a dense and complex physical structure that contains many contractile and structural proteins (Wolosewick and Porter, 1976; Korn, 1978; Hodge and Adelman, 1980; Ellisman and Porter, 1980). This sublemmal structure is called the cell cortex. Such cortex structure was found in the squid axon in a study using micro-injection of oil drops and dyes (Chambers and Kao, 1952) and confirmed by differential and electron microscopy (Metuzals and Izzard, 1969; Metuzals, 1969). Several laboratories have shown that this cortical structure is composed mainly of microtubules and neurofilaments and crossbridges (Metuzals and Tasaki, 1978; Hodge and Adelman, 1980; Bloodgood and Rosenbaum, 1976). Recently, studies of the internal surface structure of perfused squid axon using scanning electron microscope have given clear evidence that the intactness of this cortical structure is correlated with the excitability of the axon (Metuzals and Tasaki, 1978).

2. *Biochemical Evidence*

When the axon is perfused internally with proteases, (e.g., 0.1 mg/ml pronase), it rapidly develops spontaneous firing, and excitability soon is suppressed. However, the same concentration of proteases applied externally has no demonstrable effect (Tasaki *et al.*, 1965). This finding suggests that the protein structure at the internal surface of the perfused axon is critical in maintaining excitability. With careful perfusion of pronase inside, one can partially remove the inactivation mechanism of the early conductance (Armstrong *et al.*, 1973), which indicates that the internal proteins are related to the gating mechanism of the conductance pathway. Gainer *et al.* (1974), Inoue *et al.* (1976), and Takenaka *et al.* (1976) studied the chemical composition of the internal perfusate from the squid axon. They found that a 12,000 dalton protein was released when the axon was stimulated repetitively or depolarized by high external potassium. Later, release of proteins of 45,000 and 68,000 daltons in association with repetitive firing or prolonged depolarization of the axon was also observed (Pant *et al.*, 1978; Yoshioka *et al.*, 1978). These findings suggest that alteration of the sub-axolemmal protein structure is involved in the excitation process.

3. *Physiological Evidence*

It is well known that the physiological properties of the axon "membrane" are asymmetrical. For example, the internal surface of the "physiological membrane" is sensitive to different anions, while the external surface is relatively insensitive (Tasaki *et al.*, 1965). In addition, cations that produce a high conductance state when applied externally have no such effect when applied internally (Singer and Tasaki, 1968). High concentration of Ca^{++} outside does not affect the excitability of the axon, but even low $[Ca^{++}]$ inside the axon causes it to degenerate quickly. Other studies have shown that pH, heavy metals, and organic ions all affect the two sides of the "physiological membrane" differently (Singer and Tasaki, 1968). This property of asymmetry can be attributed to the existence of a cortical layer at the inner side of the "physiological membrane."

4. *Physical Evidence*

Cohen *et al.* (1968) studied birefringence changes in squid axon during the action potential and concluded that the surface structure of the axon changes during the excitation process. Similar structural changes were observed by Tasaki *et al.* (1968, 1969) in their studies of birefringence and fluorescent probes in squid axon. Electromagnetic emission at micron wavelength also was detected in crab nerve during activity (Fraser and Frey, 1968). These studies indicate that during the excitation process, molecules at the periphery of the axon undergo a certain conformational change. Since the sublemmal proteins are released during nerve activity, the molecules that undergo conformational changes as detected in the optical studies are likely to be those of the axon cortex. Indeed, recent studies by Iwasa and

Tasaki (1980) on mechanical changes in squid axon have shown that the action potential is associated with a small, quick swelling and shrinkage of the axon, which reflects a conformational change including the cortex.

In short, ample evidence supports the notion that the cortical protein structure is an inseparable part of the functional "excitable membrane." The excitation process seems to be correlated with a conformational change of this cortical protein. We feel that effects on the alteration of the cortical structure and its ion-exchange properties must be considered in the studies of the various effects of the "excitable membrane." In the following, we will use such an approach to examine some of the recent electrophysiological findings.

III. ION-SELECTIVITY OF CHANNELS AS DETERMINED BY MEASUREMENTS OF THE REVERSAL POTENTIAL

A. Determination of the Permeability Ratio Using a Reversal-Potential Method

One important problem in the study of excitation is to understand the properties of the conductance pathways through which ions move across the cell surface. Hodgkin and Huxley (1952b) originally thought that ions may be transported by means of carriers. Subsequent studies indicated that the conductance pathways of the excitable membrane are more likely to be in the form of "channels." Based on the assumption of ion-independence, it was suggested that there are separate "channels" for each ion species. For instance, the "Na^+ channel" is selectively permeable to Na^+, and the "K^+ channel" is selectively permeable to K^+.

Later studies in both frog nerve fibers and squid axons showed that the so-called "Na^+ channel" is not exclusively permeable to Na^+, as K^+ also enters it (Frankenhauser and Moore, 1963; Chandler and Meves, 1965). In fact, various ions can pass through the Na^+ channel with different degree of ease. The ion-selectivity then becomes an important property of the channel and has been studied actively in the past decade.

Working with perfused squid axon, Chandler and Meves (1965) suggested a method to determine the ion-selectivity of the Na^+ channel experimentally, which is called the

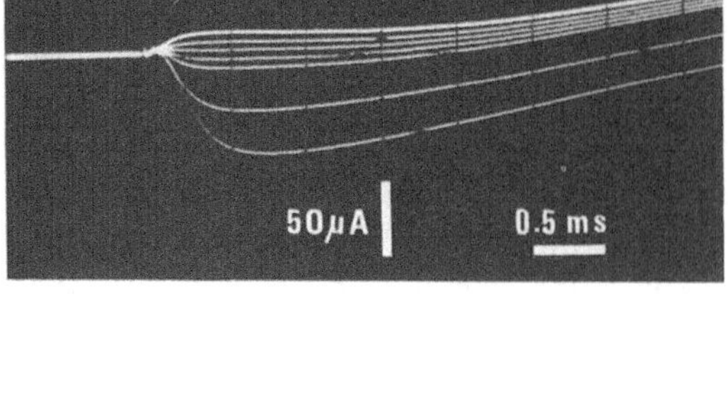

Figure 1. A sample record of reversal-potential measurement. (Top) Membrane potentials under voltage-clamping. (Bottom) Membrane currents in association with various depolarizing-potential steps. The axon was clamped at a holding potential (V_h) of −67.5 mV. A hyperpolarizing conditioning-pulse of 30 mV and 20 msec was applied before the depolarizing pulse was applied. The current traces (counting upward) correspond to depolarizing potentials of 70, 80, 90, 92, 94, 96, 98, 108 mV above V_h. The reversal potential is determined as the one giving zero early current. In this case, V_{rev} is estimated as 94 mV above V_h, or V_{rev} = +26.5 mV. Axon #81-39 $[K^+]_i$ = 200 mM, $[Na^+]_i$ = 200 mM, $[K^+]_o$ = 0, $[Na^+]_o$ = 425 mM. Temperature = 5.6°C.

"reversal potential method." As an axon is depolarized under voltage-clamping, it gives rise to an early current that varies from negative (inward flowing) to positive (outward flowing), depending on the depolarizing potential (see Fig. 1). One can determine a particular depolarizing potential at which the early current is zero. This potential is called the "reversal potential" (V_{rev}). Assuming a constant field model, Chandler and Meves (1965) proposed that the measured V_{rev} is equal to the diffusion potential as described by Goldman (1943) and Hodgkin and Katz (1949), that is, in the absence of $[Na^+]_i$ and $[K^+]_o$,

$$V_{rev} = \frac{RT}{F} \ell n \frac{P_{Na}[Na^+]_o}{P_K[K^+]_i} \tag{1}$$

In their experiment, P_{Na}/P_K was determined to be 12.7. The reversal potential method is used now widely to determine the "permeability ratio" in many membrane channels.

B. Ion-Selectivity and Channel Structure: Measurement of Reversal Potential as a Function of $[K^+]_i$

The study of ion-selectivity is important not only to test the ion-independence principle, it also can give important information on the structure and properties of the ionic conductance pathway. One popular hypothesis in the traditional membrane view assumes that the ionic channel is a rigid, narrow, aqueous pore formed by protein molecules embedded in the lipid bilayer of the cell membrane (Hille, 1975a). The ion-selectivity of the channel is explained by relating the pore size to the radius of the ion. For example, the relative permeability of the potassium channel of the squid axon to various ions was explained by assuming that the inner pore of this channel is 2.6–3.0 Å in diameter, thus, only ions that have radius less than this size can cross the channel (Armstrong, 1975). A similar view of this mechanical sieve effect is used by Hille (1971) to explain his findings of permeability ratios for monovalent cations in Na^+ channels of the frog node. He proposed that deep inside the sodium channel is a selectivity filter, which is a hole with a size of 3 × 5 Å. The large cations cannot permeate the sodium channel because they cannot pass this hole. The relative permeability to smaller ions is determined by the extent to which the partly hydrated ion can fit into this hole structure.

This rigid-pore hypothesis can be tested experimentally. Specifically, the idea of a mechanical selectivity filter implies that the ion-selectivity of a channel is fixed. It would be instructive to measure the permeability ratio at different ionic concentrations and see if the ion-selectivity of the channel is fixed or labile.

Using the method of reversal potential, Chandler and Meves (1965) had measured the permeability ratio (P_{Na}/P_K) of the early channel, i.e., Na^+ channel in squid axons, and found that it varied from 11.5 to 7.9 as the $[K^+]_i$ was changed from 300 mM to 24 mM. To verify their results, we have measured the reversal potential of the early channel as a function of $[K^+]_i$. The results are shown in Fig. 2. Based on these data, one can calculate the permeability ratio by using Eq. (1). It is found that the "permeability ratio" varies with $[K^+]_i$ (see Fig. 3). Our results confirm the earlier finding of Chandler and Meves (1965), however, our data show a much sharper change of the "permeability ratio" than they observed. We found that P_{Na}/P_K varies from 13.2 at $[K^+]_i$ = 400 mM to 3.7 at $[K^+]_i$ = 24 mM. Cahalan and Begenisich (1976) also have reported studies of the "permeability ratio" of the early channel at three different values of $[K^+]_i$, and their results agree with ours (see Fig.

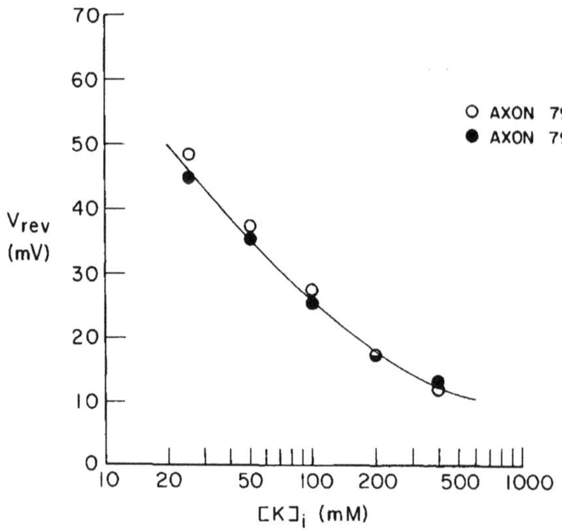

Figure 2. Change of reversal potential (V_{rev}) with internal K^+ concentration. Temperature was 5°C. $[Na^+]_i$, $[Na^+]_o$, and $[K^+]_o$ were kept constant at 0, 50, and 0 mM, respectively.

3). A change of the P_{Na}/P_K ratio also was observed by Ebert and Goldman (1976), in Myxicola axon. It is generally believed now that the "permeability ratio" of the Na^+ channel is likely to be labile.

Chandler and Meves (1965) suggested four possible causes for their observations of a variable P_{Na}/P_K: (1) a change in ionic strength, (2) a change in $[K^+]_i$, (3) a small chloride permeability, or (4) the effect of membrane potential on selectivity. Cahalan and Begenisich (1976) have reported findings that exclude possibilities (1), (3), and (4). We also have conducted a few preliminary studies and confirm their findings. At present, it seems clear that the change of P_{Na}/P_K is caused directly by the change of $[K^+]_i$.

These findings of a labile permeability ratio are not consistent with the notion that the channel is a mechanical pore and the selectivity filter is a hole of fixed size. In the single-pore model, the selectivity filter is "protected" from contact with the internal solution. The

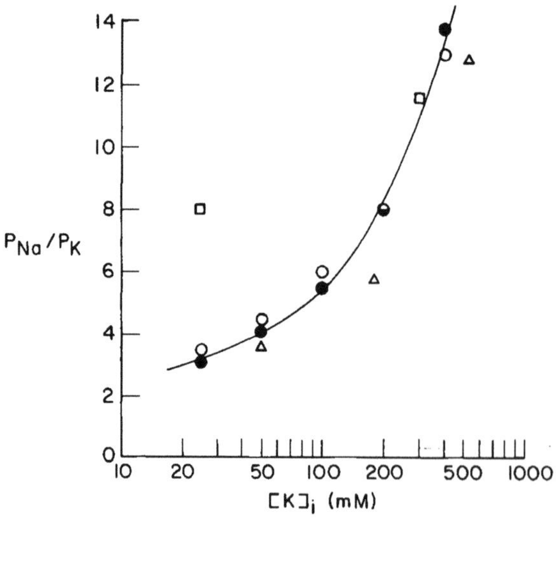

Figure 3. The permeability ratio P_{Na}/P_K, calculated from data of Fig. 2 and plotted as a function of the internal K^+ concentration (○,●). Data reported of Chandler and Meves (1965) (□) and Cahalan and Begenisich (1976) (△) are also plotted to permit comparison.

ion is supposed to move in single file, and the filter can interact with only one cation at a time. It is difficult to explain why the ion-selectivity should change when the $[K^+]_i$ is varied.

To improve its ability to fit experimental data, later versions of the single-pore model have considered more than one barrier in the channel, which permits a larger number of fitting parameters to be employed. The probability of the passage of ions over each barrier is assumed to be given by the rate equation of Eyring *et al.* (1949). These modified models often are referred to as "multi-barrier models" (Woodbury, 1971; Läuger, 1973; Hille, 1975b). Features associated with ion-selectivity in a four-barrier one-ion pore channel have been computed by Hille (1975b). His model predicts that the selectivity depends on membrane potential. However, the experimental findings of Cahalan and Begenisich (1976) and those from our laboratory show that the permeability ratio between Na^+ and K^+ in early channel remains unchanged when the reversal potential is varied by changing $[Na^+]_o$. The lack of effect of membrane potential indicates deficiency in the four-barrier model.

Another problem with some of the multi-barrier models is that they do not explain the dependence of the permeability ratio on $[K^+]_i$. Läuger (1973) and Hille (1975a,b) showed that, for one-ion pore models that feature saturable binding to sites in the channel, the permeability is concentration dependent, however, permeability ratios may remain constant for varying degree of saturation (Cahalan and Begenisich, 1976). Therefore, their models cannot explain the dependence of the permeability ratio on $[K^+]_i$. Recently, some of the one-ion pore models were extended to consider multi-ion occupancy (Hille and Schwarz, 1978; Sandblom *et al.*, 1977). Using a three-barrier, two-site single-file model, Begenisich and Cahalan (1980) reported some success in fitting the concentration-dependence of the permeability ratio of the Na^+ channel in the squid axon. The problem with these extended models is that they often involve too many adjustable parameters, which makes them extremely difficult to test directly.

C. *Alternative Interpretations of the Reversal Potential Data: Measurements of* V_{rev} *as a Function of* $[Na^+]_o$ *and* $[Na^+]_i$

A different approach to explain the results of reversal potential measurements as a function of $[K^+]_i$ is to examine causes other than changes of P_{Na}/P_K. Specifically, two problems need to be considered:

1. The method of reversal potential may not give a true "permeability ratio," because, due to violation of the ion-independence principle and the constant field assumption, the Goldman equation may not hold.
2. Because of an ion-retaining mechanism (such as a layer of ion-exchanger), the ionic concentrations at the inner surface of the membrane may be significantly different from those of the perfusion solution.

Both of these possibilities have been considered in the literature. Many investigators have expressed doubts about the validity of the constant field equation in explaining the detailed findings in excitable membranes (Cahalan and Begenisich, 1976; French and Adelman, 1976). However, the diffusion potential equation, which is based on the constant field assumption, has the advantage of being simple and easily applied to interpret experimental results. It is widely believed that a more exact treatment would not alter significantly the general form of the Goldman–Hodgkin–Katz equation. Because of its simplicity, this equation generally is regarded as a useful approximation to estimate the permeability ratio. As Cahalan and Begenisich (1976) had pointed out, however, our conception of permeability may have to be modified due to limitations of the constant field equation.

Even if the Goldman–Hodgkin–Katz equation holds for the reversal-potential measurements, one still needs to consider the possibility that the ionic concentration at the inner surface of the membrane may differ from that of the internal perfusion solution. The cortex attached to the axolemma is a protein network that, although amphoteric in nature, possesses mainly negatively charged groups at physiological pH (Jirgensons and Straumanis, 1962). These anionic groups can absorb many mobile cations, such as Na^+ and K^+, and the protein as a whole forms an ion-exchanger (Chang, 1977). Because of these ion-exchanger properties of the cortex, the ionic concentration at the inner surface of the axolemma can differ from the ionic concentration of the internal solution. In fact, Hinke (1961), using ion-selective electrodes, found that part of the ions in the axoplasm are bound. Furthermore, it is known that the cortical proteins can undergo a conformational change during excitation (see Sections II.A.2 and II.A.4): It has been suggested that they uncoil or unfold when the nerve is excited (Chang, 1979). Some of the bound ions may be released during this conformational change. Therefore, it is not unreasonable to expect that when the axon is depolarized, the concentration of free ions at the inner surface of the membrane, i.e., the cortical region, is not the same as that of the internal perfusion solution. Then the diffusion potential equation applied to the axolemma should be modified by replacing $[K^+]_i$ with $[K^+]_{\text{cortex}}$ (the effective concentration of K^+ at the cortical region), i.e.,

$$V_{\text{rev}} = \frac{RT}{F} \ln \frac{P_{\text{Na}}[\text{Na}^+]_o}{P_{\text{K}}[\text{K}^+]_{\text{cortex}}} \tag{2}$$

Let us define a parameter "β" as the "partition coefficient" of K^+ in the cortex at the excited state, i.e.,

$$\beta = \frac{[\text{K}^+]_{\text{cortex}}}{[\text{K}^+]_i}$$

Then, Eq. (2) becomes

$$V_{\text{rev}} = \frac{RT}{F} \ln \frac{P_{\text{Na}}[\text{Na}^+]_o}{\beta P_{\text{K}}[\text{K}^+]_i} \tag{3}$$

In comparison with Eq. (1) we can easily see that the observed "permeability ratio" determined from the reversal potential method actually is $P_{\text{Na}}/\beta P_{\text{K}}$ rather than $P_{\text{Na}}/P_{\text{K}}$.

Since the partition coefficient β in general is expected to vary as a function of the ionic concentration, there is no surprise that the observed "permeability ratio" should depend on $[K^+]_i$.

The idea of this ion-partition interpretation is very simple and therefore offers some attraction. We would like to test it further. If indeed the cortex functions as an ion-exchanger, one should predict that:

1. The ion-exchanger effect also must hold true for other kinds of ions, that is, the effective concentrations of other ions at the inner surface of the axolemma also should differ from those of the internal solution.
2. Since there is no counterpart of the cortex at the outer side of the axolemma, the ionic concentration seen by the membrane at the outer side would not deviate from that of the external solution.

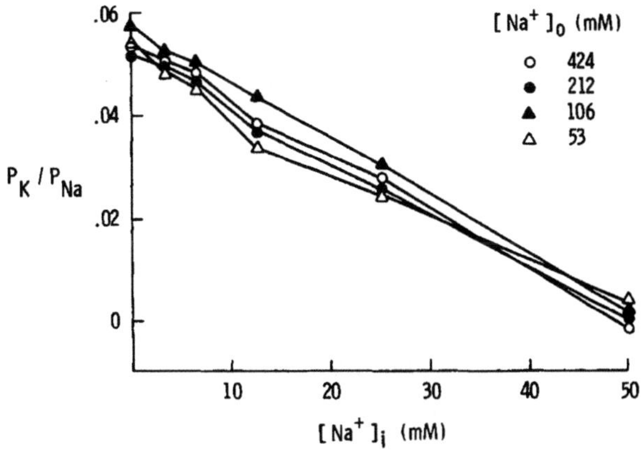

Figure 4. The permeability ratio, P_K/P_{Na}, calculated from the observed reversal potential, plotted as a function of $[Na^+]_i$ and $[Na^+]_o$. The unmodified Goldman–Hodgkin–Katz equation was used in this calculation. Concentrations were converted to activities according to the tables provided by Robinson and Stokes (1949). $[K^+]_o$ = 10 mM and $[K^+]_i$ = 400 mM. Temperature = 5°C. Axon #78-67.

These two predictions can be translated to mean that if one ignores the cortical ion-exchange property and applies the unmodified Goldman–Hodgkin–Katz equation directly to compute the observed "permeability ratio," one would find that (1) the observed "permeability ratio" varies with the concentrations of most of the internal ions, and (2) the observed "permeability ratio" does not vary with the concentration of the external ions.

We have tested these predictions in a preliminary study that used Na^+ as the test ion. Figure 4 shows the results of our measurements of the observed "permeability ratio" of Na^+ and K^+ in the early channel of squid axon using the reversal potential method. It is found that, although the observed "permeability ratio" is not sensitive to $[Na^+]_o$, it changes monotonically when $[Na^+]_i$ increases. These results agree very well with the predictions of the cortical ion-exchanger concept.

We would like to point out that earlier work by Cahalan and Begenisich (1976) also had considered an ion-retaining mechanism to explain the dependence of the observed "permeability ratio" on $[K^+]_i$. However, they rejected the idea because they did not observe a dependence of the "permeability ratio" on $[Na^+]_i$. Our measurements give results different to theirs. We recently finished a study of reversal potential under conditions of high $[Na^+]_i$. We observed that the reversal potential sometimes exceeds the Nernst potential of Na^+, which clearly indicates that the Na^+ concentration at the cortex differs from that of the internal solution (Chang, 1981).

D. Conclusion of the Reversal Potential Measurements

1. Contrary to the assumption of ion independence, the conductance pathway is permeable to more than one ion, i.e., the "Na^+ channel" is not exclusively permeable to Na^+, and one internal ion species often affects the currents carried through the membrane by other ion species.

2. It is questionable whether the ion-selectivity of the early channel can be determined accurately with the reversal potential method. First, the usefulness of the Goldman–Hodgkin–Katz equation is limited because of difficulties in the assumptions of constant field and independent movements of ions. Second, the ionic concentration at the inner surface of the axolemma could be significantly different from that of the internal solution.

3. If we assume that the observed "permeability ratio" obtained by using the reversal potential method basically is correct, the ion-selectivity would seem to change with $[K^+]_i$.

This observed lability of ion-selectivity is inconsistent with the concept of a single-pore channel that has a mechanical selectivity filter. The existing models of multi-barrier single-ion pores also cannot explain the reversal potential data. Whether a multi-barrier, multi-ion occupancy rigid-pore model can satisfactorily explain the selectivity data is not yet clear.

4. We believe that the reported $[K^+]_i$ dependence of the "permeability ratio" based on reversal potential measurements is caused mainly by the ion-exchanger property of the axon cortex. This interpretation is supported by the observation that the measured "permeability ratio" of the early channel changes with $[Na^+]_i$ but not with $[Na^+]_o$.

IV. EFFECTS OF INTERNAL IONS ON THE KINETICS OF ACTIVATION AND INACTIVATION OF THE EARLY CHANNEL

Since the "sodium channel" is not permeable exclusively to Na^+, we adopt the convention of calling it the "early channel" to avoid confusion. Similarly, the so-called "K^+ channel" will be called the "late channel."

An important property of the current-conduction mechanism, besides ion-selectivity, is the kinetics of the activation and inactivation of the conductance. To study the effects of various ions on channel kinetics, we used a method of digital TTX (tetrodotoxin) subtraction. This method has an advantage over the more commonly used ion-substitution methods, such as the choline method used by Hodgkin and Huxley (1952a) and the tris method used by Adelman and Senft (1966). We know the early current is not carried by Na^+ alone, the Na^+ current obtained by an ion-substitution method does not represent all currents that pass through the early channel. Furthermore, any substitute ion introduced into the internal solution affects the properties of the conductance mechanisms.

A. The Digital TTX-Subtraction Method

To study the time course of the current going through the early conductance channel, one must subtract the capacitive current, the leakage current, and the late current. These requirements can be satisfied by using a digital computer and the specific current-blocking property of tetrodotoxin (TTX). It is known that 0.4 μM TTX in the external solution blocks the early current completely, while the late current is not affected (Narahashi *et al.*, 1964; Kao, 1966). If one measures the membrane current without TTX and then repeats the measurement with TTX, one can obtain the early current by subtracting the current after TTX from the current before TTX.

A sample record of the TTX-subtraction method is given in Fig. 5. In this experiment, the external solution is Na^+, K^+-free artificial sea water, hence, both the Na^+ current and the K^+ current are outward. In Fig. 5A, the membrane current measured at $V = 80$ mV without TTX is shown as the top trace. The membrane current measured after the addition of TTX is shown as the bottom trace. It is clear that the early transient current is blocked completely by TTX, while the late current is not affected. In Fig. 5B, the early current is obtained by taking the difference between the two current traces in Fig. 5A.

B. Dependence of the Kinetics of the Early Current on $[K^+]_i$

A sample record of the ionic effect on the kinetics of the early current is shown in Fig. 6. The top graph (Fig. 6A), in which the time course of the early conductance is determined for $[K^+]_i = 400$ mM and $[K^+]_i = 200$ mM, shows that the conductance is inactivated

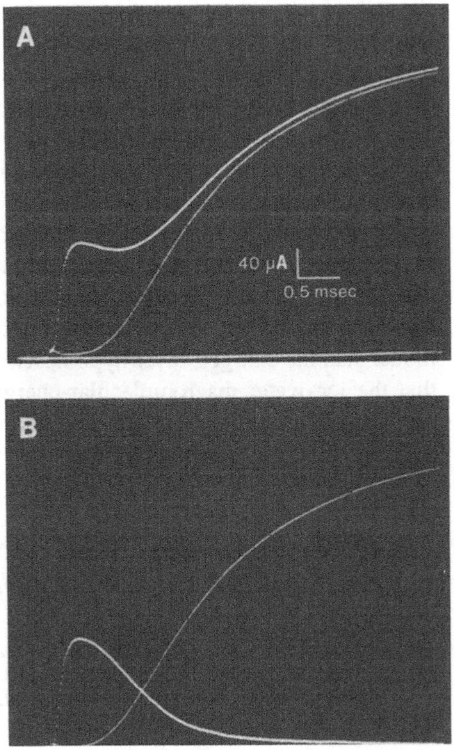

Figure 5. Measurement of early current using a digital TTX-subtraction technique. (A) The upper trace is the membrane current measured under the control condition. The lower trace is the membrane current after 0.4 μM TTX is added (late current). (B) Upper trace: the late current, from top panel. Lower trace: the early current, is taken as the difference between the two traces in the top panel. Measurement was done under conditions of reverse Na^+ gradient, i.e., $[K^+]_i$ = 400 mM, $[Na^+]_i$ = 50 mM, $[K^+]_o$ = 0, and $[Na^+]_o$ = 0. Axon #79-41. (The vertical bar represents 40 μA of membrane current which is equivalent to a current density of 0.76 mA/cm^2).

more slowly in low $[K^+]_i$. Figure 6B, which plots the expanded time course of the early conductance at $[K^+]_i$ = 400 and 200 mM, shows that activation of the early channel is also slower at low $[K^+]_i$. Thus, it is clear that *a reduction in $[K^+]_i$ affects not only the "ion-selectivity" of the early channel, but also the activation and inactivation of the "gating" mechanism.* No detailed study of the rigid-pore model has been done to explain the $[K^+]_i$ dependence of the channel kinetics. One possible explanation of the phenomenon may be the effect of ionic strength. Chandler *et al.* (1965) have shown that reducing the ionic

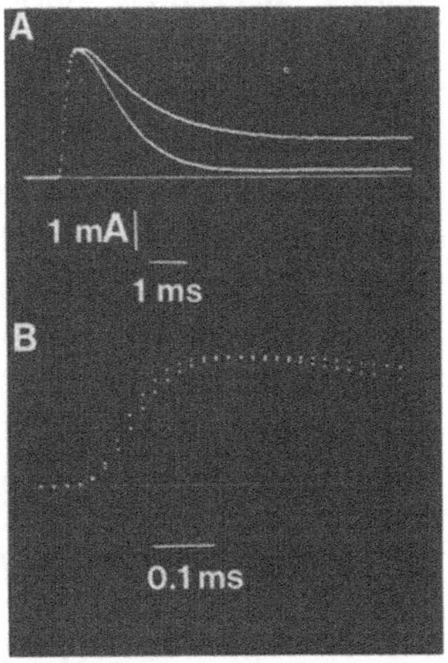

Figure 6. The time course of the early current at two different $[K^+]_i$, obtained by using a TTX-subtraction method. (A) Early currents with $[K^+]_i$ = 200 mM (upper trace) and 400 mM (lower trace). The kinetics are slower with reduced $[K^+]_i$. (B) Expanded view on the rising phase of the early currents with $[K^+]_i$ = 400 mM and 200 mM, respectively. The kinetics of activation are also slower with lower $[K^+]_i$. $[Na^+]_i$ = 50 mM, $[K^+]_o$ = $[Na^+]_o$ = 0. Temperature = 5°C. Axon #79-40. The vertical bar represents a current density of 1 mA/cm^2.

strength of the internal solution can shift the "h" (inactivation) mechanism. A larger τ_h at low $[K^+]_i$ may partially account for our observed change of kinetics; however, this interpretation is not totally satisfactory. For example, we found that adding 100 mM of Cs^+ to the internal solution actually prolongs the inactivation of the early current (see Section V.B). This finding is opposite to the prediction of the ionic-strength interpretation.

Chandler *et al.* (1965) postulated a layer of negative fixed charges at the inner surface of the axolemma to explain the effect of ionic strength. Since the protein layer at the cortex is also supposed to be negatively charged, it may be difficult to distinguish between these two notions. However, it is known that substituting D_2O for H_2O in the perfusion solution can decrease the early conductance and slow the kinetics of activation and inactivation (Conti and Palmieri, 1968). These findings are more consistent with the cortical ion-exchanger model than the charged-layer hypothesis. The isotope effect and the ionic effect both suggest that the ion-water-macromolecular-charged-site interactions, not the ionic strength, affect the "gating" mechanism of the channel. In the following section, we will see more direct evidence that ions can interact with the "gating" mechanism of the channels.

V. EFFECTS OF INTERNAL Cs^+ AND Na^+ ON THE "GATING" PROPERTIES OF EARLY AND LATE CHANNELS

A. Effect of Cs^+ on the Late Current

Earlier studies have shown that a number of ions, including Cs^+, Na^+, Rb^+, choline, and TEA, can interfere with the potassium current flowing through the late channel (Chandler and Meves, 1965; Adelman and Senft, 1968; Bergman, 1970). Bezanilla and Armstrong (1972) studied the "negative conductance" caused by Cs^+ and Na^+ ions and interpreted the data to indicate blockage of the inner pore of the late channel by Cs^+ and Na^+. Since the results of this study have an important bearing on the structure of the channel, we would like to examine the effect of internal ions very carefully and see whether alternative interpretations should be considered.

The effect of internal Cs^+ on the membrane current of a squid axon is shown in Fig. 7. Here, the external Na^+ is replaced by choline ions so that the Na^+ current and K^+ current are both outward. In this arrangement, the two currents are additive, and the data can be analyzed more simply. Four sets of data are shown in this figure, which represents current measured under different conditions: (A) control, (B) 100 mM Cs^+ inside, (C) 0.4 μM TTX outside, and (D) Cs^+ inside and TTX outside. $[K^+]_i$ and $[Na^+]_i$ are 400 mM and 50 mM, respectively. The different traces in each panel represent currents measured at various depolarizing potentials. Since the early current is blocked by TTX, only late currents, i.e., K^+ current crossing the late channel, are observed in (C) and (D). Comparing (C) and (D), one can see that late currents are not affected by the internal Cs^+ at low depolarizing potentials. At higher depolarizing potentials, however, the late currents are very sensitive to Cs^+. Comparing the top traces in (C) and (D), it is easy to see that the time courses of the late current with and without Cs^+ are very different. At time longer than a few milliseconds, the late current observed with Cs^+ is much smaller than that without Cs^+. These data indicate clearly that internal Cs^+ can suppress the K^+ current (or late current). This suppression apparently is time- and potential-dependent.

B. Effect of Cs^+ on the Early Current

An interesting property of Cs^+ is that it also affects the early current. Adelman and Senft (1966) reported that internal Cs^+ delays the turn off of sodium conductance. Our

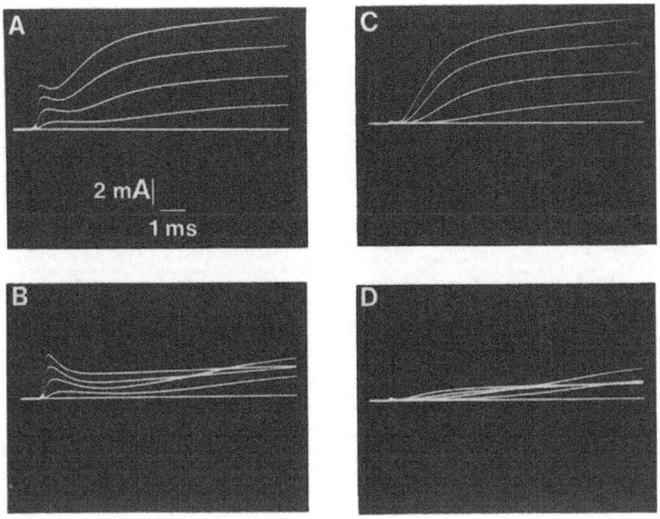

Figure 7. Membrane currents under *V*-clamp with step pulses. Current traces in each panel correspond to depolarization to V = 0, 40, 80, and 120 mV. The axon was in Na^+, K^+-free artificial sea water (ASW). $[K^+]_i$ = 400 mM and $[Na^+]_i$ = 50 mM. (A) Control. (B) 100 mM Cs^+ added inside the axon. (C) No Cs^+ inside, 0.4 μM of TTX added outside. (C) 100 mM Cs^+ inside and 0.4 μM TTX outside. Temperature = 5°C.

study suggests that the inactivation of the early conductance is partially removed in the presence of internal Cs^+. This effect also is demonstrated in Fig. 7.

Comparison of the top traces in panels (A) and (C) of Fig. 7 shows that in the absence of Cs^+, the current at large t (say, $t > 7$ msec) is practically unaffected by TTX. However, comparison of the top traces in (B) and (D) shows that in the presence of Cs^+, the current at large t is decreased significantly when TTX is added. These observations suggest that the addition of Cs^+ gives rise to a TTX-sensitive component of current. Since the early current is defined as the component of current abolishable by TTX, the data in (B) and (D) suggest that internal Cs^+ can "create" a new component of early current that is not inactivated with time. This is equivalent to saying that the internal Cs^+ partially removes the inactivation of the early conductance.

The effect of Cs^+ on early conductance is demonstrated more clearly in Fig. 8. The data in (A) were recorded without Cs^+, while those in (C) were recorded after 100 mM Cs^+ was added to the internal perfusion solution. The top traces in (A) and (C) give the membrane current without TTX, the lower traces give the membrane current with TTX. The differences between the two currents in (A) and (C) are shown in panels (B) and (D), respectively. The trace in (B) represents early current without Cs^+ and the trace in (D) represents early current with Cs^+. Comparison of (B) and (D) makes clear that Cs^+ partially removes inactivation of the early current.

By examining the current traces in Fig. 7 at different levels of depolarization, one can see that the effect of Cs^+ on the early channel also is time- and potential-dependent. This dependency will be discussed in more detail in the next subsection.

It is difficult to explain the effect of Cs^+ on the early channel on the basis of the single-pore model. The proponent of the single-pore model assumes that Cs^+ suppresses the late current by blocking the pore. Clearly, the same explanation cannot be used to describe what Cs^+ does to the early current. One may hypothesize that internal Cs^+ may interfere with the gating mechanism of the early conductance pathway by a "foot in the door" effect, which has been suggested for the action of some large organic ions, like tetraethylammonium

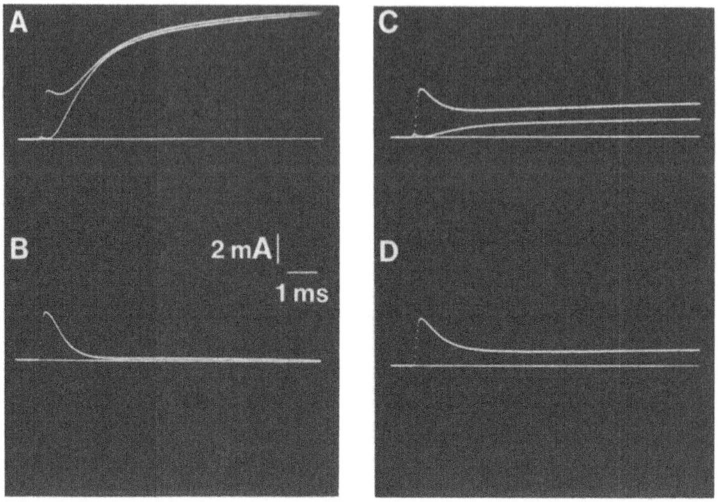

Figure 8. Effect of internal Cs^+ on early current. Data from Fig. 7. $V = 120$ mV. (A) Control. Upper trace: No TTX; lower trace: with TTX. (B) Difference between the two traces in panel A gives the early current under the control condition. (C) Same as panel A except 100 mM Cs^+ was added inside the axon. (D) Difference between the two traces in panel C gives the early current under 100 mM $[Cs^+]_i$. The vertical bar represents 2 mA/cm^2.

(Armstrong, 1971), *N*-methylstrychnine (Cahalan and Almers, 1979), or pancuronium (Yeh and Armstrong, 1978). However, we believe that this explanation is probably overly simplified. For example, Swenson and Armstrong (1981) observed that K^+ channels in squid axon close more slowly in the presence of external K^+ and Rb^+. They interpreted the data as a "foot in the door" effect. Later, Matteson and Swenson (1982) discovered that some external ions (such as Tl^+ and NH_4^+) had an opposite effect. They speeded the rate at which K^+ channels close. This observation makes the "foot in the door effect" interpretation untenable.

We suspect that the action of Cs^+ is caused by a "salt-in/salt-out" effect on the proteins that make up the conductance pathway. The conformation of protein in general is sensitive to the ionic environment. Suppose the protein structure must stay in a certain conformation for the channel to be fully functional. Interactions with Cs^+ ions will cause the protein complex to change its structure, which in turn alters its properties. This interpretation is also consistent with findings of Chandler and Meves (1970), who showed that when the axon is internally perfused with a high concentration of NaF, part of the early conductance cannot be inactivated.

C. Comparison of the Potential and Concentration Dependence of the Cs^+ Effect on the Early Current and the Late Current

To shed some light on the mechanism by which internally Cs^+ affects the early current and the late current, we have investigated the correlation between the effects of Cs^+ on the early current and the late current. If the effects of Cs^+ are caused by two different mechanisms, as suggested in the rigid-pore view, that is, if the late current is reduced by Cs^+ blocking the pore, and the sustained early current results from Cs^+ jamming the inactivation gate ("foot in the door"), the effects of Cs^+ on the early and late currents should be totally

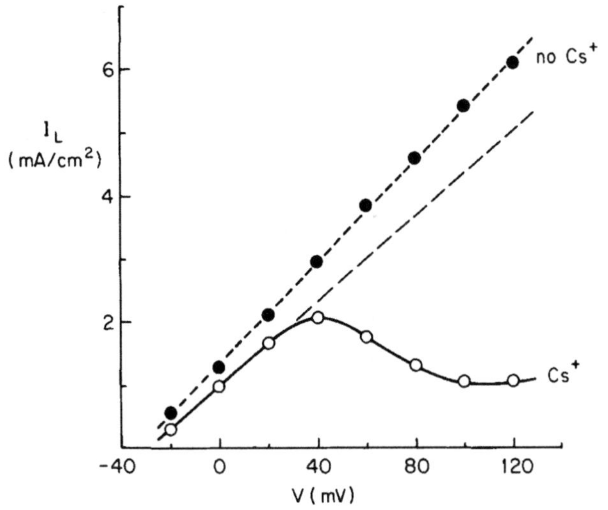

Figure 9. Effects of internal Cs^+ on the late current. I_L is the late current, which was measured at 9.6 msec after the onset of the depolarizing step, in the presence of TTX. V is the depolarizing potential. Experimental conditions same as in Fig. 7 (panels C and D).

uncorrelated. On the other hand, if the effects of Cs^+ are due to its interference with protein conformation, one may observe some correlation between the effect of Cs^+ on the early current and the effect of Cs^+ on the late current, since the proteins that control the early and late conductance pathways could be similar in nature. Therefore, we set out to examine the potential- and concentration-dependence of the effects of Cs^+ on the early and late currents.

Figure 9 plots the late current as a function of the depolarizing potential. One can see that Cs^+ begins to suppress the late current only when V is larger than +20 mV. Figure 10 plots the current vs. V under the effect of Cs^+. It can also be seen that Cs^+ gives rise to a TTX-sensitive component of current only when the depolarizing potential exceeds 20 mV. Figure 11 summarizes the effects of Cs^+ on the late current and the TTX-sensitive current, i.e., the early current. The filled circles represent the relative magnitude of the late current under the effect of internal Cs^+, and the empty circles represent the fraction of early

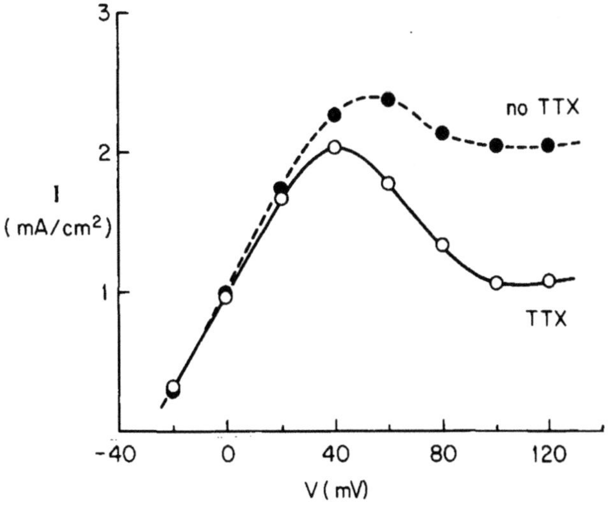

Figure 10. Potential dependence of the "Cs^+-induced" TTX-sensitive current. 100 mM Cs^+ was applied inside the axon. I is the membrane current measured at 9.6 msec after the onset of the depolarizing step. V is the depolarizing potential. Experimental conditions same as in Fig. 7 (panels B and D).

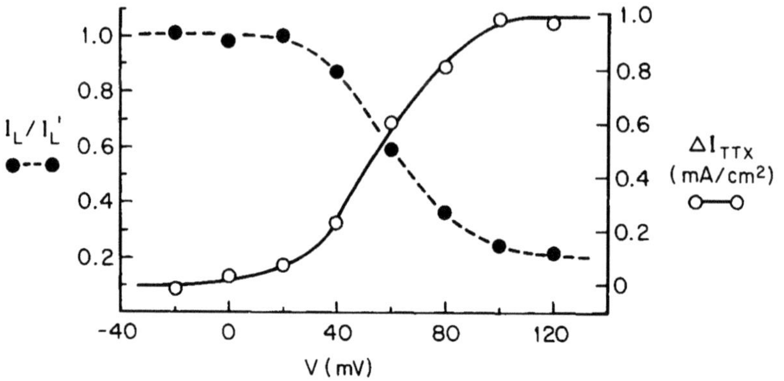

Figure 11. Potential dependence of the Cs^+ effects on early and late currents. V is the depolarizing potential. I_L is the late current measured in the presence of internal Cs^+ and I_L' is the estimated value of the late current if the potential-dependent Cs^+ effect were absent. I_L and I_L' were determined from data in Fig. 9. I_L/I_L' represents the fraction of late current not suppressed by internal Cs^+. ΔI_{TTX} represents the "Cs^+-induced" TTX-sensitive current, the values of which were determined from data in Fig. 10, i.e., subtracting data labeled "TTX" from data labeled "no-TTX".

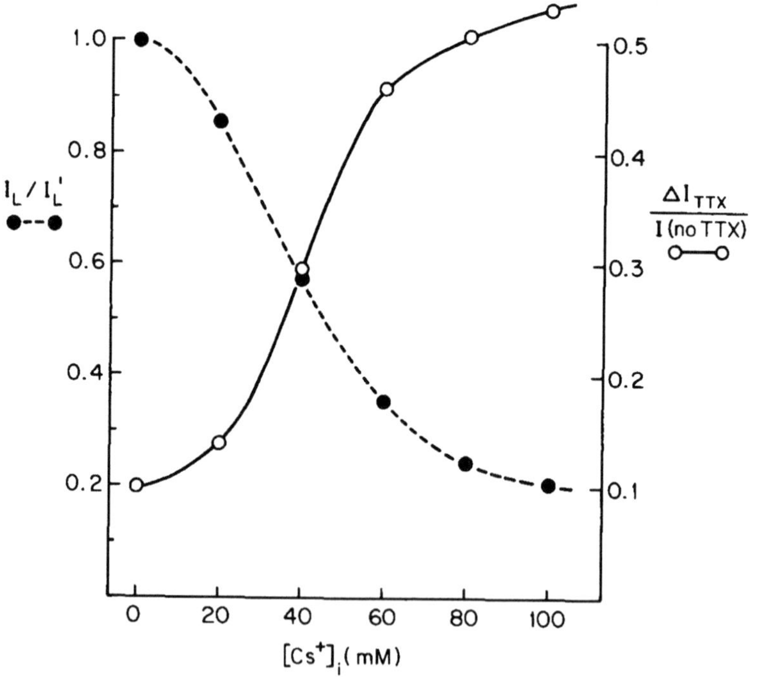

Figure 12. Concentration dependence of the Cs^+ effects on early and late currents. Filled circles represent the fraction of late current not suppressed by the internal Cs^+. Empty circles represent the relative amount of "Cs^+-induced" early current. All currents determined at V = 100 mV.

current that remains un-inactivated due to the effect of Cs^+. The data in Fig. 11 indicate that the *potential-dependence of the* Cs^+ *effects on the late current and the early current are very similar*. Both effects begin to turn on at $V = 20$ mV and become "saturated" when V reaches roughly 80 mV.

In addition to depending on potential, the effects of Cs^+ on the early conductance and the late conductance also depend on the internal concentration of Cs^+. We have studied the effects of Cs^+ on both the early current and the late current at $[Cs^+]_i = 0, 20, 40, 60, 80$, and 100 mM. We found that the effects of Cs^+ become progressively more prominent as $[Cs^+]_i$ increases. Figure 12 summarizes the concentration-dependence of the effects of Cs^+ on early and late currents. The filled circles show the reduction of the late current by Cs^+ at various concentrations. The empty circles show the fraction of "Cs^+ -induced" early current as a function of $[Cs^+]_i$. It is evident that *the concentration-dependence of the effects of* Cs^+ *on the early and late currents are very similar*. The effects of Cs^+ begin to show at $[Cs^+]_i = 20$ mM and become less sensitive to concentration when $[Cs^+]_i$ reaches 80 mM.

D. Implication of the Ionic Effects on the Structure of the Conductance Pathway

From the studies of dependence of the two experimentally controllable parameters (potential and concentration) it is evident that the effects of Cs^+ on the early current and the late current are correlated. In fact, this correlation is not limited to the effects of internal Cs^+. It is known that internal Na^+ ions also can suppress the late current (Bezanilla and Armstrong, 1972) and prolong the early current (Chandler and Meves, 1970). We have conducted a preliminary study to examine the potential dependence of the effects of internal Na^+ on the early current and the late current. The result is shown in Fig. 13. Our study shows that, as in the case of Cs^+, the potential-dependence of the Na^+ effects on the early current and the late current are very similar.

The correlation between the effects of internal ions on the early and late currents can be destroyed by internal perfusion of pronase. It is known that internal pronase can remove the inactivation of early current (Armstrong *et al.*, 1973). When Cs^+ is applied inside this pronase-treated axon, it can still cause a reduction of the late current, but the early current is no longer significantly affected.

Our findings that the effect of internal ions on the early channel and the late channel are correlated have several implications:

(1) The findings are not consistent with the single-pore view. It is highly unlikely that Cs^+ (or Na^+) affect the early channel by jamming the gate, and block the late channel by plugging up the narrow region (selectivity filter) of the channel pore. It is difficult to see why a blocking of the late channel pore should correlate with the jamming of the gate of the early channel.

(2) This observed correlation suggests that the activation of late conductance and the inactivation of early conductance are caused by similar molecular mechanisms, such as a certain change in the conformation of channel proteins. The effects of the internal ions probably resulted from interfering with this conformational change.

(3) An alternative interpretation of the findings is that the early channel and the late channel could be "coupled." The two-separate-channel scheme is a convenient assumption that makes interpretation of data simple, but there is no unequivocal evidence showing that the two pathways must be independent. There has been a debate on whether the early and late conductances are single channel or dual channel (Mullins, 1968; Narahashi and Moore, 1968; Hille, 1968). In view of the pronase effect (Armstrong *et al.*, 1973) it is more likely

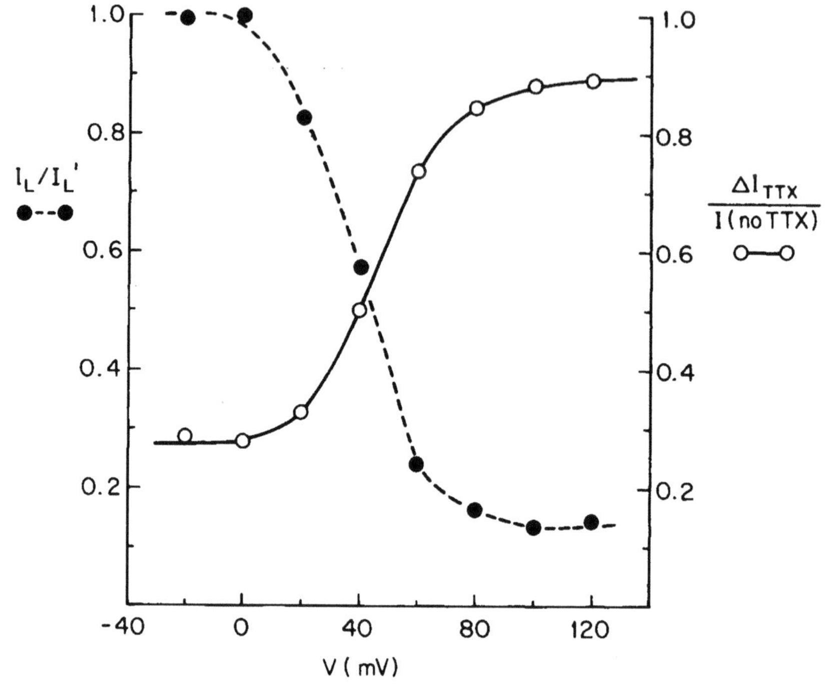

Figure 13. Potential dependence of the effects of internal Na^+ on early and late currents. Filled circles represent the fraction of late current not suppressed by the internal Na^+ ions. Empty circles represent the relative amount of early current which, due to the presence of high $[Na^+]_i$, did not inactivate. Protocols similar to those used in the Cs^+ experiment. $[K^+]_i$ = 200 mM, $[Na^+]_i$ = 400 mM, $[K^+]_o$ = $[Na^+]_o$ = 0.

that the dual channel notion will prevail, but it is not yet clear how the two pathways relate to each other. The study of chiriquitoxin effect (Kao, 1981) in muscle and nerve suggests that the early and late channels are near neighbors (their separation is estimated to be less than 15 Å). Our findings of good correlation between the effects of internal ions on the early and late currents suggest that these two pathways could be actually "coupled." Our data also suggest that this "coupling" mechanism probably is part of the sublemmal proteins on the cytoplasmic side (since it can be destroyed by internal perfusion of pronase).

VI. CONCLUSION: THE EXCITABLE MEMBRANE AS A POLYELECTROLYTE SYSTEM

A. Structure of the Nerve Membrane

Traditionally, it has been assumed that the functional "excitable membrane" is the axolemma itself. The structure of this axolemma has been described by a fluid mosaic model (Keynes, 1972; Singer and Nicolson, 1972). The partially mobile lipid bilayer provides a high-impedance boundary that separates the internal and external ions and gives rise to a large surface capacitance. The membrane proteins provide hydrophilic pathways that allow currents carried by ions to pass through.

This picture undoubtedly has a number of advantages and some of its features certainly will be retained in future models. At present, however, a significant amount of evidence indicates that this axolemmal model is oversimplified. Its major fault is its failure to recognize the physiological importance of the cytoplasmic proteins. Moreover, its conceptual structure is not consistent with the observed anatomical structure of the nerve cells. Recent morphological studies from several laboratories (see Part I of this book) have indicated that the membrane proteins are connected directly to the cytoplasmic protein network.

We think the fluid mosaic model of the nerve membrane should be significantly revised to include the anatomical and functional structures of the axon cortex. It can be said that the membrane proteins and the cortical proteins are parts of an integrated system. A more realistic picture of the nerve "membrane" may look like that drawn in Fig. 14.

Two features of this new model of the nerve "membrane" should be pointed out clearly:

1. The "membrane" is not a homogenous structure. The proteins are lumped in patches where currents will pass.
2. The membrane proteins have the properties of polyelectrolytes, i.e., a fixed-charge system. The effective conductance for a particular ionic current is determined to a large extent by the interactions between the ions and the macromolecular charged sites.

B. *Conceptual Model of "Channels": Long, Narrow Pores or Polyelectrolyte Networks?*

Based on evidence presented in the preceding sections, we think the entire cell-surface structure (cell membrane plus the attached cytoplasmic structure) is responsible for the excitation process. This surface structure is a highly organized system, the function of which depends on the interlocking of its components.

At present, we have only fragmental knowledge about the molecular structure of each of the components. The plasmalemma undoubtedly consists mainly of the lipid bilayer, and the cortical proteins are in the form of a network interwoven by microtubules, neurofilaments, and other fibrous proteins. As to the membrane proteins, the resolution of morphological techniques is not yet good enough to determine their detailed structure. Many investigators have assumed that the membrane proteins form long, narrow pores. This single-pore model of the channel has been investigated actively in the past decade, and extensive reviews have been published (Hille, 1970, 1975a; Narahashi, 1974; Armstrong, 1975; Taylor, 1974), but

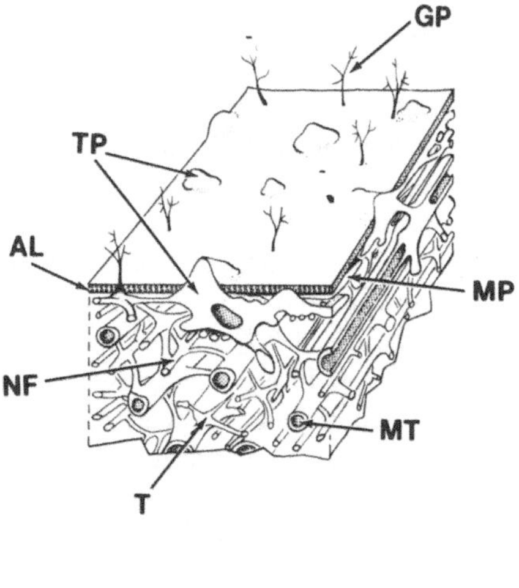

Figure 14. Conceptual picture of what an "excitable membrane" may look like. AL: axolemma, GP: glycolproteins, MP: membrane-associated proteins, MT: microtubules, MF: microfilaments, NF: neurofilaments, T: microtrabecular structure, TP: transmembrane proteins.

so far most of the evidence cited in support of it has been indirect. We feel that many of the findings in regard to channel properties can be explained by the simple fact that the membrane protein is a polyelectrolyte network. In the following, we compare the single-pore model and the notion of "polyelectrolyte network" in detail.

1. The pore model regards the channel mainly as a mechanical system. Whether an ion can pass through a channel depends on the diameter of the ion relative to that of the pore, and the change of conductance of the channel is often attributed to blocking of the pore by other ions (Cs^+, Na^+, Rb^+, etc.) or chemical "plugs," e.g., TTX, TEA. On the other hand, the polyelectrolyte concept regards the channel as a physicochemical system; the permeation of ions depends on a combined process of electrodiffusion and jumping between fixed-charge sites on the proteins. Changes of conductance in the channel can be attributed to a change either of the conformation of the ion-exchanger matrix, which would restrict the flow of counter ions in the interstices of the matrix (Eisenman *et al.*, 1967), or the association/dissociation energy between the mobile ions and the fixed-charged sites (Ling, 1962; Eisenman, 1962).

Most of the earlier theoretical treatments of the physicochemical properties of the biological membrane consider the membrane as a homogeneous planar structure (Eisenman *et al.*, 1967). In view of the large amount of evidence showing that the cell surface is highly heterogeneous (Lackie, 1980) and ions pass mainly through protein "patches," this approach should be modified. However, the earlier theoretical treatments can be borrowed to study the transport of ions in the "membrane protein patches" instead of the entire membrane. For example, one may use the approaches formulated by Nernst (1888) and Planck (1890) to study the electrodiffusion of ions within the patch, or one may utilize the theories of irreversible thermodynamics (Prigogine, 1961; deGroot, 1951) to calculate the conductance properties of the membrane proteins. Katchalsky (1964, 1967) and Teorell (1953, 1959) have carried out important studies of the physicochemical properties of physical membranes. It may be fruitful to extend their studies to examine channel properties of the "patches" formed by membrane proteins.

Recent versions of the single-pore model have upgraded the mechanical picture by assuming the existence of a local thermodynamic equilibrium within the pore so that the rate-process theory of Eyring *et al.* (1949) can be applied. This approach allows one to calculate the steady-state current-voltage relationship and conductance vs. ionic concentration (Hille, 1975b; Hille and Schwarz, 1978). The drawbacks are that a detailed structure of the channel must be assumed, and the model calculation often involves a large number of adjustable parameters; for example, a multi-occupancy, 4-site model of gramicidin A channel produces 92 parameters (Sandblom *et al.*, 1977).

2. Many known properties of the channel can be explained by both the single-pore model and the polyelectrolyte concept. For example, the sequence of ion-selectivity for the early channel in squid axon is $Li^+ \simeq Na^+ > K^+ > Rb^+ > Cs^+$, which is the same as the reverse sequence of the sizes of the unhydrated ions. This observation is frequently cited to support the pore model of the channel. However, many ion-exchangers, both biological and nonbiological, exhibit the same sequence of ion-selectivity (Diamond and Wright, 1969). This sequence and many others can be explained by the ion-exchanger theories of Eisenman (1962) or Ling (1962) without invoking the assumption of pore size inside the ion-exchanger (Reichenberg, 1969). Hence, the selectivity sequence of the early channel simply indicates that the membrane proteins are fixed-charged systems that behave like ion-exchangers; there is no need to assume critical-size pores formed by proteins. The same argument also applies to the ion-selectivity sequence of the late channel, which coincides with one of the eleven ionic sequences commonly observed in ion-exchangers (Diamond and Wright, 1969).

3. Other evidence often quoted in support of the pore model is the finding that small organic ions can cross the membrane, while large organic ions cannot (Hille, 1971; Armstrong, 1975). However, it is well known that ion-exchangers have a similar property. As ion-exchange resins are extensible networks, there is no definite pore size. Yet, the uptake of large molecules or ions decreases steadily as their size increases. For example, Kressman and Kitchener (1949) found that a phenol-sulfonate resin exchanged fully but increasingly slowly with $(CH_3)_4N^+$,$(C_2H_5)_4N^+$, $(CH_3)_3(C_5H_{11})N^+$, $PhN(CH_3)_2(C_2H_5)^+$, and $PhN(CH_3)_2CH_2Ph^+$. The same resin would not exchange fully with quinine hydrochloride even in 20 weeks. Hale *et al.* (1953), investigating the exchange of large organic cations on sulfonated polystyrene resins, found that many of the resins did not exchange with TEA ions. Hence, the properties of organic ion-uptake in the ion-exchanger are similar to the properties of permeability in the nerve membrane.

4. Do the actions of neurotoxins prove that channels are pores? The pore model assumes that neurotoxins (such as TTX) block the early channel by plugging the mouth of the pore (Hille, 1975c). However, TTX may bind to the membrane protein and change the conformation of the channel structure. In fact, a class of chemicals, known as "neurotransmitters," can modify the properties of the membrane proteins by binding to receptors. A good example is the binding of acetylcholine molecules, which makes the endplate "channel" conductive. If the "opening" of a channel can be attributed to a receptor-binding action, why should the "closing" of a channel not be explained in a similar manner? The action of neurotoxin, then, need not necessarily imply that the channel is a pore that can be plugged.

5. Is the membrane conductance controlled by physically blocking the channel pore? The pore model assumes the existance of a "gating mechanism" at the inner opening of the channel that controls the ionic conductance by blocking or unblocking the pore (Armstrong, 1975; Ehrenstein, 1976; Armstrong and Bezanilla, 1977). It is intructive to examine this particular aspect of the pore model in squid axons that have been perfused internally with pronase. In these pronase-treated axons, one often observes that the excitable current is reduced while the "leakage" current increases. The situation can be described as if part of the excitable channels are converted into "leakage channels." If the effect of pronase is to destroy the "gating mechanism" at the inner opening of the channel (while retaining the rest of the pore structure) as the single-pore model suggested (Armstrong *et al.*, 1973; Armstrong and Bezanilla, 1977), the leakage current should be blocked by TTX on the outside (if the "leakage channels" are converted from early channels), or blocked by TEA from the inside (if the "leakage channels" are converted from late channels). However, it is observed in experiments that neither TTX nor TEA can block the pronase-induced "leakage current." In fact, similar results can be observed if the leakage is induced by means other than pronase treatment. These observations clearly do not agree with the prediction of the single-pore concept of the channel.

We think the membrane conductance is controlled by changes of conformation in the membrane proteins. When the axon is weakened by pronase treatment or damaged due to other causes, the membrane proteins become partially denatured. This denaturing produces two effects: (1) the membrane proteins behave more like a kind of unfolded polyelectrolyte, which is less restrictive to ionic flow and has poor ion-selectivity, and (2) the receptor-binding mechanism of the membrane proteins no longer functions. As a result, neurotoxins, such as TTX, cannot change the conductance of the channel.

6. Have studies with artificial membranes shown that channels are pores? A positive answer has been assumed in many publications; however, the evidence is not definitive. There is no doubt that ionophores applied to lipid bilayers can induce large changes of conductance, however, none of the ionophores used in those studies (such as gramicidin, valinomycin, nonactin, or alamethicin) exist in "excitable membrane." These ionophores

may form either "ion carriers" or "ion pores," but regardless of their structure, one must bear in mind that studies of ionophores can only help to explore ways that a biological membrane may function, while the actual mechanisms may be totally different. The difference between the ionophores and the membrane proteins is tremendous, for example, the estimated values of the molecular weight of channel proteins are orders of magnitude larger than those of the gramicidin molecules, and their structures are considerably more complex.

Furthermore, artificial models have also been used to demonstrate that polyelectrolyte systems can give rise to properties of biological channels. Shashoua (1969, 1975) studied current–voltage characteristics of polymer membranes loaded with proteins, polynucleic acids, and non-biological polyelectrolytes. He found that these artificial membranes underwent spike-like electrical activity similar to that of nerve impulses. These membranes were lipid-free and had no well-defined pore structure, hence, their electrical properties must be attributed to the properties of polyelectrolytes rather than to the properties of rigid pores.

7. Another line of observations used frequently to support the rigid-pore model is the recording of noise that shows rapid changes of conductance. However, the implication of these observations on the channel structure is not clear. First, many of the systems known to give step-like noises are chemical receptors and ionophores (such as gramicidin A). They are not potential-activated biological channels. Second, in those noise measurements which recorded step-like "single-channel" currents in electrically excitable cells (Conti and Neher, 1980; Sigworth and Neher, 1980), the experimental data were often incomplete. Extensive filtering had been used to obtain those records, and some of the experimental details (such as pharmacological properties of the "channel") still are lacking. Third, even if the "single-channel" recordings are confirmed, findings of step-like current noises simply indicate that the channel has two stable conductance states and the transition time between these states is short in comparison to the dwell times in either state. These noise data may provide useful information on channel kinetics, but they give no support to the notion that the channel has the structure of single pore.

In summary, we have found no definitive evidence that the ionic pathways are mechanical pores. On the contrary, the properties of channels can be explained more simply on the basis of the fact that membrane proteins are polyelectrolyte networks that have the properties of a biological ion-exchanger. The conductance of the channel is more likely to be determined by the energy involved in the interactions between ions, water, and macromolecular charged sites, rather than by physical blockage of a narrow pore. Judging from freeze–fracture electron micrographs, channel proteins are more than one hundred Å in diameter. It is doubtful that these proteins would form a solid structure with a hole 3–5 Å wide, as the interchain distances within the proteins would have been much larger than the dimension of this hypothetical pore. Proponents of the single-pore model often explain the experimental findings from a simple mechanical point of view. They have implicitly hypothesized that a molecular mechanism can be modeled by miniaturizing a macroscopic, mechanical, structure to a scale of atomic dimension (see Fig. 15). It is questionable whether such an approach is physically sound.

C. *Functional Linkage Between the Membrane Proteins and the Cortical Proteins*

In Section II we have presented evidence that the proteins at the cortex are important in the excitation process. The electrophysiological data discussed in Sections III, IV, and V further support such a view. First, we found that the assumption of ion-independence is totally violated in regard to ions on the inside of the cell, in sharp contrast to the case when

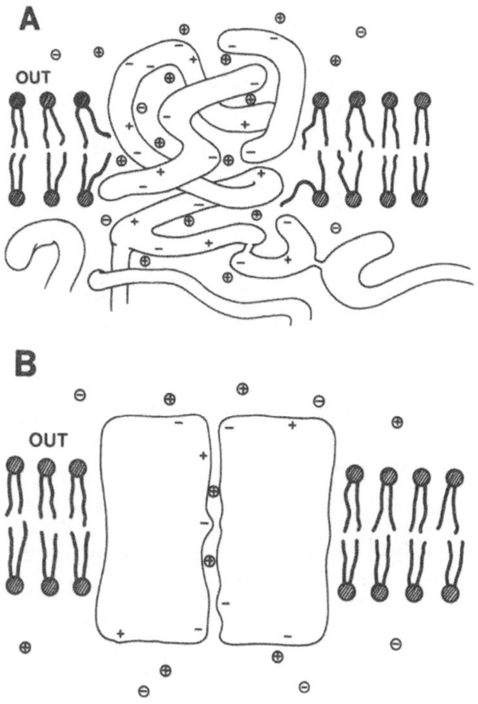

Figure 15. (A) Hypothetical structure of the membrane proteins that make up the "channel" in a membrane–cortex view. The proteins function as a polelectrolyte network. Ions pass through the "channel" by either moving through the interstices of the matrix or hopping between charged sites. (B) Hypothetical structure of a "channel" according to the rigid-pore view. The membrane proteins form a long, narrow, aqueous pore. The narrow region of the pore is the "selectivity filter," which is about 3 Å wide. Hypothetical gates (not shown) control the opening and closing of the inner mouth of the pore.

the ions are outside of the cell. All aspects of the conductance pathway, including "selectivity," kinetics, and activation and inactivation of conductance, are sensitive to the internal ionic environment. Second, the effects of the internal ions on the early conductance pathway and the late conductance pathway are correlated, and this correlation can be destroyed by the internal perfusion of pronase. These findings suggest that the "channel" probably is not a mechanical system made up of independent functional components (such as "filters" and "gates"), rather, it is likely to be a highly organized macromolecular structure in which the so-called "selectivity" and activation–inactivation are determined by the conformation of the proteins. A large portion of this macromolecular complex must extend into the cytoplasmic region, since it is highly sensitive to the internal ionic environment and is vulnerable to the attack of pronase in the internal solution.

The concept of functional linkage between the membrane proteins and cortical proteins is also supported by two of our preliminary studies on the effects of chemicals known to disrupt cytoplasmic proteins. One major component of cytoplasmic proteins are microtubules. Matsumoto and Sakai (1979) have suggested that the microtubules may be important in regulating both resting and action potentials. They showed that when axons are treated by microtubule-disrupting chemicals, the excitability is lost. We also have investigated the role of microtubules in the process of excitation. First, using an immunofluorescent technique (Brinkley *et al.*, 1980), we found evidence that the axolemma is attached to a network of microtubules at the cortical region (Chang, 1980). Second, using a voltage-clamp technique, we found that colchicine, a chemical agent that disrupts the assembly of microtubules, can reversibly and selectively suppress the early conductance of the axon (Chang, 1980; 1983). A sample voltage-clamp record showing the colchicine effect is shown in Fig. 16.

Another major component of cytoplasmic proteins are microfilaments. We have also

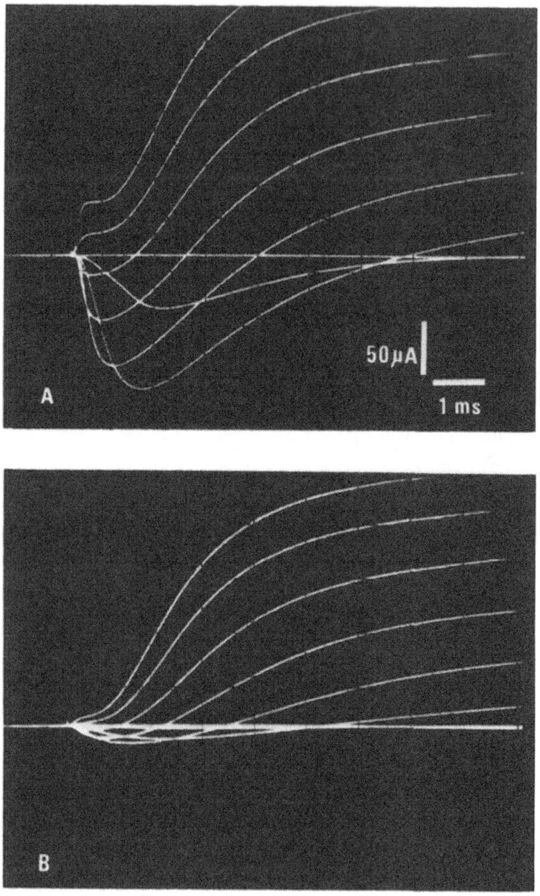

Figure 16. Sample voltage-clamp records that show the effect of colchicine. The axon was clamped from the resting potential (−60 mV) to depolarizing potentials at −40, −20, 0, 20, 40, 60, and 80 mV. (A) Control. (B) 10 mM colchicine applied inside the axon. Temperature = 5°C. $[K^+]_i$ = 400 mM, $[Na^+]_i$ = 50 mM, $[K^+]_o$ = 10 mM, and $[Na^+]_o$ = 424 mM.

studied the physiological role of microfilaments by means of the voltage-clamp technique. Microfilaments are made up mainly of actins that can be disrupted by cytochalasin B. Thus, we have examined the effect of cytochalasin B on the excitation currents. We found that cytochalasin B affects mainly the late current, although it also suppresses the early current to a smaller extent. A sample record showing the effects of cytochalasin B is shown in Fig. 17.

It seems that the cytoplasmic proteins are functionally important to the excitation properties of the nerve cells. At this point we still have to bear in mind that the results of these chemical studies are preliminary, and alternative interpretations, such as whether colchicine may behave as a local anesthetic, must be considered. In all regards, however, we feel that detailed investigations of the physiological role of the cytoplasmic proteins will bear fruitful results and should be included in future research on the "excitable membrane."

ACKNOWLEDGMENTS. This work is supported by ONR research contract N0004-76-C-0100. I thank Professor R. L. Vick for critical reading of the manuscript and Ms. Colette Dougherty and Ms. Gerrie Hazlewood for secretarial assistance. I also thank the staff of the Marine Biological Laboratory, Woods Hole, Massachusetts for support of my research.

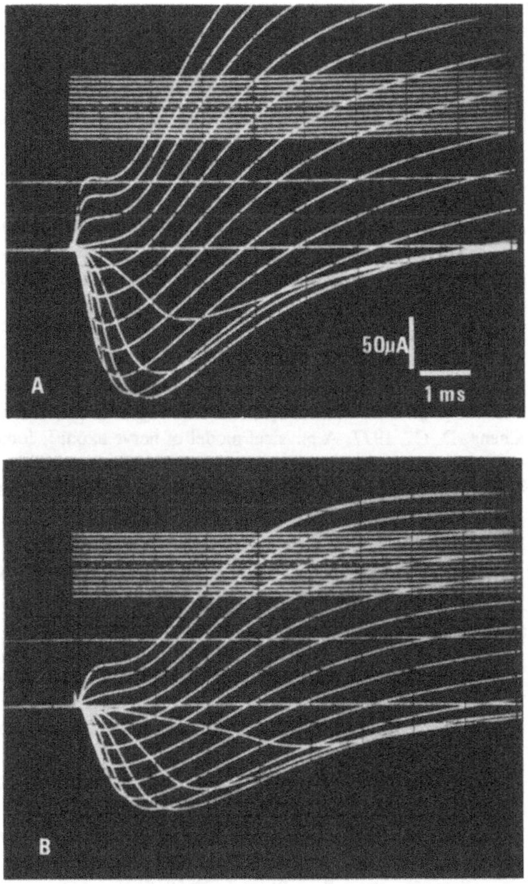

Figure 17. Sample voltage-clamp records showing the effect of cytochalasin B. The parallel lines in the upper half of each panel are records of membrane potential (in association with each of the current traces). After a hyperpolarizing conditioning pulse, the axon was clamped to depolarizing potentials at −40, −30, −20, −10, 0, 10, 20, 30, 40, 50, 60, 70, 80 mV. (A) Control. (B) 150 μg/ml of cytochalasin B (in 3% DMSO) added to the internal perfusion solution. Temperature = 5°C. $[K^+]_i$ and $[Na^+]_i$ were 400 mM and 50 mM, respectively, and $[K^+]_o$ and $[Na^+]_o$ were 10 mM and 424 mM, respectively.

REFERENCES

Adelman, W. J., Jr., and Senft, J. P., 1966, Voltage clamp studies on the effect of internal cesium ion on sodium and potassium currents in the squid giant axon, *J. Gen. Physiol.* **50:**279–293.

Adelman, W. J., Jr., and Senft, J. P., 1968, Dynamic asymmetries in the squid axon membrane, *J. Gen. Physiol.* **51:**102s–114s.

Armstrong, C. M., 1971, Interaction of tetraethyl ammonium ion derivatives with the potassium channels of giant axons, *J. Gen. Physiol.* **58:**413–437.

Armstrong, C. M., 1975, K pores of nerve and muscle membranes, in: *Membranes: A Series of Advances,* Vol. 3 (G. Eisenman, ed.), pp. 325–358, Marcel Dekker, New York.

Armstrong, C. M., and Bezanilla, F., 1977, Inactivation of the sodium channel. II. Gating current experiments, *J. Gen. Physiol.* **70:**567–590.

Armstrong, C. M., Bezanilla, F., and Rojas, E., 1973, Destruction of sodium conductance inactivation in squid axons perfused with pronase, *J. Gen. Physiol.* **62:**375–391.

Begenisich, T., and Cahalan, M., 1980, Sodium channel permeation in squid axons. I: Reversal potential experiments, *J. Physiol.* **307:**217–242.

Bergman, C., 1970, Increase of sodium concentration near the inner surface of the nodal membrane, *Pflügers Arch. (Eur. J. Physiol.)* **317:**287–302.

Bezanilla, F., and Armstrong, C. M., 1972, Negative conductance caused by the entry of sodium and cesium ions into the potassium channels of squid axons, *J. Gen. Physiol.* **60:**588–608.

Bloodgood, R. A., and Rosenbaum, J., 1976, Ultrastructural observations on the squid giant axon and associated Schwann cell sheath, *Biol. Bull.* **151**:402.

Brinkley, B. R., Fistel, S. H., and Pardue, R. L., 1980, Microtubules in cultured cells: Indirect immunofluorescent staining with tubulin antibody, *Int. Rev. Cytol.* **63**:59–95.

Cahalan, M., and Almers, W., 1979, Block sodium conductance and gating current in squid giant axons poisoned with quaternary strychnine, *Biophys. J.* **27**:57–74.

Cahalan, M., and Begenisich, T., 1976, Sodium channel selectivity: Dependence on internal permeant ion concentration, *J. Gen. Physiol.* **68**:111–125.

Chambers, R., and Kao, C. Y., 1952, The effect of electrolytes on the physical state of the nerve axon of the squid and of stentor, a protozoan, *Exp. Cell Rev.* **3**:564–573.

Chandler, W. K., and Meves, H., 1965, Voltage clamp experiments on internally perfused giant axons, *J. Physiol.* **180**:788–820.

Chandler, W. K., and Meves, H., 1970, Sodium and potassium currents in squid axons perfused with fluoride solutions, *J. Physiol.* **211**:623–652.

Chandler, W. K., Hodgkin, A. L., and Meves, H., 1965, The effect of changing the internal solution on sodium inactivation and related phenomena in giant axon, *J. Physiol.* **180**:821–836.

Chang, D. C., 1977, A physical model of nerve axon: I. Ionic distribution potential profile, and resting potential, *Bull. Math. Biol.* **39**:1–22.

Chang, D. C., 1979, A physical model of nerve axon. II. Action potential and excitation currents. Voltage-clamp studies of chemical driving forces of Na^+ and K^+ in squid giant axon, *Physiol. Chem. Phys.* **11**:263–288.

Chang, D. C., 1980, Possible role of cytoplasmic microtubule structure in the excitation properties of nerve axon, *Eur. J. Cell Biol.* **22**:304.

Chang, D. C., 1981, The effect of internal sodium ions on the action potential and reversed potential in squid axon, *Biol. Bull.* **161**:341.

Chang, D. C., 1983, A voltage-clamp study of the effects of colchicine on the squid giant axon. *J. Cell. Physiol.* **115**:260–264.

Cohen, L. B., Keynes, R. D., and Hille, B., 1968, Light scattering and birefringence changes during nerve activity, *Nature* **218**:438–441.

Conti, F., and Neher, E., 1980, Single channel recordings of K^+ currents in squid axons, *Nature* **285**:140–143.

Conti, F., and Palmieri, G., 1968, Nerve fiber behavior in heavy water under voltage-clamp, *Biophysik* **5**:71–77.

deGroot, S. R., 1951, *Thermodynamics of Irreversible Processes,* North Holland, Amsterdam.

Diamond, J. M., and Wright, E. M., 1969, Biological membranes: The physical basis of ion and nonelectrolyte selectivity, *Ann. Rev. Physiol.* **31**:581–646.

Ebert, G. A., and Goldman, L., 1976, The permeability of the sodium channel in *Myxicola* to alkali cations, *J. Gen. Physiol.* **68**:327–340.

Ehrenstein, G., 1976, Ion channels in nerve membrane, *Phys. Today* **29**(10):33–39.

Eisenman, G., 1962, Cation selective glass electrodes and their modes of operation, *Biophys. J.* **2** (suppl. 2):259–323.

Eisenman, G., Sandblom, J. P., and Walker, J. L., Jr., 1967, Membrane structure and ion permeation, *Science* **155**:965–974.

Ellisman, M. H., and Porter, K. R., 1980, Microtrabecular structure of the axoplasmic matrix: Visualization of the cross-linking structures and their distribution, *J. Cell Biol.* **87**:464–479.

Eyring, H., Lumry, R., and Woodbury, J. W., 1949, Some applications of modern rate theory to physiological systems, *Rec. Chem. Prog.* **10**:100–114.

Frankenhauser, B., and Moore, L. E., 1963, The effect of temperature on the sodium and potassium permeability changes in myelinated nerve fibres of *Xenophus laevis, J. Physiol.* **169**:431–437.

Fraser, A., and Frey, A. H., 1968, Electromagnetic emission at micron wavelength from active nerves, *Biophys. J.* **8**:731–734.

French, R. J., and Adelman, W. J., Jr., 1976, Competition, saturation, and inhibition—Ionic interactions shown by membrane ionic currents in nerve, muscle, and bilayer systems, *Current Topics in Membrane and Transport* **8**:161–207.

Gainer, H., Carbone, E., Singer, I., Sisco, K., and Tasaki, I., 1974, Depolarization-induced change in the enzymatic radio-iodination of a protein on the internal surface of the squid giant axon membrane, *Comp. Biochem. Physiol.* **47A**:477–484.

Goldman, D. E., 1943, Potential, impedance, and rectification in membranes, *J. Gen. Physiol.* **27**:37–60.

Hale, D. K., Packham, D. I., and Pepper, K. W., 1953, Properties of ion-exchange resins in relation to their structure. Part V. Exchange of organic cation, *J. Chem. Soc.* **1953**:844–851.

Hille, B., 1968, Pharmacological modifications of the sodium channels of frog nerve, *J. Gen. Physiol.* **51**:199–219.

Hille, B., 1970, Ionic channels in nerve membranes, *Prog. Biophys. Mol. Biol.* **21**:1–32.

Hille, B., 1971, The permeability of the sodium channel to organic cations in myelinated nerve, *J. Gen. Physiol.* **58**:599–619.

Hille, B., 1975a, Ionic selectivity of Na and K channels of nerve membranes, in: *Membranes: A Series of Advances*, Vol. 3 (G. Eisenman, ed.), pp. 255–323, Marcel Dekker, New York.

Hille, B., 1975b, Ionic selectivity, saturation and block in sodium channels. A four barrier model, *J. Gen. Physiol.* **66**:535–560.

Hille, B., 1975c, The receptor for tetrodotoxin and saxitoxin. A structural hypothesis, *Biophys. J.* **15**:615–619.

Hille, B., and Schwarz, W., 1978, Potassium channels as single file pores, *J. Gen. Physiol.* **72**:409–442.

Hinke, J. A. M., 1961, The measurement of sodium and potassium activities in the squid axon by means of cation-selective glass micro-electrodes, *J. Physiol.* **165**:314–335.

Hodge, A. J., and Adelman, W. J., Jr., 1980, The neuroplasmic network in *Loligo* and *Hermissenda* neurons, *J. Ultrastruct. Res.* **70**:220–241.

Hodgkin, A. L., and Huxley, A. F., 1952a, Currents carried by sodium and potassium ions through the membrane of the giant axon of *Loligo*, *J. Physiol.* **116**:449–472.

Hodgkin, A. L., and Huxley, A. F., 1952b, A quantitative description of membrane current and its application to conduction and excitation in nerve, *J. Physiol.* **117**:500–544.

Hodgkin, A. L., and Katz, B., 1949, The effect of sodium ions on the electrical activity of the giant axon of the squid, *J. Physiol.* **108**:37–77.

Inoue, I., Pant, H. C., Tasaki, I., and Gainer, H., 1976, Release of proteins from the inner surface of squid axon membrane labeled with tritiated N-ethylmaleimide, *J. Gen. Physiol.* **68**:385–395.

Iwasa, K., and Tasaki, I., 1980, Mechanical changes in squid giant axons associated with production of action potentials, *Biochem. Biophys. Res. Comm.* **95**:1328–1331.

Jirgensons, B., and Straumanis, M. E., 1962, *A Short Textbook of Colloid Chemistry*, MacMillan, New York.

Kao, C. Y., 1966, Tetrodotoxin, saxitoxin and their significance in the study of excitation phenomena, *Pharmacol. Rev.* **18**:997–1049.

Kao, C. Y., 1981, Tetrodotoxin, saxitoxin, chiriquitoxin: New perspective on ionic channels, *Fed. Proc.* **40**(1):30–35.

Katchalsky, A., 1964, Polyelectrolytes and their biological interactions, *Biophys. J.* **4**(Suppl.):9–41.

Katchalsky, A., 1967, Membrane thermodynamics, in: *The Neuroscience: A Study Program*, Rockefeller University Press, New York.

Keynes, R. D., 1972, Excitable membranes, *Nature* **239**:29–32.

Korn, E. D., 1978, Biochemistry of actomyosin-dependent cell motility (a review), *Proc. Natl. Acad. Sci. USA* **75**:588–599.

Kressman, T. R. E., and Kitchener, J. A., 1949, Cation exchange with a polysynthetic phenosulfonate resin. Part III. Equilibria with large organic ions, *J. Chem. Soc.* **1949**:1208–1210.

Lackie, J. M., 1980, The structure and organization of cell surface, in: *Membrane Structure and Function* (E. Bittar, ed.), pp. 73–102, John Wiley and Sons, New York.

Läuger, P., 1973, Ion transport through pores: A rate theory analysis, *Biochim. Biophys. Acta* **311**:423–441.

Ling, G. N., 1962, *A Physical Theory of the Living State*, Blaisdell, New York.

Matsumoto, G., and Sakai, H., 1979, Microtubules inside plasma membrane of squid giant axons and their possible physiological function, *J. Membr. Biol.* **50**:1–14.

Matteson, D. R., and Swenson, R. P., 1982, Permeant cations alter closing rate of K channels, *Biophys. J.* **37**(2) part 2:17a.

Metuzals, J., 1969, Configuration of a filamentous network in the axoplasm of the squid *(Loligo pealii L.)* giant nerve fiber, *J. Cell Biol.* **43**:480–505.

Metuzals, J., and Izzard, C. S., 1969, Spatial patterns of thread-like elements in the axoplasm of the giant nerve fiber of the squid *(Loligo pealii L.)* as disclosed by differential interference microscopy and by electron microscopy, *J. Cell Biol.* **43**:456–479.

Metuzals, J., and Tasaki, I., 1978, Subaxolemmal filamentous network in the giant nerve fiber of the squid *(Loligo pealii L.)* and its possible role in excitability, *J. Cell Biol.* **78**:597–621.

Mullins, L. J., 1968, A single channel or a dual channel mechanism for nerve excitation, *J. Gen. Physiol.* **52**:550–553.

Narahashi, T., 1974, Chemicals as tools in the study of excitable membranes, *Physiol. Rev.* **54**:813–889.

Narahashi, T., and Moore, J., 1968, A single or dual channel in nerve membrane? *J. Gen. Physiol.* **52**:553–555.

Narahashi, T., Moore, J., and Scott, W. R., 1964, Tetrodotoxin blockage of sodium conductance increase in lobster giant axons, *J. Gen. Physiol.* **47**:965–974.

Nernst, W., 1888, Zur Kinetikder in lösung befindlichen Körper, *Z. Physik. Chem.* **2**:613–637.

Pant, H. C., Terakawa, S., Baumgold, J., Tasaki, I., and Gainer, H., 1978, Protein release from the internal surface of the squid giant axon membrane during excitation and potassium depolarization, *Biochim. Biophys. Acta* **513**:132–140.

Planck, M., 1890, Ueber die Potentialdifferenz zwischen zwei verdünnten lösungen binärer electrolyte, *Ann. Phys. Chem.* **40:**561–576.

Prigogine, I., 1961, *Introduction to Thermodynamics of Irreversible Processes*, 2nd Ed., Wiley, New York.

Reichenberg, D., 1969, Ion-exchange selectivity, *Ion-Exchange* **1:**227–276.

Robinson, R. A., and Stokes, R. H., 1949, Tables of osmotic and activity coefficients of electrolytes in aqueous solution at 25°C, *Trans. Faraday Soc.* **45:**612–624.

Sandblom, J., Eisenman, G., and Neher, E., 1977, Ionic selectivity, saturation, and block in gramicidin A channels. I. Theory for the electrical properties of ion selective channels having two pairs of binding sites and multiple conductance states, *J. Membr. Biol.* **31:**383–417.

Shashoua, V., 1969, Electrically active protein and polynucleic acid membrane, in: *Molecular Basis of Membrane Function* (D. C. Tosteson, ed.), Prentice-Hall, Englewood.

Shashoua, V., 1975, Electrical oscillatory phenomena in protein membranes, *Faraday Symp. Chem. Soc.* **9:**174–181.

Sigworth, F. J., and Neher, E., 1980, Single Na^+ channel currents observed in cultured rat muscle cells, *Nature* **287:**447–449.

Singer, I., and Tasaki, I., 1968, Nerve excitability and membrane macromolecules, in: *Biological Membranes: Physical Facts and Function* (D. Chapman, ed.), pp. 347–410, Academic Press, New York.

Singer, S. J., and Nicolson, G. L., 1972, The fluid mosaic model of the structure of cell membranes, *Science*, **174:**720–731.

Swenson, R. P., and Armstrong, C. M., 1981, K^+ channels close more slowly in the presence of external K^+ and Rb^+, *Nature* **291:**427–429.

Takenaka, T., Yoshioka, T., Horie, H., and Watanabe, F., 1976, Changes in ^{125}I-labeled membrane proteins during excitation of the squid giant axon, *Comp. Biochem. Physiol.* **55B:**89–93.

Tasaki, I., Singer, I., and Takenaka, T., 1965, Effects of internal and external ionic environment on excitability of squid giant axon: A macromolecular approach, *J. Gen. Physiol.* **48:**1095–1123.

Tasaki, I., Watanabe, A., Sandlin, R., and Carnay, L., 1968, Changes in fluorescence, turbidity and birefringence associated with nerve excitation, *Proc. Natl. Acad. Sci. USA* **61:**883–888.

Tasaki, I., Carnay, L., Sandlin, R., and Watanabe, A., 1969, Fluorescence changes during conduction in nerves stained with acridine orange, *Science* **163:**683–685.

Taylor, R. E., 1974, Excitable membrane, *Annu. Rev. Phys. Chem.* **25:**387–405.

Teorell, T., 1953, Transport processes and electrical phenomena in ionic membrane, *Prog. Biophys. Biochem.* **3:**305–369.

Teorell, T., 1959, Elektrokinetic membrane processes in relation to properties of excitable tissues. 2. Some theoretical consideration, *J. Gen. Physiol.* **42:**847–863.

Wolosewick, J. J., and Porter, K. R., 1976, Stereo high-voltage electron microscopy of whole cell of human diploid line, WI-38, *Am. J. Anat.* **147:**303–324.

Woodbury, J. W., 1971, Eyring rate theory model of the current-voltage relationships of ion channels in excitable membranes, in: *Chemical Dynamics: Papers in Honor of Henry Eyring* (J. O. Hirschfelder, ed.), pp. 601–617, John Wiley & Sons, New York.

Yeh, J. Z., and Armstrong, C. M., 1978, Immobilization of gating charge by a substance that stimulates inactivation, *Nature* **273:**387–389.

Yoshioka, T., Pant, H. C., Tasaki, I., Baumgold, J., Matsumoto, G., and Gainer, H., 1978, An approach to the study of intracellular proteins related to the excitability of the squid giant axon, *Biochim. Biophys. Acta* **538:**616–626.

12

Stochastic Modeling of the Aggregation-Gating Site

Gilbert Baumann

You just have to try and imagine what the universe is like.

Paul A. M. Dirac*

I. GATING: CONFORMATIONAL CHANGE OR AGGREGATION PROCESS?

A nerve signal is usually taken to be controlled by way of a potassium- and a sodium-conductance system, each consisting of a large number of discrete sites located in the axon membrane. At each site either predominantly potassium ions or predominantly sodium ions can cross the membrane (Hodgkin and Huxley, 1952; Hille, 1975a). It is generally presumed that such a site is made up of one or several proteins forming a *permanent* pore-shaped opening in the membrane. This opening provides a channel for the flow of ions (Fig 1a). The flow of ions through the channel of this hypothetical site is controlled by specialized domains in the protein structure that can obstruct the channel and can thus function as gates. Opening and closing of the channel gates (also referred to as *gating* †) is thought to occur via *conformational change* in the protein. That is to say, the process of gating in this model can be related to the familiar experience of the opening and closing of a door.

As an alternative to this model it is proposed in this chapter that a gating site consists either of several protein subunits or of a single protein with special domains that do not form a permanent channel, but can provide a *transient* channel by collectively forming a pore-shaped opening (Fig. 1b). These hypothetical protein subunits or domains can change

* From *The Search for Solutions* by Horace Freeland Judson. Copyright © 1980 by Playback Associates. Reprinted by permission of Holt, Rinehart and Winston, Publishers.

† The term *gating* is used here without implying a specific mechanism, but in the more general sense of switching the channel conductance between *on* and *off*. A site where gating occurs is called a *gating site*.

Gilbert Baumann • Department of Physiology, Duke University Medical Center, Durham, North Carolina 27710.

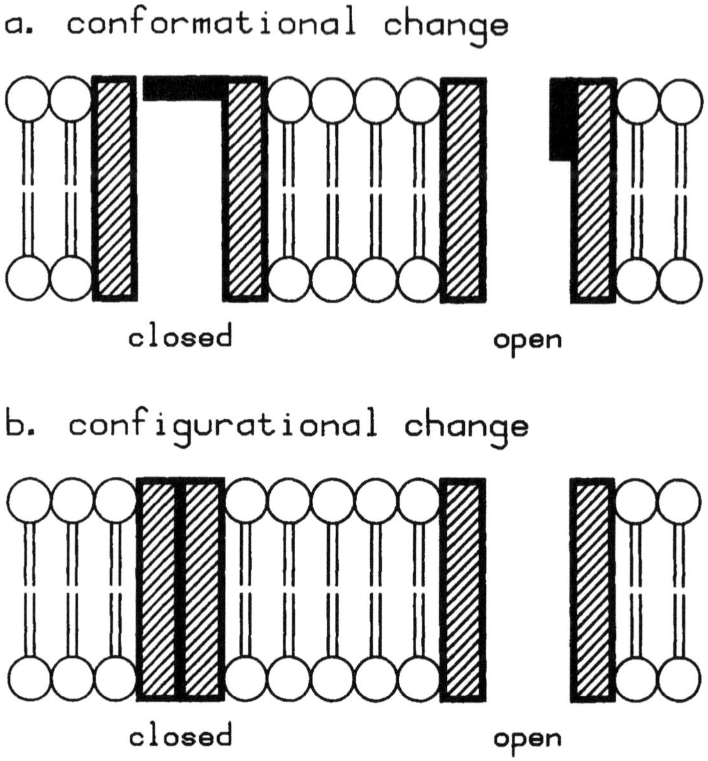

Figure 1. Two concepts of channel opening and closing (or *gating*) in a two-dimensional representation. (*a*) The conventional model of a site where channel gating occurs consists of one or several proteins forming a *permanent* pore. The ionic current is controlled by a gate (black rectangle) that opens and closes via *conformational* change in the protein and thus unblocks and blocks the channel. (*b*) The alternative model proposed here also consists of one or several proteins, but they form a *transient* pore. The ionic current in this model is controlled by change in the position of interacting domains within the protein (or in the position of interacting protein subunits) relative to each other involving an aggregation process. That is, gating occurs by way of *configurational* change in the site.

their positions relative to each other by way of a *configurational change* involving a conformational change* and an aggregation process (Baumann and Easton, 1980, 1981). The opening and closing of such a channel occurs because some site configurations create a pore and some do not.† Despite the simplicity of the assumptions in this model it takes a great deal of imagination to comprehend how the process of gating is proposed to work, because it is unlike anything one can experience in everyday life.

This chapter defines the concept of an aggregation-gating site and describes the probabilistic methods that are needed for numerically analyzing the dynamic or kinetic properties of the proposed model. Modeling results derived with such methods are then compared to data and, finally, the various approaches to modeling membrane excitability and the purpose of molecular modeling in biology are discussed.

* The assumption of conformational change is an alternative to reorientation of a protein subunit proposed earlier (Baumann and Mueller, 1974; Baumann and Easton, 1980).

† Notice that in the conventional model of a permanent channel with gate, the terms *gating site* and *channel* are synonymous and interchangeable. However, in the aggregation model, gating site and channel must be distinguished, since a channel may exist or may not exist, depending on the configuration of the gating site.

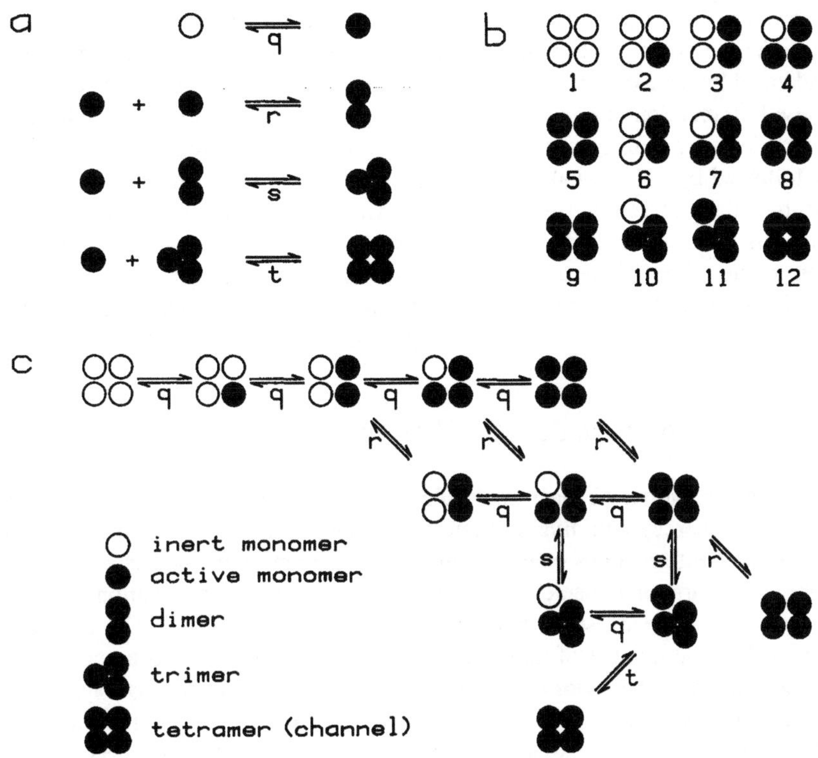

Figure 2. Model assumptions of a hypothetical aggregation-gating site consisting of four domains or subunits. (*a*) The eight elementary reactions in this model are a reversible voltage-dependent conformational change labeled *q* from inert (white disc) to active monomer (black disc) and three reversible, not voltage-dependent aggregation steps labeled *r, s,* and *t* involving dimer (two fused discs), trimer (three fused discs), and tetramer (four fused discs). (*b*) Based on these assumptions the hypothetical gating site can exist in 12 configurations. (*c*) Each of the 14 reversible transitions between the 12 configurations involves one elementary reaction. The type of reaction taking place in the transition is indicated by a letter (below arrows).

II. THE CONCEPT OF AN AGGREGATION GATING SITE

For modeling the potassium- or sodium-gating site, it is assumed that the number of subunits or domains per site is four, because this is the smallest number that gives realistic kinetic behavior* (Baumann and Easton, 1980, 1981). Furthermore, it is assumed that the subunits or domains are identical in their kinetic behavior, which does not preclude that they may be chemically different from each other. It is then postulated that each subunit or domain of the hypothetical gating site can undergo a conformational change between an inert and an active conformation. This conformational change, indicated by the letter *q* below the arrows in Fig. 2a, is voltage-dependent. A subunit or domain is called a *monomer*. It is further postulated that active monomers can aggregate. That is, an active monomer can interact with another active monomer to form a *dimer*. A dimer, in turn, can interact with an active monomer to form a *trimer,* and a trimer can interact with an active monomer to

* Assuming more than four subunits or domains per site results in qualitatively the same kinetics. For each additional subunit or domain per site but otherwise unchanged model assumptions, however, the maximum ratio of conducting to nonconducting configurations increases.

form a *tetramer*. The tetramer can function as an ionic channel because it makes a hydrophilic opening in the membrane that is big enough for ions to pass through. The three aggregation steps, indicated by the letters *r, s,* and *t* below the arrows in Fig. 2a, are voltage-independent. All of these four reactions are reversible. Thus, there are eight elementary reactions possible. The assumptions are consistent with the kinetic scheme of a voltage-dependent first-order reaction followed by voltage-independent aggregation steps (Baumann and Mueller, 1974).

The hypothetical gating site can exist in 12 possible configurations (Fig. 2b). Two configurations are connected if it takes only one elementary reaction to go from one configuration to the other. There are 14 such connections* possible between the 12 configurations. These 14 reversible transitions are the configurational changes that constitute the transition scheme of the hypothetical site (Fig. 2c).

III. STOCHASTIC MODELING METHODS

Molecules, including the macromolecules involved in the gating process, are in perpetual random motion, thereby colliding and interacting with each other. And molecules that can exist in various conformations change from one conformation to another at random. Thus, the number of a particular species of molecule in a particular conformation is not constant in a system even at steady state, but fluctuates about a mean value. The fluctuations are negligible if the total number of molecules in a system is very large. If a very large system is perturbed, its kinetic behavior can be predicted by the differential equations derived from the mass-action law of classical, *deterministic* chemical kinetics. If the total number of molecules in a system is very small, however, the random fluctuations become significant and, thus, probabilistic, *stochastic* methods must be used to describe the kinetic behavior of such a system.†

There are two stochastic techniques that allow one to analyze the kinetic behavior of a model of a gating site either from a microscopic or from a macroscopic point of view. First, with a *Monte Carlo method* one can derive individual *kinetic sample paths*‡ which are the theoretical counterpart of the experimentally obtained single-channel conductance records. The kinetic behavior of a modeled or real single site is also referred to as *microkinetic behavior*, or *microkinetics* for short. Application of this method will be called a *simulation*** to indicate that a random number generator is involved. Then, with a *Markov-process method* one can derive the theoretical or *expected kinetic behavior* of a gating-site model which is the average microkinetic behavior of an infinite number of modeled gating sites. The kinetic behavior of a large number of modeled or real sites is also referred to as *macrokinetic behavior*, or *macrokinetics* for short. Application of this method will be called a *computation*

* Notice that configuration 9 and 12 could also be connected. This configurational change, however, would violate the basic assumption of aggregation by monomer addition and deduction. Although qualitatively such a modification of the model has little effect on the kinetics (Baumann and Easton, 1981), there is no compelling reason for making this assumption.

† Furthermore, notice that in the case of a bimolecular reaction such as the dimerization step in the aggregation model, the deterministic and the stochastic approach lead to different differential equations, because the classical mass-action law does not apply to small systems. In a system of n identical reactants, each reactant can interact with the $n - 1$ other reactants, and the reaction rate is thus proportional to $n(n - 1)$. If n is very large, the rate is approximately proportional to n^2 which is the basis of the classical mass-action law. This approximation cannot be made for small systems and, thus, the time courses calculated from the two methods deviate from each other.

‡ A *sample path* is defined as a possible conductance time course resulting from a simulation of a single gating site.

** The term *simulation* is sometimes used to encompass all forms of numerical modeling.

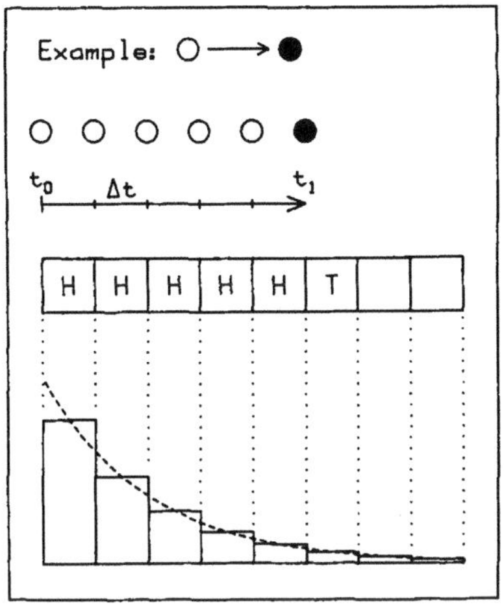

Figure 3. A chemical reaction modeled as a Poisson process is demonstrated with the example of a conformational change from inert (white disc) to active monomer (black disc). The time axis is divided up into intervals Δt. For each Δt a coin is flipped that is biased to land head up. If head (H) is up, the molecule survives the Δt in its present conformation and if tail (T) is up, it does not. If the molecule does not survive a Δt, one unit is added to the appropriate column in the histogram below that time interval. The histogram suggests that if this procedure is repeated a large number of times and as Δt approaches zero, the probability of the reaction not occurring until after time t is exponentially distributed (dashed curve).

to indicate that no random number generator is involved. The two methods are based on the same assumptions and, thus, are closely related. These fundamental assumptions concern the question of how to model a single chemical reaction and are described in more detail elsewhere (Baumann and Easton, 1982a).

IV. MONTE CARLO SIMULATION

Consider the conformational change from inert to active monomer shown in Fig. 3. It is assumed that a chemical reaction is a Poisson process. In a Poisson process the probability that an event, e.g., a reaction, occurs between time t and time $t + \Delta t$ is proportional to Δt if Δt is small. The probability distribution of times between events in a Poisson distribution is exponential (Breiman, 1969), as Fig. 3 suggests. This means that the probability of the reaction *not* occurring until after time t is

$$P(t,\infty) = e^{-\lambda t} \tag{1}$$

This is the exponential distribution* with parameter λ, which is the rate of the reaction (in the example from inert to active monomer) and is given by the standard rate equation

$$\lambda = k \prod_{m=1}^{n} N_m,$$

where k is the rate constant of the reaction and N_m is the number of molecules of the mth reactant (rather than the reactant concentration). Based on this relation, microkinetic sample paths for the gating site can be simulated† by a procedure demonstrated with an example

* It should be noted that exponential random variables exhibit the unique property of "memorylessness." This means that if the process is in configuration i at time t, the probability of transition in the time interval $t + \varepsilon$ is $1 - e^{-\lambda\varepsilon}$, independent of t.

† Such simulations are the basis of an animated 16-mm movie with soundtrack (rated G) that is available on loan from the author.

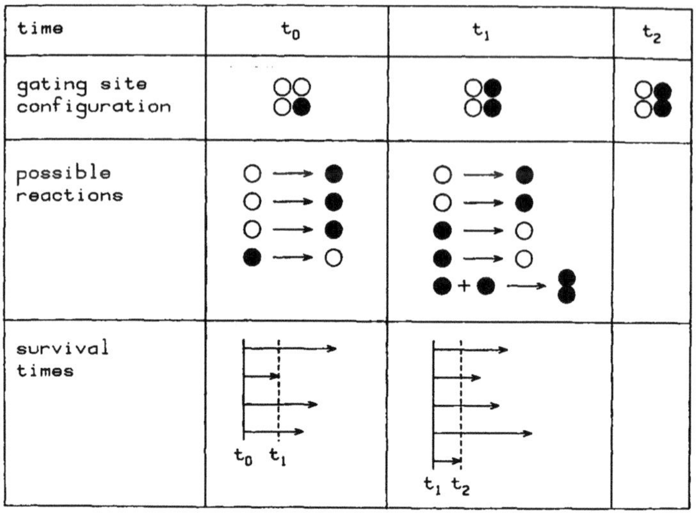

Figure 4. Monte Carlo method demonstrated by an example. Suppose that at time t_o the gating site is in configuration *2*. In this configuration there are four reactions possible. For each one a survival time, *t*, during which a reaction does not occur is calculated from Eq. (1) by generating a random number with a uniform distribution on the interval [0,1]. The reaction with the shortest survival time *t* in the example is taken to occur. Configuration *2* changes to configuration *3* and the time is advanced to $t_o + t = t_1$. The same procedure is repeated for the new configuration *3* which results in a transition to configuration *6* at time t_2. This process can be repeated as many times as desired to simulate the random behavior of the site. Whenever the site is in configuration *12*, it has unit conductance, and when it is in any of the 11 other configurations, its conductance is 0. The conductance of the site is stored at regular time intervals, the length of which is referred to as *time resolution*. This is how sample paths of single-channel conductances can be simulated for any voltage pulse sequence.

in Fig. 4. The method is described elsewhere (Baumann and Easton, 1980) in a manner that allows the reader to adapt the algorithm to any model.

The calculated macrokinetic behavior of an aggregation process is remarkably similar to the experimentally found macrokinetic behavior of gating systems. Notice that macrokinetically, an aggregation process can proceed in two different ways, depending on the stability of the dimer. If the dimer is *as stable as* two active monomers, the macrokinetics

Table I. Model Parameters

	Activation energy ($kJ \cdot mole^{-1}$)	
Reaction[a]	Monophasic	Biphasic
q_f	2	2
q_b	0	0
r_f	0	0
r_b	0	15
s_f	0	0
s_b	0	0
t_f	0	0
t_b	0	0

[a] Forward and backward reactions are indicated by the subscripts *f* and *b*, respectively.

of any aggregate beyond the dimer resemble the macrokinetics of the potassium-conductance system. If the dimer is *more stable than* two active monomers, the macrokinetics of any aggregate beyond the dimer resemble the macrokinetics of the sodium-conductance system (Baumann and Mueller, 1974; Baumann and Easton, 1981). The stability of the dimer depends on the activation energy assumed for the backward dimerization step r_b shown in Fig. 2a. Only reaction r_b of the four reversible reactions q, r, s, and t shown in Fig. 2a has such a profound influence on the macrokinetic behavior of the proposed gating-site model. Table I gives the two sets of activation energies used in all simulations and calculations in this chapter.

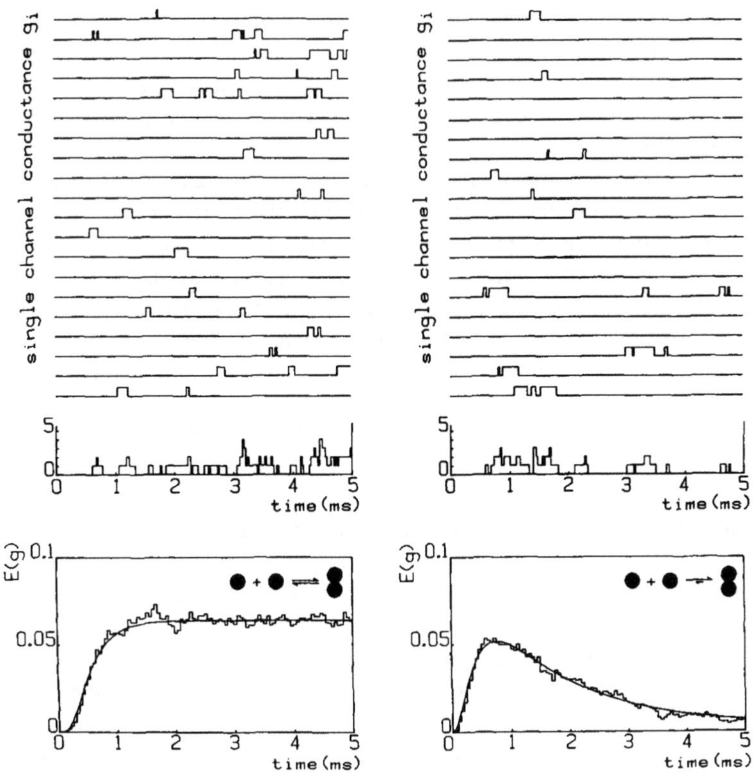

Figure 5. Relation between Monte Carlo simulations and a Markov-process computation of a single aggregation-gating site. Conductances in response to a pulse from −80 to 0 mV under voltage clamp are shown. The graphs are produced with the two parameter sets given in Table I that result in macrokinetics resembling the one of the potassium (low dimer stability, graphs to the left) and sodium-conductance system (high dimer stability, graphs to the right), respectively. In each case a series of 20 simulated sample paths is shown. The sum of the 20 microkinetic paths already reveals a different trend between the two kinetic modes, and the average of 4999 paths given in the two graphs at the bottom shows the difference clearly. There the average conductance time course (jagged curve) is compared to the expected conductance time course derived from a Markov-process computation (smooth curve) for the same two parameter sets. The initial configuration for each simulation was chosen to properly account for the steady-state distribution of the 12 configurations at the holding potential. For the potassium case (left graph) 4674 simulations were started from configuration *1*, 303 from configuration 2, seven from configuration *3*, and 15 from configuration *6*. For the sodium case (right graph) 1509 simulations were started from configuration *1*, 98 from configuration 2, two from configuration *3*, 2780 from configuration *6*, 427 from configuration *9*, 90 each from configurations *7* and *10*, and one each from configurations *8*, *11*, and *12*. Time resolution of the simulations is 50 μsec in the two bottom graphs and 25 μsec in all other graphs of Fig. 5. (Parts of this figure reproduced with permission from the *J. Theoret. Biol.*, 1981, **93**:798. Copyright by Academic Press Inc., London, Ltd.)

For the two cases of low and of high dimer stability one can simulate single microkinetic sample paths, as shown in Fig. 5. Comparison between the two sets reveals that for both assumptions of dimer stability very similar microkinetic sample paths can occur. At the microscopic level the difference between the two kinetic modes is not readily detectable but lies in the probability that channels occur during a pulse. This is demonstrated in Fig. 5 by adding or averaging many sample paths.

Now suppose one wants to determine the macrokinetic behavior of an entire hypothetical membrane patch assumed to consist of many identical and independent* aggregation-gating sites in response to a voltage pulse. Of course, one approach would be to simulate many sample paths of a single site using the above-described Monte Carlo procedure and then average these microkinetic paths. This would give one sample of the possible macrokinetic behavior of the hypothetical membrane patch. If this process were repeated enough times, a mean value of the macrokinetic time courses could be calculated at each time. This mean time course would give an *estimate* that converges to the expected macrokinetic time course of the hypothetical membrane patch as the number of sample paths gets large. This is demonstrated in Fig. 5.

The second approach, based on probability theory, would be to calculate the theoretical or expected macrokinetic time course directly from the underlying assumptions about the individual sites composing the patch. Since the patch is assumed to consist of identical and kinetically independent sites, the expected macrokinetic conductance path of a patch consisting of N sites is N times the expected macrokinetic conductance path of a single site. Thus, the problem reduces to finding the expected macrokinetic conductance time course of one site which can be calculated from a Markov-process† formulation of the transition scheme.

To illustrate the difference between averaging Monte Carlo simulations and computing a Markov process, consider the following analogy. Suppose one has a coin whose physical properties we know, and want to determine the probability, p, that it will land heads up when flipped. This can be done in two ways. The first approach is to flip the coin a large number of times, say n times, and count the number of times, s, that it lands heads up. The true probability of landing heads up can then be estimated from the ratio s/n. If n is large, then the ratio s/n will tend to be very close to p. The second approach is to consider the physical properties of the coin and the laws that govern its fall, and directly calculate p. Such a calculation could be complex and difficult, but would result in the theoretical value of p.

The two approaches are analogous to the method of averaging many Monte Carlo simulations and the Markov-process method, respectively. The Monte Carlo algorithm gives a simple way of estimating the macrokinetic behavior of a single site with any desired accuracy. This approach, however, requires a prohibitive amount of computer time due to the large number of sample simulations that must be generated. Thus, the Monte Carlo method is suitable for simulating samples of the microkinetic behavior of a site, but it is inefficient for routinely estimating the average macrokinetic behavior of a site. In contrast,

* Conceivably, sites could be tightly packed, but open and close independently of each other. The term *independent* thus refers to the non-cooperative kinetic behavior of the sites and does not imply anything about the distance between sites.

† A Markov process is a network of connected Poisson processes, and times between events in a Poisson process are exponentially distributed. In other words, a dynamic isolated system has the Markov property if the probability of the system being in a given configuration at a future time depends only on its configuration at the present time and not on how it arrived at that configuration. In other words, such a system is memoryless.

with the Markov-process method one can directly calculate the limit of an infinite number of Monte Carlo simulations. This is achieved at the cost of complexity. Although it is easy to define a Markov process (as is done in the formulation of a Monte Carlo simulation), numerical calculation of the true probability of being in a given configuration at a given time directly from the probabilistic assumptions about the site is quite difficult.

V. MARKOV-PROCESS COMPUTATION

The task of calculating the expected macrokinetic behavior of a single site in response to a voltage step by a Markov-process formulation can be broken down into two parts. First, one needs to determine the steady-state probability distribution of the 12 configurations at the initial voltage and, then, the probability of transition from any configuration to any other configuration for any time, t, after the voltage perturbation occurs. Suppose that the voltage is perturbed at time $t_0 = 0$ by a voltage-clamp step from a holding potential, V_0, to a clamp potential, V_1. The steady-state probability distribution at voltage V_0 is a vector, π, whose ith element, π_i, is the probability that the site is in configuration i. This probability distribution is given by solving the system of linear equations $H'\pi = 0$, subject to the constraint that $\Sigma\pi_i = 1$, since π is a probability distribution. The matrix H' is the transpose of H. H is called the *generating* or *infinitesimal matrix,* and has as its ijth entry the rate parameter of the reaction connecting configuration i to configuration j which may be a function of the voltage. If direct transition is not possible, the ijth entry is 0. The diagonal entries are calculated from the off-diagonal elements (Çinlar, 1975) by the equation

$$h_{ii} = -\sum_{j=1}^{12} h_{ij} \text{ for } j \neq i$$

Construction of the matrix H for an aggregation-gating site is explained in detail elsewhere (Baumann and Easton, 1981).

The probability of transition at time t after the voltage perturbation can be represented by a matrix exponential (Çinlar, 1975),

$$P(t) = e^{Ht}$$

whose ijth element is the probability that the site is in configuration j at time t given that it was in configuration i at time 0. Suppose P(t) is the vector whose jth element is the probability that the site is in configuration j at time t. Then P (t) $= \pi' P(t)$, where π' is the transpose of the vector π. That is, the probability of being in configuration j at time t is just the sum over each configuration i of the probability of being in configuration i at time t_0 times the probability of being in configuration j at time t given that the site was in configuration i at time 0.

The scalar exponential e^x is defined by the power series

$$e^x = \sum_{n=1}^{\infty} \frac{x^n}{n!}$$

The maxtrix exponential e^{Ht} is similarly defined, except that instead of a scalar argument, the power series has the matrix argument Ht. That is,

$$e^{Ht} = \sum_{n=1}^{\infty} \frac{(Ht)^n}{n!}$$

where the nth power of Ht is the matrix product of n matrices Ht. In general, determination of the matrix exponential is very difficult. In the case of small chemical systems, however, the matrix exponential can be evaluated due to a special property of the generating matrix (Easton, 1981). Evaluation of the matrix exponential for these systems shows that the probability $p_{ij}(t)$ of being in configuration j at time t given that the site was in configuration i at time 0 is given by the equation*

$$p_{ij}(t) = \sum_{l=1}^{n} a_{ijl} e^{\lambda_l t} \tag{2}$$

where λ_l is the lth eigenvalue of the generating matrix, and a_{ijl} can be determined from the eigenvectors. Determination of the probability of being in any configuration at a given time allows calculation of the expected macrokinetic conductance path and other properties of the site. The method is described elsewhere (Baumann and Easton, 1981) again in great detail to allow the reader to adapt the algorithm to any model.

VI. KINETIC PROPERTIES OF THE AGGREGATION-GATING SITE

Various kinetic and steady-state aspects of the aggregation model have been published elsewhere and are briefly summarized and referenced here.† Initial interest in aggregation stemmed from the fact that the process could account for the delay in the voltage-clamp records of potassium currents. The finding that the same process could also account for the feature of inactivation in such records of sodium currents was an unexpected result. This remarkable analogy between aggregation macrokinetics and the macrokinetics of excitable systems was only one of a long series of molecular explanations free of *ad hoc* assumptions that the aggregation concept could provide for experimental data. Some of these data had not even been recorded yet when the model was first formulated and the initial finding of the dual macrokinetic behavior was made in the fall of 1970.

The aggregation model can account for some basic steady-state features of excitable systems, such as the steep conductance-voltage relation, and provides an alternative molecular interpretation for the model of Hodgkin and Huxley (1952) of the potassium conductance system (Baumann and Bond, 1978).

In addition to the basic macrokinetic behavior of the potassium conductance system, the aggregation model with low dimer stability can account for the Cole–Moore shift and superposition (Cole and Moore, 1960) and gives a possible molecular interpretation for these phenomena (Baumann, 1979). Furthermore, the model has the potential to account for deviations from the superimposable Cole–Moore shift (Baumann and Mueller, 1974; Bau-

* Eq. (2) shows that the functional form of a model description is a sum of weighted exponentials. In the case of the proposed model, there are many exponentials. Some of their coefficients are small and some terms can be eliminated. Even so, it appears that a reasonable functional form for describing voltage-clamp data may contain about half a dozen exponentials, and attempts to fit such data with a single exponential seem problematic in this context (Baumann and Easton, unpublished results).

† Some of the findings were also reported for review purposes by Mueller (1975a,b, 1976, 1979).

mann and Easton, 1980). Such deviations have been discovered in excitable bilayers (Baumann and Mueller, 1974) and in the squid axon (Begenisich, 1979).

In addition to accounting for the basic macrokinetic behavior of the sodium conductance system, the aggregation model with high dimer stability predicted novel macrokinetic behavior (Baumann, 1981) that subsequently was detected in the sodium currents of *Myxicola* giant axons (Schauf and Baumann, 1981), as shown in Fig. 6. The novel macrokinetic

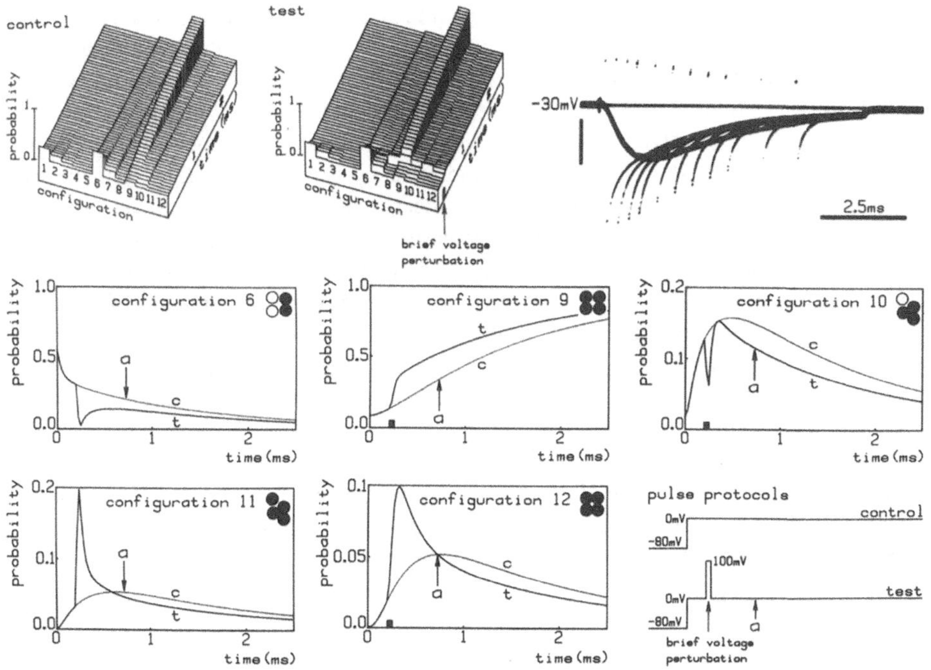

Figure 6. Novel macrokinetic behavior: Model prediction, experimental verification, and explanation in terms of the model. The three-dimensional (3-D) plots show the probability time courses of the 12 configurations for the case of high dimer stability (kinetics like sodium-conductance kinetics) in response to the control and test pulse protocols shown at bottom right of the figure. A comparison of the 3-D plots reveals that the brief voltage perturbation in the test pulse protocol causes the time course of some nonconducting configurations to deviate as shown in detail in the graphs in the lower part of the figure for configurations *6, 9, 10,* and *11*. This deviation is long-lasting and results in crossing over between the control (*c*) and test (*t*) time course of the conducting configuration *12* (graph at bottom middle). In the upper right of the figure, time courses of currents from a *Myxicola* axon show the crossing-over effect predicted by the aggregation model in response to control and test pulse protocols similar to the protocols described above. The axon was repeatedly depolarized from a holding potential of −80 mV to −30 mV with perturbations to +200 mV lasting 100 μsec beginning at 0.1, 0.4, 0.7, 1.0, 1.5, 2.0, 2.5, 3.0, 3.5, and 5.5 msec. (The current calibration is 0.05 mA · cm^{-2}.) As shown in the five graphs the crossing-over behavior can be given a molecular interpretation in terms of the aggregation model. In each graph the control time course (thin line) is marked with *c*, the test time course (thick line) with *t*, and the duration of the voltage perturbation is indicated by a thick bar on the time axis. The difference in the relative frequencies of the nonconducting configurations at time *a* results in different conductance time courses after time *a*. This occurs despite the fact that the conductance at time *a* is the same for both pulse protocols and the voltage after the time the brief voltage perturbation ends is identical for both cases. That is, the long-lasting effect of the brief voltage perturbation on the conductance time course arises from the experimentally unobservable behavior. (Parts of this figure reproduced from the *Biophys. J.*, 1981, **35:**711 by copyright permission of the Biophysical Society, and other parts reprinted by permission of the publisher from "Modeling state-dependent sodium conductance data by a memoryless random process" by G. Baumann and G. S. Easton, *Math. Biosci.* **60:**273, copyright 1982 by Elsevier Science Publishing Co., Inc.)

behavior implies "memory" or state dependence in the data. This means that the macrokinetic conductance path "remembers" a brief voltage perturbation long after it occurs. In the context of the aggregation model, such memory exhibited by the conductance path arises from the experimentally unobservable configurations, illustrating that such data do not necessarily preclude an underlying mechanism that consists of only a few components that are themselves memoryless and interact in a memoryless fashion (Baumann and Easton, 1982a).

Monte Carlo simulations of the microkinetics of an aggregation-gating site resemble the single-channel fluctuations of nerve and muscle or the more complex single-channel data of excitable lipid bilayers, depending on the simulation parameters (Baumann and Easton, 1980). That is, a simulation can result either in single-step open–closed microkinetic behavior as elucidated by direct recordings of ionic currents from individual potassium channels in the squid axon (Conti and Neher, 1980) and sodium channels in cultured rat-muscle cells (Sigworth and Neher, 1980), or it can result in records resembling the multi-step conductance bursts that are measured in lipid bilayers modified by alamethicin (Gordon and Haydon, 1972; Boheim, 1974). If one calculates the macroscopic behavior with the two sets of parameters, then the theoretical results are consistent with experimental results. The parameter set that produces nerve- and muscle-like open–closed microkinetics gives nerve- and muscle-like macrokinetics, and the parameter set that results in alamethicin-like multi-step bursts of conductance gives alamethicin-like macrokinetics featuring a fast and a slow phase in the time course. That is, the model relates the microkinetic burst behavior to the biphasic macrokinetics and gives for both phenomena a possible common molecular explanation (Baumann and Easton, 1980).

Finally, expected gating currents can be derived from a Markov-process formulation of the random behavior of a single aggregation-gating site. As shown in Fig. 7, the model can account for the phenomenon of charge immobilization in asymmetry-current data of the voltage-clamped sodium-conductance system (Baumann and Easton, 1982b).

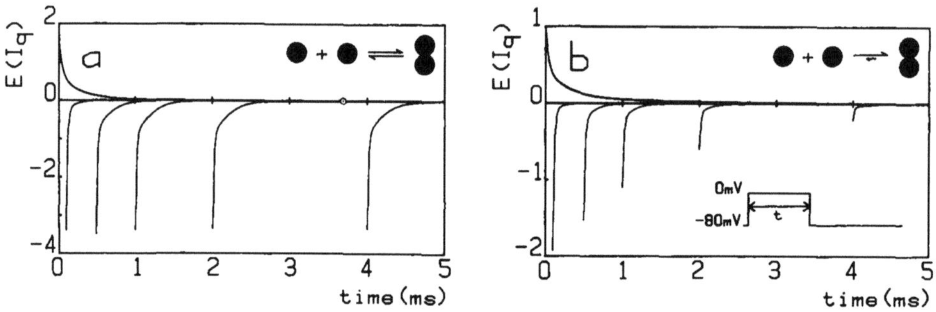

Figure 7. Macrokinetic behavior of gating currents derived from a Markov-process characterization of a single hypothetical aggregation-gating site. The expected gating current, $E(I_q)$, is assumed to be proportional to the expected net rate of conformational change of the monomers between their inert and active forms. The proportionality constant between current and rate is taken to be unity. $E(I_q)$ is calculated for low (*a*) and high (*b*) dimer stability as a function of time for the pulse sequence described in the insert of (*b*) with pulse duration t = 0.1, 0.5, 1, 2, and 4 msec. High dimer stability (*b*) results in decreasing amplitude of the gating currents upon repolarization with increasing duration of the depolarizing pulse. This feature is similar to the phenomenon of *charge immobilization* observed in gating-current data from the sodium-conductance system in nerve axons. In contrast, low dimer stability (*a*) does not lead to the charge immobilization phenomenon. In terms of the aggregation model, the cause of immobilization and inactivation are the same. Due to the high stability of the dimer, the site gets trapped in configuration *9*. The domains or subunits carry gating charge which, thus, becomes *immobilized*. The trapped site only slowly tends to return to the preferred configurations *1* and *6* upon repolarization. (Reproduced with permission from the *J. Theoret. Biol.*, 1982, **99**:254, Copyright by Academic Press Inc., London, Ltd.)

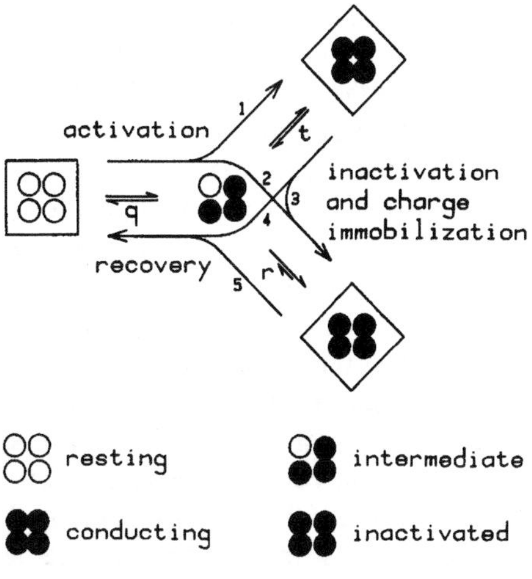

Figure 8. A greatly simplified transition scheme that does not give realistic macrokinetics, yet allows one to demonstrate some basic features of the aggregation model. At the resting potential the probability of being in the resting state is high. Upon depolarization both the conducting and inactivated configurations become more probable. This shift in probability is represented by arrows 1 and 2. The site tends to get trapped in configuration *9*, but not in configuration *12*. Thus, the probability of configuration *12* soon begins to decrease, while the probability of configuration *9* keeps rising as indicated by arrow 3. This decrease in the probability of being in the conducting configuration leads to inactivation in this model. The longer the depolarizing voltage lasts, the greater the likelihood of the site being in configuration *9*. Thus, recovery from inactivation upon repolarization (arrows 4 and 5) takes longer with increasing inactivation. The rate of transition q, which in this didactical, simplified version of the model generates the gating current, also decreases as the tendency of being trapped in configuration *9* increases. It can be demonstrated that the same principles apply to the original aggregation model. Thus, in the aggregation model, like in data from the axon, charge immobilization rises as inactivation progresses.

The probability of some of the 14 transitions occurring is very small for most pulse protocols. If all transitions that kinetically play an insignificant role are eliminated from the transition scheme in Fig. 2c assuming high dimer stability, then a simplified transition scheme is obtained that is Y-shaped with configurations 1, 9, and 12, each at the end of a branch, and configuration 7 at the intersection (Baumann and Easton, 1982b). If the other configurations of this simplified scheme are omitted, then a scheme of instructive value is obtained (Fig. 8). This didactical scheme shows how, in essence, inactivation and charge immobilization are linked in the aggregation model.

VII. DISCUSSION

Stochastic modeling of the aggregation-gating site successfully predicted novel macrokinetic behavior in the nerve axon (Fig. 6) and gives plausible explanations of such features as inactivation (Fig. 5) and gating charge immobilization (Fig. 7). Clearly, the proposed model offers a viable alternative to the conventional way of thinking about channel gating. Even so, why should one question the idea of a channel with gates which is so familiar that it appears to be the obvious truth, and consider a strange mechanism that functions like nothing one can experience? And what can be achieved by formulating models of excitability in the first place?

In biology, in contrast to other natural sciences, theoretical work such as postulating and analyzing models* is still unusual. Furthermore, the approach to theoretical modeling of excitable systems taken here is different from the approach generally taken. For these two reasons, the relation between various approaches is briefly discussed. One approach consists of fitting one or more mathematical equations to a data set. The goal in that case

* Still, in the biological literature, molecular models are sometimes erroneously referred to as cartoons. A cartoon is defined as a deliberately *exaggerated* representation of a well-known fact, while a model in science is a *simplified, abstract* representation of an often poorly understood natural phenomenon.

is to summarize the data in a concise form. Another approach consists of fitting a kinetic scheme of least complexity to a data set. There the goal is to eliminate schemes that cannot account for the data, with the aim of developing a scheme that can generate the experimentally observed behavior and can be interpreted in terms of a realistic underlying molecular mechanism. A third approach consists of guessing and postulating a molecular mechanism such as, for example, cooperativity, conformational change, or aggregation, and then finding by kinetic analysis which of these processes, if any, can account for excitability data. Clearly, the main aim in that case is to postulate a molecular framework for interpreting the data.

While these modeling approaches are obviously different from each other, they are nevertheless related and have important features in common. Notice that a molecular mechanism implies a kinetic scheme, and a kinetic scheme implies a set of equations that may account for a set of data. The converse is not necessarily the case. There may be many equations that can adequately account for the same data. Similarly, there may be many kinetic schemes that are described by the identical set of equations. Finally, there may be many molecular mechanisms that are described by a given kinetic scheme. In other words, a model may not be unique, whether it consists of an equation, a scheme, or a molecular mechanism. Therefore, the only conclusion one can draw from a model that can account for a data set is that it cannot be eliminated from the list of possible models. Agreement between model and data does not prove that a model is right.

The viability of a model must be established by submitting it to a set of modeling criteria. This set of criteria is different for each type of model. For establishing the viability of models based on mathematical equations or kinetic schemes, agreement between model prediction and data is the main (and often the only) criterion. In the case of models based on a molecular mechanism, however, agreement between model and data is but one of a number of *modeling criteria,* such as, for example:

1. *Physical consistency.* A model based on a molecular mechanism must be compatible with the laws of chemistry and physics. That is, such a model must be consistent with existing theories. For example, if the model contains a reaction loop, the sum of all forward activation energies must be equal to all backward activation energies around that loop in order not to violate the law of conservation of energy.
2. *Self-consistency.* There should be consistency between, for example, microscopic and macroscopic, steady-state and kinetic, or ionic and gating-current properties of the model.
3. *Analyzability.* A molecular model must be mathematically analyzable. That is, one must be able to formulate the model in mathematical terms, and the mathematics must be such that analysis is feasible. Notice that if this condition cannot be met, the model may still be right, but one cannot compute its consequences and, therefore, the model is not useful.
4. *Explanatory power.* Within the framework of a conceptual model it should be possible to interpret steady-state and kinetic features in the data in terms of molecular properties and events. For example, solvent-substitution experiments (Schauf, 1982) can be interpreted within the framework of the aggregation model in terms of a change in the rate of channel closing (Baumann and Schauf, unpublished results).
5. *Plausibility.* Molecular descriptions derived from a model should not be construed as being explanations, unless they are supported by independent evidence. For example, the assumption of aggregation gating appears to be a plausible assumption because excitability-inducing molecules such as alamethicin or EIM are known to aggregate (McMullen and Stirrup, 1971; Bukovsky, 1977).

6. *Predictive power.** A conceptual model should predict unexpected behavior that is a direct consequence of the underlying molecular mechanism itself. For example, crossing-over of the sodium currents with and without brief voltage perturbation (Fig. 6) was calculated from the aggregation concept before it was observed experimentally.

7. *Simplicity.* A conceptual model should be as simple as possible. Given two models that predict the same phenomena, the simpler of the two is the preferable one. It should be of least physical complexity and free of *ad hoc* mathematical assumptions made in order to facilitate data fitting, because given enough parameters, any data can be fit. It should be remembered that a model is always a simplification of reality and that the ability to perform meaningful analysis is inversely related to complexity.

8. *Universality.* No model, no matter how good, ever agrees with all the data. Given a choice, however, a conceptual model should rather account for a large variety of data in a qualitative sense than fit only a small set of data with high precision.† A model for membrane excitability needs to account for the wide range of voltage-clamp data including steady-state, ionic, gating, and single-channel-current data, and should apply equally to nerve, muscle, and the excitable lipid bilayer.

Three decades ago Hodgkin and Huxley (1952) suggested among other possible molecular interpretations of their model that an aggregate could be the pore-providing structure in the nerve membrane. Later, it was found that with excitability-inducing molecules such as EIM or alamethicin, a bilayer can be made to behave kinetically like a nerve membrane, and again it was proposed that aggregation may be involved in the gating process (Mueller and Rudin, 1969). It is now well known that excitability-inducing molecules have a strong tendency to aggregate (McMullen and Stirrup, 1971; Bukovsky, 1977). Furthermore, as demonstrated in this chapter, the macrokinetics of aggregation coincide with the macrokinetics of nerve gating. Yet, neither the concept of aggregation gating nor the concept of aggregation†† itself is well established, perhaps indicating that they are difficult concepts to grasp. This is probably because in the macroscopic world there is no analog to a gate that functions by subunit or domain assembly and disassembly.

The principle of building a channel-forming structure from subunits or domains (sub-

* Ultimately, it is irrelevant whether a phenomenon is first predicted by a model and then discovered, or first discovered and then accounted for by a model. Predictions verified, however, tend to have greater psychological impact than phenomena explained.

† It should be remembered that the collection of data is always motivated by a question formulated within the context and, therefore, within the limitations of a certain way of thinking. A new way of thinking, i.e., a new theoretical bias, can lead to novel data, as demonstrated here with the crossing-over data of Fig. 6. During development of a new model, therefore, it is more important to qualitatively rather than quantitatively fit data, with emphasis on logically connecting different types of data, e.g., ionic- and gating-current data, rather than on exactly matching one type of data.

†† Etymologically, to *aggregate* refers to the process of coming together by *addition,* e.g., $1 + 1 = 2$, $2 + 1 = 3$, $3 + 1 = 4$. It derives from the Latin stem *greg*-which refers to anything that comes together and the prefix *ag*-(from the Latin *ad*-) which stands for *addition to.* Aggregation is different from congregation. To *congregate* derives from the same stem and the prefix *con*- (from the Latin *com*-) meaning *together,* and refers to the process of coming together *simultaneously,* e.g., $1 + 1 + 1 + 1 = 4$. In the scientific literature this subtle but important distinction in meaning is frequently not made, and any process of coming together may be referred to as aggregation, which may contribute to the widespread unfamiliarity with the process of aggregation in the strict etymological sense of the word. Moreover, notice that aggregation can formally be likened to addition polymerization. In both cases monomers combine in a step-by-step fashion. Yet, the two processes differ. Polymerization refers to an *irreversible* process that results in a stable product involving sometimes thousands of monomers. Aggregation, however, implies a *reversible* process usually involving just a few monomers that does not yield a stable product.

assembly principle) eliminates the need for a special molecular structure acting as a gate, since an aggregate of subunits or domains has the potential of changing between conducting and nonconducting configurations and, thus, inherently provides a mechanism for the regulation of ion flow through the membrane. From an evolutionary point of view, the sub-assembly principle, as opposed to the principle of building a gated channel-forming structure in one piece, seems advantageous in at least two more respects. In general, the technique of sub-assembly reduces the number of assembly errors compared to the technique of assembly of a complete structure directly from elementary building blocks. Furthermore, coding a protein made of n identical subunits or domains requires $1/n$ times as much nucleic acid as coding a protein of the same size made of one piece.

The aggregation model can account for many features in the steady state and kinetic data of gating systems and explain them in molecular terms. So far, there is no experimental evidence against the model and we cannot exclude it as a possible mechanism of channel gating. It should be emphasized, however, that every model has its limitations. For example, the aggregation model (like the gate model) does not give an explanation for the fact that channels may prefer the passing of one type of ion over another. There exist models of ion selectivity in membrane channels (Eisenman, 1962; Hille, 1975b) and they are compatible with both the model of a channel with gate and the aggregation model. A comprehensive model of membrane excitability will consist of several models, each covering one aspect of excitability.

In conclusion, the value of a model lies ultimately in its ability to describe the essential properties of a system in a manner that enables prediction of observable behavior and, consequently, increases our understanding of the underlying mechanism. Yet, the fact remains that successful predictions, plausible explanations, or even precise data fits cannot prove a model. They just make it more viable and acceptable and increase its heuristic value. A model is a tool much like a language. It provides a common basis for thinking and communicating about experimental findings. When a large number of the available puzzle pieces of experimental data logically fit together within the framework of a model, the model becomes useful to the experimenter. The evidence presented and cited in this chapter indicates that the aggregation model may have reached this stage and may be a good guess for how membrane excitability works.

ACKNOWLEDGMENTS. I thank George Easton and Drs. Andrés Manring and Kaspar Zürcher for reading the manuscript and making valuable suggestions, and William Boyarsky, David E. Lupo, and R. Dennis Rockwell for technical assistance. Funding for this research has been provided by NSF grant PCM78-02802 and NIH grant GM27260.

REFERENCES

Baumann, G., 1979, Modular gating channel and Cole-Moore effect: Next-neighbor (Hill-Chen) versus non-restricted aggregation, *Math. Biosci.* **46:**107–115.

Baumann, G., 1981, Novel kinetics in the sodium conductance system predicted by the aggregation model of channel gating, *Biophys. J.* **35:**699–705.

Baumann, G., and Bond, J. D., 1978, A novel molecular interpretation of the Hodgkin–Huxley model of electrical excitability, *Math. Biosci.* **39:**291–297.

Baumann, G., and Easton, G. S., 1980, Micro- and macrokinetic behavior of the subunit gating channel, *J. Membr. Biol.* **52:**237–243.

Baumann, G., and Easton, G. S., 1981, Markov process characterization of a single membrane gating site, *J. Theoret. Biol.* **93:**785–804.

Baumann, G., and Easton, G. S., 1982a, Modeling state-dependent sodium conductance data by a memoryless random process, *Math. Biosci.* **60**:265–276.

Baumann, G., and Easton, G. S., 1982b, Charge immobilization linked to inactivation in the aggregation model of channel gating, *J. Theoret. Biol.* **99**:249–261.

Baumann, G., and Mueller, P., 1974, A molecular model of membrane excitability, *J. Supramol. Struct.* **2**:538–557.

Begenisich, T., 1979, Conditioning hyperpolarization-induced delays in the potassium channels of myelinated nerve, *Biophys. J.* **27**:257–265.

Boheim, G., 1974, Statistical analysis of alamethicin channels in black lipid membranes, *J. Membr. Biol.* **19**:277–303.

Breiman, L., 1969, *Probability and Stochastic Processes. With a View Toward Applications,* Houghton Mifflin, Boston, Massachusetts.

Bukovsky, J., 1977, Production, purification, and characterization of excitability-inducing molecule, *J. Biol. Chem.* **252**:8884–8889.

Çinlar, E., 1975, *Introduction to Stochastic Processes,* Prentice-Hall, Englewood Cliffs, New Jersey.

Cole, K. S., and Moore, J. W., 1960, Potassium ion current in the squid giant axon: Dynamic characteristics, *Biophys. J.* **1**:1–14.

Conti, F., and Neher, E., 1980, Single channel recordings of K^+ currents in squid axons, *Nature* **285**:140–143.

Easton, G. S., 1981, The calculation of the transient behavior of Markov processes with an application to membrane excitability, Masters Thesis, University of North Carolina, Chapel Hill.

Eisenman, G., 1962, Cation-selective glass electrodes and their mode of operation, *Biophys. J.* **2** (2)Part 2:259–323.

Gordon, L. G. M., and Haydon, D. A., 1972, The unit conductance channel of alamethicin, *Biochim. Biophys. Acta.* **255**:1014–1018.

Hille, B., 1975a, Ionic selectivity of Na and K channels of nerve membranes, in: *Membranes: Lipid Bilayers and Biological Membranes: Dynamic Properties,* Vol. 3 (G. Eisenman, ed.), pp. 255–323, Dekker, New York.

Hille, B., 1975b, Ionic selectivity, saturation, and block in sodium channels. A four-barrier model, *J. Gen. Physiol.* **66**:535–560.

Hodgkin, A. L., and Huxley, A. F., 1952, A quantitative description of membrane current and its application to conduction and excitation in nerve, *J. Physiol. (London)* **117**:500–544.

McMullen, A.I., and Stirrup, J.A. 1971, The aggregation of alamethicin, *Biochim. Biophys. Acta* **241**:807–814.

Mueller, P., 1975a, Electrical excitability in bilayers and cell membranes, in: *Energy Transducing Mechanisms,* Vol. 3 (E. Racker, ed.), MTP Int. Rev. Sci., Biochem. Ser. 1, pp. 75–120, Butterworth, London.

Mueller, P., 1975b, Membrane excitation through voltage induced aggregation of channel precursors, *Ann. N.Y. Acad. Sci.* **264**:247–264.

Mueller, P., 1976, Molecular aspects of electrical excitation in lipid bilayers and cell membranes, in: *Horizons Biochem. Biophys.,* Vol. 2 (E. Quagliariello, F. Palmieri, and T. Singer, eds.), pp. 230–284, Addison-Wesley, Reading, Massachusetts.

Mueller, P., 1979, The mechanism of electrical excitation in lipid bilayers and cell membranes, in: *Neurosciences, Fourth Study Program.* (F. O. Schmitt, ed.), MIT Press, Cambridge, Massachusetts.

Mueller, P., and Rudin, D. O., 1969. Translocators in bimolecular lipid membranes: Their role in dissipative and conservative bioenergy transductions, in: *Current Topics in Bioenergetics* (R. Sanadi, ed.) Academic Press, New York, **3**:157–249.

Schauf, C. L., 1982, Solvent substitution as a probe of gating processes in voltage-dependent ion channels, in: *Structure and Function in Excitable Cells* (D. C. Chang, I. Tasaki, W. J. Adelman, Jr., and H. R. Leuchtag, eds.), Plenum Press, New York.

Schauf, C. L., and Baumann, G., 1981, Experimental evidence consistent with aggregation kinetics in sodium currents in *Myxicola* giant axons, *Biophys. J.* **35**:707–714.

Sigworth, C. F., and Neher, E., 1980, Single Na^+ channel currents observed in cultured rat muscle cells, *Nature* **287**:447–449.

13

Effect of Scorpion Toxins on Sodium Channels

Hans Meves

I. INTRODUCTION

Neurotoxins have become an important tool for studying the function of nerve fibers (for review see Catterall, 1980). Among the various neurotoxins, scorpion venoms have particularly attracted the attention of electrophysiologists.

II. LEIURUS VENOM

Koppenhöfer and Schmidt (1968a) studied the effect of venom from the scorpion *Leiurus quinquestriatus* on the membrane currents of the node of Ranvier. Normally, a 50 mV pulse produces a transient Na inward current followed by a delayed K outward current (Fig. 1A). After application of 1 μg/ml *Leiurus quinquestriatus* venom, the decay of the Na inward current is markedly slowed (Fig. 1B). Even at the end of a 98 msec pulse there is still a considerable Na inward current (Fig. 1C). The transient and the maintained inward current disappear in Na-free Ringer's (Fig. 1D) or after addition of tetrodotoxin, indicating that they are both carried by Na ions through the Na channel. Fig. 1 suggests that scorpion venom makes inactivation of the Na permeability slow and incomplete.

As shown by Koppenhöfer and Schmidt (1968b), scorpion venom also has a pronounced effect on the steady-state Na inactivation curve $h_\infty(V)$. The curve is obtained by using long prepulses of variable height followed by a constant test pulse. Plotting test-pulse current vs. prepulse amplitude normally yields a curve that decays steeply for depolarizing prepulses and reaches zero for prepulses larger than 45 mV (curve A in Fig. 2). In the venom-treated node the curve is flatter, does not reach zero, and turns upward again at strong depolarizing prepulses (curves B and C in Fig. 2).

Subsequent work, using venom and isolated toxins from *Leiurus quinquestriatus* and other scorpion species, on squid giant axons (Narahashi *et al.*, 1972; Gillespie and Meves, 1980), other nonmyelinated axons (Romey *et al.*, 1975), neuroblastoma cells (Bernard *et*

Hans Meves • I. Physiologisches Institut der Universität des Saarlandes, 6650 Homburg, Saar, West Germany.

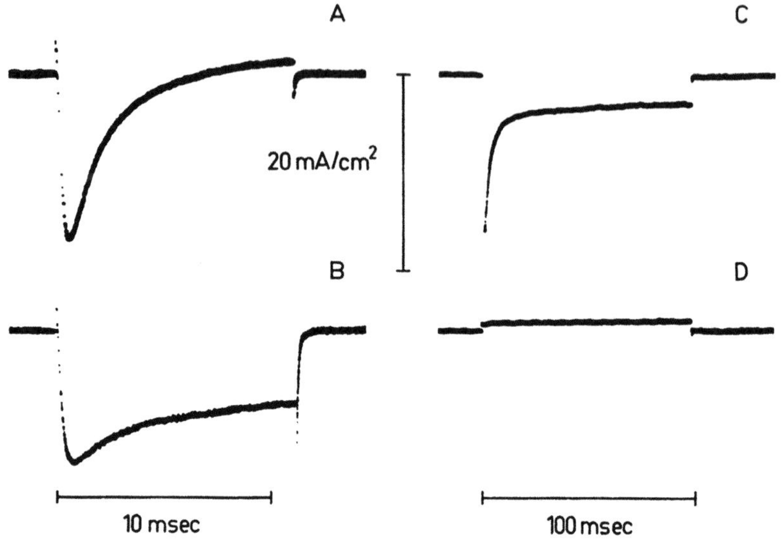

Figure 1. Effect of *Leiurus quinquestriatus* venom on the voltage clamp currents of the node of Ranvier. A, normal Ringer; B—C, Ringer + 1 μg/ml venom; D, Na-free Ringer + 1 μg/ml venom. The records show membrane currents associated with 50 mV pulses of 11 msec (A, B) or 98 msec (C, D) duration. Records A, B and C, D from two different fibres. Temperature 18 °C. (From Koppenhöfer and Schmidt, 1968a).

al., 1977), and nodes of Ranvier (Conti *et al.*, 1976; Nonner, 1979; Mozhayeva *et al.*, 1980) confirmed these findings.

Figures 3 and 4 illustrate the effect of scorpion venom on the intracellularly perfused squid giant axon, which has been investigated by Gillespie and Meves (1980). Normally, the Na inward current of a squid axon decays to a small maintained inward current (Fig. 3A). External application of scorpion venom in a relatively high concentration (60 μg/ml) increases the maintained inward current, that is, makes inactivation less complete (Fig. 3B). (Intracellular application of the venom is ineffective.) The peak current and the maintained current reverse sign at the same potential and both of these are blocked by tetrodotoxin. However, the voltage dependence of the maintained conductance is different from that of

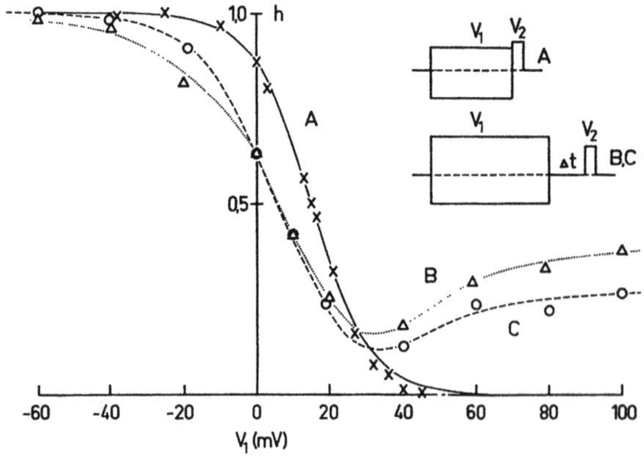

Figure 2. Effect of *Leiurus quinquestriatus* venom on the steady-state Na inactivation curve of a node of Ranvier. A, normal Ringer; B and C, Ringer + 1 μg/ml venom. Pulse program (see inset) consists of a variable prepulse V_1 and a constant test pulse V_2 which follows the prepulse immediately (for A) or after a variable interval (Δt) (for B and C). Prepulse duration 50 msec for A and B, 500 msec for C. Ordinate: current produced by test pulse V_2 (expressed as fraction of the maximum current obtained after a −60 mV prepulse). Abscissa: prepulse potential V_1 (referred to the normal resting potential). Temperature 18 °C. (From Koppenhöfer and Schmidt, 1968b).

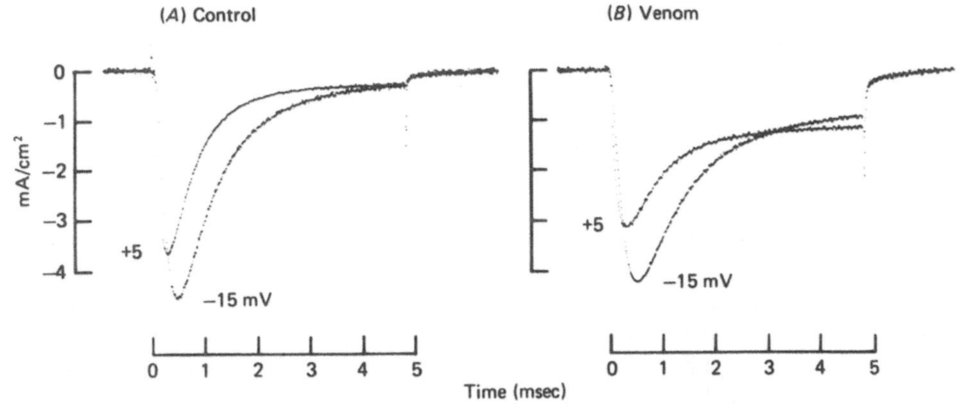

Figure 3. Effect of *Leiurus quinquestriatus* venom on the Na currents of an internally perfused squid giant axon. A, control records; B, 15 min after application of sea water with 60 μg/ml venom. Resting potential −55 mV in A, −56 mV in B. Holding potential −60 mV in A and B. 50 msec prepulse to −85 mV in A and B. Potential during clamp pulse −15 and 5 mV. K currents blocked by perfusion with 218 mM KF + 54 mM tetraethylammonium chloride (TEA) + sucrose. Tetrodotoxin-insensitive currents, measured after the records in B, have been subtracted. Temperature 15.5 °C. (From Gillespie and Meves, 1980).

the peak conductance. As shown in Fig. 3B, an increase in the pulse potential from −15 to 5 mV reduces the peak current but increases the maintained current.

Gillespie and Meves (1980) determined the voltage dependence of the maintained conductance by fitting a modified Hodgkin–Huxley equation with four adjustable parameters (I'_{Na}, τ_m, h_∞, τ_h) to the Na currents. Here, I'_{Na} is the value which the Na current would attain if inactivation remained at its resting level; τ_m and τ_h are the time constants of activation and inactivation, respectively; h_∞ is the steady state value of inactivation. The h_∞ values for control (○) and venom (●) found in a typical experiment are plotted in Fig. 4 against pulse potential. The control values (○) show a minimum of h_∞ at small negative potentials (see Fig. 11 in Chandler and Meves, 1970b). Scorpion venom (●) substantially increases h_∞ without altering its voltage dependence. (In addition, the curve fits showed a decrease of I'_{Na} in scorpion venom, no significant effect on τ_m and a mild increase of τ_h.) This observation is compatible with the idea that scorpion venom increases the number of channels that go

Figure 4. Inactivation variable h_∞ as a function of pulse potential in a perfused squid giant axon before (○, control) and after (●, venom) treatment with scorpion venom. The variable h_∞ was obtained by fitting a modified Hodgkin and Huxley equation to the Na currents associated with 10 msec pulses to different potentials; values near reversal potential (=38 mV) are not plotted. Currents in venom were recorded 10 min after application of sea water with 200 μg/ml *Centruroides sculpturatus* venom. Resting potential = holding potential = −67 mV. The axon was perfused with 218 mM KF + 54 mM TEA + sucrose. Temperature 10.5 °C. (From Gillespie and Meves, 1980).

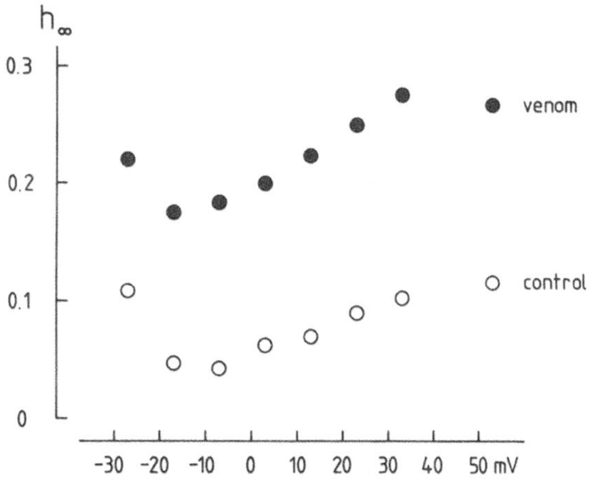

from the peak conductance state into the maintained conductance state (the open ⇆ closed ⇆ open transition of the inactivation gate; see Chandler and Meves, 1970a,b).

III. CENTRUROIDES VENOM

Figures 1–4 illustrate the typical scorpion venom effect; a slow and incomplete Na inactivation, which becomes more incomplete with increasing depolarization. However, a completely different effect has been described for venom from the scorpion species *Centruroides sculpturatus* on nodes of Ranvier. As shown by Cahalan (1975), this venom affects mainly Na activation rather than Na inactivation. Fig. 5 illustrates a family of Na currents in normal Ringer's (top) and after a brief treatment with Ringer + 1 μg/ml *Centruroides sculpturatus* venom (bottom). After the venom treatment the Na currents are smaller and an inward current can be seen to develop after the end of each depolarizing pulse, the size of the inward current increasing with increasing height of the preceding pulse. The inward current developing after the pulse can be blocked by tetrodotoxin. Records on slower time base show that the inward current grows to a maximum and then declines very slowly. According to Cahalan (1975), the phenomenon is due to a transient, depolarization-induced shift of the Na activation curve $m_\infty(V)$ to more negative potentials. The amount of shift depends on the size of the depolarizing pulse. Because of the shift of the $m_\infty(V)$ curve, Na activation is *not* turned off when the potential is returned to the holding potential at the end of the depolarizing pulse; the Na current slowly increases in size as Na inactivation is removed.

Scorpion venoms are mixtures of 30–40 small polypeptides with different amino acid sequences. From the venom of *Centruroides sculpturatus*, Watt *et al.* (1978) have isolated by chromatography several fractions, named toxin I, III, IV, B140-1 (now named toxin V) and variant 1,2,3. The amino acid sequences of some of these toxins (toxin I, variant 1,2,3) have been analyzed (Babin *et al.*, 1974, 1975; Fontecilla-Camps *et al.*, 1980; see Fig. 6). In addition, the three-dimensional structure of variant 3 has been determined (Fontecilla-

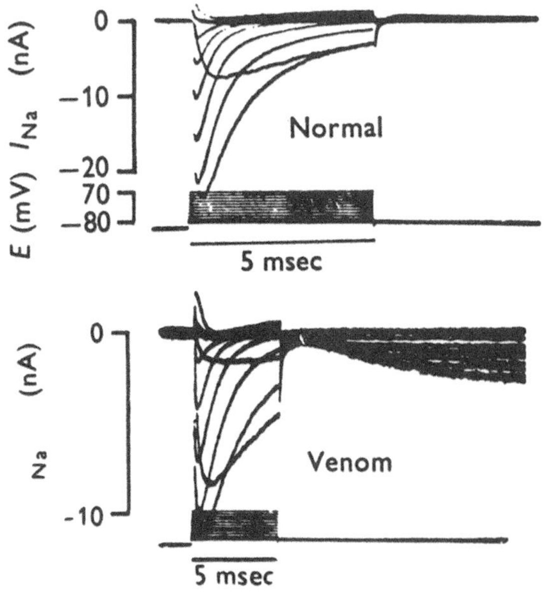

Figure 5. Effect of *Centruroides sculpturatus* venom on the Na currents of a node of Ranvier. Eleven 5 msec pulses to different potentials up to 70 mV, holding potential −80 mV. Top: normal; bottom: after treatment with 1 μg/ml *Centruroides* venom for 1 min. Note inward current developing after end of pulses in bottom records. K currents were blocked by Ringer with 12 mM TEA and by cutting the ends of the fiber in 108 mM CsF + 12 mM TEA. Leakage currents were subtracted. Temperature 15 °C. (From Cahalan, 1975).

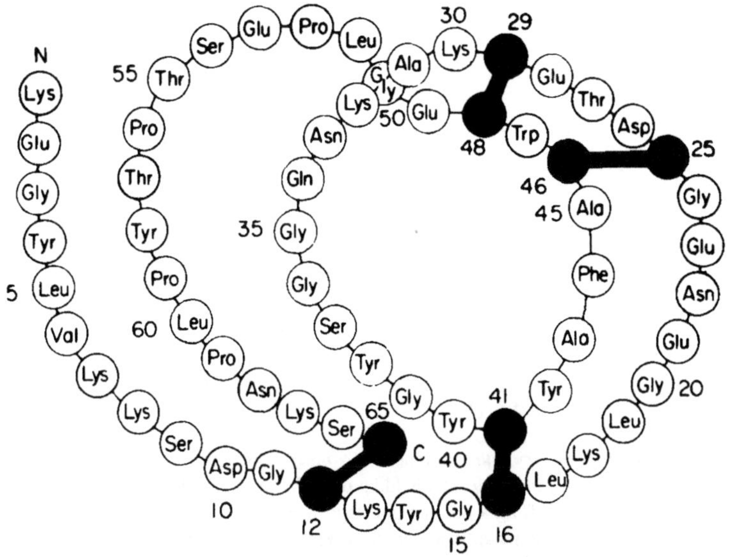

Figure 6. Amino acid sequence for variant 3 of *Centruroides sculpturatus*. Blackened portions represent cysteine residues in disulfide bridges. (From Fontecilla-Camps *et al.*, 1980). The first published complete amino acid sequence of a scorpion toxin was that of neurotoxin I of the North African scorpion *Androctonus australis* (Rochat *et al.*, 1970).

Camps *et al.*, 1980). In order to find out which of these fractions is responsible for the effect described by Cahalan (1975), Meves and Rubly (1981) and Meves *et al.* (1982) investigated the effect of the various *Centruroides sculpturatus* toxins on the Na current of the node of Ranvier.

IV. TESTING THE FRACTIONS

The first step was to design experiments for investigating the effects of the various chromatographic fractions derived from the venom. The two types of experiment illustrated in Fig. 7A and B were found most suitable for demonstrating the characteristic effect of *Centruroides sculpturatus* venom and for testing the fractions. In a node treated with *Centruroides sculpturatus* venom, depolarizing clamp pulses of 500 msec duration produce a transient Na inward current (barely visible on the slow time base employed) followed by a slowly increasing inward current; after the end of the pulse a large tail of inward current occurs (Fig. 7A). A 15 msec pulse to 38 mV is followed by an inward current that increases to a peak and then declines gradually (Fig. 7B, record a). In both cases the depolarization-induced long-lasting inward current is blocked by tetrodotoxin. These observations can be explained by assuming that a depolarization of sufficient strength and duration induces a Na permeability which does not completely inactivate and does not immediately shut off when the potential is returned to the holding potential. In other words, depolarization of a node treated with *Centruroides* venom causes a transient modification in the gating of Na channels that manifests itself in both the inactivation and the activation process. The modification of the activation process consists in a shift of the Na activation curve to more negative potentials. Because of this shift some of the Na channels conduct inward current at the holding potential (−72 mV in Fig. 7A, −92 mV in Fig. 7B). The gradual decay of this current reflects the

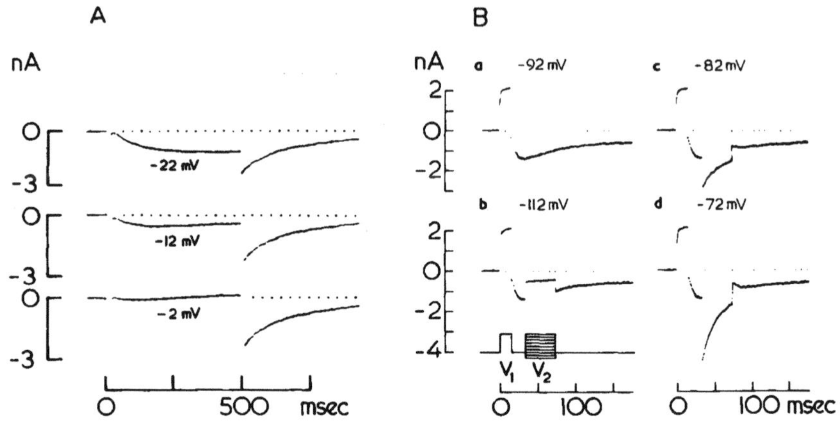

Figure 7. Effect of *Centruroides sculpturatus* venom on voltage clamp currents. A, slowly developing Na inward currents associated with 500 msec pulses from a holding potential of −72 mV to the potentials indicated, 10 μg/ml venom. B, a 15 msec pulse V_1 from a holding potential of −92 mV to 38 mV is followed by a long-lasting Na inward current (record **a**) whose height was rapidly altered by a second pulse V_2 to the potentials indicated (records **b,c,d**), 3.3 μg/ml venom. K currents were blocked as in Fig. 5. Temperature 15 °C (From Meves and Rubly, 1981).

slow return of the Na activation curve into its normal position. A second depolarizing pulse applied 20 msec after the end of the first pulse transiently increases the inward current (Fig. 7B, records c and d), a hyperpolarizing pulse decreases it (Fig. 7B, record b). After the end of the second pulse the inward current resumes its original time course.

The two types of experiment illustrated by Fig. 7A and B were performed with the various toxin fractions. Only toxin III (1.1–3.3 μg/ml), toxin IV (0.83–3.3 μg/ml), and toxin I (10 μg/ml) reproduced the effects seen with the whole *Centruroides* venom; toxins III and IV were more potent than toxin I. The other fractions (toxin V, variants 1,2,3) merely inhibited Na inactivation, that is, acted like *Leiurus quinquestriatus* venom. In a node treated with toxin V (3.3 μg/ml) or variants 1,2,3 (10–50 μg/ml) and subjected to a 500 msec depolarizing pulse, the transient Na inward current was followed by a slowly *decreasing* residual inward current (rather than by an *increasing* inward current as in Fig. 7A). At the end of the depolarizing pulses the membrane current rapidly returned to zero, indicating that the Na activation curve is *not* shifted.

V. CONDUCTANCE-PARAMETER SHIFTS WITH VOLTAGE

The next step was to study the transient depolarization-induced shift of the Na activation curve in nodes treated with toxins III or IV quantitatively. The $I_{Na}(V)$ curve was measured with two different pulse programs (see insets in Fig. 8A and B), test pulses of varying height with or without a strong depolarizing prepulse. As shown in Fig. 8A, the $I_{Na}(V)$ curve measured *with* a prepulse (filled symbols) is different from that measured *without* a prepulse (empty symbols). Its negative-resistance branch is shifted by 40 mV to more negative potentials, the maximum Na inward current $I_{Na\ max}$ is increased, and the potential at which $I_{Na\ max}$ occurs is shifted −15 mV. The amount of shift depended on the interval between prepulse and test pulse which was 20 msec in Fig. 8. Toxin IV concentrations of 0.83–1.33 μg/ml were sufficient to obtain the transient depolarization-induced shift, but the size of the

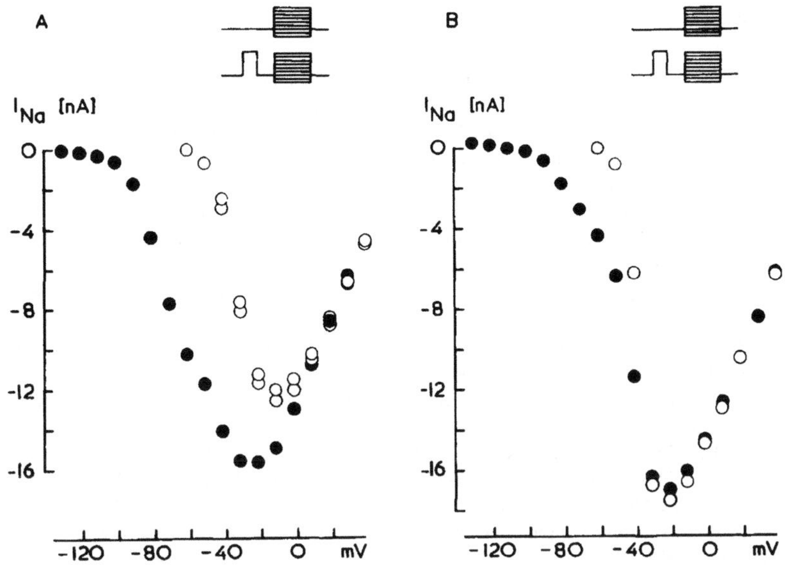

Figure 8. Depolarization-induced shift of the negative-resistance branch of the $I_{Na}(V)$ curve in a node treated with 3.3 μg/ml (A) or 0.33 μg/ml (B) toxin IV of *Centruroides sculpturatus*. The pulse program (see insets) consisted of test pulses of varying height with or without a 15 msec, 130 mV prepulse, followed by a 20 msec interval. Holding potential −92 mV. Ordinate: peak I_{Na} (corrected for tetrodotoxin-insensitive current) during test pulse; abscissa: test-pulse potential. Filled and empty symbols indicate peak I_{Na} with and without prepulse, respectively. Temperature 15 °C (From Meves *et al.*, 1982).

shift varied between fibers. In 26 experiments with 0.83–1.33 μg/ml toxin IV, the shift (measured with a pulse interval of 20 msec) was − 10 to − 60 mV. Even in 0.33 μg/ml toxin IV a small shift of the foot of the I_{Na}(V) curve could be induced (Fig. 8B). Presumably, in Fig. 8B only *some* of the Na channels are modified, whereas the shift in Fig. 8A concerns *all* Na channels. The shift of the negative-resistance branch of the I_{Na}(V) curve in toxin IV-treated nodes was not affected by the simultaneous presence of toxin V in a four times higher concentration, suggesting that the two toxins bind to different receptors (for details see Meves, Rubly, and Watt, 1982).

Further experiments have shown that the transient shift of the negative-resistance branch of the I_{Na}(V) curve in toxin III or IV-treated nodes is accompanied by a transient shift of the time to peak, the time constant of Na activation (τ_m), and the time constant of Na inactivation (τ_h). There is, however, no marked change in the ion selectivity of the Na channels. Details of these experiments are described in a recent publication by Hu *et al.* 1983.

VI. CONCLUSION

The overall conclusion is that the venom of the scorpion *Centruroides sculpturatus* contains toxins of different biological action. Toxins III, IV and I produce a transient shift of the I_{Na}(V) curve, toxin V and variants 1,2,3 merely inhibit inactivation, that is, act like other scorpion toxins. It remains to be seen which differences in chemical structure are responsible for the differences in biological action.

REFERENCES

Babin, D. R., Watt, D. D., Goos, S. M., and Mlejnek, R. V., 1974, Amino acid sequences of neurotoxic protein variants from the venom of *Centruroides sculpturatus* Ewing, *Arch. Biochem. Biophys.* **164**:694–706.

Babin, D. R., Watt, D. D., Goos, S. M., and Mlejnek, R. V., 1975, Amino acid sequence of neurotoxin I from *Centruroides sculpturatus* Ewing, *Arch. Biochem. Biophys.* **166**:125–134.

Bernard, P., Couraud, F., and Lissitzky, S., 1977, Effects of a scorpion toxin from *Androctonus australis* venom on action potential of neuroblastoma cells in culture, *Biochem. Biophys. Res. Commun.* **77**:782–788.

Cahalan, M. D., 1975, Modification of sodium channel gating in frog myelinated nerve fibres by *Centruroides sculpturatus* scorpion venom, *J. Physiol.* **244**:511–534.

Catterall, W. A., 1980, Neurotoxins that act on voltage-sensitive sodium channels in excitable membranes, *Annu. Rev. Pharmacol. Toxicol.* **20**:15–43.

Chandler, W. K., and Meves, H., 1970a, Evidence for two types of sodium conductance in axons perfused with fluoride solutions, *J. Physiol.* **211**:653–678.

Chandler, W. K., and Meves, H., 1970b, Rate constants associated with changes in sodium conductance in axons perfused with sodium fluoride, *J. Physiol.* **211**:679–705.

Conti, F., Hille, B., Neumcke, B., Nonner, W., and Stämpfli, R., 1976, Conductance of the sodium channel in myelinated nerve fibres with modified sodium inactivation, *J. Physiol.* **262**:729–742.

Fontecilla-Camps, J. C., Almassy, R. J., Suddath, F. L., Watt, D. D., and Bugg, C. E., 1980, Three-dimensional structure of a protein from scorpion venom: A new structural class of neurotoxins, *Proc. Natl. Acad. Sci. USA* **77**:6496–6500.

Gillespie, J. I., and Meves, H., 1980, The effect of scorpion venoms on the sodium currents of the squid giant axon, *J. Physiol.* **308**:479–499.

Hu, S. L., Meves, H., Rubly, N., and Watt, D. D., 1983, A quantitative study of the action of *Centruroides sculpturatus* toxins III and IV on the Na currents of the node of Ranvier, *Pflügers Arch.* **397**:90–99.

Koppenhöfer, E., and Schmidt, H., 1968a, Die Wirkung von Skorpiongift auf die Ionenströme des Ranvierschen Schnürrings. I. Die Permeabilitäten P_{Na} and P_K, *Pflügers Arch.* **303**:133–149.

Koppenhöfer, E., and Schmidt, H., 1968b, Die Wirkung von Skorpiongift auf die Ionenströme des Ranvierschen Schnürrings. II. Unvollständige Natrium-Inaktivierung, *Pflügers Arch.* **303**:150–161.

Meves, H., and Rubly, N., 1981, Effect of various fractions of *Centruroides sculpturatus* venom on the frog node of Ranvier, *J. Physiol.* **320**:116.

Meves, H., Rubly, N., and Watt, D. D., 1982, Effect of toxins isolated from the venom of the scorpion *Centruroides sculpturatus* on the Na currents of the node of Ranvier, *Pflügers Arch.* **393**:56–62.

Mozhayeva, G. N., Naumov, A. P., Nosyreva, E. D., and Grishin, E. V., 1980, Potential-dependent interaction of toxin from venom of the scorpion *Buthus eupeus* with sodium channels in myelinated fibre. Voltage clamp experiments, *Biochim. Biophys. Acta* **597**:587–602.

Narahashi, T., Shapiro, B. I., Deguchi, T., Scuka, M., and Wang, C. M., 1972, Effects of scorpion venom on squid axon membranes, *Am. J. Physiol.* **222**:850–857.

Nonner, W., 1979, Effects of *Leiurus* scorpion venom on the "gating" current in myelinated nerve, in: *Advances in Cytopharmacology,* Vol. 3 (B. Ceccarelli and F. Clementi, eds.), Raven Press, New York.

Rochat, H., Rochat, C., Miranda, F., Lissitzky, S., and Edman, P., 1970, The amino acid sequence of neurotoxin I of *Androctonus australis* Hector, *Eur. J. Biochem.* **17**:262–266.

Romey, G., Chicheportiche, R., Lazdunski, M., Rochat, H., Miranda, F., and Lissitzky, S., 1975, Scorpion neurotoxin—a presynaptic toxin which affects both Na^+ and K^+ channels in axons, *Biochem. Biophys. Res. Commun.* **64**:115–121.

Watt, D. D., Simard, J. M., Babin, D. R., and Mlejnek, R. V., 1978, Physiological characterization of toxins isolated from scorpion venom, in: *Toxins: Animal, Plant and Microbial* (P. Rosenberg, ed.), Pergamon Press, Oxford and New York.

14

Modification of Voltage-Sensitive Sodium Channels by Batrachotoxin

B. I. Khodorov

I. INTRODUCTION

Batrachotoxin (BTX; Fig. 1) is a steroidal alkaloid contained in the skin secretion of the Colombian arrow poison frog, *Phylobates aurotaenia* and related species (see Albuquerque and Daly, 1976; Myers *et al.*, 1978, for reviews). It belongs to a class of neurotoxins inducing a steady depolarization of nerve and striated muscle membranes. Since this depolarization can be eliminated effectively by tetrodotoxin (TTX) or by removal of Na^+ from the external solution the conclusion was made that BTX increases the resting sodium permeability of excitable membranes (Narahashi *et al.*, 1971; Albuquerque, 1972; Catterall, 1975).

Voltage-clamp experiments carried out on frog myelinated nerves (Khodorov *et al.*, 1975; Peganov *et al.*, 1976; Revenko and Khodorov, 1977; Khodorov and Revenko, 1979) clarified the origin of the effects of BTX. It has been established that BTX produces dramatic and practically irreversible changes in activation, inactivation, ion selectivity, and sensitivity of Na channels to the blocking action of local anesthetics (for a review see Khodorov, 1981).

Further investigations of the effects of BTX on frog node of Ranvier (Zaborovskaja and Khodorov, 1980a, 1981; Mozhayeva *et al.*, 1981a; Guselnikova *et al.*, 1981; Khodorov *et al.*, 1981; Revenko *et al.*, 1980, 1981a,b; Dubois and Khodorov, 1982) and neuroblastoma cells (Zubov *et al.*, 1981, 1983a,b) revealed new remarkable properties of BTX-modified Na channels.

The aim of the present chapter is to review briefly some of the most essential data obtained in this field.

II. INTERACTION OF BTX WITH OPEN Na CHANNELS

BTX interacts with open Na channels on a one-to-one stoichiometric basis (Revenko, 1977; Khodorov and Revenko, 1979). This conclusion was derived from the study of the

B. I. Khodorov • Vishnevsky Surgery Institute, Moscow 113093, U.S.S.R.

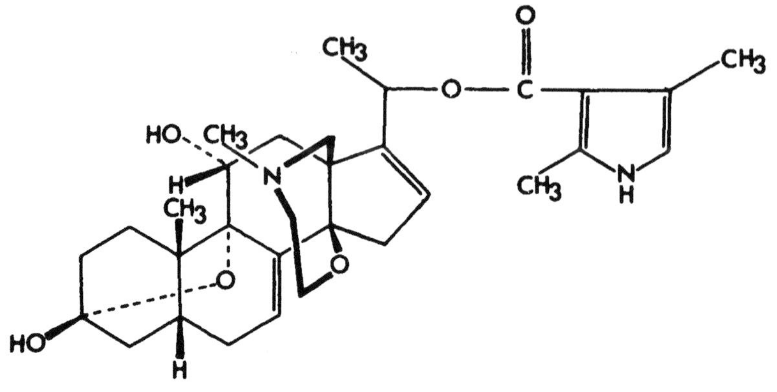

Figure 1. The structure of batrachotoxin.

effects of repetitive membrane stimulation of voltage-clamped frog nodes of Ranvier treated with BTX at various concentrations (in the range 2×10^{-9}–5×10^{-5} M). The acceleration of Na channels modification by repetitive pulsing was also observed in the electric-eel electroplax (Bartels-Bernal *et al.*, 1977) and voltage-clamped clonal neuroblastoma cells (Zubov *et al.*, 1981). It is necessary, however, to stress that the number of channels modified during each depolarizing pulse (even when it is very strong, e.g., + 80 mV) is relatively small. Therefore, about 6000 short and strong pulses (+ 80 mV, 5 ms duration, 10 Hz) are required to modify almost all the Na channels in myelinated nerve or neuroblastoma cell treated with 10^{-5} M BTX (Mozhayeva *et al.*, 1981a, 1982a; Zubov *et al.*, 1981; Dubois and Khodorov, 1982).

A study of the structure–activity relation (see Warnick *et al.*, 1975; Albuquerque and Daly, 1976, for reviews) suggests the existence of a specific receptor for BTX in the sodium channel. The BTX binding to this "recognizing site" is very tight and, in some membranes (such as the nodal membrane) it seems to be practically irreversible. Therefore, it is possible to remove BTX from the bathing solution at any stage of membrane treatment, and to acquire thereby a membrane with two stable populations of normal and modified Na channels. The relative size of these populations may be varied within wide limits by changing the BTX concentration, the duration of its application, and the parameters of repetitive pulsing. Recent experiments (Mozhayeva *et al.*, 1981a, 1982a; Zubov *et al.*, 1981; Dubois and Khodorov, 1982) dealing with the effects of BTX on Na inactivation, gating charge movements, ion selectivity, and the effects of H^+ and Ca^{++} ions have been carried out under the condition of almost total modification of Na channels.

III. THE ACTIVATION PROCESS

Binding to open Na channels BTX hinders their deactivation upon membrane repolarization to the normal resting potential ($E = -70$ mV). To close the BTX-modified channels it is necessary to hyperpolarize the membrane to about $-$ 120 mV. Starting from such holding potentials it is possible to elicit modified Na currents at negative potentials (− 100−− 70 mV), where almost all the normal Na channels are still in the resting state. These modified currents rise slowly and have practically exponential kinetics (instead of being sigmoid as in the normal channels; Fig. 2).

It was revealed (Khodorov and Revenko, 1979) that at the first approximation the modified Na permeability P^*_{Na} may be described by modified Hodgkin–Huxley equations (Hodgkin and Huxley, 1952; Frankenhaeuser and Huxley, 1964):

$$P^*_{\mathrm{Na}} = \overline{P}^*_{\mathrm{Na}}\, m \tag{1}$$

$$dm/dt = \alpha_m(1-m) - \beta_m m \tag{2}$$

$$m_\infty = \frac{\alpha_m}{\alpha_m + \beta_m} \tag{3}$$

$$\tau_m = \frac{1}{\alpha_m + \beta_m} \tag{4}$$

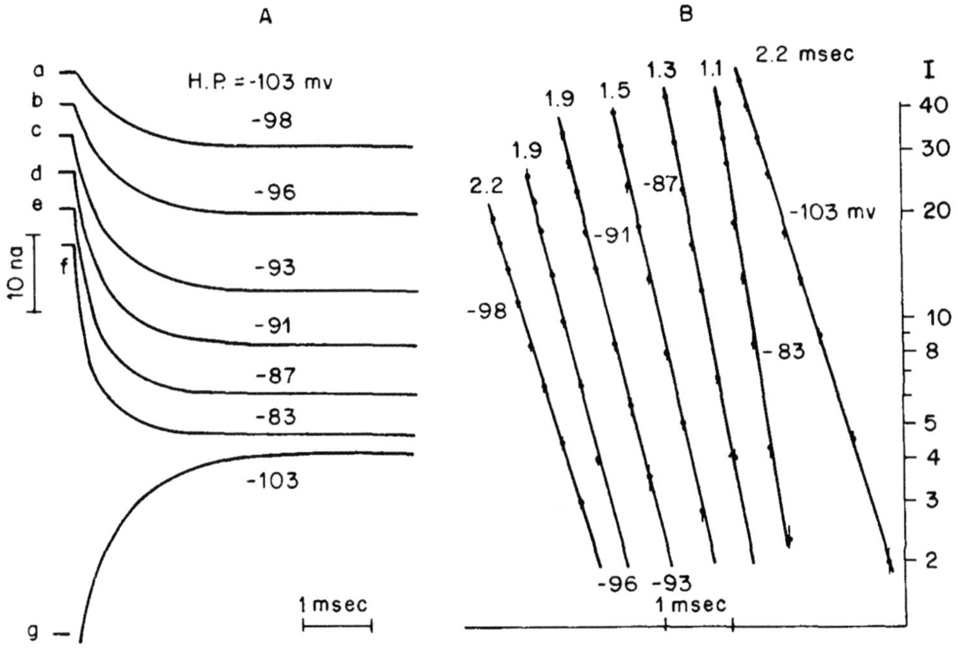

Figure 2. The kinetics of BTX-modified sodium currents. (A) Modified currents in the node treated with 10^{-5} M BTX. Numbers attached to the traces indicate the test potential (E) in mV. The lower record is off-current obtained after a 5 ms voltage step to -83 mV. (B) The same currents in logarithmic scale. The upper numbers are the time constants and the lower ones are the membrane potentials in mV. Temperature 20°C (Khodorov and Revenko, 1979).

Here, $\bar{P}^*_{Na}$ is the maximum modified Na permeability, m is the fraction of open Na channels, α_m and β_m are forward (opening) and backward (closing) rate-constants, respectively.

Figure 3A shows that half of the modified Na channels ($m_\infty = 0.5$) is open at E about $-$ 90 mV, while the midpoint potential for normal P_{Na} is about $-$ 25 mV (Dodge, 1963; Hille, 1967). Thus, the modified permeability curve is shifted by about 65 mV to more negative E.

The voltage-dependence of the activation time-constant τ^*_m is presented in Fig. 3A. The curve is bell-shaped and symmetrical. The maximum $\tau^*_m(\tau_{max})$ varies from fiber to fiber; at 20°C its mean $\pm$SD value is 1.5 $\pm$ 0.5 ms (n = 12).

Table I presents the calculated values of α's and β's for modified (α^*_m, β^*_m) and normal (α_m, β_m) Na channels.

It can be seen that BTX produces a drastic reduction of the closing rate-constant β^*_m, shifting its voltage-dependence by about 60 mV to more negative E. The α_m–E relation is also shifted towards negative E but to a much lesser extent ($\approx$ 20 mV) (Fig. 3B).

Reduction of β^*_m leads to an increase in the maximum time-constant τ_{max} and is the main cause of the negative shift of P_{Na}–E relation. Indeed, the fact that at $E = -$ 90 mV about half of the modified channels become open is caused not by an increase in the activation rate-constant α_m (its value at $-$ 90 mV is even smaller than in the normal node; see Table I) but results from a drastic reduction of β_m. Here, $\beta^*_m/\alpha^*_m = 1$, while in normal Na channels β_m/α_m at $E = -$ 90 mV is about 40.

The sigmoid time-course of normal Na activation described formally (Hodgkin and Huxley, 1952) by a power function of m (m^2 or m^3) is interpreted at present as a result of a multistep channel transition from resting to the open state (Moore and Cox, 1976; Armstrong and Bezanilla, 1977; Bezanilla and Armstrong, 1977). If this interpretation is correct, the exponential kinetics of the modified Na activation indicate that BTX somehow reduces the number of such energetic barriers to only one, separating the closed- and open-channel conformations.

Therefore a simple two-state (one-barrier) system (see Rojas and Keynes, 1975; Keynes

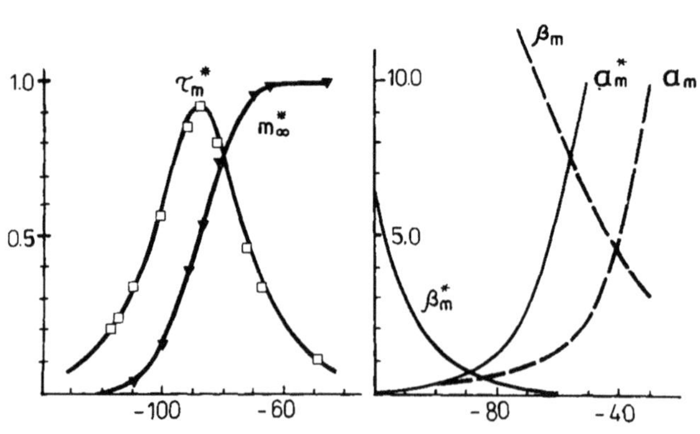

Figure 3. The voltage-dependent parameters of modified sodium activation. (A) Steady-state activation m^*_∞ (E) and the time-constant of activation $\tau^*_m(E)$, in msec. Theoretical curves were drawn in accordance with Eqs. (5) and (6). $E' = -88$ mV; $z = -4$; τ_{max} = 0.9 msec. Holding potential = -100 mV (Khodorov and Revenko, 1979). (B) The calculated rate-constants for BTX-modified (α^*_m and β^*_m) and normal (α_m and β_m) in Na channels (see Table I). BTX concentration 2 $\times$ 10^{-5} M; temperature 19°C.

Table I. Empirical Rate Constants for Na-Channels Activation in Frog Node of Ranvier[a]

	Normal Na channels			BTX-modified sodium channels			Normal/modified	
E	α_m	: B_m	:β_m/α_m	α_m^* :	β_m^*	:β_m^*/α_m^*	β_m/β_m^*	: α_m/α_{m^*}
−120				0	6.25			
−110				0.12	2.90	24.2		
−100	0.33	19.67	59.6	0.26	1.37	5.27	14.35	1.27
−90	0.43	16.52	38.4	0.50	0.61	1.22	27.08	0.86
−80	0.59	13.70	23.2	1.08	0.32	0.80	42.81	0.55
−70	0.82	10.93	13.3	2.44	0.15	0.06	72.87	0.34
−60	1.29	8.51	6.6	5.33	0.05	0.01	170.2	0.24
−50	2.26	6.29	2.8	10	0.00			0.23
−40	4.70	4.47	0.95					
−30	9.30	3.03	0.3					

[a] E: membrane potential (internal minus external) in millivolts; α's and β's: opening and closing rate constants, respectively, in ms^{-1}, calculated according to equations

$$\alpha_m = m_\infty/\tau_m \text{ and } \beta_m = (1 - m_\infty)/\tau_m.$$

Values of m_∞ and τ_m for normal channels are borrowed from Hille (1967); m_∞^* and τ_m^* are from the experiment presented in Fig. 3.

and Rojas, 1974; Neher and Stevens, 1979) may be used in the analysis of BTX-induced changes in Na activation.

The molecular mechanism of the striking change in the closing rate-constant β_m under the action of BTX requires further theoretical and experimental investigation.

Modification of the activation process produced by BTX in clonal neuroblastoma cells proved to be similar to that observed in nerve fibers. However, the voltage shift of the Na activation in these cells is smaller; it varies between 20 (Zubov *et al.*, 1981) and 40 mV (Huang *et al.*, 1981).

It is worthwhile to compare the effects of BTX on Na activation with those produced by the alkaloid aconitine. When applied to the node of Ranvier, aconitine causes a shift of P_{Na}–E curve by about 50 mV to more negative E (Schmidt and Schmidt, 1974; Mozhayeva *et al.*, 1976). The kinetics of Na activation become approximately exponential with an increased maximum τ_{max}. This increase, however, is not greater than 2–3 times in contrast to the 5–10-fold increase of τ_{max} caused by BTX.

Such a difference means that in aconitine-treated node α_m (= β_m) at the midpoint potential is larger than in the case of BTX action. This suggests that aconitine produces a smaller shift in the β_m–E and a larger shift in the α_m–E relations than BTX. Unfortunately, a detailed study of the effects of aconitine on Na activation kinetics has not been performed as yet.

IV. INTRAMEMBRANE CHARGE MOVEMENT

It was shown earlier (Khodorov and Revenko, 1979) that m_∞–E and τ_m–E relations for modified Na channels may be described by the equations (Keynes and Rojas, 1974):

$$m_\infty(E) = \frac{1}{1+\exp\left(-z\frac{E-E'}{RT/F}\right)} \quad (5)$$

$$\tau_m(E) = \frac{\tau_{max}}{\cosh\frac{(zE-E')}{2RT/F}} \quad (6)$$

Here, z is the effective valency of an activating charged particle in elementary units, multiplied by the fraction of the electrical field acting on it. E' is the midpoint potential (at which charges are equally distributed between resting and open configurations), and R, T, and F have their usual meanings.

The theoretical curves of m_∞ and τ_m in Fig. 3 were drawn in accordance with Eq. (5) and (6), using $E' = -88$ mV and $z = -4$. The calculated average value of $z = -3.5 \pm 0.5$ ($n = 12$) proved to be the same as that derived from the maximum slope of P_{Na}–E curve of a normal fiber (Fishman *et al.*, 1971).

This led us to assume that BTX does not change the maximum charge displaced in the Na channel (Khodorov and Revenko, 1979). Recently, this suggestion has been confirmed by direct measurement of the effect of BTX on intramembrane charge movement in the node of Ranvier (Dubois and Khodorov, 1982).

Figure 4 presents the records of integral gating currents before (left) and after (middle) treatment of the nodal membrane with 10^{-5} M BTX.

Modification of channels was produced by 5–8 min repetitive stimulation of the BTX-treated node with strong (to +80 mV) and short (5 ms) pulses at a rate of 10 Hz. After the end of such a pulsing, the peak of normal I_{Na} became very small or practically disappeared. In spite of this the maximum charge displaced remained almost unchanged. Several determinations (n) of the maximum charge movement for E between +20 and +80 mV gave the following mean ± SEM values: control, 123 ± 11 fC($n = 5$); BTX, 122 ± 7 fC ($n = 9$).

It is also seen from Fig. 4 that at the depolarizing potentials used (−20 mV and +20 mV) BTX did not change appreciably the kinetics of the ON-response. By contrast, the OFF-

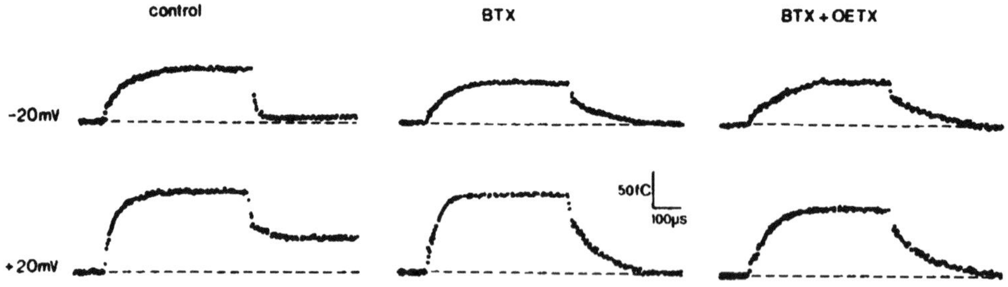

Figure 4. Intramembrane charge movement (integral of the asymmetrical capacity current) elicited by depolarizations to −20 and +20 mV in control conditions, after a treatment consisting of the total modification of sodium channels by 10^{-5} M BTX (middle) and after 10 min application to BTX-treated node of 40 μM OETX. The averaged integral gating current was obtained in response to 12 depolarizing and 24 half-amplitude hyperpolarizing pulses from a holding potential of −120 mV. No channels were blocked by 5×10^{-7} M TTX. Several determinations (n) of the maximum charge movement for depolarizations between +20 and +80 mV gave the following mean ±SEM values: control 123 ± 11 fC ($n = 5$); BTX 122 ± 7 fC ($n = 9$); OETX 111- ± 9 fC ($n = 3$). Temperature 16°C (Dubois and Khodorov, 1982).

charge movement associated with membrane repolarization to the holding potential -120 mV was slowed down and became equal to the ON-charge movement, which indicates a removal of the depolarization-induced charge immobilization.

This result is consistent with the above data concerned with the action of BTX on I_{Na} kinetics: the activation time-constant τ_m of modified channels, very large at $E < -70$ mV becomes at $E > -50$ mV equal or even smaller than in intact Na channels (Khodorov and Revenko, 1979). The removal of charge immobilization during membrane depolarization reflects the inhibitory action of BTX on Na inactivation.

V. INSTANTANEOUS CURRENTS THROUGH MODIFIED NA CHANNELS: EFFECTS OF CA^{++} IONS

The voltage-dependence of instantaneous Na currents (I_{Na} vs. E) was measured in frog myelinated fibers under the condition of a complete modification of Na channels by 10^{-5} M BTX (Mozhayeva *et al.*, 1982a). Figure 5 presents I'_{Na}–E relations observed in Ringer's solutions containing 2 or 20 mM Ca^{++}. In the first case (2 mM Ca^{++}) the I'_{Na} curve saturated at $E > 60$ mV, and revealed a negative slope at $E < -100$ mV. An increase in the external Ca^{++} concentration to 20 mM produced a moderate decrease of I'_{Na} in the potential range -40–$+100$ mV and a drastic inhibition of I'_{Na} at $E < -70$ mV. Thus the potential shift of the maximum I'_{Na} proved to be about 30 mV.

The conclusion was made that the negative slope of I'_{Na} curve reflects the voltage-dependent block of modified Na channels by external Ca^{++}. According to the data obtained Ca^{++} binds to a site located about halfway across the membrane ($\delta \sim 0.45$) with a dissociation constant of 200 ± 54 ($n = 5$). Judging by recent estimations of the inner acid group location in the Na channel (Mozhayeva *et al.*, 1981b), the Ca^{++}-binding site is a part of the channel selectivity filter.

VI. SELECTIVITY IN MODIFIED NA CHANNELS

BTX decreases the selectivity of Na channels both in nerve fibers (Revenko and Khodorov, 1977; Khodorov and Revenko, 1979) and neuroblastoma cells (Huang *et al.*, 1979;

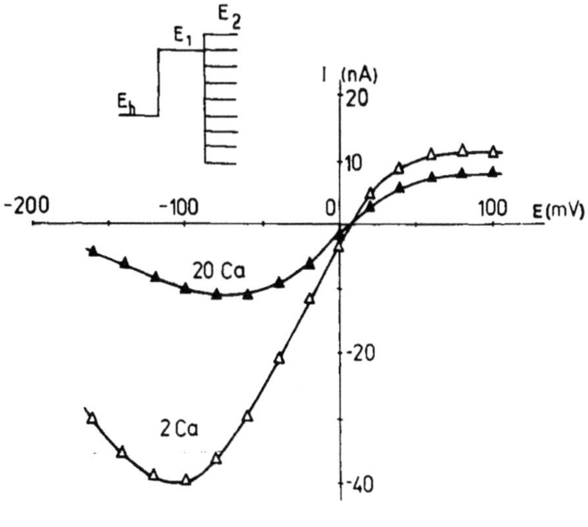

Figure 5. Instantaneous current IR^{*}_{Na}, through BTX-modified Na channels of the node of Ranvier as a function of membrane potential, E_2 and external Ca^{++} ion concentration (2 and 20 mM Ca). Complete modification of Na channels in the node was achieved by 5-min BTX (10^{-5} M) treatment combined with repetitive membrane pulsing (10 Hz). The pulse procedure is shown in the inset. Holding potential, $E_h = -120$ mV; during the first pulse membrane was depolarized to $+80$ mV (Mozhayeva *et al.*, 1982a).

Table II. Effects of BTX and Aconitine on the Relative Permeability of Sodium Channels and pK_{eff} of This Acid Group[a]

	Frog nerve		
	Normal (1)	BTX (2)	Aconitine (3)
P_K/P_{Na}	0.09	0.20	0.60
P_{NA4}/P_{Na}	0.16	0.50	1.35
ΔpK_{eff}		−0.41	−0.74 (4)
	Neuroblastoma cell		
	Normal	BTX (5)	Aconitine (6)
P_K/P_{Na}	0.11	0.29	
P_{NH4}/P_{Na}	0.35	0.76	1.37

[a] Permeability ratios were calculated by means of equation

$$R_{r,s} - E_{r,Na} = \frac{RT}{F}\ln\frac{P_s(S)}{P_{Na}(NA)}$$

where $E_{r,Na}$ and $E_{r,s}$ are the reversal potentials in control Na Ringer's and Na substitute (S) Ringer's solutions respectively, P_S and P_{Na} are permeabilities of the Na channel for Na and S-ions; ΔpK_{eff}: change in the effective pK of the hypothetical acid group in the Na channel, calculated according to a simplified model (Woodhull, 1973); References: Hille (1972); Mozhayeva *et al.* (1981a, 1982c); Mozhayeva *et al.* (1977); Naumov *et al.* (1979); Zubov *et al.* (1981); Grishchenko *et al.* (1981).

Zubov *et al.*, 1981, 1983a). The modified channels in the nodal membrane proved to be measurably permeable to Rb^+, Cs^+, Ca^{++}, and even to methylammonium. Table II presents some recent estimations of P_K/P_{Na} and P_{NH4}/P_{Na} in frog node of Ranvier (Mozhayeva *et al.*, 1981a, 1982c) and neuroblastoma cells (Zubov *et al.*, 1981, 1983a). The table also compares the changes in ion selectivity of Na channels caused by BTX and aconitine. It can be seen that aconitine is more effective in its action on the selectivity mechanism of the channel than BTX. Note that the opposite is true for the action of these alkaloids on channel gating: BTX is a stronger modifier of the gating mechanism than aconitine.

VII. BLOCK BY H^+ IONS

A decrease in external pH from 7.7 to 5.12 or 4.4 exerts a dual effect on BTX-modified Na channels. The voltage-dependence of Na activation is shifted to more positive potentials, and the maximum of sodium conductance in the region of E_{rev} is reduced (Mozhayeva *et al.*, 1981a, 1982c). The first effect is believed to be a result of the interaction of H^+ ions with negatively charged groups of the membrane surface, whereas the second one reflects a direct blocking action of H^+ ion on Na channels. The shift of the sodium-conductance curve does not differ appreciably from that observed in the normal node of Ranvier at corresponding pH (see Hille, 1968a). On the contrary, the blocking action of H^+ on I_{Na} is noticeably decreased under the action of BTX. Calculation according to simplified model (Woodhull, 1973) with a single titrable acid group shows that its effective pK (pK_{eff}) at $E = 0$ mV decreases under the action of BTX by 0.37–0.45 at pH 5.12 and 4.8 respectively.

According to Eisenman's theory (1962) this indicates a decrease in the effective "field strength" of the ion-binding site in the selectivity filter.

The voltage-dependence of the H^+ block of modified I_{Na} proved to be the same as that of the normal I_{Na} (Woodhull, 1973). This apparently means that BTX does not affect the fraction (δ) of the total potential difference across the membrane that acts upon the H^+-ion binding site in the channel.

Aconitine causes much more pronounced change of K_{eff}; K_{eff} drops by about 0.74 (Naumov *et al.*, 1979). Simultaneously, a considerable decrease of δ takes place (Mozhayeva *et al.*, 1981b). The results obtained are consistent with the fact mentioned above that BTX produces a lesser change in Na-channel selectivity than aconitine.

VIII. SINGLE Na CHANNEL CONDUCTANCE

It has been noted above that BTX-modified Na channels are measurably permeable to such relatively big cations as Rb^+, Cs^+, Ca^{++}, and methylammonium, which cannot pass through normal Na^+ pores (Hille, 1971, 1972). This suggests that BTX alters the structure of the selectivity filter such that its minimum dimensions are increased to 3.8 × 6.0 Å as compared to a pore size of 3.0 × 5.0 Å postulated by Hille (1972) for a normal Na channel (Khodorov, 1978; Khodorov and Revenko, 1979; Huang *et al.*, 1979).

Thus, one could expect that a single BTX-modified channel has a higher ionic conductance compared to one of normal channels.

To estimate the conductance and some other properties of a single modified channel, Na current fluctuations have been recorded and analyzed in frog myelinated fiber treated with BTX (Khodorov *et al.*, 1981). The spectral density of I_{Na} fluctuations was fitted by the sum of a $1/f$ component and a Lorentzian function. The time-constant $\tau_c = 1/(2\pi f_c)$ obtained from the corner frequency f_c of the Lorentzian function approximately agreed with the activation time-constant τ_m^* of the macroscopic currents.

The conductance of a single modified Na channel, γ, was calculated from the integral of the Lorentzian function and the steady-state I_{Na}. The values found, between 1.6–3.45 pS, proved to be significantly lower than the values 6.4–8.45 pS reported in the literature for normal Na channels (Conti *et al.*, 1976, 1980).

The reason for the marked decrease of γ under the action of BTX is not clear as yet.

One possibility is that a slow passage of big Ca^{++} ions decreases the flux of Na^+ through modified Na channels. Ca^{++} block of modified I_{Na}, shown above (see Fig. 5), increases as membrane hyperpolarization becomes larger. Thus, one could expect a larger reduction of γ at more negative E. Preliminary data obtained by Khodorov *et al.* (1981) seem to reveal the opposite tendency in the γ–E relation. A study of the effects of Ca^{++} ions on γ may clarify this problem.

The other explanation of γ-reduction in BTX-modified Na channels could be that BTX, having bound to the receptor in the channel wall, "protrudes" its positively charged nitrogen into the channel lumen not far from the selectivity filter. This may decrease both the pK of the inner acid group and the Na^+ ion flux through the modified channels.

IX. INACTIVATION PROCESS AND "THE SECOND OPEN STATE"

When the fraction of BTX-modified Na channels in the node of Ranvier is small, the modified current as a rule seems to be practically constant during a maintained membrane depolarization, even when such a depolarization lasts many hundreds of milliseconds (Fig. 6). This led us to conclude that BTX removes Na inactivation completely (Khodorov *et al.*, 1975; Khodorov, 1978; Khodorov and Revenko, 1979).

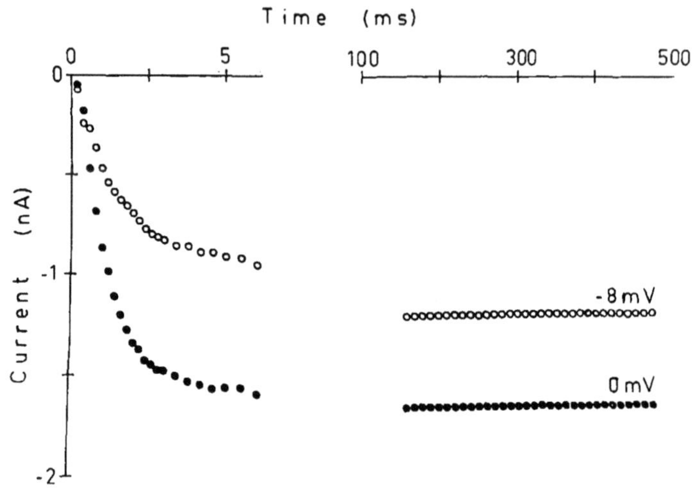

Figure 6. Current relaxations and steady-state currents through the BTX-modified Na channels at $V = -8$ mV (open symbols) and 0 mV (filled symbols). Note different time scaling during relaxations and in the steady state. The activation time-constants τ_m^* and the steady-state Na currents I, are 1.60 ms, -1.19 nA (-8 mV), 0.85 ms, and -1.64 nA (0 mV). Holding potential $E_h = -32$ mV. Exp. B 2, motor fiber. Temperature 15°C. V denotes the displacement of the membrane potential from its resting value, E_o (where 30% of normal Na channels were inactivated). $E_o = -71.6 \pm 1.2$ mV ($n = 9$; Khodorov *et al.*, 1981).

Recently, however, under the condition of a total modification of Na channels by BTX (see Section I) it has been found that in some fibers a small but distinct decay of I_{Na} to a steady-state level during membrane depolarizing does occur. The detailed analysis of this effect (Mozhayeva *et al.*, 1981a) showed that it reflects a partial inactivation of modified Na channels.

Inactivation was also revealed by the use of conditioning depolarizing prepulses, E_c. The best way to estimate quantitatively the steady-state inactivation of modified channels proved to be a study of the effects of E_c on the instantaneous value of Na current, I_{Na}, measured at the end of a short (0.5 ms) and strong (to + 80 mV) depolarizing pulse. (Fig. 7, inset). The conditioning depolarization, E_c, of 50 ms duration, changes the steady-state channel distribution between resting, open, and inactivated states. The intermediate short and strong depolarizing pulse, E_2, quickly opens all the channels that remained free from

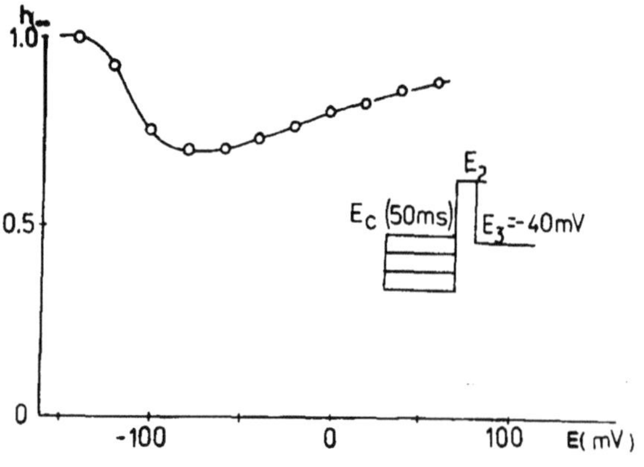

Figure 7. Steady-state inactivation of BTX-modified Na channels measured by means of pulse procedure shown in the inset. h_∞ is the fraction of modified channels free from inactivation at the end of conditioning 50-ms voltage step of various magnitude E_c; Temperature 10°C. Explanation in the text (Mozhayeva *et al.*, unpublished).

inactivation, and the third (E_3) repolarizing pulse, only produced an instantaneous inward I_{Na} proportional to the fraction of noninactivated Na channels. The h_∞–E_c curve obtained in this way (Fig. 7) has both descending and ascending branches. h_∞ decreases from 1 to about 0.7 as E_c rises to -70 (-60) mV, and then grows to 0.85 (0.9) at $E_c \geqslant +60$ mV. In some cases, h_∞ increases at $E_c > +80$ mV up to 1.0.

Thus, at moderate depolarization, about 30% of Na channels can undergo inactivation. The fact that during a test depolarizing pulse the inward I_{Na} very often does not reveal a marked decrease can be explained by "unfavorable" relation between the rates of activation and inactivation. Na channels modified by BTX open at high negative E very slowly, and there is no excess in the rate of activation over that of inactivation appropriate to form the peak of I_{Na} (Mozhayeva *et al.*, 1981a).

In neuroblastoma cells a clear-cut partial inactivation of modified channels reveals readily even without conditioning prepulses (Zubov *et al.*, 1981, 1983c).

During a depolarizing step from the holding potential $E = -130$ mV to $E_c < -60$

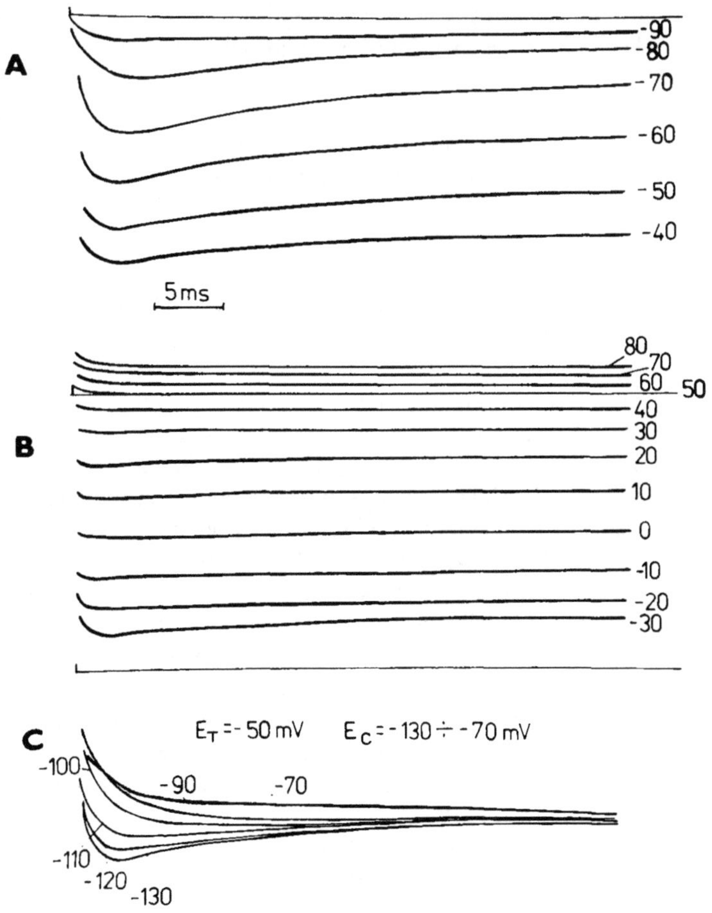

Figure 8. Records of modified Na currents in the voltage-clamped internally perfused neuroblastoma cell. (A) and (B) are families of current traces associated with membrane depolarization from $E = -130$ mV to various potentials indicated beside each current trace. (C) Effect of conditioning 50 ms membrane depolarization E_c on the time-course and I_{Na} maximum elicited by the test pulse E_T to -50 mV. $E_h = -130$ mV; values of E_c in mV are indicated at each trace (Zubov *et al.*, 1983c).

(− 50) mV, the inward I_{Na} first rises (time to peak 2–10 ms, 20°C) then slowly decays to a steady-state level. This partial inactivation usually does not exceed 10–20% but sometimes, as for example in the case shown in Fig. 8A, amounts to 50%. At larger depolarizations ($E_c > -50$ mV; Fig. 8B), the inactivation of modified channels diminishes, and at $E \geqslant 0$, mV becomes negligible. The conditioning depolarizing prepulse of 50 ms duration to $E < -50$ mV decreases the rate of rise and the peak value of I_{Na} elicited by the test pulse. At $E_c > 0$ mV the steady-state inactivation practically vanishes (Fig. 8C).

We have interpreted this increase of h_∞ at large E as a result of conformational transition of the modified channels to a new, "second" open state, 0_2. Judging by available data, this transition can occur both from the first open (0_1) and inactivated (I) states. However, the detailed kinetics scheme of all these transitions is still in progress.

Recently, it has been found that Na inactivation is incomplete, and h_∞ has a tendency to rise at positive E, not only in the internally perfused axons (Adelman and Senft, 1966a,b; Chandler and Meves, 1970; Bezanilla and Armstrong, 1977) or scorpion-toxin-treated nerve fibers (Koppenhöfer and Schmidt, 1968; Mozhayeva *et al.*, 1980; Gillespie and Meves, 1980), but also in normal unperfused axons (Shoukimas and French, 1980).

However, in all the above cases, the secondary increase of h at positive E was relatively small (to 0.2–0.4). BTX is the first drug known as yet that allows h_∞ to approach 1.0 at large E.

A pronounced rectification of macroscopic currents at positive E suggests that individual Na channels in the 0_2 state have lower conductance (γ) and larger rectification than in the 0_1 state. The outward currents through these channels saturate at $E \geqslant +60$ (+80) mV (see Fig. 5).

X. BLOCK OF MODIFIED Na CHANNELS BY TERTIARY AND QUATERNARY AMMONIUM DRUGS

Local anesthetics, antiarrythmics, and related amine drugs are known to cause two phenomenologically different types of Na-current inhibition, the so-called "tonic" (steady-state) block (Strichartz, 1973), a decline in I_{Na} without conditioning or test pulses, and "use-dependent" (variable) block which develops during repetitive or prolonged membrane depolarization (Strichartz, 1973; Courtney, 1975, 1980; Khodorov *et al.*, 1976; Hille, 1977; Cahalan, 1978; see also Khodorov, 1979a,b, 1981, for reviews).

Experiments on nerve fibers have shown that BTX removes completely the use-dependent inhibition of I_{Na} caused by all these drugs and decreases drastically their tonic-blocking action. At concentrations sufficient to cause practically complete inhibition of normal I_{Na} these drugs produce only small suppression of modified currents.

Partial inhibition of modified I_{Na} by tertiary and quaternary local anesthetics is time- and voltage-dependent. It rises as the amplitude and duration of the depolarizing pulse increase (Khodorov *et al.*, 1977; Khodorov, 1978; Cahalan, 1978; Zaborovskaja, 1979; Zaborovskaja and Khodorov, 1982).

To produce a complete tonic inhibition of modified I_{Na} by local anesthetics it is necessary to raise their concentrations by about 10 times as compared to those sufficient to inhibit the normal I_{Na}.

Figure 9 presents an example of a selective inhibition of unmodified I_{Na} by quaternary antiarrhythmic N-propyl ajmaline (NPA) applied to the node of Ranvier treated previously with BTX.

The curves show the voltage-dependence of modified (open symbols) quasisteady-state

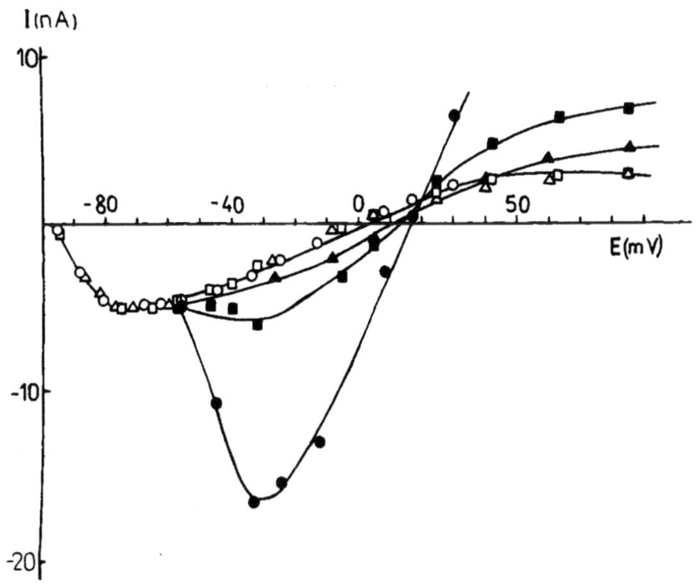

Figure 9. Resistance of BTX-modified Na channels in the node of Ranvier to the blocking action of the quaternary antiarrhythmic, N-propyl ajmaline (NPA). Sodium currents after 2 min application of 2 × 10^{-5} M BTX (○ ●), and during a subsequent application of 10^{-4} M NPA combined with repetitive (10 Hz) membrane depolarization to $E = -10$ mV, (□ ■) and to $E = 60$ (△ ▲). Open symbols denote currents through modified Na channels; filled symbols denote peak currents through the normal (unmodified) Na channels; temperature 15°C (Zaborovskaja and Khodorov, 1982).

I_{Na}, and the peak I_{Na} (filled symbols) flowing through the channels that retained their normal activation–inactivation kinetics ("normal channels"). It is seen that application of NPA (10^{-4} M) combined with repetitive membrane depolarization caused a strong inhibition of the normal I_{Na} without a noticeable decrease of the modified one.

Similar results have been obtained in experiments with 10^{-5} M yohimbine (Revenko *et al.*, 1981b), 10^{-5} ethmozine (Revenko *et al.*, 1980; Bolotina *et al.*, 1981), 5 × 10^{-5} M quinidine (Revenko *et al.*, 1981b), or 5 × 10^{-6} M strychnine (Zaborovskaja and Khodorov, 1982).

Block of modified I_{Na} by large concentrations of amine drugs proved to be reversible. Washing of the node with drug-free solution leads to almost complete restoration of modified Na currents (Zaborovskaja and Khodorov, 1982). This apparently indicates that BTX and amine blockers have different binding sites in the Na channel.

Of special interest is the fact that recovery of modified Na channels from the tonic block produced by large concentrations of amine drugs proceeds much faster than normal channels unblocking in the same node (Zaborovskaja and Khodorov, 1981). Figure 10 illustrates this observation. It is seen that after 30-min washing of the BTX-treated nodal membrane from 2 × 10^{-3} M NPA, the modified I_{Na} recovered almost completely while the normal one remained still strongly inhibited.

Thus BTX, bound to the channel, increases the dissociation rate-constant of the NPA-channel complex. This may be the major cause of a low sensitivity of modified Na channels to amine blockers.

Similar effects have been observed in experiments on voltage-clamped internally perfused neuroblastoma cells (Zubov *et al.*, 1981, 1983a). So under the action of BTX the

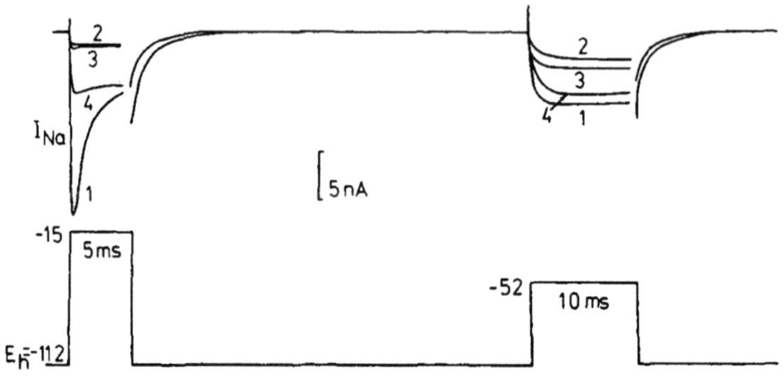

Figure 10. Difference in the rate of normal and BTX-modified channels unblocking during washing of the node from large concentration (2.10^{-3} M) NPA. The node was treated by 2.10^{-5} M BTX during 2 min, then washed out with toxin-free solution and stimulated by twin depolarizing pulses of different magnitude: to −15 and −52 mV separated by 40 ms interval. Traces: 1 and 2, before and during the action of 2 × 10^{-3} M NPA; 3 and 4, after 3- and 30-min washing of the node with drug-free solution. Note the striking difference in the rate of modified and normal Na channels unblocking (Zaborovskaja and Khodorov, 1982a,b).

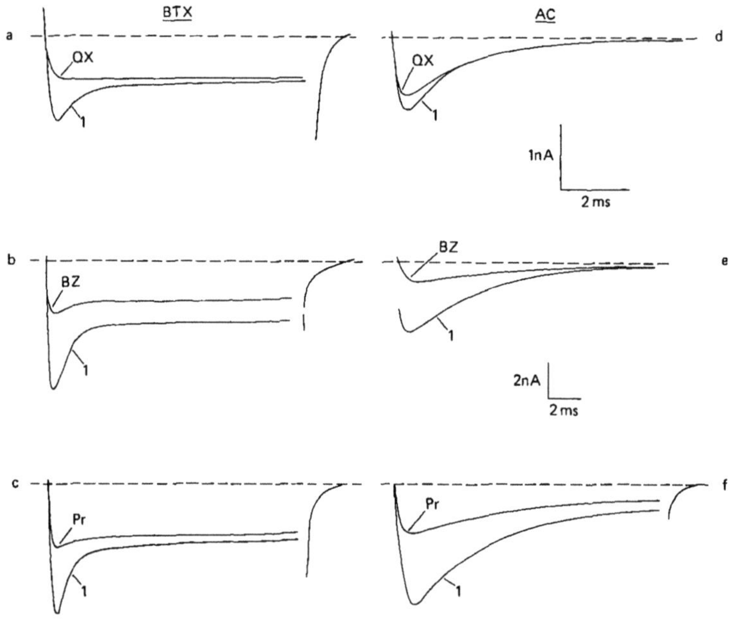

Figure 11. Effects of BTX and aconitine (AC) on the tonic blocking action of QX-314, procaine (Pr) and benzocaine (Bz) on Na channels in neuroblastoma cells. Left column: partial modification of Na channels by 4 × 10^{-5}M BTX, Trace I, two-components currents through normal (peak) and modified (steady-state) Na channels before the action of anesthetics. QX, Bz and Pr = effects of 1.7 mM QX-314 (internal application), 1 mM Bz and 1 mM Pr (external application), respectively. Test voltage; +20 mV (upper row) and −30 mV (middle and lower rows). Effects of Bz and Pr were examined on the same cell. Right column: complete modification of Na channels by 3 × 10^{-4}M AC. Trace I = ionic currents through AC-modified Na channels before the action of anesthetics. Traces QX, Bz, Pr = ionic currents during application of 1.7 mM QX-314 (internal application), 1 mM Bz and 1 mM Pr (external application), respectively. Test voltage: −10 mV (upper row), −50 mV (middle and lower rows) Solutions: *a* and *d* (in mM): 140 KF, 5 Tris −HCl (internal); 132.5 NaCl, 10 $CaCl_2$, 7.5 tetraethylammonium (TEA) Cl, Tris-HCL (external). *b* −*f* (in mM): 130 KF, 10 TEA Cl, 5 Tris-HCl (internal); 140 NaCl, 2 $CaCl_2$, 5 Tris-HCl (external). Holding potential for all records −130 mV, pH 7.6. Temperature about 20°C (Zubov et al. 1983b).

dissociation constant K_D for the tonic blocking action of lidocaine was increased from 1.21 ± 0.13 mM (n = 12) to 5.66 ± 0.68 mM (n = 5). Lipid insoluble quaternary derivative of lidocaine, compound QX 314, is known to block I_{Na} in nerve fibers only by its intracellular application (Narahashi *et al.*, 1972; Strichartz, 1973). The same result was obtained in neuroblastoma cells. At concentration 1.7 mM, QX 314 applied intracellularly blocked the normal peak I_{Na} but caused only small reduction of the modified currents (Fig. 11).

Aconitine resembles BTX in its effect on the sensitivity of Na channels of neuroblastoma cells to the blocking action of QX-314 and lidocaine (Fig. 11) (Zubov *et al.*, 1981, 1983b), in spite of the fact that aconitine, unlike BTX, does not inhibit sodium channels inactivation (Grishchenko *et al.*, 1981). Under the action of aconitine K_D for the tonic blockage of Na channels by lidocaine is increased up to 9.10 ± 2.30 mM (n = 7). However, an unexpected difference between effects of aconitine and BTX was revealed in studying the tonic blocking action of procaine. In nerve fibre aconitine increased K_D for procaine from 0.27 ± 0.03 mM to 1.32 ± 0.05 mM (Negulayev and Nosyreva, 1979). By contrast, in neuroblastoma cells aconitine did not affect appreciably K_D for procaine (0.48 ± 0.13 and 0.62 ± 0.08 mM before and after treatment of the cell by aconitine, respectively) (Zubov *et al.*, 1983a).

Aconitine significantly reduced, and in a number of experiments, completely abolished the cumulative ("use-dependent") inhibition of sodium currents both by lidocaine and procaine (Zubov *et al.*, 1983a).

XI. BLOCK BY NEUTRAL DRUGS

Unlike tertiary and quaternary amines, the neutral (at physiological pH) local anesthetic benzocaine (BZ) proved to be able to block equally effectively both normal and BTX-modified Na channels in nerve fiber (Khodorov, 1978; Zaborovskaja, 1979; Guselnikova *et al.*, 1981; Zaborovskaja and Khodorov, 1982a,b) and neuroblastoma cells. Figure 12A compares the effect of BZ on normal and modified I_{Na} in the node of Ranvier pretreated with BTX. Figure 12B shows the normalized kinetics of currents inhibition by 1.5 mM BZ, and the time-course of their subsequent restoration during washing of the node. It can be seen that curves of normal and modified I_{Na} changes coincide.

Since BZ has no terminal amine group, it has been suggested that BTX affects primarily the "anionic site" of a common local anesthetic receptor (Hille, 1977) but does not change appreciably the properties of its other binding sites interacting with hydrophobic part of the blocker molecule and its highly polarized carbonyl (or related) groups (Khodorov, 1981).

An alternative hypothesis may be that BZ has a separate receptor in the Na channel (Mrose and Ritchie, 1978) which is not affected by BTX. Such a suggestion seems to be, however, at variance with the data indicating that BZ competes with lidocaine (Schmidmayer and Ulbricht, 1980), QX 314 (Venitz and Schwarz, 1980), or QX 572 (Guselnikova *et al.*, 1981; Huang and Ehrenstein, 1981) for the Na channel.

In neuroblastoma cells, as distinct from nerve fibers, BTX increases greatly the dissociation constant for the blocking action of BZ (from 0.48 to 5.6 mM^{-1}) suggesting a total modification of the local anesthetics receptor under the action of BTX (Zubov *et al.*, 1983a). Of interest is that aconitine, unlike BTX, even increases the sensitivity of Na channels to BZ: the dissociation constant falls from 0.68 to 0.35 mM^{-1} (Zubov *et al.*, 1983b). Such a difference in the effects of BTX and aconitine can be explained if one assumes that their multisites receptors overlap only partially.

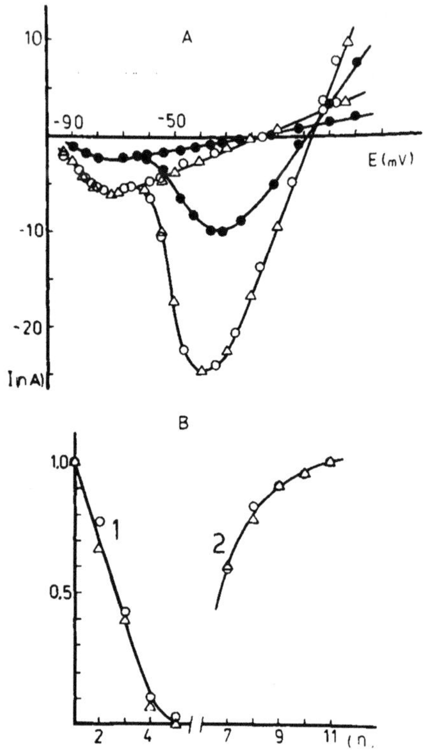

Figure 12. Equally effective blockage of normal and BTX-modified Na currents by benzocaine. (A) I_{Na}–E relations in the node of Ranvier treated with 2 × 10^{-5} M BTX. The symbol ○ denotes modified quasi-steady-state (first hump of the curve) and normal (peak) currents (second hump) before application of benzocaine; ●, the same currents during the action of 0.8 mM benzocaine; △, after 1-min washing of the node with drug-free solution. (B) 1, the time-course of Na-currents inhibition through normal (○) and BTX-modified (△) channels under the action of 1.5 mM benzocaine. 2, restoration of these currents during washing of the node by drug-free solution. Ordinate: relative value of modified I_{Na}; the original value of I_{Na} is taken for unity. Abscissa: the number of the test pulse; 0.2 Hz (Zaborovskaja and Khodorov, 1981).

In neuroblastoma cells, as distinct from nerve fibers, BTX greatly increases the dissociation constant (K_D) for the blocking action of BZ (from 0.48–5.6 mM; Zubov *et al.*, 1982b), suggesting the total modification of local-anesthetics receptor(s) under the action of BTX. Of special interest is the fact that aconitine, unlike BTX, even increases the sensitivity of Na channels to BZ; K_D for BZ block decreases from 0.68 to 0.35 mM (Zubov *et al.*, 1982b). Such a difference in the effects of BTX and aconitine can be explained if one assumes that their complex (multisite) receptors overlap only partially.

Recently, a new neutral effective blocker of Na and K channels has been found, the plant alkaloid oenanthotoxin (OETX; Dubois and Schneider, 1981a,b). Since this toxin also has no terminal amine group one could expect that, like BZ, it will be capable of effectively blocking BTX-modified Na channels.

However, the experiments (Dubois and Khodorov, 1982) showed that at concentrations of 2 × 10^{-5} M, OETX effectively inhibits Na^+ currents through normal Na channels, but produces only relatively small suppression of the modified I_{Na} (Fig. 13).

OETX-induced blockage of normal I_{Na} in nerve is associated with parallel inhibition of intramembrane charge movement (immobilization of gating charges; Dubois and Schneider, 1981a,b). BTX protects the gating currents from this blocking action of OETX. Superfusion of the BTX-treated node with 4 × 10^{-5} M OETX for 10–20 min did not significantly suppress the charge movement (see Fig. 4; Dubois and Khodorov, 1982). This observation suggests that the gating subunits of the Na channel are a common target for BTX and OETX.

The antagonism between BTX and OETX proved to be bilateral. Application of BTX to the node previously treated with 4 × 10^{-5} M OETX did not relieve the OETX-induced block of Na channels.

It has been mentioned above that in the normal Ranvier node OETX blocks effectively not only Na, but also K channels, the latter block at a given OETX concentration being even stronger (Dubois and Schneider, 1981b). The same effect was observed in experiments on BTX-treated node. BTX itself did not change the potassium currents I_K and was unable to protect it from the blocking action of OETX (Dubois and Khodorov, 1982). Thus, the antagonism between BTX and OETX in their action on Na channels seems to be a specific one.

The molecular mechanism of OETX blocking action on Na and K channels is not clear

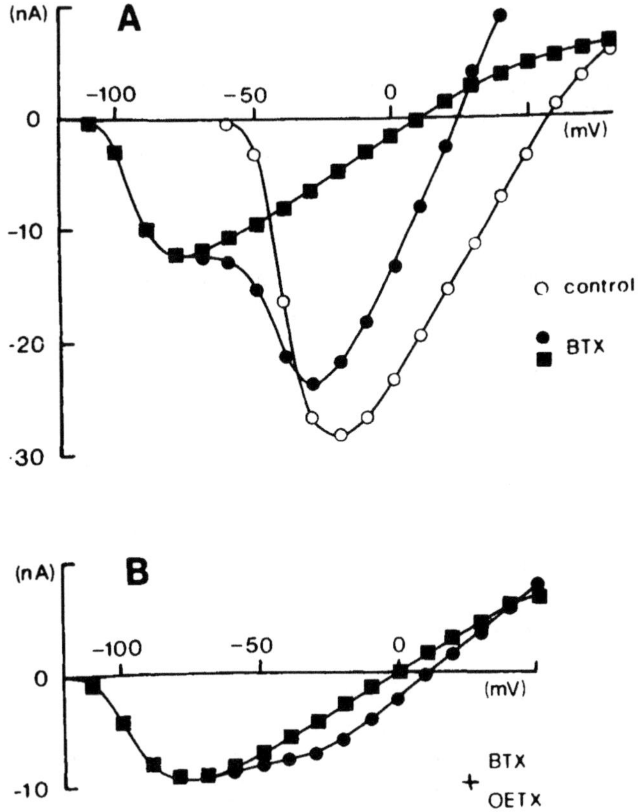

Figure 13. Sodium current-voltage curves in control conditions (A) open circles, 5 min after application of 10^{-5} M BTX combined with repetitive depolarizations (A) filled circles and squares, and 8 min after application of 4.10^{-5} M OETX to the BTX-treated node (B). In BTX, circles denote the peak current and squares denote the quasi-steady-state current measured after 80 msec of depolarization. Temperature 14°C (Dubois and Khodorov, 1982).

as yet. The molecule of OETX bears two alcohol functions and five unsaturated aliphatic (ethylenic and acetylenic) bonds. Since alcohols are known to block Na current and gating charge movement in axons (Swenson and Oxford, 1980) we have examined the effect of butanol on BTX-treated node of Ranvier (Dubois and Khodorov, 1982). Experiments have shown that butanol, (0.1–0.5 M) unlike OETX, does not distinguish normal and modified Na channels. Both currents became reversibly blocked to almost the same degree. This indicates that the blocking effect of OETX cannot be directly related to its alcohol functions. More likely, the unsaturated functions are responsible for OETX-effects under consideration. One can suppose that OETX penetrating in the lipid matrix of the membrane binds to the hydrophobic parts of the channel-gating dipoles due to its ability to form π bonds. This results in channel immobilization in its resting configuration. Unlike OETX, BTX binds to a specific receptor in the open Na channel and hinders the channel transition to the resting state. Such a preference of BTX and OETX to opposite channel conformations may underlie its antagonism on the molecular level.

The similarity of BZ and butanol action on BTX-modified channels is of great interest and deserves further special investigations since it has been shown that butanol, unlike BZ, does not compete with local anesthetics for the Na channel (Venitz and Schwarz, 1980).

XII. CONCLUDING REMARKS

Over the last few years, a certain amount of progress has been made in the study of the mechanism of BTX action on excitable membranes.

Voltage-clamp experiments on nerve fibers and neuroblastoma cells, as well as experiments performed by the use of ion flux method showed that BTX induces simultaneous modification of activation, inactivation, selectivity, and sensitivity of Na channels to many drugs and some toxins. All these effects result from the interaction of BTX with a single receptor ("recognizing site"), which, judging by the effects observed, must be structurally linked with all the major entities of the Na channel.

The alkaloids aconitine, veratridine, and grayanotoxin compete with BTX for a common receptor, which in turn is coupled allosterically to the receptor for scorpion and anemone toxins.

The only channel structure that did not change noticeably in BTX-modified Na channel proved to be the tetrodotoxin receptor. In voltage-clamped node of Ranvier K_D for TTX-block appeared to be ($\pm$SD) 4.3 $\pm$ 1.3 nM (n = 7) and 3.6 $\pm$ 1.3 nM (n = 9) for normal and BTX-modified channels, respectively (Mozhayeva *et al.*, 1982b).

The use-dependent modification of sodium currents by BTX suggests its interaction with open sodium channels.

What might be the reason for such a requirement of BTX for the open state of the Na channel?

Many other amine compounds are known to interact only or mainly with open channels. Most of them, however, are the channel blockers—local anesthetics, antiarryhthmics and related drugs. In the charged form, they enter the open axoplasmic channel mouth, bind to corresponding receptor(s), and inhibit both ionic and gating currents.

The big BTX molecule is unlikely to enter the open channel mouth. Even if this were possible, it would block the currents.

More probable therefore is the location of BTX-receptor in the hydrophilic part of the channel. Then two possible explanations of the gate-dependency of BTX-receptor interaction are worth considering.

One possibility is that in the resting Na channel, the BTX molecule cannot reach the receptor because of some steric or other hindrance. Conformational change of the channel underlying its activation makes the receptor accessible. In the frame of such a hypothesis, the BTX receptor is considered as a preformed voltage-independent structure which does not undergo conformational change during channel gating.

The other possibility, which we believe is more realistic, is that the BTX receptor acquires its active conformation or even is formed ("assembled") in the process of channel transition to the open state.

It is tempting, for example, to propose that multiple discrete binding sites of the complex receptor, interacting with different functions of the BTX molecule, are the structural constituents of the channel gating subunits. In the resting state, the arrangement of these subunits and the corresponding binding sites does not favor effective BTX binding. The latter becomes possible only when the channel activities and its subunits acquire an appropriate "open" arrangement. If it is correct, then the tight binding of BTX to an "assembled" receptor should stabilize this open channel conformation. This is precisely what is really observed in the experiments. The probability of channel closing upon membrane repolarization (deactivation) as well as during a maintained depolarization (inactivation) is greatly decreased, and the OFF-charge movement reflecting the return of the gating subunits, their closed arrangement during membrane hyperpolarization, is slowed down considerably (see Figs. 2, 3, and 4).

The hypothesis of "receptor assembling" during channel gating is relative to Hille's "modulated receptor hypothesis" (Hille, 1977) and may be considered as its particular case. Hille suggests that a "universal receptor" of local anesthetics is modulated by Na inactivation. The role of activation is reduced only to operation of the gate that opens the "hydrophilic pathway" to and from the receptor for charged drugs that have penetrated inside the cell.

By contrast, in the present paper it is proposed that at least in case of BTX action the conformational change of the Na channel creates (assembles) the receptor, which does not preexist as such in the resting channel.

It is very likely that similar mechanisms may underlie the modification of Na activation produced by some other drugs, as for example, DDT (Hille, 1968b) or *Centruroides sculpturatus* toxins (Cahalan, 1975; Meves, 1982) interacting with open Na channels.

The fact that under the action of *Centruroides* toxin the negative shift of the P_{Na}–E relation is not stable (as in case of BTX treatment) but dissipates gradually after the end of a depolarizing pulse may be explained by a loose binding of this toxin to the activating subunits.

There is little doubt that a further comparative study of the effects of BTX and other neurotoxins on the properties of Na channels will provide considerable insight into the molecular organization of the Na channel.

Note Added In Proof

A further publication about the effect of BTX on gating currents in myelinated nerves is currently in press. Dubois, Schneider and Khodorov (*J. Gen. Physiol.*, in press) have found that BTX when applied to the node of Ranvier, caused an approximately three-fold increase in steepness of the Q (intramembrane gating charge movement) vs. voltage relationship and a 50 mV negative shift in its midpoint. Q_{max} was virtually identical before and after BTX treatment. BTX treatment eliminated the charge immobilization even after a 20-ms depolarizing pulse. After BTX treatment the voltage dependence of charge movement was the same as the steady state voltage dependence of sodium conductance activation. The observations are consistent with the hypothesis (see Khodorov and Revenko, 1979) that BTX induces an aggregation of the charged gating particles associated with each Na channel and causes them to move as a unit having approximately three times the average valence of the individual particle. Movement of this aggregated unit would open the BTX-modified Na channel.

Quandt and Narahashi (1982, *Proc. Natl. Acad. Sci. USA,* **79:**6732–6736) have studied the properties of BTX-modified Na channels of neuroblastoma cells (NIE-115) by using the patch clamp method for measuring single channel currents from excised membranes. It has been established that BTX decreases the single Na channel conductance and diminishes drastically the probability of channels closing both during a maintained membrane depolarization and after its termination. BTX-modified Na channels open spontaneously at much more negative potentials than the normal channels.

REFERENCES

Adelman, W. J., and Senft, J. P., 1966a, Effects of internal sodium on ionic conductance of internally perfused axons, *Nature (London)* **212:**614–616.

Adelman, W. J., and Senft, J. P., 1966b, Voltage-clamp studies on the effects of internal cesium ion on sodium and potassium currents in the squid giant axon, *J. Gen. Physiol.* **50:**279–293.

Albuquerque, E. X., 1972, The mode of action of batrachotoxin, *Fed. Proc.* **31:**1133–1138.

Albuquerque, E. X., and Daly, J. W., 1976, Batrachotoxin, a selective probe for channels modulating conductances in elentrogenic membranes, in: *The Specificity and Action of Animal, Bacterial, and Plant Toxins. Receptors and Recognition, Series B,* Vol. 1 (P. Cuatrecasas, ed.), pp. 297–338. Chapman and Hall, London.

Armstrong, C. M., and Bezanilla, F., 1977, Inactivation of the sodium channel. II. Gating current experiments, *J. Ger. Physiol.* **50:**279–293.

Bartels-Bernal, E., Rosenberry, T. L., and Daly, J. M., 1977, Effect of batrachotoxin on the electroplax of electric eel: Evidence for voltage-dependent interaction with sodium channels, *Proc. Natl. Acad. Sci. USA* **74:**951–955.

Bezanilla, F., and Armstrong, C. M., 1977, Inactivation of the sodium channel. I. Sodium currents experiments, *J. Gen. Physiol.* **70:**549–566.

Bolotina, V. M., Revenko, S. V., and Khodorov, B. I., 1981, Use-dependent blockage of sodium channels in the node of Ranvier by ethmozine, *Neurophysiologia, Kiev, USSR* **13:**380–390 (in Russian).

Cahalan, M. D., 1975, Modification of sodium channel gating in frog myelinated nerve fibres by *Centruroides sculpturatus* scorpion venom, *J. Physiol (London)* **244:**511–534.

Cahalan, M. D., 1978, Local anesthetic block of sodium channels in normal and pronase-treated squid giant axons, *Biophys. J.* **23:**285–311.

Catterall, W. A., 1975, Cooperative activation of action potential Na^+ ionophore by neurotoxins, *Proc. Natl. Acad. Sci. USA* **72:**1782–1786.

Catterall, W. A., 1980, Neurotoxins that act on voltage-sensitive sodium channels in excitable membranes, *Annu. Rev. Pharmacol. Toxicol.* **20:**15–43.

Chandler, W. K., and Meves, H., 1970, Evidence for two types of sodium conductance in axon perfused with sodium fluoride solution, *J. Physiol.* **211:**653–678.

Conti, F., Hille, B., Neumcke, B., Nonner, W., and Stämpfli, R., 1976, Measurement of the conductance of the sodium channel from current fluctuations at the node of Ranvier, *J. Physiol. (London)* **262:**699–727.

Conti, F., Neumcke, B., Nonner, W., and Stämpfli, R., 1980, Conductance fluctuations from the inactivation process of sodium channels in myelinated nerve fibres, *J. Physiol.* **308:**217–239.

Courtney, R. K., 1975, Mechanism of frequency-dependent inhibition of sodium currents in frog myelinated nerve by the lidocaine derivative GEA 968, *J. Pharmacol. Exp. Ther.* **195B:**225–236.

Courtney, K., 1980, Structure-activity relationship for frequency-dependent sodium channel block in nerve by local anesthetics, *J. Pharm. Exp. Ther.* **213:**114–119.

Dodge, F. A., 1963, A study of ionic permeability changes underlying excitation in myelinated nerve fibres of the frog, Thesis of Rockefeller Inst., New York.

Dubois, J. M., and Khodorov, B. I., 1982, Batrachotoxin protects sodium channels from the blocking action of oenanthotoxin, *Pflügers Arch.* **395:**55–58.

Dubois, J. M., and Schneider, M. F., 1981a, Block of sodium current and intramembrane charge movement in myelinated nerve fibres poisoned with toxin from *Oenathe crocate*, *Nature (London)* **289**:685–689.

Dubois, J. M., and Schneider, M. F., 1981b, Block of ionic and gating currents in node of Ranvier with oenanthotoxin, *Adv. Physiol. Sci.* **4**:79–87.

Eisenman, G., 1962, Cation selective glass electrodes and their mode of operation, *Biophys. J.* **2**(2, Part 2):259–323.

Fishman, S. N., Khodorov, B. I., and Volkenstein, M. V., 1971, Molecular mechanisms of membrane ionic permeability changes, *Biochim. Biophys. Acta* **225**:1–10.

Frankenhaeuser, B., and Huxley, A., 1964, The action potential in the myelinated nerve fibre of *Xenopus laevis* as computed on the basis of voltage clamp data. *J. Physiol. (Lond.)* **171**:302–315.

Gillespie, J. I., and Meves, H., 1980, The effect of scorpion venoms on the sodium currents of the squid giant axon, *J. Physiol.* **308**:479–499.

Grischenko, I. I., Zubov, A. N., and Naumov, A. P., 1981, Activation, inactivation and selectivity of aconitine-modified sodium channels in neuroblastoma cells, in: *USSR-Sweden III Symposium Physico-Chemical Biology, Abstracts*, pp. 206–207, Tbilisi.

Guselnikova, G. G., Zoborovskaya, L. D., and Khodorov, B. I., 1981, Study of the properties of local anesthetics binding sites in sodium channel of nerve fibres, in: *USSR-Sweden III Symposium Physico-Chemical Biology, Abstracts*, pp. 210–211, Tbilisi.

Hille, B., 1967, A pharmacological analysis of the ionic channels of nerve, Thesis of Rockefeller University, New York.

Hille, B., 1968a, Pharmacological modifications of the sodium channels of frog nerve, *J. Gen. Physiol.* **51**:196–219.

Hille, B., 1968b, Charges and potentials at the nerve surface, divalent ions and pH, *J. Gen. Physiol.* **51**:221–236.

Hille, B., 1971, The permeability of the sodium channel to organic cations in myelinated nerve, *J. Gen. Physiol.* **58**:599–619.

Hille, B., 1972, The permeability of the sodium channel to metal cations in myelinated nerve, *J. Gen. Physiol.* **59**:637–658.

Hille, B., 1977, Local anesthetics: Hydrophilic and hydrophobic pathways for the drug-receptor reaction, *J. Gen. Physiol.* **69**:497–515.

Hodgkin, A. L., and Huxley, A. F., 1952, A quantitative description of membrane current and its application to conduction and excitation in nerve, *J. Physiol. (London)* **117**:500–544.

Huang, L. Y. M., and Ehrenstein, G., 1981, Local anesthetics QX 572 and benzocaine act at separate sites on the batrachotoxin-activated sodium channel, *J. Gen. Physiol.* **77**:137–153.

Huang, L.-Y., Ehrenstein, G., and Catterall, W. A., 1978, Interaction between batrachotoxin and yohimbine, *Biophys. J.* **23**:219–231.

Huang, L. Y. M., Catterall, W., and Ehrenstein, G., 1979, Comparison of ionic selectivity of batrachotoxin-activated channels with different tetrodotoxin dissociation constants, *J. Gen. Physiol.* **73**:839–854.

Huang, L. Y. M., Moran, N., and Ehrenstein, G., 1981, The action of batrachotoxin on sodium current in internally perfused neuroblastoma cells, *Biophys. J.* **33**:120a.

Keynes, R. D., and Rojas, E., 1974, Kinetics and steady-state properties of the charged system controlling sodium conductance in the squid giant axon, *J. Physiol. (London)* **239**:393–434.

Khodorov, B. I., 1978, Chemicals as tools to study nerve fiber sodium channels. Effects of batrachotoxin and some local anesthetics, in: *Membrane Transport Process*, Vol. 2 (D. Tosteson, Yu. Ovchinnikov, and R. Latorre, eds.), pp. 153–174, Raven Press, New York.

Khodorov, B. I., 1979a, Inactivation of the sodium gating current, *Neuroscience* **4**:865–876.

Khodorov, B. I., 1979b, Some aspects of the pharmacology of sodium channels in nerve membrane, Process of inactivation, *Biochem. Pharmacol.* **28**:1451–1459.

Khodorov, B. I., 1981, Sodium inactivation and drug-induced immobilization of the gating charge in nerve membrane, *Prog. Biophys. Mol. Biol.* **37**:49–89.

Khodorov, B. I., and Revenko, S. V., 1979, Further analysis of the mechanisms of action of batrachotoxin on the membrane of myelinated nerve, *Neuroscience* **4**:1315–1340.

Khodorov, B. I., Peganov, E. M., Revenko, S. V., and Shishkova, L. D., 1975, Sodium currents in voltage clamped nerve fiber of frog under the combined action of batrachotoxin and procaine, *Brain Res.* **84**:541–546.

Khodorov, B. I., Shishkova, L. D., Peganov, E. M., and Revenko, S. V., 1976, Inhibition of sodium currents in frog Ranvier node treated with local anesthetics. Role of slow sodium inactivation, *Biochim. Biophys. Acta* **443**:409–435.

Khodorov, B. I., Peganov, E. M., Revenko, S. V., and Shishkova, L. D., 1977, Gating and selectivity of sodium channels in Ranvier node treated with batrachotoxin. Effects of local anesthetics, in: *Proceeding of the International Union of Physiological Science*, Vol. 12, p. 195, Paris.

Khodorov, B. I., Neumcke, B., Schwarz, W., and Stämpfli, R., 1981, Fluctuation analysis of the Na channels modified by batrachotoxin in myelinated nerve, *Biochim. Biophys. Acta* **648(1)**:93–99.

Koppenhöfer, E., and Schmidt, H., 1968, Die Wirkung von Skorpiongift auf die Ioninströme des Ranvierschen Schürrings, *Pflügers Arch.* **303:**133–149.
Meves, H., 1983, Effects of scorpion toxins on sodium channels, in: *Structure and Function in Excitable Cells* (D. Chang, I. Tasaki, W. Adelman, Jr., and H. Leuchtag, eds.), Plenum, New York.
Moore, J. W., and Cox, E. B., 1976, A kinetic model for the sodium conductance system in squid axon, *Biophys. J.* **16:**171–192.
Mozhayeva, G. N., Naumov, A. P., and Negulyaev, Yu. A., 1976, Effect of aconitine on some properties of sodium channels in the Ranvier node membrane, *Neurophysiologia* **8:**152–160 (in Russian).
Mozhayeva, G. N., Naumov, A. P., Negulayev, Yu. A., and Nosyreva, E., 1977, The permeability of aconitine-modified sodium channels to univalent cations in myelinated nerve, *Biochim. Biophys. Acta* **466:**461–463.
Mozhayeva, G. N., Naumov, A. P., Nosyreva, E., and Grishin, N. M., 1980, Potential-dependent interaction of toxin from venom of scorpion *Buthus supeus* with sodium channels in myelinated fibre, *Biochim. Biophys. Acta* **597:**687–702.
Mozhayeva, G. N., Naumov, A. P., and Khodorov, B. I., 1981a, Changes in properties of sodium channels of the nodal membrane treated with batrachotoxin (BTX), in: *USSR-Sweden III Symposium Physico-Chemical Biology, Abstracts,* pp. 221–222, Tbilisi.
Mozhayeva, G. N., Naumov, A. P., and Negulayev, Yu. A., 1981b, Block of normal and aconitine-modified sodium channels with hydrogen ions, in: *USSR-Sweden III Symposium Physico-Chemical Biology, Abstracts,* pp. 208–209, Tbilisi.
Mozhayeva, G. N., Naumov, A. P., and Khodorov, B. I., 1982a, Potential-dependent blockage of batrachotoxin-modified sodium channels in frog node of Ranvier by calcium ions, *Gen. Physiol. Biophys.* **1:**281–282.
Mozhayeva, G. N., Naumov, A. P., and Khodorov, B. I., 1982b, Tetrodotoxin changes the activation kinetics of batrachotoxin-modified sodium channels, *Gen. Physiol. Biophys.* **1:**221–223.
Mozhayeva, G. N., Naumov, A. P., and Khodorov, B. I., 1982c, Ion selectivity and properties of the acid group in Na channels modified by batrachotoxin (BTX) in nerve membrane, *Gen. Physiol. Biophys.* **1:**453–455.
Mozhayeva, G. N., Naumov, A. P., and Khodorov, B. I., Proton permeability of sodium channels modified by batrachotoxin, *Gen. Physiol. Biophys.* **1:**463–464.
Mrose, H. E., and Ritchie, M., 1978, Local anesthetics: Do benzocaine and lidocaine act as the same single site? *J. Gen. Physiol.* **71:**223–225.
Myers, Ch., Daly, J., and Malkin, B., 1978, A dangerously toxic new frog *(Phyllobates)* used by Embers Indians of Western Colombia, *Bull. American Museum of Natural History,* **161:**311–365.
Narahashi, T., Albuquerque, E. X., and Deguchi, T., 1971, Effects of batrachotoxin on membrane potential and conductance of solid giant axon, *J. Gen. Physiol.* **58:**54–70.
Narahashi, T., Frazier, D., and Moor, J., 1972, Comparison of tertiary amine local anesthetics in their ability to depress membrane ionic conductances, *J. Neurobiol.* **3:**267–276.
Naumov, A. P., Negulyaev, Yu. A., and Nosyreva, E. D., 1979, Change in the affinity of the sodium channel acid group to hydrogenium ions under the action of aconitine, *Dokl. Akad. Nauk SSSR* **1:**229–232.
Negulayev, Yu., and Nosyreva, E. D., 1979, A comparative study of procaine and benzocaine effect on normal and aconitine modified sodium channels, *Zytologia* **21:**697–701 (in Russian).
Neher, E., and Stevens, C. F., 1979, Voltage-driven conformational changes in intrinsic membrane proteins, in: *The Neurosciences. Fourth Study Program* (F. O. Schmidt and F. G. Worden, eds.), pp. 623–629, MIT Press, Cambridge, Massachusetts, and London, England.
Neumcke, B., Schwarz, W., and Stämpfli, R., 1980, Modification of sodium inactivation in myelinated nerve by anemonia toxin, *Biochim. Biophys. Acta* **600:**456–466.
Peganov, E. M., Revenko, S. V., Khodorov, B. I., and Shishkova, L. D., 1976, Batrachotoxin and aconitine, the modificators of fast sodium channels in nerve fibre membrane, in: *Molecular Biology,* Vol. 15, pp. 42–56, Naukova Dumka, Kiev, USSR (in Russian).
Revenko, S. V., 1977, Effect of electric stimulation on the rate of sodium channel modification in voltage clamped Ranvier node, *Neurophysiologia, Kiev, USSR* **9:**544–547 (in Russian).
Revenko, S. V., and Khodorov, B. I., 1977, Effect of batrachotoxin on the selectivity of sodium channels in myelinated nerve fiber, *Neurophysiologia, Kiev, USSR* **9:**313–316 (in Russian).
Revenko, S. V., Khodorov, B. I., and Shapovalova, L. M., 1979, Blockage of Ranvier node sodium channels by alkaloid yohimbine, *Dokl. Akad. Nauk SSSR* **248:**494–498 (in Russian).
Revenko, S. V., Bolotina, V. M., and Khodorov, B. I., 1980, Tonic and frequency-dependent block of Na channels by ethmozine, *J. Mol. Cell. Cardiol.* (Suppl. 1) **12:**135.
Revenko, S. V., Khodorov, B. I., and Shapovalova, L. M., 1982a, Quinidine blockage of sodium and potassium channels in myelinated nerve fiber, *Neurophysiologia* **14:**324–330 (in Russian).
Revenko, S. V., Khodorov, B. I., and Shapovalova, L. M., 1982b, The effects of yohimbine on sodium and gating currents in frog Ranvier node membrane, *Neuroscience* **7:**1377–1387.

Rojas, E., and Keynes, R., 1975, On the relation between displacement currents and activation of the sodium conductance in the squid giant axon, *Phil. Trans. R. Soc. Lond. B* **270**:459–482.

Schmidmayer, J., and Ulbricht, W., 1980, Interaction of lidocaine and benzocaine in blocking sodium channels, *Pflügers Arch.* **387**:47–54.

Schmidt, H., and Schmidt, O., 1974, Effect of aconitine on the sodium permeability of the node of Ranvier, *Pflügers Arch.* **349**:133–148.

Shoukimas, J. J., and French, R., 1980, Incomplete inactivation of sodium currents in nonperfused squid axon, *Biophys. J.* **32**:857–862.

Strichartz, G. R., 1973, The inhibition of sodium currents in myelinated nerve by quaternary derivatives of lidocaine, *J. Gen. Physiol.* **62**:37–57.

Swenson, R. P., and Oxford, G. S., 1980, Modification of sodium channel gating by long chain alcohols: Ionic and gating current measurements, in: *Mol. Mechanisms of Anesthesia (Prog. in Anesthesiology* Vol. 20 (R. Fink, ed.), pp. 7–16, Raven Press, New York.

Venitz, E., and Schwarz, W., 1980, Block of Na currents in frog muscle fibres by local anesthetics and n-alcanols, *Pflügers Arch.* (Suppl.) **334**:R-18.

Warnick, J. E., Albuquerque, E. X., Onur, R., Jansson, S.-E., Daly, I., Tokuyama, T., and Witkop, B., 1975, The pharmacology of batrachotoxin. VII. Structure-activity relationships and the effects of pH, *J. Pharmacol. Exp. Ther.* **193**:232–245.

Woodhull, A. M., 1973, Ionic blockage of sodium channels in nerve, *J. Gen. Physiol.* **61**:687–708.

Yeh, J., and Narahashi, T., 1977, Kinetic analysis of pancuronium interaction with sodium channels in squid axons membrane, *J. Gen. Physiol.* **69**:293–323.

Zaborovskaja, L. M., 1979, The modes of action of local anesthetics on excitable nerve membrane, Thesis, Moscow, USSR (in Russian).

Zaborovskaja, L. D., and Khodorov, B. I., 1980a, Stimulus-dependent blockage of sodium channels in the node of Ranvier by the quaternary antiarrhythmic agent N-propylaimaline (Neo-gilurytmal), *Bull. Exp. Biol. Med. USSR* **5**:578–580 (in Russian).

Zaborovskaja, L. D., and Khodorov, B. I., 1980b, Stimulus-dependent inhibition of sodium currents by an antiarrhythmic Neo-gilurytmal, *J. Mol. Cell. Cardiol.* (Suppl.) **12(1)**:185.

Zaborovskaja, L. D., and Khodorov, B. I., 1982a, Differences in the blocking action of benzocaine and amine drugs on normal and modified sodium channels in the node of Ranvier treated with batrachotoxin, *Neurophysiologia,* **14**:636–643. (in Russian).

Zaborovskaja, L. D., and Khodorov, B. I., 1982b, Reversible blockage of batrachotoxin-modified sodium channels by amine compounds and benzocaine in frog node of Ranvier, *Gen. Physiol. Biophys.* **1**:283–285.

Zubov, A. N., Naumov, A. P., and Khodorov, B. I., 1981, Effects of batrachotoxin on gating and ion selectivity of sodium channels in neuroblastoma cells, in: *USSR-Sweden III Symposium Physico-Chemical Biology,* pp. 219–220, Tbilisi.

Zubov, A. N., Naumov, A. P., and Khodorov, B. I., 1983a, Effect of batrachotoxin (BTX) on activation, inactivation and ion selectivity of Na channels in clonal neuroblastoma cells, *Gen. Physiol. Biophys.* **2**:75–77.

Zubov, A. N., Naumov, A. P., and Khodorov, B. I., 1983b, Modification of the binding sites for local anesthetics in sodium channels by batrachotoxin (BTX) and aconitine (AC), *Gen. Physiol. Biophys.* **2**:125–127.

Zubov, A. N., Naumov, A. P., and Khodorov, B. I., 1983c, Effect of batrachotoxin on Na channels in neuroblastoma cell membrane. *Zytologia* **2,** in press (in Russian).

III

Electrochemistry and Electrophysics

15

Axolemma–Ectoplasm Complex and Mechanical Responses of the Axon Membrane

I. Tasaki and K. Iwasa

I. INTRODUCTION

The layer of axoplasm located directly underneath the axolemma is rich in fibrillar elements, such as microtubules, neurofilaments, and actin filaments (Metuzals and Izzard, 1969; Metuzals and Tasaki, 1978; Hodge and Adelman, 1980). An extensive destruction of this layer by intracellular perfusion with a solution containing chaotropic anions or proteolytic enzymes is known to lead to a loss of excitability. Therefore, we believe that this layer, which is called the *ectoplasm*, is an essential part of the excitable membrane.

The macromolecular organization of the ectoplasm is strongly affected by various types of physiological manipulations of the axon. Depolarization of the membrane by external application of a potassium-rich medium, for example, enhances the solubility of the subaxolemmal protein molecules in the internal perfusion fluid (Gainer *et al.*, 1974). Repetitive stimulation of the axon also increases the rate of release of various proteins into the perfusion solution (Pant *et al.*, 1978). Conversely, chemical modification of the ectoplasm by means of various reagents known to modify certain amino acids, or, simply by an alkaline shift of the pH of the internal perfusion fluid, leads to repetitive firing of action potential (Tasaki, 1982). The excitable sites in the axolemma are connected, mechanically and biochemically, to the ectoplasmic layer (Metuzals and Tasaki, 1978).

In this article, we present a variety of experimental facts indicating that the ectoplasm is directly involved in the production of mechanical responses of the axon. By mechanical responses, we mean rapid mechanical changes in the axon associated with the action potential (Tasaki and Iwasa, 1980; Iwasa and Tasaki, 1980; Tasaki *et al.*, 1981). We postulate that the intrinsic protein molecules in the axolemma are capable of undergoing a conformational transition from a compact, Ca-ion rich state to a swollen, univalent-cation rich state in response to a stimulating current pulse. This transition is expected to enhance the rate of invasion of extracellular cations into the ectoplasm. We propose that the pronounced me-

I. Tasaki and K. Iwasa • Laboratory of Neurobiology, NIMH, Bethesda, Maryland 20205, and Marine Biological Laboratory, Woods Hole, Massachusetts 02543.

chanical movements of the axon surface during the later phase of the action potential are attributable to this invasion of Na- and Ca-ions into the ectoplasm.

II. A NEW METHOD FOR DETECTING MECHANICAL RESPONSES OF THE SQUID GIANT AXON

Quite recently, one of us (I. T.) received a gift of a novel type of piezoelectric material from Kureha Chemical Industry Co., in Tokyo. This material is made of a synthetic polymer, polyvinylidene fluoride (abbreviated to PVDF). When a piece of PVDF film is stretched at a high temperature in the presence of a strong electric field, it becomes highly piezoelectric (Kawai, 1969; Murayama *et al.*, 1976; Robinson, 1978). We examined mechanical responses of crab nerve fibers and of squid giant axon by using a pressure sensor constructed with Kureha PVDF film (Tasaki and Iwasa, in preparation) and found that the new sensor gives a considerable improvement of the signal-to-noise ratio over the pressure sensors employed in our previous studies. In fact, with this new pressure sensor, we could record mechanical responses of the crab nerve trunk without the help of a signal-averager, namely, by connecting the output of the sensor to an oscilloscope via a voltage follower (see Fig. 1).

In most of the experiments described below, we employed squid giant axons for studying mechanical changes in the axon associated with the production of an action potential. Ordinarily, the signals had to be averaged over 500–2000 trials in order to lower the noise level sufficiently below the signal.

III. CHARACTERISTICS OF THE MECHANICAL RESPONSES OF THE SQUID AXON

The existence of a rapid swelling of the axon membrane associated with nerve excitation was predicted by Teorell (1962) on the basis of analyses of his hydraulic model. Later on, Williams (1970) speculated that the influx of Ca-ions associated with a propagated action potential might cause a contraction (or shrinkage) of the axon membrane and that the

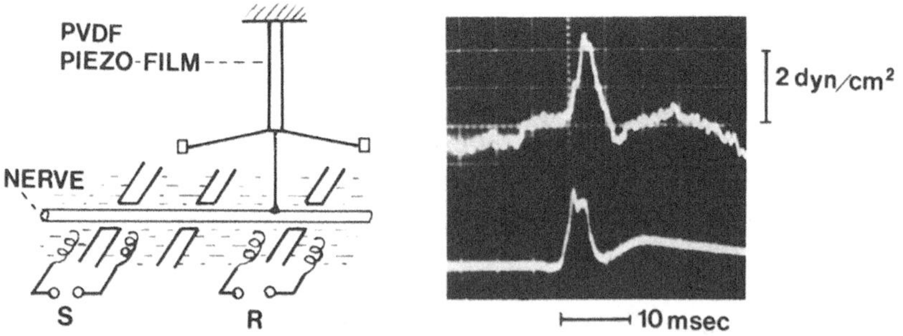

Figure 1. (Left) Schematic diagram illustrating the setup used for detecting mechanical changes in a crab nerve associated with a propagated action potential. A narrow sheet of PVDF piezofilm was kept under tension by means of a nylon fiber attached horizontally at the end. A stylus made with a few bristles (fixed vertically at the end of the film) was used to transmit changes in pressure developed by the nerve to the PVDF film. The surface electrodes on the film were connected to oscilloscope via a voltage-follower. (Right) Record of a transient rise in pressure (upper trace) produced by an electric shock (delivered to the nerve through electrodes S). The lower trace shows the action potential of the nerve (recorded simultaneously with electrodes R). Temperature 20°C.

subsequent gradual disappearance of the Ca-ions from the inner side of the membrane might be related to the recovery of excitability during the refractory period.

When we succeeded in experimentally demonstrating the existence of swelling of the squid axon during the action potential, we immediately noted that the swelling of the axon is followed by a pronounced shrinkage of the axon. The temporal relationship between the mechanical and electrical responses was examined by recording the mechanical response with a pressure sensor placed directly above the recording tip of the internal wire used for recording propagated action potentials (see Fig. 2).

We see in the figure that the pressure exerted by the axon upon the stylus of the pressure sensor starts to rise as soon as the membrane potential begins to rise. The peak of the pressure rise coincides, within about 0.05 msec, with the peak of the action potential. The peak value of the pressure rise is usually in the range between 1–3 μg. Since the surface of the stylus making contact with the axon surface was roughly 0.3 mm^2, the pressure reached at the peak is about 1 dyn/cm^2 (0.1 Pa) or slightly less.

In the figure, we also see that there is a large downward deflection of the upper trace, representing a shrinkage of the axon, at the time when the axon is in the state of after-hyperpolarization. In most axons examined, the amplitude of the shrinkage was 1.5–4 times as large as the amplitude of the swelling.

By using optical devices, it has been shown that there is an outward displacement of the axon surface during the depolarizing phase of the action potential. The amplitude of the displacement is 1–3 nm (10–30 Å), which is less than the thickness of axolemma (10 nm). The hyperpolarizing phase of the action potential is accompanied by a large inward displacement of the axon surface. We thus find that the mechanical responses obtained by the use of the optical method are nothing but isotonic counterparts of the responses recorded isometrically with a PVDF film.

Effects of various biochemical and pharmacological agents upon the mechanical responses of the axon were examined. External application of trypsin (1 mg/ml for 60 min) is known to break the proteinaceous connection between the Schwann cells and the axolemma. This treatment of the axon had no effect on the mechanical response of the axon. Internally applied proteinases rapidly digest and dissolve the endoplasm. Nearly complete removal of the endoplasm by internal perfusion with a solution containing pronase (0.1 mg/ml for 1 min) did not affect the amplitude of the mechanical response. External application of ouabain (1 mM) is expected to alter intracellular electrolyte composition of the Schwann cells very

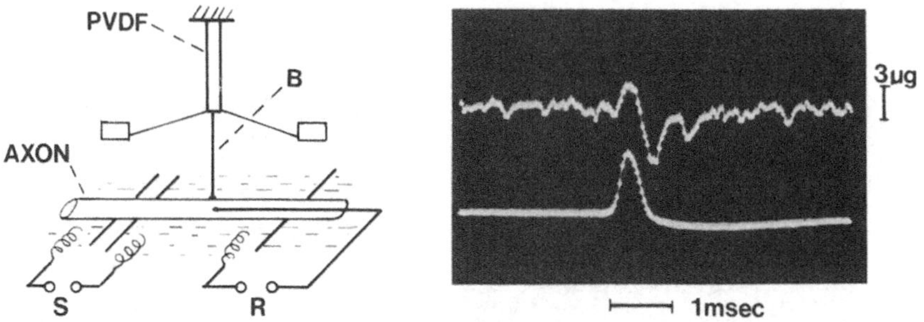

Figure 2. (Left) Schematic diagram of the setup used for recording mechanical responses of a squid giant axon. B represents a stylus for transmitting pressure changes to the PVDF piezofilm. (Right) The top trace shows an initial rise and a subsequent fall in the pressure of the axon associated with a propagated action potential. The bottom trace indicates the action potential of the axon (about 110 mV above the resting potential level at the peak). Temperature 20°C.

quickly and that of the giant axon very slowly (see White *et al.*, 1974). No clear effect was observed up to 90 min after application of this drug.

Electrolytes in the external medium exert strong influence on the mechanical response of the axon. An increase in the external Ca-ion concentration to 100–200 mM immediately lowers the amplitude of the first, i.e., swelling, phase and enhances the amplitude of the second (shrinkage) phase. A decrease in the concentration of $CaCl_2$ to a level of 2–3 mM by addition of an isotonic NaCl solution produces a slight enhancement of the swelling phase followed by initiation of repetitive firing of full-sized responses. Complete substitution of Li-ions for the Na-ions in the external seawater brings about a distinct increase of both the swelling and shrinkage phases of the mechanical response.

IV. MECHANICAL RESPONSES OF TEA-TREATED AXONS

Intracellular injection of tetraethylammonium (TEA) salt into a squid giant axon is known to bring about an enormous prolongation of the action potential duration (Tasaki and Hagiwara, 1957). The properties of the mechanical responses of such TEA-treated axons deserve special attention, because they clearly indicate that the mechanical responses of the axon is not a mere reflection of changes in the membrane potential (Iwasa and Tasaki, 1981).

Figure 3 shows two examples of the records of the mechanical responses taken from axons treated mildly with TEA. The lower trace in the figure represents the action potentials recorded *externally*. In these records, we see that the action potential duration was prolonged by TEA injection from the normal value of about 0.6 msec to 3–6 msec. (Note that the depolarization phase of the internally recorded action potential of a TEA-treated axon has a long plateau.) The upper trace in the figure indicates that the swelling phase of the response is relatively short in duration and small in amplitude. In heavily TEA-treated axons, the amplitude of the swelling is so small that it is frequently reduced to the level of the random noise of the recording system.

A significant fact revealed by using these TEA-treated axons is that, during the entire depolarization phase of the prolonged action potential, there is a continuous increase in the degree of shrinkage of the axon. This finding strongly suggests that the shrinkage in a normal axon, i.e., an axon not treated with TEA, is not directly related to the membrane hyperpolarization.

Now, the question arises: Can the effect of TEA on the mechanical response of the axon be explained on a physicochemical basis?

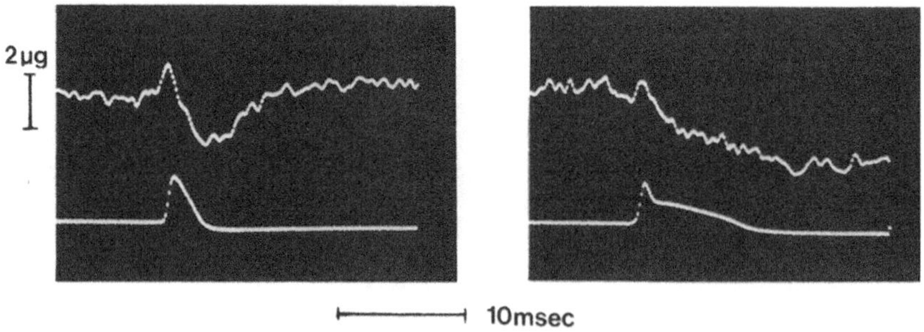

Figure 3. (Upper trace) Mechanical responses of squid giant axons into which a dilute TEA phosphate solution (made isotonic with addition of glycerol) had been injected. (Lower trace) Extracellularly recorded action potentials of the axons treated mildly with TEA.

TEA is a spherical univalent cation with a relatively large ion radius. The effective radius, estimated from the internuclear distances and angles, is approximately 4 Å (see Robinson and Stokes, 1959). This value is considerably larger than the crystal radii of K^+ (1.33 Å) and Na^+ (0.99 Å) and is close to the radius of hydrated Na^+ (3.3 Å). It is known that TEA-ions $[(C_2H_5)_4N^+]$ are selectively taken up by synthetic cation-exchangers by virtue of the Van der Waals force between the hydrophobic side-groups of TEA and the cation-exchanger matrix (Kressman and Kitchner, 1949). Exchange of TEA for Na^+ or K^+ in cation-exchangers increase the volume of the exchangers (Gregor, 1951).

From these physicochemical properties of TEA, it is expected that TEA-ions injected into a squid giant axon are preferentially bound to the anionic protein sites surrounded by hydrophobic amino acids. These anionic protein sites in the ectoplasm are expected to be occupied normally by K^+, because, in the resting axon, the axolemma constitutes the major diffusion barrier between the axon interior and the external medium. K^+ are known to enter readily into hydrophobic sites, losing their hydration (Williams, 1970, p. 343 and 362). There is a continuous influx of Ca-ions during the plateau of a prolonged action potential (see Pant *et al.*, 1978). It is well known that replacement of divalent cations in polyelectrolyte gels for univalent cations brings about a pronounced shrinkage of the gel (see Katchalaky and Zwick, 1955). It seems reasonable, therefore, to explain the progressive shrinkage of the axon during the plateau of the action potential as being caused by the continuous influx of Ca-ions into the ectoplasm. Incorporation of TEA appears to make the axoplasm more compact and reduces the amplitude of rapid movements of the ectoplasm.

V. CONFORMATIONAL TRANSITION OF MEMBRANE MACROMOLECULES

There are a variety of experimental findings that suggest the existence of conformational transitions of the macromolecules in the nerve membrane. The phenomenon of "abrupt depolarization" illustrated in Fig. 4 is one of these findings (Tasaki, 1982, p. 243). Here, squid giant axons immersed in a rapidly circulating solution of $CaCl_2$ (made isotonic with glycerol), is intracellularly perfused with a dilute NaF solution. When the pH and osmolarity of the internal and external solutions are in a proper range, the axons remain electrically excitable under these (bi-ionic) conditions. A brief pulse of outwardly directed current through the axon membrane can produce a prolonged (all or none) action potential. In the following observations, however, no electric stimuli are employed. Only a recording electrode (of the Ag–AgCl–agar type) is introduced into the axon interior (see top of Fig. 4).

The procedure of demonstrating abrupt depolarization is to dilute the external Ca-ion concentration smoothly and continuously by adding an isotonic solution of salt of one of alkali metal ions. While the external Ca-ion concentration is falling uniformly over the entire axon surface, the potential difference across the axon membrane is recorded continuously.

By this simple procedure, it is shown that the membrane potential jumps upward at a critical (or threshold) univalent–divalent cation-concentration ratio. The critical concentration ratio varies widely among different alkali metal ions. Nevertheless, one (and only one) potential jump can be observed with any one of the alkali metal ions. By introducing the axon under study into an A.C. impedance bridge, it has been shown that the potential jump is accompanied by a simultaneous fall in the membrane impedance.

As has been hinted by the discoverers of abrupt depolarization (Hill and Osterhout, 1938), the observed potential jump represents a transition of the axon membrane from its resting state conformation to its excited state conformation. We thus conclude that the axon

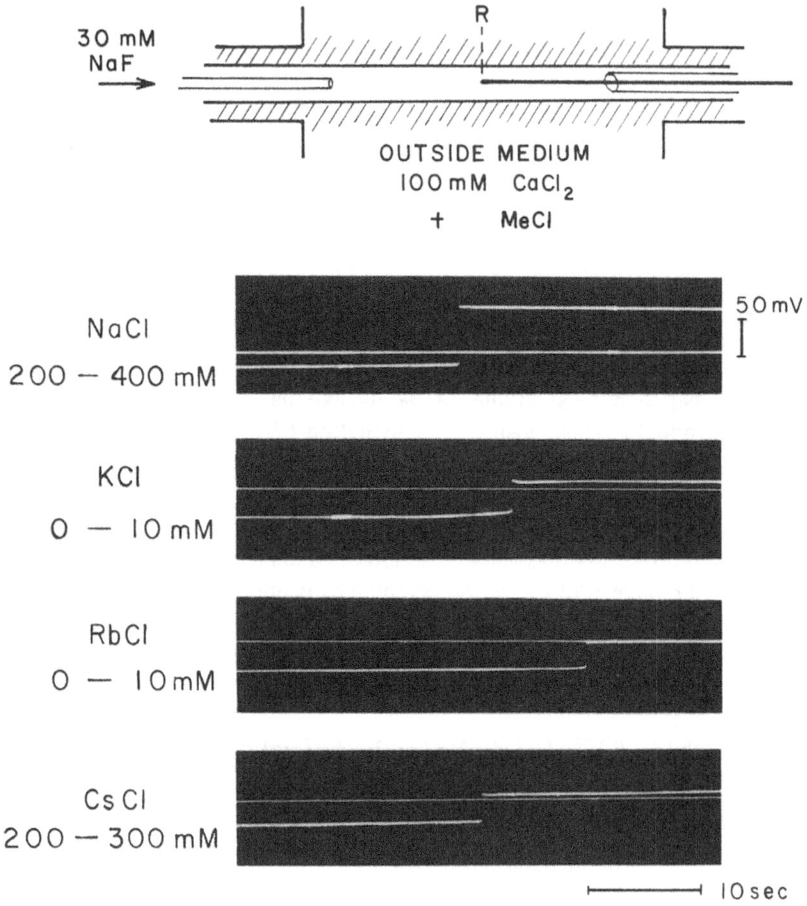

Figure 4. Abrupt depolarization of the squid axon membrane produced by addition of the salts of alkali metal ions to the 100 mM $CaCl_2$ solution outside the axons. Prior to each observation, the endoplasm of the axon used was removed by mild pronase digestion. The concentration of the univalent cation salt was raised smoothly and uniformly over the axon surface in the range indicated. (From Tasaki, 1982).

membrane is "stable" both in the Ca-rich (resting) state and Ca-deficient (excited) state, and that the transition from one state to the other is discontinuous in nature. Other experiments that support our conclusion are found in a recent monograph (Tasaki, 1982).

(Note that abrupt depolarization cannot be demonstrated unless the external electrolyte composition can be altered more or less uniformly over the entire membrane surface within the time comparable to the action potential duration. In axons which develop brief action potentials, the spatial nonuniformity of the axon membrane complicates the results.)

VI. A MODEL OF TWO DISCRETE STATES OF THE MEMBRANE MACROMOLECULES

There are several inanimate systems that are capable of undergoing transitions between two stable states. In the Ostwald–Lillie iron-wire analog (see Franck, 1956), the oxidized (passive) state of the iron surface is one of the stable states; the reduced state is the other. In Teorell's nerve analog (Teorell 1959; Meares and Page, 1972; Franck, 1978b), there are two stable concentration profiles in the membrane. In EIM molecules incorporated in a lipid bilayer, there are two discrete conductance states (Bean *et al.*, 1969).

In the nerve membrane, the divalent cations in the membrane are considered to play a crucial role in determining the state of the membrane. It is possible to construct a physicomathematical model that is capable of undergoing transitions between two states in response to a continuous change in the univalent–divalent cation-concentration ratio in the medium (Tasaki, 1982, p. 269). A brief description of this model is given below.

We postulate that a class of intrinsic protein molecules in the axolemma (see Fig. 5) is made of a long polypeptide chain along which anionic sites are distributed in a more or less regular manner. When immersed in a Ca-rich medium, the chain is considered to be crosslinked by Ca-ions at many paired sites in the molecule. A heavily crosslinked state of the molecule is stable. When the salt of univalent cation is added to the external medium, the crosslinks are broken. This breakage involves replacement of two univalent cations for one Ca-ion.

We denote the ratio of the number of crosslinked pairs of anionic sites to the total number of pairs available for crosslinking by f. We assume that the free energy required to form a single crosslink is dependent on f:

$$(a - bf)kT \tag{1}$$

where kT has the usual physicochemical significance and a and b are regarded as being independent of f. The constant b, which we assume to be positive, is a measure of cooperativity. When f is small, namely, when the majority of the pairs are in their unlinked state, much work is required, because the polypeptide chains are widely separated by electrostatic repulsion ($b > 0$). The work done when the fraction increases from 0 to f is then

$$(a - \tfrac{1}{2}bf)fNkT \tag{2}$$

when N is the number of pairs of sites available for the formation of Ca-bridges. The free energy of the whole molecule, F, may then be written in the following form:

$$\begin{aligned}(F - F_o) / (NkT) = {} & (a - \tfrac{1}{2}bf)f \\ & + f\ln f + (1 - f)\ln(1 - f) \\ & + 2f\ln\lambda_1 - f\ln\lambda_2\end{aligned} \tag{3}$$

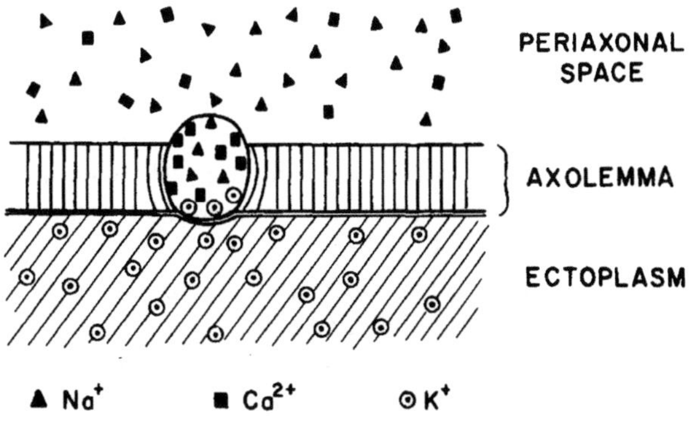

Figure 5. Schematic diagram illustrating the cation distribution in the axolemma–ectoplasm complex in the resting state of the axon.

where F_o represents the portion of the free energy which is independent of f. Quantities λ_1 and λ_2 represent the absolute activities of the univalent and the divalent cations, respectively. When the protein molecules are in their compact state, cation fluxes through the molecules are small. The absolute activities of the cations in the superficial layer of the macromolecules are not different from their values in the external medium. The second and third terms on the right-hand side of Eq. (3) represent the mixing entropy of the pairs of sites available for crosslinking. The last two terms denote the free energy change associated with replacement of univalent cations with divalent cations.

When f is varied from 0 to 1, we find one minimum or two minima in the value of F, depending on a, b, and C_1^2/C_2. In equilibrium, only those states corresponding to the minima of F are realized. The equilibrium values of f are determined by the condition $\partial F/\partial f = 0$, namely by

$$a - bf + \ln [f/(1 - f)] + \ln [KC_1^2/C_2] = 0 \tag{4}$$

where the absolute activities were replaced with the external concentrations, $\lambda_1^2/\lambda_2 = KC_1^2/C_2$, where K is a constant. This is the desired equation relating the equilibrium value of f to the variable of the system C_1^2/C_2.

The mode of dependence of the fraction of crosslinked pairs, f, on the value of C_1^2/C_2 in the medium varies with the choice of b and $(a + \ln K)$. When b is larger than 4, the relationship between f and C_1^2/C_2 is represented by an S-shaped curve. Figure 6 shows the relationship between the value of C_1^2/C_2 in the external medium and the fraction of broken Ca-bridges, $1 - f$, calculated by the use of Eq. (4). In calculation, b is chosen to be equal to 5 and $[5 - a - \ln K]$ to be 12. This diagram indicates that, when C_1 (univalent cation concentration) is 500 mM, transition $P - P^*$ takes place as C_2 (divalent cation concentration) is reduced to 12 mM. The difference in behavior among different alkali metal ions is represented by different sets of constants. In this manner, the experimental results shown in Fig. 4 can be explained by the use of the diagram in Fig. 6. This diagram also explains the origin of the hysteresis loop known to appear when the ratio of C_1 to C_2 is varied in a cyclic manner.

VII. A TENTATIVE INTERPRETATION OF THE ORIGIN OF THE MECHANICAL RESPONSES

We now consider the origin of the mechanical responses of an intact squid axon immersed in a solution containing 30–50 mM $CaCl_2$ and about 500 mM NaCl. We postulate, at the outset, that the intrinsic protein molecules in the axolemma (see Fig. 5) possess the properties of our model of two discrete states illustrated in Fig. 6. In the resting state, the membrane resistance is relatively high (0.1–$0.2\Omega \cdot m^2$). From this fact, we infer that the protein molecules in the axolemma of a resting axon are highly crosslinked with Ca-ions and that the axolemma is playing the role of being the sole diffusion barrier between the axon interior and the external medium. The anionic sites in the ectoplasm are expected to be occupied by K^+ under these conditions (see Fig. 5).

Now, we consider the effect of a pulse of outwardly directed current delivered to the axon membrane. When the current is strong enough to produce a potential change of about 25 mV, i.e., the familiar electrochemical quantity, RT/F, a finite number of crosslinks are broken by invasion of K^+ into the protein molecules. As soon as the value of f in Fig. 6 reaches the critical level for a conformational transition (represented by point P in the figure),

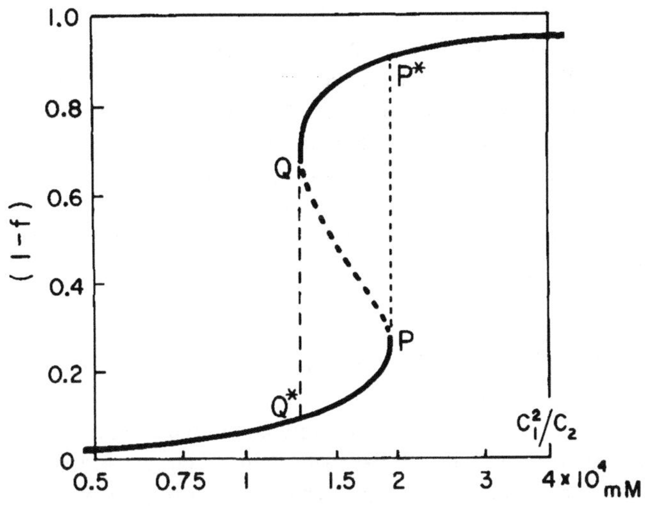

Figure 6. A theoretical cation-exchange isotherm illustrating the origin of two stable states of the axon membrane. The ordinate, $1-f$, represents the fraction of unlinked pairs of anionic sites in the intrinsic protein molecule. The abscissa represents the ratio of the square of the concentration of the univalent cation salt to the concentration of the divalent cation salt. See Eq. (4) in the text. (From Tasaki, 1982.)

a cooperative process sets in and the major portion of the remaining crosslinks are rapidly broken. The state of the molecule immediately after the conformational transition is represented by P* in the figure. Since the cooperative process involves replacement of two Na^+ for each Ca-ion, there is a sudden rise in the osmotic pressure within the molecules. Consequently, the molecules swell immediately. This swelling brings about a profound increase in the mobilities of the cations in the molecule. It is also probable that the arrangement of the lipid bilayer around the intrinsic protein molecules is strongly disturbed when the conformational transition takes place.

The increase in the cation mobilities (hence, in the membrane conductance) brings about a profound enhancement of cation interdiffusion across the membrane. Since the mobility of Na^+ in the membrane proteins is much greater than that of Ca-ions (see Williams, 1970), a rapid increase in the Na^+ concentration is expected in the superficial layer of the ectoplasm. In many biocolloids, replacement of Na^+ for K^+ enhances swelling (Hermans, 1949, p. 566). Therefore, the ectoplasm is expected to swell also as soon as the intrinsic protein molecules undergo a transition.

It is well known that, in swollen gels, the diffusion constants of small molecules or ions are not significantly different from those in water (Hermans, 1949, p. 580). We assume, therefore, that the diffusion constants, D, of Na^+ and K^+ are of the order of $10^{-9} m^2 s^{-1}$. In the ectoplasm, these ions spread by diffusion to a distance of the order of $(Dt)^{-2}$ at time t after the transition. When $t = 0.1$ msec, this distance is roughly 0.3 μm. Therefore, we postulate that the effects of Na^+ invasion at different membrane sites overlap and the ectoplasm swells more or less uniformly over the entire axon surface.

We assume that invasion of Ca-ions into the ectoplasm is slow because of the strong electrostatic interaction with the anionic sites of protein molecules with multivalent cations. As Williams has suggested, a marked shrinkage of the ectoplasm is produced during the later phase of the action potential by the invasion of Ca-ions (see Figs. 2 and 3).

The invasion of Ca-ions into the ectoplasm lowers the overall membrane conductance (measured by using a high-frequency A.C.). At about the time when the membrane conductance returns to normal, we see a small downward deflection of the mechanical trace (see Fig. 2). The small shrinkage of the axon indicated by this deflection may be interpreted as the onset of the relative refractoriness associated with the transition of the intrinsic protein molecules from their swollen state to the compact state.

The state of the ectoplasm returns gradually to normal, as the Ca-ions are removed initially by diffusion and subsequently by binding to various Ca-sequestering elements in the axon. The recovery of the axon excitability during the relatively refractory state may be related to the gradual change in the state of the ectoplasm.

VIII. PSEUDOCAPACITY OF THE AXON MEMBRANE

A long pulse of electric current can be delivered to the axon membrane by using either extracellular electrodes or an intracellular metal wire electrode. Mechanical changes in the axon induced by transmembrane currents have been examined by employing these two methods of delivering current pulses. Figure 7 shows an example of the records obtained by using an intracellular current electrode. Again, a piece of Kureha piezoelectric film was employed for detection of mechanical changes. Along the axis of the axon under study, a 12-mm long platinized platinum electrode was introduced. The potential variations induced by the applied current were recorded with another platinum electrode located in the middle of the current electrode.

The mechanical changes shown in the left-hand record in the figure were taken by using a pulse of inwardly directed, i.e., hyperpolarizing, current that produced a change in the membrane potential of about 120 mV (see the lower trace in the figure). It is seen that a distinct shrinkage of the axon was produced by the current. The significant feature of the mechanical change observed under these conditions is that the time-course of the mechanical change is very similar to that of the potential change.

By using an optical method, it can be shown that an inwardly directed membrane current produces an inward displacement of the axon surface. The amplitude of the displacement induced by 120 mV hyperpolarization was found to be 2–5 nm.

The following experimental findings indicate that the mechanical changes induced by a transmembrane current are not simple physical changes of the dielectric material of the membrane induced by the applied potential changes: (1) there is a sign of gradual rise in the membrane resistance during the passage of an inward current, (2) when the axon is excited (by delivering a brief pulse of outwardly directed current) during the passage of an inward current, a mechanical response with a very large amplitude is generated, and (3)

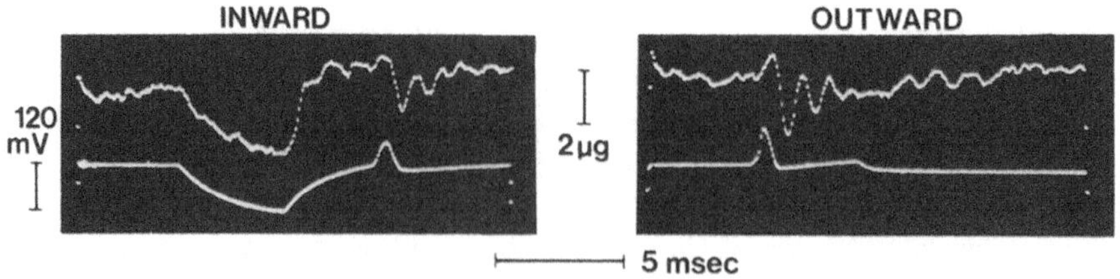

Figure 7. Records showing the effects of a 3-msec long transmembrane current pulse (approximately 10 µA) applied to the axon membrane using a 12-mm long internal metal electrode.

substitution of Mg-ion for Ca-ion in the external medium decreases the amplitude of the mechanical change associated with the potential change of the same amplitude.

The mechanical change generated by an inwardly directed membrane current may be interpreted in the following manner. The mobility of Ca-ion in the external medium is not very different from that of Na^+. In the intrinsic protein molecules and in the ectoplasm, however, the mobility of Ca-ions is far smaller than that of the univalent cations (see Tasaki *et al.*, 1967). The discontinuous change in the mobility ratio at the axon surface brings about accumulation of Ca-ion in the ectoplasm when an inwardly directed current transverses the membrane. The observed shrinkage of the axon may be attributed to the action of Ca-ion accumulated by this process.

We emphasize in this connection that the interpretation of the effect of an inward current mentioned above is nothing new. A long time ago, Albrecht Bethe concluded that, in the region of a nerve (trunk) traversed by an inward current, the nerve membrane becomes "compact" as the consequence of Ca-ion accumulation (Bethe, 1920, p. 294). He also pointed out that the pH near the axon surface is also affected by the current.

The hydroxyl ion concentration in the external medium is only about 1 μM. In the axon interior, the OH-ion concentration is even lower. Nevertheless, it is probable that the OH-ion concentration falls at and near the axon surface by the passage of a current. As Bethe and Toropoff (1914) have shown previously, the relative change in the H-ion concentration depends on the mobilities and the concentrations of other mobile ion species and is usually much larger than the relative concentration changes of other ions.

The shrinkage of the axon membrane brought about by the accumulated Ca-ions and H-ions is responsible for the gradual rise in the membrane resistance. The IR-drop, i.e., the potential drop across an ohmic resister, must then gradually increase during the passage of an inward current. Franck (1978a) named the component of the membrane capacity attributable to a progressive change in the membrane resistance a "pseudo-capacity." The similarity in the time-course between the mechanical changes and the potential changes in Fig. 7 may therefore be explained in terms of the pseudocapacity of the axon membrane.

The right-hand record in Fig. 7 shows that mechanical changes of the axon membrane are produced by a pulse of outwardly directed current of the same intensity (taken from the same axon). It is seen that an action potential was generated at the onset of the current pulse. The mechanical response associated with the action potential was followed by a small swelling of the axon. During the passage of an outward current, we do not expect a large mechanical change because it is unlikely that an outward current alters the cation composition in the ectoplasm except when it generates an action potential.

IX. A GENERAL REMARK CONCERNING THE HISTORICAL BACKGROUND OF THE PRESENT INTERPRETATION

The importance of Ca-ion as the initiator of structural changes in the nerve was fairly well recognized by Jacques Loeb a long time ago. He remarked, "it is not impossible that a substitution of K for Ca, or *vice-versa*, in ion colloid actually occurs at the cathode, while a constant current flows through the nerve" (Loeb, 1906, p. 102). "The normal irritability of animal tissues depends upon the presence of Na, K, Ca, and Mg ions in the right proportion" (p. 95). He emphasized that *a substitution of Ca for K or Na in colloids favors the formation of more solid or insoluble compounds* (p. 87), and suggested that the ion-substitution (or ion-exchange) process involving Ca-ion is at the base of irritability, i.e., excitability, phenomena of animal cells. Obviously, our interpretation of the mechanical changes in the axon is, in general, consistent with Loeb's concept.

Loeb's theory never gained popularity among electrophysiologists during his lifetime. Apparently, Loeb was far ahead of his contemporaries. Many physiologists at that time were fascinated by Nernst's apparently rigorous physicomathematical theory of nerve excitation (Nernst, 1908) and, for many decades, Nernst's approach and Hill's extension of Nernst's theory (Hill, 1910) dominated physiologists' way of thinking. Ironically, it was found later that the strength–frequency relation in threshold excitation, with which Nernst's theory is concerned, deals only with the physical properties of the (physiologically inert) myelin sheath and not with the excitable part, i.e., nodal membrane, of the nerve fiber (Tasaki, 1982, p. 110).

The cardinal difficulty of Nernst and Hill's mathematical approach lies in the fact that the theory does not tell anything about the mode of action of specific ions on various structures of the protoplasm (Heilbrunn, 1938, pp. 430–431). The same difficulty exists in the modern mathematical theory of nerve excitation proposed by Hodgkin and Huxley (1952). Without being supplemented by an explicit treatment of the relationship between structure and function of excitable cells, mathematical theories proposed in the past do not offer any explanation as to the origin of the mechanical responses of the nerve fiber.

Loeb's approach was adopted by a number of subsequent investigators. Heilbrunn (1938), Brink (1954), Tobias (1964), and others emphasized the importance of physicochemical changes in the axon membrane produced by changes in the Ca-ion concentration. The shortcoming of this approach lies in the fact that the "regenerative" (or cooperative) behavior of the excitation process was not explicitly taken into consideration.

In the present article, we have shown that, when supplemented by the concept of two discrete states of the intrinsic protein molecules, recent experimental findings concerning mechanical changes in the squid giant axon can be explained on the basis of Loeb's theory.

ACKNOWLEDGMENTS. We thank Mr. T. Araki and Mr. H. Obara of Kureha Chemical Co. for supporting this research project.

REFERENCES

Bean, R. C., Shepherd, W. C., Chan, H., and Eichner, J., 1969, Discrete conductance fluctuations in lipid bilayer protein membranes, *J. Gen. Physiol.* **53**:741–757.

Bethe, A., 1920, Nervenpolarisationsbilder und Erregungstheorie, *Pflüger's Arch. Ges. Physiol.* **183**:289–302.

Bethe, A., and Toropoff, T., 1914, Über elektrolytische Vorgänge an Diaphragmen, *Z. Phys. Chem.* **88**:686–742.

Brink, F., 1954, The role of calcium ions in neural processes, *Pharmacol. Rev.* **6**:243–298.

Franck, U. F., 1956, Models for biological excitation, *Prog. Biophys.* **6**:171–206.

Franck, U. F., 1978a, Chemical oscillations, *Angew. Chem. Int. Ed. Engl.* **17**:1–15.

Franck, U. F., 1978b, A quantitative treatment of oscillatory phenomena in coarse-grained ion-exchanger membranes, *Electrochimica Acta* **23**:1081–1091.

Gainer, H., Carbone, E., Singer, I., Sisco, K., and Tasaki, I., 1974, Depolarization induced change in the enzymatic radio-iodination of protein on the internal surface of the squid giant axon membrane, *Comp. Biochem. Physiol.* **47A**:477–484.

Gregor, H., 1951, Gibbs Donnan equilibria in ion exchange resin systems, *J. Am. Chem. Soc.* **73**:642–652.

Heilbrunn, L. V., 1938, *An Outline of General Physiology*, W. B. Saunders, Philadelphia.

Hermans, P. H., 1949, Chapter XII Gels, in: *Colloid Science* (H. R. Kruyt, ed.), Vol. 2, pp. 483–650, Elsevier, New York, Amsterdam.

Hill, A. V., 1910, A new mathematical treatment of changes of ionic concentration in muscle and nerve under the action of electric currents, with a theory as to their mode of excitation, *J. Physiol. (London)* **40**:190–224.

Hill, S. E., and Osterhout, W. J. V., 1938, Calculation of bioelectric potentials, *J. Gen. Physiol.* **21**:541–556.

Hodge, A. J., and Adelman, W. J., Jr., 1980, The neuroplasmic network in Loligo and Hermissenda neurons, *J. Ultrastruct. Res.* **70**:220–241.

Hodgkin, A. L., and Huxley, A. F., 1952, A quantitative description of membrane current and its application to conduction and excitation in nerve, *J. Physiol. (London)* **117**:500–544.

Iwasa, K., and Tasaki, I., 1980, Mechanical changes in squid giant axons associated with production of action potentials, *Biochem. Biophys. Res. Commun.* **95**:1328–1331.

Iwasa, K., and Tasaki, I., 1981, Rapid pressure changes of the squid giant axon accompanied by prolonged action potentials, *Biol. Bull.*, **161**:346.

Katchalsky, A., and Zwick, M., 1955, Mechanochemistry and ion exchange, *J. Polymer Sci.* **17**:221–234.

Kawai, H., 1969, The piezoelectricity of poly (vinylidene fluoride), *Jpn. J. Appl. Phys.* **8**:975–976.

Kressman, T. R. E., and Kitchner, J. A., 1949, Cation exchange with a synthetic phenosulphonate resin. Part III. Equilibria with large organic cations, *J. Chem. Soc.* **1949**:1208–1210.

Loeb, J., 1906, *The Dynamics of Living Matter*, Columbia University Press, New York.

Meares, P., and Page, K. R., 1972, Rapid force-flux transitions in highly porous membranes, *Phil. Trans.* **272**:1–46.

Metuzals, J., and Izzard, C. S., 1969, Spatial patterns of threadlike elements in the axoplasm of the giant nerve fiber of the squid (*Loligo pealii* L.) as disclosed by differential interference microscopy and by electron microscopy, *J. Cell Biol.* **43**:456–479.

Metuzals, J., and Tasaki, I., 1978, Subaxolemma filamentous network in the giant nerve fiber of the squid *(Loligo pealei L.)* and its possible role in excitability, *J. Cell Biol.* **78**:597–621.

Murayama, N., Nakamura, K., Obara, H., and Segawa, M., 1976, The strong piezoelectricity in polyvinylidine fluoride (PVDF), *Ultrasonics* (January) 15–23.

Nernst, W., 1908, Zur Theorie des elektrischen Reizes, *Pflüger's Arch. Ges. Physiol.* **122**:275–314.

Pant, H. C., Terakawa, S., Baumgold, J., Tasaki, I., and Gainer, H., 1978, Protein release from the internal surface of the squid giant axon membrane during excitation and potassium depolarization, *Biochim. Biophys. Acta* **513**:132–140.

Robinson, A. L., 1978, Flexible PVF_2 film: An exceptional polymer for transducer, *Science* **200**:1371–1374.

Robinson, R. A., and Stokes, R. H., 1959, *Electrolyte Solutions*, Butterworths, London.

Tasaki, I., 1982, *Physiology and Electrochemistry of Nerve Fibers*, Academic Press, New York.

Tasaki, I., and Hagiwara, S., 1957, Demonstration of two stable potential states in the squid giant axon under tetraethylammonium chloride, *J. Gen. Physiol.* **40**:859–885.

Tasaki, I., and Iwasa, K., 1980, Swelling of nerve fibers during action potentials, *Upsala J. Med. Sci.* **85**:211–215.

Tasaki, I., Watanabe, A., and Lerman, L., 1967, Role of divalent cations in excitation of squid giant axons, *Am. J. Physiol.* **213**:1465–1474.

Tasaki, I., Iwasa, K., and Gibbons, R. C., 1981, Mechanical changes in crab nerve fibers during action potentials, *Jpn. J. Physiol.* **30**:897–905.

Teorell, T., 1959, Electrokinetic membrane processes in relation to properties of excitable tissues, II, Some theoretical considerations, *J. Gen. Physiol.* **42**:847–862.

Teorell, T., 1962, Excitability phenomena in artificial membranes, *Biophys. J.* (Suppl.) **2**:27–52.

Tobias, J. M., 1964, A chemically specified molecular mechanism underlying excitation in nerve: A hypothesis, *Nature (London)* **203**:13–17.

White, G. L., Schellhase, H. U., and Hawthorne, J. N., 1974, Phosphoinositide metabolism in rat superior cervical ganglion, vagal and phrenic nerve; effects of electrical stimulation and various blocking agents, *J. Neurochem.* **22**:149–158.

Williams, R. J. P., 1970, The biochemistry of sodium, potassium, magnesium and calcium, *Quart. Rev. Chem. Soc. (Lond.)* **24**:331–365.

16

History of the Physical Chemistry of Charged Membranes

Torsten Teorell

An important principle in any department of knowledge is to find the approach from which a problem appears in its greatest simplicity.

Quotation attributed to J. Willard Gibbs (1839–1903)

I. INTRODUCTION

One has to look for the roots of what now has been called membranology far back in time and search into many different fields of human activity. My task in this chapter must necessarily be more limited as indicated by the title "Charged Membranes."* Seen in retrospect, the primary problems and speculations were concerned with the riddle of many life processes. The search began with experiments on frog muscle contractions by Galvani about 1780 (Fig. 1). Incidentally, it can be said that the ensuing controversy between Galvani and Volta and others, about the actual nature of "animal electricity" also led to the development of electromagnetism, which gave rise to our first electrical measuring instruments (galvanometers and voltmeters). The debates on muscle electricity and muscle contractions enhanced old speculations on the mechanism of bodily excretions and secretions, such as urine, bile, and gastric juice. These are more "visible" problems, which inspired "model making" with pig bladder, parchment, and other "diaphragms." Dialysis and osmosis were discovered, and this was the beginning of colloid chemistry. It is of particular interest to recall that the "electrified osmosis" called electroendosmosis was discovered already in 1803 by Reuss. Later, Dutrochet, Quincke, and particularly Wiedemann (1893) were able to set up the

* This brief and very incomplete historical review is mainly based on papers by K. Sollner (1976) and T. Teorell (1976). Sollner's paper is a rich source for the literature before 1930. Rothschuh's books (1953, 1973) have also been consulted. The reference list added to the present paper is limited only to works which have been concerned with sectors of charged membranes research in which the present author has taken some part.

Torsten Teorell • Department of Physiology and Medical Biophysics, Biomedical Center, Uppsala University, S-751 23 Uppsala, Sweden.

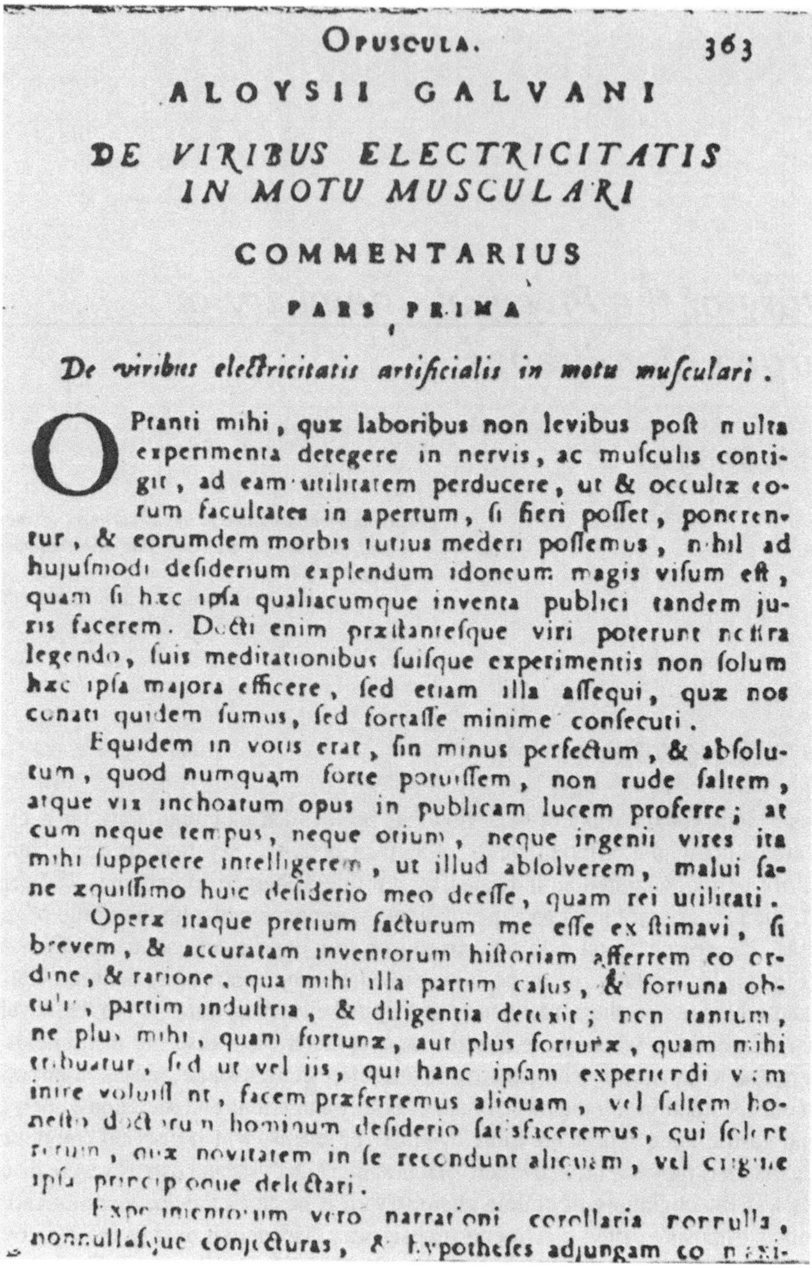

OPUSCULA. 363

ALOYSII GALVANI

DE VIRIBUS ELECTRICITATIS IN MOTU MUSCULARI

COMMENTARIUS

PARS PRIMA

De viribus electricitatis artificialis in motu musculari.

OPtanti mihi, quæ laboribus non levibus post multa experimenta detegere in nervis, ac musculis contigit, ad eam utilitatem perducere, ut & occultæ eorum facultates in apertum, si fieri posset, ponerentur, & eorumdem morbis tutius mederi possemus, nihil ad hujusmodi desiderium explendum idoneum magis visum est, quam si hæc ipsa qualiacumque inventa publici tandem juris facerem. Docti enim præstantesque viri poterunt nostra legendo, suis meditationibus suisque experimentis non solum hæc ipsa majora efficere, sed etiam illa assequi, quæ nos conati quidem sumus, sed fortasse minime consecuti.

Equidem in votis erat, sin minus perfectum, & absolutum, quod numquam forte potuissem, non rude saltem, atque vix inchoatum opus in publicam lucem proferre; at cum neque tempus, neque otium, neque ingenii vires ita mihi suppetere intelligerem, ut illud absolverem, malui sane æquissimo huic desiderio meo deesse, quam rei utilitati.

Operæ itaque pretium facturum me esse existimavi, si brevem, & accuratam inventorum historiam afferrem eo ordine, & ratione, qua mihi illa partim casus, & fortuna obtulit, partim industria, & diligentia detexit; non tantum, ne plus mihi, quam fortunæ, aut plus fortunæ, quam mihi tribuatur, sed ut vel iis, qui hanc ipsam experiendi viam inire voluissent, facem præferremus aliquam, vel saltem honesto doctorum hominum desiderio satisfaceremus, qui solent rerum, quæ novitatem in se recondunt aliquam, vel cognitione ipsa principioque delectari.

Experimentorum vero narrationi corollaria nonnulla, nonnullasque conjecturas, & hypotheses adjungam eo maxi-

Figure 1. Galvani, Aloisio (1737–1798), Italian natural scientist and physician, is regarded as the founding father of electrobiology. Here is a reproduction of Galvani's first "reprint" dealing with muscle contractions.

quantitative laws for electroosmosis which are valid even today, notwithstanding the advent of a new powerful tool, irreversible thermodynamics. The author's own work on excitability models rests on "Wiedemann's law" and on the important "hydrodiffusion" law of A. Fick (Figs. 2 and 3). Fick (1866) has an excellent chapter on diffusion in his book *Medizinische Physik,* probably the first text on this subject. (Fick was a physiologist!) In the workshops

Figure 2. Fick, Adolf (1829–1901), German physiologist, was a pupil of Ludvig in Zürich and later a professor in Würtzburg. He was one of the leading physiologists of the nineteenth century, striving to apply exact sciences to the life sciences. His work included, besides diffusion, important contributions to muscle, circulation, and sensory physiology. There is a lineage to modern membrane science through Jaques Loeb (1859–1924), who early worked as an assistant to Fick. (Portrait from *Rothschuh,* 1953.)

of living "membranes" the state was much more confused, although many "laws" were proposed for electrical action and injury currents in muscle and other preparations. Unfortunately, we have to leave these interesting debates between the strong personalities who were the founders of modern physiology.

From a physical–chemical point of view the real leap forward came in 1884 when the Swede, Svante Arrhenius, published his first paper on electrolytic dissociation into free ions of salts in aqueous solutions (Fig. 4). Arrhenius was then only 25 years of age and his paper was received coolly in Sweden. Abroad, particularly in Germany, Arrhenius' papers, combined with Fick's work on diffusion led to one of the most remarkable booms in physical chemistry, which continued roughly through the last decade of the nineteenth century. Now appeared the fundamental contributions by Nernst and Planck on diffusion and electrode potentials (Fig. 5). Many believed optimistically that the riddles of bioelectrogenesis were now solved; the key was in the different ion mobilities and different ion concentrations across the excitable nerve and tissue membranes. Neither Nernst nor Planck were directly involved in biological work. The application of their theoretical ideas came through Ostwald and Overton. Overton recognized the selectivity differences between sodium and potassium,

Figure 3. Wiedemann, Gustav (1826–1899) was a professor of physical chemistry, and later of physics at Leipzig in Germany. He worked in several areas other than electroendomosis and was also a editor of journals in his fields. His extensive "handbooks" were standard works for a long time.

Figure 4. Arrhenius, Svante (1859–1927), Swedish physicist. The idea of charge separation (Helmholtz's "double layers") existed before Arrhenius but his theory of preexisting, free ions was epochmaking (Nobel Prize in 1903). He was a versatile person in many respects. He applied the mass action law to immunology, advanced theories on astrophysics and wrote excellent popularized science.

Figure 5. Planck, Max (1858–1947), one of the greatest scientists of modern times, the founder of the quantum theory. Planck was the leading figure in German natural science. Like Helmholtz, great honors were bestowed on him. His interest in electrochemistry was maintained throughout his life and new papers were published as late as the 1930s.

to become the foundation of the famous ionic theory of Hodgkin and Huxley, about 50 years later. The portal figure in this advancement of electrobiology was Bernstein, who clearly stated "the membrane theory" of ion migration and water movement in 1902 (Fig. 6). Bernstein seems to have been a remarkably productive scientist. In 1860, and subsequently, he published work on experimental devices called "rheotomes" for the analysis of the time events of action currents of various organs. He also kept abreast of the theoretical concepts of Helmholtz and Gibbs as to the thermodynamics of bioelectric currents. Bernstein's (1902) paper in *Pflügers Arch.* leaned heavily on Ostwald, who in turn referred to the artificial precipitation membranes introduced by Traube and the botanist Pfeffer. Ostwald looked upon these "semipermeable" diaphragms as "Ionensiebe" (ion sieves). This expression could still be a keyword for the membrane research related to the excitability of living tissues. The early terms were "Scheidewänden," septa or diaphragms. It was remarkable that a comprehensive presentation of Bernstein's (1912) work did not appear until the publication of his book *Elektrobiologie*. In a foreword Bernstein excused himself by alluding to the "constant flow" in the area. In particular, he referred to the many new results that he and others obtained with the new instruments, the "capillary electrometer" and Eindhoven's

Figure 6. Bernstein, Julius (1839–1917), German physiologist, was one of the few pupils of Helmholtz. In Halle, he demonstrated his great experimental skill as a biophysicist (propagation rate of contraction and electric "waves" in muscles and nerves). His "membrane theory" (1902, 1912) has greatly influenced modern developments in electrophysiology.

"string galvanometer." These instruments were for this period a revolution equivalent to the introduction of the vacuum-tube amplifier in the late 1920s. Bernstein's book is entertaining reading in many respects. Somewhere among the pages there is the remark that a cell phase boundary potential might be of the order of 0.02 volt, and perhaps one might anticipate up to 0.08 volts! He did not mention any sign-reversal of the action potentials, "the overshoot." That belongs to another era of development. In Bernstein's time, depolarization was a decline towards zero voltage.

II. THE FIRST YEARS OF THE TWENTIETH CENTURY

Apparently, the first decades of this century were a time of reconciliation between the conflicting, almost hectic developments in physical chemistry that had occurred about 1890. Of course, this does not mean that progress was not constantly made. "Oil chains," phase-boundary potential differences between immiscible organic liquids and water, were investigated and interpreted (Haber and others, particularly Beutner (1920, 1933), summarized this work). During this period, the glass electrode was conceived, although it had hardly any practical implications at that time. Another 20 years were to pass before the first commercial glass electrode was introduced by the Cambridge Instrument Company in Eng-

The second decade, comprising the first world war, spawned new fundamental views. Although "colloid chemistry" was much investigated in the early 1900s, it was scarcely coupled with bioelectrogenesis until Donnan in 1911, and his follower, J. Loeb, in theory and experiments emphasized the importance of the living "colloid-electrolytes." By custom, colloids were defined as materials that could not pass through semipermeable membranes. Proteins belong to this category. It should be remembered that the molecular-weight classification of our time did not become available until the Swede, The(odor) Svedberg, and his collaborators in Uppsala, introduced the ultracentrifuge in the 1930s. In 1911, Donnan presented a clear theory of the influence charged colloids had upon the distribution equilibria and potentials when common salt ions were present in a semipermeable system. Less stringent but very elaborate papers appeared in 1916 by Bethe and Toropoff. They studied the neutrality changes in porous material when a dc current was applied, in other words, the electroosmosis effect. They clearly invoked the existence of membrane-bound ions. In accordance with dominating views from colloid chemistry they considered the presence of charges in the pores as due to "adsorption," following Freundlich and others. Thus, these papers became a forerunner to Michaelis' concepts and the "TMS" theory, which will be presented later in this paper.

The well-known Donnan effects did not immediately attract much attention among the electrobiologists. Donnan himself, once stated to the present author: "I woke up one morning in 1922 and found myself famous, due to Jaques Loeb in New York." Jaques Loeb (1922) made a real breakthrough through his book, *Proteins and the Theory of Colloidal Behaviour*, which rapidly became a "bible," electrobiologists were given a new push forward. (In passing, it may be mentioned that Donnan's very early papers mentioned the formation of hydrochloric acid in gastric juice as a biological application of his views. In fact, this was the impetus that started the present author on an investigation of the gastric-juice electrolytes and brought him into the membrane realm.)

There were two other "bibles" for me and those of my generation who had started in biomedical research. The most comprehensive was R. Höber's already classical *Physicalische Chemie der Zelle und der Gewebe* and L. Michaelis' (1922) volume *Die Wasserstoffionen-Konzentration*.

After my dissertation on gastric juice I wanted to learn more about "permeability," the new fashionable term, and was accepted as a research fellow at the famous Rockefeller Institute for Medical Research in New York (The Rockefeller University), a Mecca for the bioscientists. In 1934, I came to J. Loeb's old department, then headed by the famous plant electrophysiologist W. J. V. Osterhout. Another of the twelve distinguished members who ran the Institute was L. Michaelis, and the young W. Wilbrandt was a contemporary fellow (Wilbrandt, 1938). The prevailing atmosphere was one of free communication, which was extremely stimulating for us youngsters. Osterhout's and Michaelis' departments had membrane electrogenesis in common, but with different approaches. Osterhout's group had attracted me because they worked on "ion accumulation," which was near to my own interest in hydrogen-ion secretion in the stomach (the gastric HCl acid has a pH of ~1, while the blood is at pH 7, a million-fold ion accumulation!). Michaelis and collaborators repeated some of his work from the 1920s on the potentials of dried collodion, which served as possible models for cell membranes.

My first term in New York was a thorough training in basic electrochemistry guided by Osterhout's staff, which included the well known electrochemists McInnes, Shedlovsky, and Longsworth. It resulted in one paper that was essentially a generalization of the Donnan effects, in which a steady supply of one ion influenced the redistribution of the other ions (the "passive ions") into a final steady state formally following the Donnan requirements. This was called "the diffusion effect" (Teorell, 1935a, 1937a). Another paper dealt with an application of McInnes' method of determining of the Hittorf "transference numbers" in a new type of membrane, cellophane, which was far less dense than the Michaelis' material

(Teorell, 1936a). The essential finding was that the cation transference number of several 1–1 valent salts was greatly enhanced. According to views in both Osterhout's and Michaelis' groups, this was interpreted in terms of the Nernst–Planck potential equation such that the *relative* mobility, i.e., the ratio cation/anion mobility, was changed [implying a change of the actual friction encountered by the ions during the passage through the membrane (Osterhout, 1956)]. I dared to suggest that the membrane itself was an (indiffusible) electrolyte and that the diffusibilities might not at all be different from those in "free" water, because I recalled Arrhenius' (1892) early work that an "added" salt could greatly accelerate the diffusion rate in salt mixtures. The "masters" Osterhout and Michaelis were somewhat reluctant to accept to my "heretic" views. Then I pieced the evidence together to give a picture of a "mixed potential," that is, *three* potentials in series summing up to the total measurable potential difference consisting of two opposing potential jumps at the membrane interfaces (simplified to two Donnan P.D.) *plus* one interior membrane potential difference, according to the Henderson (Planck) formula. The ensuing formula yielded a quite clear cut prediction of the old widely discussed "concentration effect," i.e., that a "diaphragm" potential was highly dependent on the bathing concentrations, even if these were kept in

TABLE II

X or membrane 'activity' = 1. Membrane negative. Mobility relation *u*:*v* in the membrane the same as in water. (Signs refer to the dilute solution in the external circuit.)

a_1	a_2	Partial E.M.F. Boundary mv.	Diffusion mv.	Total E.M.F. mv.
100	10	+ 1.1	− 13.2	− 12.1
10	1	+ 10.9	− 12.1	− 1.3
5	0.5	+ 20.5	− 12.0	+ 8.5
1	0.1	+ 46.2	− 5.4	+ 41.8

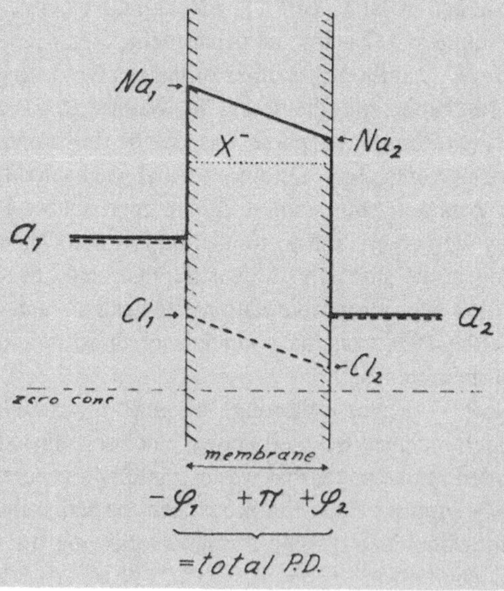

Figure 7. The Table demonstrates the "concentration effect," while the diagram depicts probably the first "concentration profiles" at a fixed charge membrane. (From Teorell, 1976).

the same ratio "inside" to "outside." This was a problem which Loeb and Beutner had already studied in 1912, but mainly in terms of partition-solubility effects (their object of study was apple peel!). When I finally presented my attempt to explain the "concentration effect" to Loeb's successor, Osterhout, he recommended a short paper. I also remember Michaelis' reaction; he nodded kindly his approval for the publication (Fig. 7). Osterhout introduced the three-page communication to the *Soc. Exp. Biol. Med.* in the fall (Teorell, 1935b). The title was "An attempt to formulate a quantitative theory of membrane permeability." I then had to go back to Sweden for a new job.

The following year, K. H. Meyer and J. F. Sievers (1936) independently published the same kind of theory at much greater length in a Swiss journal. They also considered the partition effects as due to solubility. Both versions of this "fixed-charge theory," as it was named, were directed towards the origin of the membrane potential. Still, it was no theory of the *kinetics* of the ion transport; this required further elaboration. At a Faraday Society meeting (Teorell, 1937b) the first rough scale drawing of the important concentration profile was presented. I realized that the transport kinetics had to rely on the Planck concepts and was aware of the work by Planck's pupil Behn, back in 1897. Behn's very clear paper was concerned with "neutral," convection-free membranes. In fact, it contained predictions between potentials, transport rate, and ionic equilibria, which were later rediscovered and used by H. Ussing in Copenhagen in his now famous work on frog skin. The darkening situation in Europe and World War II greatly hampered continued work and international communications. It was not until 1949 I discovered that David Goldman (1943) in his brilliant work had attacked the general electrical phenomena of membranes. He also leaned on Planck's theory. In fact, he touched upon the fixed-charge effect but he did not pursue it fully. When the war was over, I met K. H. Meyer and we agreed on the abbreviation "TMS" (Teorell, Meyer, Sievers) for our early theories of 1935–1936, which seems to have become an accepted designation.

III. THE YEARS AFTER 1950

Peace regained, I felt compelled to extend the Planck–Behn ideas to include the effect of the membrane fixed charge. At a conference in Göttingen in 1951, I presented a more complete analysis of "mixed potentials," membrane conductance and rectification effects. Formulas for concentration profiles were also given (reviewed in *Prog. Biophys.* in 1953; Teorell, 1951, 1953). R. Schlögl (1964), who had translated my Göttingen manuscript into proper German, had the skill and insight to extend our generalized membrane theory. In a series of beautiful works he and Helfferich (1959) reexamined the fixed-charge membranes, now with the new proper name, "ion-exchange membranes," which by this time had found industrial applications. R. Schlögl, G. Schmid, and other workers put the formulations into the new language of irreversible thermodynamics. Hereby, the phenomena of anomalous osmosis were rationalized. This was a phenomenon which Freundlich's pupil, Karl Sollner, had treated extensively ever since the early 1930s. Sollner is one of the major figures of modern membranology, and we owe him much. He introduced the "permselective" membranes, which furnished excellent experimental material. Sollner also dealt with ionic mixtures, the so-called "bi-ionic" systems, heterogenous membranes of mosaic and sandwich type, as well as with studies of liquid nonaqueous current generators.

The literature on charged membranes grew immensely in the 1960s. The concepts were widened to include "liquid fixed charges" and "mobile charges." Valid criticism on the TMS theories were also expressed.* Lakshminarayanaiah's (1969) book, *Transport Phenomena in Membranes*, gives an impressive account (in about 500 pages!) of the past and the present. The advances were also collected in several volumes with the title *Membranes* (edited by G. Eisenman, 1972) and other similar series. Several new journals specializing in membrane research were launched. The field had become really fertile!

IV. THE DEVELOPMENT OF THE MEMBRANE OSCILLATOR

In the early 1950s we used the porous Sollner membranes in attempts to verify the extended Planck theory which I published in 1951 (Teorell, 1951; see also Teorell, 1953). I had a notion that it might be possible to construct an artificial membrane system, which would give *oscillatory* membrane potentials and transport phenomena, simulating those of nerves and other excitable cells, which had been extensively studied with the aid of microelectrodes and modern electronics. The introduction of radioactive labeled ions had shown intense dynamic transport events in nerves and other tissues. In 1953, after many failures, I stumbled on an artificial system of fritted glass pores which showed remarkable similarity with many features of excitability experiments. The whole rhythmic display of the model could be traced back to electroosmotic water streaming across the membranes, which rearranged the concentration profiles and hence the conductance (Teorell, 1955; 1959a,b; 1961; 1971). It was essentially a feedback system involving something mysterious called "negative conductance." This is a property which goes back to some very important work of Cole and Curtis in the late 1930s. Cole had demonstrated that the action potentials in the alga *Nitella* and in squid axons were accompanied by great changes in the membrane conductance and impedance. Cole's (1968) impressive book, *Membranes, Ions and Impulses*, should be consulted as a standard work on the boundary fields between physics and electrobiology.

U. Franck in Germany has thoroughly investigated the experimental aspect of the membrane oscillator ("MO") against the general background of other unstable processes, from 1963 and later (see Franck, 1980). Of special interest is the demonstration that a system of fixed-charge membranes can be arranged to also simulate the *propagation* of potential waves, the most conspicuous property of the living nerve system. Mauro and Israeli workers have also analyzed the MO system. P. Meares (Meares, 1980; Meares and Page, 1972) and collaborators have thoroughly extended the MO concepts both experimentally and by aid of a concise examination of the irreversible thermodynamics. He presents his recent results in another chapter of this book. Over many years I have had a fruitful collaboration with I. Tasaki. As early as 1968 he had adopted the membrane oscillator as a nerve model, later extending and modifying it in several aspects (Fig. 8).

During the last ten years I have pursued the MO work. The main motivation has been its possible relevance to the mechanism of "mechano-electrical transduction," which is exhibited by the physiological "pressure receptors," where mechanical force is coded into electric nerve impulses (hearing, touch, and stretch; Teorell, 1971). The last effort along

* The very name "fixed charge" gave rise to some vague interpretations. Dean (1947) and Teorell (1953) tried to introduce the more proper term "ionic membranes" to emphasize the electrolyte character of the membrane matrix. In particular, G. Ling (1962) in his large, important treatise advanced an "association–induction" hypothesis and abandoned the fixed-charge name to forestall "further confusion" (see Ling, 1962, p. 214). Compare G. Ling's chapter in this volume. References should also be given to the chapters by D. C. Chang, H. R. Leuchtag, and others in this book, as well as to A. Strickholm (1981), who has discussed the possible chemical groups giving rise to "charge."

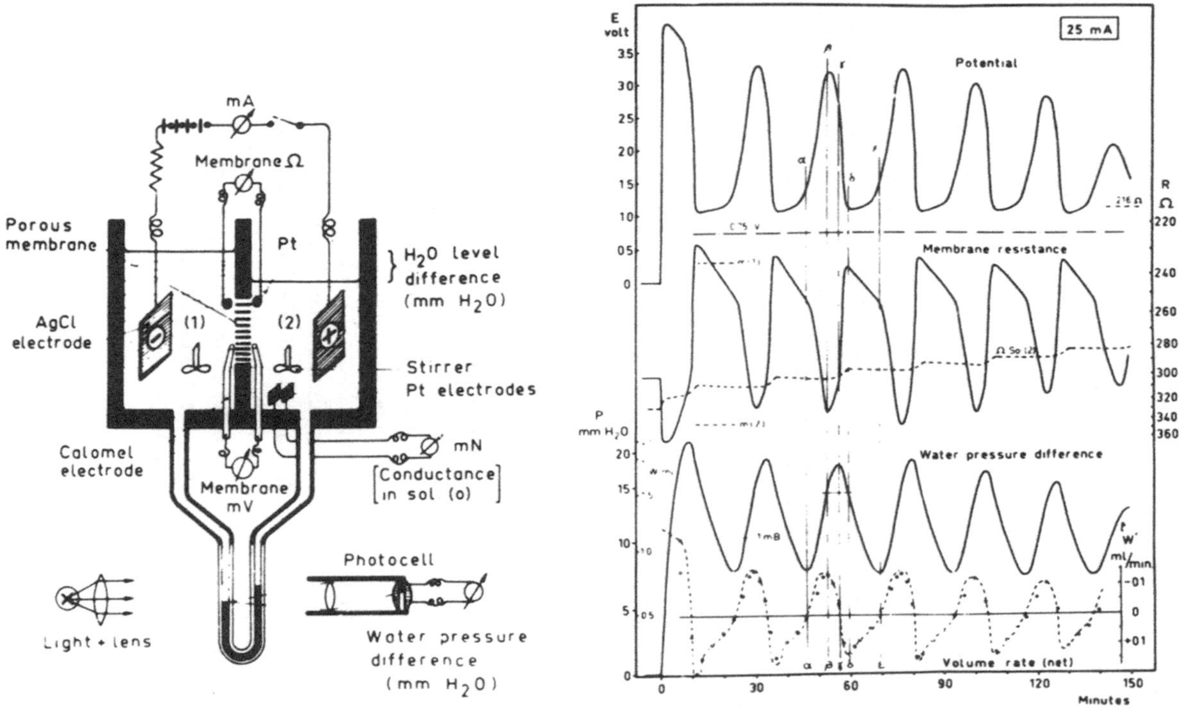

Figure 8. The first experimental set up of the membrane oscillator. The membrane was a porous glass filter or unglazed porcelain. Solutions were 0.1 N and 0.01 N NaCl. The graphs show the time variations of potential, resistance, and hydrostatic pressure as directly recorded. The volume rates were obtained as slopes of the pressure curves. (From Teorell, 1955).

this line deals with the "intrinsic excitation–contraction" coupling in smooth muscles and the heart (Teorell, 1980).

A salient feature of the MO concept is that *flow of water accompanies the ion transport process* (due to the presence of *charged* groups in membranes). Therefore, it is of greatest interest that Tasaki and Iwasa (1980), and in another chapter of this book, have reported swelling and shrinkage processes and pressure variations in isolated squid ions (due to water shifts?), which may conform with the MO model (Teorell, 1980). We realize, of course, that although the similarities between the axon observations and the MO predictions are ambiguous, they may hopefully be a challenge for further search for electroosmotic events in excitable tissues.

V. THE DAWN OF THE NEW AGE OF MEMBRANE RESEARCH

When drawing analogies between models and electrophysiology, caution is necessary, because several other oscillatory membrane systems have been observed in recent years. Mueller *et al.* (1962) discovered action-potential-like phenomena in thin protein-lipid films. A. Monnier has observed potential perturbation in oxidized oil layers. These findings seem to be the start of a new avenue of membrane research, that of the *bimolecular layers* ("bilayers," BLM = black lipid membrane). The status of BLM is covered by M. Jain (1972; with over 1300 references!) and more has come. The subject is treated in part by G. Eisenman in another chapter of this book. Also, this line of research can be traced back to the mid-1930s when Danielli and Davson, pursuing "surface chemistry," studied "bimolecular films" and suggested that they were the basic architecture of living cell components. These studies seem to have opened the search for the "infrastructures." They have shifted the interest from "macro" pores to submicroscopic "channels," "gates," and "carriers." Eisenman and collaborators have used the so-called transition-state theory of Eyring and thereby brought the membrane investigation to an almost atomic level. This has been of particular value for the understanding of ion selectivity. It seems, however, that there remains a great deal to do before the oscillatory events and signal propagation of the living tissues can be understood. The new "channel" concepts do not necessarily mean that charged-pore theories such as the TMS theory are obsolete. Some way of unification has to be found.

In spite of the enormous progress over almost a century it may be safely stated that the nature of bioelectricity still is much more of a riddle than the old pioneers could have ever anticipated. A satisfactory solution of this riddle should, among other things, comprise a complete description of *the major problems of excitation:* (1) the origin of the resting and action potential (and current), the nature of the driving forces and ion selectivity; (2) propagation of the impulse; (3) the "coding" of the nerve signals (oscillation); and (4) the energy sources.

ACKNOWLEDGMENTS. My thanks are due to Prof. K. Sollner for discussion and for permission to reproduce the portrait of Wiedemann and Bernstein (from his private collection), to Inga Ström, librarian, for valuable aid in the literature search, and to Ebon Arnelund and Ragnhild Westerberg for their excellent secretarial work.

REFERENCES

Arrhenius, S., 1892, Untersuchungen über Diffusion von in Wasser gelösten Stoffen, *Zeitschr. Physik. Ch.* **10**:51–95.

Bernstein, J., 1902, Untersuchungen zur Thermodynamik der bioelektrischen Ströme, I, *Pflügers Archiv.* **92**:521–540.

Bernstein, J., 1912, *Elektrobiologie*, Fr. Vieweg & Sohn, Braunschweig.

Beutner, R., 1920, *Die Entstehung Elektrischer Ströme in Lebenden Geweben*, Enke, Stuttgart.

Beutner, R., 1933, *Physical Chemistry of Living Tissues and Life Processes*, Williams & Williams, Baltimore.

Cole. K. S., 1968, *Membranes, Ions and Impulses*, University of California Press, Berkeley and Los Angeles.

Dean, R.B., 1947, The effects produced by diffusion in aqueous systems. *Chem Reviews*, **41**:503–521.

Fick, A., 1855, Über Diffusion, *Peendorfs Ann.* **10**:337.

Fick, A., 1866, *Die Medizinische Physik*, Verlag Fr. Veiweg & Sohn, Braunschweig.

Franck, U., 1980, The Teorell membrane oscillator—a complete nerve model, *Upsala J. Med. Sci.* **85**(No. 3):265–282.

Goldman, D., 1943, Potential, impedance and rectification in membranes, *J. Gen. Physiol.* **27**:37–60.

Helfferich, F., 1959, *Ionenaustauscher, Bd. I*, Verlag Chemie, Wiedheim (English Edition, 1962, *Ion Exchange*, McGraw-Hill, New York).

Jain, M. K., 1972, *The Bimolecular Lipid Membrane*, van Nostrand Reinhold, New York, Toronto, London.

Lakshminarayanaiah, N., 1969, *Transport Phenomena in Membranes*, Academic Press, New York and London.

Ling, G. N., 1962, *A Physical Theory of the Living State: The Association-Induction Hypothesis*, Blaisdale, New York and London.

Loeb, J., 1922, *Proteins and the Theory of Colloidal Behavior*, McGraw-Hill, New York.

Loeb, J., and Beutner, R., 1912, Über die Potentialdifferenzen an der unverletzten und verletzten Oberfläche pflanzlicher und tierischer Organe, *Biochem. Zeitschr.* **41**:1–26 (also cited in Beutner, 1933, p. 196).

Meares, P., 1980, Coupling of ion and water fluxes in synthetic membranes, *Upsala J. Med. Sci.* **85**(No. 3):259–264.

Meares, P., and Page, K. R., 1972, Rapid force-flux transitions in highly porous membranes, *Phil. Trans. R. Soc. A (London)* **272**:1–46.

Meyer, K. H., and Sievers, J. F., 1936, several articles in *Helv. Chim. Acta* **19**:649, 665, 987.

Michaelis, L., 1922, *Die Wasserstoffionen-Konzentration*, 2nd ed., Springer Verlag, Berlin.

Monnier, A. M., 1980, The possible role of dielectric constant variation and of electro-osmosis in excitable natural and artificial membranes. An extension of Teorell's "membrane oscillator," *Upsala J. Med. Sci.* **85**(No. 3):237–246.

Mueller, P., Rudin, D., Tien, H. T., and Wescott, W., 1962, Reconstitution of cell membrane structure in vitro and its transformation into an excitable system, *Nature* **194**:979–980.

Osterhout, W. J. V., 1956, The role of water in protoplasmic permeability and in antagonism, *J. Gen. Physiol.* **39**:963–976.

Rothschuh, K. E., 1953, *Geschichte der Physiologie*, Springer-Verlag, Berlin, Göttingen, Heidelberg (English version, 1973, *History of Physiology*, R. Krieger, Huntington, New York).

Schlögl, R., 1964, *Stofftransport durch Membranen*, Steinkopff Verlag, Stuttgart.

Sollner, K., 1976, The early development of the electrochemistry of polymer membranes, in: *Charged Gels and Membranes*, Part 1 (E. Sélégny, ed.), pp. 3 –55, Reidel, Dordrecht-Holland.

Strickholm, A., 1981, Control of ionic permeability by membrane charged groups: Dependency of pH, depolarization, tetrodotoxin and procaine, *Upsala J. Med. Sci.* **86**:9–21.

Tasaki, I., 1968, *Nerve Excitation, A Macromolecular Approach*, Charles C. Thomas, Springfield, Illinois.

Tasaki, I., and Iwasa, K., 1980, Swelling of nerve fibers during action potential, *Upsala J. Med. Sci.* **85**(No. 3):211–215.

Teorell, T., 1935a, Studies on the "diffusion effect" upon ionic distribution. I. Some theoretical considerations, *Proc. Natl. Acad. Sci. USA* **21**:152–161.

Teorell, T., 1935b, An attempt to formulate a quantitative theory of membrane permeability, *Proc. Soc. Exp. Biol. Med.* **33**:282–285.

Teorell, T., 1936a, Ionic transference numbers in cellophane membranes, *J. Gen. Physiol.* **19**:917–927.

Teorell, T., 1936b, A method of studying conditions within diffusion layers, *J. Biol. Chem.* **113**:735–748.

Teorell, T., 1937a, Studies on the "diffusion effect" upon ionic distribution. II. Experiments on ionic accumulation, *J. Gen. Physiol.* **21**:107–132.

Teorell, T., 1937b, The properties and functions of membranes. Natural and artificial, *Trans. Faraday Soc.* **33**:939, 983, 1053, 1086.

Teorell, T., 1948, Membrane electrophoresis in relation to biological polarization effects, *Nature* **162**:961.

Teorell, T., 1949a, Permeability, *Annu. Rev. Physiol.* **11**:545 –564.

Teorell, T., 1949b, Membrane electrophoresis in relation to biological polarization effects, *Arch. Sci. Physiol.* **3**:205–220.

Teorell, T., 1951, Zur quantitativen Behandlung der Membranpermeabilität, *Zeitschr. Elektrochemie. u. Angew. Physik. Ch.* **55**:460 –469.

Teorell, T., 1953, Transport processes and electrical phenomena in ionic membranes, *Prog. Biophys.* (and *Biophys. Chem.*) **3**:305–369.

Teorell, T., 1955, A contribution to the knowledge of rhythmical transport processes of water and salt, *Exp. Cell Res.* (Suppl.) **3**:339–345.

Teorell, T., 1956, Transport phenomena in membranes, Eighth Spiers memorial lecture, *Faraday Soc. Disc.* **21**:9–26.

Teorell, T., 1958, Transport processes in membranes in relation to the nerve mechanism, *Exp. Cell Res.* (Suppl.) **5**:83–100.

Teorell, T., 1959a, Electrokinetic membrane processes in relation to properties of excitable tissues. I. Experiments on oscillatory transport phenomenon in artificial membranes, *J. Gen. Physiol.* **42**:831–845.

Teorell, T., 1959b, Electro-kinetic membrane processes in relation to the properties of excitable tissues. II. Some theoretical considerations, *J. Gen. Physiol.* **42**:847–863.

Teorell, T., 1961, Oscillatory electrophoresis in ion exchange membranes, *Arkiv. Kemi. (R. Acad. Sci. Stockholm)* **22**(18):401–408.

Teorell, T., 1971, A biophysical analysis of mechano-electrical transduction, in: *Handbook of Sensory Physiology, Vol. 1. Principles of Receptor Physiology* (W. R. Loewenstein, ed.), pp. 292–339, Springer Verlag, Berlin, Heidelberg, New York.

Teorell, T., 1976, The development of the modern membrane concepts in relation to biological phenomena, in: *Charged Gels and Membranes,* Part I (E. Sélégny, ed.), pp. 57–69, Reidel, Dordrecht-Holland.

Teorell, T., 1980, A fixed charge system as a formal model for the behaviour of smooth muscles and the heart, *Upsala J. Med. Sci.* **85**(No. 3):201–209.

Wiedemann, G., 1893, *Die Lehre von der Elektrizität,* 2nd Ed., Vol. 1, pp. 982–1023, Fr. Vieweg & Braunschweig.

Wilbrandt, W., 1938, Review of current membrane research, *Ergebn. Physiol.* **40**:204.

17

Flux Coupling and Nonlinear Membrane Phenomena

Patrick Meares

I. THE TEORELL OSCILLATOR

A well studied and understood artificial system that shows affinity with biological excitability is the Teorell membrane oscillator (Teorell, 1959). In addition to showing rhythmical oscillations in potential and pressure, the membrane oscillator can, over certain ranges of external variables, exist in two stable steady states. These are a high-conductance state and a low-conductance state between which it can be triggered to switch. Although the membranes that have been used so far in the membrane oscillator bear no resemblance, chemical or structural, to biological membranes, it is possible that a thorough physicochemical understanding of the membrane oscillator may help to reveal some of the constraints governing excitable biological systems.

The configuration of the oscillator is shown in Fig. 1. It consists of a membrane situated between two solutions differing in composition or concentration so that when the membrane is in equilibrium with one its conductance is lower than when it is in equilibrium with the other. For convenience we shall consider only two solutions of a single electrolyte such that the solution in compartment I is more dilute than that in compartment II. Without wishing to stress analogy with biological systems, compartment I may be regarded as the exterior and compartment II as the interior of an excitable cell.

Electrodes in compartments I and II permit an ion current to be driven electrically through the membrane. If it is less permeable to anions than to cations the direction of polarization is as shown, otherwise it is reversed. Means are provided also to maintain and record a negative pressure differential between I and II. Probe electrodes and a transducer are used to record continuously the potential difference and pressure difference across the membrane.

The arrangement shown is suitable for studying the behavior of the system when all variables except the fluxes can be fixed or varied at will. One can fix either current or

Patrick Meares • Chemistry Department, University of Aberdeen, Old Aberdeen AB9 2UE, Scotland.

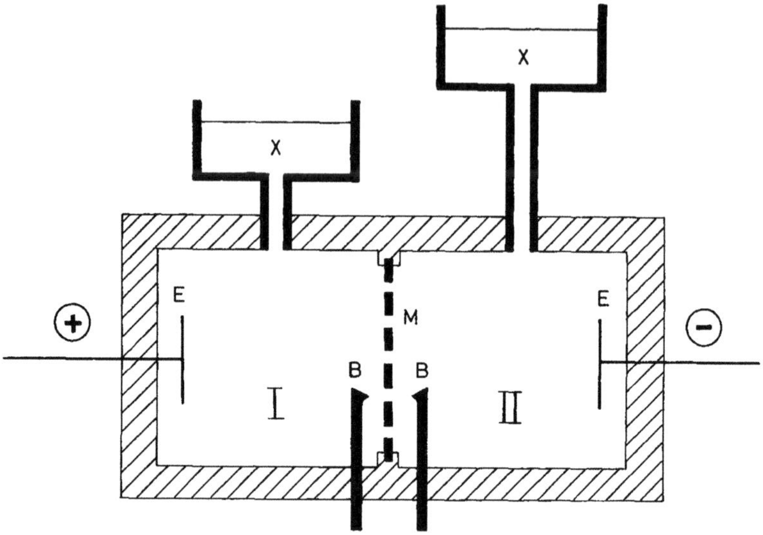

Figure 1. Membrane oscillator (diagrammatic) M membrane, E current electrodes, B probe electrodes, X solution reservoirs, I dilute compartment, II concentrated compartment. (Reproduced with permission from Meares and Page, *Ion Transport in Plants*, W. P. Anderson ed., Copyright Academic Press, London, 1973).

potential but not both. In this way nonlinear connections between fluxes and forces can be examined and the ranges where bistability occurs noted.

Oscillations are observed only when two forces, commonly pressure and potential, are unrestrained. Oscillating regimes are always possible in regions of bistable stationary states. It is sufficient to discuss only the latter to explore the role of flux coupling in the behavior of such a membrane system.

II. FLIP-FLOP TRANSITIONS

Figure 2 is a typical plot of voltage vs. current. It shows nonlinearity and two stable states, a low resistance state R_1 and a high resistance state R_2. The experiment shown was carried out galvanostatically by adjusting the current and observing the potential (Meares and Page, 1972). Transition from the low- to the high-resistance state occurs at high current I_1 and the reverse transition at a low current I_2. The membrane in this experiment was a Nucleopore filter of 0.5 μm pore diameter separating 0.10 and 0.01 M NaCl at a pressure difference of -589 Pa (1 pascal = 1 newton/m^2).

The physical explanation of Fig. 2 is as follows. At zero current the pressure drives the concentrated solution from II to I; the pores being occupied by this solution puts the membrane into its high-conductance, i.e., low-resistance, state R_1. As current is increased, the principal ion flux is of cations directed from I to II. A parallel flow of solution is induced electroosmotically in the pores from I to II. At low currents, electroosmosis reduces the pressure-driven efflux. Interdiffusion of solutions I and II in the pores establishes a concentration profile in the membrane which is dependent on the net volume flow (Fig. 3). As the current is increased, electroosmosis increases and the concentration profile is modified.

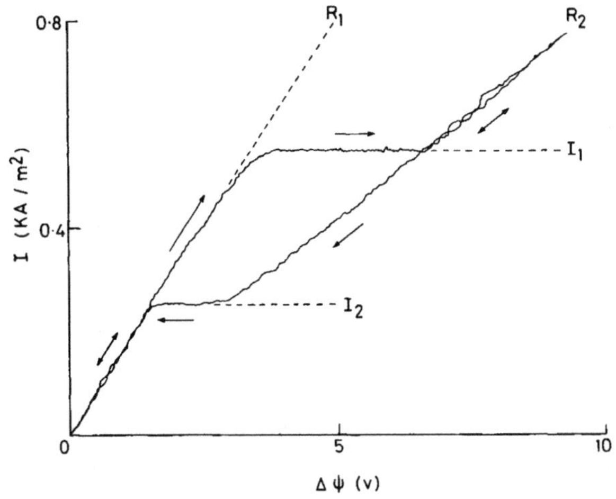

Figure 2. Galvanostatic voltage vs. current plot showing flip-flop. Nucleopore membrane 0.5 μm pore diameter, 0.01 and 0.10 M NaCl, $\Delta p = -589$ Pa. (Reproduced with permission from Meares and Page, *Phil. Trans. R. Soc. A* **272,** 1972).

Curvature of the voltage–current plot results. At a critical value of the current the electroosmotic influx exceeds the hydrodynamic efflux and the pores fill with the dilute solution. At this point either the membrane potential increases sharply or, under potentiostatic conditions, the current falls.

Electroosmosis increases monotonically, though not always linearly, with membrane potential. Once the system is in its high-resistance high-potential state, it will remain there while the current is reduced below I_1 until at some lower current I_2 the hydrostatic pressure once again takes control and refills the pores with solution II. The larger the pressure difference the larger I_1 and the larger the hysteresis loop. At sufficiently low pressures only a monostable but curved voltage–current plot is obtained because interdiffusion of I and II across the membrane acts to reduce the sharpness of these so-called "flip-flop" transitions.

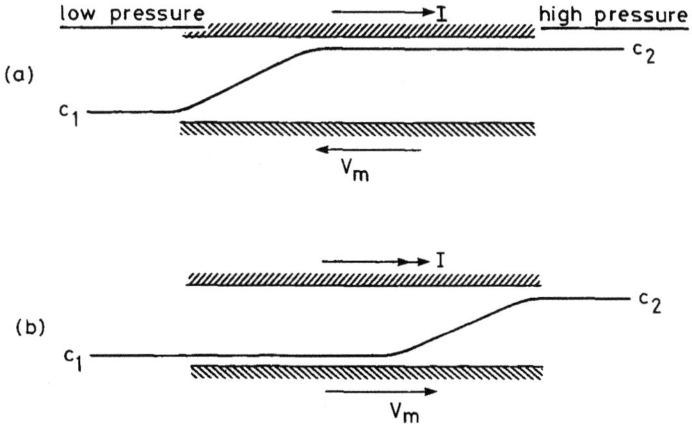

Figure 3. Schematic concentration profiles in membrane capillary at (a) low current, (b) high current.

III. ION–WATER FLUX COUPLING IN POROUS MEMBRANES

The electroosmosis opposing the hydrodynamic flow arises as a result of coupling between ion and water fluxes in the membrane. This coupling, moderated slightly by ion–ion coupling, determines quantitatively the behavior of the Teorell membrane system. To understand the connections between excitability-reminiscent phenomena and structure in various artificial membranes we must examine the nature of flux coupling in them.

The first membranes examined were highly porous and included glass sinters, powdered quartz, and aluminum oxide and plugs of compressed, fine-mesh ion-exchange beads (Teorell, 1959; Franck, 1963, 1967; Drouin, 1969). Although some of these materials have a high surface charge density their high porosity makes it easy for hydrodynamic flow to predominate over electroosmotic flow.

Theories developed to describe the behavior of such membranes concentrated on their porous nature (Kobatake and Fujita, 1964; Mikulecky and Caplan, 1966; Meares and Page, 1972). The membrane volume flux can be expressed by

$$V_m = -\pi a^2 N \left[\frac{a^2}{8\eta} \left(\frac{dp}{dx} \right) - \frac{\varepsilon \phi_o}{4\pi\eta} \left(\frac{d\psi}{dx} \right) \right] \tag{1}$$

where a is the radius of the membrane pores and N their number per unit area. η and ε are the viscosity and dielectric constant of the pore fluid. dp/dx and $d\psi/dx$ are the pressure and potential gradients at distance x along a membrane pore. ϕ_o is the electric potential at the pore walls. It is a function of x because the surface charge density, which results from ion adsorption or the ionization of membrane bound weakly basic or acidic groups, is a function of the concentration at x.

Equation (1) is a superposition of two terms: (1) a viscous flow given by Poiseuille's law, and (2) an electroosmotic flow described by classical electrokinetic theory in which the Gouy–Chapman theory may be used to describe the diffuse electric double layer at the pore walls. This is a good approximation for broad pores such that $\kappa a \gg 1$ (κ is the Debye–Hückel reciprocal length parameter) and provided the membrane thickness greatly exceeds a.

To be useful, Eq. (1) has to be integrated across the membrane subject to boundary conditions of concentrations and pressure and to the imposed electric current. The integrated form of V_m is very complex (Meares and Page, 1972).

Figure 4 shows values of V_m computed as a function of membrane potential at several pressures. The associated current–voltage and resistance–voltage plots are also shown. It can be seen that the switch from low- to high-resistance states begins close to the potential at which the volume flux changes sign. Notice also that the V_m–$\Delta\psi$ plot becomes less straight as Δp is increased.

When V_m is plotted against current, an S-shaped curve, such as that in Fig. 5, is obtained. The system can have three solutions between the transition currents I_1 and I_2, but only the highest and lowest volume fluxes are accessible at constant current.

Once the volume flux is known, the cation and anion fluxes can also be calculated by using the Nernst–Planck equation to describe the diffusional and electrically driven ion fluxes relative to the local center of mass and then adding to these a convective term made up of the product of the volume flux and the local ion concentration. Figure 5 shows that for these fluxes multivalued solutions are also found between I_1 and I_2.

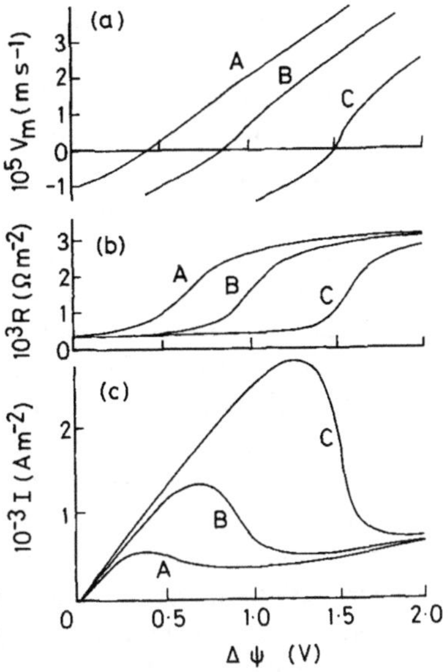

Figure 4. (a) Volume flow V_m, (b) membrane resistance R and (c) current density I each plotted against membrane potential. Curve A, $\Delta p = -196$ Pa; curve B, $\Delta p = -392$ Pa; curve C, $\Delta p = -687$ Pa.

Meares and Page (1972) used Nucleopore filter membranes to test the theory. These membranes have a uniform circular pore structure, which enabled reliable values to be assigned to a, N, and the membrane thickness. No such definite values could be given to the electroosmotic characteristics. They were determined by observing electroosmotic flow in the absence of pressure and concentration gradients, and by assuming the Helmholtz–Smoluchowski theory of electrokinetic processes. Electroosmosis was therefore idealized as plug flow in large capillaries and no specific information on the coupling between water and ion flows was required. The quantitative agreement obtained between the observed and predicted behaviors of the Teorell system owed much to the predictions being independent of a molecular scale model of electroosmosis.

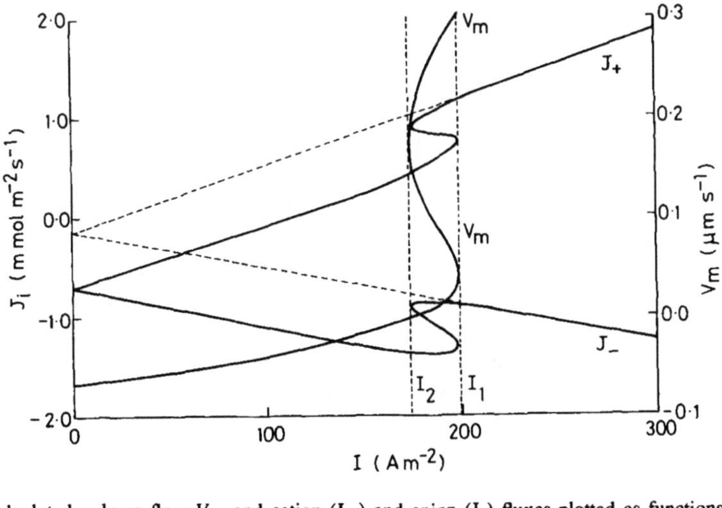

Figure 5. Calculated volume flow V_m, and cation (J_+) and anion (J_-) fluxes plotted as functions of current I. Membrane and solutions as Fig. 2, $\Delta p221 = -294$ Pa. (Reproduced with permission from Meares and Page, *Phil. Trans. R. Soc. A* **272,** 1972.)

IV. THE FINE-PORE MEMBRANE OSCILLATOR

Equations developed to describe long pores that are wide compared with the electric double layer thickness cannot be extended to describe the majority of excitable biological membranes. To discuss these it is desirable to avoid the concept of discrete pores, and especially to avoid describing solution in any pores as a viscous continuum. The fine-pore theory of charged membranes introduced by Schmid (1950, 1951) and extended and expounded by Schlögl (1964) forms the starting point for an alternative treatment. Despite its name, the existence and structure of membrane pores does not form part of this theory. The space charge due to the free ions counter to the membrane charges is assumed to be uniformly distributed either through the membrane or through the liquid swelling it. (Authors have not always made clear which assumption was being made. The difference is significant when the ions and water are not both uniformly distributed.) In the fine-pore theory there is no meaning to be attached to the different velocity profiles of plug flow and Poiseuille flow in cylindrical pores. Convective movement of the membrane liquid is treated as being directly proportional to the net force applied uniformly throughout its volume. It is not affected by the individual contributions that go to make up this force.

The mathematical treatment of the membrane oscillator fitted with a fine-pore membrane uses the Nernst–Planck flux equation, supplemented for convective flow, to deal with the ion fluxes (Mackie and Meares, 1955; Schlögl and Schödel, 1955). The convective flow is driven by pressure, osmotic pressure, and potential differences. Franck (1963) first combined these into a theory of the oscillator.

Very recently Langer *et al.* (1981) have re-examined the problem of the fine-pore membrane oscillator on the basis of three equations. For the anion and cation fluxes they used

$$J_i = -u_i c_i \,(RTd \ln c_i/dx + z_i F \, d\psi/dx) + c_i v \qquad (2)$$

where J_i is the flux of ions i whose charge number z_i includes the appropriate sign. u_i and c_i are the mobility and concentration in the membrane, and v is the local barycentric velocity and is essentially equivalent to the volume flux V_m of Eq. (1). It was expressed by Langer *et al.* (1981) by using the equation (Schlögl, 1955)

$$v = -d_h \left[\left(\frac{dp}{dx} \right) - F\omega X \left(\frac{d\psi}{dx} \right) \right] \qquad (3)$$

where ω is the charge number and X the concentration of fixed ions in the membrane expressed in terms of the volume of solution absorbed by the membrane.

The analogy between Eqs. (1) and (3) is obvious but, whereas in Eq. (1) the potential ϕ_o can be estimated only from electrokinetic theory, X can be measured analytically. Thus, according to Eq. (1), the absolute value of the hydrodynamic flow is fixed but its ratio with the electroosmotic flow is not, whereas in Eq. (3), the ratio of hydrodynamic and electroosmotic flows is fixed but their absolute magnitudes are not. They can be determined only by experimental measurement of the hydrodynamic permeability d_h.

There is a further important difference between Eq. (1) and (3). In the porous membrane c_i, p, and ψ may be regarded as continuous across the membrane–solution interfaces. In the fine-pore membrane, these symbols refer to quantities inside the membrane, and they change sharply across its boundaries. In the theory, the boundary changes in concentration and potential were treated as ideal Donnan equilibria. The concentrations so calculated were used with the Van't Hoff equation to estimate the osmotic pressure differences across the boundaries.

The integration of the two ion flux equations [Eq. (2)] and the volume-flux Eq. (3) subject to these boundary conditions gives a complicated algebraic result. It can be handled only by the application of numerical methods. Langer *et al.* (1981) have applied the results to and made measurements of voltage–current plots on several cellulose nitrate and polyamide membranes with equivalent pore radii in the range 8–51 nm.

They have observed flip-flop behavior in membranes down to 17 nm equivalent pore radius with fixed-ion concentrations of the order 0.1 mol m^{-3} when using pressures up to 10^4 Pa. The computations were in accord with the experimental findings. Langer *et al.* (1981) concluded there was a lower limit on the hydrodynamic permeability and an upper limit on the fixed-ion concentration beyond which flip-flop and oscillatory behavior would not be found at these pressures.

Further computations indicated that at higher pressures and in thinner membranes these limits could be extended, especially where there was a large difference between the mobilities of the anions and cations in the membrane. They concluded that under such conditions, flip-flops might be found at potentials in the region of 100 mV in membranes with permeabilities and fixed ion concentrations not unlike those of some biological tissues and plant cell walls. Thus, they left open the question as to whether the Teorell oscillator mechanism might be of direct relevance in biology.

V. THE EFFECTS OF IDEALIZATIONS AND SIMPLIFICATIONS

We shall now examine whether the idealizations and simplifications required by Langer *et al.* (1981) when solving and applying their equations make it more or less likely that the Teorell oscillator may be of biological significance. We shall examine in particular how flux couplings are handled in their treatment which uses Eq. (2), the Nernst–Planck equation, extended to include solvent drag. It has been tested in detail and found to be satisfactory provided certain precautions are observed (Meares, 1973, 1981). These include the necessity to measure directly the ion mobilities in the membrane. This can be done conveniently and with little error by using radiotracers. Although these mobilities often vary with concentration, it is not practicable to take this into account when integrating the flux equations. It is not satisfactory quantitatively to estimate the u_i from ionic mobilities in free solution. Interaction of the mobile ions with the electric fields around the fixed charges slows the counter-ions relative to the co-ions to an important extent. The effect is concentration-dependent and is greatest when the membrane is in contact with dilute solutions. The effect is even greater when the counter-ions are of valency higher than 1. Data are shown in Table I for two cation-selective membranes. The ratio of the co- and counter-ion mobilities in the membrane is clearly greater than in solution. According to Langer *et al.* (1981) this effect would increase the probability of flip-flop behavior.

The ion concentrations at and the potential differences across the membrane–solution interfaces are related to the properties of the membrane and solutions by the Gibbs–Donnan equilibrium. This must be written in terms of ion activities a_i in the membrane and a_i° in the solution

$$a_i = a_i^\circ \exp\left[-(z_i\psi_D + p_D\overline{V}_i)/RT\right] \tag{4}$$

Table I. Relative Mobilities of Co-Ions and Counter-Ions in Membranes and in Free Solution at 25°C

	Concentration			$u_{\text{co-ion}}/u_{\text{counter-ion}}$	
	mol m^{-3}	Counter-ions	Co-ions	Membrane	Solution
Zeo-Karb 315	10	Na^+	Br^-	2.22	1.56
cation	100	Na^+	Br^-	2.00	1.56
exchanger membrane,	10	Cs^+	Br^-	2.70	1.01
X = 745 mol m^{-3}	100	Cs^+	Br^-	1.96	1.01
	5	Sr^{2+}	Br^-	9.00	2.63
	50	Sr^{2+}	Br^-	4.35	2.63
Eastman–Kodak 398.3	25–200	Na^+	Cl^-	1.79	1.52
cellulose	25–200	K^+	Cl^-	1.54	1.04
acetate membrane,	12–100	Ca^{2+}	Cl^-	4.17	2.56
X = 5.5 mol m^{-3}					

Here ψ_D is the potential and p_D the pressure just inside the membrane relative to that just outside, and $\overline{V}_i$ is the partial molar volume of the ions. Usually, the pressure term is omitted but the validity of this omission may be questioned in view of the way in which the pressure and potential terms in the volume flow Eq. (3) have to be estimated.

When p_D is ignored and the Donnan equations for counter-ions and co-ions are combined to eliminate ψ_D there results (Glueckauf, 1962)

$$(c_c/\nu_c)^{\nu_c}\,(c_c/\nu_c + X/\nu_g)^{\nu_g} = (\alpha c^o)^{\nu} \tag{5}$$

where c_c is the molarity of co-ions in the membrane, c^o the molarity of the salt in the solution, ν_c and ν_g the number of co-ions and counter-ions per mole of the electrolyte, and α a ratio of mean activity coefficients. The ideal Donnan equation is obtained by letting $\alpha = 1$. When c^o is small it has been found many times that the ideal Donnan equation seriously underestimates c_c. For example, in the cation exchanger Zeo-Karb 315, α falls below 0.7 for NaBr at 0.01 M. It is inconvenient to have α varying with concentration and it has been found empirically (Meares, 1973) that

$$(c_c/\nu_c)^{\nu_c}\,(c_c/\nu_c + X/\nu_g)^{\nu_g} = (\alpha c^o)^{\beta\nu} \tag{6}$$

where α and β are constants. This equation represents the data very well over wide concentration ranges. For NaBr in Zeo-Karb 315 α is then close to 1 and β is about 0.8.

Because the co-ions are not attracted by the fixed charges, the Donnan potentials can be estimated from the ratio of the inside and outside co-ion concentrations

$$\psi_D = \frac{RT}{F}\ln(c_c/\alpha\nu_c c^o) \tag{7}$$

where c_c is obtained from Eq. (6).

If one uses the ideal Donnan expression, the concentration difference across the interior of the membrane is exaggerated and, hence, so too are the diffusional ion fluxes. In the flip-flop theory, these fluxes tend to obliterate the steep concentration profiles needed to create the transitions. Thus, by following Langer *et al.* (1981) and using the ideal Donnan distribution the likelihood of flip-flop phenomena tends to be underestimated.

The Nernst–Planck equation neglects coupling between ion fluxes. Careful measurements have shown these to be small in the well-hydrated cation exchanger Zeo-Karb 315 (Meares, 1976). There are no direct measurements on other membranes and it must be recognized that ion–ion coupling may be significant in membranes of low dielectric constant in which ion clustering tends to occur.

The term $c_i v$ in the ion-flux equation accounts for the coupling between ion and water fluxes. It implies that tight coupling occurs in all cases. In Zeo-Karb 315 the coupling coefficients q_{i3}, where 3 indicates water, have been found to vary between 0, i.e., no coupling, and unity, i.e., tight coupling (Meares, 1976). In fact q_{i3} was found to lie between 0.0 and 0.4 for the co-ions and between 0.4 and 0.7 for the counter-ions. Thus, the use of $c_i v$ without a coupling factor less than unity overestimates the effect of coupling between ion and water fluxes. Coupling is essential to the occurrence of flip-flop transitions, but overestimating it probably overestimates the pressure region in which the flip-flops would be found. There does not appear to be any comparable information on other membranes to indicate whether these observations are of wider significance.

VI. THE VOLUME FLUX EQUATION

It is important also to check the volume flux Eq. (3) against experimental data but there are no data on fine-pore membranes where p and ψ have been varied systematically. Volume fluxes in Zeo-Karb 315 have been measured and analyzed by using the formalism of nonequilibrium thermodynamics (Foley *et al.* 1974). The appropriate equation is

$$-V_m = L_{p\pi}[\pi(1 + \bar{c}^o \bar{V}_s)/v_g\bar{c}^o] + L_p(p^o - \pi) + L_{pE}E \tag{8}$$

Here p^o, E, and π are the pressure, reversible co-ion potential, and osmotic pressure differences between the external solutions, $\bar{c}^o$ is a thermodynamically defined mean of the external solution concentrations, and $\bar{V}_s$ the partial molar volume of the salt. L_p, L_{pE}, and $L_{p\pi}$ are thermodynamic permeability coefficients. L_p is concerned with hydrodynamic flow, L_{pE} with electroosmosis, and $L_{p\pi}$ with the coupling of solute and water fluxes.

When measurements are made with no concentration difference across the membrane, Eq. (8) becomes

$$-V_m = L_p p^o + L_{PE}E; \; \pi = 0 \tag{9}$$

Under short-circuit conditions E is zero and L_p can be determined in a pressure flow experiment. It is identical to d_h/δ, where δ is the membrane thickness. Similarly, L_{pE} can be determined from the electroosmotic flow in the absence of a pressure difference. Comparison of Eq. (3) with the nonequilibrium thermodynamic scheme shows that

$$L_{pE} = -F\omega X d_h/\delta = FXL_p \tag{10}$$

should hold.

Table II shows that L_{pE}/FL_p is a good constant over a range of concentrations for three salts in Zeo-Karb 315. In each case, the ratio was considerably less than X determined analytically from the exchange capacity divided by the water content of Zeo-Karb 315. The difference is greater for Cs^+ than for Na^+ and is greatest for Sr^{2+}. This is the order of increasing binding of these counter-ions by the membrane charges.

Table II. Electroosmotic Permeability/Hydrodynamic Permeability in mol m^{-3} for Several Salts in Zeo-Karb 315

c^0	NaBr		CsBr		$SrBr_2$	
mol m^{-3}	L_{pE}/FL_p	X	L_{pE}/FL_p	X	L_{pE}/FL_p	X
10	328	727	290	762	195	744
30	315	733	280	764	194	746
90	275	740	290	772	204	750
500	349	750			183	759

More limited measurements (S. Gashi, unpublished data) have been made on membranes of polysulphone sulphonic acid, which absorb far less water than Zeo-Karb 315. They act like ionomer membranes in that the ions and water in them are clustered rather than homogeneously distributed. From the data below, it can be concluded that the electroosmotic effect in these membranes relative to their hydrodynamic permeability is far less than their ionic content would lead one to expect.

L_{pE}/FL_p	X
mol m^{-3}	mol m^{-3}
8	1560

These observations show that in membranes of low water content and dielectric constant, which physically resemble biological membranes more closely than do typical synthetic ion-exchange membranes, electroosmosis declines in importance relative to hydrodynamic flow. Whereas the equation for V_m used by Langer *et al.* (1981) would predict that electroosmosis would overwhelm hydrodynamic flow at moderate values of X, our results show that this may not happen and that flip-flops might easily be found in the pressure and potential ranges of interest.

It should be mentioned that Langer *et al.* (1981) determined X for their membranes from measurements of L_p and L_{pE}. Hence, they found an "electroosmotically effective" value of X. Unfortunately there is no reason to believe that this is equal to the "ideal Donnan effective" value of X needed to describe the boundary distributions in terms of the ideal Donnan equation.

It must also be considered whether Eq. (3) handles properly the effect of an osmotic pressure difference across the membrane. This might be tested by comparing the observed and predicted volume flows in the absence of an external pressure difference. Attempts to carry out this test, while setting the osmotic and ion-activity coefficients at unity, have produced meaningless results because they depend upon small differences between large terms. Indeed the neglect of the pressure term in the Donnan distribution Eq. (5) has a significant effect on these small differences.

The osmotic flow is directed against the hydrodynamic flow in the fine-pore membrane oscillator. If Eq. (3) exaggerates the importance of the osmotic pressure difference due to its neglect of coupling between the efflux of salt and the osmotic influx, flip-flops would be expected at somewhat lower pressures than the calculations of Langer *et al.* (1981) suggest.

VII. CONCLUSIONS

The outcome of this analysis has been to show that the quantitative errors introduced by the use of idealizations in the calculations of Langer *et al.* (1981) have tended to underplay the probability of flip-flop and oscillatory phenomena occurring in fine-pore membranes by the Teorell mechanism. Their new theory is greatly to be welcomed and one hopes it will be further refined. Langer *et al.* (1981) have already reported that the inclusion in the theory of unstirred layers at the membrane faces enhances the probability that flip-flops will be found. This was demonstrated earlier also by Meares and Page (1972) with porous membranes. Experimental research on these nonlinear processes should be continued and extended to membranes more closely related in structure and properties to biological types because there is a real possibility that hydraulically coupled mechanoelectric transduction processes may be more widespread in biology than has previously been believed.

ACKNOWLEDGMENT. The author expresses his thanks to Drs. Langer, Page, and Wiedner for showing him their paper before publication.

REFERENCES

Drouin, H., 1969, Experiments with the Teorell membrane oscillator, *Ber. Bunsenges. Phys. Chem.* **73**:223–229.
Foley, T., Klinowski, J., and Meares, P., 1974, Differential conductance coefficients in a cation exchange membrane, *Proc. R. Soc. A* **336**:327–354.
Franck, U. F., 1963, On the electrochemical properties of porous ion-exchange membranes, *Ber. Bunsenges. Phys. Chem.* **67**:657–671.
Franck, U. F., 1967, Phenomena in biological and artificial membranes, *Ber. Bunsenges. Phys. Chem.* **71**:789–799.
Glueckauf, E., 1962, A new approach to ion exchange polymers, *Proc. R. Soc. A* **268**:350–370.
Kobatake, Y., and Fujita, H., 1964, Flows through charged membranes I. Flip-flop current versus voltage relation, *J. Chem. Phys.* **40**:2212–2218.
Langer, P., Page, K. R., and Wiedner, G., 1981, A Teorell oscillator system with fine pore membranes, *Biophys. J.* **36**:93–107.
Mackie, J. S., and Meares, P., 1955, The diffusion of electrolytes in a cation-exchange resin membrane, *Proc. R. Soc. A* **232**:498–509.
Meares, P., 1973, The permeability of charged membranes, in: *Transport Mechanisms in Epithelia* (H. H. Ussing and N. A. Thorn, eds.), pp. 51–71, Munksgaard, Copenhagen.
Meares, P., 1976, Some uses for membrane transport coefficients, in: *Charged and Reactive Polymers* (E. Selegny, ed.), pp. 123–146, Reidel, Dordrecht.
Meares, P., 1981, Coupling of ion and water fluxes in synthetic membranes, *J. Membr. Sci.* **8**:295–307.
Meares, P., and Page, K. R., 1972, Rapid force-flux transitions in highly porous membranes, *Phil. Trans. R. Soc. A* **272**:1–46.
Mikulecky, D. C., and Caplan, S. R., 1966, The choice of reference frame in the treatment of membrane transport by non-equilibrium thermodynamics, *J. Phys. Chem.* **70**:3049–3056.
Page, K. R., and Meares, P., 1973, Solute–water interactions in the Teorell oscillator membrane model, in: *Ion Transport in Plants* (W. P. Anderson, ed.), pp. 65–75, Academic Press, London.
Schlögl, R., 1955, On the theory of anomalous osmosis, *Z. Phys. Chem. N.F.* **3**:73–102.
Schlögl, R., 1964, *Material Transport Through Membranes*, Chapters V and VI, Steinkopf Verlag, Darmstadt.
Schlögl, R., and Schödel, U., 1955, On the transport properties of a porous charged membrane, *Z. Phys. Chem. N.F.* **5**:372–397.
Schmid, G., 1950, On the electrochemistry of fine-pore capillary systems I. Survey, *Z. Electrochem.* **54**:424–430.
Schmid, G., 1951, On the electrochemistry of fine-pore capillary systems II. Electro-osmosis, *Z. Electrochem.* **55**:229–237.
Teorell, T. T., 1959, Electro-kinetic membrane processes in relation to properties of excitable tissues I. Experiments on oscillatory transport phenomena in artificial membranes, *J. Gen. Physiol.* **42**:831–846.

18

Solvent Substitution as a Probe of Gating Processes in Voltage-Dependent Ion Channels

Charles L. Schauf

I. SUBSTITUTION OF D_2O FOR H_2O AS A PHYSIOLOGICAL TOOL

Although it is widely accepted that ion transport across excitable membranes occurs via voltage-dependent ion-selective channels, and much has been learned about the microscopic events which occur prior to, during, and following activation of such channels, many important questions remain unresolved. For example, does the transient conductivity of channels arise from a relatively small alteration in the conformation of a preexisting structure (a "gate"), or is it due to a field-dependent activation of channel precursors, which must then interact locally to form a conducting pathway? Because the transmembrane electric field is the independent variable that controls permeability, activation of channels is likely to be triggered by some intramembrane charge movement, but what is the molecular origin of this displacement current and what relationship does it have to the creation of a conducting channel? Once opened, what is the nature of the pathway an ion sees in traversing the membrane? Does physiological inactivation of the Na^+ channel depend on the existence of another specialized gate-like structure, or on the interaction of a blocking particle with the conducting channel? Is inactivation perhaps just a consequence of the relative stability of one of several intermediate, nonconducting protein conformations, so that the term "inactivation" becomes a misnomer to the extent that it is taken to imply the existence of a separate kinetic process? In an attempt to answer such questions investigators have drugged, irradiated, and digested axons, have grossly altered internal and external electrolyte compositions, and have subjected axons to increasingly more complex voltage-clamp protocols.

Relatively little attention has, however, been directed toward using the solvent itself as a means of revealing the nature of microscopic events, despite the intense use of such techniques in elucidating reaction mechanisms of chemical systems. In recent studies (Schauf and Bullock 1979, 1980), we have demonstrated the feasibility of using the substitution of heavy water (D_2O) for H_2O as such a probe in voltage-clamped *Myxicola* giant axons.

Charles L. Schauf • Department of Physiology, Rush University, Chicago, Illinois 60612.

Two levels of sophistication need to be considered. On the one hand, solvent substitution, because of its differential effects on various aspects of gating and ion translocation, has been a useful pharmacological tool to elucidate the relationships between microscopic events. It has, for example, been possible to fully dissociate effects on the rate of channel opening from any changes in intramembrane ("gating") charge movement, thus ruling out those classes of kinetic models in which the rate limiting step leading to a conducting channel has a significant gating current associated with it. In addition, we have suggested that sodium channels that have already become conducting might not inactivate by the same mechanism as resting channels which have not yet opened, and that channel closing on repolarization may not simply be the reverse of opening. On the other hand, the effect of making an isotopic substitution can also provide other molecular insights. In particular, knowledge of both the magnitude of an isotope effect and, more importantly, its temperature-dependence, can be used to infer the degree to which the rate limiting step in a particular molecular process may involve changes in solvent structure.

It is the aim of this article to review and summarize both these approaches as they have been carried out in our laboratory over the past several years. The basic observations will be outlined first, followed by a detailed discussion of the potential molecular implications, and some ideas for future experimental approaches.

II. EXPERIMENTAL METHODS

Methods for the combined dialysis and voltage-clamp of *Myxicola* giant axons have been previously detailed (Bullock and Schauf, 1978; Schauf and Bullock, 1980). Briefly, this technique involves inserting a length of cellulose acetate dialysis tubing 250 microns in diameter longitudinally into the giant axon. The internal voltage and current electrodes are then placed inside this tubing. The tubing is freely permeable to compounds with molecular weights of 6000 or less, and does not significantly increase membrane series resistance. It offers the ability to control the composition of the internal solution without increasing membrane leakage conductance, and with no tendency for flow to be occluded by residual axoplasm. When potassium currents were measured, the internal solution was normally composed of 450 mM K^+ glutamate, 50 mM KF, 30 mM K_2HPO_4, and 1 mM Hepes buffer, and the external solution was isosmotic Tris buffer containing 10^{-6} M tetrodotoxin. For sodium current experiments the dialyzate contained 600 mM Cs^+ glutamate and 1 mM Hepes, and the external solution was K^+-free artificial seawater (430 mM NaCl, 10 mM $CaCl_2$, 50 mM $MgCl_2$, 20 mM Tris). Solutions were adjusted to pH 7.3 ± 0.1 both externally and internally. Intramembrane charge movements were recorded with a P/4 pulse protocol (Armstrong and Bezanilla, 1977) in Cs^+ dialyzed axons bathed in isosmotic Tris. Series resistance was routinely compensated, and leakage and capacity currents eliminated by appropriate analog circuitry (Schauf and Bullock, 1980).

Several features of solutions prepared in D_2O require special consideration. Commercial deuterium oxide (99.8% D_2O, Sigma Chemical) occasionally contains some impurities, and it is necessary to redistill it prior to use. The limiting equivalent conductivity of electrolytes is approximately 20% lower in D_2O (Swain and Evans, 1966). Thus, it is essential to recompensate series resistance following solvent substitution. Ionic equilibria differ in D_2O because the self-ionization of D_2O is an order of magnitude smaller than for H_2O. As a consequence the pD is higher than the measured pH by 0.41 units, and pH readings for glass electrodes must be appropriately corrected (Katz and Crespi, 1970). In addition, D_2O behaves as a stronger acid than H_2O, thus effectively raising the pK_a of titratable weak acids by about 0.5 pH units (Covington and Jones, 1968). Fortunately, extensive data concerning the effecting of pH changes are already available in *Myxicola* (Schauf and Davis, 1976), and no significant effects are expected to result from a change of 0.5 pH units around neutrality. However, this difficulty would need to be dealt with, if solvent substitution experiments were to be designed with pH as a variable.

III. DIFFERENTIAL EFFECTS OF D_2O ON NORMAL CHANNELS

When H_2O is replaced by D_2O (both externally and internally), the steady-state voltage-dependence of both the sodium and potassium conductances [$g_{Na}(V)$ and $g_K(V)$], and the steady-state Na^+ inactivation [h_∞] curves remain the same (Schauf and Bullock, 1979, 1980). Sodium channel selectivity ratios for both alkali metal and organic cations are also insensitive to D_2O substitution. In H_2O, the P_X/P_{Na} ratios calculated in *Myxicola* from reversal-potential measurements (using the methods of Hille, 1971, 1972) are 1.05 ± 0.10 for Li^+, 0.45 ± 0.05 for guanidine, 0.41 ± 0.02 for $NH_4{}^+$, 0.077 ± 0.008 for K^+, and 0.037 ± 0.003 for Cs^+. In D_2O, the respective values are 1.07 ± 0.09, 0.50 ± 0.06, 0.42 ± 0.06, 0.77 ± 0.009, and 0.037 ± 0.002. In different experiments the P_K/P_{Na} ratio was determined as a function of internal [K^+] in a manner similar to that used by Begenisich and Cahalan (1980). Although we also found that P_K/P_{Na} decreases with increasing internal [K^+], there is no change in this dependence following solvent substitution. Thus, within the accuracy of such measurements, selectivity does not depend on the nature of the solvent. On the basis of current theories (Eisenman, 1962; Hille, 1975), this result would not be expected if D_2O substitution produces significant changes in the structure of the Na^+ channel.

We have also examined the effects of D_2O on tetrodotoxin (TTX) binding in *Myxicola* using standard procedures (Ulbricht, 1979). The dissociation constant for equilibrium block is 3.8 ± 0.5 nM, in agreement with data in other tissues, and was not altered by D_2O (4.0 ± 0.8 nM). Recently, a similar negative result for TTX was reported by Hahin and Strichartz (1981) using frog nerve. However, they were also in a position to measure saxitoxin (STX) block. Unlike TTX, STX block was enhanced by D_2O. Kinetic studies showed this was due to a selective decrease in the rate of unbinding of STX from its receptor, with little or no change in the binding step. They concluded that hydrogen bonding was more significant for STX's interaction with the Na^+ channel than that of TTX.

The only nonkinetic parameters affected by D_2O are the maximum conductances for the sodium and potassium channels. In experiments in which $\bar{g}_{Na}$ and $\bar{g}_K$ are both determined simultaneously in the same axons, D_2O decreases $\bar{g}_{Na}$ by 26 ± 2% and $\bar{g}_K$ by 32 ± 2% (significant with $p < .002$). This decrease is independent of the identity of the current-carrying species. Thus, when Li^+ is the sole external monovalent cation, the Na^+ channel conductance is decreased in D_2O by the same proportion as when the experiments are done in normal sea water. The magnitude of the D_2O effect is slightly greater than that seen in bulk electrolyte solutions (Krishnan and Friedman, 1969; Tronstad and Stokland, 1937). Within the context of kinetic schemes in which the majority of Na^+ channels are conducting at the time of peak Na^+ conductance, changes in the maximum channel conductances can only arise from a decrease in either the number of available channels or single-channel conductance. However, D_2O substitution does not affect the magnitude of intramembrane charge movement (see below). If the number of conducting channels had decreased, total charge movement should have been reduced as well, suggesting that single-channel conductance is the parameter decreased by D_2O substitution (but see later discussion for another interpretation).

In contrast to the lack of a D_2O effect on most equilibrium properties, for temperatures at which the solvent properties of H_2O and D_2O are quite different (see Discussion), the kinetics of the ionic currents are generally slowed by D_2O (Conti and Palmieri, 1968; Meves, 1974; Schauf and Bullock, 1979, 1980). This is shown in records (a) and (b) of Fig. 1 and in the first portion of Table I (which serves to conveniently summarize all the solvent-substitution results discussed in this chapter). However, these kinetic isotope effects are significantly different for various aspects of channel gating. For example, at 5°C Na^+ activation and K^+ activation are slowed by 43–47% and 40% respectively. The decline in I_{Na} during a maintained depolarization (time-constant $_s\tau_h$, describing the inactivation of channels that have become conducting) is slowed by approximately the same amount (52%). However, at 5°C the time-constants for inactivation produced by a depolarizing prepulse (here abbreviated $_p\tau_h$), and for reactivation of inactivated channels are increased by 2 1/2 times in D_2O (Table I). This difference is so large as to preclude interpretation as a minor effect.

At higher temperatures (13–14°C), where the solvent properties of H_2O and D_2O become comparable, the kinetic effects of D_2O are generally less, although the magnitude of the decrease with temperature and thus the Q_{10}'s vary for the different processes (Table I). Both Na^+ and K^+ activation are now slowed by 17–20%, while $_s\tau_h$ and $_p\tau_h$ are increased by 37% and 43% respectively. These data yield Q_{10}'s for the solvent effect of 0.72 for Na^+ activation,

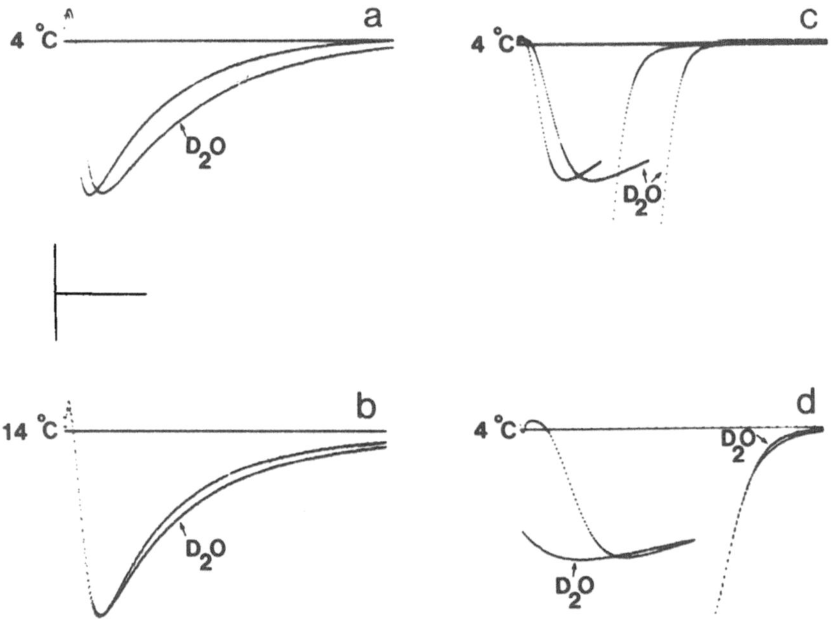

Figure 1. Sodium currents in *Myxicola* axons in H_2O and D_2O. Records in (a) were obtained during a depolarization to 0 mV in H_2O (unmarked) and D_2O (arrow) at a temperature of 4°C. Both activation and inactivation are slowed by about 50%. In contrast, the records in (b) (note the different time scale) were obtained at 14°C where D_2O substitution has a much smaller effect. In the experiment shown in (c), the axon was repolarized just after maximum inward current at 4°C in H_2O and D_2O. The pulse duration was longer in D_2O so that repolarization occurred at a similar point relative to the time at which inward current was a maximum. Again Na^+ activation and inactivation are slowed by D_2O, but there is no effect on the tail currents observed following repolarization. In (d) the tail current records have been translated and expanded for comparison. The time scale bar coresonds to 2 msec in (a), 0.8 msec in (b), 1.0 msec in (c), 0.5 msec in (D). Current calibration bar is 0.8 mA/cm^2 in (a) and (b), and 0.15 mA/cm^2 in (c) and (d).

Table I. Summary of Isotope Effects on Myxicola

	Parameter	Ratio of parameter in D_2O to that in H_2O		
		5°C	13–14°C	Q_{10}
Untreated axons	Na^+ activation	1.47 ± 0.02	1.18 ± 0.03	0.72
	$_s\tau_h$	1.52 ± 0.03	1.37 ± 0.03	0.84
	$_p\tau_h$	2.61 ± 0.21	1.43 ± 0.05	0.47
	Na^+ tails	No effect	No effect	
	Gating current	No effect	No effect	
	K^+ activation	1.40 ± 0.03	1.17 ± 0.02	0.79
Gallamine-treated axons	Na^+ activation	1.42 ± 0.09	1.10 ± 0.07	0.75
	$_s\tau_{h1}$	1.42 ± 0.15	0.94 ± 0.43	0.60
	$_s\tau_{h2}$	2.00 ± 0.20	1.88 ± 0.09	0.92
	Na^+ tails	1.69 ± 0.08	1.32 ± 0.06	0.75
C_9—treated axons	K^+ activation	1.46 ± 0.05	1.16 ± 0.08	0.75
	τ_{C9}	1.74 ± 0.07	1.14 ± 0.05	0.61
	$_p\tau_{C9}$	2.20 ± 0.17		
	$f(V)$	No effect	No effect	
Ba^{++}—treated axons	K^+ activation	1.51 ± 0.13		
	$\tau_{Ba}(0\ K^+)$	1.56 ± 0.14		
	$\tau_{Ba}(215\ K^+)$	1.12 ± 0.08		
	$f(V)$	No effect		

0.84 for $_s\tau_h$, and 0.47 for $_p\tau_h$. Measured in a single group of axons, these differences in temperature-dependence are all statistically significant.

The only aspect of ionic current kinetics that is not slowed by D_2O is the closing of Na^+ channels following repolarization (Na^+ tail currents). The records in (c) of Fig. 1 show Na^+ tail currents in H_2O and D_2O with the latter scaled so that the maximum inward currents are equal, and in (d) the current in D_2O has been shifted so that the repolarization times superimpose. As noted above, the time to peak inward current is increased by D_2O and inactivation is slowed. In contrast, solvent substitution does not change the rapid decrease in Na^+ currents following repolarization. The slower components of Na^+ tail currents can be relatively labile and the slight decrease in the magnitude of the slow component(s) in this example in the D_2O solution does not appear to be significant. Measurements of tail currents are exceedingly sensitive to series resistance compensation and/or the presence of any spatial nonuniformity (Schauf *et al.*, 1977). However, we do not feel that this is relevant here, since the observation is that D_2O is without an effect on tail currents, even though activation is delayed. We have attempted to perform comparable experiments on potassium currents, however, at present the interpretation of our results is complicated by the presence of significant loading of the periaxonal space, and the fact that washout is, as might be expected, D_2O sensitive. It is our impression that, here again K^+ tail currents are not altered by D_2O, however, more experiments and analysis are necessary.

IV. D_2O AND INTRAMEMBRANE CHARGE MOVEMENT

Since voltage is the independent variable controlling permeability, it is apparent that whatever the gating mechanism is, it must be sensitive to electric field changes. A reasonable assumption is that a structure with a permanent or induced dipole moment is linked to the activation of ion transport. Changes in the average orientation of this structure should be detectable as a nonlinear component of total displacement current, provided the charge involved is sufficiently large. Predicted originally by Hodgkin and Huxley (1952), such "gating currents" were first detected by Armstrong and Bezanilla (1973, 1974), and their existence has since been confirmed in many systems (Almers, 1978). However, when examined in sufficient detail (Neumcke *et al.*, 1976; Schauf and Bullock, 1981; Keynes *et al.*, 1981), the time course of charge movement does not correspond in a simple way to opening of Na^+ channels.

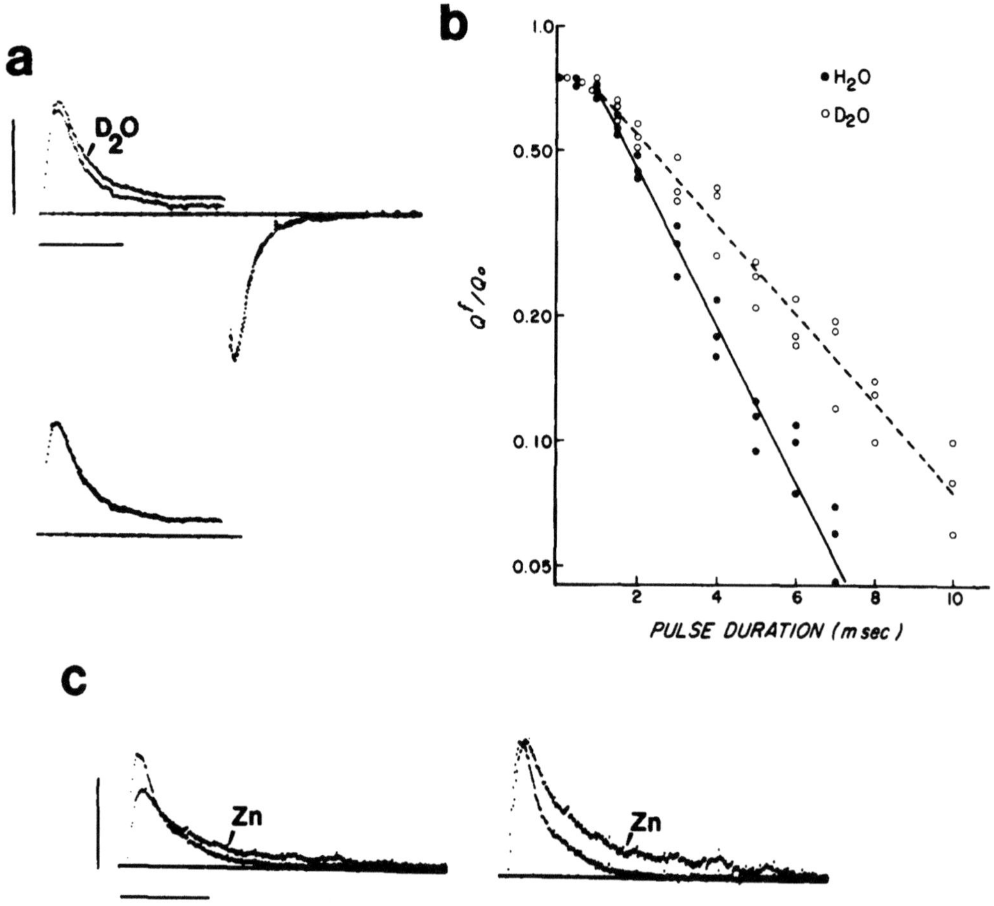

Figure 2. Effects of D_2O substitution and 30 mM external Zn^{++} on gating currents in *Myxicola*. In (a) we show gating currents (P/4 protocol, holding potential −80 mV, reference potential −140 mV) during and following a 2 msec pulse to +20 mV in H_2O (unmarked) and D_2O (arrow). The leakage pedestal is slightly higher in D_2O because this record was obtained about 30 min later. The OFF responses clearly superimpose. The ON responses also superimpose (lower records in part a) when they are shifted vertically to account for the increased nonlinear leakage current. The time course of charge immobilization in H_2O and D_2O is illustrated in part (b). The ratio of the charge movement in the fast component of the OFF response to the total charge movement in all components of the OFF response, or to the total charge in the ON response, is plotted as a function of pulse duration. Immobilization proceeds more slowly in D_2O. The gating currents in part (c) illustrate the effects of external Zn^{++} in slowing the ON response for purposes of comparison with the lack of effect of D_2O. The current calibration bar in all records is 40 $\mu A/cm^2$, time scale is 1 msec.

Another feature associated with gating current measurements has been the inability to detect a separate current corresponding to inactivation (Armstrong and Bezanilla, 1977). If activation and inactivation involved distinct field sensitive "gates" not coupled to each other in any way, one would expect some charge movement associated with each. Instead, it appears that the inactivation process involves a change in channel structure which, although not itself associated with charge redistribution, tends to retard or "immobilize" those charges that originally activated the channel. Thus, at least to some extent, inactivation seems dependent on the same charge movement that causes activation, as was suggested much earlier from studies of the kinetic properties of the ionic currents themselves (Goldman and Schauf, 1972). The exact nature of this coupling remains unclear, however. Studies of single Na^+ channels suggest it is probably more complex than that expected if a channel was simply obliged to become conducting before it could be inactivated by some voltage-independent process (Horn *et al.*, 1981).

In contrast to its effect on ionic currents, heavy water substitution does not alter the asymmetry currents during or following short (1–2 msec) pulses in *Myxicola* (Q_{on} or Q_{off}, P/4 procedure; see Schauf and Bullock, 1981; Fig. 2a). In ten axons, τ_{on} averages 288 ± 23 μsec (range 240–370 μsec) in H_2O, and 304 ± 30 μsec (range 263–380 μsec) in D_2O at a potential of 0 mV. Similarly, τ_{off} is 177 ± 6 μsec (range 160–208 μsec) in H_2O, and 178 ± 10 μsec (range 153–215 μsec) in D_2O. No effects on Q_{on} are seen at any other voltages either (−40–+100 mV). Total charge is also not changed and averages 11.2 ± 2.2 nC/cm^2 in H_2O, and 10.9 ± 2.8 nC/cm^2 in D_2O. The position of the $Q(V)$ curve along the voltage axis is not altered by solvent substitution. However, marked effects on the gating current are seen with 30 mM external Zn^{++} (shown in Fig. 2c for comparison). As in squid (Armstrong and Gilly, 1979), Zn^{++} slows both Na^+ activation and Q_{on}, though it has little effect on Na^+ tail currents, total charge moved, or Q_{off}. The ability to detect clear modifications of gating current with Zn^{++} (as well as Ca^{++}), but not with D_2O suggests that the negative result is real and not due to poor resolution.

Although D_2O does not affect gating currents during or following short pulses, it does seem to slow immobilization. As a depolarizing pulse is made longer, the asymmetry current following pulse termination ceases to be a single exponential. Some charge is retarded and the OFF asymmetry current declines biphasically. The extent of this immobilization can be quantified by measuring the decrease in the fraction (Q_f) of the total intramembrane charge (Q_o) which appears in the fast component. This is a difficult experiment, but Fig. 2b shows that the time-constant for immobilization seems to be slowed by about 70%. Considering the scatter in the data, this is consistent with the effect of D_2O on Na^+ inactivation and the hypothesis that Na^+ inactivation indeed retards intramembrane charge movements (Armstrong and Bezanilla, 1977).

V. D_2O EFFECTS ON DRUGS OCCLUDING Na^+ CHANNELS

As stated earlier, sodium activation and inactivation are not completely independent and a detailed model has postulated a physiological "inactivating particle" that can enter and block an open channel from the interior (Armstrong and Bezanilla, 1977; Armstrong and Gilly, 1979). While probably not correct in all its features, this model has stimulated interest in substances that can block Na^+ currents in ways consistent with a voltage- and/or time-dependent occlusion, and thus possibly mimic physiological inactivation (Strichartz, 1973; Cahalan, 1978; Yeh and Narahashi, 1977; Cahalan and Almers, 1979a,b; Kirsch *et al.*, 1980; Morello *et al.*, 1980). Such drugs should have solvent-sensitive kinetics. We therefore attempted to evaluate the effects of D_2O substitution on occlusion of Na^+ channels.

The drug chosen for initial examination was gallamine triethiodide, a neuromuscular blocking agent with actions similar to those of pancuronium (Yeh and Narahashi, 1977). The details of the effects of gallamine have been reported (Schauf and Smith, 1982), but are briefly reviewed here for convenience. Gallamine does not affect g_K in *Myxicola,* and does not alter the Na^+ channel when applied externally. However, when it is added internally, the decline in I_{Na} during step depolarizations is altered, and as the membrane potential increases, Na^+ channels fail to inactivate (Fig. 3).

For membrane potentials less than -10 mV, gallamine simply slows inactivation. However, inactivation remains complete, and for long pulses, e.g., 15 msec, no Na^+ tails are seen. At membrane potentials greater than -10 mV, inactivation is biphasic, with an initial voltage-dependent decrease in I_{Na} (time-constant abbreviated as ${}_s\tau_{h1}$) that is much faster than normal inactivation, followed by a later decline (${}_s\tau_{h2}$) that is slower than seen in unmodified Na^+ channels. As the membrane potential increases, the fraction of channels rapidly shutting off increases and, simultaneously, the Na^+ tail conductance following 15 msec pulses increases by an equivalent amount. These tail currents (not illustrated) have a "hook" and decline more slowly than normal. For membrane potentials positive to E_{Na} a current-dependent block of outward Na^+ current is also evident. Gallamine does not affect steady-state prepulse inactivation (h_∞), the time-constants for recovery from and development of inactivation for membrane potentials less than -30 mV, or Na^+ activation.

Thus, gallamine appears to have a dual effect on the Na^+ channel. First, it is similar to pancuronium (Yeh and Narahashi, 1977) in that it occludes sodium channels in a voltage-dependent fashion. This occlusion by gallamine also prevents inactivation. On repolarization, channels cannot close until gallamine dissociates, resulting in slow, hooked tails. However, unlike pancuronium, gallamine and Na^+ are apparently able to compete for occupancy of

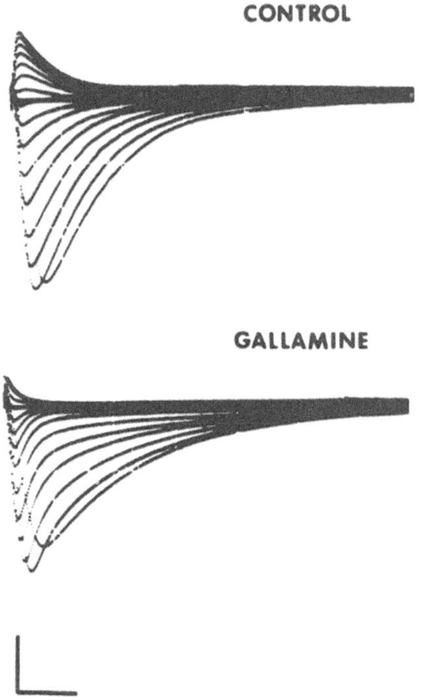

Figure 3. Effects of gallamine triethiodide on membrane currents of *Myxicola* axons in H_2O. Currents were obtained during step depolarizations of 50–200 mV from a holding potential of -80 mV before and during internal application of 10 mM gallamine. Note the presence of three effects: for the two lowest voltages inactivation is slowed, for larger depolarizations inactivation is biphasic, for depolarizations beyond E_{Na} outward currents are blocked. Calibration bars are 0.75 mA/cm^2 and 2 msec. Temperature 5°C.

a binding site, so that a current-dependent block of outward I_{Na} is seen. A second effect of gallamine is to slow inactivation of nonoccluded channels without preventing their ultimate inactivation.

However, a major difference between gallamine and pancuronium (as well as other quaternary ammonium compounds) is the effect on asymmetry current. In axons with intact inactivation, pancuronium eliminates the fast OFF charge which is resistant to immobilization, and in pronase-treated axons pancuronium induces immobilization even in the absence of Na^+ inactivation (Yeh and Armstrong, 1978). Thus, only slow OFF charge movement can be detected. In contrast, we found that gallamine almost completely abolished immobilization, so that the OFF responses following long pulses are as rapid as those following short pulses (Schauf and Smith, 1982). These responses were recorded at a voltage at which 80% of the Na^+ channels still inactivated in the presence of gallamine, suggesting that inactivation does not invariably result in charge immobilization. A possible explanation could be that gallamine inhibits the interaction between the inactivation particle and its binding site, without substituting for the particle and thereby inducing immobilization.

Gallamine thus appears to interact with Na^+ channels in several ways, only one of which is channel occlusion. We thus felt that an examination of the effects of D_2O on gallamine-modified channels might provide insight into the ways different processes are altered by solvent substitution. Again these are summarized in Table I (see also Schauf and Bullock, 1982). The rapid, voltage-dependent occlusion of sodium channels evident for membrane potentials greater than -10 mV is slowed by D_2O substitution with an average ${}_s\tau_{h1}(D_2O)/{}_s\tau_{h2}(H_2O)$ ratio of 1.42 ± 0.15 at 5°C. The D_2O-induced slowing disappears when the temperature is increased to 13°C with the time-constant ratio averaging 0.94 ± 0.43. Thus the Q_{10} is 0.60. The voltage-independent, slow decline, seen both immediately following the rapid gallamine-occlusion at potentials above -10 mV, and in isolation for lower voltages, is also slowed by D_2O substitution. The ratio ${}_s\tau_{h2}(D_2O)/{}_s\tau_{h2}(H_2O)$ averages 2.00 ± 0.20 at 5°C, significantly larger than the effect on ${}_s\tau_{h1}$ at this temperature. Moreover, the kinetic effect on ${}_s\tau_{h2}$, persists at 13°C with a ${}_s\tau_{h2}(D_2O)/{}_s\tau_{h2}(H_2O)$ ratio of 1.88 ± 0.09. Thus the Q_{10} for the isotope effect on ${}_s\tau_{h2}$ is 0.92. There is no effect of D_2O on the fraction of Na^+ channels that fail to inactivate. Gallamine tails are also slowed by D_2O substitution. The average ratio of tail time-constants in gallamine-treated axons in D_2O compared to H_2O is 1.69 ± 0.08 at 5°C and 1.32 ± 0.06 at 13.5°C, giving a Q_{10} of 0.75. This is a particularly interesting result because, as we showed above, Na^+ tail currents in untreated axons are not sensitive to D_2O substitution. The presence of a solvent effect is, however, consistent with the notion that the gallamine tail currents represent the dissociation of drug molecules from their binding sites.

VI. EFFECTS OF D_2O ON K^+ CHANNEL BLOCKERS

In addition to Na^+ channel blockers, we also felt it would be useful to look at some agents known to occlude K^+ channels. One of these was the TEA^+ derivative nonyltriethylammonium (C_9). The effects of internal C_9 in *Myxicola* giant axons resemble those in squid axons (Armstrong, 1966, 1969, 1971). The potassium current increases to a peak and then decreases exponentially (Fig. 4). As $[C_9]$ increases, the peak K^+ currents become smaller, and inactivation is faster and more complete. C_9 does not affect activation of potassium channels. In K^+-free artificial sea water the fraction of blocked K^+ channels, $f(V)$, increases with increasing $[C_9]$, but is not voltage-dependent. The K^+ inactivation time-constant, $\tau_{C9}(V)$, decreases from 17 msec at 0 mV to 7 msec at $+100$ mV. There is little

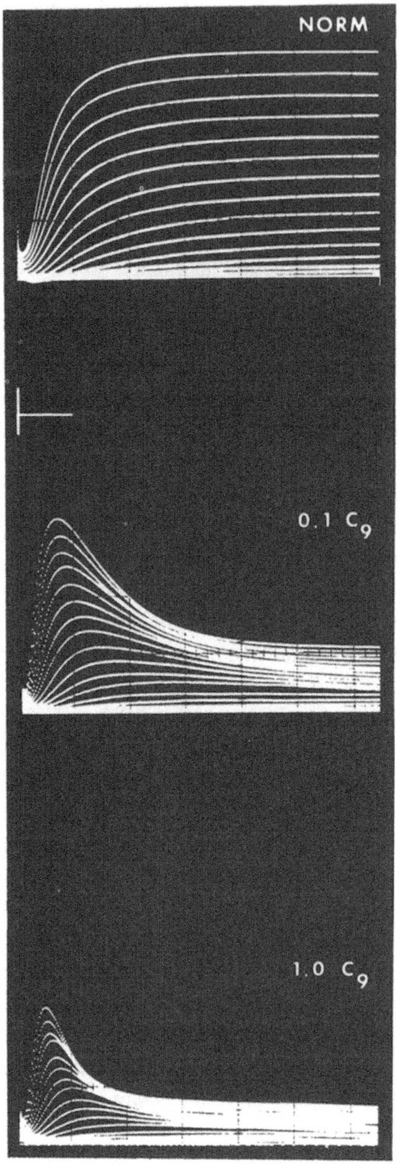

Figure 4. Effects of internal nonyltriethylammonium on membrane potassium currents of *Myxicola* in H_2O. Currents are shown for step depolarizations of 30–180 mV (10 mV increments) from a holding potential of −80 mV in an axon bathed in K^+-free sea water and dialyzed with control K^+-glutamate (top), and K^+-glutamate plus 0.1 mM C_9 (middle) or 1.0 mM C_9 (lower records). Calibration bars are 3 msec and 0.75 mA/cm². Temperature 5°C.

effect of external $[K^+]$ on C_9 block. In prepulse experiments peak g_K decreases after an initial delay, allowing one to define the fraction of g_K that is inactivated in a way analogous to prepulse inactivation of g_{Na}. Values for what we will call ${}_p\tau_{C9}$ are 45 msec at −40 mV, 34 msec at −20 mV, 17 msec at 0 mV, and 13.5 msec at +20 mV. These time-constants are comparable to the values of τ_{C9} measured during maintained depolarizations. A similar protocol allows calculation of the recovery time-constant. This averages 180 msec at −80 mV.

The effects of D_2O on C_9-modified K^+ channels are summarized in Table I. At 5°C, D_2O increases τ_{C9} at all voltages by 74 ± 7%. At 14°C, the D_2O-induced slowing averages 14 ± 5%, resulting in a Q_{10} of 0.61. At neither temperature does D_2O change $f(V)$. In prepulse experiments D_2O increases ${}_p\tau_{C9}$ by 120 ± 17% at 5°C, a significantly greater effect than that seen during maintained depolarizations. In recovery experiments, the time-constant is increased by 64% at 5°C and is unaffected at 14°C. In the same axons, D_2O slows K^+ activation by 46 ± 5%, significantly less than the D_2O effect on τ_{C9}.

We also examined internal Ba^{++} (Fig. 5). In contrast to C_9, for membrane potentials less than +20 mV little effect of Ba^{++} is seen. However, for more positive potentials, K^+ currents inactivate in the presence of Ba^{++}, and furthermore, unlike C_9, the steady-state K^+ inactivation is strongly voltage-dependent. In fact, $f(V)$ increases so rapidly that the $I(V)$ curve has a negative slope conductance. The Ba^{++} block is antagonized by increasing $[K^+]_o$. These effects qualitatively resemble those seen in squid axons (Armstrong and Taylor, 1980; Eaton and Brodwick, 1980).

In axons bathed in K^+ free ASW, substitution of D_2O slows the rate of K^+ inactivation in the presence of internal Ba^{++}. The ratio $\tau_{Ba}(D_2O)/\tau_{Ba}(H_2O)$, describing the inactivation of Ba^{++}-modified K^+ channels, is voltage-independent in K^+ free ASW and averages 1.58 ± 0.06 at 5°C. At higher $[K^+]_o$ the behavior is very different, however. Although the isotope effect on K^+ activation is unchanged in 215 mM $[K^+]_o$, the ratio $\tau_{Ba}(D_2O)/\tau_{Ba}(H_2O)$ is only 1.12 ± 0.08. Thus, increased external K^+ seems to antagonize not only the rate and degree of Ba^{++} block, but also the magnitude of the isotope effect itself.

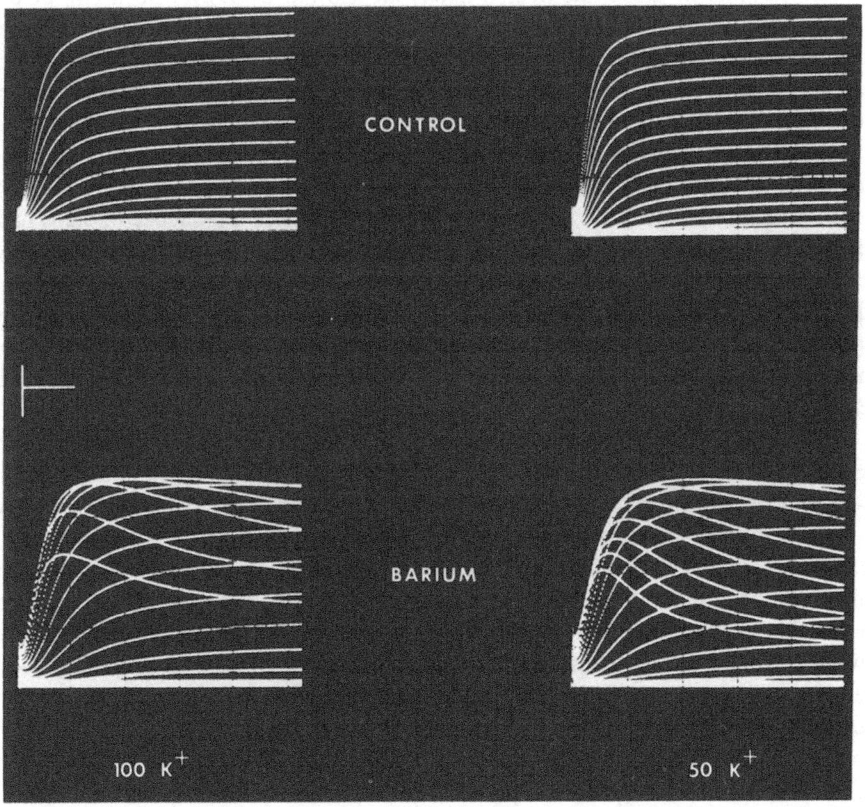

Figure 5. Effects of barium on K^+ currents of *Myxicola* axons in H_2O. Membrane currents are shown for depolarizatons of 30–180 mV from a holding potential of −80 mV before and during dialysis with 5 mM Ba^{++}. The external potassium concentration was 50 mM (right-hand records) or 100 mM (left-hand records). Calibration bars are 0.75 mA/cm² (upper records) or 0.4 mA/cm² (lower records), and 3 msec. Temperature 5°C.

VII. SOME IMPLICATIONS FOR SODIUM CHANNEL MODELS

What are the immediate implications of these results? Concerning the ionic conductances themselves, the differential effects of solvent substitution allow various molecular events to be separated, thus constraining the number of possible reaction schemes. In the Cs^+ dialyzed *Myxicola* axons used here, the time-constants for prepulse inactivation and inactivation during a maintained depolarization are equal when compared at the same membrane potential, while in intact (Goldman and Schauf, 1973) or K^+-dialyzed (Schauf and Bullock, 1979) axons, prepulse inactivation is significantly slower than inactivation of conducting channels. At 5°C, prepulse inactivation is slowed by D_2O by a much larger amount than the rate of inactivation of conducting channels, so that these two measures of Na^+ inactivation again no longer correspond to one another. The fact that D_2O selectively increases the time-constant of prepulse inactivation supports the concept that inactivation produced by changes in initial conditions is not the same as that occurring when an open channel inactivates during a maintained depolarization. This might be expected if the inactive state of the sodium channel could be reached either via the conducting state, or via an alternate pathway not requiring opening of the channel. Originally suggested a decade ago (Goldman and Schauf, 1973), this view has recently received independent support from experiments on single Na^+ channels in excised membrane patches (Horn *et al.*, 1981).

Tail currents in gallamine-treated axons are slowed by D_2O with a Q_{10} of 0.75. In contrast, Na^+ tail currents in untreated axons are completely insensitive to solvent substitution, despite the fact that Na^+ activation and inactivation are markedly slowed by D_2O. Thus, while the creation of an open channel normally involves solvent interactions, closing is solvent-independent. This differential effect may well be an important clue to understanding the underlying mechanisms of channel gating. At a minimum, the rate constant(s) for going from closed to open channels must be D_2O-sensitive, while those for going from open to closed channels are not. This might occur if only the last of a sequence of steps had to be reversed for a channel to close, and this particular step was solvent-insensitive. However, such an interpretation is complicated by the fact that D_2O slows activation of Na^+ channels with no effect on the intramembrane charge movement presumably associated with transitions among preceding closed states (Schauf and Bullock, 1979, 1981). Perhaps the process by which a channel closes upon repolarization is not simply the reverse of that by which it opened.

The ability to completely dissociate the time course of the Na^+ conductance from that of gating currents by D_2O substitution is perhaps the most striking result of our experiments. Most kinetic models to date have proposed linear sequences of reversible first-order reactions with purely voltage-dependent rate constants (Moore and Cox, 1976; Armstrong and Gilly, 1979). The fact that D_2O slows ionic but not asymmetry currents severely constrains such models. Provided that the transition sensitive to D_2O is that leading to the conducting state, and that this step involves little or no charge movement (the charge movement must occur between preceding closed states), the D_2O results can in principle be accounted for. However, whether a satisfactory quantitative fit can be achieved given this constraint has not yet been investigated for any such model. For example, it is not obvious how one would achieve tail currents with the proper voltage-dependence. It is thus of interest to explore alternative models that assumed a specific separation between the initial response to electric field changes, and the ultimate creation of a conducting channel. In particular, one wishes to know whether a self-consistent kinetic representation can be achieved.

Baumann has, for example, discussed one model in which a reversible, first-order, voltage-dependent reaction step, converting inert to active subunits, is followed by a reversible, voltage-independent, step-by-step assembly of those active subunits into a channel-forming aggregate (Baumann and Mueller, 1964; Baumann, 1981; see also the chapter by Baumann in this volume). One can easily imagine that the voltage-dependent creation of active monomers may not involve interaction with the solvent, while their voltage-independent aggregation to form a transmembrane aqueous pathway does. Thus, intramembrane charge movement (presumably associated with the voltage-dependent monomer activation step) would not be D_2O sensitive, while the ionic currents, reflecting occupancy of the conducting state, are slowed. It is important to stress that such an inherent separability would exist in this type of model even if the details of the precursor-production vs. assembly processes were more complex than the simple aggregation scheme assumed by Baumann (for example, the subunits may not be identical). Another feature of such a model is that the number of channels in the conducting state as a function of membrane potential depends, in part, on the proportion of activated subunits. Subunit activation is voltage-dependent, so that the rate of rise of the Na^+ conductance can become more rapid with increasing depolarization in the absence of any charge movement in the transition to the conducting state.

We have been able to account for the observed D_2O effect with such a model by assuming that only a single rate-constant is altered (Baumann and Schauf, in preparation). Sodium activation and inactivation are slowed, while gating and tail currents are unaffected. In addition, the voltage-dependence of both sodium activation and tail currents is in line with that actually observed in H_2O. While curve fitting does not constitute proof for any model, it does indicate that a different physical formulation is feasible. Moreover, it becomes incumbent upon those suggesting more traditional kinetic schemes to demonstrate that ionic currents can indeed be slowed in the absence of any effect on gating current. In this context, it is important to note that the essential experimental features of the D_2O results, slowing of Na^+ activation and inactivation with no effects on tail currents or charge movement, have been confirmed in squid giant axons (Bullock and Schauf, in preparation). We should note that the aggregation model predicts that, under some circumstances, the maximum Na^+ conductance may decrease in D_2O due to a change in the fraction of available Na^+ channels that are conducting at the time of peak I_{Na}, without any change in the conductance of a single open channel. Thus, determination of single-channel conductances in D_2O may provide an experimental means of further deciding between classes of molecular schemes.

The effects of D_2O on the occlusion of sodium and potassium channels by drugs and ions can be compared with solvent effects on normal channels. The blocking effects of gallamine, C_9, and Ba^{++} are all D_2O-sensitive at 5°C but the effects disappear when the temperature is increased. This temperature-dependence is larger than that seen for activation of either modified or normal channels, but comparable to that for the isotope effect on normal prepulse inactivation. In contrast, the gallamine-induced slowing of sodium inactivation remains D_2O-sensitive at all temperatures, suggesting a different origin for this phenomenon.

VIII. PHYSICAL CHEMISTRY OF ISOTOPE EFFECTS

Aside from the previous considerations, in which D_2O is essentially regarded as a pharmacological probe, albeit a very gentle one, additional conclusions are possible, based on what is known concerning the physical chemistry of isotope effects. The theoretical foundations of isotope effects are in general well understood (Laidler, 1969; Melander, 1960; Thornton and Thornton, 1970). Isotope effects arise fundamentally as the quantum mechanical result of changes in nuclear mass on the energy levels of a molecule. The central function of a statistical mechanical treatment of a molecular system is the molecular partition function. The Born–Oppenheimer approximation considers the motion of electrons to be determined by a field produced by nuclei at fixed positions. The equilibrium and rate-constants for a reaction can then be expressed in terms of the partition functions. Consider the simple reaction:

$$\alpha A + \beta B \rightleftharpoons \gamma C + \delta D$$

The equilibrium constant K_{eq} is given by

$$K_{eq} = \frac{Q_C^\gamma \, Q_D^\delta}{Q_A^\alpha \, Q_B^\beta} \, \frac{\exp(-E_0/kT)}{v \, N_A \, \Delta N}$$

where E_0 is the difference in zero point energies of products from reactants, v is the volume of the system, N_A is Avogadro's number, $\Delta N = \gamma + \delta - \alpha - \beta$, and the Q_i are the molecular partition functions of the various species. For the forward kinetic rate-constant of the above elementary reaction, transition state theory yields an expression similar in form to that shown above (Thornton and Thornton, 1970).

We now wish to consider the effects of substituting a heavier nuclear isotope for a lighter one on the partition functions, and thus on the equilibrium and rate-constants. Because of the use of the Born–Oppenheimer approximation the internuclear potential functions are identical for the two isotopic molecules. Thus, the major contribution to isotope effects on partition functions simply arises from the difference in zero point energy. In particular, the ratio of the equilibrium or rate-constants of the unsubstituted system (K) to that of the deuterium substituted system (K^*) is generally given by $K/K^* = I \exp(E_0/RT)$, where E_0 is the isotopically-induced difference in the separation of the zero point vibrational energies of products and reactants and I is an inertial factor (ranging between 1 and 2) that involves the ratio of the masses and moments of inertia of both systems. In general, there is expected to be a net loss in zero point energy in the transition state so that the formation of an activated complex is favored by the lighter isotope.

If only the ground vibrational state is significantly occupied, the temperature-dependence of the isotope effect is directly determined by the zero point energy term. For example, the fairly strong C-D and C-H bonds have $E_0 \sim 1$ kcal/mole. Assuming a large molecule for which I would be near unity, the maximum K/K^* ratio will be about 7.0. Thus between 5–15°C the value of K/K^* varies by a factor (Q_{10}) of 0.93. The differences in zero point energies for hydrogen and deuterium "hydrogen" bonds is much smaller (perhaps only 0.14 kcal/mole; see Nemethy and Scheraga, 1964) and the expected Q_{10} is unity.

When isotope effects are measured by replacing the water with D_2O, not only will exchangeable hydrogens be replaced by deuterium, thus producing the primary and/or secondary equilibrium and kinetic effects discussed above, but the nature of the solvent and its interactions with solutes will also be altered. Liquid H_2O is a solvent with anomalously high melting point, boiling point, and heat capacity. It has long been appreciated that such behavior has its origin in extensive intermolecular hydrogen bonding which confers some sort of "structure" to H_2O. However, the exact molecular definition of this structure varies from one model to another. Although similar to H_2O in some properties (molecular geometry, dielectric constant, and dipole moment), D_2O has a higher viscosity, melting point, and temperature of maximum density. Thermodynamic quantities, such as heat capacity, are also elevated in D_2O. These data are usually taken to mean that liquid D_2O is more structured than H_2O at the same temperature, or alternatively that D_2O behaves like H_2O at a lower temperature. However, the significant feature of interest to us is that, whatever the molecular details, the structure of D_2O breaks down more rapidly than that of H_2O with increasing temperature from 0–15°C, so that at higher temperatures the two liquids have similar properties (Nemethy and Scheraga, 1964; Heppolette and Robertson, 1960). This gives rise to phenomena that are much more temperature sensitive than the hydrogen-deuterium exchange effects discussed above. For example, Kresheck *et al.* (1965) found a Q_{10} of 0.7 for the enthalpy of transfer of propane, and Heppolette and Robertson (1960) found similar Q_{10}'s for several hydrolysis reactions in D_2O.

We observe K/K^* ratios from 1.5 (for Na^+ and K^+ activation, inactivation of open Na^+ channels, gallamine occlusion, and gallamine tail currents) to approximately 2.5 (for prepulse inactivation of Na^+ channels) with Q_{10}'s between 0.5–0.8. Such a temperature-dependence, if it were interpreted as a primary or secondary (nonsolvent) isotope effect, would correspond to a zero point energy difference of 4–8 kcal/mole. Not only is such a difference in zero point energy unreasonable even for a primary effect involving the strongest covalent bonds, but even if several bonds were broken simultaneously, the magnitude of the isotope effect (K/K^*) would be some 1000 times larger than that actually observed. Thus, our tentative conclusion is that for these processes, rather than a H-D exchange leading to alterations in channel structure, changes in solvent ordering may be of primary importance. This could be the case if the activation energy of channel opening was higher in D_2O because a transition state involved an energetically unfavorable interaction with the solvent. For example, part of the "gate" might need to swing out into the solution, or significant solvent reordering might occur during formation of a transmembrane aqueous pore. In contrast, the temperature-independent D_2O effect on the gallamine-induced slowing of Na^+ inactivation is consistent with a primary or secondary isotope effect in which solvent interaction plays no significant role.

Clearly, the interpretation in terms of microscopic events of the solvent effects we have observed represents only a beginning step. More detailed experiments, performed over a wider range of temperatures, with better resolution, and perhaps tied to a more specific working model will no doubt yield more information.

ACKNOWLEDGMENT. Supported by USPHS Research Grant NS15741, National Multiple Sclerosis Society Grant 1313A2, and by the Morris Multiple Sclerosis Research Fund.

REFERENCES

Almers, W., 1978, Gating currents and charge movements in excitable membranes, *Rev. Physiol. Biochem. Pharmacol.* **82**:97–190.

Armstrong, C. M., 1966, Time course of TEA^+-induced anomalous rectification in squid giant axon, *J. Gen. Physiol.* **50**:491–503.

Armstrong, C. M., 1969, Inactivation of the potassium conductance and related phenomena caused by quaternary ammonium ion injection in squid axons, *J. Gen. Physiol.*, **54**:553–574.

Armstrong, C. M., 1971, Interaction of tetraethylammonium ion derivatives with the potassium channels of giant axons, *J. Gen. Physiol.* **58**:413–437.

Armstrong, C. M., and Bezanilla, F., 1973, Currents related to the movements of the gating particles of the sodium channel, *Nature* **242**:459–461.

Armstrong, C. M., and Bezanilla, F., 1974, Charge movement associated with the opening and closing of the activation gates on the sodium channel, *J. Gen. Physiol.* **63**:533–552.

Armstrong, C. M., and Bezanilla, F., 1977, Inactivation of the sodium channel. II. Gating current experiments, *J. Gen. Physiol.* **70**:567–590.

Armstrong, C. M., and Gilly, W. F., 1979, Fast and slow steps in the activation of sodium channels, *J. Gen. Physiol.* **74**:691–711.

Armstrong, C. M., and Taylor, S. R., 1980, Interaction of barium ions with potassium channels in squid giant axons, *Biophys. J.* **30:**473–488.

Baumann, G., 1981, Novel kinetics in the sodium conductance system predicted by the aggregation model of channel gating, *Biophys. J.* **35:**699–705.

Baumann, G., and Mueller, P., 1964, A molecular model of membrane excitability, *J. Supramol. Struct.* **2:**538–557.

Begenisich, T. B., and Cahalan, M. D., 1980, Sodium channel permeation in squid axons. I. Reversal potential experiments, *J. Physiol.* **307:**217–242.

Bullock, J. O., and Schauf, C. L., 1978, Combined voltage-clamp and dialysis of *Myxicola* axons: Behavior of membrane asymmetry currents, *J. Physiol.* **278:**309–324.

Cahalan, M. D., 1978, Local anesthetic block of sodium channels in normal and pronase-treated squid giant axons, *Biophys. J.* **23:**285–311.

Cahalan, M. D., and Almers, W., 1979a, Interactions between quaternary lidocaine, the sodium channel gates, and tetrodotoxin, *Biophys. J.* **27:**39–56.

Cahalan, M. D., and Almers, W., 1979b, Block of sodium conductance and gating current in squid giant axons poisoned with quaternary strychnine, *Biophys. J.* **27:**57–74.

Conti, F., and Palmieri, G., 1968, Nerve fiber behavior in heavy water under voltage clamp, *Biphysik* **5:**71–79.

Covington, A. K., and Jones, S. J., 1968, *Hydrogen Bonded Solvent Systems,* Taylor and Francis Publishers, London.

Eaton, D. C., and Brodwick, M. S., 1980, Effects of barium on the potassium conductance of squid axon, *J. Gen. Physiol.* **75:**727–750.

Eisenman, G., 1962, Cation-sensitive glass electrodes and their mode of operation, *Biophys. J.* **2**(Part 2):259–295.

Goldman, L., and Schauf, C. L., 1972, Inactivation of the sodium current in *Myxicola* giant axons: Evidence for coupling to activation, *J. Gen. Physiol.* **59:**659–675.

Goldman, L., and Schauf, C. L., 1973, Quantitative description of sodium and potassium currents and computed action potentials in *Myxicola* giant axons, *J. Gen. Physiol.* **61:**261–284.

Hahin, R., and Strichartz, G., 1981, Effects of deuterium oxide on the rate and dissociation constants for saxitoxin and tetrodotoxin action. Voltage clamp studies on frog myelinated nerve, *J. Gen. Physiol.* **78:**113–140.

Heppolette, R. L., and Robertson, R. E., 1960, The temperature dependence of the solvent isotope effect, *J. Am. Chem. Soc.* **83:**1834–1838.

Hille, B., 1971, The permeability of the sodium channel to organic cations in myelinated nerve, *J. Gen. Physiol.* **58:**599–620.

Hille, B., 1972, The permeability of the sodium channel to metal cations in myelinated nerve, *J. Gen. Physiol.* **59:**637–658.

Hille, B., 1975, Ionic selectivity, saturation, and block in sodium channels; a four-barrier model, *J. Gen. Physiol.* **66:**535–560.

Hodgkin, A. L., and Huxley, A. F., 1952, A quantitative description of membrane current and its application to conduction and excitation in nerve, *J. Physiol. (Lond.)* **117:**500–544.

Horn, R., Patlak, J., and Stevens, C. F., 1981, Sodium channels need not open before they inactivate, *Nature* **291:**426–427.

Katz, J. L., and Crespi, H. L., 1970, Isotope effects in biological systems, in: *Isotope Effects in Chemical Reactions* (C. J. Collins and N. S. Bowman, eds.), pp. 286–363, Van Nostrand-Reinhold, New York.

Keynes, R. D., Malachowski, G. C., Van Helden, D. F., and Greeff, N. G., 1981, Components of the asymmetry current in the squid giant axon, in: *Advances in Pysiological Sciences,* Vol. 4 (J. Salanki, ed.), pp. 37–49, Pergamon Press, Budapest.

Kirsch, G. E., Yeh, J. Z., Farley, J. M., and Narahashi, T., 1980, Interaction of n-alkylguanidine with the sodium channel of squid axon membrane, *J. Gen. Physiol.* **76:**315–336.

Kresheck, G. C., Schneider, H., and Scheraga, H. A., 1965, The effect of D_2O on the thermal stability of proteins. Thermodynamics parameters for the transfer of model compounds from H_2O to D_2O, *J. Phys. Chem.* **69:**3132–3144.

Krishnan, C. V., and Friedman, H. L., 1969, Solvation enthalpies of various nonelectrolytes in water, propylene carbonate, and dimethyl sulfoxide, *J. Phys. Chem.* **73:**1572–1580.

Laidler, K. J., 1969, *Theories of Chemical Reaction Rates,* McGraw Hill, New York.

Melander, L., 1960, *Isotope Effects on Reaction Rates,* Ronald Press, New York.

Meves, H., 1974, The effect of holding potential on the asymmetry currents in squid giant axons, *J. Physiol* **243:**847–867.

Moore, J. W., and Cox, G., 1976, A kinetic model for the sodium conductance system in squid axon, *Biophys J.* **16:**171–191.

Morello, R., Begenisich, T., Trzos, W., and Reed, J. K., 1980, Interaction of nonylguanidine with the sodium channel, *Biophys. J.* **31:**435–440.

Nemethy, G., and Scheraga, H. A., 1964, Structure of water and hydrophobic bonding in proteins IV. The thermodynamic properties of liquid deuterium oxide, *J. Chem. Phys.* **41**:680–687.

Neumcke, B., Nonner, W., and Stämpfli, R., 1976, Asymmetrical displacement current and its relation with activation of the sodium current in the membrane of frog myelinated nerve, *Pflügers Arch.* **363**:193–203.

Schauf, C. L., and Bullock, J. O., 1979, Modification of sodium channel gating in *Myxicola* giant axons by deuterium oxide, temperature and internal cations, *Biophys. J.* **27**:193–208.

Schauf, C. L., and Bullock, J. O., 1980, Solvent substitution as a probe of channel gating in *Myxicola:* differential effects of D_2O on some components of membrane conductance, *Biophys J.* **30**:295–306.

Schauf, C. L., and Bullock, J. O., 1981, Isotope effects on ionic currents and intramembrane charge movements in *Myxicola* axons: Implications for models of sodium channel gating, in: *Advances in Physiological Sciences,* Vol. 4 (J. Salanki, ed.), pp. 51–66, Pergamon Press, Budapest.

Schauf, C. L., and Bullock, J. O., 1982, Solvent substitution as a probe of channel gating in *Myxicola:* Effects of D_2O on kinetic properties of drugs that occlude channels, *Biophys. J.* **37**:441–452.

Schauf, C. L., and Davis, F. A., 1976, Sensitivity of the sodium and potassium channels of *Myxicola* giant axons to changes in external pH, *J. Gen. Physiol.* **67**:185–195.

Schauf, C. L., and Smith, K. J., 1982, Gallamine triethiodide-induced modifications of the sodium conductance in *Myxicola* giant axons, *J. Physiol.* **323**:157–171.

Schauf, C. L., Bullock, J. O., and Pencek, T. L., 1977, Characteristics of sodium tail currents in *Myxicola* axons: Comparison with membrane asymmetry currents, *Biophys. J.* **19**:7–28.

Strichartz, G. R., 1973, The inhibition of sodium current in myelinated nerve by quaternary derivatives of lidocaine, *J. Gen. Physiol.* **62**:37–57.

Swain, C. G., and Evans, D. F., 1966, Conductances of ions in light and heavy water at 25 °C, *J. Am. Chem. Soc.* **88**:383–390.

Thornton, E. L., and Thornton, E. R., 1970, Origin and intepretation of isotope effects, in: *Isotope Effects in Chemical Reactions* (C. J. Collins and N. S. Bowman, eds.), pp. 286–363, Van Nostrand-Reinhold, New York.

Tronstad, L., and Stokland, K., 1937, Electrical conductivities and ion mobilities in heavy water, *K. Nor. Vidensk. Selsk. Forh.* **10**:141–144.

Ulbricht, W., 1979, Kinetics of tetrodotoxin and saxitoxin action at the node of Ranvier, *Adv. Cytopharmacol.* **3**:363–372.

Yeh, J., and Armstrong, C. M., 1978, Immobilization of gating charge by a substance that stimulates inactivation, *Nature* **273**:387–389.

Yeh, J. Z., and Narahashi, T., 1977, Kinetic analysis of pancuronium interaction with sodium channels in squid axon membranes, *J. Gen. Physiol.* **69**:293–323.

19

The Molecular Mechanisms of Cellular Potentials

Gilbert N. Ling

I. INTRACELLULAR POTASSIUM: PUMPED OR ADSORBED?

Water, proteins, and potassium ions are the three most abundant components of the living cells. According to the membrane-pump theory, intracellular water and K^+ exist largely in the free state. This theory has provided the theoretical framework for a great number of brilliant achievements, including the theory of cellular potentials of Bernstein (1912), Hodgkin and Huxley (1952), and many others.

According to the alternative "bulk-phase theories" as championed by Moore and Roaf (1908), Fisher and Suer (1935), Gortner (1930), and others, a substantial part of the cells' K^+ and water may exist in a bound state. This view all but vanished in the early 1940s. The membrane-pump concept became almost universally accepted and taught. But not everyone was so convinced; among the doubters were Ernst (1963), Troshin (1966), and myself.

Having discovered that iodoacetate plus anoxia failed to slow down the rate of Na^+ efflux in frog muscle (Ling, 1951, 1952, 1962; which in the membrane-pump theory largely represents outward Na^+ pumping rate), a finding subsequently confirmed by Keynes, Conway and co-workers (Keynes and Maisel, 1954; Conway *et al.*, 1961), I reached the conclusion that the postulated Na pump would consume more energy than the cell commands. As an alternative, I suggested in 1951 and 1952 a new theory that includes a molecular mechanism for the selective accumulation of K^+ over Na^+ in living cells as well as in (sulfonate type) cation-exchange resins. This theory comprises four postulates: (1) fixed anionic (and cationic) sites exist throughout the living cell and not limited to the cell membrane as previously suggested (see below), (2) these fixed ionic sites belong primarily to proteins, e.g., β- and γ-carboxyl groups, (3) in living cells and some model systems there is a high degree of counter-ion (for example, K^+) association with fixed anionic sites, and (4) electrostatic adsorption on fixed anionic sites favors the smaller hydrated K^+ over Na^+ (Ling, 1951, 1952).

Gilbert N. Ling • Department of Molecular Biology, Pennsylvania Hospital, Philadelphia, Pennsylvania 19107.

A variety of experimental studies were carried out to test whether intracellular K^+ is free or "bound," including the vapor pressure measurement of Hill and co-workers (Hill, 1930; Hill and Kupalov, 1930), the measurement of K^+ mobility in nerve and muscle cells (Hodgkin and Keynes, 1953; Kushmerick and Podolsky, 1969), the measurement of intracellular K^+ activity with a K^+-sensitive microelectrode (Hinke, 1961), the demonstration of an ability of pure natural cell membranes (red cell ghosts; Freedman, 1976), or synthetic membrane vesicles (phospholipid vesicles containing K-Na activated ATPase) selectively to accumulate K^+ or Na^+ (Hilden and Hokin, 1975; Goldin and Tong, 1974). These findings, in addition to the *apparent* contradiction (Berendsen and Edzes, 1973; Cooke and Kuntz, 1974) of early claims of demonstration of K^+ and Na^+ binding (Cope, 1965; Ling and Cope, 1969) as well as H_2O binding by NMR (Hazlewood *et al.*, 1969; Cope, 1969) led many scientists to the conclusion that the bulk of intracellular K^+ as well as water is in a free state.

More careful scrutiny of the evidence, however, has revealed that this conclusion was incorrect (a full in-depth review will be given in my forthcoming book (Ling, 1983), for a briefer and less complete one, see Ling and Negendank, 1980).

On the other hand, the pure membrane "vesicles," par excellence, i.e., the squid axon membrane sacs free of axoplasm do not pump K^+ or Na^+ (Ling and Negendank, 1980), whereas an effectively membraneless open-ended (EMOC) preparation of muscle cells does preferentially accumulate K^+ and extrude Na^+ (Ling, 1978a).

II. EVIDENCE THAT SUPPORTS BINDING OF INTRACELLULAR K^+

The most definitive experimental evidence that the bulk of intracellular K^+ exists in an adsorbed state has come in the last five years.

According to the association–induction (AI) hypothesis, the bulk of intracellular K^+ is adsorbed on β- and γ-carboxyl groups of cellular proteins (Ling, 1952). Because in muscle over 60% of these anionic groups belong to myosin (Ling and Ochsenfeld, 1966), and myosin is found only in the A band of the muscle myofibrils (Hanson and Huxley, 1953), the theory predicts the localization of K^+ in the A bands. There is also belief that the sites that bind uranium in electron-microscope sections of tissues are also the β- and γ-carboxyl groups (Ling, 1977a). Accepting this view, the AI Hypothesis further predicts that if one can in some way visualize K^+ as it exists in the living cell, the pattern of its distribution would resemble a conventional electron-microscope picture, in which the darkly stained area (by uranium) would be where most K^+ is found. To test this hypothesis, I used autoradiography. Unfortunately, the two radioactive isotopes of K^+ were not suitable because one has too short a half life (^{42}K) and the other is much too expensive (^{40}K). For this reason, I settled for two "surrogates": ^{134}Cs and ^{208}Tl, both long-lived and inexpensive.

With a technique for preserving isolated frog muscle in room temperature for up to eight days (Ling and Bohr, 1969), most of the cell K^+ can be replaced physiologically and reversibly by ^{134}Cs-labeled Cs^+ or ^{208}Tl-labeled thallium (Tl^+; Ling, 1977b). Single frog muscle fibers, whose K^+ had been completely or largely replaced by labeled Cs^+ or Tl^+ were isolated and rapidly air dried. After coating with photoemulsion and an exposure of about two weeks, the developed pictures appear as shown in Fig. 1, where the partial coverage of the emulsion allows identification of the lines along which silver granules congregate to be the A (dark) band.

This work was completely corroborated and extended by Edelmann (1980a), who used frozen (rather than dried) muscle fibers loaded with, for example, ^{134}Cs-labeled Cs^+, and carried out the exposure in liquid nitrogen and separated the emulsion film from the underlying

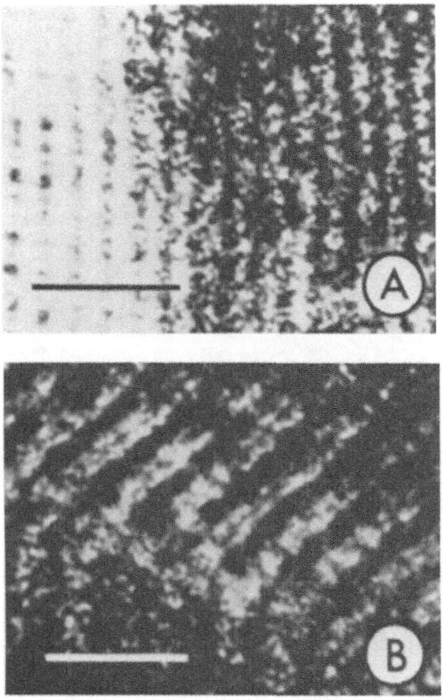

Figure 1. Autoradiographs of frog muscle fibers. Single ^{134}Cs-loaded fiber partially covered with photographic emulsion.(From Ling, 1977a, by permission of *Physiol. Chem. Phys.*).

muscle fiber before taking light-microscopic or electron-microscopic pictures. Besides autoradiography, Edelmann carried out no less than three additional types of experimental testing of the predictions of the AI hypothesis, including transmission electron microscopy. In this work, Edelmann (1977) first replaced electron-light K^+ (at. wt. 39) in frog muscle cells with electron dense Cs^+ (at. wt. 133) and Tl^+ (at. wt. 204). These Cs^+- or Tl^+-loaded muscles were then frozen dried, and dry cut. With neither chemical fixation nor staining, he obtained electron-micrograph plates (Figs. 2B and C) that reveal a pattern of distribution of these K^+-surrogates strikingly similar to the glutaraldehyde-fixed and uranium-stained preparation (Fig. 2A). When the sections were exposed to water (Fig. 2E) and in normal "K^+-loaded" muscle (Fig. 2F) only faintly dark areas show.

Edelmann (1978) then used another method to test the hypothesis, dispersive x-ray microprobe analysis. In this, an electron beam was focused on either the A band or I band of Cs^+- or Tl^+-loaded as well as normal "K^+-loaded" muscle thin sections. The x-ray spectrum revealed that the concentrations of all three elements were higher in the A band than in the I band. This work has been fully confirmed by Trombitas and Tigyi-Sebes (1979), who demonstrated much higher K^+ concentration in the A bands than in the I bands of isolated myofibrils of honeybee thorax muscle.

The observations described above have established that in voluntary muscle K^+ is localized in distribution. Other evidence shows that the accumulated K^+ can be displaced by other univalent cations; the effectiveness varies greatly with different ions that have the same long-range attributes (that is, univalency) but differ by short-range attributes only. The fact that this ion-specific effect is fully preserved in muscle cells whose membrane (plus postulated pumps) were made functionally ineffective (EMOC preparation; see Ling, 1977b) suggests that the localized K^+ is specifically adsorbed, one K^+ to one anionic site. Additional evidence of the *adsorbed* state of K^+ and Ca^{++} distribution in the A band is derived from another new technique Edelmann developed.

Edelmann's (1980b) fourth effort to test the association–induction hypothesis was even

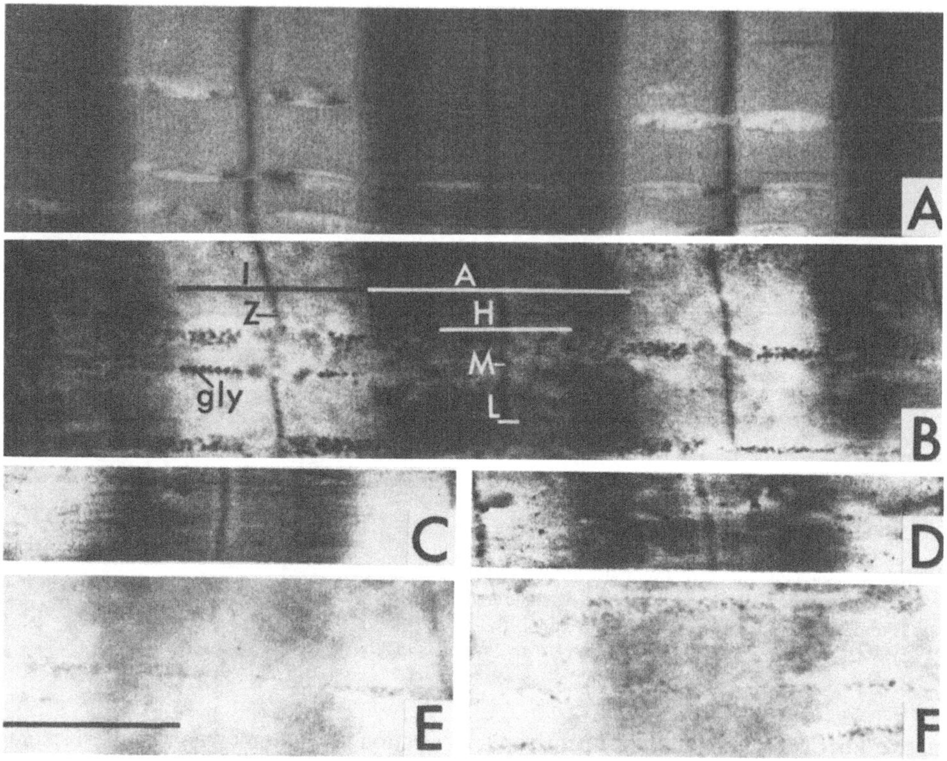

Figure 2. Electron micrographs of frog sartorius muscle. (A) Muscle fixed in glutaraldehyde only and stained with uranium by conventional procedure. (B) EM of section of freeze-dried Cs^+-loaded muscle, without chemical fixation or staining. (C) Tl^+-loaded muscle without chemical fixation or staining. (D) Same as (C) after exposure of section to moist air, which causes the hitherto even distribution of thallium to form granular deposits in the A band. (E) Section of central portion of (B) after leaching in distilled water. (F) Normal "K-loaded" muscle. (From Edelmann, 1977, by permission of *Physiol. Chem. Phys.*)

more striking. He used the new technology called laser-microprobe mass spectrometric analysis (LAMMA). In this, he used freeze–dried and dry cut regular normal frog muscle cells. The thin sections obtained were dipped in a solution containing 50 mM K^+, 50 mM Na^+, and 10 mM Cs^+. Freed of adhering fluid, the A band was vaporized by a focused laser beam and the vaporized atoms analyzed quantitatively. Peak heights, when compared to a control gelatin film containing known quantities of 50 mM K^+, 50 mM Na^+, and 10 mM Cs^+, revealed preferential uptake of K^+ and Cs^+ over Na^+. Here, there is no question of membrane or membrane pumps, as we are dealing with a thin section less than 1 μm thick from a single muscle cell 60 μm in diameter. Yet selective K^+ and Cs^+ over Na^+ adsorption can be demonstrated *in vitro*, yielding convincing evidence for a fundamental issue of century-old debate. The bulk of intracellular K^+ is adsorbed on anionic protein sites.

III. OSMOTIC ACTIVITY

The establishment of the adsorbed state of the major cation, K^+, demands a "new" source of osmotic activity inside the cell to balance that of free Na^+ in the external medium

(Ling, 1980b). Remembering that osmotic activity in fact is a measure of the decrease of water activity, this "new" source, in the AI hypothesis, is the third major component of the cells, the proteins. More specifically, it is the extended polypeptide chain of a "matrix protein" postulated to exist throughout the whole cell in all living cells. Although, not yet clearly established, actin, tubulin, and other cytoskeletal proteins are likely candidates, though in the AI hypothesis, they must exist, in the resting living cells, in a finer state of dispersion than those usually seen in electronmicroscopic pictures.

The matrix protein or proteins polarize in multilayers and hence reduce the osmotic activity of the bulk of cell water amounting to, on an average, about ten layers of water molecules between two chains. It is the reduced solubility of water in this state for large hydrated ions, e.g., Na^+, and molecules, e.g., sugar and amino acids, that functions to maintain the low level of these solutes in muscle, nerve, and other cells. Since there are no other osmotically active molecules or ions in the cell which match in concentration that of K^+, the establishment of the adsorbed state of K^+ adds an important reason for the polarized multilayer theory of cell water. (For recent experimental evidence supporting this view, see Ling *et al.*, 1980a,b; Ling, 1981a.)

One major concept of the association–induction hypothesis, a high degree of association between the major components of the cell (proteins, water, and K^+), is now reasonably well established by the experiments described above. The second major concept is yet far from being established, it concerns the control of coherent behavior through the propagated short-range inductive effect, or more completely the indirect F-effect, where the F-effect refers to the combined inductive or I-effects mediated through intervening atoms and D-effect mediated through space. A propagated F-effect is called an indirect F-effect (Ling, 1962). This provides a possible mechanistic basis for what later Monod *et al.* (1965) called "allosteric" effect (Ling, 1962, 1977d, 1981b). In the AI hypothesis this inductive effect lends coherence to the associated protein–K^+-water system as an autocooperative assembly by providing a mechanism for near-neighbor interactions among adsorption sites for ions as well as H_2O. Just as important, the inductive effect provides the theoretical basis for the control of many binding sites for K^+, H_2O, and other molecules on the protein chains by a small number of "cardinal adsorbents," including ATP and Ca^{++} which adsorb on specific controlling sites of the proteins, called "cardinal sites." Cardinal sites include what are commonly known as receptor sites (Ling, 1977d, 1981b).

IV. SOLUTE DISTRIBUTION

A general equation for solute distribution in living cells and model systems was introduced in 1965 (Ling, 1965). Choosing K^+ (and Na^+) as examples, and assuming the existence of only one type of adsorption site, the equation reads:

$$[K^+]_{cell} = \alpha q_K [K^+]_{ex} + \frac{[f]}{2}\left(1 + \frac{\frac{[K^+]_{ex}}{[Na^+]_{ex}} \cdot K^{oo}_{Na\to K} - 1}{\sqrt{\left(\frac{[K^+]_{ex}}{[Na^+]_{ex}} K^{oo}_{Na\to K} - 1\right)^2 + 4\frac{[K^+]_{ex}}{[Na^+]_{ex}} \cdot K^{oo}_{Na\to K} \exp\left(\frac{\gamma}{RT}\right)}}\right) \quad (1)$$

The first term on the right-hand side of Eq. (1) represents free K^+ in the cell water, and the second term represents adsorbed K^+. The concentration of free K^+ in the cell depends on α, the water content, $[K^+]_{ex}$, the external K^+ concentration, and q_K, the (average) equilibrium distribution coefficient of K^+ in the cell water. The value of q is near unity for small molecules and molecules that can fit in the polarized water lattice but are low for hydrated ions ($q_{Na} \approx 0.1$; $q_K \approx 0.3$) and other larger molecules (Ling, 1977c; Ling *et al.*, 1980a,b; Negendank and Shaller, 1979). The term representing adsorbed K^+ is an adsorption isotherm that was introduced in 1964 (Ling, 1964). Here $[f]$ is the concentration of the adsorption sites for K^+ in moles per kilogram of fresh cells. $K^{oo}_{Na \to K}$ is the intrinsic equilibrium constant for Na → K exchange. $-\gamma/2$ is the nearest neighbor interaction energy between K^+ and Na^+ (for more details, see Ling, 1970, 1980a; Ling and Bohr, 1970). When $-\gamma/2$ is zero, the isotherm becomes in essence the usual Langmuir type. However, when $-\gamma/2$ is larger than zero, the isotherm is autocooperative. The equilibrium concentration of K^+ adsorbed ($[K^+]_{ad}$) by the cells when plotted against increasing external K^+ concentration (at a constant $[Na^+]_{ex}$) is S-shaped or sigmoid, showing lower increment of uptake per unit increment of external K^+ concentration at low external K^+ concentration than at a higher external concentration. When the same data are plotted in a double logarithmic plot, i.e., when log $[K^+]_{ad}$ or log ($[K^+]_{ad}/[Na^+]_{ex}$) is plotted against either log $[K^+]_{ex}$ (at constant $[Na^+]_{ex}$) or log ($[K^+]_{ex}/[Na^+]_{ex}$), the slope at the locus where $[K^+]_{ad}$ equals $[Na^+]_{ad}$ is greater than unity. Indeed, the straight line with this slope and passing through the locus of equal K^+ and Na^+ occupancy is one described by the equation introduced by A. V. Hill (1910) to represent the oxygen uptake of hemoglobin. The slope of this double-log plot is the Hill's coefficient n. n, so far an empirical parameter, has long been suspected to be related to cooperativity. It is, in fact, equal to $\exp\left(-\frac{\gamma}{2RT}\right)$ where R and T have the usual meanings (Ling, 1964, 1980a).

The remarkable feature of an autocooperative adsorption is its "all-or-none" nature. That is, with a minor change in the ratio of K^+ and Na^+ concentration in the medium, the adsorption shifts from all K^+ to all Na^+, as diagrammatically illustrated in Fig. 3A, where the *i*th and *j*th solute may represent Na^+ and K^+ respectively.

Experimental studies showed that in a variety of living cells, the K^+ and Na^+ distribution follows Eq. (1) well. These include frog muscles (Ling, 1966; Ling and Bohr, 1970), rabbit uterine muscle (Jones, 1970), canine carotid arteries (Jones, 1973), guinea pig taenia coli (Karreman, 1973; Gulati, 1973), and human lymphocytes (Negendank and Shaller, 1979). Furthermore, in agreement with the association–induction hypothesis, a number of agents including drugs such as ouabain and ions such as Ca^{++}, which collectively are referred to as cardinal adsorbents, act on the K^+ and Na^+ distribution by changing primarily $K^{oo}_{Na \to K}$ as diagrammatically illustrated in Fig. 3B. Thus, exposure of frog muscle to a low concentration of ouabain (3.26×10^{-7} M) causes virtually all intracellular K^+, to be replaced stoichiometrically by Na^+ by merely changing the value of $K^{oo}_{Na \to K}$ from 200 to about 10, as shown in Fig. 4. The fact that K^+ is adsorbed to begin with and that the ouabain effect is entirely intact in an EMOC preparation (Ling, 1978a) shows that ouabain does not act on membrane-pumping mechanism, but diminishes the selectivity of the adsorption sites of K^+ over Na^+. Ca^{++}, on the other hand, acts to preserve the high $K^{oo}_{Na \to K}$ value, so that removal of Ca^{++} then acts like ouabain (Jones, 1973).

How a small number of bound Ca ions or ouabain molecules can affect many more K^+ and Na^+ adsorption sites is a subject of central interest in the AI hypothesis. As I have already mentioned, the propagated inductive effect or indirect *F*-effect offers the basic

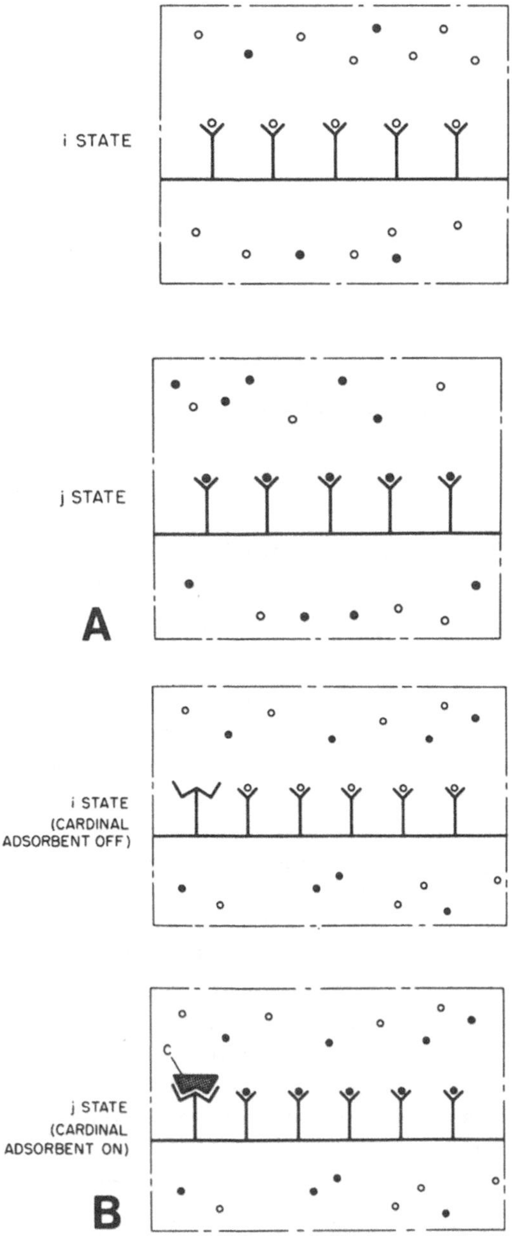

Figure 3. (A) Cooperative shifts between i and j states due to a change in the relative concentration of the i and j solutes in the environment. (B) Cooperative shifts between i and j states due to adsorption-desorption of cardinal adsorbent in an environment with unchanging i and j concentrations. (From Ling, 1977d, by permission of *Mol. Cell Biochem.*).

information- and energy-transmitting medium. I shall now review the mechanism of shift in $K^{oo}_{Na \to K}$ as a result of the propagated inductive effect.

The facts that sulfonate type of ion exchange resin with low pK_a selects K^+ over Na^+ and phosphoric and carboxyl types of ion exchange resin with high pK_a values select Na^+ over K^+ (Bregman, 1953), and the theory of Teunissen and Bungenberg de Jong (1938) that similar rank order of selectivity changes reflect differences of the field strength of the anionic groups, as well as the work of Eisenman *et al.* (1957; see also Ling, 1960) provided the stimulation to extend the earlier (1952), more restrictive model of selective adsorption

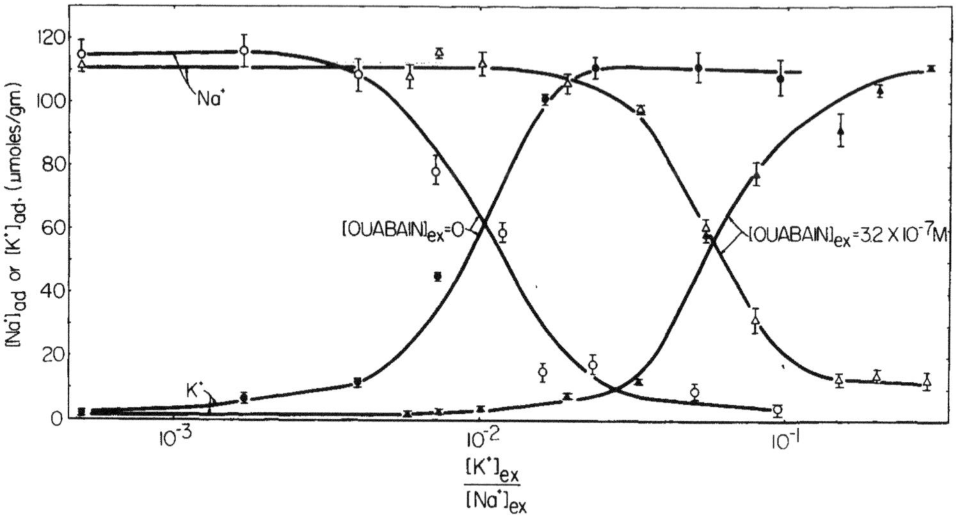

Figure 4. Effect of ouabain (3.2×10^{-7} M) on the equilibrium distribution of K^+ and Na^+ ion. Curves with open (Na^+) and filled (K^+) circles were equilibrium-distribution data from muscles not treated with ouabain. The point of intersection gives a $K^{\infty}_{Na\to K}$ of 100. In muscles treated with ouabain (3.2×10^{-7} M), $K^{\infty}_{Na\to K}$ has shifted to 21.7. (From Ling and Bohr, 1971, by permission of *Physiol. Chem. Phys.*)

of K^+ over Na^+ to a more general model in which selectivity of K^+ and Na^+ as well as other monovalent ions are variable (Ling, 1957, 1960). Indeed, in this microscopic model, the selectivity of K^+ and Na^+ varies with a small change in the electron density of an anionic oxygen atom, measured as a *c*-value. Rigorously defined elsewhere (Ling, 1962, p. 58), the *c*-value may be more simply described as a way to quantitatively simulate the aggregate effects of the remaining atoms of an oxyacid on the interaction of a hypothetical, prototype, singly-charged oxygen atom with a cation, as a displacement (in angstrom units) of the unit electric charge on the oxygen atom from its original prototype location at the center of the oxygen atom. Thus, if the aggregate effect produces an overall displacement of electrons in the system toward the oxyacid oxygen, it can be exactly matched by a specific displacement of the unit charge toward the cation, represented as a positive *c*-value, e.g., +1.0 Å. On the other hand, if the aggregate effect is to produce the opposite effect, it would be represented as a negative *c*-value, e.g., −1.0 Å. Thus, high *c*-value corresponds to high pK_a as in acetic acid, and low *c*-value corresponds to a low pK_a as in trichloracetic acid.

In this theory, it is the propagated change of the *c*-value of cooperatively linked β- and γ-carboxyl groups that gives rise to the observed alteration of $K^{00}_{Na\to K}$ (Ling, 1957, 1958, 1962, 1969).

V. SELECTIVE IONIC PERMEATION

In 1953, I suggested that the basic mechanism of selective K^+ adsorption on the β- and γ-carboxyl groups on the cytoplasmic protein may be extended to β- and γ-carboxyl groups of proteins on the cell surface (Ling, 1953). That the cell surface is endowed with fixed negative charges has long ago been considered by Bethe and Toropoff (1915), Michaelis (1925), Teorell (1935–1936), Meyer and Sievers (1936), Sollner *et al.* (1941), and others. *It was new, however, to postulate a high degree of counter-cation association as the basis* *of selective K^+ permeability* (Ling 1953, 1960, 1962, 1969). In 1965, Ling and Ochsenfeld provided positive evidence that the surface anionic sites may well be β- and γ-carboxyl groups since the rate of K^+ permeation shows an inflection at pH 4.6, the characteristic pK_a of β- and γ-carboxyl groups (Ling and Ochsenfeld, 1965). Entry of K^+ by asssociation of the ion with surface β- and γ-carboxyl groups is followed by libration of the ion near the fixed anion. Subsequent dissociation and entry into the cell shows "saturation" and "competition." This adsorption–desorption route is one way by which a charged ion of opposite sign to the surface fixed ions may gain entrance into the cell, as is indicated by route 2 in Fig. 5. A second route of entry is via the interstices among the fixed ionic sites, called the saltatory route (route 1).

It has been widely accepted that the cell surface is covered with a continuous lipid layer (Overton, 1899) punctured occasionally with narrow water-filled pores. Yet, in the last 20 years or so, new evidence has been put forth to suggest that this assumption might be true only for certain highly specialized cells, e.g., human erythrocytes (for review, see Ling, 1981b). Among the reasons cited for this view are the following: (1) with few exceptions, e.g., human erythrocytes, most cell membranes do not contain enough lipids to form a continuous bilayer (Ling, 1981b; Jain, 1972, Table 9-2), (2) removal of virtually all lipids did not change the overall thickness or the trilaminar layer spacings of the membranes (Fleischer *et al.*, 1967; Morowitz and Terry, 1969), (3) in the prototypical egg cell (of the frog) the rate of diffusion of labeled water through the cell membrane is equal to the rate of diffusion in the egg cytoplasm (Ling *et al.*, 1967), and (4) at 10^{-7} M the K^+-ionophore valinomycin reduces the resistance of artificial bilayers of lipids, extracted from erythrocytes, from $10^8 \Omega$ cm^2 to 2×10^4 Ω cm^2 in the presence of 30 mM KCl (Andreoli *et al.*, 1967). The membrane resistance of frog muscle cells is 4000 Ω cm^2 (Katz, 1966, Table 1) of which more than half is due to Cl^--conductance (Hutter and Padsha, 1959). Thus, the K^+ conductance of frog muscle cells is also in the realm of 10^4 Ω cm^2. If a substantial part of the muscle membrane barrier is lipid, exposure of frog muscle to 10^{-7} M valinomycin in the presence of 30 mM KCl should cause a substantial increase of inward K^+ flux rate. In fact, none was observed, nor was any change observed in frog ovarian eggs (Ling and Ochsenfeld, unpublished), in squid axon (Stillman *et al.*, 1970), or in the inner membrane of mouse liver mitochondria (Maloff *et al.*, 1978). However, a two-fold increase of K^+ flux rate was observed for human erythrocytes (Ling and Ochsenfeld, unpublished), in agreement with the unusually high lipid content of the red cell membranes (Jain, 1972, Table 9-2).

According to the AI hypothesis, the *physiological barrier* at the cell surface is multi-

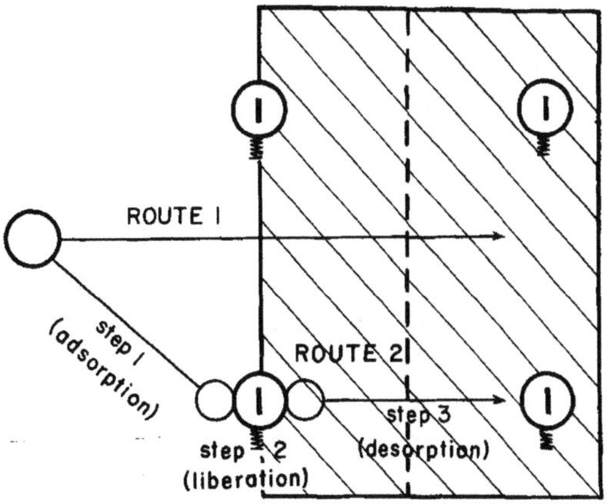

Figure 5. Diagrammatic illustrations of the two routes of ion entry into a fixed-charge system. Shaded area represents a microscopic portion of the surface of a fixed-charge system in which four fixed anions are represented. Route 1 is the saltatory route. Route 2, the adsorption–desorption route, involves a sequence of three steps, adsorption, libration near the fixed anion, and desorption. This adsorption–desorption route corresponds to the double type, since two ions are involved (the free cation and the fixed anion). (From Ling and Ochsenfeld, 1965, by permission of *Biophys. J.*)

layered water polarized by cell-surface matrix proteins with extended polypeptide chains (Ling, 1981a). Membrane lipids serve the role of stabilizing the protein–water system and of simple insulation (Ling, 1977e). This model readily explains why some model semipermeable membranes possess "pores" many times the diameter of solutes to which it is virtually impermeable. Thus, the copper ferrocyanide membrane consists of a network of crystalline particles with an average diameter of 150 Å, and the interstices are expected to be of similar dimensions (see Glasstone, 1946, p. 656). The average pore size of the active layer of cellulose acetate is about 44 Å (Schultz and Asumaa, 1969). Both the copper ferrocyanide and cellulose acetate membrane are semipermeable, highly permeable to water but virtually impermeable to sucrose which has a diameter of only 9.4 Å (Ling, 1973). An additional advantage of the polarized-water model over the conventional lipid layer-rigid pore model is that the cell surface permeability is amenable to the control by cardinal adsorbents (such as Ca^{++} or ATP), which are known to react with cardinal sites (receptor sites) on cell surface proteins.

This model of surface permeation was described by an equation that is a direct extension of Eq. (1), except that, for simplicity, nearest-neighbor interaction was neglected. Under this simplifying condition, the rate of entry of an ith monovalent cation into normal muscle cells is described by Eq. (2) (Ling and Ochsenfeld, 1965):

$$V_i = \underbrace{A[p_i]_{\text{ex}}}_{\text{Route 1}} + \underbrace{\frac{V_i^{\max}\,[p_i]_{\text{ex}}\,\tilde{K}_i}{1 + [p_i]_{\text{ex}}\,\tilde{K}_i + [p_j]_{\text{ex}}\,\tilde{K}_j}}_{\text{Route 2}} \tag{2}$$

where the first and second term on the right-hand side of the equation represent the saltatory route (route 1 of Fig. 5) and the adsorption–desorption route (route 2). $[p_i]_{\text{ex}}$ and $[p_j]_{\text{ex}}$ are the concentration of the ith and jth cation in the external medium. $\tilde{K}_i$ and $\tilde{K}_j$ are their respective adsorption constants on the cell surface. A and $V_i^{\max}$ are both constants. Strong support for this model came from the demonstration that ion entry into living cells as well as model systems of cation-exchange resin sheets, sheeps wool, and even a layer of isolated actomyosin gel are adequately described by this basic equation (Ling, 1960; Ling and Ochsenfeld, 1965, 1970).

VI. MOLECULAR MECHANISMS OF CELLULAR POTENTIALS

The verification of a localized and adsorbed state of K^+ makes a fundamental change of the cellular electrical potential inevitable. Indeed, a new theory of cellular electrical potential based on the early version of the AI hypothesis was already introduced in 1955 (Ling, 1955, 1959, 1960). According to this theory, the cellular resting potential is not a "membrane potential," but is a surface-adsorption potential as was one time suggested in general principle by Baur (see Baur and Kronmann, 1917). However Baur's theory, as a variant of the membrane theory, envisages two such potentials, one on each side of the cell membrane. The AI hypothesis suggests that the entire cell constitutes a proteinaceous fixed-charge system sharing similar attributes due to ion binding and water polarization. Therefore, there is only one surface adsorption potential at the outer cell boundary. It is the density and polarity of the surface β- and γ-carboxyl groups, which also determine the rate of ion permeation, as well as the temperature and the concentration of the external cation that adsorbs onto these anionic sites that determine the magnitude of the resting potential. For resting frog muscle, the simple equation was given by Ling (1959, 1960, 1962, 1967, 1982):

$$\psi = \text{constant} - \frac{RT}{F} \ln \left(\sum_{i=1}^{n} \tilde{K}_i [p_i^+]_{\text{ex}} \right) \tag{3}$$

where $[p_i^+]_{ex}$ represents the concentration of the *i*th monovalent cation in the external medium and $\tilde{K}_i$ is the adsorption constant of the *i*th monovalent cation on the surface anionic sites. In a more explicit form, Eq. (3) may be expressed as

$$\psi = \text{constant} - \frac{RT}{F} \ln (\tilde{K}_K [K^+]_{\text{ex}} + \tilde{K}_{\text{Na}} [Na^+]_{\text{ex}}) \tag{4}$$

where $\tilde{K}_K$ and $\tilde{K}_{Na}$ are the adsorption constants of K^+ and Na^+, respectively, and should be the same as those in Eq. (2) ($\tilde{K}_i$, $\tilde{K}_j$).

When the surface contains no fixed cation (as I believe is the case in resting muscle), there is no sensitivity to external Cl^- (Hodgkin and Horowicz, 1960), even though muscle has a very high Cl^- permeability (Hutter and Padsha, 1959). Indeed, after eliminating the Cl^- terms from the Hodgkin–Katz equation (Katz, 1966; a procedure I feel not easily defensible), the modified Hodgkin–Katz equation may, for short-term experiment involving little change of internal ion concentration, be reduced to a form entirely analogous to Eq. (3), with permeability constants P_i instead of adsorption constants $\tilde{K}_i$,

$$\psi = \text{constant} - \frac{RT}{F} \ln \left(\sum_{i=1}^{n} P_i [P_i]_{\text{ex}} \right) \tag{5}$$

Edelmann *et al.* (1971) have determined both the permeability constants and surface-adsorption constants of guinea pig heart with the help of Eq. (2) and concluded that ψ followed Eq. (3) but not Eq. (5) (Edelmann, 1973).

The surface-adsorption model offers an explanation for the data of a large number of reports, including those of Tobias (1950), Falk and Gerard (1954), Kao (1956), Shaw and Simon (1955), Shaw *et al.* (1956), Koketsu and Kimura (1960), and Tasaki and his co-workers (1964, 1965), which indicate that the resting potential does not vary with changes of intracellular K^+ concentration. Other data of Adrian (1956), Baker *et al.*(1961), and Hagiwara *et al.*(1964), show that some dependency of ψ on internal K^+ can sometimes be demonstrated (Ling, 1978b), on the basis of variation of one or more of the following factors: (1) change in surface fixed-ion density (Adrian, 1956), and (2) creation of new amphoteric interface between perfusing or injected fluids and the exposed intracellular protoplasm as well as the degree of ionization of fixed anions (compared to fixed cations).

In 1979, I further revised Eq. (3) to take into account nearest-neighbor interaction among the cell-surface sites (Ling, 1979). The equation for the resting potential now takes the form:

$$\psi = \text{constant}_1 - \frac{RT}{F} \ln \left[\frac{1}{[K^+]_{\text{ex}}} \left(1 + \frac{\xi - 1}{\sqrt{(\xi - 1)^2 + 4\xi \exp(\gamma/RT)}} \right) \right] \tag{6}$$

where,

$$\xi = \frac{[K^+]_{\text{ex}}}{[Na^+]_{\text{ex}}} \cdot \tilde{K}^{\infty}_{\text{Na} \to \text{K}} \tag{7}$$

$\tilde{K}^{\infty}_{Na \to K}$ and γ have the same significance as in Eq. (1) but refer only to anionic sites on a microscopically thin cell surface. Eq. (6) can account for the gradual decrease of ψ with decreasing external K^+ concentration (and high Na^+ concentration) below that of the normal Ringer's solution (Fig. 6). This theoretical equation is capable of explaining, for example, changes of ψ described by Ruzyllo and Vick (1974) for canine Purkinje cells.

The broader usefulness of Eq. (6) is demonstrated by its ability to explain the experimental data of Maloff *et al.* (1978) who measured the resting potential of the giant mitochondria of cuprizone-fed mouse liver. This potential was virtually indifferent to external K^+ concentration before 10^{-7} M valinomycin was added. Contrary to the expectation based on the membrane theory, in which the expected effect of valinomycin is to increase K^+ permeability of the mitochondrial inner membrane, the absence of change in K^+ conductance was demonstrated (see also Ling, 1981a). On the other hand, these observations can be described by Eq. (6), with the assumption that the affinity for K^+ on the surface anionic sites is increased by valinomycin by a factor of three (for further discussion, see Ling, 1981b).

A more general equation for the resting and action potential would include a diffusion potential term,

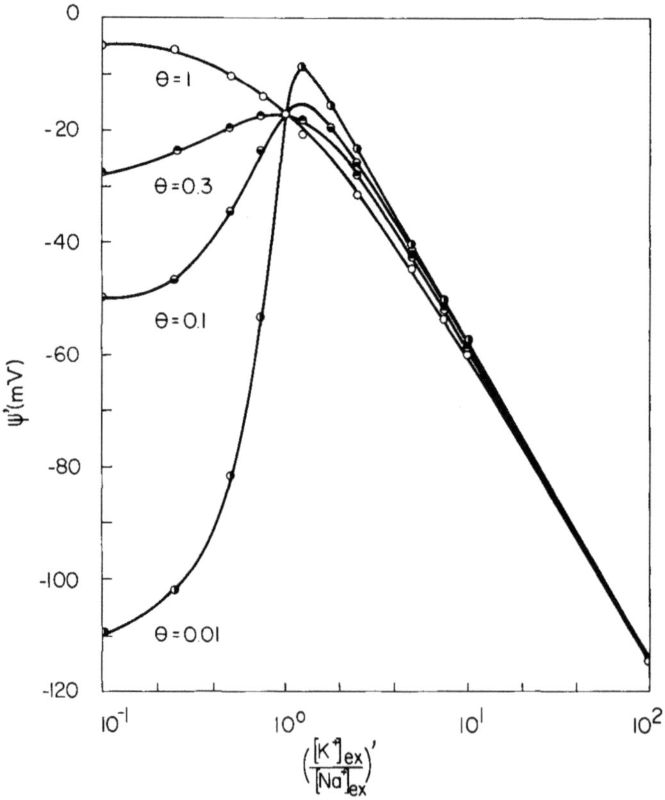

Figure 6. A plot of the resting potential against external K^+ and Na^+ concentration ratio. According to Eq. (6) ordinate represents ψ' which is equal to ψ—constant, abscissa represents $([K^+]_{ex}/[Na^+]_{ex})'$ which is $([K^+]_{ex}/[Na^+]_{ex} + K^{\infty}_{Na \to L})$. For experiments carried out in the presence of a constant concentration of Na^+, e.g., 100 mM, the abscissa is then $[K]_{ex} \cdot (K^{\infty}_{Na \to K}/0.1)$. Symbols are computed points, presented to mark out the separate curves. (From Ling, 1979, by permission of *Physiol. Chem. Phys.*)

$$\psi = \text{constant}_1 - \frac{RT}{F} ln \left\{ \frac{1}{[K^+]_{ex}} \left(1 + \frac{\xi - 1}{\sqrt{(\xi - 1)^2 + 4\xi \exp(\gamma/RT)}} \right) \right\} \tag{8}$$
$$- \frac{RT}{F} ln \frac{\gamma_{in}^{Na} [Na^+]_{in}^{free} + \gamma_{in}^{K} [K^+]_{in}^{free}}{\gamma_{ex}^{Na} [Na^+]_{ex} + \gamma_{ex}^{K} [K^+]_{ex}}$$

where γ_{in}^{Na}, γ_{in}^{K}, γ_{ex}^{Na}, and γ_{ex}^{K} are the activity coefficients of Na^+ and K^+ in the cell-surface layer water and external solution respectively. $[Na^+]_{in}^{free}$ and $[K^+]_{in}^{free}$ are the free Na^+ and K^+ concentration in the cell-surface water, respectively. The second term on the right-hand

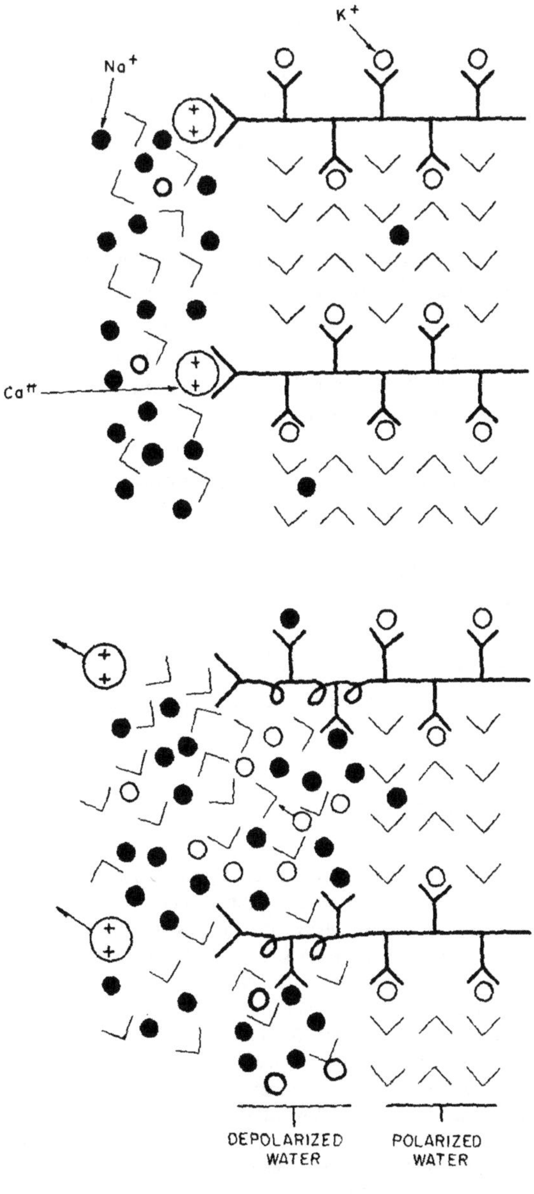

Figure 7. Schematic illustration of the events occurring at the cell surface at rest and during action potential, according to the association–induction hypothesis. The upper figure shows the resting condition, in which K^+ is preferentially adsorbed at the cell surface. Water in the state of polarized multilayers, represented as ordered *V*'s, has low solubility and permeability for both K^+ and Na^+. Resting condition of preferential K^+ adsorption and polarized water depends on a cardinal adsorbent, Ca^{++} occupying the cardinal site. During activity, Ca^{++} removal causes a time-dependent autocooperative change of the electronic conformation of the surface proteins, causing a decrease of K^+/Na^+ preference in response to a *c*-value increase and the depolarization of water, with a local increase of specific permeability to Na^+ via the adsorption–desorption route and nonspecific permeability to Na^+ and other solutes via the depolarized water. (From Ling, 1981c, by permission of Wissenschaftlich Verlagsgesesellschaft, Stuttgart).

side of Eq. (8), a diffusion potential term, vanishes when the cell is at rest because under this condition, $\gamma_{in}^{Na}/\gamma_{ex}^{Na} = [Na^+]_{ex}/[Na^+]_{in}$ and $\gamma_{in}^{K}/\gamma_{ex}^{K} = [K^+]_{ex}/[K^+]_{in}$. Now

$$q_{Na} = \frac{\gamma_{ex}^{Na}}{\gamma_{in}^{Na}}, \quad q_K = \frac{\gamma_{ex}^{K}}{\gamma_{in}^{K}} \tag{9}$$

where q_{Na} and q_K are the equilibrium distribution coefficients of Na^+ and K^+ between the surface-cell water and the external solution, respectively. Substituting Eq. (9) into (8), and taking into account of the fact that γ_{ex}^{Na} $[Na^+]_{ex}$ + γ_{ex}^{K} $[K^+]_{ex}$ is a constant and that $\gamma_{ex}^{Na} = \gamma_{ex}^{K}$, Eq. (8) simplified to:

$$\psi = \text{constant}_2 - \frac{RT}{F} \ln \left\{ \frac{1}{[K^+]_{ex}} \left(1 + \frac{\xi - 1}{\sqrt{(\xi - 1)^2 + 4\xi \exp(\gamma/RT)}} \right) \right\} - \frac{RT}{F} \ln \left\{ \frac{[Na^+]_{in}^{free}}{q_{Na}} + \frac{[K^+]_{in}^{free}}{q_K} \right\} \tag{10}$$

During excitation, $K_{Na\to K}^{oo}$ falls due to the *c*-value increase of the surface anionic sites, q_{Na} (and q_K) concomitantly increases. The inrush of Na^+ brings about not only a cancellation of the resting potential but an overshoot. The adsorption of Na^+ displaces K^+ from adsorption sites at and near the cell surface, causing an increase of free K^+, i.e., $[K^+]_{in}^{free}$, and hence the delayed outward K^+ current (Fig. 7).

To be noted is the fact that the fixed-surface anionic site would function as a K^+ channel at rest and following activation. But the same sites with a transient *c*-value change provide an apparent Na^+ channel. Yet in both cases, there is both a noncompetitive ("independent") component through the cell-surface water and a competitive component through the adsorption–desorption route.

The delayed outward current more likely reflects not an opening of a specific "gate" but the diffusion of K^+ after its liberation from adsorption sites at and near the cell surface.

ACKNOWLEDGMENT. The foregoing work was supported by NIH Grants 2-R01-CA16301-03 and 2-R01-GM11422-13, and by Office of Naval Research Contract N00014-79-C-0126.

REFERENCES

Adrian, R. H., 1956, The effect of internal and external potassium concentration on the membrane potential of frog muscle, *J. Physiol.* **133**:631–658.

Andreoli, T. E., Tiffenberg, M., and Tosteson, D. C., 1967, The effect of valinomycin on the ionic permeability of thin lipid membranes, *J. Gen. Physiol.* **50**:2527–2545.

Baker, P. E., Hodgkin, A. L., and Shaw, T. I., 1961, Replacement of the protoplasm of a giant nerve fiber with artificial solutions, *Nature* **190**:885–887.

Baur, E., and Kronmann, S., 1917, Über die Ionenadsorptionspotentiale, *Z. Physik. Chem.* **92**:81–97.

Berendsen, H. J. C., and Edzes, H. T., 1973, The observation and general interpretation of sodium magnetic resonance in biological material, *Ann N.Y. Acad. Sci.* **205**:459–480.

Bernstein, J., 1912, *Elektrobiologie*, F. Vieweg und Sohn, Braunschweig.

Bethe, A., and Toropoff, T., 1915, Über elektrolytische vorgänge and diaphragmen teil I die neutralitätsstörung, *Zeitsch. Physik. Chem.* **88**:686–742.

Bregman, J. I., 1953, Cation exchange processes, *Ann. N.Y. Acad. Sci.* **57**:125.

Conway, E. J., Kernan, R. P., and Zadunarsky, J. A., 1961, The sodium pump in skeletal muscle in relation to energy barriers, *J. Physiol.* **155**:263–279.

Cooke, R., and Kuntz, I. D., 1974, The properties of water in biological systems, *Annu. Rev. Biophys. Bioeng.* **3**:95–126.

Cope, F. W., 1965, Nuclear magnetic resonance evidence for complexing of sodium ions in muscle, *Proc. Natl. Acad. Sci. USA* **54**:225–227.

Cope, F. W., 1969, Nuclear magnetic resonance evidence using D_2O for structured water in muscle and brain, *Biophys. J.* **9**:303–319.

Edelmann, L., 1973, The influence of rubidium, cesium ions on the resting potential of guinea pig heart muscle cells as predicted by the association–induction hypothesis, *Ann. N.Y. Acad. Sci.* **204**:534–537.

Edelmann, L., 1977, Potassium adsorption sites in frog muscle visualized by cesium and thallium under the transmission electron microscope, *Physiol. Chem. Phys.* **9**:303–319.

Edelmann, L., 1978, Visualization and x-ray microanalysis of potassium tracers in freeze-dried and plastic imbedded frog muscle, *Microscop. Acta Suppl. Microprobe Anal. Biol. Med.* **2**:166.

Edelmann, L., 1980a, Potassium binding sites in muscle: Electron microscopic visualization of K, Rb, and Cs in freeze-dried preparations and autoradiography at liquid nitrogen temperature using ^{86}Rb and ^{134}Cs, *Histochemistry* **67**:233–242.

Edelmann, L., 1980b, Preferential localized uptake of K^+ and Cs^+ over Na^+ in the A-band of freeze-dried embedded muscle section detected by x-ray microanalysis and laser microprobe mass analysis, *Physiol. Chem. Phys.* **12**:509–514.

Edelmann, L., Pfleger, K., and Matt, K. H., 1971, Untersuchungen der in Vergleish zur Kalium-Anreichung bevorzugsten Anreicherung von Cäsium 137 im Saugetierorganismus. I. Kalium, Rubudium and Cäsium Anreicherung im perfundierten Meerschweinschenherzen, *Biophysik* **7**:181–199.

Eisenman, G., Rudin, D. O., and Casby, J., 1957, Paper verbally prsented at the 10th annual Conference on Electrical Techniques in Medicine and Biology of the A.I.E.E., I.S.A., and I.R.E., Boston.

Ernst, E., 1963, *Biophysics of the Striated Muscle*, Hungarian Academy of Sciences, Budapest, Hungary.

Falk, G., and Gerard, R. W., 1954, Effects of microinjected salts and ATP in the membrane potential and mechanical response of muscle, *J. Cell. Comp. Physiol.* **43**:393–404.

Fisher, M. H., and Suer, W. J., 1935, Base-protein-acid compounds, *Arch. Pathol.* **20**:683–689.

Fleischer, S., Fleischer, B., and Stoeckenius, W., 1967, Fine structure of lipid-depleted mitochondria, *J. Cell. Biol.* **32**:193–203.

Freedman, J. C., 1976, Partial restoration of sodium and potassium gradients by human erythrocyte membrane, *Biochim. Biophys. Acta* **455**:989–992.

Glasstone, S., 1946, *Textbook of Physical Chemistry*, 2nd Ed., Van Nostrand, New York.

Goldin, S. M., and Tong, S. W., 1974, Reconstitution of active transport catalyzed by the purified sodium and potassium ion-stimulated adenosine triphosphatase from canine renal medulla, *J. Biol. Chem.* **249**:5907–5915.

Gortner, R. A., 1930, The state of water in colloidal and living systems, *Trans. Faraday Soc.* **26**:678–704.

Gulati, J., 1973, Cooperative interaction of external calcium, sodium and ouabain with the cellular potassium in smooth muscle, *Ann. N.Y. Acad. Sci.* **204**:337–357.

Hagiwara, S., Chichibu, S., and Naka, K. I., 1964, The effects of variouis ions on resting and spike potential of barnacle muscle fibers, *J. Gen. Physiol.* **48**:163–179.

Hanson, J., and Huxley, H. E., 1953, Structural basis of the cross-striation in muscle, *Nature* **172**:530–532.

Hazlewood, C. F., Nichols, B. F., and Chamberlain, N. F., 1969, Evidence for the existence of a minimum of two phases of ordered water in skeletal muscle, *Nature* **222**:746–750.

Hilden, S., and Hokin, L. E., 1975, Active potassium transport coupled to active sodium transport in vesicles reconstituted from purified sodium and potassium ion-activated adenosine triphosphate from the rectal gland of Squalus acanthias, *J. Biol. Chem.* **250**:6296–6303.

Hill, A. V., 1910, The possible effects of the aggregation of molecules of hemoglobin on its dissociation curves, *J. Physiol. (London)* **40**:iv.

Hill, A. V., 1930, The state of water in muscle and blood and the osmotic behavior of muscle, *Proc. R. Soc. B* **106**:477–505.

Hill, A. V., and Kupalov, P. S., 1930, The vapor pressure of muscle, *Proc. R. Soc. B* **106**:445–476.

Hinke, J. A. M., 1961, The measurement of Na and K activities in the squid axon by means of cation selective glass microelectrodes, *J. Physiol.* **156**:134–335.

Hodgkin, A. L., and Horowicz, P., 1960, The effect of sudden changes in ionic concentration on the membrane potential of single muscle fibers, *J. Physiol.* **153**:370–375.

Hodgkin, A. L., and Huxley, A. F., 1952, A quantitative description of membrane current and its application to conduction and excitation in nerve, *J. Physiol. (London)* **117**:500–544.

Hodgkin, A. L., and Keynes, R. D., 1953, The mobility and diffusion coefficient of potassium in giant axons from Sepia, *J. Physiol.* **119**:513–528.

Hutter, O. F., and Padsha, S. M., 1959, Effect of nitrate, and other anions on the membrane resistance of frog skeletal muscle, *J. Physiol.* **146:**117–132.

Jain, M. K., 1972, *The Bimolecular Lipid Membrane*, Van Nostrand Rheinhold, New York.

Jones, A. W., 1970, Effects of progesterone treatment on potassium accumulation and permeations in the rabbit myometrium, *Physiol. Chem. Phys.* **2:**151–167.

Jones, A. W., 1973, Control of cooperative K accumulation in smooth muscle by divalent ions, *Ann. N.Y. Acad. Sci.* **204:**379–392.

Kao, C. Y., 1956, Absence of membrane potential in presence of asymmetrical ion distribution in the Fundulus egg, *Biol. Bull.* **111:**292.

Karreman, G., 1973, Cooperative specific adsorption, *Ann. N.Y. Acad. Sci.* **204:**393–409.

Katz, B., 1966, *Nerve, Muscle, and Synapse*, McGraw-Hill, New York.

Keynes, R. D., and Maisel, G. W., 1954, The energy requirement for sodium extrusion from a frog muscle, *Proc. R. Soc. B* **142:**383–392.

Koketsu, K., and Kimura, Y., 1960, The resting potential and intracellular potassium of skeletal muscle in frogs, *J. Cell. Comp. Physiol.* **55:**239–244.

Kushmerick, M. J., and Podolsky, R. J., 1969, Ionic mobility in muscle cells, *Science* **166:**1297–1298.

Ling, G. N., 1951, Tentative hypothesis for selective ionic accumulation in muscle cells, *Am. J. Physiol.* **167:**806.

Ling, G. N., 1952, The role of phosphate in the maintenance of the resting potential and selective ionic accumulation in frog muscle cells, in: *Phosphorous Metabolism*, Vol. II (W. D. McElroy and B. Glass, eds.), pp. 748–795, Johns Hopkins University Press, Baltimore, Maryland.

Ling, G. N., 1953, Ionic permeability according to the fixed charge hypothesis, *Proc. 19th Int. Physiol. Cong.*, Montreal, Canada, p. 566.

Ling, G. N., 1955, New hypothesis for the mechanism of cellular resting potential, *Fed. Proc.* **14:**93.

Ling, G. N., 1957, Fixed charge hypothesis for a key mechanism in change of Na-K permeability during excitation, *Fed. Proc.* **16:**81.

Ling, G. N., 1958, Fixed charge induction hypothesis for biological transmitter amplifier and mixer at the molecular level, *Fed. Proc.* **17:**98.

Ling, G. N., 1959, On the mechanism of cell potential, *Fed. Proc.* **18:**371.

Ling, G. N., 1960, The interpretation of selective ionic permeability and cellular potentials in terms of the fixed charge-induction hypothesis, *J. Gen. Physiol.* **43:**149–174.

Ling, G. N., 1962, *A Physical Theory of the Living State: The Association-Induction Hypothesis*, Blaisdell, Waltham, Massachusetts.

Ling, G. N., 1964, Role of inductive effect in cooperative phenomena of proteins, *J. Biopolymers* **1:**91–116.

Ling, G. N., 1965, The membrane theory and other views for solute permeability, distribution, and transport in living cells, *Perspect. Biol. Med.* **9:**87–106.

Ling, G. N., 1966, All-or-none adsorption by living cells and model protein-water systems: Discussion of the problem of permease-induction and determination of secondary and tertiary structures of proteins, *Fed. Proc.* **25:**958–970.

Ling, G. N., 1967, Anion-specific and cation-specific properties of the collodion-coated glass electrode and a modification, in: *Glass Electrodes for Hydrogen and Other Cations*, Chapter 10 (G. Eisenman, ed.), pp. 284–292, Marcel Dekker, New York.

Ling, G. N., 1969, A new model for the living cell: A summary of the theory and recent experimental evidence in its support, *Int. Rev. Cytol.* **26:**1–61.

Ling, G. N., 1970, Diphosphoglycerate and inosine hexaphosphate control of oxygen binding by hemoglobin: A theoretical interpretation of experimental data, *Proc. Natl. Acad. Sci. USA* **67:**296–301.

Ling, G. N., 1973, What component of the living cell is responsible for its semipermeable properties? Polarized water or lipids? *Biophys. J.* **13:**807–816.

Ling, G. N., 1977a, K^+ localization in muscle cells by autoradiography, and identification of K^+ adsorbing sites in living muscle cells with uranium binding sites in electron micrographs of fixed cell preparations, *Physiol. Chem. Phys.* **9:**319–328.

Ling, G. N., 1977b, Thallium and cesium in muscle cells compete for the adsorption sites normally occupied by K^+, *Physiol. Chem. Phys.* **9:**217–225.

Ling, G. N., 1977c, Potassium accumulation in frog muscle: The association-induction hypothesis versus the membrane theory, *Science* **198:**1281–1283.

Ling, G. N., 1977d, The physical state of water and ions in living cells and a new theory of the energization of biological work performance by ATP, *J. Mol. Cell Biochem.* **15:**159–172.

Ling, G. N., 1977e, The functions of polarized water and membrane lipids: A rebuttal, *Physiol. Chem. Phys.* **9:**301–311.

Ling, G. N., 1978a, Maintenance of low sodium and high potassium levels in resting muscle cells, *J. Physiol.* **280:**105–123.

Ling, G. N., 1978b, Two opposing theories of the cellular electrical potential: A quarter of a century of experimental testing, *Bioelectrochem. and Bioenerget.* **5:**411–419.

Ling, G. N., 1979, The equations for cellular resting potentials according to the surface adsorption theory, a corollary of the association-induction hypothesis, *Physiol. Chem. Phys.* **11:**59–64.

Ling, G. N., 1980a, The theory of the allosteric control of cooperative adsorption and conformation changes: A molecular model for physiological activities according to the association-induction hypothesis, in: *Cooperative Phenomena in Biology* (G. Karreman, ed.), pp. 39–69, Pergamon Press, New York.

Ling, G. N., 1980b, The role of multilayer polarization of cell water in the swelling and shrinkage of living cells, *Physiol. Chem. Phys.* **12:**383–384.

Ling, G. N., 1981a, Water and the living cell as seen from the viewpoint of a new paradigm, in: *International Cell Biology 1980–1981* (H. G. Schweiger, ed.), pp. 904–914, Springer-Verlag, Berlin.

Ling, G. N., 1981b, Oxidative phosphorylation and mitochondrial physiology: A critical review of chemiosmotic theory, and reinterpretation by the association-induction hypothesis, *Physiol. Chem. Phys.* **13:**29–96.

Ling, G. N., 1981c, Elektrische Potentiale lebender Zellen, in: *Die Zelle: Struktur und Funktion,* 3rd Ed. (H. Metzner, ed.), pp. 356–389, Wissenschaftlich Verlagsgesellschaft mbH, Stuttgart, Germany.

Ling, G. N., 1982, The cellular resting and action potentials: Interpretation based on the association-induction hypothesis, *Physiol. Chem. Phys.* **14:**47–96.

Ling, G. N., 1983, *In Search of the Physical Basis of Life,* Plenum, New York, in preparation.

Ling, G. N., and Bohr, G., 1969, Studies on ionic distribution in living cells. I. Long-term preservation of isolated frog muscles. *Physiol. Chem. Phys.* **1:**591–599.

Ling, G. N., and Bohr, G., 1970, Studies on ion distribution in living cells II. Cooperative interaction between intracellular K^+ and Na^+ Ions, *Biophys. J.* **10:**519–538.

Ling, G. N., and Bohr, G., 1971, Studies on ion distribution in living cells. III. Cooperative control of electrolyte accumulation by ouabain in the frog muscle, *Physiol. Chem. Phys.* **3:**431–447.

Ling, G. N., and Cope, F. W., 1969, Potassium ion: Is the bulk of intracellular K^+ adsorbed? *Science* **163:**1335–1336.

Ling, G. N., and Negendank, W., 1980, Do isolated membranes and purified vesicles pump sodium? A critical review and reinterpretation, *Perspsect. Biol. Med.* **23:**215–239.

Ling, G. N., and Ochsenfeld, M. M., 1965, Studies on the ionic permeability of muscle cells and their models, *Biophys. J.* **5:**777–807.

Ling, G. N., and Ochsenfeld, M. M., 1966, Studies on ion accumulation in muscle cells, *J. Gen. Physiol.* **49:**819–843.

Ling, G. N., and Ochsenfeld, M. M., 1970, Demonstration of saturability and competition in ion transport into a membraneless protein-water system, *Physiol. Chem. Phys.* **2:**189–194.

Ling, G. N., Ochsenfeld, M. M., and Karreman, G., 1967, Is the cell membrane a universal rate-limiting barrier to the movement of water between the living cell and its surrounding medium? *J. Gen. Physiol.* **50:**1807–1820.

Ling, G. N., Ochsenfeld, M. M., Walton, C., and Bersinger, T. J., 1980a, Mechanism of solute exclusion from cells: The role of protein-water inteaction, *Physiol. Chem. Phys.* **12:**3–10.

Ling, G. N., Walton, C., and Bersinger, T. J., 1980b, Reduced solubility of polymer-oriented water for sodium salts, amino acids, and other solutes normally maintained at low levels in living cells, *Physiol. Chem. Phys.* **12:**111–138.

Maloff, B. L., Scordillis, S. P., Reynolds, C., and Tedeshi, H., 1978, Membrane potentials and resistance of giant mitochondria, *J. Cell. Biol.* **78:**199–213.

Meyer, K. H., and Sievers, J. F., 1936, La Permeabilité des membranes. II. Essais avec des membranes selectives artificielles, *Helv. Chim. Acta* **19:**665–677.

Michaelis, L., 1925, Contributions to the theory of permeability of membranes for electrolytes, *J. Gen. Physiol.* **8:**33–59.

Monod, J., Wyman, J., and Changeux, J., 1965, On the nature of allosteric transitions: A plausible model, *J. Mol. Biol.* **12:**88–118.

Moore, B., and Roaf, H. E., 1908, On the equilibrium between the cell and its environment in regard to soluble constituents, with special reference to the osmotic equilibrium of the red blood corpuscles, *Biochem. J.* **3:**55–81.

Morowitz, H. J., and Terry, T. M., 1969, Characterization of the plasma membrane of mycoplasma Laidlawii, V. Effect of selective removal of proteins and lipids, *Biochim. Biophys. Acta* **183:**276–294.

Negendank, W., and Shaller, C., 1979, K-Na distribution in human lymphocytes: Description by the association-induction hypothesis, *J. Cell. Physiol.* **98:**95–105.

Overton, E., 1899, Über die allgemeinen osmotischen eigenschaften der zellen ihre vermultchen ursachen und ihre bedeutung für physiologie, *Vierteljahrschrift Naturf. Ges. Zürich* **44**:88–136.

Ruzyllo, W., and Vick, R. L., 1974, Cellular resting and diastolic potentials in canine cells: Effects of external [K^+] and purkinje repetitive excitation, *Mol. Cell Cardiol.* **6**:27–37.

Schultz, R. D., and Asumaa, S. K., 1969, Ordered water and the ultrastructure of the cellular plasma membrane, *Rec. Prog. Surf. Sci.* **3**:291–295.

Shaw, F. H., and Simon, S. E., 1955, The nature of the sodium and potassium balance in nerve and muscle, *Aust. J. Exp. Biol. Med.* **33**:153–177.

Shaw, F. H., Simon, S. E., Johnstone, B. M., and Holman, M. E., 1956, The effects of the changes of environment on the electric and ionic pattern of muscle, *J. Gen. Physiol.* **40**:263–288.

Sollner, K., Abrams, I., and Carr, C. W., 1941, The structure of the collodion membrane with its electrical behavior. II. The activated collodion membrane, *J. Gen. Physiol.* **25**:7–27.

Stillman, I. M., Gilbert, D. L., and Robbins, J., 1970, Monactin does not influence potassium permeability in the squid axonal membrane, *Biochim. Biophys. Acta* **203**:338.

Tasaki, I., and Takenaka, T., 1964, Effects of various potassium salts and proteases upon excitability of intracellular perfused squid giant axons, *Proc. Natl. Acad. Sci. USA* **52**:804–840.

Tasaki, I., Luxuro, M., and Ruarte, A., 1965, Electrophysiological studies of Chilean squid axons under internal perfusion with sodium rich medium, *Science* **150**:899–901.

Teorell, T., 1935–1936, An attempt to formulate a quantitative theory of membrane permeability, *Proc. Soc. Exp. Biol. Med.* **33**:282–285.

Teunissen, P. H., and Bungenberg de Jong, H. G., 1938, Negative, nicht amphotere biokolloide alf hochmolekulare elektrolyte, II. Reihenfolgen der kationen bei der umladung mit neutralsalzen. Analogien mit den reihenfolgen der löslichkeit von entsprechenden kleinmolekularen elektrolyten, *Kolloidbeihefte* **48**:33–92.

Tobias, J. M., 1950, Injury and membrane potentials in frog muscle after depleting potassium and producing other changes by soaking in potassium-free salt solution and distilled water, *J. Cell. Comp. Physiol.* **36**:1–13.

Trombitas, K., and Tigyi-Sebes, A., 1979, X-ray microanalysis studies on native myofibrils and mitochondria isolated by microdissection from honey-bee flight muscle, *Acta Biochem. Biophys. Acad., Hung.* **14**:271–277.

Troshin, A. S., 1966, *Problems of Cell Permeability*, Pergamon Press, London.

20

Electrical Behavior of Single-Filing Channels

George Eisenman, John Sandblom, and Jarl Hagglund

I. BARRIER-MODEL STUDIES

Several biological channels, as well as the model peptide channel formed by gramicidin, are believed to be capable of multiple occupancy by permeant ions (Hodgkin and Keynes, 1955; Hladky, 1972; Hille and Schwarz, 1978; Begenisich, 1979; Eisenman *et al.*, 1978; Urban *et al.*, 1980). Some of the peptide channels, such as those formed by Gramicidin A, have been estimated to be so narrow that ions and water molecules cannot pass each other (Schagina *et al.*, 1978; Procopio and Andersen, 1979; Levitt *et al.*, 1978; Finkelstein and Andersen, 1981). That such channels also have multiple barriers with different locations in the potential field is inferred from the finding of flux-ratio exponents greater than 1 and from the existence of an ion-concentration dependence of the current–voltage characteristic, which in the gramicidin channel was recognized quite early (Hladky, 1972; Läuger, 1973) to be concave toward the voltage axis at low salt concentrations ("sublinear") and convex at high concentrations ("supralinear").

This chapter is concerned with the way structure and function of single-filing ionic channels are inferred from their electrical and flux behaviors. We will present experimental and theoretical work that is still in progress, directed toward understanding the neutral peptide channel formed in lipid bilayers by Gramicidin A. This molecule is emphasized because its primary and higher structures are better known than any other channel molecule (Sarges and Witkop, 1965; Urry, 1971; Koeppe *et al.*, 1978). We will use strict Eyring barrier models but will also scrutinize certain deficiencies of such models recently pointed out by Andersen (1982a,b,c), who has examined two effects in the aqueous solutions external to the channel mouth, diffusion limitations and "interfacial polarization" (also see Andersen and Procopio, 1980; Finkelstein and Andersen, 1981). It should also be noted that other extensions of barrier models are under way, including a general procedure for generating equivalent circuits

George Eisenman • Department of Physiology, UCLA Medical School, Los Angeles, California 90024. *John Sandblom* • Department of Physiology and Medical Biophysics, University of Uppsala, S-751 23 Uppsala, Sweden. *Jarl Hagglund* • Department of Neurology, University Hospital, S-751 23 Uppsala, Sweden.

(Sandblom *et al.*, 1982) and a study of the effects of nonstatic energy profiles (Läuger *et al.*, 1980).

The present situation as to theory can be summarized with the help of Fig. 1. At the top, a reasonably realistic energy profile for the gramicidin channel in the absence of an applied voltage is indicated schematically. This profile results from the superposition of an unfavorable electrostatic image energy (Läuger, 1973; Levitt, 1978) upon favorable interaction energies with the ligands of the channel. This produces a large number of local potential-energy minima, "sites," separated from each other by maxima, "barriers," which are less favorable to occupancy by ions. The profile in Fig. 1A is seen to contain *N* barriers "B" and *N*-1 sites "S" and is therefore designated "*N*B(*N*-1)S." Such a complicated energy profile has so far been treated only for the case of single-ion occupancy (Woodbury, 1971; Läuger, 1973; Hille, 1975a,b), and is not analytically tractable for occupancy by more than one cation. It can, however, be approximated by simpler profiles shown (B–D), which have already been at least partially analyzed. The 3-barrier 2-site case (3B2S) has been applied to the K channel (Chizmadjev and Aityan, 1977; Hille and Schwarz, 1978) and the gramicidin channel (Urban and Hladky, 1979), Hille and Schwarz have obtained numerical solutions for certain properties of the 4-barrier 3-site case (4B3S), and the 1-barrier 4-site model (1B4S) of Sandblom *et al.* (1977) was the first multi-occupancy model applied to explain the concentration-dependent permeability ratios for gramicidin and related blocking effects.

Probably, a completely satisfactory approximation to Fig. 1A can be made by the general 5-barrier 4-site case (5B4S) shown in (E); we have solved the two 3-barrier 4-site subcases 3B4S′ and 3B4S″ illustrated in (F) and (G), assuming that no more than 3 of the 5 barriers are ever significantly rate-determining. In the 3B4S′ case (Hagglund *et al.*, 1982) the barrier between the sites in each half channel is assumed never to be as important as the barrier either at the mouth of the channel or at the center. In the 3B4S″ case (Sandblom *et al.*, 1983), the barrier between the outermost site and the solution is considered to be small compared to the barrier between it and the next site in the channel. For the neutral gramicidin channel it is necessary to extend the analysis to 4 sites rather than 3 sites, because a 3-site channel would require the electrostatically unlikely (Levitt, 1978) existence of a binding site in the middle of the channel.

Though existing barrier models are successful in many respects in accounting for the oberved properties of the gramicidin channel, they have shown certain inadequacies. In

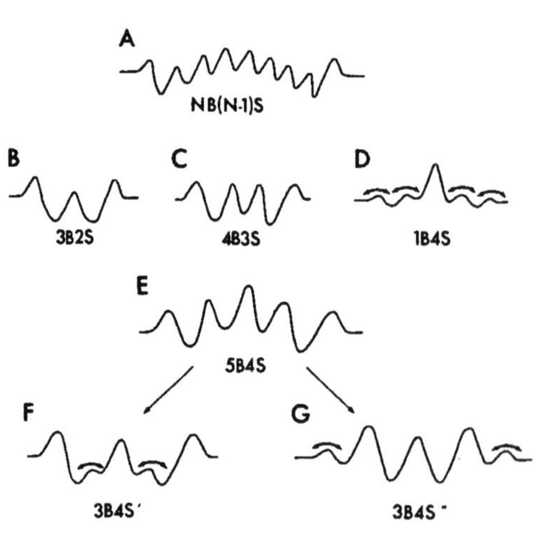

Figure 1. Principal categories of energy profiles for ion transfer through an ionic channel like gramicidin A. The code symbolizes the number of barriers, B, across which there is no equilibrium, and the number of sites, S, e.g., 3B2S means 3 barriers 2 sites. A double-arrow across a barrier indicates that neighboring sites (or a site and the adjacent external solution) are in equilibrium with each other. The top profile "*n*B(*n*-1)S" presents a realistic profile for the gramicidin A channel with a large number of sites and barriers (Eisenman *et al.*, 1978; Levitt, 1978). The remaining profiles give various approximations to this, certain of which have been analyzed in the literature. See text for details.

particular, the 1B4S model cannot account for the observed flux-ratio exponent, wnich exceeds unity (Schagina *et al.*, 1978; Procopio and Andersen, 1979), nor can it yield the observed concentration-dependence of the *I–V* behavior (Läuger, 1973; Hladky, 1972; Eisenman *et al.*, 1980a,b,c). On the other hand, models with more barriers but fewer sites, for example, 3 barriers and 2 sites (Urban and Hladky, 1979), while being able to account for the flux-ratio behavior, have been shown (Hagglund *et al.*, 1979; Eisenman *et al.*, 1980a) to have problems in simultaneously describing Lev and co-workers' flux-ratio data for Rb and the conductance data for this ion. More recently, we have shown (Eisenman *et al.*, 1980b) that the 3-barrier 2-site model of Urban and Hladky (1979) can be ruled out on the basis of the observed *I–V* shape in the limit of zero-occupancy, specifically, the *I–V* curve is much less sublinear than is predicted by this model from a fit to the conductance data. By contrast, both the 3B4S′ and 3B4S″ models are capable of simultaneously predicting the limiting shape at low concentration of the *I–V* relationship and the concentration-dependence of conductance (see Eisenman *et al.*, 1980b, Fig. 8). Whereas both of these models are compatible with the electrical data, the flux-ratio data of Procopio and Andersen (1979) favor the 3B4S″ case.

The models presented here are all based on conventional rate theory (Läuger, 1973; Hille and Schwarz, 1978). Such models describe any process from diffusion to chemical reaction in terms of elementary jumps over energy barriers and can be used to represent the process of permeation in as much detail or with as much accuracy as desired, subject to the restriction discussed by Läuger *et al.* (1980) that the time needed for the adjustment of ligands after an ion jump is small compared to the dwelling time in the potential well. Further assumptions are examined in the discussion under "Effects Not Considered in Strict Barrier Models," but it should be noted that inherent limitations of rate theory are not fundamental vis a vis alternative continuum, e.g., Nernst–Planck, formulations, but have to do with the "graininess" of description. Since these relate to the sharpness of the barriers and the consequent length of time spent in crossing them, such limitations can always be circumvented by proliferating the number of barriers used to represent a given real transition.

II. THEORETICAL EXPECTATIONS

Figure 2 shows the kinds of concentration-dependences expected for the electrically measurable properties of a 4-site channel, exemplified by the 3B4S′ model (Hagglund *et al.*, 1982) here. The solid curves are for a channel with a highly asymmetrical outer barrier (see diagram at top left, with $f_1 = 0$ and $f_2 = 0.2$). The solid curves are drawn for a negligible central barrier; the dashed curves show the effects expected when this barrier becomes significant. For simplicity, both sites in each channel half have been assumed to be at the same position in the potential field ($f_3 = f_4 = 0$). Besides the electrically measureable properties of single-channel conductance at low voltage (G_0) and "normalized chord conductance" (G_c/G_o; Eq. (1) below gives the theoretical voltage-dependence of this quantity), Fig. 2 presents the flux-ratio exponent n and the probability P for occupancy by 1, 2, etc., ions. A 3B2S model would exhibit the behavior seen for log concenteration (log C) more negative than -4, and the only essential differences that would be observed in a plot for the alternative 4-site model 3B4S″, would be a maximum value of 2 for n and the possibility that the high-concentration drops of the dashed curves might not occur because the outer sites have a finite maximum occupancy even when the aqueous concentration becomes infinite.

Notice that characteristic "half-saturations" of G_o occur at half occupancy (indicated

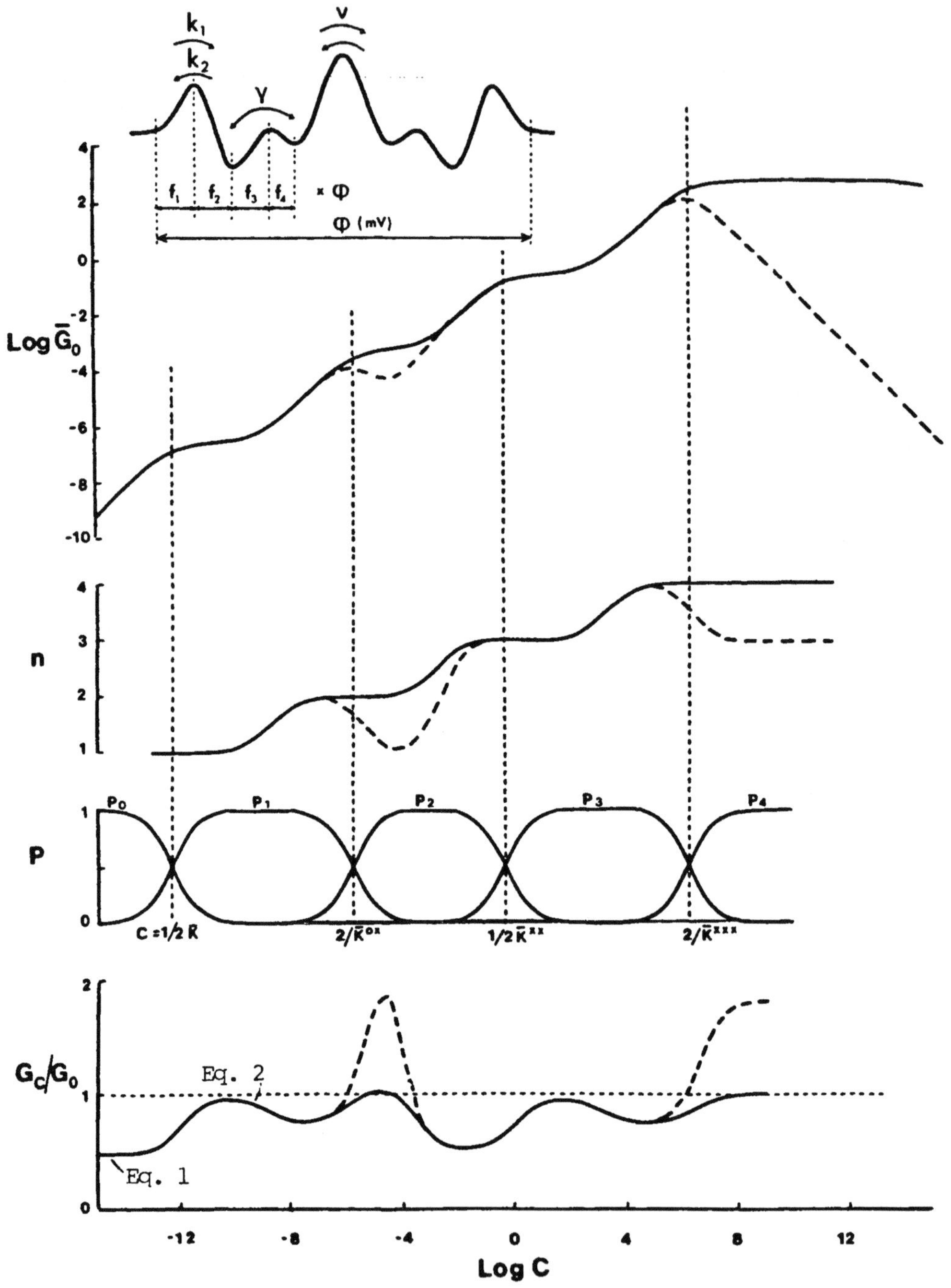

Figure 2. Summary of the principal properties of the 4-site model (as calculated from the double-diamond approximation of Hagglund *et al.*, 1982), which neglects the states with double occupancy in each half channel. Insert: Energy vs. potential diagram of the 3B4S′ model. k_1 and k_2 are the on- and off-rate constants for ion transfer across external barriers. ν is the rate-constant across the central barrier. γ stands for the ratio of binding constant between inner and outer sites, which are assumed to be in equilibrium with each other. ϕ is the total potential drop across the channel in mV; f_1—f_4 are the fractions of ϕ indicated between the dashed vertical lines. Main figure, from top to bottom: Conductance, flux-ratio exponent, probability of occupancy, and chord conductance, as functions of concentration. The curves are given for negligible potential across the channel, except the chord

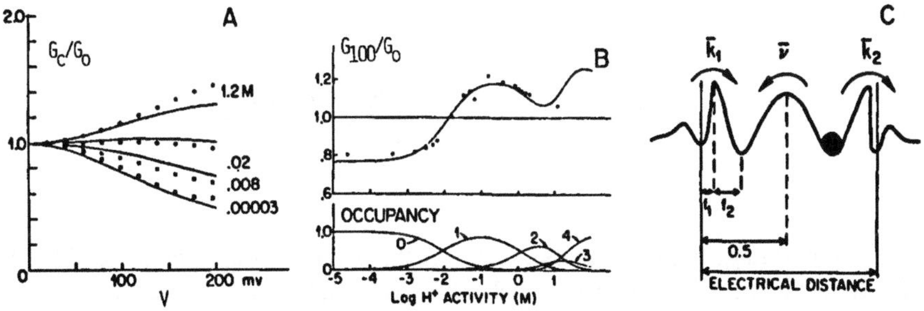

Figure 3. Changes of *G–V* shape with increasing occupancy for H. For illustrative purposes, previously published polynomial data for gramicidin in GMO/Dec and GMO/Hxdec bilayers in symmetrical H^+ solutions (reproduced with permission after Eisenman *et al.*, 1980b, Fig. 1) are presented. These differ slightly at the lowest concentrations from the more refined spline data in Fig. 9, but not sufficiently to alter the point of the figure. The theoretical curves have been drawn according to Eqs. (3) and (6), for $f_3 = 0$ and $f_4/2 = .2553$, with the following values of the *k* and *P* parameters: $k_{yo} = 0.6$ M^{-1}, $k_{yi} = 53.6$ M^{-1}, $k_{yoyo} = 25$ M^{-2}, $k_{yoyoyi} = .083$ M^{-3}, $k_{yoyiyi} = 2$ M^{-3}, $k_{yoyiyiyo} = .355$ M^{-4}, $P_1 = 19{,}300$, $P_2 = 527{,}000$, $P_3 = 38{,}800$, $P_4 = 9440$, $P_5 = 10{,}800$, $P_6 = 665{,}000$. (Since the conductances in the theoretical expressions are normalized with respect to *F/RT*, all *P* parameters should be multiplied by 1.568×10^5 to express these in conventional units of $M^{-1}S^{-1}$, $M^{-2}S^{-2}$, etc.)

by the vertical dotted lines drawn at the halfway points of the levelling-off processes) as each successive ion is loaded. Also note in Fig. 2 (bottom) that the *I–V* shape, as reflected in the behavior of the normalized chord conductance at 100 mV (G_c/G_o), changes with changing occupancy. Unfortunately, the binding constants of the successive loading states in real channels are not as widely separated as in this example, so it is not always easy to distinguish these changes. However, when the situation is favorable, those at the lowest concentrations are clearly visible, as is shown in Fig. 3 for H^+ and as has been shown elsewhere for Tl^+ (Eisenman *et al.*, 1980a, Fig. 6). The correspondence between the change of shape and occupancy by the first ion is clearly seen for H^+ in Fig. 3B (also see Fig. 10 for Cs^+).

A. Low Occupancy Limit

This shape change, which corresponds to the one at the lowest concentrations in Fig. 2, represents the change in the *I–V* characteristic when proceeding from the simple expression (Eisenman *et al.*, 1980b) derived elsewhere (Hagglund *et al.*, 1982), which applies as ion occupancy approaches zero,

$$\left(\frac{G_c}{G_o}\right)_{C\to 0} = \frac{1 + \overline{k}_2/2\overline{\nu}}{\cosh\,(0.5-f_1)\,U + \overline{k}_2/2\overline{\nu}} \cdot \frac{\sinh U/2}{U/2} \qquad (1)$$

conductance, which was calculated for 100 mV channel potential with an unsymmetric external barrier ($f_1 = f_3 = f_4 = 0$ and $f_2 = 0.2$). Solid curves represent an insignificant central barrier and dashed curves correspond to a central barrier which at intermediate and high concentrations becomes significant, causing blocking with decrease of conductance and flux-ratio exponent and change of *I–V* behavior to become supralinear. Note that change of occupancy is reflected not only in the flux-ratio exponent, but also in the single-channel conductance and in the normalized chord conductance. Up to 2-ion occupancy (P_2) the curves also illustrate the general behavior of the 2-site model. The rate constants used were assigned for illustrative purposes, and in an actual data for the gramicidin channel the step-wise changes are less distinct. The parameters used will be found in Hagglund *et al.*, 1982, Fig. 10.

to that at 1-ion occupancy,

$$\left(\frac{G_c}{G_o}\right)_{C=\text{low}} = \frac{1 + \bar{k}_2/2\bar{\nu}}{a + b \cdot \bar{k}_2/2\bar{\nu}} \cdot \frac{\sinh U/2}{U/2} \tag{2}$$

where U is the voltage (in RT/F units), G_c/G_o is the chord conductance divided by the zero current conductance, $a = \cosh f_1U \cdot \cosh (0.5 - f_1 - f_2)U$, and $b = \cosh (f_1 + f_2)U$.

In Eq. (1), G_c/G_o as a function of U is seen to depend on only two parameters: f_1, the voltage-dependence of the entry step, and $\bar{k}_2/\bar{\nu}$, the ratio of the rate-constant for leaving the channel ($\bar{k}_2$) to that for crossing its center ($\bar{\nu}$), as defined in Fig. 3 or Fig. 2 (when $f_3 = f_4 = 0$). In Eq. (2) the voltage-dependence of the exit step (f_2) is now also involved in the shape. The transition between the I–V shapes of Eqs. (1) and (2) occurs at half occupancy of the first site. These equations, which hold for the low-occupancy states of all barrier models (Läuger, 1973), define the "plateaus" labeled "Eq. (1)" and "Eq. (2)" of G_c/G_o in Fig. 2.

B. Concentration-Dependent Behavior for the 3B4S″ Model

We will use the 3B4S″ model (Sandblom *et al.*, 1983) to exemplify the theoretical expectations for higher concentrations C because this variant of the 4-site models is compatible with the flux-ratio exponent measurements of Procopio and Andersen (1979). For this model, the zero-voltage conductance (G_o) is given by Eq. (3) in terms of six "peak parameters," $P_1, P_2, \ldots, P_6$, and four lumped equilibrium binding constants, $K_1, \ldots, K_4$, which define the occupancy of the channel, n_o,

$$G_o = n_o/[2(P_1c + P_2c^2 + P_3c^3)^{-1} + (P_4c + 2P_5c^2 + P_6c^3)^{-1}] \tag{3a}$$

$$n_o = [1 + 2K_1c + K_2c^2 + 2K_3c^3 + K_4c^4]^{-1} = \text{fraction of sites empty} \tag{3b}$$

Note that Eq. (3) can be reduced to the 3B2S model by assuming no occupancy states higher than 2 ($K_3 = K_4 = 0$, $P_3 = P_5 = P_6 = 0$), to the 2B4S″ model by assuming the central barrier to be negligible ($P_4 = P_5 = P_6 = \infty$), and to the 1B4S model when the outer barriers are negligible ($P_1 = P_2 = P_3 = \infty$).

The ten lumped parameters of Eq. (3) are defined in Fig. 4 in terms of the 16 intrinsic parameters of the 3B4S″ model. Figure 4 also shows schematically the energies of the ionic configurations corresponding to these intrinsic parameters. For example, the first term in the brackets of Eq. (3a) depends on the outer barrier only and corresponds to the definitions of P_1, P_2, and P_3 in the energy diagrams where the ion sits at the top of the outer barrier. The second term gives the contribution due to the central barrier (P_4, P_5, and P_6), corresponding to the ion sitting at the top of the central barrier. Thus, the first term corresponds to the 2B4S″ model and the second term to the 1B4S model.

Equation (4) (Sandblom *et al.*, 1983) gives the flux-ratio exponent, n, for the complete 3B4S″ case

$$n = \frac{[P_1 + P_2c + P_3c^2] \cdot [P_1 + 2P_4 + (4P_5 + 0.5P_2 + 0.5P_{\bar{y}oyo})c + (2P_6 + 0.5P_3)c^2]}{[P_1 + 0.5(P_2 + P_{\bar{y}oyo})c + 0.5P_3c^2] \cdot [P_1 + 2P_4 + (4P_5 + P_2)c + (2P_6 + P_3)c^2]} \tag{4}$$

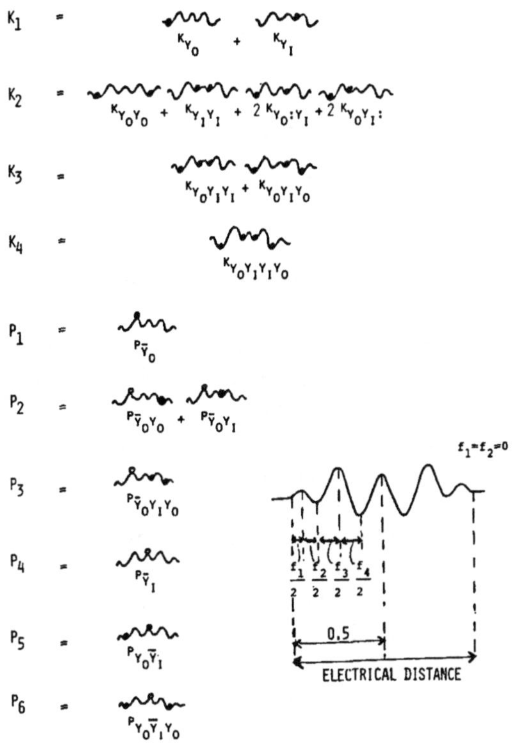

Figure 4. Definition of the binding constants (K) and peak parameters (P) of the 3B4S″ model in terms of the intrinsic parameters (k, p) whose occupancy states are indicated by the little energy diagrams. Note that, in Eq. (6), $2k_{yoyi} = k_{yo:yi} + k_{yoyi:}$ (these double occupancy states are lumped). Also note as in Fig. 3 that the conductance in the theoretical expressions is normalized with respect to F/RT, so that all P parameters and rate-constants should be multiplied by 1.586×10^5 in order to express these in the conventional units $M^{-1}S^{-1}$, $M^{-2}S^{-2}$, etc.

and Eq. (5) gives it for the 2B4S″ limit when the central barrier is unimportant

$$n = \frac{P_1 + P_2c + P_3c^2}{P_1 + 0.5\,P_2c + 0.5\,[(P_2 - P_{\bar{y}oyi}) + P_3c]c} \tag{5}$$

Equations (4) and (5) depend solely on the 3 P parameters for each barrier of Eq. (3) as well as on the additional intrinsic parameter ($P_{\bar{y}oyo}$ or $P_{\bar{y}oyi}$) defined in Fig. 4, which explains the meaning of the various subscripts.

Equation (6) expresses the I–V behavior as a function of concentration and voltage in terms of the normalized chord conductance and depends upon four additional intrinsic parameters, as well as the voltage-dependence of the forward and backward jump at the mouth of the channel (f_3, f_4) defined in Fig. 4 (note that f_1 and f_2 have been taken as zero here for simplicity).

$$\frac{G_c}{G_o} = \frac{c}{G_o}\,\frac{\sinh u/2}{u/2}\,\frac{1}{A/B + D/E}$$

$$\begin{aligned} A = {} & 2[1 + 2\,(K_1 - k_{yi})c + (K_2 - k_{yiyi} - 2k_{yoyi})c^2]\cosh((1 - f_3)u/2) + 4c\,(k_{yi} \\ & + k_{yoyi}c + (K_3 - k_{yiyiyo})c^2)\,(\cosh(f_3u/2))(\cosh(1 - (f_3 + f_4)u/2) + 2c^2\,(k_{yiyi} \\ & + 2k_{yiyiyo}c + K_4c^2)(\cosh f_3u/2)(\cosh(1 - f_4)u/2)/\cosh(f_4u/2) \end{aligned} \tag{6}$$

$$B = P_1 + P_{\bar{y}oyo}c + (P_3c + (P_2 - P_{\bar{y}oyo}))(\cosh(f_3u/2))c/\cosh(f_4u/2)$$

$$D = 1 + 2(K_1 - k_{yi})c + (K_2 - k_{yiyi} - 2k_{yoyi})c^2 + $$
$$+ 2c(k_{yi} + k_{yoyi}c + (K_3 - k_{yiyiyo})c^2)\cosh(f_3 + f_4)\,u/2) +$$
$$+ c^2(k_{yiyi} + 2k_{yiyiyo}c + K_4c^2)$$
$$E = P_4 + 2P_5c + P_6c^2$$

III. EXPERIMENTAL METHODS

The general methods were those we have used previously (Neher *et al.*, 1978; Eisenman *et al.*, 1980b). Except where specifically noted, all measurements have been made on glycerol monooleate (GMO) bilayers with either n-decane (Dec) or n-hexadecane (Hxdec) as the solvent. In agreement with Hladky and Haydon (1972), we found no significant differences for the open-channel properties G_c/G_o and G_{50}, the single-channel conductance at 50 mV, between these two solvents. Ionic strength has been maintained with 9mM $MgSO_4$ or $MgCl_2$ as an "indifferent" electrolyte. This is important not only to maintain a sufficiently low solution resistance but also to screen any surface-charge contaminants and to minimize interfacial-polarization effects for the lowest concentrations of permeant ions. We have also carried out selected studies in 1.0 M tetramethylammonium chloride (TMACl) and 1.0 M $MgCl_2$ and $CaCl_2$. We find that 9 mM $MgSO_4$ (or $MgCl_2$ when sulfate is undesirable for solubility reasons) is without effect on single-channel conductances (the agreement between "*" and "o" data in Figs. 13–14), as well as on many-channel *I–V* shapes at 10 mM permeant-ion concentration (data not shown). Consistent with these results, McBride (1981) reports that 9 mM BaF_2 is without effect on the single-channel conductances of 1.0 mM AgF or KF. We settled on Mg salts to maintain ionic strength since Bamberg and Läuger (1977) had previously found this to be the least blocking of the divalent cations, and we had previously found that adding $MgSO_4$ to maintain ionic strength constant at 5.1 mM (Neher *et al.*, 1978, Fig. 4) did not alter the single-channel conductances for Tl^+.

We have made two technical refinements in our many-channel triangular-wave technique (Eisenman, *et al.*, 1980b,c). We now correct for "capacitive currents" (due to electrocompression, ion-injection, etc.)* by subtracting the currents recorded from bare-bilayer controls at comparable salt concenterations, frequencies, and voltages, and we have used a spline (at 10 mV intervals) instead of a polynomial approximation to fit the data. Typical records showing the effects of these corrections are shown in Figs. 5 and 6, for low ionic strength and high ionic strength, respectively. The inserts compare Lissajous plots of *I* vs. *V* for a typical cycle of raw data for a bare bilayer (left inset) and for a bilayer to which gramicidin has been added (right inset). The data points and solid connecting curves give the spline representations for I/G_o, the "normalized currents," for the positive-going limbs

* Currents flowing through the bilayer between channels can distort the many-channel *I–Vs*, particularly at the lowest ion concentrations. Such currents have been ascribed to a variety of causes, e.g., voltage- and time-dependent changes of membrane capacitance (Schoch *et al.*, 1979; Benz *et al.*, 1975) due to electrocompression (White and Thompson, 1973; Requena *et al.*, 1975; Alvarez and Latorre, 1978), as well as ion-injection (Benz and Janko, 1976; Andersen, 1982b). We have found the empirical procedure of subtracting appropriate bare bilayer control currents to satisfactorily reduce such distortions (Figs. 5 and 6). In addition, area changes most recently discussed by White and Chang (1981), could affect our bare-bilayer subtraction procedure. This last effect is most important for large torus to bilayer ratios. In the present experiments the radius of the bilayer was usually 90% of the radius of the aperture, although occasionally it was as small as 20% without serious differences seen in the results.

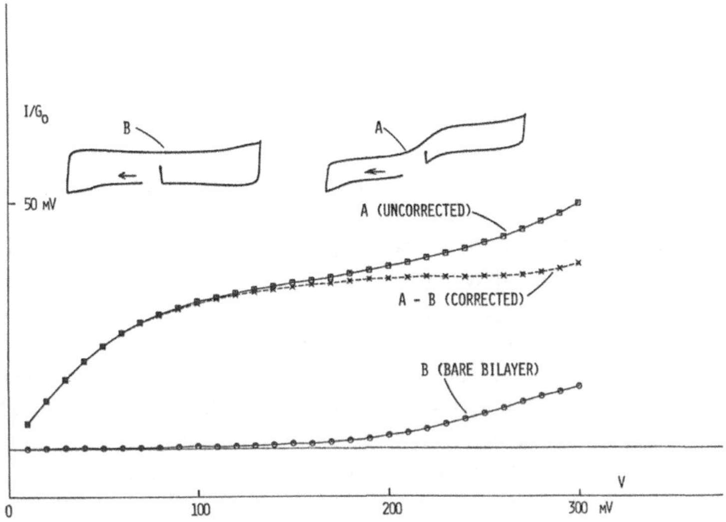

Figure 5. Illustration of the way a typical *I–V* record at low permeant-ion concentration and low-ionic strength is corrected for "bare bilayer" currents. The Lissajous figures (with current as the Y-axis, voltage as the X-axis, the arrow indicating the direction of time) plot the raw data for a single triangular wave at 100 Hz from a holding potential near zero to −32 mV, then to +320 mV and then to zero. (A) is for 0.2 mM HCl in 9 mM $MgCl_2$ in the presence of gramicidin. (B) is a bare-bilayer control measured in 10 mM CsCl in the absence of gramicidin. The data points plot the spline representation of the data from 0–+300 mV isolated from the −320 to +320 mV limb.

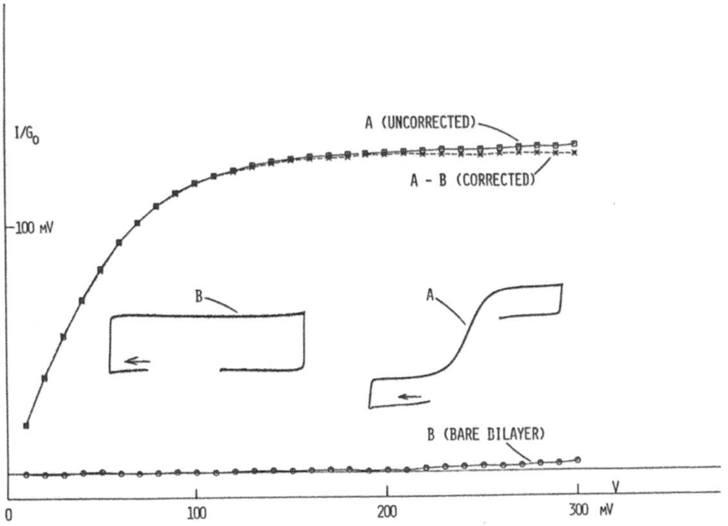

Figure 6. Similar to Fig. 5 but illustrating the corrections at high ionic strength (1.0M $Ca(NO_3)_2$ for a permeant-ion concentration of 1 mM HCl in both A and B). In this case, the Lissajous figures are for traces at 200 Hz and for a 320 millivolt triangular wave.

from 0–300 mV, and the dashed curves give the gramicidin curves corrected for the bare-bilayer currents. Notice that the upward "tailing" is most severe at low ionic strength and above 200 mV and that the subtraction removes virtually all of it. These refinements have negligible effects on our results except for the lowest concentrations, but the increased resolution in the low concentration limit has enabled the trends seen previously (Eisenman, *et al.*, 1980b, Fig. 7) to be reliably magnified so that the species differences are now much more striking, as will be discussed when we come to Fig. 9.

Previous multichannel *I–V* measurements appear to have been largely restricted to the use of brief pulses from which "instantaneous" current–voltage curves have been measured (Bamberg and Benz, 1976; Bamberg and Läuger, 1977) although the ramp technique has been used as a standard procedure with carriers (Laprade *et al.*, 1975). In our method, a single triangular ramp of voltage is applied through a pair of chlorided Ag plates and the resulting current was measured with a fairly fast current amplifier (Keithley Model 427, rise time 0.01 msec). The transmembrane potential difference between a second pair of chlorided Ag wires close to the membrane was measured with a floating differential amplifier (Keithley Model 604). A magnetic reed switch enabled the current amplifier to be removed from the circuit for measuring the open circuit membrane potential (V_o). Voltage and current were digitized and stored on magnetic tape with a Nicolet Model 1090A digital oscilloscope and a Kennedy Model 9700 tape recorder. Typical *I–V* records, written directly out from the tape onto an *X–Y* plotter, can be seen in the inserts to Figs. 5 and 6. The tapes were read into an IBM 3033 computer, using a Fortran program written by Bruce Enos, in which the data were filtered with five iterations of a Hanning filter at intervals of 1/2048 of the sweep duration, after which the positive-going *I–V* limb (usually from −220–+220 mV, but on occasion for a wider voltage range) was isolated and fitted to a spline at 10 mV intervals for convenient storage. The midpoint of the spline was chosen from the measured value of V_o and the corresponding current at zero p.d. Particular care was taken to add lipid symmetrically and to keep fluid levels as symmetrical as possible to avoid asymmetries in the *I–V* curve. This procedure was almost always successful, as judged by agreement between positive and negative limbs, which were always computed although the only results presented here are for the positive limb.

The data sets so created were stored on a disk and later read into an APL file, upon which the following further computations were carried out. First, the currents at zero potential were calculated, using a third-degree polynomial, both for the appropriate bare-bilayer control and for the records in the presence of gramicidin. Then, the currents from the bare bilayer

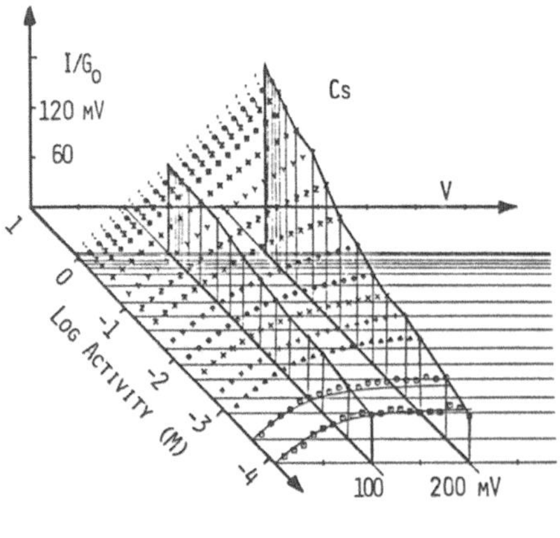

Figure 7. Experimentally observed currents for Cs plotted as a function of voltage and log Cs activity. The data presented are for GMO/Dec bilayers at 100 Hz. 9mM $MgSO_4$ was present for all plots except the circles and squares which were done in the presence of 9mM $MgCl_2$.

were subtracted from those in the presence of gramicidin, after scaling these in proportion to the measured capacitances (obtained directly from the "jumps" at the ends of the Lissajous figures (see Figs. 5 and 6)). The data from positive and negative limbs, as well as their arithmetic mean, were stored separately. This completed the processing of the experimental data, but it was found to be convenient to "normalize" all data by dividing the measured currents by G_o, which was calculated using the lowest 70 mV of data from an equation of the form of Eq. (1) (a polynomial was found to be unsatisfactory) with a noninteractive Levenberg–Marquardt nonlinear least-squares curve-fitting routine developed by Chris Clausen. Typical I/G_o vs. V curves so computed are shown in Figs. 5 and 6.

The usual useful range of frequency-independence for GMO/Hxdec was 5–200 Hz. For GMO/Dec and PE/Dec 50–200 Hz proved satisfactory. An important control of the correctness of single-channel I–V shape inferred from the many-channel technique is, of course, the direct comparison with single-channel measurements at comparable ionic concentrations. We have previously presented data demonstrating satisfactory agreement between our many-channel measurements and published single channel data of others (Eisenman *et al.*, 1980b), although not, of course, at ultra-low ionic concentrations where single-channel data are just beginning to be measured (McBride, 1981).

IV. EXPERIMENTAL FINDINGS

A. Current vs. Voltage and Activity of Permeant Ion

1. Without Blocking Ions

Figures 7 and 8 plot the experimentally observed currents for Cs and Li (in the presence of 9mM $MgSO_4$ or $MgCl_2$ as supporting electrolyte) as 3-D plots of I/G_o as a function of voltage and log ionic activity a.* Looking at the dependence of I/G_o vs. log a on the 100 mV and 200 mV planes, the changes of I–V shape with ionic activity are apparent, as well as the fact that it is possible to obtain a limiting shape that no longer depends upon ionic activity by going to sufficiently low permeant-ion concentration (for comparison with the experimental data points in Figs. 7 and 8 a theoretical limiting curve (see below) is also shown at the two lowest concentrations). A limiting shape has been obtainable for all permeant species studied so far (data not shown were obtained for Na, K, Rb, T1, CH_3NH_3,

* Although full details will be given in a forthcoming experimental paper (Eisenman, Sandblom, and Hagglund, in preparation), a comment about reliability and reproducibility of the I–V curves is in order. Except of the very lowest concentrations, the precision and repeatability of the individual data points are so good that all significant trends can be seen in plots of randomly selected individual traces, without need to concatenate multiple sweeps. This can be seen by the internal consistency of the plots in Figs. 7 and 8, which present families of such individual records. Actually, about ten records have been taken at each concentration, none of which differ from the data shown by more than the size of the symbols. Indeed, it is possible to remove the "jitter" apparent at the two lowest Cs concentrations by averaging several records, but the additional data processing that this would entail has not been felt worthwhile for the present. Some feeling for the kind of reproducibility of the data can be gained by comparing the records for 30 mM LiCl in Fig. 8 at 10 Hz (triangles) and 100 Hz (circles), as well as in the presence of 9 mM $MgCl_2$ (crosses) and in its absence (diamonds). This not to say that there cannot be variability in I–V behavior introduced by very large variations in concentration of gramicidin in the bilayer, but these have been avoided in the present studies which were generally done at the lowest density of channels that produced an adequate membrane conductance. In this region, changes of channel density by factors of 2–4 were found to be without detectable effect on the I–V shape.

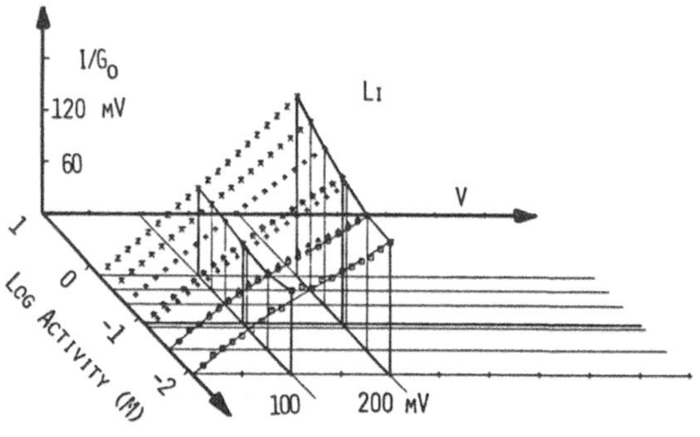

Figure 8. Same as Fig. 7 but for LiCl in GMO/Dec. 9 mM $MgCl_2$ was present except where noted. Data are for 100 Hz except as noted below. Comparison of the x's with the diamonds obtained in the absence of $MgCl_2$ illustrates both the kinds of reproducibility seen between different experiments and the inertness of $MgCl_2$ at this concentration. The frequency-independence of the *I–V* shape is seen by comparing the triangles and circles obtained at 10 Hz vs. 100 Hz, respectively.

and H) and can be seen more clearly for Cs in the solid curve of Fig. 10. This curve plots the normalized chord conductance at 100 mV (G_{100}/G_o, corresponding to the "normalized currents" (I/G_o) for the 100 mV planes in Fig. 7 multiplied by *V*. The 0-ion plateau and the 1-ion plateau (recall Fig. 3) are detectable and indicated on the figure by arrows labeled "0-ion" and "1-ion."

The limiting shapes for each of the permeant species are given by the data points in Fig. 9 where they can be compared with theoretical curves drawn according to Eq. (1) for the parameters in Table I. These data will be analyzed and interpreted below. Table I also contains values for the voltage-dependence of the exit step, where known from analysis of data at higher concentrations.

2. *The Effects of Ca and Mg As Voltage-Independent Blockers*

Figure 10 illustrates the effects of 1.0 M of the blocking (Bamberg and Läuger, 1977) cations, Ca and Mg on the *I–V* shape for Cs, as represented by G_{100}/G_o. It can be seen that Mg and Ca displace the curve to the right to varying degrees without, however, altering the value of the low-concentration limit, as previously reported (Eisenman, 1981). Since the

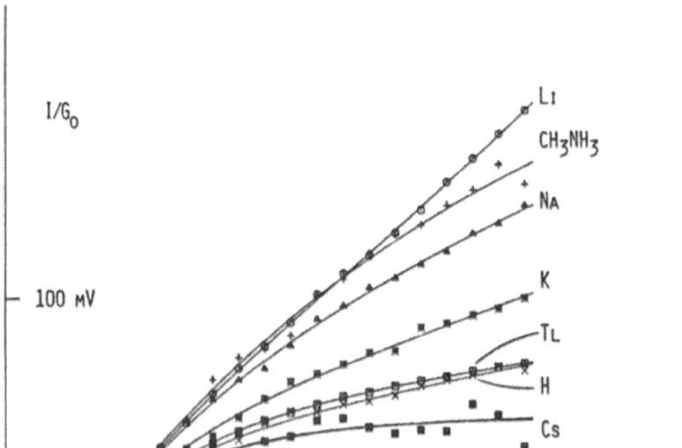

Figure 9. Low-concentration limiting shapes for *I–V* behavior. Theoretical curves are drawn according to Eq. (1) for the parameters listed in Table I.

Table I

	Voltage-dependence of entry	Voltage-dependence of exit	Ratio of rate-constants of leaving to crossing
Li	10.8%		0.7
Na	5.7%		1.31
K	6.5%		0.41
Cs	0.4%	23%[a]	0.5
Tl	3.3%	20%[b]	0.5
H	4.8%	26%[c]	0.19
CH_3NH_3	4.6%		2.0

[a] From 2B4S" fit of Figs. 19–20.
[b] From Eisenman *et al.* (1980a).
[c] From 3B4S" fit of Fig. 3.

voltage-dependences of the entrance step are the same (< 1%) for all three curves (data not shown), this implies that the ratio of rate-constants for leaving vs. crossing are unaltered by either Ca or Mg. Moreover, the apparent binding constant for the first Cs ion to occupy the channel (as judged by the concentration at which the normalized chord conductance changes its value from its low-concentration limit) is lowered to a differing degree by Mg and Ca, as can be seen by the different displacements of the curves to the right (half occupancy for the 1st Cs ion to bind occurs at about $10^{-3.5}$ M in "pure" CaCl, at $10^{-1.5}$ M in 1.0 M $MgCl_2$, and at $10^{-0.8}$ M in 1.0 M $CaCl_2$). The effects of the divalent ions, therefore, are as if they simply lowered the effective concentration of permeant ion in the channel without altering the relative barrier heights at the middle vs. the mouth, perhaps simply by decreasing the occupancy through competitive binding with Cs for a common binding site.

Because there is no sign of block at voltages as high as 300 mV for Cs^+ (the data are not shown but they are similar to Fig. 6 for H^+), in sharp contrast to what we will show

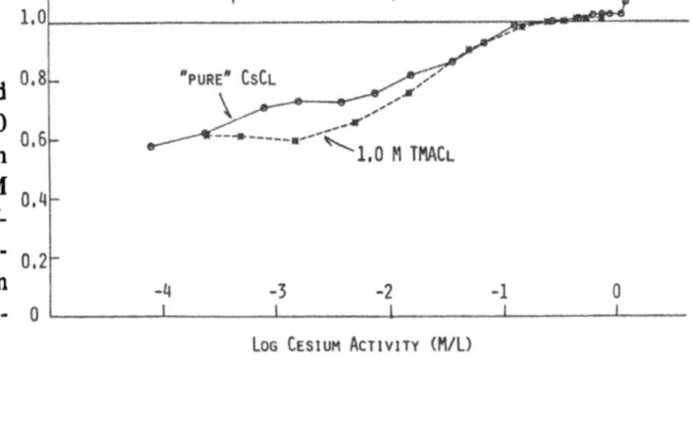

Figure 10. *G–V* shape for Cs, exemplified by the "normalized chord conductance" at 100 mV, and the effects of divalent cations upon this shape. "Pure" CsCl actually contains 9 mM $MgCl_2$. The curves have no theoretical significance and merely connect data points, but plateaus expected for 0-ion occupancy and 1-ion occupancy from Fig. 2 are visible and are indicated by arrows.

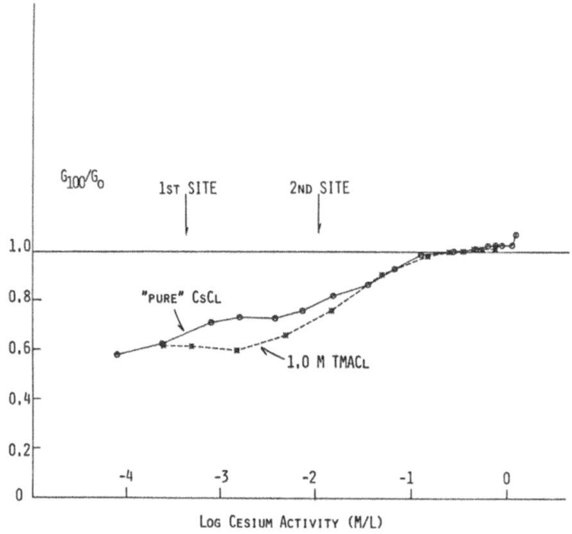

Figure 11. Effect of TMACl on the concentration-dependence of the *I–V* shape for Cs, as exemplified by the "normalized chord conductance" at 100 mV. The curves merely connect data points and are without theoretical significance, but from the shape changes in "pure" CsCl (in the presence of 9 mM $MgCl_2$) the midpoints of two "titrations" can be detected and are indicated by arrows labeled "1st site" and "2nd site."

for TMA in the next section (Fig. 12), it appears that the binding site for Mg and Ca is electrically external to the channel, probably being the outer site of the 3B4S″ model.*

3. *The Effects of TMA as a Voltage-Dependent Competitive Blocker*

Although it has been reported (Urban, 1978; Andersen, 1982b) that various impermeant methylated ammonium ions (including TMA and TEA) are inert. Urban's measurements were confined to permeant-ion concentrations higher than 0.01 M, and Andersen's only published data at lower concentrations (0.002 M CsCl) are for TEACl. The data of Figs. 11 and 12, which present typical *I–V* data for the alkali cations in the presence of 1.0 M TMACl, agree with their findings at higher concentrations but indicate that TMA is not inert at lower permeant-ion concentrations. From the the congruence between the dashed and solid curves for G_{100}/G_o in Fig. 11 it is apparent that the shape of the *I–V* curves is indistinguishable from that in the absence of TMA for concentrations higher than 0.05 M (see the data of Fig. 7). However, at lower concentrations, the values of G_{100}/G_o indicate that the *I–V* curve is more "saturating" in the presence of TMA than in its absence, as can be seen by the displacement of the dashed curve to the right.

Formally, the effect in Fig. 11 is as if TMA markedly decreased the strength of the first binding site, for example, by competing for the site rather weakly so that low concentrations of permeant ions displace it. The alternative interpretation, if TMA is inert, would be that the first binding site is a "spurious" consequence of interfacial polarization. Against this interpretation is the downturn in *I–V* shape seen at high voltages (above 200 mV) in Fig. 12. Such behavior is just what would be expected if TMA were a competitive blocker because it could bind competitively but could not cross the channel (Horn *et al.*, 1981), for this would lead to the appearance of an apparent voltage-dependent block, exactly as has

* In seeming contradiction to our conclusion for the effects of divalent cations on Cs is the conclusion of McBride (1981) from the voltage-dependence of the effects of 9 mM BaF_2 on 1 mM KF or AgF that BaF_2 produces a voltage-dependent block because he finds it to decrease the conductance at 250 mV without decreasing G_o. We also find a similar effect for K (which, in contrast to Cs, has a significant voltage-dependence of entry), but presently attribute it to a decrease in the voltage-dependence of the entry step produced by divalent cations.

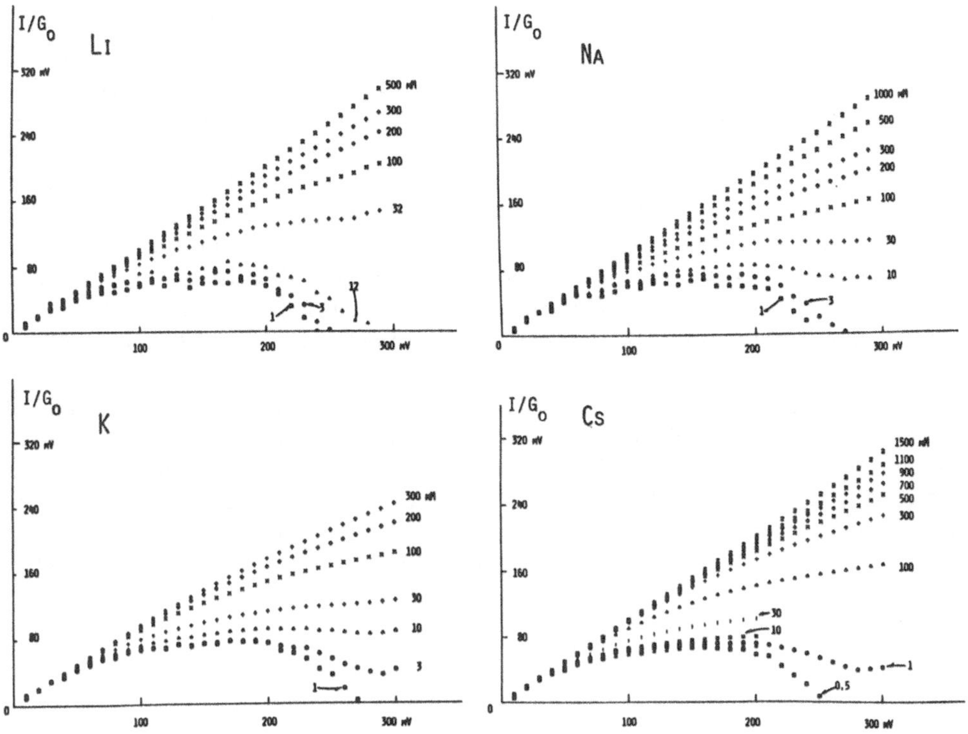

Figure 12. *I–V* behavior of representative monovalent cations in the presence of 1.0 M TMACl, suggesting voltage–dependent block by TMA at low cation concentrations (below 30 millimolar for the least permeant species (Na, Li) and below 10 millimolar for the most permeant species (Cs, K)). The concentrations of permeant cation are indicated on each subfigure and in all cases the data were obtained at 200 Hz. No Mg was present in these experiments.

been postulated occurs for TMA in the Na channel by Horn *et al.* (1981). A detail worth noting in Fig. 12 consistent with this interpretation is the finding that the concentrations at which the *I–V* curves turn down depend upon the permeant-ion species, with TMA producing its effects at the highest permeant-ion concentration for just those species, e.g., Li, Na, that are thought (Eisenman *et al.*, 1978) to bind the least strongly. These results indicate that TMA is not inert and, because of the voltage-dependence of the TMA effect, suggest that TMA competes for a binding site within the channel, e.g., to the inner site of the 3B4S″ model.*

B. G_o vs. Activity of Permeant Ion

Even though a rather wide range of data already exists for the single-channel conductances (G_o, usually approximated by measurements at 50 mV, G_{50}) in GMO bilayers (Neher *et al.*, 1978; Eisenman *et al.*, 1978; Urban *et al.*, 1980; McBride, 1981; Urry *et*

* This effect may be specific for TMA since Andersen finds (personal communication) that there is no sign of a negative-slope resistance up to 400 mV in 0.005 M CsCl plus 0.495 M TEACl. Unfortunately, data were not obtained for 0.005 M CsCl at voltages above 200 mV in Fig. 12, so it is impossible to state with certainty whether such a negative-slope resistence would have been expected even for TMA, although it clearly should be seen for TEACl at Cs concentrations below 0.001 M if TEA blocks similarly to TMA.

al., 1980), we have extended the low-concentration measurements by several orders of magnitude for comparison with our ultralow-concentration *I–V* data by using "noise" methods (Zingsheim and Neher, 1974; Kolb *et al.*, 1975) in collaboration with Erwin Neher. The measurements were done in GMO/Dec bilayers, after controls on single channels showed that the amplitudes were indistinguishable from those previously found in GMO/Hxdec (Neher *et al.*, 1978). The measurements were ultimately limited not by the inability to resolve channel amplitudes but by the unavoidable channel conductance of hydrogen ions, which in neutral solutions sets a "mud" level of two femptosiemens at pH 6.5. (We do not trust measurements at more alkaline pH owing to an apparent titratable change of behavior of unclear origin which we observe around this pH.) Although this work is still in progress, we present selected measurements on the higher permeant species Tl, Cs, and Rb in Figs. 13 and 14.

Figure 13 presents data for Tl, combining our previously published results (○) for TlF (Eisenman *et al.*, 1978) with our unpublished noise measurements with Neher (*) for TlCl. Figure 14 presents our data for CsCl similarly plotted. For comparison, our noise data for Rb at ultralow concentrations are plotted as small dots on both figures. (The deviation from the slope of 1 at the lowest concentrations of Rb is due to the "mud" level conductance due to H. The bracketed dots for Rb were obtained at a slightly more alkaline pH (6.5–6.8) than

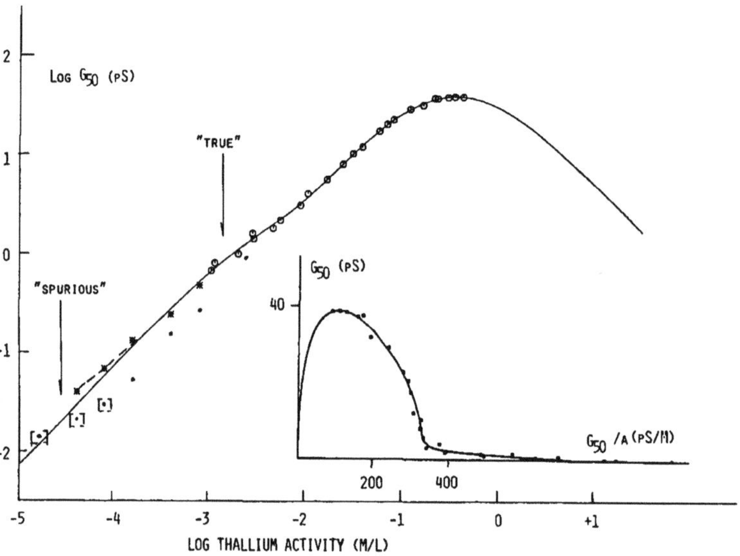

Figure 13. Concentration-dependence of single-channel conductances measured at low voltage (50 mV). Previously published data (Eisenman *et al.*, 1978) are presented for TlF in GMO/Hxdec bilayers (circles) as well as values (Neher, Eisenman, and Sandblom, unpublished) in GMO/Dec bilayers. For comparison with the Cs data, our values by noise methods at ultra-low concentrations of RbCl under the same conditions, are also presented (small dots). All noise data were obtained in the presence of 9mM $MgSO_4$; the single-channel data were obtained in its absence. Except for the bracketed data points, obtained at pH 6.5–6.8, the pH was 5.4–5.8. The insert gives an Eadie–Hofstee-type (Eisenman *et al.*, 1977) representation. The arrows indicate the approximate location of "true" and "spurious" binding sites. The theoretical curve has been drawn, using the following parameters, according to the "double diamond" approximation of the 3B4S′ model (Hagglund *et al.*, 1982) using the nomenclature of this reference and conventional units: $\bar{k}_1 = 4.85 \times 10^8$ $M^{-1}sec^{-1}$, $\bar{k}_2 = 3.14 \times 10^6$ sec^{-1}, $\bar{k}_1^{oa} = 1.03 \times 10^9$ $M^{-2}sec^{-1}$, $\bar{k}_2^{oa} = 4.23 \times 10^6$ sec^{-1}, $\bar{k}_1^{aa} = 9.05 \times 10^8$ $M^{-3}sec^{-1}$, $\bar{k}_2^{aa} = 2.94 \times 10^8$ sec^{-1}, $\bar{k}_1^{aaa} = 9.56 \times 10^8$ $M^{-4}sec^{-1}$, $\bar{k}_2^{aaa} = 3.03 \times 10^8$ sec^{-1}, $\bar{\nu} = 1.44 \times 10^6$ sec^{-1}, $\bar{\nu}^{aa} = 2.69 \times 10^7$ sec^{-1}. The nomenclature follows that in Fig. 2, with the superscripts indicating the loading state of the channel.

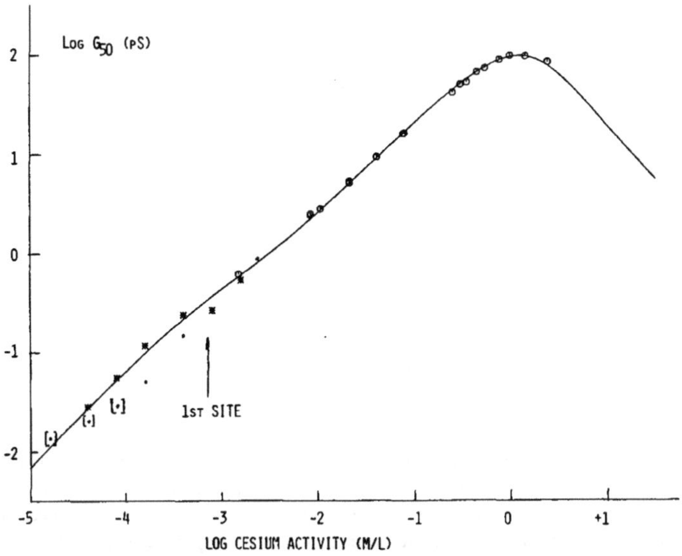

Figure 14. Same as Fig. 13 but for CsCl. The approximate location of the 1st binding site is indicated by the arrow. The parameters for the theoretical curve drawn according to Eq. (3) are: K_1 = 662.4 M^{-1}, K_2 = 2323 M^{-2}, K_3 = 97.9 M^{-3}, K_4 = 897.3 M^{-4}, P_1 = 1416 pSM^{-1}, P_2 = 6 − 1300 pSM^{-2}, P_3 = 289000 pSM^{-3}, P_4 = 862.9 pSM^{-1}, P_5 = 1.58×10^{10} pSM^{-2}, P_6 = 1.34×10^{9} pSM^{-3}. (See legend to Fig. 4 for conversion factor to conventional units.)

the remaining data of the figure (pH = 5.5–5.8) to reduce H-ion concentration.) The single-channel data were measured without supporting electrolyte, but all the noise data were obtained in the presence of 9 mM $MgSO_4$ (ionic strength = 36 mM). Theoretical curves are drawn for Tl in Fig. 13 using the 3B4S′ model and the parameters of Eisenman *et al.* (1980b, Fig. 8), and for Cs in Fig. 14 using the 3B4S″ model and the preliminary parameter set given in the figure legend. The main figures are log–log plots of conductance vs. ionic activity, while the insert to Fig. 13 shows an Eadie–Hofstee-type plot of conductance vs. the ratio of conductance to activity. Figure 15 shows the interrelations between these two kinds of plots and the way the phenomenological binding constants and rate-constants (Eisenman *et al.*, 1978; also defined in the figure) manifest themselves.

This figure compares a log–log representation of conductance vs. concentration at the left with an Eadie–Hofstee (E–H) type plot (Eisenman *et al.*, 1977) of conductance vs. conductance divided by concentration at the right. The inflections on the log–log plot correspond to the slopes on the E–H plot, and represent the binding constants K for the indicated occupancy states (denoted by the subscripts 1, 2, etc.). The four plateaus labeled $G/2$ on the log–log plot correspond to the Y-intercepts of the E–H plot and represent the maximal limiting conductances for each occupancy state. Lastly, the Y-values, labelled $G \times K$, at the conductances corresponding to the intersections of the diagonal slopes with the vertical line at C = 1 (log C = 0), are the counterparts of the X-intercepts on the E–H plot and reflect the "permeability" products, G × K.

Notice in Figs. 13 and 14 the adequate correspondence between channel amplitudes measured by noise and by conventional methods (theoretically expected differences (Kolb, 1980) are too small to see on the log–log plots). Also notice that the "foot" of the Eadie–Hofstee plot (insert in Fig. 13), which shows up as an inflection on a log–log plot (Fig. 15), is much more pronounced for both Tl and Cs than for Rb. The significance of this will be discussed later in relation to interfacial polarization effects.

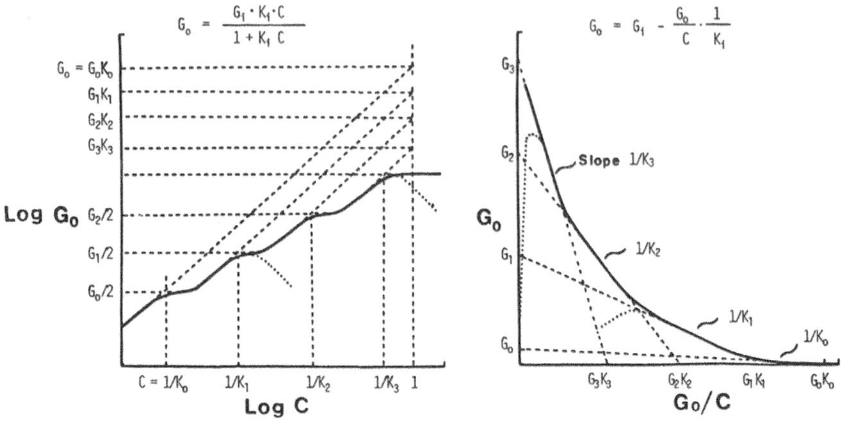

Figure 15. Comparison between log–log plot (left) and Eadie–Hofstee (Eisenman *et al.*, 1977) representation (right) of the conductance at zero voltage (G^o). The "phenomenological variables" of the Eadie–Hofstee plot, K (inverse slope) and G (Y-intercept), also have a simple interpretation in the log–log plot, as indicated by the dashed lines. The subscripts 1, 2, etc., refer to successive occupancy states. (Reproduced from Hagglund *et al.*, 1982, Fig. 16).

C. Monoglycerides vs. Phospholipids

Comparing the I–V shapes of gramicidin channels in GMO vs. bacterial phosphatidyl ethanolamine (PE) is of relevance because interfacial polarization and diffusional access limitations seem to be more prominent in phospholipids than in monoglycerides (Andersen, 1982b,c). This is done in Figs. 16–18. Figure 16 summarizes the behavior of the normalized current as a function of V and log a_K in PE/Dec bilayers, presenting the corresponding data for GMO-Hxdec for comparison as small dots. Figure 17 compares the limiting shapes in these two lipids, drawn according to Eq. (1) with f_1 and $k_2/\bar{\nu}$ being 0.0174 and 1.64 in PE in comparison with 0.065 and 0.41 in GMO. These values indicate that the central barrier is four times more important in PE than in GMO, which is consistent with what Andersen (personal communication) has predicted from considerations of dipole potential differences between these two lipids (Szabo and McBride, 1978). The voltage-dependence of the entry

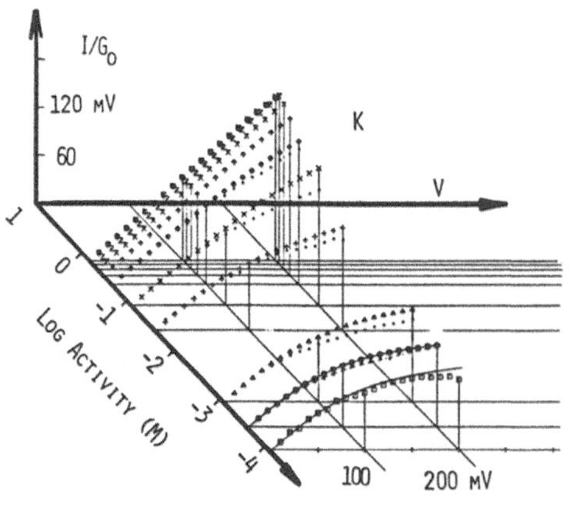

Figure 16. Comparison of I–V shape for KCl in phospholipid (PE) vs. monoglyceride (GMO) bilayers. The same type of plot is used as in Figs. 7 and 8, but the large symbols present the data for PE/Dec bilayers at 50–100 Hz. and the small symbols for GMO/Hxdec bilayers at 5 Hz. 9 mM $MgSO_4$ was present in all experiments. The two solid curves show the theoretical limiting curve for PE/Dec at the low-ion concentration, while the dashed curve is the theoretical limiting curve for GMO/Hxdec. The difference in shape of these two curves can be seen more clearly in Fig. 17.

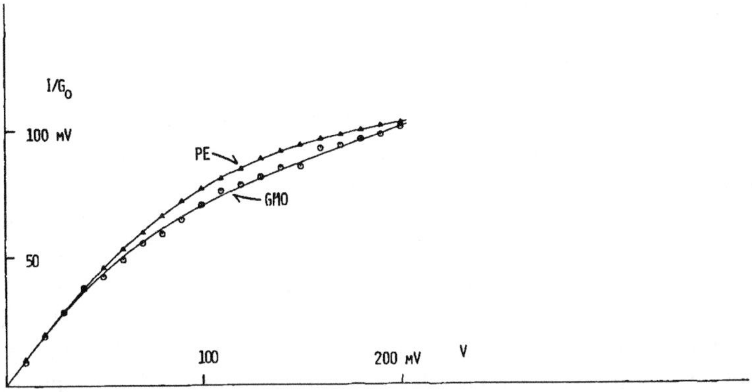

Figure 17. Comparison of low concentration limiting shapes in monoglycerides (GMO) vs. phospholipid (PE). Data points are for 0.3 mM KCl in the presence of 9 mM $MGSO_4$. Theoretical curves have been drawn according to Eq. (1) for the values of parameters given in the text.

step also appears to be considerably smaller in PE than in GMO, although this could be an effect of a more pronounced interfacial polarization in the phospholipid, as discussed later.

As far as the concentration dependence of the *I–V* shape is concerned, this manifests itself in the normalized chord conductance of Fig. 18 as an upward and leftward displacement in PE relative to GMO, with the theoretically most important difference being the concentration-dependence seen in PE below 0.001 M. Naively interpreted, this would imply a much stronger first binding constant for the gramicidin channel in a PE vs. GMO bilayer. This is intuitively unreasonable, and a more likely explanation will be given in the discussion.

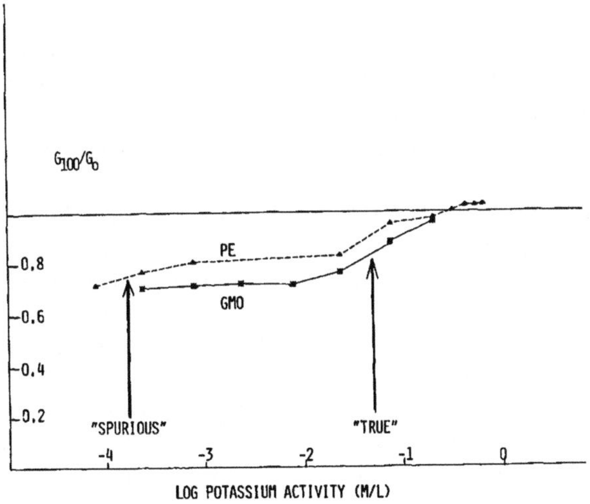

Figure 18. Comparison of concentration-dependence of *G–V* shape for phospholipid (PE/Dec) vs. monoglyceride (GMO/Hxdec) bilayers. It is of interest that if the changes in shape seen in PE bilayers below 0.001 M are due to interfacial polarization, as suggested in the text, then the true low concentration limiting shape for gramicidin in PE, undistorted by interfacial polarization, should be obtained from the shape observed on the "plateau" seen around 0.001–0.01 M. When the shapes are compared in this concentration region, the voltage-dependence of the entry step appears to be quite similar for the two lipids, and their chief difference appears to be that the central barrier is relatively larger in PE than in GMO, which is intuitively reasonable (see Szabo and McBride, 1978).

V. DATA ANALYSIS AND PHYSICAL INTERPRETATION

A. Low-Occupancy Limit

We begin with the relatively simple data in the limit of low-channel occupancy. The data of Fig. 9 and Table I show a number of interesting species-dependent trends. The most obvious is that the limiting shape depends strongly on the species of permeant ion. This is a consequence of differences in the two parameters of Eq. (1). Considering the voltage-dependence of entry (f_1) first, we see that for the alkali cations it decreases with increasing size, with virtually no voltage-dependence for Cs (0.5%) but a substantial one (11%) for Li. Physically, the differences between Li and Cs would be understandable if Cs enters without having to "strip" tightly-bound water, while Li has to "shed" some to enter. The voltage-dependence of the entry step would then reflect the voltage-dependence of the partial dehydration process or, alternatively, of channel deformation to accomodate a more highly hydrated species. Interestingly, CH_3NH_3 and H (presumably H_3O) both have normal voltage-dependences (5%) similar to the intermediate-sized alkali cations, while Tl may really be a little less dependent (3.3%) because it is a more easily deformable (polarizable) cation.

Considering the trends in the ratio $k_2/\bar{\nu}$ next, it can be seen that there are important species differences. For example, the present data indicate that a major part of the "selectivity filter" for Na is in the middle of the channel while for K it is at the mouth. Such species differences have important consequences for describing the "selectivity" of a channel generally. Thus, the observed differences in relative barrier heights, as well as which barrier is rate-determining, will have to be considered to interpret selectivity correctly for gramicidin; and a similar situation may well hold for any multibarrier biological channel. It also makes physical sense that the relative ease of leaving vs. crossing the channel ($k_2/\bar{\nu}$) is anomalously low for H relative to the other permeant species since this would be consistent with its not having to displace a water plug (Levitt *et al.*, 1978; Finkelstein and Rosenberg, 1979). It is also interesting that the bulkiest cation, CH_3NH_3, finds more difficulty in crossing than in leaving.

The relative ease of leaving vs. crossing must be taken into account in assessing the physical implications of differences in flux-ratio exponents for different species. In particular, it may not be correct to infer low occupancy from low-flux coupling for a species such as Na (as has been done by Procopio and Andersen, 1979) for which Fig. 9 indicates that the central barrier is larger than the exit barrier.

Besides the information in Table I from the low-occupancy limit, it is possible to use combined electrical and flux measurements over the full range of concentration to assess the intrinsic parameters of the 3B4S″ model and to examine their physical implications. An example will be given below for Cs, and we have also included in Table I the voltage-dependences of the exit step for this ion calculated from such data, as well as for H for which we have also carried out such an analysis (Eisenman *et al.*, 1980c). These additional data indicate that the exit step is quite voltage-dependent (25%), which implies that the barriers at the channel mouth are highly asymmetrical. Since the barrier shape is quite similar for H and Cs, it would seem that the only novelty in the behavior of H relative to the other monovalent cations is in its ability to cross the channel without having to entrain a plug of water.

B. A Fit to Cs in the 2-Barrier Approximation (2B4S″)

To illustrate how, in principle, one can use the 3B4S″ model to interpret the combined electrical and flux behavior over the full range of concentrations we present a preliminary analysis for Cs, for which species single-channel conductance (Neher *et al.*, 1978), flux-ratio exponents (Procopio and Andersen, 1979), and *I–V* data (Hladky *et al.*, 1979; Eisenman *et al.*, 1980b) are available in the literature. These are plotted as points in Fig. 19 as a function of the log of Cs activity. The conductances include our unpublished noise data (*x*'s), previously presented in Fig. 14. The *I–V* behavior is from Figs. 7 and 10, together with the data of Hladky *et al.* (1979) shown as Z's.

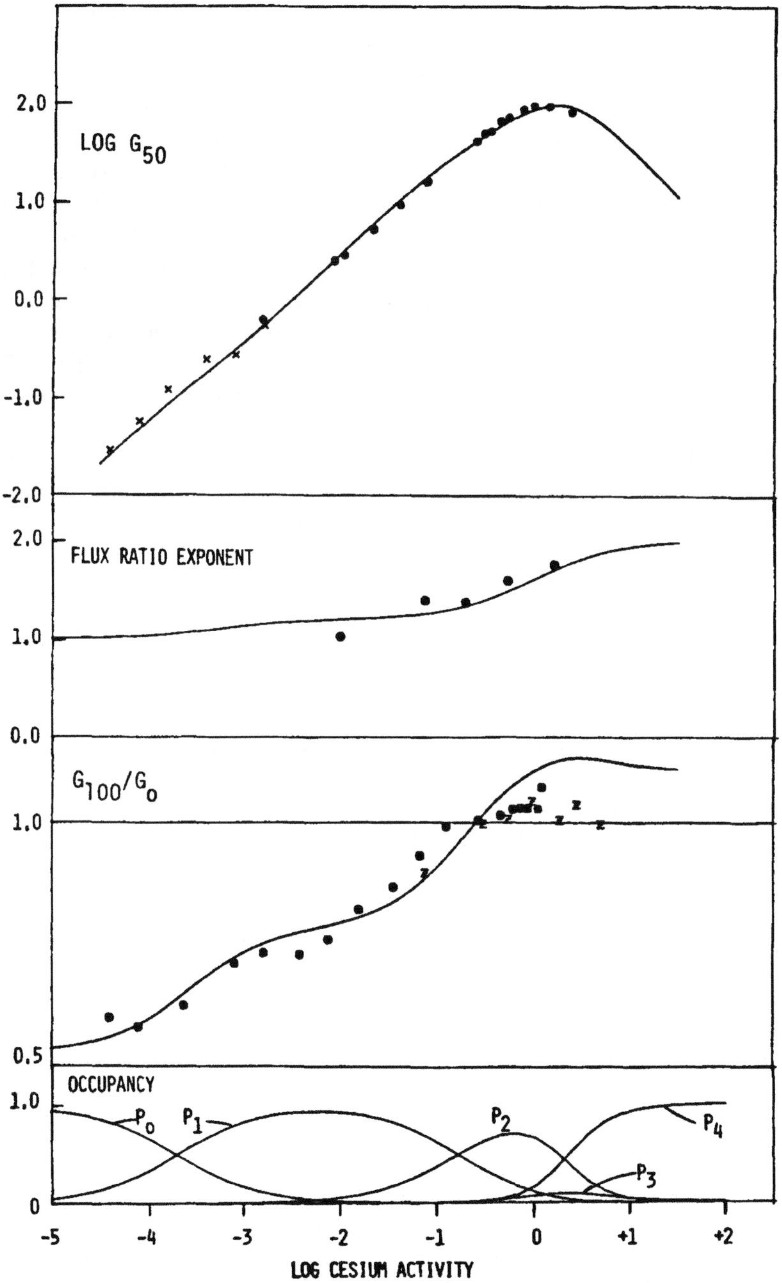

Figure 19. Comparison of concentration-dependences of experimentally observed electrical and flux behaviors with those expected from the 2B4S″ approximation of the 3B4S″ model, which considers the central barrier to be insignificant at all concentrations.

The experimental values of the flux-ratio exponent are seen to approach a level of two at high concentrations. From the lack of a downturn at concentrations where the conductance shows a maximum, we can conclude (see Sandblom *et al.*, 1983) that the middle barrier is small compared to the outer barriers at high concenterations. Since the middle barrier is also smaller than the outer barrier at low occupancy (recall Table I), it seems reasonable to neglect it at all concentrations. For illustrative purposes we will therefore use the 2B4S″ approximation of the 3B4S″ model, i.e., to set $P_4 = P_5 = P_6 = \infty$, reducing the number of lumped parameters to seven instead of ten (recall Fig. 4).

The conductance data at low voltage (50 mV, G_{50}) are first used to define seven of these parameters according to Eq. (3). We next use the flux-ratio exponent data to fix the value of the intrinsic peak parameter (P_{yoyo}) of Eq. (5) for the theoretical curve for the flux-ratio exponent. Note that we are comparing electrical data in a monoglyceride (GMO) with flux data in a phospholipid (diphytanoyl lecithin, DPPC). Because Andersen finds (personal communication) that the maximum in conductance occurs at about the same concentration in DPPC and GMO, this comparison is probably valid at higher concentrations, but its validity at the lowest concentrations will need to be studied because there are important differences in behavior of gramicidin channels in typical phospholipids vs. GMO (Figs. 16–18). Finally, we use the concentration dependence of the I–V relationship of Eq. (6) and the chord conductance at 100 mV to fix the four additional intrinsic binding constants as well as the two voltage-dependences (f_3 and f_4).*
The parameters extracted from Fig. 19 are summarized in Fig. 20 under their corresponding energy diagrams, and the occupancies corresponding to the K parameters, which are calculated quite simply according to Eq. (7) have also been indicated in the figure.

$$\begin{aligned} n_1 &= 2K_1 c\, n_0 = \text{fraction singly occupied} \\ n_2 &= K_2 c^2\, n_0 = \text{fraction doubly occupied} \\ n_3 &= 2K_3 c^3\, n_0 = \text{fraction triply occupied} \\ n_4 &= K_4 c^4\, n_0 = \text{fraction quadruply occupied} \end{aligned} \tag{7}$$

Do the extracted parameters make physical sense? First note that, except for the doubly inner occupied state to be discussed below, each loading step reduces the binding constants in Fig. 20 from those expected from independently reaching the loaded state (for example, $k_{yoyo} < k_{yo}{}^2$ and $k_{yoyiyiyo} < k_{yiyi} \cdot k_{yoyo}$).

In these two examples, the binding constant for a channel occupied by two ions, with one ion in each outer site (k_{yoyo}) is observed to be smaller than the product of the binding constants for loading each outer site independently ($k_{yo} \cdot k_{yo}$). Similarly, the binding constant for loading a channel with four ions ($k_{yoyiyiyo}$) is seen to be smaller than the product of the binding constant for loading two ions on the inner sites (k_{yiyi}) times the binding constant for loading two ions on the outer sites (k_{yoyo}).

* The fit to the G_{100}/G_o data is *not* good. We believe this is because the central barrier has been neglected in the approximation we are using for illustrative purposes here. Although we have successfully fitted the full 3B4S″ model to data for H^+ by iteratively fitting the G_{50} data and then the G_{100}/G_o data in a self-consistent way (Eisenman *et al.*, 1980c; recall Fig. 3), the lesser detail in the conductance data for Cs^+ has made this procedure too time-consuming for this species, and a fit to the full 3B4S″ model awaits our developing a procedure for simultaneously fitting the G_{50} and I/G_o data, now in progress. Also, note that the two double-occupancy states symbolized by $k_{yoyi:}$ and $k_{yo:yi}$ cannot be separated for a negligible central barrier since they are in equilibrium. Their *sum* is given as a lumped binding constant by our data.

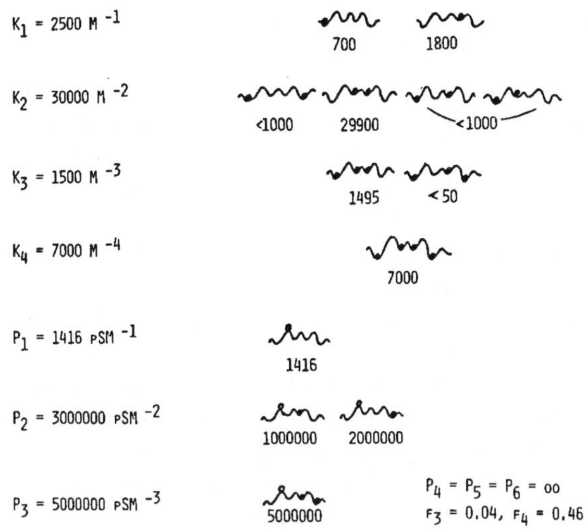

Figure 20. Parameters of the 3B4S″ model extracted from Fig. 19. The intrinsic parameters, according to the definitions in Fig. 4, are given under their corresponding energy diagrams; and the lumped parameters are given at the left. Note that the voltage-dependences given at the bottom right imply a 2% dependence of the entry step and a 23% voltage-dependence of the exit step. Note that the voltage-dependence of the entry step differs from that in Table I, where the effect of a small but significant central barrier was also included.

This is consistent with electrostatic repulsion between ions which requires that the binding constant for loading several ions simultaneously in a given state should be less than the product of individual constants for loading them independently (e.g., that $k_{yoyiyiyo} \ll k_{yo} \cdot k_{yi} \cdot k_{yi} \cdot k_{yo}$).

Second, comparing the values for the doubly-occupied states, one observes that the doubly inner occupied states are more stable than the doubly outer occupied states, and even that $k_{yiyi} > k_{yi}^2$! This implies that there must be an allosteric conformational stabilization of the doubly inner occupied configurations since it is contrary to electrostatic expectations. Physically, double occupancy would be favored for reasons of compatibility between comparable conformations of each half channel, and a further favoring of the inner sites could come about if binding within the channel induced a greater change in the orientation of the carbonyl oxygens and water molecules in each half channel than binding at the channel mouth. This interpretation is compatible with the finding (Sandblom, Neher, and Eisenman, unpublished) that the channel lifetime increases markedly for the doubly-occupied state.

VI. DISCUSSION

A. Effects Not Considered in Strict Barrier Models

This paper has been concerned with strict Eyring single-filing barrier models. Our strategy has been, within the context of such models, to find the simplest one that can characterize the observed data without introducing unnecessary special physical assumptions or restrictions, i.e., interactions are not restricted to electrostatic repulsion, so that we can then let the data tell us what the underlying physical interactions are. In this sense, treatments such as that of Urry *et al.* (1980), which introduce specific physical assumptions very early in order to simplify the mathematics, represent a complementary approach. We have, however, assumed that the location of barriers and wells is independent of the occupancy state of the channel. (Lest this statement cause confusion concerning our previous conclusion of an allosteric conformational change, we point out that such a change need not change significantly the locations of the barriers and wells, since it might merely involve rotations of ligands and changing tilts of water molecules and backbone H-bonding structure.) This restriction can easily be relaxed if the data warrant it.* We have also assumed that the water in the system need not be considered as a molecular species but can be included as if it were a part of the channel structure, i.e., hydration water. This allows for the effects of water on ions, but not the converse.

An alternative strategy, initiated recently by Andersen (1982a,b,c), starts from the point of view that strict barrier models are inadequate since they do not take into account phenomena that can occur in the aqueous solutions external to the channel mouth. This bears on two further assumptions made here, and in all strict barrier models to date: (1) that diffusion up to the mouth of the channel is not rate-determining, and (2) that interfacial-polarization effects are negligible. We have done no experiments bearing on the former, but some of our present results do bear on the latter, and will be discussed below.

Many aspects of the behavior of "hybrid" models incorporating continuum representations of diffusion external to the channel with barrier representations within it are likely to be approximatable by a 1-barrier 1-site model for diffusion. The extensions of 3-barrier 2-site models by adding additional sites external to the barriers (as in our 3B4S″ treatment) or by including aqueous-solution effects, as done by Andersen, lead to quite similar expectations regarding what will be observed experimentally. The differences between these two approaches, which for the present should be regarded as complementary,† center on such questions of interpretation as whether the outer sites of the 3B4S″ model are "true" sites belonging to the channel or only apparent ones reflecting the presence of phenomena occurring in the aqueous solutions external to the channel.

Interfacial-polarization effects, according to Andersen (1982b) arise because some of the potential applied across a membrane is distributed across the aqueous double layers, giving rise to an accumulation of cations at the positive interface and of anions at the negative interface. Andersen, through the use of certain approximations, extends the analysis of Walz *et al.* (1969) for lipid bilayers to include the effects of single-channel currents, obtaining analytical expressions for several important distortions that can arise from interfacial polarization. The most important such distortion for the present paper is the possibility of a spurious high-affinity site, which would mimic many of the properties we attribute to the outer sites of the 3B4S″ model. Although we have not specifically directed our experiments to distinguish between "true" outer sites and "spurious" ones, being content at this stage to establish that additional sites are demanded beyond those of the 3-barrier 2-site model, we believe for the reasons given below that our data favor the notion that, at least in GMO bilayers, the additional sites are actual sites of the gramicidin channel and not manifestations of behavior in the aqueous solutions near the mouth of the channel.

* McBride (1981) has removed this restriction in his considerations of the 3B2S model by allowing locations of sites and barriers to move outward with increasing occupancy. Even with this additional freedom he finds the 3B2S model to be inadequate to describe his new measurements for AgF and KF (McBride and Szabo, 1978; McBride, 1981), for which he finds the same inability to reconcile G_0 and G_c/G_0 data as we reported previously for other species (Eisenman *et al.*, 1980b).

† It should be noted that we emphasize the *shape* of the I–V behavior, whereas Andersen emphasizes the *magnitude* of the currents in reaching those interpretations of the data which are presently divergent. Another difference is that we rely on data from monoglycerides whereas he emphasizes phospholipids, and it appears that interfacial-polarization effects may be less prominent in the former.

A crucial piece of evidence for the importance of interfacial polarization is the residual linear dependence of current on voltage seen at high voltages and reduced by adding 400 mM TEACl, which should not exist for a diffusion-controlled entry step and which Andersen (1982b) interprets as due to a voltage-dependence of the entry process expected at low ionic strength from interfacial polarization. An alternative interpretation, if the concentrations were insufficiently low for occupancy of the channel to be negligible, would be that this voltage-dependence is due to the exit step, rather than to the entry step.* For example, a voltage-dependent exit step is expected to give the same behavior as a voltage-dependent entry step, as can be seen in Figs. 21A and B which show the dependence of the the *I–V* shape on the voltage-dependence of the entry step (f_1) or exit step (f_2) according to Eqs. (1) or (2) for 0-ion occupancy or 1-ion occupancy, respectively, for the particularly simple case where the central barrier is negligible ($\bar{k}_2/\bar{\nu} = 0$). It should be apparent that the same kind of a linear dependence of *I* on *V* is observed in both situations for a small but significant, i.e., 5–10%, voltage-dependence. Since this might not be seen in an actual channel, owing to the voltage-dependence of occupancy, Fig. 22 illustrates how it would occur for a 25% 1-ion occupancy, using the parameters and occupancy for H from Fig. 3. Clearly, a linear dependence of *I* on *V* is observed for a 2.5–10% voltage-dependence of the exit step if the ionic concentration is not lowered sufficiently to be certain that 1-ion occupancy is less than 25%. Our estimates of channel occupancy for Cs (recall Fig. 20) indicate that Andersen's measurements for Cs were made at an insufficiently low concentration to be certain that he was not observing the voltage-dependence of the exit step. In addition, it should be pointed out that Andersen finds, and considers it disturbing, that the effective capacitance of the channel and surrounding bilayer must be three times larger than that of the unmodified bilayer for interfacial-polarization effects to be large enough to be the sole cause of the linear dependence of *I* on *V* at high voltages.

The use of TEA to increase ionic strength rests on the assumption that this ion is inert and exerts its effects purely through the changes in ionic strength. We will argue here that the effect of TEA to remove the asymptotic voltage-dependence might, alternatively, be due to a lowering of the effective concentration of permeant ion in the channel by a competitive binding. The data of Fig. 12 indicate that the related species TMA is not inert, and indeed competes for the binding site. If TEACl had a similar effect, which has not been excluded, then the shape changes produced by increasing ionic strength with TEACl, which Andersen uses as evidence for interfacial polarization, could alternatively be interpreted as a sign that for his (probably insufficiently low) concentrations of permeant species, it is only in the presence of TEACl that the occupancy of the channel has been reduced to a sufficiently low level to see the voltage-dependence of the entry step. It is this interpretation that we now favor. Depending on which interpretation is correct, the dissociation constant of about 0.6 mM indicated by the arrow in Fig. 11 will be either "true" or "spurious." In this regard, it should be noted that this value is quite consistent with the value of around 0.7 mM suggested by the conductance data of Fig. 14 (see arrow labeled "1st site"), which we will argue below is not distorted by interfacial polarization.

Several further points should be noted in regard to the possibility that interfacial polarization is distorting our *I–V* shapes. First, consider the results in Fig. 9, where a linear

* It should be pointed out that Andersen argues (see Andersen, 1982b, Figs. 5–7) that the linear-dependence of intercept currents on aqueous Cs concentration is evidence that the *V*-dependence is due to the entrance step rather than the exit step and states (personal communication) that such a V-dependence on entry is expected even if there is a significant 1-ion occupancy, provided that the first ion binds so strongly that it will only exit when an additional ion enters the channel.

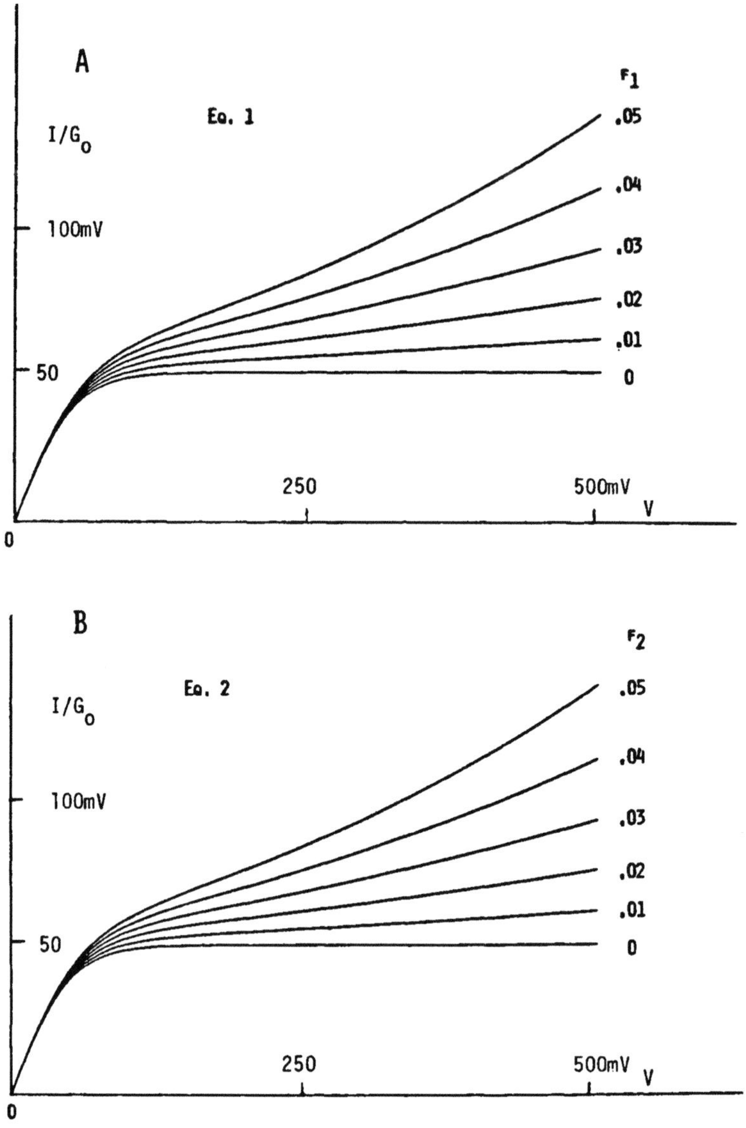

Figure 21. Theoretical illustration of the way a voltage-dependent entry step (A) or exit step (B) can lead to the appearance of a linear dependence of current on voltage at high voltages. (A) shows the expectations of Eq. (1) for the dependence of current upon voltage as a function of the voltage-dependence of the entry step (f_1). (B) shows the expectations of Eq. (2) for the dependence of current upon voltage as a function of the voltage-dependence of the exit step (f_2). In both (A) and (B) the central barrier has assumed to be negligible ($\bar{k}_2/\bar{\nu} = 0$). Note that the shape of the curve is indistinguishable between these two situations.

dependence of I on V at high voltages is apparent for Na and K but where *no such V-dependence* is seen for Cs. Thus, we find that there is no sign of interfacial polarization for Cs, provided a sufficiently low occupancy is achieved by going to lower concenterations than Andersen did. This suggests that the linear V-dependence of I for the other species is not due to interfacial polarization, a conclusion supported by the large species differences seen in the limiting I–V shapes in Fig. 9 (interfacial polarization effects should be species-independent). Since the data in Fig. 9 are well described by the strict barrier model (see the data points and the theoretical curves), we favor the interpretation that the species differences represent true differences in the voltage-dependences of the entry step and in the ratios of rate constants for leaving vs. crossing the channel.

Second, Fig. 10 illustrates, using G_{100}/G_o as an indication of *I–V* shape, that the limiting shape at low permeant-ion concentration is independent of large changes in ionic strength. This indicates that interfacial polarization effects are negligible, at least for GMO bilayers, for if such effects were present, the addition of 1.0 M $CaCl_2$ or $MgCl_2$ would have reduced them markedly and thereby changed the limiting value of G_{100}/G_o at low Cs concentration. (The full *I–V* curves corresponding to the chord conductance for the lowest concentration points on all three curves of Fig. 10 are virtually indistinguishable, data not shown.)

In view of the absence of indications of interfacial-polarization effects in our experiments in GMO bilayers, it is comforting to find signs of such effects in PE bilayers, for this signifies that we can see such effects in our experiments and gives us some idea of their magnitude. Scrutiny of the data of Fig. 18 suggests that whereas interfacial-polarization effects are negligible in GMO bilayers, they are indeed distorting the data in phospholipid bilayers. The argument is as follows. In GMO bilayers, the data in Figs. 16 and 18 indicate a concentration-independence of shape below 10 mM, but a large change in shape around 0.1 M, which would imply a "true" phenomenological first dissociation constant between 10–100 mM, quite consistent with the value of 69 mM found by Dani and Levitt (1981) from water permeability and the value of 25 mM previously inferred by us (Neher *et al.*, 1978; Eisenman *et al.*, 1978) from conductance. The G_{100}/G_o data for PE also show signs of a "true" dissociation constant in the same concentration range, but they also show an additional apparent dissociation constant at an unreasonably low value of less than 0.1 mM. This has been indicated by the arrow labeled "spurious"; we suggest that this change is a consequence of interfacial polarization for it seems unlikely that the first binding constant would be 100 times stronger in PE than in GMO. An alternative possibility that the "spurious" binding is due to a negative surface charge seems unlikely in view of our previous findings with PE which has a negligible surface charge at the pH (5.6) of these measurements (McLaughlin, *et al.*, 1970, Fig. 3), which would be further screened by the presence of 9 mM $MgSO_4$.* In addition, the concentration-dependence of the difference between G_{100}/G_o for PE vs. GMO is not the monotonically decreasing displacement with increasing ionic strength that would be expected if the differences between these lipids reflected solely the presence of a negative charge on the PE.

The most important consequence of interfacial polarization for the present work, as well as for our previous studies (Eisenman *et al.*, 1978; Neher *et al.*, 1978) would be the possibility that the highest affinity sites are "spurious" in that they are not due to binding to the channel but reflect aqueous-solution effects, for, as described by Andersen (1982b), such effects can manifest themselves in such "ohmic" properties as the conductance at low voltage. The data of Figs. 13 and 14 allow us to examine this possibility.† With the Rb data in these figures taken as a control, the existence of clear species differences, manifested in the "feet" for Tl and Cs, but not seen for Rb, indicates that the first binding constants for Tl and Cs are not simply consequences of interfacial polarization, which should produce the same effect for all species at the same ionic strength. However, interfacial polarization

* The batch of PE used in the present experiments, although from the same manufacturer (Supelco), has not been verified as having as low a surface charge as that in our previous studies (McLaughlin *et al.*, 1970).

† Even if the conductances from our noise measurements were not known with better than 30% accuracy, this would be more than sufficient for these conclusions.

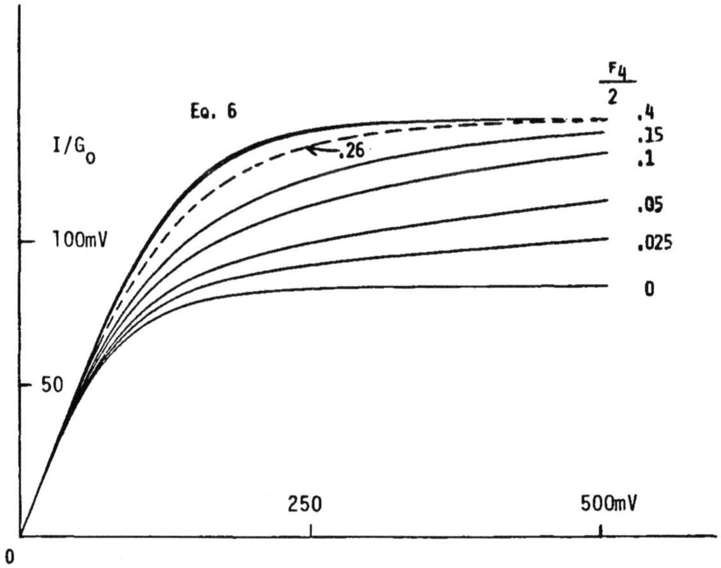

Figure 22. Effects of voltage-dependence of exit step at 25% 1-ion occupancy. This fugure illustrates the kind of V-dependence of I that would be observed in a real channel having parameters like those of H^+ in Fig. 3. The dashed theoretical curve, corresponding to the expectations for 0.003 M H^+, was drawn according to Eq. (6) according to the parameters in the legend of Fig. 3. The remaining theoretical curves were drawn for the indicated voltage-dependences of the exit step ($f_4/2$) with no voltage-dependence of the entry step ($f_3/2 = 0$). Note that the curves corresponding to 2.5–10% voltage-dependence of the exit step ($f_4/2 = 0.025$–0.1) show a distinct region where I depends linearly upon V. This shows how a small voltage-dependence of the exit step at rather low occupancy can produce a linear dependence of I on V at high V in a partially occupied channel.

could well be the cause of the "toe" for Tl (dashed line and arrow labeled "spurious" in Fig. 13), which is seen at concentrations well below the true "foot," indicated by the arrow labeled "true." (The phenomenological binding constants (recall Fig. 15) corresponding to these two arrows would be approximately 500 M^{-1} for the "true" and 20,000 M^{-1} for the "spurious.")

ACKNOWLEDGMENTS. We wish to thank Bruce Enos for writing the Fortran programs for analysis of the digital tapes, Chris Clausen for setting up the APL procedures for data analysis, and Olaf Andersen and Dick Horn for valuable discussions. We also thank Sally Krasne and Dick Horn for their helpful comments on the manuscript and gratefully acknowledge the support of the NSF (PCM 76-20605), the USPHS (GM 24749 and F05 TW2468-01), and the Swedish Medical Research Council (4238).

REFERENCES

Alvarez, O., and Latorre, R., 1978, Voltage-dependent capacitance in lipid bilayers made from monolayers, *Biophys. J.* **21**:1–17.

Andersen, O. S., 1982a, Ion movement through gramicidin A channels. Single channel measurements at very high potentials, *Biophys. J.*, submitted.

Andersen, O. S., 1982b, Ion movement through gramicidin *A* channels. Interfacial polarization effects on single-channel current measurements, *Biophys. J.*, submitted.

Andersen, O. S., 1982c, Ion movement through gramicidin *A* channels. Studies on the diffusion-controlled association step, *Biophys. J.*, submitted.

Andersen, O. S., and Procopio, J., 1980, Ion movement through gramicidin *A* channels. On the importance of the aqueous diffusion resistance and ion-water interactions, *Acta Physiol. Scand.* **481**:27–35.

Bamberg, E., and Benz, R., 1976, Voltage-induced thickness changes of lipid bilayer membranes and the effect of an electric field on gramicidin *A* channel formation, *Biochem. Biophys. Acta* **426**:570–580.

Bamberg, E., and Läuger, P., 1977, Blocking of the gramicidin *A* channel by divalent cations, *J. Membr. Biol.* **35**:351–375.

Begenisich, T., 1979, Ionic interactions in potassium channels, in: *Membrane Transport Processes,* Vol. 3 (C. F. Stevens and R. W. Tsien, eds.), pp. 105–111, Raven Press, New York.

Benz, R., and Janko, K., 1976, Voltage-induced capacitance relaxation of lipid bilayer membranes. Effects of membrane composition, *Biochim. Biophys. Acta* **455**:721–738.

Benz, R., Forlich, O., Lauger, P., and Montal, M., 1975, Electrical capacity of black lipid films and of lipid bilayers made from monolayers, *Biochem. Biophys. Acta* **394**:323–334.

Chizmadjev, Yu. A., and Aityan, Kh., 1977, Ion transport across sodium channels in biological membranes, *J. Theoret. Biol.* **64**:429–453.

Dani, J. A., and Levitt, D. G., 1981, Binding constants of Li^+, K^+, and Tl^+ in the gramicidin channel determined from water permeability measurements, *Biophys. J.* **35**:485–500.

Eisenman, G., 1981, Ca changes the I-V characteristic of the gramicidin channel to that seen in the limit of low permeant ion concentration, *Biophys. J.* **33**:65a.

Eisenman, G., Sandblom, J., and Neher, E., 1977, Ionic selectivity, saturation, binding, and block in the gramicidin A channel: A preliminary report, in: *Metal-Ligand Interactions in Organic Chemistry and Biochemistry,* Part 2 (B. Pullman and N. D. Goldblum, eds.), pp. 1–36, Reidel, Dordrecht, Holland.

Eisenman, G., Sandblom, J., and Neher, E., 1978, Interactions in cation permeation through the gramicidin channel Cs, Rb, K, Na, Li, Tl, H, and effects of anion binding, *Biophys. J.* **22**:307–340.

Eisenman, G., Enos, B., Hagglund, J., and Sandblom, J., 1980a, Gramicidin as an example of a single-filing ionic channel, *Ann. N.Y. Acad. Sci.* **339**:8–20.

Eisenman, G., Hagglund, J., Sandblom, J., and Enos, B., 1980b, The current-voltage behavior of ion channels: Important features of the energy profile of the gramicidin channel induced from the conductance-voltage characteristic in the limit of low ion concentrations, *Upsala J. Med. Sci.* **85**:247–257.

Eisenman, G., Sandblom, J., Enos, B., and Hagglund, J., 1980c, H^+ permeation of the gramicidin A channel, *Fed. Proc.* **39**:1707.

Finkelstein, A., and Andersen, O. S., 1981, The gramicidin *A* channel: A review of its permeability characteristics with special reference to the single-file aspect of transport, *J. Membr. Biol.* **59**:155–171.

Finkelstein, A., and Rosenberg, P. A., 1979, Single-file transport: Implications for ion and water movement through gramicidin *A* channels, in: *Membrane Transport Processes,* Vol. 3 (C. F. Stevens and R. W. Tsien, eds.), pp. 73–88, Raven Press, New York.

Hagglund, J., Enos, B., and Eisenman, G., 1979, Multi-site, multi-barrier, multi-occupancy models for the electrical behavior of single filing channels like those of gramicidin, *Brain Res. Bull.* **4**:154–158.

Hagglund, J., Eisenman, G., and Sandblom, J., 1982, Single salt behavior of a 4-site single-filing channel with barriers at its middle and ends, *Bull. Math. Biol.* (in press).

Hille, B., 1975a, Ionic selectivity of Na and K channels in nerve membranes, in: *Membranes, Vol. 3, Lipid Bilayers and Biological Membranes: Dynamic Properties* (G. Eisenman, ed.), pp. 255–323, Marcel Dekker, New York.

Hille, B., 1975b, Ion selectivity, saturation, and block in sodium channels, *J. Gen. Physiol.* **66**:*535–560*.

Hille, B., and Schwarz, W., 1978, Potassium channels as multi-ion single-file pores, *J. Gen. Physiol.* **72**:409–442.

Hladky, S. B., 1972, The two-site lattice made for the pore, Ph.D. Dissertation, Cambridge University, England.

Hladky, S. B., and Haydon, D. A., 1972, Ion transfer across lipid membranes in the presence of gramicidin *A*. Studies of the unit conductance channel, *Biochim. Biophys. Acta* **274**:294–312.

Hladky, S. B., Urban, B. W., and Haydon, D. A., 1979, Ion movements in the gramicidin pore, in: *Membrane Transport Processes,* Vol. 3 (C. F. Stevens and R. W. Tsien, eds.), pp. 89–103, Raven Press, New York.

Hodgkin, A. L., and Keynes, R. D., 1955, The potassium permeability of a giant nerve fibre, *J. Physiol.* **128**:61–88.

Horn, R., Patlak, J., and Stevens, C. J., 1981, The effect of tetramethylammonium on single sodium channel currents, *Biophys. J.* **36**:321–327.

Koeppe, R. E., Hodgson, K. O., and Stryer, L., 1978, Helical channels in crystals of gramicidin *A* and of cesium-gramicidin *A* complex: An X-ray diffraction study, *J. Mol. Biol.* **121**:41–54.

Kolb, H. A., 1980, Determination of single-pore conductance from noise analysis, *Biochim. Biophys. Acta* **600**:986–992.

Kolb, H. A., Läuger, P., and Bamberg, E., 1975, Correlation analysis of electrical noise in lipid bilayer membranes: Kinetics of gramicidin A channels, *J. Membr. Biol.* **20**:133–154.

Laprade, R., Ciani, S., Eisenman, G., and Szabo, G., 1975, The kinetics of carrier-mediated ion permeation in lipid bilayers and its theoretical interpretation, in: *Membranes,* Vol. 3 (G. Eisenman, ed.), pp. 127–214, Dekker, New York.

Läuger, P., 1973, Ion transport through pores: A rate-theory analysis, *Biochim. Biophys. Acta* **311**:423–441.

Läuger, P., Stephan, W., and Frehland, E., 1980, Fluctuations of barrier structure in ionic channels, *Biochim. Biophys. Acta* **602**:167–180.

Levitt, D. G., 1978, Electrostatic calculations for an ion channel, II. Kinetic behavior of the Gramicidin A channel, *Biophys. J.* **22**:221–248.

Levitt, D. G., Elias, S. R., and Hautman, J. M., 1978, Number of water molecules coupled to the transport of sodium, potassium, and hydrogen ions via gramicidin, nonactin, or valinomycin, *Biochim. Biophys. Acta* **512**:436–451.

McBride, D. W., 1981, Anomalous mole fraction behavior, momentary block, and lifetimes of gramicidin A in silver and potassium fluoride solutions, Ph.D. Dissertation, UCLA.

McBride, D. W., and Szabo, G., 1978, Blocking of the gramicidin channel conductance by Ag, *Biophys. J.* **21**:25a.

McLaughlin, S., Szabo, G., Eisenman, G., and Ciani, S., 1970, Surface charge and the conductance of phospholipid membranes, *Proc. Natl. Acad. Sci USA* **67**:1268–1275.

Neher, E., Sandblom, J., and Eisenman, G., 1978, Ionic selectivity, saturation, and block in gramicidin *A* channels. II. Saturation behavior of single-channel conductances and evidence for the existence of multiple binding sites in the channel, *J. Membr. Biol.* **40**:97–116.

Procopio, J., and Andersen, O. S., 1979, Ion tracer fluxes through gramicidin *A* modified lipid bilayers, *Biophys. J.* **25**:8a.

Requena, J., Haydon, D. A., and Hladky, S. B., 1975, Lenses and the compression of black lipid membranes by an electric field, *Biophys. J.* **15**:77–81.

Sandblom, J., Eisenman, G., and Neher, E., 1977, Ionic selectivity, saturation and block in gramicidin *A* channels: I. Theory for the electrical properties of ion selective channels having two pairs of binding sites and multiple conductance states, *J. Membr. Biol.* **31**:383–417.

Sandblom, J., Ring, A., and Eisenman, G., 1982, Linear network representation of multistate models of transport, *Biophys. J.* **38**:93–104.

Sandblom, J., Eisenman, G., and Hagglund, J. V., 1983, Multi-occupancy models for single-filing ionic channels. Theoretical behavior of a 4-site channel with 3 barriers separating the sites, *J. Membr. Biol.*, **71**:61–78.

Sarges, R., and Witkop, B., 1965, Gramicidin *V*, The structure of valine- and isoleucine-gramicidin *A*, *J. Am. Chem. Soc.* **87**:2011–2020.

Schagina, L. V., Grinfeldt, A. E., and Lev, A. A., 1978, Interaction of cation fluxes in gramicidin *A* channels in lipid bilayer membranes, *Nature* **273**:243–245.

Schoch, P., Sargent, D. F., and Schwyzer, R., 1979, Capacitance and conductance as tools for measurement of asymmetric surface potentials and energy barriers of lipid bilayer membranes, *J. Membr. Biol.* **46**:71–89.

Szabo, G., and McBride, D. W., 1978, Influence of double-layer and dipolar surface potentials on ionic conductance of gramicidin channels, *Biophys. J.* **21**:25a.

Urban, B. W., 1978, The kinetics of ion movements in the gramicidin channel, Doctoral Thesis, University of Cambridge, England.

Urban, B. W., and Hladky, S. B., 1979, Ion transport in the simplest single-file pore, *Biochim. Biophys. Acta* **554**:410–429.

Urban, B. W., Hladky, S. B., and Haydon, D. A., 1980, Ion movements in gramicidin pores. An example of single-file transport, *Biochim. Biophys. Acta* **602**:331–354.

Urry, D. W., 1971, The gramicidin *A* transmembrane channel: A proposed helix, *Proc. Natl. Acad. Sci. USA* **68**:672–676.

Urry, D. W., Venkatachalam, C. M., Spisni, A., Bradley, R. J., Trapane, T. L., and Prasad, K. U., 1980, The malonyl gramicidin channel: NMR-derived rate constants and comparison of calculated and experimental single-channel current, *J. Membr. Biol.* **55**:29–51.

Walz, D., Bamberg, E., and Läuger, P., 1969, Nonlinear electrical effects in lipid bilayer membranes. I. Ion injection, *Biophys. J.* **9**:1150–1159.

White, S. H., and Chang, W., 1981, Voltage dependence of the capacitance and area of black lipid membranes, *Biophys. J.* **36**:449–453.

White, S. H., and Thompson, T. E., 1973, Capacitance, area, thickness variations in thin lipid films, *Biochim. Biophys. Acta* **323**:7–22.

Woodbury, J. W., 1971, Eyring rate theory model of the current-voltage relationships of ion channel in excitable membranes, in: *Chemical Dynamics: Papers in Honor of Henry Eyring* (J. O. Hirschfelder, ed.), pp. 601–661, New York, Wiley.

Zingsheim, H. P., and Neher, E., 1974, The equivalence of fluctuation analysis and chemical relaxation measurements: A kinetic study of ion pore formation in thin lipid membranes, *Biophys. Chem.* **2**:197–207.

21

The Nonlinear Kinetics of an Electrodiffusion Membrane

H. R. Leuchtag and H. M. Fishman

I. WHY ELECTRODIFFUSION?

Two approaches that have had a dominant influence on the study of excitability are macroscopic electrodiffusion, based on the Nernst–Planck equation, and the channel approach, based on the phenomenological formulation of Hodgkin and Huxley. Although electrodiffusion served as the historical starting point of electrophysiological theory and was once described by K. S. Cole as "the most complete, most attractive, and the most widely used model for the electrical properties of living cell membranes," the focus of attention in recent years has shifted to the channel approach. The opening sections of this chapter will present some of the reasons for which the study of electrodiffusion remains an important, indeed a necessary, endeavor toward obtaining a full understanding of excitability.

The goal of this chapter is to present an introduction to the problem of time-dependent, nonlinear electrodiffusion. To this end we formulate a simple model involving a single ion species, which nevertheless leads to equations requiring some mathematical sophistication. Because the resulting nonlinear wave equations have been studied in other contexts, the problem can be viewed by comparison to analog systems such as shock waves and wind-driven water waves. An approximate analysis allows examination of a simple current-clamp case.

A. Historical Background

The equations of electrodiffusion were first formulated by Nernst and Planck in the 1880s for electrolyte solutions, for which the condition of local electroneutrality is valid. Bernstein's 1902 membrane hypothesis, which applied Nernst's work to excitable mem-

H. R. Leuchtag • Department of Physiology and Biophysics, University of Texas Medical Branch, Galveston, Texas 77550; and Department of Biology, Texas Southern University, Houston, Texas 77004. *H.M. Fishman* • Department of Physiology and Biophysics, University of Texas Medical Branch, Galveston, Texas 77550.

branes, was supported by Fricke's 1923 measurement of membrane capacitance. K. S. Cole (1947) provided a link between electrodiffusion and a linear circuit model of a membrane and, with Curtis (1939), demonstrated the decrease of membrane impedance during the passage of an action potential. Following an analysis of Mott (1939), Goldman (1943) introduced the constant-field approximation, which was extended by Hodgkin and Katz (1949) and supplemented by the sodium hypothesis to become the bridge to the description by Hodgkin *et al.* (1952) and Hodgkin and Huxley (1952a,b,c,d) of squid-axon behavior under voltage clamp and during an action potential. Ussing (1949) extended the application of electrodiffusion to isotopic tracers and, under idealized boundary conditions, obtained a simple relation for the ratio between inward and outward fluxes; a more general form of this ratio was derived by Behn (1897) and Teorell (1949). Much of the analytical work done on electrodiffusion has been devoted to its steady-state aspects. This body of work will not be reviewed here, as reviews of various aspects of it are available (see Cole, 1965, 1968; Schwartz, 1971; Arndt and Roper, 1972; Mackey 1975a; Vaidhyanathan, 1976; Finkelstein and Mauro, 1977; Leuchtag and Swihart, 1977; Sten-Knudsen, 1978; deLevie, 1978; Teorell, this volume).

Time-dependent analyses, with which this chapter deals, include both transient and spectral studies. Cohen and Cooley (1965) studied the case in which the displacement current was neglected and the electroneutrality approximation holds, which gave an almost entirely resistive response. McGillivray (1970) investigated the response of an electroneutral membrane to a voltage-step perturbation. Hagglund (1972) studied a convection–diffusion model of the early current, equivalent to time-dependent electrodiffusion under the constant-field approximation, and found a series solution exhibiting similarities to the *m* kinetics of Hodgkin and Huxley. This solution, earlier obtained by FitzHugh (Cole, 1968, p. 188), was compared with multisite Eyring models by Hays *et al.* (1978). Sandblom (1972) and Hagglund and Sandblom (1972), extending the method to the frequency domain, obtained a Cole–Cole plot of the impedance locus with both inductive and capacitive reactance. Meyer and Kostin (1974) computed the onset of the transient potential response to a step change in the sodium partition coefficient in the presence of potassium. Small-signal admittance analyses of an electrodiffusion membrane were carried out by deLevie *et al.* (1974) and Mackey (1975b). Leuchtag (1974a,b) showed that the nonlinear one-ion electrodiffusion equations yield Burgers's equation (Burgers, 1950). Kazantsev (1978) also obtained Burgers's equation in an approximate analysis of a two-state electrodiffusion membrane. Mikulinsky and Mikulinsky-Fishman (1976) based a theory of $1/f$ noise on electrodiffusion. Brumleve and Buck (1978, 1981a,b) have obtained transient and impedance-frequency responses numerically.

B. Critiques of Electrodiffusion

Despite the considerable achievements gained up to the 1960s, the results of the electrodiffusion approach remained far short of explaining the growing body of data, and in 1965 K. S. Cole, in an influential review, expressed a pessimistic outlook toward the electrodiffusion analyses that had been carried out. There were many difficulties; the rectifications obtained from constant-field analysis were far too gentle to account for squid-axon data, there seemed to be no way to accommodate negative resistance associated with the sodium currents into the electrodiffusion formalism, the Ussing equation appeared to require an arbitrary exponent n', equal to about 2.5, to fit squid-axon data, and so on. These conclusions were based on analyses that, as Cole himself put it, were simple: (1) They were linear analyses. The electroneutrality analysis for two (but not more) ions produces a linear system; the constant-field approximation likewise linearizes the mathematical system and separates it into independent sets of equations. (2) They neglected a possible field dependence of the dielectric constant. (3) In some analyses the effects of ion partition between membranes and aqueous phases were ignored. (4) The effects of extra-axolemmal layers and structures, both outside (Frankenhaeuser and Hodgkin, 1955; Villegas and Villegas, 1960; Adelman and FitzHugh, 1975; Grisell and Fishman, 1979; Taylor *et al.*, 1980) and inside (Metuzals and Tasaki, 1978; Fishman, 1981) the axon, were not considered. (5) Possible discrepancies stemming from the special boundary-value assumptions* in the Ussing version of the flux-ratio equation, rather than the more general Behn–Teorell version, were not examined.

The concern about negative resistance is ameliorated by the fact that Hodgkin and Huxley (1952d) were able to describe the ion-conduction phenomena with only positive quantities, by recognizing that the concentration gradient of the sodium ions sets up an inwardly directed driving force. Cole, who was aware of many of the limitations of the analyses he cited, proposed two choices for researchers in the field: either correct and improve upon the electrodiffusion approach or look for a better mechanism of ion permeability. The present work is pursuing the former path.

As has been pointed out (Hille, 1978) a number of phenomena have not been explained by one-ion electrodiffusion. (This critique likewise applies to other current approaches, since formalisms based on special assumptions cannot be considered to be explanations.) It is true that a definitive theory of membrane electrodiffusion will have to include at least two ion species. In the introduction of new generalizations (such as that below), however, the simpler one-ion case may well have to precede more general ones.

It is important that the goal of an explanation from first principles not be abandoned. Only with continued efforts to "correct and improve upon the electrodiffusion approach," can a realistic assessment of this approach be made.

II. CONCEPTUAL AND EXPERIMENTAL FOUNDATIONS OF THE CHANNEL APPROACH

The acceptance of the channel approach (Hille, 1970) is essentially based on three lines of evidence:

1. The use of the Hodgkin–Huxley conductances, $g_{Na}(V,t)$ and $g_K(V,t)$ to fit voltage-clamp data accurately for a great diversity of preparations under many ionic and pharmacological conditions.

2. The measurement, in electrically isolated small patches of membrane, of random, rectangular electrical pulses, known as "single-channel currents."

3. The separation of protein macromolecules from excitable cells that, when reconstituted in artificial lipid membranes, cause them to exhibit some properties similar to those in excitable membranes.

These lines of evidence will be briefly considered in order to examine whether their combination justifies an exclusive reliance on the channel approach.

One barrier to the effective consideration of this question is the lack of a formal definition of the term "channel." The minimal components of the term "channel" as it is used in the context of excitable cells appear to be: the hypothesis of separability of Na and K currents, the assumption of a voltage- and time-dependent gating action on ion conductance, and the

* The derivation of the Ussing equation assumes steady fluxes of isotopic ions between regions of finite and zero electrochemical activities. In Ussing's (1949) terminology, $\bar{a}_1 = \bar{a}_2 = 0$. While these may be valid assumptions for large well-mixed compartments, they are dubious for potassium efflux from squid axon, which is bounded by the Schwann-cell layer. This difficulty was not taken into consideration by Hodgkin and Keynes (1955) and others.

implicit assumption that the V- and t-dependence is inherent in the membrane itself and not the result of interactions of the membrane with its boundary layers and neighboring structures. Therefore, for the purposes of this discussion, a "voltage- and time-dependent excitability channel" or, briefly, a "channel," will be defined as a macromolecular entity embedded in the plasma membrane that alone and directly controls ("gates") the passage of specific individual ions through its interior in response to the potential difference across the membrane and to the time. Let us consider whether such channels can be said to have been observed.

A. Hodgkin–Huxley Conductances

The Hodgkin–Huxley formulation divides the voltage-clamp current, after subtracting a dielectric contribution assumed to be due to a lumped linear capacitance, into three components, taken to be due to independent ionic current contributions,

$$I(V,t) = I_{\mathrm{Na}}(V,t) + I_{\mathrm{K}}(V,t) + I_{\mathrm{L}}(V).$$

The third, leakage, is subtracted by assuming that its voltage-dependence is ohmic and that the Na and K conductances are zero for hyperpolarization. The K current is assumed to be unchanged when the external sodium is replaced with choline (assumed impermeant), so that the current remaining after subtraction of the K and L currents is taken to be the Na current. Ion (chord) conductances are then introduced by the equations

$$I_{\mathrm{Na}} = g_{\mathrm{Na}}(V,t)\,(V - \mathscr{E}_{\mathrm{Na}}) \text{ and } I_{\mathrm{K}} = g_{\mathrm{K}}(V,t)\,(V - \mathscr{E}_{\mathrm{K}}) \qquad (1)$$

where $\mathscr{E}_{\mathrm{Na}}$ and $\mathscr{E}_{\mathrm{K}}$ are the experimentally determined reversal potentials (now considered not necessarily equal to the Nernst potentials, as originally assumed by Hodgkin and Huxley). It is these nonlinear "ion conductances" that form the basis for the search for "channels." Their time- and voltage-dependence was analyzed by Hodgkin and Huxley by assuming that

$$g_{\mathrm{Na}} = \bar{g}_{\mathrm{Na}} m^3 h, \qquad g_{\mathrm{K}} = \bar{g}_{\mathrm{K}} n^4,$$

where the sodium activation and inactivation m and h, and the potassium activation n, are normalized to one, and taken to obey linear kinetic equations with parameters the voltage-dependence of which was obtained by curve fitting.

The usefulness of this formulation cannot be questioned. From it, Hodgkin and Huxley were able to reconstruct the action potential, and it has been used successfully to describe the responses from a large number of preparations. It has been supported by a number of findings: the pufferfish poison tetrodotoxin at micromolar concentrations has an effect similar to the substitution of an impermeant ion for sodium; in mechanistic language, it "blocks the Na channel." (A significant exception (Kao and Fuhrman, 1967; Kidokoro *et al.*, 1974) is the tetrodotoxin-insensitive response of the axons of the pufferfish itself. Another complication is the discovery of tetrodotoxin sensitivity in nonexcitable cells, including Schwann cells (Villegas and Villegas, 1981)). Tetraethylammonium, internally applied, and 4-aminopyridine similarly suppress the K current, but are not as selective. Asymmetry currents have been measured, which are consistent with the interpretation of voltage-dependent movements of Na channel "gates." The action of pronase, which eliminates "Na inactivation" while only slightly reducing the "Na activation," supports the m^3h formulation.

Given that the Hodgkin–Huxley formulation remains satisfactory as a quantitative description of axon currents, can we assume that it therefore forms the basis for a physical theory? There are good grounds for doubting that we can:

1. The formulation is not derived directly from the laws of physics. It is basically an *inductive* approach, based on manipulating the voltage-clamp data in various ways. It needs to be complemented by a *deductive* approach grounded in the laws of electrodynamics and statistical physics.

2. The explicit dependence of the "conductances" on time is a particularly objectionable feature if these are to be interpreted as independent physical entities rather than as parameters in a phenomenological description. Figure 1, adapted from Cole (1968), illustrates this point by showing Nat and Kal, Cole's personified controllers of the Na and K conductances, consulting their watches as well as sensing the membrane potential difference. A closed mechanical system must be characterizable by a function, the Lagrangian, that does not depend explicitly on time. It is from this principle, the homogeneity of time, that the law of energy conservation follows (Landau and Lifshitz, 1960). The fact that the function that is taken to define the quantitative properties of a Hodgkin–Huxley channel, $g_{Na,K}(V,t)$, does depend explicitly on time indicates that such a channel is not an independent system but part of a larger system with which it exchanges energy. An analysis of an excitable system can therefore not be complete if it limits itself to the study of time-dependent channels.

3. It uses the assumption of independent ionic currents to justify the subtraction of the K current (and leakage) from the net ionic current to obtain the Na current; yet according to more recent reports (Baker *et al.*, 1962; Tasaki *et al.*, 1962; Chandler and Meves, 1965; Meves and Chandler, 1965; Moore *et al.*, 1966) some K "goes through the Na channel" and vice versa. From an operational point of view, this contradicts the supposed independence of the two channels and makes the subtraction process from which I_{Na} is obtained questionable.

4. One of the reasons Hodgkin and Huxley(1952b) gave in support of the linear forms in Eq. (1) is their observation that the instantaneous *I–V* curves are linear. More recent data in various preparations show nonlinear *I–V* curves (see Dodge and Frankenhaeuser, 1959; Sigworth and Spalding, 1979; Swenson and Armstrong, 1981). These findings indicate that the use of a chord conductance introduces uncertainties in the determination of the temporal variation of ion conductances and driving forces.

5. In ramp-clamp experiments (Fishman, 1969, 1970; Palti and Adelman, 1969), a continuous sequence of *I–V* curves ranging from those typical of the K system to those

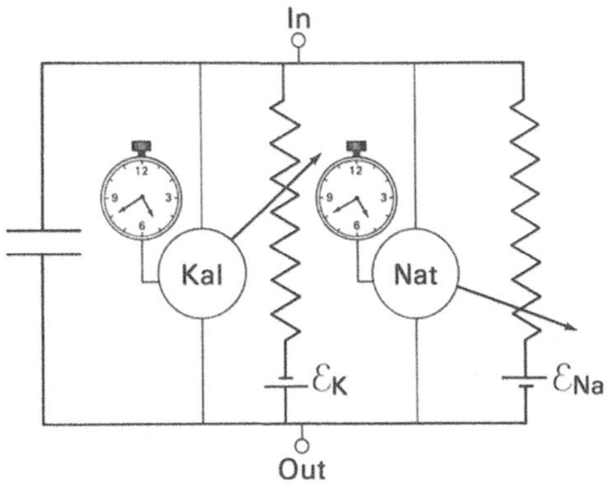

Figure 1. Schematic drawing of fundamental concepts in the Hodgkin–Huxley formulation. This figure is adapted from Cole (1968) to exhibit the explicit time-dependence of the K and Na conductances, personified by Cole's Kal and Nat, who are consulting their watches and sensing the membrane potential difference to set the K and Na conductances.

typical of the Na system can be obtained by simply varying the ramp rate, as shown in Fig. 2. Thus, while the terms "Na current" and "K current" have become conventional, the nature of the response appears to be less a property of specific ion species than of the kinetic range of the stimulus. The terms "early" and "late" current have been suggested as more appropriate. However, if we are dealing with different kinetic properties rather than with different ion species, there is no reason to assume that they reside at distinct sites, and the concept of selectivity becomes meaningless.

6. While many preparations have been successfully described within the Hodgkin–Huxley formulation, phenomena have been observed that required extension of its concepts far beyond its original scope, such as Ca-activated K currents (Junge, 1981), excitation currents of Ca ions (Hagiwara and Byerly, 1981), and others. And even in the squid axon there is now evidence that ion-channel kinetics are both ion-species and ion-concentration dependent (Adelman and French, 1978; Oxford and Adams, 1981; Swenson and Armstrong, 1981; Matteson and Swenson, 1982). The Hodgkin–Huxley phenomenology thus has become an elastic scheme subject to emendation as needed to fit experimental data, rather than a theory, which must be capable of being falsified by comparison with experiment.

7. Calculations by Hodgkin and Huxley (1952d) predicted a threefold increase in the average influx and efflux of sodium ions when the temperature is lowered by 10°C, but tracer experiments (Landowne, 1973; Cohen and Landowne, 1974) show a much smaller dependence on temperature. Since the duration of the nerve impulse is much longer at the lower temperature, these data represent a serious challenge to any model that accounts for the sodium current by a simple permeability change (Landowne, 1973).

8. Studies with new analogs of tetrodotoxin and saxitoxin (chiriquitoxin, neosaxitoxin, decarbamylsaxitoxin, and reduced saxitoxin) have shown inconsistencies in the "plug-in-the-channel" model of these toxins (Hille, 1975), so that this model is no longer believed to be tenable (Kao *et al.*, 1981; Kao and Walker, 1982).

9. The Hodgkin–Huxley description is not unique in its ability to fit voltage-clamp data, as several authors have demonstrated (Cole and Moore, 1960; Hoyt, 1968; Moore and

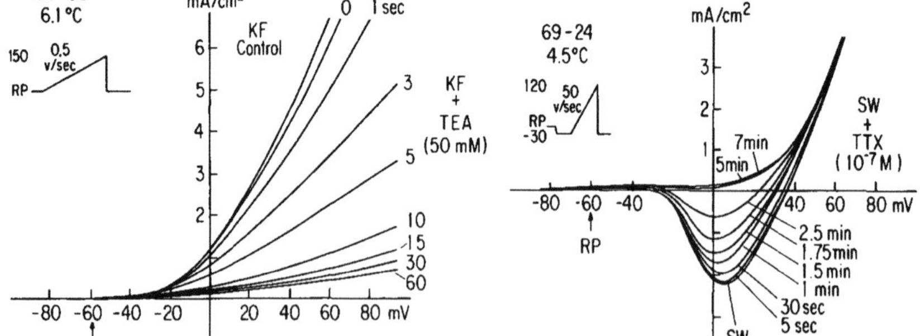

Figure 2. Current–voltage curves obtained from perfused squid axons during applied ramp voltage clamp. In (A), the clamp voltage is increasing at the rate of 0.5 V/sec on an axon in seawater. The upper curve is a control with internal KF; the lower curves were recorded at the times indicated after addition of 50 mM tetraethylammonium. In (B), the ramp rate is increased to 50 V/sec; note the appearance of negative differential conductance in the control curve (SW). The upper curves show the time sequence of responses to the addition of tetrodotoxin. The appearance of typical "I_K" and "I_{Na}" curves, together with their characteristic pharmaceutical sensitivities, as a function of ramp rate suggests that these curves represent dynamical properties of the membrane system. (Reproduced by permission from Fishman, 1970).

Jakobsson, 1971; Starzak, 1973a,b; Easton, 1978), even though it undoubtedly covers more of the known phenomena than any other existing description.

The separation of currents into Na, K, and leakage, and the further separation of Na currents into activation and inactivation factors has led to a fragmentation of the excitation problem which, while yielding temporary advantages in fitting data from specific experiments, may be leading away from a global solution of the problem. The concept of a channel, which is so closely dependent on the Hodgkin–Huxley formalism, is far from established as a physical entity.

B. Rectangular Spontaneous Electrical Pulses

In the early reports of voltage-dependent, ion-channel current fluctuations (Fishman, 1973; Conti *et al.*, 1975; Siebenga *et al.*, 1974) in isolated, macroscopic areas of nerve membrane, the assumption of discrete conductances was invariably used to interpret data. Although these and subsequent measurements dealt with the statistical properties of current fluctuations, they were macroscopic measurements due to the relatively large areas involved. Consequently, these measurements of membrane current fluctuations were not proof of the existence of channels. More recently, advances in the isolation of a very small patch of membrane have yielded random, pulse-like current waveforms at constant potential, which are thought to reflect the elementary current through single channels (Conti and Neher, 1980; Sigworth and Neher, 1980; see Figs. 3A and B). Do these observations constitute proof of membrane channels?

There exists a substantial literature outside the realm of membrane biophysics that is pertinent to this question. In semiconductor physics a type of current noise similar to the random, discrete-level, pulse-like fluctuations found in membranes is known as "burst" or "popcorn" noise. Figure 3C shows burst noise current in a reverse-biased germanium p-n junction and Fig. 3D, that in a 200 kΩ carbon composition resistor (Card and Chaudhari, 1965). Although the magnitudes of the currents differ widely, the qualitative appearance of these bursts is remarkably similar to the membrane fluctuations. Explanation of these phenomena do not require channels in the materials of these devices (Hsu *et al.*, 1970). A similar argument can be made with respect to spontaneous pulses exhibited by ferroelectric materials, called Barkhausen pulses and believed to be due to movement of dipolar domains in the electric field (Chynoweth, 1958). Clearly, the observation of random, pulse-like waveforms in isolated patches of membrane does not resolve the issue of whether channels of the type specified exist.

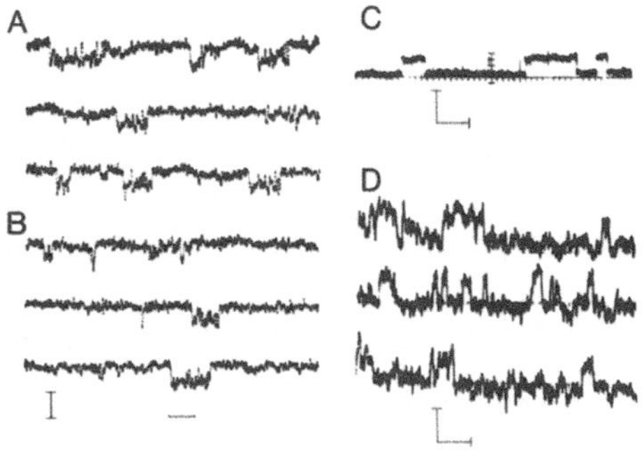

Figure 3. Comparison of current-noise records from a squid-axon membrane patch (A and B) with those from nonbiological conductors (C and D). Membrane potential was −25 mV in A, −35 mV in B. Horizontal scale bar represents 50 msec; vertical, 3 pA. The waveform in C was produced by a germanium junction biased at 24 V; horizontal bar, 2 msec; vertical bar, 50 nA. The three traces in D were recorded from a 200 kΩ, 2 W, carbon composition resistor with a dc potential difference of 9 V; horizontal bar, 2 msec; vertical bar, 270 nA. (A and B are reproduced by permission from Conti and Neher, 1980 and C and D from Card and Chaudhari, 1965.)

C. Isolation and Reconstitution Experiments

An important line of investigation consists of the solubilization, isolation, and reconstitution of membrane constituents necessary to excitability. Although the molecules found to bind neurotoxins that affect the early current have been referred to as "the Na channel," closer consideration of the actual results of these studies appears to raise questions regarding the premises of the channel approach.

There are three types of neurotoxins: (1) guanidinium-containing water-soluble heterocyclic compounds, (2) polypeptides (sea anemone and scorpion toxins), and (3) lipid-soluble polycyclic compounds (veratridine, grayanotoxin, batrachotoxin, and aconitine). Analysis of the tetrodotoxin–saxitoxin receptor has yielded peptide fractions ranging from 46,000 to 300,000 daltons and a protein fraction of about 700,000 daltons. The receptor for the lipid-soluble toxin veratridine, however, has been found to bind mainly on the lipid phase of the membranes or on a lipoidal protein. The receptors that bind polypeptide neurotoxins yield polypeptides of molecular weights in the same range as those of the tetrodotoxin–saxitoxin receptor (Villegas and Villegas, 1981; Villegas *et al.*, this volume). Although the findings that lipid-soluble neurotoxins bind at different membrane sites from neurotoxins of the other groups can be incorporated into the channel hypothesis by assuming a sufficiently complex channel structure to include both types of binding sites, a less localized conduction process would also be consistent with these findings.

Vesicles formed by plasma membrane fragments isolated from lobster nerves exhibit a slow (0.08 pmol/cm^2 sec) steady efflux of labeled sodium ions. The flow rate is increased in the presence of veratridine, but diminished by tetrodotoxin (Barnola and Villegas, 1976). Beyond this simple criterion of functional activity, properties characteristic of excitable cells have not been reported in these vesicles.

A series of studies on synaptosomes shows evidence for membrane potentials, potassium-ion accumulation, and sensitivity to neurotoxins. Enhancement of sodium uptake by veratridine, grayanotoxin 1, or batrachotoxin was diminished by 50% when tetrodotoxin was also present (Villegas and Villegas, 1981).

Liposomes with incorporated solubilized membrane proteins exhibited a similar neurotoxin sensitivity. Permeability ratios for K or Rb with respect to Na for tetrodotoxin-sensitive fluxes in veratridine-activated liposomes were K/Na = 0.47 and Rb/Na = 0.55. These selectivities are lower than those reported for untreated squid axons or nodes, but similar to nodes treated with batrachotoxin or aconitine. These results, obtained with nerve membrane fragments, were confirmed with liposomes incorporating protein or peptide particles (Villegas and Villegas, 1981).

The fact that preparations of supposedly purified "Na channels" are not only permeable to potassium ions, but lack any enhancement in selectivity for sodium ions over that of an intact axon membrane carries significant implications for the channel approach. Thus, if K and Na channels exist as separate entities, as the channel hypothesis requires, they have not been separated in this series of experiments. On the other hand, if it is assumed that a successful isolation of the protein molecules supporting excitability (Na channel) has been attained, the claim of the existence of separate structures selectively conducting K and Na currents has been disproved. Rather than supporting the channel hypothesis, the results of these experiments therefore appear to represent a serious challenge to its basic tenets.

D. Have Specialized Conductance Structures Been Demonstrated?

The conceptual and experimental foundations of the channel approach have been briefly considered in this section. We have pointed out that the Hodgkin–Huxley formalism on which it is based contains serious difficulties if it is to be regarded as a physical theory. The term "channel" is ambiguous and ill-defined. Clearly, there must exist channels in the sense that ions must traverse certain pathways through the membrane. What is not clear is whether these pathways are of the three types, Na, K, and leakage, corresponding to the Hodgkin–Huxley formulation, whether they are specialized conductance structures corresponding to the Hodgkin–Huxley conductances, whether they are independent, and whether a proper mathematical description should make them functions of membrane potential difference and time. The observations of spontaneous pulses, while of great importance, neither confirm nor rule out the existence of the hypothesized voltage- and time-dependent excitability channels. The results of the protein isolation and reconstitution experiments are consistent with the interpretation that, while certain proteins play key roles in ion conduction, the electrical response of an excitable cell is an integral expression of the behavior of the entire cell. Therefore, a rigid and exclusive reliance on the channel paradigm is not warranted by the current state of knowledge of excitable cells.

III. THE ONE-ION ELECTRODIFFUSION MODEL

The characteristic features of the electrodiffusion model are that it is macroscopic and that it represents an idealization.

1. It is a macroscopic formulation, in the sense that the distinct components of the mosaic membrane are averaged, over sufficiently large membrane area, to a uniform region supporting continuous measurable changes in the electric field and the ion concentration. Experiments suggest that this area need not be very large, and that the response of a patch 50 μm in diameter differs little from that of a square millimeter of membrane.*

2. It is an idealized model, focusing on the irreducible minimum formulation of the physical problem to permit an analysis by rigorous mathematical methods. Against such a background of analytical knowledge, more general hypotheses can then be imposed and tested.

A. Fundamental Equations

The transient electrodiffusion model has three major components: the Nernst–Planck equation, the equation of continuity, and Gauss's law of electrostatics.

The Nernst–Planck equation states that the ion flux, and hence the ionic current density J_{ion}, is the sum of two terms: a diffusional one given by Fick's law, and a migrational term proportional to the electrostatic force on the ion times the ion density N.

$$J_{\text{ion}} = -q\mu\left(kT\frac{\partial N}{\partial x} - qEN\right) \tag{2}$$

Here q is the ion charge, μ its mechanical mobility (velocity per unit force), and E ($= -\partial V/\partial x$) the electric field. As usual, x is transverse displacement, k is Boltzmann's constant, and T the Kelvin temperature. When the ion current is zero, these two effects balance each other

* It is sometimes proposed (Hille, 1978; Chang, this volume) to apply the Nernst–Planck equations separately to each of the various types of membrane channels or regions. This is not done here because, in the absence of specific knowledge about the dimensions, intrinsic properties, and interactions of these regions, their inclusion would be a purely formal device, which would result only in greatly increasing the number of adjustable parameters.

exactly to give the Nernst potential. As the temperature approaches absolute zero, the diffusional term becomes very small relative to the migrational term; this fact provides the basis for a convenient simplification below.

The equation of continuity accounts for the ions by assuming that none are created or destroyed. For a single ion species,

$$\frac{\partial J_{\text{ion}}}{\partial x} = -q \frac{\partial N}{\partial t} \tag{3}$$

Gauss's law, the continuum equivalent of the inverse square law for point charges, describes the coupling of the ions to the electric field.

$$\frac{\partial E}{\partial x} = \frac{1}{\kappa \varepsilon_0} qN \tag{4}$$

Here ε_0 is the permittivity of vacuum and, as in other electrodiffusion treatments, the assumption is made that the relative permittivity κ is a constant, i.e., that the polarization is proportional to the field. It should be pointed out, however, that in many classes of materials, such as the ferroelectrics, this assumption does not hold.

Combining the last two equations and integrating gives us Maxwell's well-known statement that the total current $J(t)$ is equal to the current due to the movement of ions plus the displacement current.

$$J_{\text{ion}} + \kappa \varepsilon_0 \frac{\partial E}{\partial t} = J(t) \tag{5}$$

The membrane will be assumed to be current clamped, so $J(t)$ is a known function.

B. Boundary and Initial Conditions

It is necessary at this point to introduce the boundary conditions. We will assume given ion concentrations just inside the membrane, N_{I} and N_{II}, given by

$$\begin{aligned} N(0,t) &= N_{\text{I}} = \beta_{\text{I}} N_{\text{in}} \\ N(L,t) &= N_{\text{II}} = \beta_{\text{II}} N_{\text{out}} \end{aligned} \tag{6}$$

These are related to the external and axoplasmic concentrations N_{out} and N_{in} by the partition coefficients β_{I} and β_{II}, considered constant. The values of the field at the boundaries $E_{\text{I}}(t)$ and $E_{\text{II}}(t)$, are not zero because the interfaces are shielded from the bulk axoplasm and external solution by charged extramembrane boundary layers. The initial concentration profile is

$$N(x,0) = N_0(x) \tag{7}$$

in the range $x = 0$ to $x = L$.

IV. THE PARTIAL DIFFERENTIAL EQUATION GOVERNING ONE-ION ELECTRODIFFUSION

It is now simple to combine the Maxwell equation with the Nernst–Planck equation and Gauss's law to obtain the following partial differential equation in the field E:

$$\frac{\partial E}{\partial t} = \mu kT \frac{\partial^2 E}{\partial x^2} - q\mu E \frac{\partial E}{\partial x} + \frac{1}{\kappa \varepsilon_0} J(t) \tag{8}$$

This equation may be described as a parabolic quasilinear partial differential equation. (It is parabolic because it belongs to the class of equations the prototype of which is the diffusion equation, rather than that of linear wave equation, which is hyperbolic, or that of Laplace's equation, which is elliptic. It is quasilinear because its nonlinearity is not in the dominant second-order term, but in the first-order term.)

It is convenient at this point to strip away the constant parameters in our equations by transforming to new variables, which will be dimensionless. Consider N_1 to be an arbitrary unit ion concentration and λ the Debye length corresponding to it,

$$\lambda = \left(\frac{\kappa\varepsilon_0\, kT}{q^2\, N_1}\right)^{1/2} \tag{9}$$

Then define new variables as follows:

$$\begin{aligned} &\xi = x/\lambda,\ l = L/\lambda, && \tau = (\mu kT/\lambda^2)t \\ &u = (q\lambda/kT)E, && n = N/N_1 \\ &f(\tau) = (\lambda/\mu q N_1 kT)J(t), && v = (q/kT)V \end{aligned} \tag{10}$$

A. Burgers's Equation

Consider now our parabolic quasilinear equation with the current term set to zero, expressed in the dimensionless variables,

$$\frac{\partial u}{\partial \tau} + u\frac{\partial u}{\partial \xi} = \frac{\partial^2 u}{\partial \xi^2} \tag{11}$$

This equation, well known in hydrodynamics and acoustics, is called Burgers's (1950) equation, despite the fact that it first appeared in a 1915 paper by Bateman. While the major fields of application of Burgers's equation have been the theory of shock waves and the study of turbulence, it serves more generally as a simple model equation for nonlinear dissipative phenomena (Witham, 1974; Leibovich and Seebass, 1974; Karpman, 1975). The properties of Burgers's equation have been extensively studied (see the comprehensive review by Benton and Platzman, 1972). A transformation discovered by Hopf (1950) and J. D. Cole (1951) permits the Burgers equation to be converted into the diffusion equation, which makes it in principle exactly soluble. A number of explicit solutions are known (Benton and Platzman, 1972), including periodic solutions applicable to the two-point boundary-value problem and the spectral solutions of Fay (1931) and of Kochina (1961).

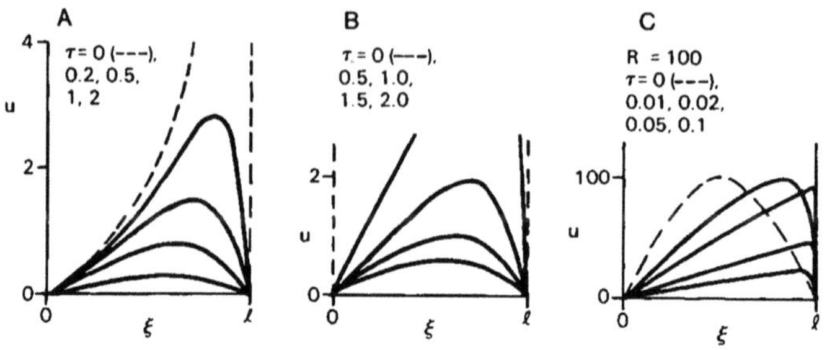

Figure 4. Solutions to the Burgers equation, adapted from Benton and Platzman (1972). (A) Eq. (12). (B) Fay's series, Eq. (13). (C) Eq. (14). The boundary conditions used in these figures are suitable for certain hydrodynamic problems, but are not applicable to membrane electrodiffusion.

Three solutions that have been developed for Burgers's equation in the context of the shock-wave problem, adapted from Benton and Platzman (1972), are shown in Fig. 4. Figure 4A corresponds to the solution

$$u(\xi,\tau) = \frac{2 \sin(\pi\xi/l)}{\cos(\pi\xi/l) - \exp(\tau)} \tag{12}$$

Figure 4B is Fay's series,

$$u(\xi,\tau) = -2 \sum_{n=1}^{\infty} (-1)^n \frac{\sin(n\pi\xi/l)}{\sinh(n\tau)} \tag{13}$$

and Fig. 4C shows the behavior of the solution

$$u(\xi,\tau) = \frac{4 \sum_{n=1}^{\infty} n a_n \exp(-n^2\tau) \sin(n\pi\xi/l)}{a_0 + 2 \sum_{n=1}^{\infty} a_n \exp(-n^2\tau) \cos(n\pi\xi/l)} \tag{14}$$

where $a^n = (-1)^n I_n(R/2)$, I_n is a Bessel function of the second kind, and R is a numerical parameter, the Reynolds number. The latter solution exhibits the typical leading-edge sharpening of a profile that begins at $\tau = 0$ as a half sine wave.

B. The Inhomogeneous Burgers Equation

Let us return now to the problem of ion conduction in membranes. First, these solutions cannot be taken over directly to the membrane case, because the boundary conditions are different. In hydrodynamic problems, the dependent variable (pressure or fluid velocity) is usually specified at the boundaries; this is referred to as a Dirichlet condition. Substitution of Eq. (4) into Eq. (6) shows that in the present case it is the first spatial derivative of the dependent variable (the electric field) that is specified; this is a Neumann boundary condition. Second, in general there will be a finite current term J, or $f(\tau)$ in dimensionless notation. The full equation, with $f(\tau)$ included, is known as the inhomogeneous Burgers equation, or as the driven Burgers equation, because $f(\tau)$ represents a forcing function on the system.

$$\frac{\partial u}{\partial \tau} + u\frac{\partial u}{\partial \xi} = \frac{\partial^2 u}{\partial \xi^2} + f(\tau) \tag{15}$$

This equation has been applied to wind-driven water waves (Jeng and Meecham, 1972). The inhomogeneous Burgers equation is likewise exactly soluble by the Hopf–Cole transformation; furthermore, Rodin (1970) has derived an explicit Riccati solution for a forcing function of the form $f(\tau) = a\tau^{-3/2}$. This discussion will focus on the simpler Burgers equation, corresponding to the turn-off phase of a current-clamp pulse.

V. LOW-TEMPERATURE APPROXIMATION

A. A Remarkable Hyperbolic Equation

The simplest case is one in which the contribution of diffusion is negligible; this occurs at low temperature, as mentioned above. In this case, the resting Nernst potential, which arises as a consequence of the balance between diffusion and migration, vanishes. Since transformation of Eq. (10) does not permit the important limiting case where $T = 0$, we return to Eq. (8). For this case, and with $J(t) = 0$ as well,

$$\frac{\partial E}{\partial t} + q\,\mu\,E\frac{\partial E}{\partial x} = 0 \tag{16}$$

This equation is remarkable in that it is a wave equation in which E plays the roles both of the magnitude of the wave and (multiplied by the constant $q\mu$) its velocity. This equation has the simple general solution

$$E = F\,(x - q\,\mu\,E\,t) \tag{17}$$

which represents a wave of arbitrary shape described by the function F, propagating with the velocity $q\mu E$. Equation (17) resembles the forward-propagating solution of the linear wave equation $F(x - ct)$, with one important difference; the wave velocity c is not constant, but is proportional to the dependent variable, the electric field! This means that the wave is dispersive, so that different parts of a waveform travel at different speeds; the larger the field, the greater the speed. Thus, the waveform will tend to sharpen as it travels. Furthermore, this wave will move either forward or backward, depending on whether the field is positive or negative.

B. A Simple Approximate Solution to the Boundary-Value Problem

A particular solution of Eq. (16),

$$E = \frac{1}{q\mu}\frac{x - x_0}{t + \sigma} \tag{18}$$

(where x_0 and σ are constants), illustrates some features of the system. The quantity $q\mu E$ has to be of positive slope, since $q\dfrac{\partial E}{\partial x}$ is, by Eq. (4), positive definite (even for anions). An approximate solution for the electric-field profile in a membrane can be made of three joined line segments. The outer segments, of thickness Δx, satisfy the boundary conditions by having slopes proportional to N_{I} and N_{II}, respectively, and the middle segment is given by Eq. (18). Thus, the field is given by

$$E(x) = \begin{cases} E(0) + (qN_{\mathrm{I}}/\kappa\varepsilon_0)x, & 0 \leqslant x \leqslant \Delta x \\ (q\mu)^{-1}(x - x_0)/(t + \sigma), & \Delta x \leqslant x \leqslant L - \Delta x \\ E(L) - (qN_{\mathrm{II}}/\kappa\varepsilon_0)(L - x), & L - \Delta x \leqslant x \leqslant L \end{cases} \tag{19}$$

When $E(0)$ and $E(L)$ are evaluated by matching at Δx and $L - \Delta x$, the membrane potential difference is found, by integration, to equal

$$V_{\mathrm{m}} = \frac{q}{2\kappa\varepsilon_0}(N_{\mathrm{II}} - N_{\mathrm{I}})\,\Delta x^2 + \frac{L}{2q\mu}\frac{L - 2x_0}{t + \sigma} \tag{20}$$

This relation may be written

$$V'_{\mathrm{M}} = \frac{1}{2}(N'_{\mathrm{II}} - N'_{\mathrm{I}})(\Delta x')^2 + \frac{L'}{2}\frac{L' - 2x'_0}{t' + \sigma'} \tag{21}$$

where $V_{\mathrm{M}}/V'_{\mathrm{M}} = V_0 = 1$ mV, $N'/N = \kappa\varepsilon_0 V_0 L'^2/qL^2$, $t/t' = \sigma/\sigma' = L^2/q\mu V_0 L'^2 = 1$ msec; also $E' = q\mu E$.

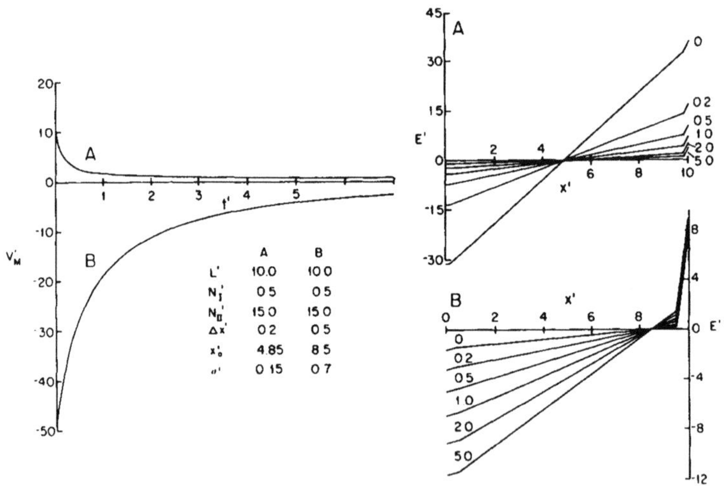

Fig. 5. Voltage–time responses (left) and field profiles (right) for an approximate solution to the membrane electrodiffusion problem. The diffusion term has been neglected by assuming zero temperature. The long central line segments in the profiles correspond to Eq. (18), while the outer segments, of width Δx, have constant slopes proportional to the corresponding boundary concentrations. Field profiles are shown at current turnoff ($t' = 0$) and subsequent (dimensionless) times. Two conditions, A and B, are considered, with parameters given in the inset. The potential difference decays from a positive value in A, with $x_0' = 4.85$, and from a negative value in B, with $x_0' = 8.5$, illustrating the threshold role of x_0' relative to the midpoint of the membrane, $x' = 5$. The solutions are exact only in the limit as $\Delta x' \to 0$.

A set of these approximate solutions to the membrane boundary-value problem is shown in Fig. 5. The corresponding ion-concentration profiles are, from Eqs. (4) and (19), double-step functions; the outer ion concentration, at $x = L$, is taken much greater than that at $x = 0$, corresponding to the sodium case. Because the outer segments do not satisfy Eq. (16), this solution is exact only in the limit $\Delta x \to 0$.

Transient voltage responses for two sets of parameters, A and B, are given in Fig. 5 (left). The parameters were chosen to roughly simulate relaxations from positive and negative current pulses (see Cole, 1968, p. 250). Closer examination shows that the data contain second-order features that the mathematical curves lack, but that is to be expected from the simplicity of the model and the restrictive assumptions that were made.

More realistic solutions are obtained when the diffusion term is "switched on" again. As the temperature is brought up from zero, the diffusion term begins to round off the sharp corners of the field profile, and the step functions in the concentration profile are converted into hyperbolic-tangent-like regions (Burgers, 1974). The thickness of the rounding layer grows as the temperature, and hence the diffusivity, increases. A more interesting case to model would be the part of the curve during which the current pulse is on. This will require an appropriate solution of the inhomogeneous Burgers equation with a constant forcing term.

VI. TOWARD AN ELECTRODIFFUSION THEORY OF EXCITABLE CELLS

It is significant that the electrodiffusion system of equations, which has been associated with the study of excitability since the early years of this century, is now seen to predict the existence of nonlinear waves analogous to shock waves and wind-driven waves. However, it is not sufficient merely to develop this analogy; it is necessary to develop solutions that can yield quantitative predictions. The analyses carried out so far require further elaboration and generalization. In addition to a thorough analysis of the homogeneous and inhomogeneous Burgers equations and associated two-point boundary-value problems, it will be necessary to generalize the hypotheses to cover multiple ions, access structures, and a field-dependent dielectric constant. Although this will be a substantial task, the above calculations show that even simple approximations can yield significant insights.

What is proposed here is that the functional mechanism of excitability may not be exactly separable into distinct submechanisms responsible for voltage sensing, gating, and selectivity in the way that the channel hypothesis assumes, but that the ion distributions within the membrane respond in a nonlinear fashion to the nonuniform electric field and the oppositely directed K and Na concentration gradients, so that under certain conditions a transverse nonlinear wave is generated which carries ions transiently across the membrane. In so doing, it increases the ion density in the membrane, thus bringing about the decrease in the membrane impedance found by Cole and Curtis (1939). The facts about selectivity may then follow from the differences in the mobilities, the ion concentrations present at the boundaries, and other physical parameters. The explanations for the action of specific drugs could be sought in a realm of phenomena wider than the current concept of mechanical "blocking." The investigation of stochastic pulses could be linked with studies on similar phenomena in nonbiological systems, such as burst noise and ferroelectric Barkhausen pulses.

The dielectric permittivity describes the composite electric properties of the components of the membrane, phospholipid and protein. Although for many substances this quantity is independent of the electric field over a wide range of field strengths, this is a special assumption that cannot be taken for granted in membranes of excitable cells. Some components of the membrane, particularly the proteins, may contribute a field dependence to the membrane dielectric permittivity, possibly making it ferroelectric. Possibilities of this nature have already been proposed by Wei (1969) and Matthias (1973); see also the discussion of domains in Tasaki, (1982). The assumption of the constancy of the dielectric permittivity, which is made here as a convenient approximation, must therefore be viewed as a restriction to be removed in future analyses.

The present results suggest that a more extensive examination of the nonlinear equations arising in electrodiffusion theory may help resolve some of the difficulties that have plagued prior applications of electrodiffusion to membrane data (Cole, 1965; Hille, 1978). Other suggested approaches include consideration of the effects of boundary ion accumulation and depletion (Attwell, 1979) and the modification of Eq. (2) to account for the limited density of available sites (Agin, 1969). It is difficult to compare the electrodiffusion approach with the more conventional "channel" approaches, which relate more directly to the obtainable data. Voltage clamping is the preferred measurement technique because it allows (at least within the lumped-parameter model) separation of ionic from capacitive current. Yet for a continuum approach the voltage clamp represents a difficult integral condition on the membrane and its boundary layers. Again, while the maintenance of a viable membrane requires minimal amounts of divalent cations, the inclusion of these in the nonlinear analysis greatly complicates the problems. For the steady-state case, it has been shown that each additional valency present raises the order of the governing differential equation by one (Leuchtag, 1981).

While the discussion has been limited to a one-dimensional case, it is significant that equations describing the space and time dependence of the action potential (FitzHugh, 1967; Nagumo *et al.*, 1962; McKean, 1970) are of the same general type (reaction-diffusion) as Eqs. (8), (11), and (15), suggesting the possibility of a more general oblique wave in the two-dimensional case. Although the present electrodiffusion model is a simple idealization, its analysis indicates that electrodiffusion membrane systems exhibit time-dependent behavior qualitatively similar in some significant respects to the observed transient responses of excitable cells.

ACKNOWLEDGMENTS. This work was supported in part by NIH grant NS 13778. H.R.L. gratefully acknowledges helpful discussions of this work with Dr. James C. Swihart. We thank Dr. L. E. Moore for his useful comments and Ms. Sarah Adams for her careful typing of the manuscript.

REFERENCES

Adelman, W. J., Jr., and FitzHugh, R., 1975, Solutions of the Hodgkin–Huxley equations modified for potassium accumulation in a periaxonal space, *Fed. Proc.* **34**:1322–1329.

Adelman, W. J., Jr., and French, R. J., 1978, Blocking of the squid axon potassium channel by external caesium ions, *J. Physiol.* **276**:13–25.

Agin, D., 1969, An approach to the physical basis of negative conductance in squid axon, *Biophys. J.* **9**:209–221.

Arndt, R. A., and Roper, L. D., 1972, *Simple Membrane Electrodiffusion Theory*, Physical Biological Sciences Misc., Blacksburg, Virginia.

Attwell, D., 1979, Problems in the interpretation of membrane current-voltage relations, in: *Membrane Transport Processes*, Vol. 3 (C. F. Stevens and R. W. Tsien, eds.), pp. 29–41, Raven Press, New York.

Baker, P. F., Hodgkin, A. L., and Shaw, T. I., 1962, The effects of changes in internal ionic concentrations on the electrical properties of perfused giant axons, *J. Physiol.* **164**:355–374.

Barnola, F. V., and Villegas, R., 1976, Sodium flux through the sodium channels of axon membrane fragments isolated from lobster nerve, *J. Gen. Physiol.* **67**:81–90.

Behn, U., 1897, Ueber wechselseitige Diffusion von Electrolyten in verdünnten wässerigen Lösungen, insbesondere über Diffusion gegen das Concentrationsgefälle, *Ann. Physik u. Chemie N.F.* **62**:54–67.

Benton, E. R., and Platzman, G. W., 1972, A table of solutions of one-dimensional Burgers equation, *Appl. Math.* **30**:195–212.

Brumleve, T. R., and Buck, R. P., 1978, Numerical solution of the Nernst–Planck and Poisson equation system with applications to membrane electrochemistry and solid state physics, *J. Electroanal. Chem.* **90**:1–31.

Brumleve, T. R., and Buck, R. P., 1981a, Potential reversals across site-free passive membranes: A simulation analysis, *J. Electroanal. Chem.* **126**:55–71.

Brumleve, T. R., and Buck, R. P., 1981b, Transmission line equivalent circuit models for electrochemical impedances, *J. Electroanal. Chem.* **126**:73–104.

Burgers, J. M., 1950, Correlation problems in a one-dimensional model of turbulence. I., *Proc. Kon. Ned. Akad. Wet.* **53**:247–261.

Burgers, J. M., 1974, *The Nonlinear Diffusion Equation. Asymptotic Solutions and Statistical Problems*, Chapter I, D. Reidel, Dordrecht-Holland, Boston.

Card, W. H., and Chaudhari, P. K., 1965, Characteristics of burst noise, *Proc. IEEE* **53**:652–653.

Chandler, W. K., and Meves, H., 1965, Voltage clamp experiments on internally perfused giant axons, *J. Physiol.* **180**:788–820.

Chynoweth, A. G., 1958, Barkhausen pulses in barium titanate, *Phys. Rev.* **110**:1316–1332.

Cohen, H., and Cooley, J. W., 1965, The numerical solution of the time-dependent Nernst–Planck equations, *Biophys. J.* **5**:145–162.

Cohen, L. B., and Landowne, D., 1974, The temperature dependence of the movement of sodium ions associated with nerve impulses. *J. Physiol.* **236**:95–111.

Cole, J. D., 1951, On a quasi-linear parabolic equation occurring in aerodynamics, *Quart. Appl. Math.* **9**:225–236.

Cole, K. S., 1947, *Four Lectures on Biophysics*, pp. 45–57, Institute of Biophysics, University of Brazil, Rio de Janeiro.

Cole, K. S., 1965, Electrodiffusion models for the membrane of squid giant axon, *Physiol. Rev.* **45**:340–379.

Cole, K. S., 1968, *Membranes, Ions and Impulses*, University of California Press, Berkeley, California.

Cole, K. S., and Curtis, H. J., 1939, Electric impedance of the squid giant axon during activity, *J. Gen. Physiol.* **22**:649–670.

Cole, K. S., and Moore, J. W., 1960, Potassium ion current in the squid giant axon: Dynamic characteristic, *Biophys. J.* **1**:1–14.

Conti, F., and Neher, E., 1980, Single channel recordings of K^+-currents in squid axons, *Nature* **285**:140–143.

Conti, F., DeFelice, L. J., and Wanke, E., 1975, Potassium and sodium ion current noise in the membrane of the squid giant axon, *J. Physiol.* **248**:45–82.

deLevie, R., 1978, Mathematical modeling of transport of lipid-soluble ions and ion-carrier complexes through lipid bilayer membranes, in: *Advances in Chemical Physics* (I. Prigogine and S. A. Rice, eds.), pp. 99–137, Wiley, New York.

deLevie, R., Seidah, N. G., and Moreira, H., 1974, Transport of ions of one kind through thin membranes. IV. Admittance for membrane-soluble ions, *J. Membr. Biol.* **16**:17–42.

Dodge, F. A., and Frankenhaeuser, B., 1959, Sodium currents in the myelinated nerve fibre of *Xenopus laevis* investigated with the voltage clamp technique, *J. Physiol.* **148**:188–200.

Easton, D. M., 1978, Exponentiated exponential model (Gompertz kinetics) of Na^+ and K^+ conductance changes in squid giant axon, *Biophys. J.* **22**:15–28.

Fay, R. D., 1931, Plane sound waves of finite amplitude, *J. Acoust. Soc. Am.* **3**:222–241.

Finkelstein, A., and Mauro, A., 1977, Physical principles and formalisms of electrical excitability, in: *Handbook of Physiology*, Section 1, Vol. I (E. R. Kandel, ed.), pp. 161–213, Am. Physiol. Soc., Bethesda.

Fishman, H. M., 1969, Direct recording of K and Na current-potential characteristics of squid axon membrane, *Nature* **224**:1116–1118.

Fishman, H. M., 1970, Direct and rapid description of the individual ionic currents of squid axon membrane by ramp potential control, *Biophys. J.* **10**:799–817.

Fishman, H. M., 1973, Relaxation spectra of potassium channel noise from squid axon membrane, *Proc. Natl. Acad. Sci. USA* **70**:876–879.

Fishman, H. M., 1981, Material from the internal surface of squid axon exhibits excess noise, *Biophys. J.* **35**:249–255.

FitzHugh, R., 1967, Mathematical models of excitation and propagation in nerve, *National Institutes of Health Technical Report*.

Frankenhaeuser, B., and Hodgkin, A. L., 1955, The after-effects of impulses in the giant nerve fibres of *Loligo*, *J. Physiol.* **131**:341–376.

Goldman, D. E., 1943, Potential impedance and rectification in membranes, *J. Gen. Physiol.* **27**:37–60.

Grisell, R. D., and Fishman, H. M., 1979, K^+ conduction phenomena applicable to the low frequency impedance of squid axon, *J. Membr. Biol.* **46**:1–25.

Hagglund, J. V., 1972, Convection-diffusion as a model of the early current in the giant axon, *Upsala J. Med.* **77**:77–90.

Hagglund, J. V., and Sandblom, J., 1972, The kinetic behavior of the potassium channel in nerve membrane: A single-ion electrodiffusion process, *T.-I.-T. J. Life Sci.* **2**:107–119.

Hagiwara, S., and Byerly, L., 1981, Calcium channel, *Annu. Rev. Neurosci.* **4**:69–125.

Hays, T. R., Buckwalter, C. Q., Lin, S. H., and Eyring, H., 1978, Ion flow through a membrane: Concentration and current responses to a step potential change, *Proc. Natl. Acad. Sci. USA* **75**:1612–1615.

Hille, B., 1970, Ionic channels in nerve membranes, in: *Progress in Biophysics and Molecular Biology*, Vol. 21 (J. A. V. Butler and D. Noble, eds.) pp. 1–32, Pergamon, New York.

Hille, B., 1975, The receptor for tetrodotoxin: A structural hypothesis, *Biophys. J.* **15**:615–619.

Hille, B., 1978, Ionic channels in excitable membranes. Current problems and biophysical approaches, *Biophys. J.* **22**:283–294.

Hodgkin, A. L., and Huxley, A. F., 1952a, Currents carried by sodium and potassium ions through the membrane of the giant axon of *Loligo*, *J. Physiol.* **116**:449–472.

Hodgkin, A. L., and Huxley, A. F., 1952b, The components of membrane conduction in the giant axon of *Loligo*, *J. Physiol.* **116**:473–496.

Hodgkin, A. L., and Huxley, A. F., 1952c, The dual effect of membrane potential on sodium conductance in the giant axon of *Loligo*, *J. Physiol.* **116**:497–506.

Hodgkin, A. L., and Huxley, A. F., 1952d, A quantitative description of membrane current and its application to conduction and excitation in nerve, *J. Physiol.* **116**:500–544.

Hodgkin, A. L., and Katz, B., 1949, The effect of sodium ions on the electrical activity of the giant axon of the squid, *J. Physiol.* **108**:37–77.

Hodgkin, A. L., and Keynes, R. D., 1955, The potassium permeability of a giant nerve fiber, *J. Physiol.* **128**:61–88.

Hodgkin, A. L., Huxley, A. F., and Katz, B., 1952, Measurement of current–voltage relations in the membrane of the giant axon of *Loligo*, *J. Physiol.* **116**:424–448.

Hopf, E., 1950, The partial differential equation $u_t + uu_x = \mu u_{xx}$, *Comm. Pure Appl. Math.* **3**:201–230.

Hoyt, R. C., 1968, Sodium inactivation in nerve fibers, *Biophys. J.* **8**:1074–1097.

Hsu, S. T., Whittier, R. J., and Mead, C. A., 1970, Physical model for burst noise in semiconductor devices, *Solid-State Electron* **13**:1055–1071.

Jeng, D. T., and Meecham, W. C., 1972, Solution of forced Burgers equation, *Phys. Fluids* **15**:504–506.

Junge, D., 1981, *Nerve and Muscle Excitation*, 2nd Ed., pp. 109–111, Sinauer Associates, Sunderland, Massachusetts.

Kao, C. Y., and Fuhrman, F. A., 1967, Differentiation of the actions of tetrodotoxin and saxitoxin, *Toxicon* **5**:25–34.

Kao, C. Y., and Walker, S. E., 1982, Active groups of saxitoxin and tetrodotoxin as deduced from actions of saxitoxin analogues on frog muscle and squid axon, *J. Physiol.* **323**:619–637.

Kao, C. Y., Yeoh, P. N., Goldfinger, M. D., Fuhrman, F. A., and Mosher, H. S., 1981, Chiriquitoxin, a new tool for mapping ionic channels, *J. Pharmacol. Exp. Ther.* **217**:416–429.

Karpman, V. I., 1975, *Non-Linear Waves in Dispersive Media*, Chapter 4, Pergamon, New York.

Kazantsev, E. F., 1978, Electro-diffusion model of the nerve impulse, *Biophysics* **23**:303–309.

Kidokoro, Y., Grinnell, A. D., and Eaton, D. C., 1974, Tetrodotoxin sensitivity of muscle action potentials in pufferfishes and related fishes, *J. Comp. Physiol.* **89**:59–72.

Kochina, N. N., 1961, On periodic solutions of Burgers' equation, *J. Appl. Math. Mech.* **25**:1597–1607.

Landau, L. D., and Lifshitz, E. M., 1960, *Mechanics*, pp. 13–14, Pergamon, Oxford, and Addison-Wesley, Reading Massachusetts.

Landowne, D., 1973, Movement of sodium ions associated with the nerve impulse, *Nature* **242**:457–459.

Leibovich, S., and Seebass, A. R., 1974, Examples of dissipative and dispersive systems leading to the Burgers and the Korteweg–deVries equations, in: *Nonlinear Waves* (S. Leibovich and A. R. Seebass, eds.), pp. 103–138, Cornell University Press, Ithaca.

Leuchtag, H. R., 1974a, On the theory of electrodiffusion and its application to the electrical properties of squid axon membrane, Ph.D. Dissertation, Indiana University, Bloomington.

Leuchtag, H. R., 1974b, On the theory of electrodiffusion and its application to the electrical properties of squid axon membrane, *Diss. Abst. Int.* **35**:1520-B.

Leuchtag, H. R., 1981, A family of differential equations arising from multi-ion electrodiffusion, *J. Math. Phys.* **22**:1317–1320.

Leuchtag, H. R., and Swihart, J. C., 1977, Steady-state electrodiffusion scaling, exact solution for ions of one charge, and the phase plane, *Biophys. J.* **17**:27–46.

Mackey, M. C., 1975a, *Ion Transport Through Biological Membranes*, Springer-Verlag, New York.

Mackey, M. C., 1975b, Admittance properties of electrodiffusion membrane models, *Math. Biosci.* **25**:67–80.

Matteson, D. R., and Swenson, R. P., 1982, Permeant cations alter closing rates of K channels, *Biophys. J.* **37**:17a.

Matthias, B. T., 1973, Organic ferroelectricity, in: *Proc. 3rd Int. Conf. From Theoretical Physics to Biology* (M. Marois, ed.), pp. 12–20, Karger, Basel.

McGillivray, A. D., 1970, Asymptotic solutions of the time-dependent Nernst–Planck equations, *J. Chem. Phys.* **52**:3126–3132.

McKean, H. P., 1970, Nagumo's equation, *Adv. Math.* **4**:209–223.

Metuzals, J., and Tasaki, I., 1978, Subaxolemmal filamentous network in the giant nerve fiber of the squid (*Loligo pealei* L.) and its possible role in excitability, *J. Cell Biol.* **78**:597–621.

Meves, H., and Chandler, W. K., 1965, Ionic selectivity in perfused giant axons, *J. Gen. Physiol.* **48**:31–33.

Meyer, J. P., and Kostin, M. D., 1974, Time-dependent Nernst–Planck and Poisson equations: Initial phase of action potential, *J. Chem. Phys.* **61**:4067–4069.

Mikulinsky, M. A., and Mikulinsky-Fishman, S. N., 1976, The theory of 1/f noise, *Phys. Lett.* **58A**:46–48.

Moore, J. W., Anderson, N., Blaustein, M., Takata, M., Lettvin, J. Y., Pickard, W. F., Bernstein, T., and Pooler, J., 1966, Alkali cation selectivity of squid axon membrane, *Ann. N.Y. Acad. Sci.* **137**:818–829.

Moore, L. E., and Jakobsson, E., 1971, Interpretation of the sodium permeability changes of myelinated nerve in terms of linear relaxation theory, *J. Theoret. Biol.* **33**:77–89.

Mott, N. F., 1939, The theory of crystal rectifiers, *Proc. R. Soc. A* **171**:27–38.

Nagumo, J., Arimoto, S., and Yoshizawa, S., 1962, An active pulse transmission line simulating nerve axon, *Proc. IRE* **50**:2061–2070.

Oxford, G. S., and Adams, D. J., 1981, Permeant cations alter K channel kinetics and permeability, *Biophys. J.* **33**:70a.

Palti, Y., and Adelman, W. J., Jr., 1969, Measurement of axonal membrane conductances and capacity by means of a varying potential control voltage clamp, *J. Membr. Biol.* **1**:431–458.

Rodin, E. Y., 1970, A Riccati solution for Burgers' equation, *Quart. Appl. Math.* **27**:541–545.

Sandblom, J., 1972, Anomalous reactances in electrodiffusion systems, *Biophys. J.* **12**:1118–1131.

Schwartz, T. L., 1971, The thermodynamic foundations of membrane physiology, in: *Biophysics and Physiology of Excitable Membranes* (W. J. Adelman, Jr., ed.), pp. 47–95, Van Nostrand Reinhold, New York.

Siebenga, E., De Goede, J., and Verveen, A. A., 1974, The influence of TTX, DNP and TEA on membrane flicker noise and shot effect noise of the frog node of Ranvier, *Pflügers Arch.* **351**:25–34.

Sigworth, F. J., and Neher, E., 1980, Single Na^+ channel currents observed in cultured rat muscle cells, *Nature* **287**:447–449.

Sigworth, F., and Spalding, B., 1979, Toxin-resistant sodium channels have low conductance, *Biophys. J.* **25**:194a.

Starzak, M. E., 1973a, A model for conductance changes in the squid giant axon. I. Interactive relaxation, *J. Theoret. Biol.* **39**:487–504.

Starzak, M. E., 1973b, A model for conductance changes in the squid giant axon. II. Channel kinetics, *J. Theoret. Biol.* **39**:505–522.

Sten-Knudsen, O., 1978, Passive transport processes, in: *Membrane Transport in Biology*, Vol. 1 (G. Giebisch and D. C. Tosteson, eds.), pp. 5–113, Springer-Verlag, Berlin.

Swenson, R. P., Jr., and Armstrong, C. M., 1981, K^+ channels close more slowly in the presence of external K^+ and Rb^+, *Nature* **291**:427–429.

Tasaki, I., 1982, *Physiology and Electrochemistry of Nerve Fibers*, Academic, New York.

Tasaki, I., Watanabe, A., and Takenaka, T., 1962, Resting and action potential of intracellularly perfused squid giant axon, *Proc. Natl. Acad. Sci. USA* **48**:1177–1184.

Taylor, R. E., Bezanilla, F., and Rojas, E., 1980, Diffusion models for the squid axon Schwann cell layer, *Biophys. J.* **29**:95–117.

Teorell, T., 1949, Membrane electrophoresis in relation to bio-electrical polarization effects, *Arch. Sci. Physiol.* **3**:205–219.

Ussing, H. H., 1949, The distinction by means of tracers between active transport and diffusion, *Acta Chem. Scand.* **2**:43–56.

Vaidhyanathan, V. S., 1976, Philosophy and phenomenology of ion transport and chemical reactions in membrane systems, in: *Topics in Bioelectrochemistry and Bioenergetics*, Vol. I, (G. Milazzo, ed.), Wiley, New York.

Villegas, G. M., and Villegas, R., 1960, The ultrastructure of the giant nerve fibre of the squid: Axon-Schwann cell relationship, *J. Ultrastruct. Res.* **3**:362–373.
Villegas, R., and Villegas, G. M., 1981, Nerve sodium channel incorporation in vesicles, *Annu. Rev. Biophys. Bioeng.* **10**:387–419.
Wei, L. Y., 1969, Molecular mechanisms of nerve excitation and conduction, *Bull. Math. Biophys.* **31**:39–58.
Witham, G. B., 1974, *Linear and Nonlinear Waves,* Chapter 4, Wiley, New York.

22

The Effects of Surface Compartments on Ion Transport across Membranes

Martin Blank

I. THE ROLE OF SURFACE PROCESSES IN MEMBRANE SYSTEMS

Physiologists have generally sought explanations of the phenomena they study in terms of the chemistry and physics of living systems. This has been especially evident in the area of excitable membranes where, starting in the late eighteenth century with the controversy between Galvani and Volta about animal electricity, physical chemistry has played a central role in the elucidation of the mechanism. The Bernstein (1902) hypothesis at the beginning of this century used the emerging ideas about diffusion potentials and selective permeability of membranes to provide us with the first explanation of excitation, and this has been the basis of our current understanding. The ideas of electrical double layer theory, formulated about a decade later,* suggested that the surfaces of membranes had special properties (such as surface potentials, ion concentrations, and electrical capacitance) that could influence ion fluxes. However, it was not until 1935 that Teorell first introduced the idea that the membrane potential included the effects of two phase boundary potentials in addition to a diffusion potential. When Goldman (1943) derived a steady-state equation relating the transmembrane potential to the ionic mobilities and the concentrations in the two solutions, it was found to be useful in steady-state applications without the surface potentials. If one includes the surface potentials (Goulden, 1976) there are additional exponential terms, but the equation is essentially similar to the Goldman equation. The difference between surface and bulk concentrations is balanced by the difference between surface and bulk electrical potentials, so the neglect of surface properties is not critical in steady states.

The role of surface processes in dynamic systems is much more complex. In 1957, Teorell described the "membrane oscillator" which was able to develop oscillating membrane

* A good summary of the electrochemistry of the double layer has been given by Overbeek (1952).

Martin Blank • Department of Physiology, College of Physicians and Surgeons, Columbia University, New York, New York 10032.

potentials by the interplay of fluxes resulting from concentration, electrical, and pressure differences, and in 1971, Goldman outlined the types of equations required for a theory of excitation that is compatible with physiological properties as well as physical chemistry. These papers (Teorell, 1957; Goldman, 1971) described essential elements of a dynamic theory, including the effects of surface potentials but neglecting ionic processes in the surface region. However, if ions are bound at surfaces and ion binding, release and exchange reactions are possible, the changes in ionic concentrations in this region may be significant. Furthermore, in transient states the rate-constants for changes of surface concentrations and surface potentials may be quite different and this could lead to the oscillatory effects described by Teorell (1957).

To study these possibilities and to see if the effects of surface processes could account for the unusual ionic fluxes during excitation, we have described the ionic events in membranes during transients, including the electrical double-layer region. The initial results, reported at the Sixth International Symposium on Bioelectrochemistry, June 1981 (Blank and Kavanaugh, 1982; Blank et al., 1982), suggest that nonspecific increases in cation permeability can cause significant changes in the surface compartments and lead to an apparently selective inward sodium flux such as occurs in excitable membranes. Furthermore, our studies suggest ranges of certain physical properties that would contribute to this unusual ion flux.

II. SURFACE PROCESSES

The surface of a phase differs in many ways from the bulk region. (For example, at the surfaces of solutions there are significant concentration changes at the expense of the surface-free energy, along with molecular orientation effects.) The special properties of surfaces often lead to unusual effects, especially in dynamic systems. In this section, we shall review some surface processes that have been studied in model systems and emphasize those aspects that are relevant to the problems of membrane transport.

A. The Structural Basis of Membrane Permeability

Several aspects of membrane properties can be understood in terms of the physical chemistry of systems of the same size and composition as the natural membrane (Blank, 1980a). The membrane that surrounds the nerve cell is composed of two molecular layers of lipids which make up the hydrophobic core of the membrane, as well as polymers (for example, proteins and glycoproteins) on the two surfaces and traversing the membrane. This membrane is essentially the permeability barrier that excludes ions and prevents the free intermixing of intra- and extracellular ions. The hydrocarbon interior is an important part of the permeability barrier, as we have shown from transport studies with lipid monolayers (Blank, 1979).

Monolayers show porous behavior similar to that associated with membrane structures. Generally, smaller molecules move through more rapidly than larger ones, and there is interference between fluxes in opposing directions. (see Fig. 1). We can explain this behavior on the basis of a theoretical model where the monolayer molecules are in constant motion, and the motion leads to transient holes which allow molecules to pass through. Obviously, small molecules pass more frequently than larger ones. The equation derived for monolayer permeability also explains the way molecules diffuse within an oriented lipid layer. (Diffusion within a layer is related to permeation through the layer because a hole not only allows a

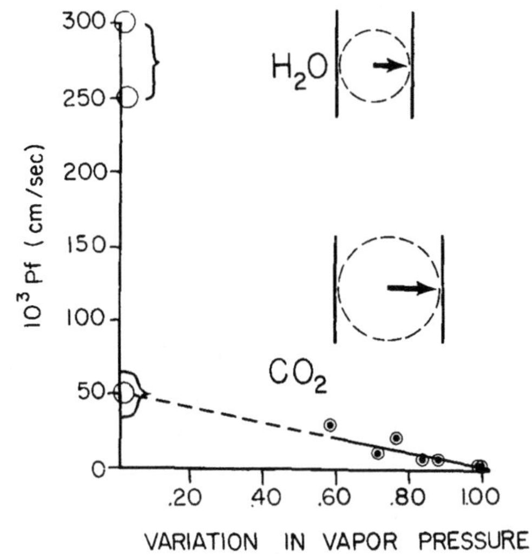

Figure 1. The permeabilities *Pf* of an octadecanol monolayer to H_2O and to CO_2 are indicated on the ordinate, and the approximate relative cross-sectional areas of the two molecules are also shown. The permeability of an octadecanol monolayer to CO_2 varies with the water vapor pressure of the aqueous phase upon which it is spread, indicating interference between CO_2 and H_2O (the two gases that are crossing the monolayer), and these data are plotted on the lower part of the figure. The permeabilities are determined by the sizes of the gases and not by their solubilities. For more details see Blank (1979). (Reproduced with permission from the *Journal of General Physiology*).

small molecule to pass through, but an adjacent monolayer molecule can also move over and occupy that hole.) The predicted value for the diffusion coefficient of lipid molecules in membranes, which has been confirmed, is on the order of 10^{-8} cm^2/sec (Blank and Britten, 1965). It appears that the model to explain movement through a monolayer is reasonable for small uncharged solutes, and that it also accounts for molecular diffusion in membrane structures.

The factors that influence the rate at which a charged permeant moves through a membrane structure have been studied in an experimental system using charged monolayers adsorbed at an electrode surface. In a polarograph, the current across the interface between mercury and an aqueous solution is measured as one varies the potential, and there is a classic Ohm's law relation between current and polarization unless something in the solution reacts at the electrode. When a cation is reduced at a particular potential there is a large increase in the current, known as the diffusion current, that is proportional to the concentration of the cation. The potential at which this reaction occurs, the half-wave potential, is characteristic of the electrode reaction.

In our experiments, we used materials similar to the monolayer materials in the earlier studies and measured the transport of ions through these layers at the electrode interface. Using divalent copper ions as permeants and decylammonium ions as the monolayer-forming material, we were able to determine the parameters of the ion transport process (Blank and Miller, 1968; Miller and Blank, 1968). With no monolayer there is a normal diffusion current, but as the monolayer at the electrode surface becomes more and more concentrated, there is a decrease in the current. Since the current is due to the reduction of cupric ions when they reach the mercury surface, it appears that the charged monolayer (cationic) surface repels cations, and there are fewer cations to move through the monolayer. The essential difference between the permeability of uncharged and charged monolayers is that the rate of movement of ions through a charged layer is also a function of the surface charge. There is actually a linear relation between the logarithm of the rate-constant k and the surface charge σ so that an order of magnitude change in the rate is caused by a change of one charge per hundred square angstroms (1 nm^2). As expected from electrical double-layer theory, the rate is also a function of the ionic strength (see Fig. 2).

The explanation for the observed variation of permeability with surface charge and

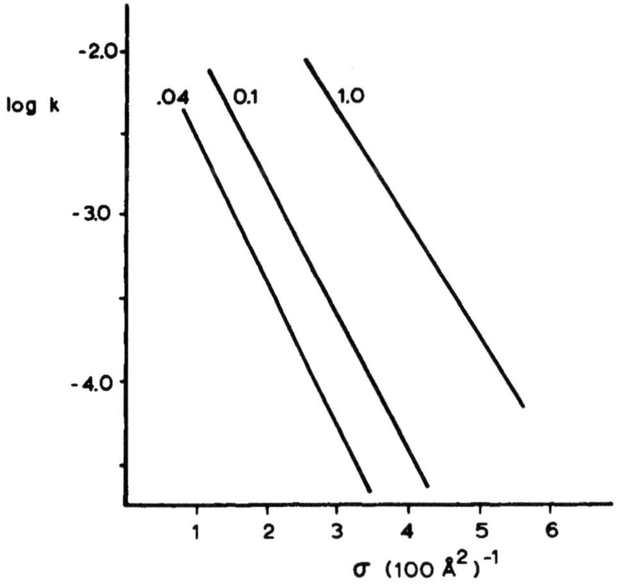

Figure 2. The logarithm of the ion transport rate-constant k (in cm/s) of Cu^{2+} ion diffusing through a decylammonium monolayer at a mercury–water interface, plotted against σ, the surface-charge density (in charges/100 Å²). The numbers on the lines indicate the ionic strength of the aqueous phase containing 10^{-3} M Cu^{2+} ions. (Reproduced with permission from *Bioelectrochemistry and Bioenergetics*).

ionic strength has been given in terms of electrical double-layer theory (Britten and Blank, 1977). The concentration of ions at an interface depends upon the electrical potential at that surface relative to the electrical potential of the solution, and the flux of an ion across an interface due to an electrochemical potential difference is proportional to the surface concentration of ions. If we assume that the transport of ions does not disturb the electrical double layer, i.e., the reestablishment of the electrical double-layer distribution is much faster than the ion-transport process, we predict that the logarithm of the permeability varies linearly with the surface-charge density, as demonstrated (see Fig. 2) in a monolayer study (Miller and Blank, 1968), and also in measurements on bilayers (Sweeney and Blank, 1973). Therefore, the surface potential or charge density should control the surface concentration of ions in natural membranes, and we can use the estimated values of the surface charge for excitable membranes to calculate the expected ion concentrations.

Without going into detail, some of the ideas of electrical double-layer theory that we have found useful for understanding membrane properties are as follows:

1. At charged surfaces, the concentrations of adsorbed counter-ions is higher, and that of co-ions, lower than in solution. The surface concentration is related to the bulk concentration by an exponential factor that is a function of the electrical potential (or the charge) at the surface.

2. Since the surface can store charge, i.e., counter-ions adsorb when the surface is charged, it has a capacitance that is generally very large.

3. The thickness of the electrical double-layer region, i.e., the distance where the potential falls to $1/e$ of the value at the charged surface, in physiological saline is less than 10 Å thick.

B. Ion-Concentration Changes Due to Current Flow

An action potential in an *in vitro* experiment is initiated by passing a cathodal current, i.e., when the cathode is on the outside, across the membrane. To understand the effect of passing a current due to an electrical potential gradient, let us first consider interfacial transfer of charged and uncharged substances under a concentration gradient. When a substance is transported between two liquid phases, the concentrations on the two sides of the interface are in partition equilibrium. This factor applies to charged as well as uncharged substances, but in the case of ions there is an additional complication that can lead to significant changes in the interfacial concentration.

When an electrical field is applied across an ionic solution, the cations move toward the cathode, the anions toward the anode, and there are concentration changes in the vicinity of the electrodes depending upon the ion transport numbers. When current flows across a liquid–liquid system, there are concentration changes at the interface as well as at the electrodes, since ions generally have different transport numbers in the two liquids. The increase or decrease in the interfacial concentration depends upon the difference in transport numbers between the two phases, the rate of flow of ions through the interface, the direction of the current and the rate of diffusion of the ions to or from the interface. Blank and Feig (1963) have explored some of these effects in a liquid–liquid junction of water and nitrobenzene. Using platinum electrodes, current was passed between these two phases, each of which contained hexadecylammonium ions. These ions are: (1) charged under the conditions of the experiments, (2) distributed between both phases with a partition coefficient close to one, and (3) surface-active. (There is a linear relationship between the surface tension and the surface concentrations of ions, so when we measure the surface tension we are in effect measuring the number of ions at the interface.) When we pass current in one direction across this interface, we find that there is a slow increase in the number of ions that is proportional to the magnitude of the current (see Fig. 3). When we pass current in the reverse direction, we deplete the interface of ions as shown by the change in surface tension in the opposite direction. The change of ion concentration as a result of differences in ion-transport numbers was first observed in the laboratory of Nernst and reported in terms of changes in bulk composition (Nernst and Riesenfeld, 1902). In our study, we measured the concentration change right at the surface and were able to determine its time course (Blank, 1967).

Applying the results of this experiment to the problem of excitation, let us consider the case of a cation-selective membrane, where almost all the current is carried by cations in the membrane region, while in the aqueous phase the current is carried by both cations and anions, that is, only about half is due to cations. In the case of a cathodal current, cations move from a region where they carry the full current to a region where they carry only half the current. As a result, there is an excess of cation in the aqueous phase adjacent to the outer face of the membrane. By the same token there is a depletion of cation at the

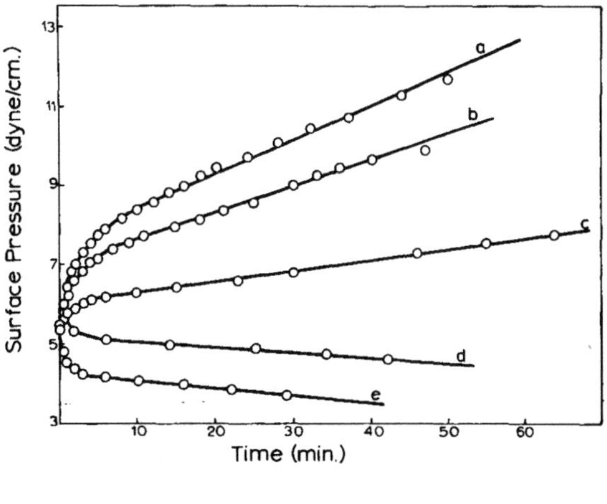

Figure 3. The variation in surface tension with time at a water–nitrobenzene interface when different currents are passed through the interface. The values of the current in microamperes are: (a) 150, (b) 106, (c) 52, (d) −93, and (e) −208, and the long-chain cationic surfactant, CTAB, is initially 0.96×10^{-4} M in water and 1.55×10^{-4} M in nitrobenzene, the equilibrium partition at 25°C. (Reproduced with permission from *Science*).

inner face of the membrane. In both cases there is a corresponding change in the concentration of anion. Therefore, in the natural membrane, we expect cathodal (outward) currents to give rise to a decrease in the ionic concentration at the inner and an increase at the outer surface of the membrane. Furthermore, there is also a change of ionic species. The outside has sodium as the major cation, but the current brings potassium ions into this region. An exchange of bound cations in the surface regions could cause additional changes if the membrane has adsorbed (bound) cations in the surface.

C. Surface-Ion Exchange Reactions

If cations are adsorbed on both surfaces of a membrane nonselectively, and potassium comes into a region where sodium is adsorbed, we have the possibility of ion exchange. An ion-exchange reaction releases sodium in this region, and the resulting higher concentration leads to a higher gradient driving it into the cell. Similar processes tend to decrease the concentration gradient driving the potassium across the membrane. These changes promote the conditions for a reversal of the ion flow. The gradient driving the original flow of potassium is decreased and the gradient driving sodium is elevated.

We have demonstrated ion exchange reactions in surface films, using a mercury–water interface having adsorbed serum albumin at pH 8 where it is negatively charged, and bound monovalent thallium ions, which are a frequent substitute for potassium in biological systems (Britten and Blank, 1968). Using standard polarographic measurements, it can be shown that both thallium and serum albumin are adsorbed on this electrode surface (Blank *et al.*, 1976). If the mercury cathode is polarized to above the potential where there is a current due to the reduction of thallium, and spontaneously reverse the potential so that the mercury is now an anode, the mercury dissolves mercurous ions primarily. (Under these conditions, mercurous ion is over 90% of the product of the electrolytic reaction.) Since mercury ions are now in a region where thallium ions are bound to the serum albumin, the mercury ions displace thallium ions. We can measure the thallium concentration above a polarization of -0.4 volts. The current below -0.4 volts picks up only the mercury, while above -0.4 volts, it picks up both the mercury and the thallium. In this case, the determined thallium concentration includes additional ions that have arisen as a result of the ion-exchange reaction. (The passage of the anodic pulse not only brings mercurous ions into the region of bound thallium ions, but also causes a desorption of the whole complex. Therefore, the measurements must be made within a confined range.)

III. THE SURFACE COMPARTMENT MODEL (SCM)

A. Development of the SCM

The surface compartment model (SCM) was introduced (Blank, 1965) to study the role of surface processes in ion transport across membranes. Many simplifying assumptions were made in the original paper, but it was shown that ionic concentration changes at membrane surfaces after current flow could contribute to the unusual fluxes in excitable membranes. The original SCM has been expanded to include the effects of surface charge, membrane potential, and ion-binding equilibria in the steady state (Blank and Britten, 1978) and during transients (Blank and Kavanaugh, 1982), so we can now study changes in surface regions under dynamic conditions. The SCM incorporates the ideas presented in the previous section: (1) ion transport is strongly dependent upon surface charge, (2) surface layers accumulate or are depleted of ions as a result of current flow across regions having different ion mobilities, and (3) ion-exchange processes can occur in surface regions. In addition, the generally accepted ideas of conservation (of mass and charge) or chemical kinetics (applied to ionic reactions) and of current flow in electrical circuits (composed of resistances and capacitances) have been used in the equations given below.

B. Equations of SCM for Transients

The SCM equations for transient states under voltage clamp are 14, highly coupled, independent differential equations, plus one nondifferential equation for the current. (In addition there are ten flux equations for the three ions and the mobile membrane charge, and three pump-flux equations to balance the steady-state ion fluxes across the membrane.) The 15 basic equations are given in terms of a membrane that consists of the discrete regions shown in Table I. Six mass balance equations describe the time variation of the ionic concentrations in the two surface compartments:

$$\dot{N}_2 = (1/L_2)\,(J_{N1} - J_N - P_N - \dot{N}_{22}) \tag{1}$$

$$\dot{K}_2 = (1/L_2)\,(J_{K1} - J_K - P_K - \dot{K}_{22}) \tag{2}$$

$$\dot{A}_2 = (1/L_2)\,(J_{A1} - J_A - P_A) \tag{3}$$

$$\dot{N}_3 = (1/L_3)\,(J_N + P_N - J_{N3} - \dot{N}_{33}) \tag{4}$$

$$\dot{K}_3 = (1/L_3)\,(J_K + P_K - J_{K3} - \dot{K}_{33}) \tag{5}$$

$$\dot{A}_3 = (1/L_3)\,(J_A + P_A - J_{A3}) \tag{6}$$

In the equations *J*s are fluxes given by Nernst–Planck equations and the *P*s are pump fluxes.

Table I. Surface Compartment Model—Symbols[a]

Compartment	Outside reservoir	Outside surface	Membrane	Inside surface	Inside reservoir
Thickness		L_2		L_3	
Electrical potential	$E_1 = 0$	E_2		E_3	E_4
Capacitance		C_1	C_2	C_3	
Sodium concentration	N_1	N_2		N_3	N_4
Passive flux	J_{N1} →		→ J_N →	J_{N3} →	
Pump flux			→ P_N →		
Potassium concentration	K_1	K_2		K_3	K_4
Passive flux	J_{K1} →		→ J_K →	J_{K3} →	
Pump flux			→ P_K →		
Anion concentration	A_1	A_2		A_3	A_4
Passive flux	J_{A1} →		→ J_A →	J_{A3} →	
Current	I →		→ I →	I →	
Surface sites					
Charge density		X_2	→ J_X →	X_3	
Bound Na density		N_{22}		N_{33}	
Bound K density		K_{22}		K_{33}	

[a] Reproduced with some modifications from *Bioelectrochemistry and Bioenergetics*, with permission.

The four chemical kinetic equations that give the changes in the bound cations at the two surfaces are given by:

$$\dot{N}_{;2} = B_F X_2 N_2 - B_R N_{22} \tag{7}$$

$$\dot{K}_{22} = B_F X_2 K_2 - B_R K_{22} \tag{8}$$

$$\dot{N}_{33} = B_F X_3 N_3 - B_R N_{33} \tag{9}$$

$$\dot{K}_{33} = B_F X_3 K_3 - B_R K_{33} \tag{10}$$

where ion binding is a bimolecular process, ion dissociation a monomolecular process, and the ratio of the kinetic constants, $B_F/B_R = K_{EQ}$, is the binding equilibrium constant.

Changes in the negative surface charge on the two sides of the membrane are given by the following equations based on charge balance:

$$\dot{X}_2 = -(\dot{N}_{22} + \dot{K}_{22}) - J_X \tag{11}$$

$$\dot{X}_3 = -(\dot{N}_{33} + \dot{K}_{33}) + J_X \tag{12}$$

The currents in the surface compartments and the membrane during voltage clamp to a step pulse ($\dot{E}_4 = 0$) are given by equations for parallel resistance–capacitance elements:

$$I = F(J_{N1} + J_{K1} - J_{A1}) - C_1\dot{E}_2 \tag{13}$$

$$I = F(J_{N3} + J_{K3} - J_{A3}) + C_3(\dot{E}_3 - \dot{E}_4) \tag{14}$$

$$I = F(J_N + J_K - J_A - J_X + P_N + P_K - P_A) + C_2(\dot{E}_2 - \dot{E}_3), \tag{15}$$

where F is the Faraday. (The $\dot{E}_4$ term is necessary for the initial current upon imposing the clamp.) Equations (13–15) can be solved for the state variables $\dot{E}_2$ and $\dot{E}_3$ as well as the current I.

C. *Initial Values and Parameters*

To study the behavior of the SCM equations (Blank *et al.*, 1982), we have used initial conditions based on published values for the squid axon. These are shown in Table II along with other steady-state values calculated by assuming ranges for the parameters, again based on experimental values when available, but including some simplifying assumptions. The SCM parameters are:

1. Two of the binding parameters, the binding rate-constant (B_F), the dissociation rate-constant (B_R) and their ratio, the equilibrium constant (K_{EQ}) must be set. We have set K_{EQ} and B_R assuming that they are the same for sodium and potassium on the two sides of the membrane. K_{EQ} was chosen to give the maximum ion binding consistent with the known dimensions of the membrane and B_R has been varied over a range of magnitudes.

2. The three ionic mobilities in the surface compartments were assumed equal (M) as a first approximation, and have been varied over a range of magnitudes.

3. The four conductances in the membrane-sodium (G_N), potassium (G_K), anion (G_A), and mobile charge associated with gating currents (G_X) are fixed by the magnitudes of the steady-state ionic fluxes and the gating current. These values are known for squid axon.

4. The three capacitances, two at the surfaces and one for the dielectric are also fixed by observations, since the total capacitance must be about 0.8 $\mu F/cm^2$. We have chosen values that are consistent with the properties of surfaces and dielectric materials.

Although 12 parameters must be set, we see that most of these values are fixed by experimental observations. Only two values cannot be estimated from the literature, B_R and M. However, we shall see that it is possible to determine these magnitudes from their effects on the sodium influx.

D. Conditions for Transient Sodium Influx

The equations were solved for a voltage clamp where E_4 was changed from -65 mV to -20 mV using a very rapid (time-constant ~20 μsec) exponential function, and a small fraction of the surface charge moved rapidly across the membrane, in line with the observed gating current. In all cases the mobilities (M) and conductances across the membrane (G) were held constant during a solution. Using the best values for the parameters, the solutions showed a steadily decreasing negative current until the establishment of a new steady state, with gradual and slight changes in the variables of the system. However, when the G_N and G_K were increased, the SCM equations responded with a positive, transient sodium flux. In Fig. 4 the permeability of the membrane to sodium and potassium is increased several orders of magnitude above resting values.

The transient reversal of J_N, i.e., an inwardly directed sodium flux, is accompanied by increases in N_2, the sodium at the outer surface, and decreases in N_3, the sodium at the

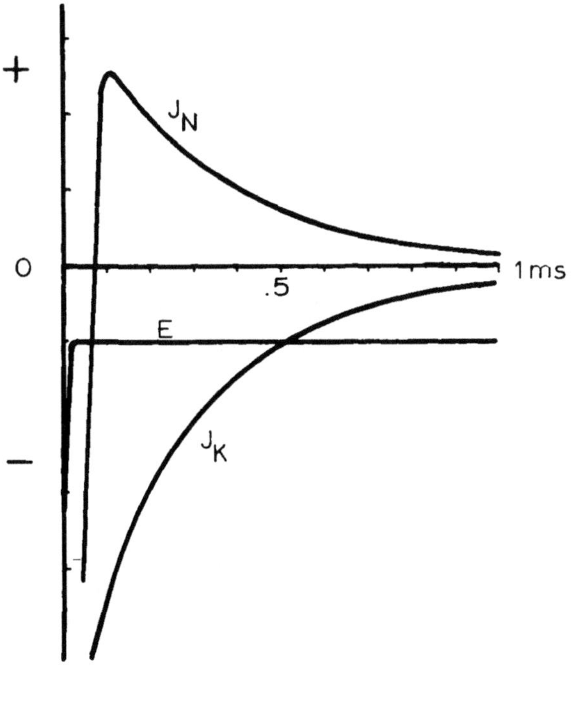

Figure 4. The net sodium flux, J_N, the net potassium flux, J_K, and the voltage-clamp potential, E, calculated from the surface-compartment model as functions of time when the membrane permeabilities to *both* sodium and potassium are raised several orders of magnitude. The transient sodium flux is positive, i.e. inward, while the potassium flux is negative. (Reproduced with permission from *Bioelectrochemistry and Bioenergetics*).

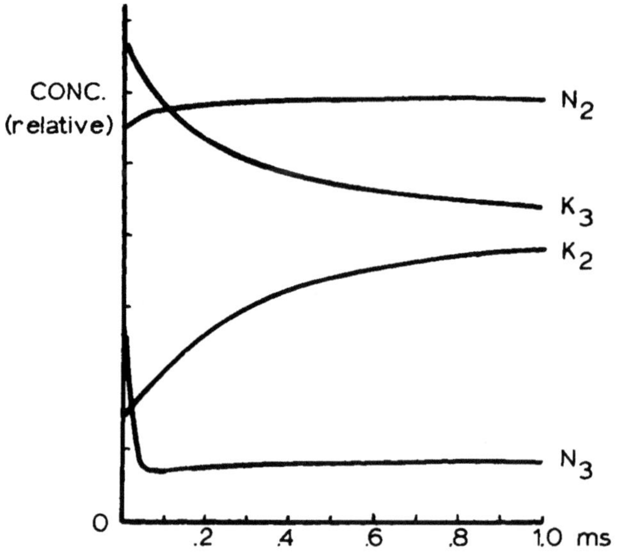

Figure 5. The changes in the cation concentrations in the surface compartments are shown under the same conditions as in Fig. 4. (Reproduced with permission from *Bioelectrochemistry and Bioenergetics*).

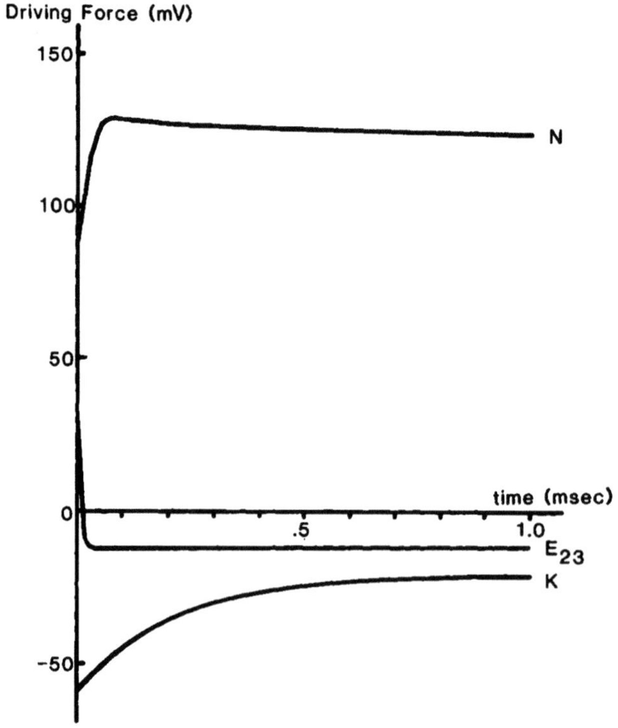

Figure 6. The components of the driving forces for sodium and potassium across the membrane as a function of time under the same conditions as in Fig. 4. The chemical potential differences are labeled N for sodium and K for potassium. The electrical potential difference is labeled E_{23}.

inner surface, as shown in Fig. 5. Accumulation and depletion of ions on opposite sides of the membrane occurs for both sodium and potassium ions, but because of the opposite directions of the concentration gradients, only the effect on the sodium flux is positive. The change in the chemical potential difference compensates for the unfavorable electrical driving force, as shown in Fig. 6. The surface compartments act as buffers for the electrochemical potential of sodium ions. The changes in bound ions, shown in Fig. 7, reflect the changes in the ionic concentrations shown in Fig. 5, because there are relatively slight changes in X_2 and X_3.

The transient sodium influx shown in Fig. 4 is seen when the permeability of the membrane to both sodium and potassium is raised several orders of magnitude. Increasing G_N or G_A separately does not give the influx, while increasing G_K separately gives less than 1% of the influx shown in Fig. 4. To observe a transient influx of sodium ions of approximately the right magnitude and with a peak at the right time, both G_N and G_K must be increased.

The transient sodium influx is quite sensitive to the magnitudes of some of the parameters. The effect of the permeability increase, the basis for the influx, is shown in Fig. 8. From this result, it appears that the membrane permeabilities (G_N, G_K) must be about 5×10^{-6} (at the fixed levels of the other quantities in Table II) to have approximately the right magnitudes of fluxes and times. The results with the other parameters are summarized in Table III, which gives the effects of increases in the parameters on the sodium influx magnitude and peak time. The parameter associated with the gating current, the conductance, G_X, has no influence on the transient positive J_N, and is probably associated with the mechanism that is responsible for the increase in membrane permeability.

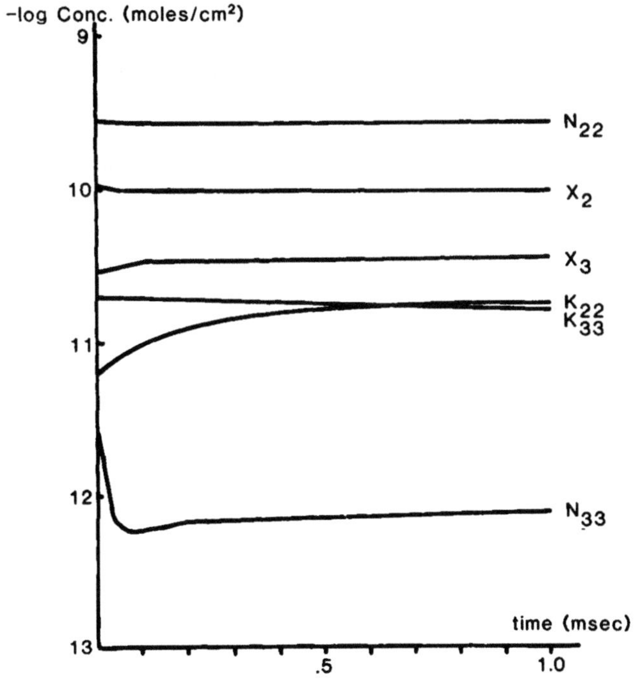

Figure 7. The variation of the membrane-bound species and the surface charge as functions of time under the same conditions as in Fig. 4. The concentrations are in moles/cm^2 on a logarithmic scale.

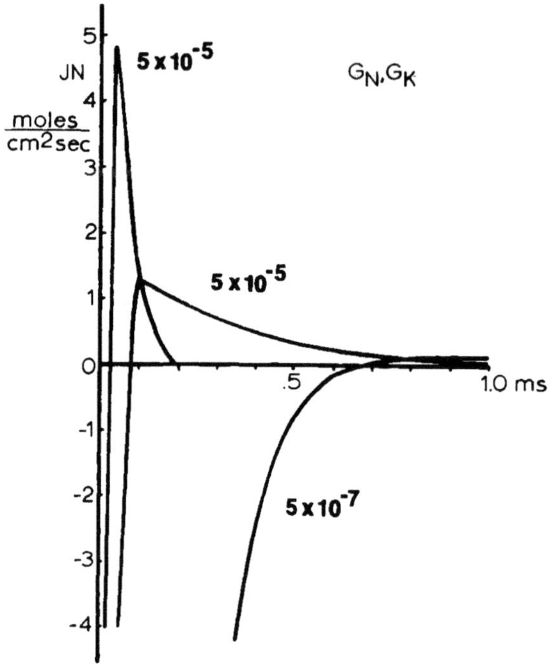

Figure 8. The variation of the sodium flux with time as a function of G_N, G_K the membrane permeability of the two cations. The magnitudes of $G_N = G_K$ in cm/(sec mV) are shown on the curves. (Reproduced with permission from *Bioelectrochemistry and Bioenergetics*).

The quantitative aspects of the parameter study enable us to draw some conclusions about the physical properties of the molecules involved in the mechanism that gives rise to the increased permeability to cations. For example, the ionic mobility in the surface compartments must be at least an order of magnitude smaller than the G_N, G_K in order to sustain the concentration gradients shown in Fig. 5 and give rise to a transient sodium influx. The magnitude of M must therefore lie between the low and high permeability values of G_N and G_K. Of the binding equilibrium parameters, K_{EQ} must be over 0.1 M^{-1} and B_R about 5×10^4 sec^{-1} or greater to give an appreciable sodium influx at reasonable values of time.

IV. CONCLUSIONS

When an excitable membrane (such as squid axon) is stimulated electrically *in vitro*, an unusual flow of ions across the membrane gives rise to the action potential. While the resting potential is determined primarily by the diffusion of potassium ions out of the cell, during the transient period, the inward diffusion of sodium ions increases. This ion flow is followed by a transient increase in the outward flow of potassium ions leading to a recovery to the resting potential. It is possible to characterize these changes in terms of (1) a reversal of selectivity, that is, a membrane which allowed potassium to go through selectively at first is now allowing sodium to go through, and (2) the ion fluxes are undergoing kinetics of the activation–inactivation type. The action potential may therefore be regarded as a problem of a change in ion selectivity and a change in the ion-flow kinetics initiated by a cathodal stimulation.

However, the SCM has enabled us to demonstrate that ionic processes in the electrical double-layer regions at the surfaces of membranes can give rise to an apparent change in ion selectivity and a transient inward sodium flux *if there is a nonselective increase in*

Table II. Surface Compartment Model—Steady-State Values[a]

Compartment	Units	Outside reservoir	Outside surface	Membrane	Inside surface	Inside reservoir
Thickness	cm		3.88×10^{-8}		6.09×10^{-8}	
Electrical potential	mV	0	−46		−78	−65
Capacitance	$\mu F\ cm^{-2}$		100	0.8	100	
Sodium concentration	$Moles\ cm^{-3}$	4.4×10^{-4}	2.73×10^{-3}		8.38×10^{-5}	5.0×10^{-5}
Passive flux	$Moles\ cm^{-2}\ sec^{-1}$		0	8.4×10^{-11}	0	
Pump flux	$Moles\ cm^{-2}\ sec^{-1}$		0	-8.4×10^{-11}	0	
Potassium concentration	$Moles\ cm^{-3}$	1.0×10^{-5}	6.21×10^{-5}		6.70×10^{-4}	4.0×10^{-4}
Passive flux	$Moles\ cm^{-2}\ sec^{-1}$		0	-5.1×10^{-11}	0	
Pump flux	$Moles\ cm^{-2}\ sec^{-1}$		0	5.1×10^{-11}	0	
Anion concentration	$Moles\ cm^{-3}$	4.5×10^{-4}	7.25×10^{-5}		2.69×10^{-4}	4.5×10^{-4}
Passive flux	$Moles\ cm^{-2}\ sec^{-1}$		0	-1.1×10^{-11}	0	
Pump flux	$Moles\ cm^{-2}\ sec^{-1}$		0	1.1×10^{-11}	0	
Current density	$amp\ cm^{-2}$		0	0	0	
Surface sites						
Charge	$Moles\ cm^{-2}$		1.06×10^{-10}		2.95×10^{-11}	
Bound Na	$Moles\ cm^{-2}$		2.89×10^{-10}		2.47×10^{-12}	
Bound K	$Moles\ cm^{-2}$		6.56×10^{-12}		1.97×10^{-11}	

[a] Reproduced with some modifications from *Bioelectrochemistry and Bioenergetics*, with permission.

Table III. Effects of Increases in the SCM Parameters on the Transient Sodium Influx[a]

Parameter	Influx magnitude	Influx peak time
G_N, G_K	+	−
B_R	+	−
K_{EQ}	+	No effect
M	−	No effect
G_X	No effect	No effect

[a] Reproduced with some modifications from *Bioelectrochemistry and Bioenergetics*, with permission.

membrane permeability to cations. In the SCM, the ion selectivity arises from the normal asymmetry of the ionic concentrations across membranes. The transient ion-flux kinetics are due to the changes in the gradients between the compartments and the complex interrelations between the variables of the SCM equations.

The transient sodium influx of Fig. 4 resembles but is quantitatively unlike the flux observed experimentally. It is obvious that one should not expect agreement where there have been many approximations in the derivation of the system and in the setting of the parameters. (What one aims for is an insight into the effects of certain processes on the ionic fluxes across membranes.) But the SCM has a built-in simplification, the assumption of constant values of the permeabilities during a solution, that should be removed to achieve a better approximation. For example, if the permeabilities were set at the best estimates based on the resting fluxes and the membrane permeabilities to sodium and potassium were increased as a result of the gating current, the magnitude of the transient sodium influx would be enhanced for lower values of the permeabilities and there would be smoother changes with variations of the voltage clamp. In Fig. 8, when the permeabilities (G_N, G_K) are high, the sodium influx starts earlier, but also shuts off earlier. If the permeability were to start at a low value and change gradually to the higher value and then back, one would sweep out a much smoother curve and the flux would not shut off as abruptly. The potassium flux would also change. This can be done by assuming a model for the permeability increase linked to the gating currents. Aggregation–disaggregation reactions in oligomeric proteins, which involve changes in surface charge and ambient ion concentrations, especially pH (Blank, 1980b), are reasonable possibilities for the basis of a molecular mechanism.

At the beginning of this paper, we referred to the early ideas of Bernstein (1902) about the breakdown in selective permeability to ions leading to an action potential, and in view of the results reviewed here, it may be worth reexamining this hypothesis. According to the SCM results, it appears that a *nonselective* increase in cation permeability to both sodium and potassium gives rise to an *apparently selective* transient sodium influx. The molecular mechanism that would underlie such a change is more in keeping with Bernstein's ideas than with the currently prevailing ideas of a selective increase in membrane permeability to sodium.

Finally, it is important to bear in mind that ions other than sodium and potassium can take part in the same kinds of processes, and that similar transient fluxes can occur when there are changes in the electrical double-layer regions at the surfaces of membranes. In this paper, the changes are due to imposed currents, but they can be due to the release of ions by the action of light (Blank *et al.*, 1981) or redox reactions due to membrane-associated enzymes. If the effects of electrical double layers are taken into consideration in these cases, some of the unusual biological properties may turn out to be the physical consequences of specialized biological reactions at surfaces.

Note Added in Proof

Since this manuscript was completed, two of the problems discussed in the last section have been successfully treated. A voltage sensitive ion channel model based on oligomer aggregation has been developed (Blank, Bioelectrochem. Bioenerg. **9**:615–624, 1983), and used in a voltage clamp computation (Blank, Bioelectrochem. Bioenerg. **10** (4), in press). The results show that a nonselective increase in permeability to cations leads to a peak inward current followed by a steady state outward current, and that both currents depend upon the clamp voltage approximately as in the squid axon membrane.

ACKNOWLEDGMENT. This work has been supported by research contract N00014-80-C-0027 from the Office of Naval Research.

REFERENCES

Bernstein, J., 1902, Untersuchungen zur Thermodynamik der bioelektrischen Ströme, *Pflugers Arch.* **92**:521–562.

Blank, M., 1965, A physical interpretation of the ionic fluxes in excitable membranes, *J. Colloid Sci.* **20**:933–949.

Blank, M., 1967, The accumulation of ions at water nitrobenzene interfaces during transference, in: *Physics and Physical Chemistry of Surface Active Substances*, Vol. II (J. Th. G. Overbeek, ed.), pp. 233–243, Gordon and Breach, University Press, Belfast.

Blank, M., 1979, Monolayer permeability, *Prog. Surface Membr. Sci.* **13**:87–139.

Blank, M., 1980a, The thickness dependence of properties of membrane protein multilayers, *J. Colloid Interface Sci.* **75**:435–440.

Blank, M., 1980b, A surface free energy model for protein structure in solution: Hemoglobin equilibria, *Colloids and Surfaces* **1**:139–149.

Blank, M., and Britten, J. S., 1965, Transport properties of condensed monolayers, *J. Colloid Sci.* **20**:789–800.

Blank, M., and Britten, J. S., 1978, The surface compartment model of the steady state excitable membrane, *Bioelectrochem. Bioenerg.* **5**:528–540.

Blank, M., and Feig, S., 1963, Electric fields across water-nitrobenzene interfaces, *Science* **141**:1173–1174.

Blank, M., and Kavanaugh, W. P., 1982, The surface compartment model (SCM) during transients, *Bioelectrochem. Bioenerg.*, **9**:427–438.

Blank, M., and Miller, I. R., 1968, Transport of ions across lipid monolayers: The structure of decylammonium monolayers at the polarized mercury water interface, *J. Colloid Interface Sci.* **26**:26–33.

Blank, M., Eisenberg, W., and Britten, J. S., 1976, Ion exchange kinetics in an adsorbed protein film, *Bioelectrochem. Bioenerg.* **3**:15–27.

Blank, M., Soo, L., Wasserman, N. H., and Erlanger, B., 1981, Photoregulated ion binding, *Science* **214**:70–72.

Blank, M., Kavanaugh, W. P., and Cerf, G., 1982, The surface compartment model-voltage clamp, *Bioelectrochem. Bioenerg.*, **9**:439–458.

Britten, J. S., and Blank, M., 1968, Thallium activation of the (Na^+-K^+)-activated adenosine triphosphatase of rabbit kidney, *Biochim. Biophys. Acta* **159**:160–166.

Britten, J. S., and Blank, M., 1977, The effect of surface charge on interfacial ion transport, *Bioelectrochem. Bioenerg.* **4**:209–216.

Goldman, D. E., 1943, Potential, impedance and rectification in membranes, *J. Gen. Physiol.* **27**:37–60.

Goldman, D. E., 1971, Excitability models, in: *Biophysics and Physiology of Excitable Membranes* (W. J. Adelman, Jr., ed.), pp. 337–358, Van Nostrand Reinhold, New York.

Goulden, P. T., 1976, The biological membrane potential, some theoretical considerations, *J. Theoret. Biol.* **58**:425–438.

Miller, I. R., and Blank, M., 1968, Transport of ions across lipid monolayers: Reduction of polarographic currents of Cu^{++} by decylammonium monolayers, *J. Colloid Interface Sci.* **26**:34–40.

Nernst, W., and Riesenfeld, E. H., 1902, Über elektrolytische erscheinungen an der grenzfläche zweier lösungsmittel, *Ann. Phys. Lpz.* **8**:600–624.

Overbeek, J. Th. G., 1952, Electrochemistry of the double layer, in: *Colloid Science*, Vol. I, (H. R. Kruyt, ed.), pp. 115–193, Elsevier, Amsterdam.

Sweeney, G. D., and Blank, M., 1973, Some electrical properties of thin lipid films formed from cholesterol and cetyltrimethylammonium bromide, *J. Colloid Interface Sci.* **42**:410–417.

Teorell, T., 1935, An attempt to formulate a quantitative theory of membrane permeability, *Proc. Soc. Exp. Biol. Med.* **33**:282–285.

Teorell, T., 1957, On oscillatory transport of fluid across membranes, *Acta Soc. Med. Upsalien,* **62**:60–66.

IV

Proteins in Excitation

23

Reconstitution of Nerve Membrane Sodium Channels

Channel Proteins

Raimundo Villegas, Gloria M. Villegas, Zadila Suárez-Mata, and Francisco Rodríguez

I. THE Na CHANNEL AND ITS PROPERTIES

Excitation is associated with a transient and highly selective increase of the Na conductance of the neuronal plasma membrane. The structure to which this Na-conductance change is generally attributed is known as the Na channel. The Na channel mechanism has unique properties: it can be activated by changes in the electric field or by the action of some chemical agents, it preferentially allows Na to cross the membrane following its electrochemical potential gradient, and it can be specifically blocked by tetrodotoxin or saxitoxin. For reviews on the Na channel and its properties in nerve cells, axons, and artificial vesicles, see Armstrong (1975 and 1981), Catterall (1980), Cole (1968), Hille (1976, 1978), Hodgkin (1964), Narahashi (1974), and Villegas and Villegas (1981).

The sensing gating processes responsible for the transient opening of the channel during a voltage-clamp pulse or an action potential are thought to involve conformational changes of membrane macromolecules. The same macromolecules should be involved in the pharmacological response to neurotoxins which is characteristic of the Na channel. Three groups of neurotoxins are ordinarily used to study the Na channel, and for each group a specialized membrane receptor has been proposed. These groups of neurotoxins are: the *lipid-soluble polycyclic compounds* veratridine, grayanotoxin, batrachotoxin, and aconitine, which alter both the activation and the inactivation processes of the Na channel, producing an increase in the membrane permeability to Na (Albuquerque, 1972; Albuquerque *et al.*, 1971; Barnola and Villegas, 1976; Herzog *et al.*, 1964; Hille, 1968; Mozhayeva *et al.*, 1977; Munson *et al.*, 1979; Narahashi, 1979; Narahashi and Seyama, 1974; Ohta *et al.*, 1973; Ulbricht, 1965, 1972), the *polypeptide toxins* from sea anemones and scorpions which bind to a common

Raimundo Villegas, Gloria M. Villegas, Zadila Suárez-Mata, and Francisco Rodríguez • Centro de Biofísica y Bioquímica, Instituto Venezolano de Investigaciones Científicas, Caracas 1010A, Venezuela, and Instituto International de Estudios Avanzados, Caracas 1015A, Venezuela.

receptor different from that of the lipid-soluble polycyclic compounds (Bergman *et al.*, 1976; Conti *et al.*, 1976; Nakamura *et al.*, 1965; Romey *et al.*, 1976, 1975), and the *water-soluble heterocyclic compounds* containing guanidinium moieties, tetrodotoxin, and saxitoxin, which specifically abolish the ionic permeability of the channel activated electrically or chemically (Hille, 1966, 1968; Kao, 1966; Nakamura *et al.*, 1965; Narahashi, 1974; Narahashi *et al.*, 1960, 1964; Ritchie *et al.*, 1976; Sevcik, 1976). The inhibition by tetrodotoxin and saxitoxin of the effect of the lipid-soluble polycyclic neurotoxins and of the polypeptide toxins is noncompetitive, indicating that they act at different receptor sites (Catterall, 1975a,b, 1977a; Henderson *et al.*, 1973; Ray and Catterall, 1978).

In addition to the pharmacological properties, the ionic selectivity exhibited by the channel when its gates are open is another characteristic that allows its identification. Selectivity is attributed to the nature, size, and charge of the Na channel pore (Eisenman, 1962; Hille, 1971, 1972, 1975; Mullins, 1959, 1968, 1975). Whatever the mechanism for ion transport and selectivity may be, the fact is that the channel is also permeable to other monovalent cations. The permeability of the Na channel to these cations is in the order Li > Na > K > Rb > Cs (Chandler and Meves, 1965; Hille, 1972; Moore *et al.*, 1976).

II. THE Na-CHANNEL PROBLEM

Is the Na channel a single macromolecule with different functional sites, or is it an assembly of molecules, each serving one or more of the Na-channel functions, such as electric field sensor, neurotoxin receptor, gating system, and/or ion selective pore? In order to answer this question the isolation of the Na channel is required. The interest in isolating neurotoxin receptors lies in the assumption that each receptor is the channel constituent responsible for the function modified by the neurotoxin. Different approaches for establishing correlations between structure and function are to study Na-channel activity in isolated nerve membranes, in membranes after one or more of its constituents have been removed, or to study the function of removed membrane components after their incorporation into lipid bilayers. At the membrane level, these experimental approaches are complicated by the small dimensions and especially by the instability of the membrane and its components. Even though mild physical or chemical manipulations of the membrane can cause irreversible loss of biological activity, some progress has been recently made in the isolation of Na channels from nerve plasma membranes.

This chapter summarizes some of the work carried out to isolate lobster nerve membranes with Na-channel activity, to incorporate Na channels present in isolated lobster nerve membranes into soybean liposomes, and to reconstitute in soybean liposomes the Na-channel activity of partially purified nerve membrane material obtained by detergent treatment followed by gel exclusion chromatography fractionation.

III. SODIUM-CHANNEL ACTIVITY OF LOBSTER NERVE MEMBRANE VESICLES

A. Isolation of Nerve Plasma Membranes

The presence of Na channels in the vesicles spontaneously formed by the plasma membrane fragments isolated from walking-leg nerves of the lobster *Panulirus argus* was first demonstrated in our laboratory (Barnola and Villegas, 1976). Electron micrographs of

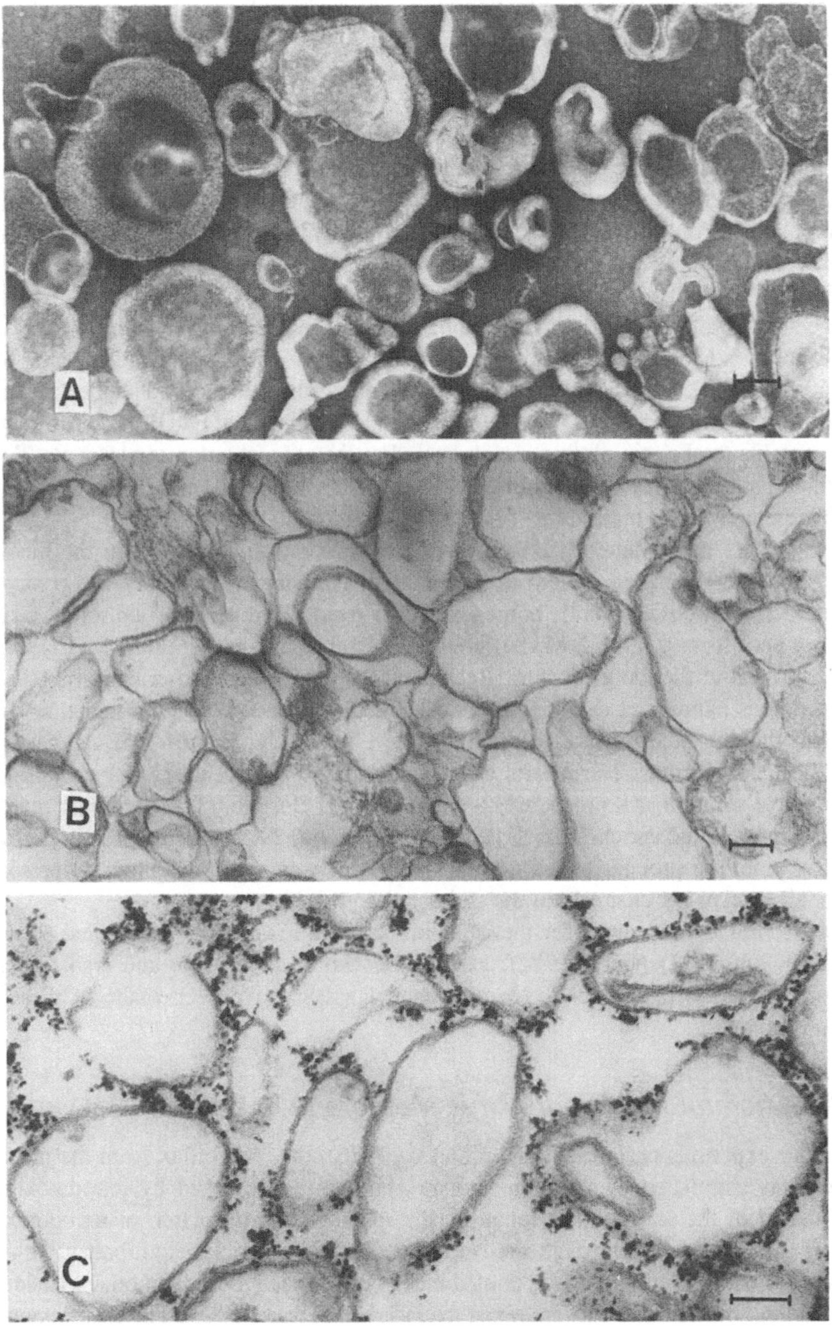

Figure 1. Electron micrographs of the original lobster nerve membrane preparations. (A) fresh membrane vesicles, 0.1–0.4 μm in diameter, negatively stained with phosphotungstic acid. (B) ultrathin section of the same membrane preparation, fixed in OsO_4 and embedded in Epon. (C) thin section of a membrane preparation incubated during 2 hr in isotonic medium containing thorium dioxide. As observed, the thorium micelles do not penetrate the vesicle lumina. The solid mark represents 0.1 μm.

the membrane vesicles are shown in Fig. 1. The method used to isolate plasma membranes from lobster nerves was described by Barnola *et al.* (1972) and is similar to the one developed by Camejo *et al.* (1969) to isolate nerve plasma membranes from squid nerves. A simplified version of the method, designed by Dr. F. V. Barnola, has been recently published (Villegas and Villegas, 1981).

Nerves (20 g) dissected from walking legs of living *Panulirus argus* lobster were placed in 40 ml of 0.33 M sucrose solution containing 2 mM $MgCl_2$ and buffered with 10 mM Tris-Cl to pH 7.5. The tissue was minced with a Sorvall blender (12,000 rpm for 2 min) and homogenized (ten strokes in a teflon-pestle homogenizer at 9,700 rpm). The homogenate was diluted to 10% (wt/vol) with the same sucrose solution and centrifuged at 65,000 g for 40 min at 4°C, and the pellet was suspended in the same sucrose solution (90 ml) and homogenized again. Portions of the homogenate (30 ml) were layered on top of 1.12 M sucrose solution (30 ml) buffered to pH 7.5 with 10 mM Tris-Cl and centrifuged at 65,000 g for 60 min (Beckmann SW-25.2 rotor). The band at the interface was collected and was diluted three-fold with 10 mM Tris-Cl pH 7.5 and centrifuged at 65,000 g for 40 min. The pellet contains total nerve plasma membranes. For the reconstitution experiments, the total membrane pellet was resuspended in 0.78 M sucrose solution, 10 mM Tris-Cl, pH 7.5 and kept frozen at − 70°C until used.

For further enrichment in axonal membrane (taking an increment in the binding of $[^3H]$tetrodotoxin per mg of protein as indication of enrichment) the pellet was resuspended in 10 mM Tris-Cl buffer (6 ml), homogenized, layered (2 ml portions) on top of a sucrose linear gradient from 0.66–1.2 M, buffered with 10 mM Tris-HCl, 2 mM $MgCl_2$, pH 7.5, and centrifuged at 65,000 g for 3 hr (Beckmann SW-25.2 rotor). After centrifugation, two cloudy regions banding at densities of 1.072 (fraction I) and 1.124 g/ml (fraction II) were collected, diluted three- to four-fold with 10 mM Tris-Cl buffer, pH 7.5, and centrifuged at 65,000 g for 40 min. The pellets were collected and resuspended separately in lobster physiological solution. All procedures were carried out close to 0°C. The lighter membrane (fraction I) was found enriched in $[^3H]$tetrodotoxin binding (9.5 pmol/mg protein for fraction I and 3.4 pmol/mg protein for fraction II). Fraction I was used to explore the presence of functionally active Na channels in the isolated membranes.

Other similar procedures for the isolation of nerve membranes from squid (Fischer *et al.*, 1970), lobster (Denburg, 1972), garfish (Chacko *et al.*, 1974b) and crab (Balerna *et al.*, 1975) have been described. No Na-channel assays have been made in these latter preparations.

B. Na-Channel Activity of Nerve Membrane Vesicles

In our experiments (Barnola and Villegas, 1976), the ^{22}Na efflux from the membrane vesicles was stimulated by veratrine or batrachotoxin and blocked by tetrodotoxin. The identification of the Na channels relied on the specificity of the action of tetrodotoxin for blocking the flux of ^{22}Na through the Na channels. When vesicles equilibrated with ^{22}Na-labeled physiological solution were diluted into a nonradioactive medium otherwise identical to the loading solution, the time course of the ^{22}Na efflux was well fitted by two exponentials: a fast one, corresponding to the first 1–10 min period, and a slow one, to the 10–60 min term. The fact that the ^{22}Na efflux is increased by veratridine or batrachotoxin and such increase is specifically abolished by tetrodotoxin indicates that the neurotoxin-sensitive flux occurs through the Na channel, since the same neurotoxins produce the same effects in intact axons. As it is shown in Fig. 2, the fast and slow rates were increased by veratrine (0.5 mg/ml) and the increments were abolished by the additional applications of tetrodotoxin

Figure 2. Efflux of ^{22}Na from axolemma vesicles. C corresponds to controls, V to vesicles treated with veratrine (0.50 mg/ml), and V + TTX to those treated with tetrodotoxin (100 nM) in the presence of veratrine (0.50 mg/ml). The vesicles were previously equilibrated with ^{22}Na lobster physiological containing 1 mM ouabain, pH 7.5, and the efflux was carried out in a nonradioactive solution that otherwise was identical to the loading solution. The values are relative to the 1-min measurement. Efflux measured at 20–22°C. (From Barnola and Villegas, 1976).

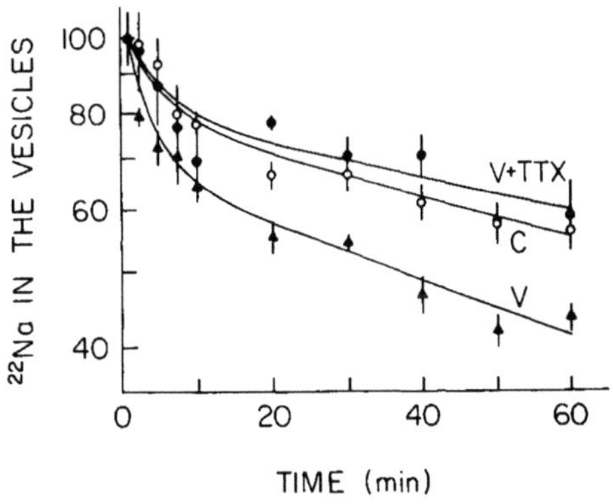

(100 nM). Similar results were observed when batrachotoxin (0.5 μM) was used instead of veratrine to activate the Na channels.

The effect of different tetrodotoxin concentrations on the slow-phase efflux rate of ^{22}Na from lobster nerve membrane vesicles in the presence of veratrine (0.5 mg/ml) is shown in Fig. 3. The data were well fitted by a dose-response curve corresponding to a one-to-one stoichiometric reaction, i.e., a model in which one tetrodotoxin molecule binds to a single specific site in the membrane and blocks the flux through a single Na channel. Complete blocking occurs when all receptors are occupied by tetrodotoxin. The value of K_i, the concentration of tetrodotoxin required to produce 50% of its maximum blocking effect, is about 12 nM (Barnola and Villegas, 1976). In intact axons of the same lobster, the K_i value is 5 nM (Villegas, *et al.*, 1976), and in isolated nerve membranes, the K_d for the specific binding of [^{3}H]tetrodotoxin to the isolated nerve membranes is 4.5 nM (Barnola *et al.*, 1972). Similar K_i and K_d values have been also determined in intact axons, and isolated nerve plasma membranes obtained from other animals (Balerna *et al.*, 1975; Benzer and Raftery, 1972; Colquhoun *et al.*, 1972; Chacko *et al.*, 1974a; Henderson and Strichartz, 1974; Ritchie *et al.*, 1976).

Experimental results of work actually in progress in our laboratory (Ana María Correa, unpublished results) have revealed the pharmacological response of the lobster nerve mem-

Figure 3. Rates of ^{22}Na efflux from axolemma vesicles treated with different tetrodotoxin concentrations in the presence of 0.50 mg of veratrine per milliliter. The rates correspond to the slow phase (10–60-min period) of the efflux curves. The continuous line is a least-squares fit to the data, assuming that the diminution by tetrodotoxin of the rate in the presence of veratrine is a rectangular hyperbolic function of the toxin concentration. K_i = 11.9 nM. (From Barnola and Villegas, 1976).

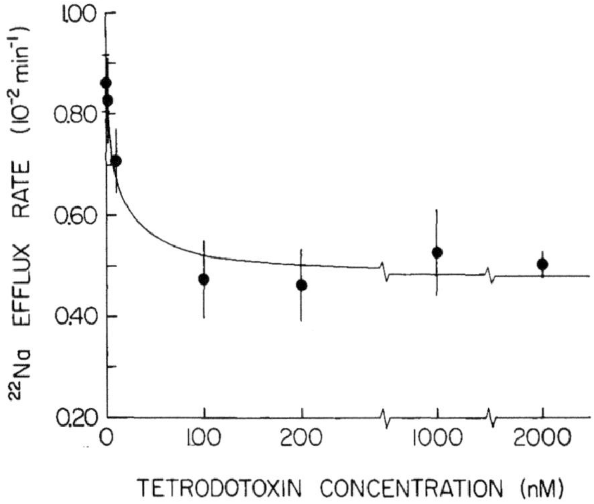

brane vesicles to the polypeptide Sea Anemone toxin II. The action of this toxin on the vesicles was found to depend, as in nerve cells and axons (Catterall, 1977a; Catterall *et al.*, 1976; Ray and Catterall, 1978; Ray *et al.*, 1978), on membrane potential. This finding supports the proposition that the receptor for the polypeptide toxins may be related to the primary electric field sensor of the channel (Catterall, 1977a, 1979; Ray and Catterall, 1978).

IV. SODIUM CHANNEL ACTIVITY OF RECONSTITUTED VESICLES

A. Incorporation of Na Channels into Soybean Liposomes

In 1977, we were able to incorporate Na channels of crude nerve membrane preparations of the lobster *Panulirus* into soybean liposomes (Villegas *et al.*, 1977) following the freeze-thaw-sonication procedure of Kasahara and Hinkle (1976). Nerve plasma membrane fragments, isolated as described above from walking-leg nerves of living lobsters and kept frozen at − 70°C in sucrose solution (0.78 M sucrose, 10 mM Tris-HCl, pH 7.5), were employed as a source of Na channels. In some experiments, nerve membranes of the lobster *Homarus americanus* were used.

For reconstitution, frozen nerve membranes were thawed and added to sonicated soybean liposomes, usually to a final concentration per ml of 0.5 mg of membrane protein and 40 mg of soybean lipids (Associated Concentrates, Woodside, N.Y.). The composition of the soybean lipids, as determined by Miller and Racker (1976), is 40% phosphatidyl choline, 33% phosphatidyl ethanolamine, 14% phosphatidyl inositol, 5% lysophosphatidyl choline, and 4% cardiolipin. The soybean liposomes were ordinarily suspended in a phosphate solution (150 mM NaPi, 150 KPi, pH 7.5, Pi stands for the phosphate formed by the orthophosphoric acid titration of NaOH and KOH to pH 7.5). Aliquots of the protein–lipid mixture were placed in different test tubes: one without drugs as a control, the others containing veratridine or veratridine plus tetrodotoxin. In some experiments veratridine was replaced by grayanotoxin-I, batrachotoxin, or aconitine. The air of the glass tubes containing the mixture was replaced with nitrogen, and the tubes were capped with parafilm. The membrane fragments–lipid mixture was then frozen by immersion in dry ice acetone for 1 min, thawed by agitating the tubes in cold water for 1–1.5 min, and then sonicated in a bath type sonicator for 30 sec. For sonication, a bath-type sonicator was used (Model T 80-80-1-RS, Laboratory Supplies, Hicksville, N.Y.), set at its maximum sonication intensity (80 Hz and 80 W) and with its bath filled with iced distilled water. The biological assay of the Na channel was carried out as usual, by the addition of ^{22}Na-labeled solution to the suspension of reconstituted vesicles kept close to 0°C. After 15 s the ^{22}Na influx was measured by the method of Gasko *et al.* (1976). All procedures were performed close to 0°C in order to preserve the membrane proteins and the reconstituted vesicles. In the experiments to be described below, membrane particles or membrane proteins obtained by detergent treatment of lobster nerve membranes were used as Na-channel sources.

The functional activity of the Na channels incorporated into the liposomes was explored by measuring the flux of ^{22}Na into the vesicles reconstituted in the absence or presence of neurotoxins. Experimental results indicated that the ^{22}Na flux into the reconstituted vesicles was increased by veratridine and the other lipid soluble neurotoxins, and the increment was abolished by the addition of tetrodotoxin. The tetrodotoxin-sensitive ^{22}Na influx caused by veratridine is considered to express the Na-channel activity. In the absence of veratridine no significant change in ^{22}Na influx was produced by tetrodotoxin. The ^{22}Na influx was observed to be time-dependent.

B. Protein Nature of the Na Channel

No response to veratridine and tetrodotoxin was observed when the addition of nerve membrane fragments to the soybean liposomes was omitted or when only nerve membrane lipids or nerve membrane fragments heated at 50°C for 15 min were added to the soybean liposomes for reconstitution (Villegas *et al.*, 1977, 1979). Treatment of the membrane protein–soybean phospholipid mixture with dicyclohexylcarbodiimide before reconstitution also abolished the response to the neurotoxin (Villegas *et al.*, 1977). We have demonstrated in the same preparation of reconstituted vesicles (using a constant amount of liposomes) that both the amount of membrane protein incorporated and the magnitude of the tetrodotoxin-sensitive ^{22}Na influx induced by veratridine increased linearly as functions of the amount of membrane protein used for reconstitution (Villegas *et al.*, 1979). These concomitant linear dependences give further support to the notion that the Na-channel activity of the reconstituted vesicles is due to the incorporation of functionally active Na channels.

C. Effect of the Neurotoxins

The effects of different concentrations of veratridine, grayanotoxin-I, batrachotoxin, and aconitine on the ^{22}Na flux into the reconstituted vesicles have been studied (Villegas *et al.*, 1977, 1980) and the results are shown in Fig. 4. The increase approaches a plateau at

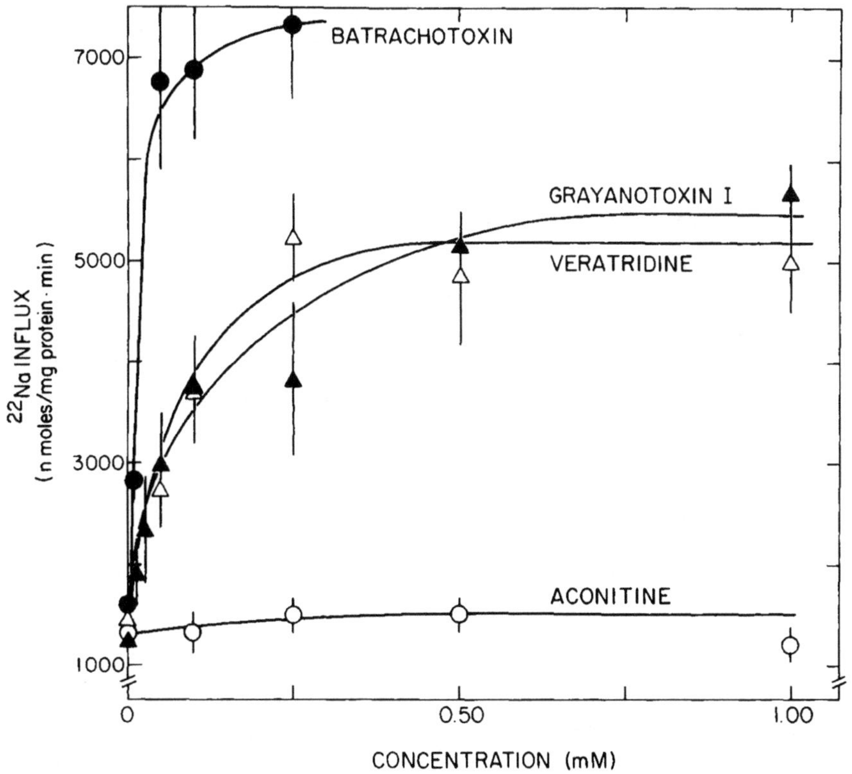

Figure 4. Effect of different concentrations of veratridine (△), grayanotoxin-I (▲), batrachotoxin (●), and aconitine (○) on the ^{22}Na flux into vesicles reconstituted with crude nerve membrane preparations (0.5 mg/ml or membrane protein) and soybean lipids (40 mg/ml). The influx measurements were made 30 sec after addition of ^{22}Na-labeled solution to the suspensions of reconstituted vesicles. (From Villegas *et al.*, 1980).

about 250 μM for veratridine and grayanotoxin-I and 100 μM for batrachotoxin, while up to 500 μM aconitine produces a slight effect. The increase in ^{22}Na flux caused by these neurotoxins was abolished by tetrodotoxin, indicating that the increment in the flux is due to activation of Na channels. The high lipid–protein ratio in reconstituted vesicles might explain why the concentrations of lipid-soluble toxins required to activate the Na channel are higher in reconstituted vesicles than in natural nerve cell membranes (Villegas and Villegas, 1981).

The blocking effect of different tetrodotoxin concentrations on the increases in the ^{22}Na

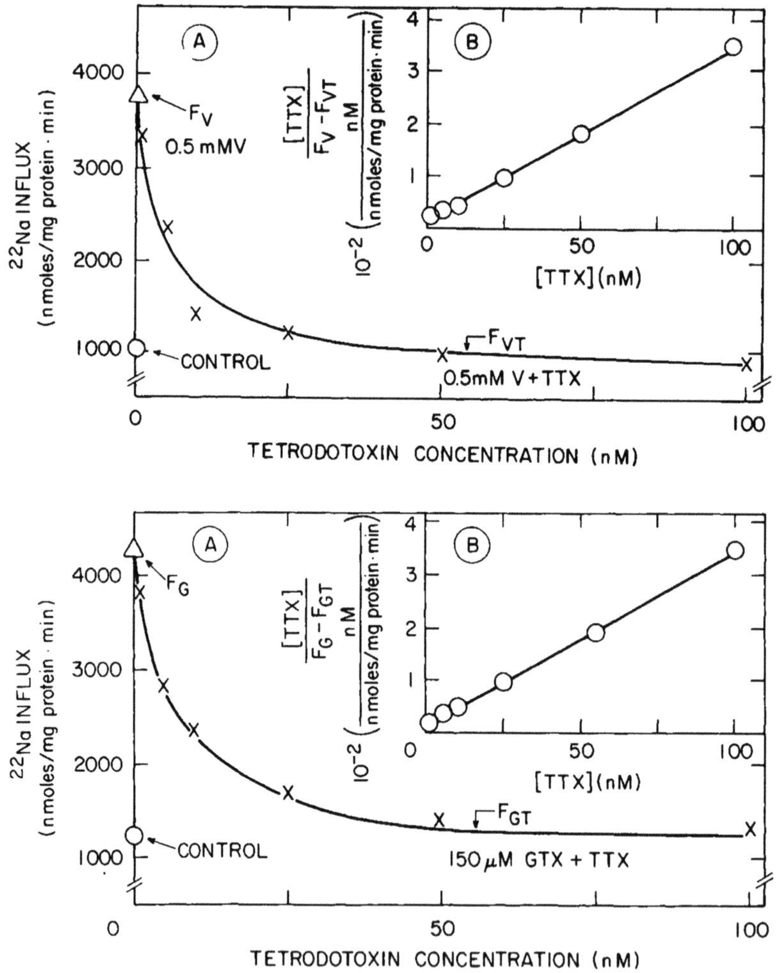

Figure 5. Effect of different tetrodotoxin concentrations on the increment of the ^{22}Na flux caused by 0.5 mM veratridine (top) or 150 μM grayanotoxin-I (bottom). The vesicles were made with crude nerve membrane preparations (0.5 mg/ml of membrane protein) and soybean lipids (40 mg/ml). (Top) (A) Values of the influx in the absence of drugs (control, O), in the presence of veratridine (F_V, Δ) and in the presence of veratridine plus different tetrodotoxin concentrations (V_{VT},X). (B) Modified Lineweaver–Burke plot of the data in (A), the intersection of Y-axis multiplied by the slope of the line is equal to $K_i \cdot K_i = 5$ nM (Bottom) (A) Influx in the absence of drugs (control, O), in the presence of grayanotoxin-I (F_G, Δ), and in the presence of grayanotoxin-I plus different tetrodotoxin concentrations (F_{GT}, X). (B) Modified Lineweaver–Burke plot of the data in (A). $K_i = 4$ nM (From Villegas *et al.*, 1980).

influx caused by veratridine and grayanotoxin-I is illustrated in Fig. 5 (Villegas *et al.*, 1980). The K_i values are 5 nM and 4 nM tetrodotoxin respectively. These values are similar to those reported for other preparations of reconstituted vesicles (Villegas *et al.*, 1977), nerve membrane vesicles (Barnola and Villegas, 1976), and intact axons (Villegas *et al.*, 1976) of the same lobster *Panulirus argus*. They are also similar to the values of K_d for the specific binding of [^{3}H]tetrodotoxin to nerve membranes isolated from the same lobster species (Barnola *et al.*, 1972), crab (Balerna *et al.*, 1975), and garfish (Benzer and Raftery, 1972; Chacko *et al.*, 1974a), and to intact nerves of lobster *(Homarus americanus)* and garfish (Colquhoun *et al.*, 1972; Ritchie *et al.*, 1976).

The response of the reconstituted vesicles to the lipid-soluble polycyclic neurotoxins, as well as to tetrodotoxin, suggests that a large part of the Na channel is preserved during the reconstitution procedure. Catterall (1975a,b, 1977b, 1979) has proposed that, at least in neuroblastoma cells, the lipid-soluble polycyclic neurotoxins share a common receptor and that tetrodotoxin and saxitoxin share another receptor. The third receptor, which was proposed for the binding of the polypeptide neurotoxins, has not yet been explored in reconstituted vesicles.

D. *Ionic Selectivity*

The ionic selectivity of the incorporated Na channel has been recently studied (Condrescu and Villegas, 1982). The relative permeability to ^{22}Na, ^{42}K, and ^{86}Rb of the incorporated channel activated by veratridine or grayanotoxin-I has been determined by measuring in the same preparation of reconstituted vesicles the influxes of these cations in the absence of drugs (control), in the presence of veratridine (0.5 mM) or grayanotoxin-I (150 μM), and in the presence of veratridine or grayanotoxin-I plus tetrodotoxin (100 nM). In these experiments phosphate was used as anion. In the experiments in which the channels were activated with veratridine, the tetrodotoxin-sensitive influxes of ^{42}K and ^{86}Rb were 0.47 and 0.55 times the ^{22}Na influx. In the experiments in which the Na channels were activated with grayanotoxin-I, the tetrodotoxin-sensitive influxes of ^{42}K and ^{86}Rb were 0.40 and 0.38 times the ^{22}Na influx.

Since cesium phosphate precipitates, the relative permeability of the Na channel to ^{22}Na, ^{42}K, ^{86}Rb, and ^{137}Cs was made in vesicles reconstituted in a medium in which sulfate instead of phosphate was used as anion. As in the preceding experiments the influxes were measured in the absence of drugs (control), in the presence of veratridine (0.5 mM) or grayanotoxin-I (150 μM), and in the presence of veratridine or grayanotoxin-I plus tetrodotoxin (100 nM). In the experiments in which the Na channels were activated with veratridine, the tetrodotoxin-sensitive influxes of ^{42}K, ^{86}Rb, and ^{137}Cs were respectively, 0.44, 0.22, and 0.07 times the ^{22}Na influx. When activated with grayanotoxin-I the tetrodotoxin-sensitive influxes of ^{42}K, ^{86}Rb, and ^{187}Cs were 0.46, 0.25, and 0.07 times the ^{22}Na influx. The results revealed that the Na channels incorporated into soybean liposomes do exhibit cationic selectivity. This is illustrated in Fig. 6. Moreover, experiments with [^{32}P]phosphate, [^{35}S]sulfate, and ^{36}Cl revealed that the anions do not move through the incorporated Na channels.

Although no information is available about intact axons of the lobster *Panulirus argus*, a comparison with selectivity data from other sources reveals that the selectivity of the reconstituted vesicles is lower than that of nontreated, voltage-clamped squid axons (Chandler and Meves, 1965) and frog Ranvier nodes (Hille, 1972), but similar to that of voltage-clamped frog Ranvier nodes treated with batrachotoxin (Khodorov, 1978) or aconitine (Mozhayeva *et al.*, 1977). Therefore, as we have noticed previously (Villegas and Villegas, 1981;

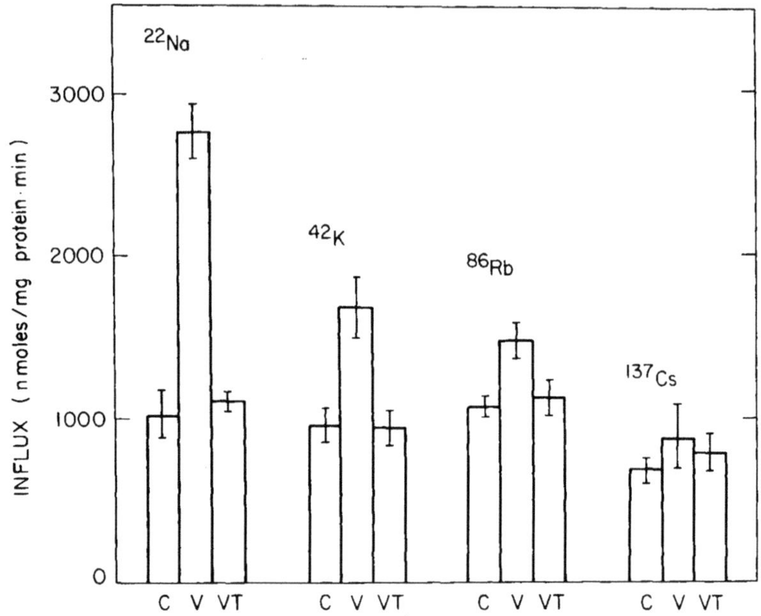

Figure 6. Tetrodotoxin-sensitive fluxes of different cations induced by veratridine. ^{22}Na, ^{42}K, ^{86}Rb, and ^{137}Cs fluxes into vesicles prepared in a solution containing 300 mEq/l of sulfate (pH 7.5) and 75 mEq/l of each Na, K, Rb, and Cs. The vesicles were reconstituted with nerve membrane (0.5 mg/ml of membrane protein) and soybean lipids (40 mg/ml). For each cation, the influx was measured in the absence of drugs (control, C), in the presence of 0.5 mM veratridine (V), and in the presence of 0.5 mM veratridine plus 100 nM tetrodotoxin (VT). The influxes were measured 30 sec after addition of the respective radioactive-labeled solution to the suspension of reconstituted vesicles. (From Condrescu and Villegas, 1982).

Condrescu and Villegas, 1982) the occupancy of the binding site of the lipid-soluble neurotoxins appears to affect the selectivity filter of the Na channels.

V. RECONSTITUTION OF THE SODIUM CHANNEL WITH PARTIALLY PURIFIED NERVE MEMBRANE MATERIAL

Reconstitution experiments carried out with particles obtained from lobster nerve plasma membrane preparations by detergent treatment, differential centrifugation, and ammonium sulfate fractionation revealed that the fractions obtained have different Na-channel activities and different proportions of the same peptides present in the original membrane (Fig. 7). The activity of the fractions appeared to be related to a large component which does not enter the 9% sodium dodecyl sulfate polyacrylamide gel and to peptides with apparent molecular weights of 220,000 and 110,000 daltons (Villegas *et al.*, 1980). Those results led us to explore the possibility of obtaining a further purified fraction with Na-channel activity. In the following paragraphs a description of the procedure used and the preliminary results obtained are described.

A. Preparation of a Partially Purified Membrane Material

Membrane (8–12 mg protein/ml) suspended in the 0.78 M sucrose solution was diluted 24-fold with 0.5 M NaCl, 1 mM EDTA (ethylenediamine tetraacetic acid), 10 mM Tris-Cl

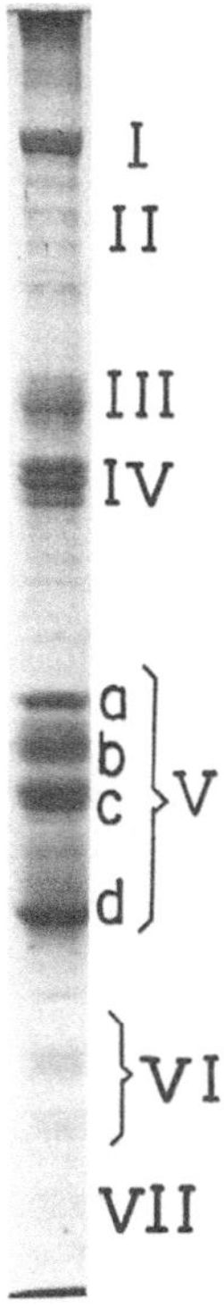

Figure 7. Electrophoretic pattern of the original nerve membrane preparation on a SDS-9% polyacrylamide gel carried out as described in the text and stained with Coomassie blue. The molecular weights of the peptides, estimated from six different experiments, are: band I, 200,000 daltons (250,000 when measured in a 3–15% gradient); bands II, 170,000–140,000; band III, 120,000; band IV, 90,000; band Va, 66,000; band Vb, 60,000; band Vc, 56,000; band Vd, 48,000; bands VI, 40,000–36,000; band VII, 30,000.

solution, pH 7.5, and centrifuged at 150,000 *g* for 30 min. The pelleted membrane was resuspended to a concentration ranging from 10 to 12 mg of protein/ml in 0.35 M sucrose, 10 mM Tris-Cl solution pH 7.5, containing 25 mg of soybean lipids/ml. The membrane was then treated with 0.5% cholic acid (neutralized with NaOH) and 30 mM octyl glucoside for 15 min. After adding the detergents the final sucrose concentration was 0.3 M. The suspension was then centrifuged at 100,000 *g* for 60 min. The supernatant was collected and 1.5 ml containing 7–8 mg of protein were delivered to the Sepharose 6B chromatography column (1.1 × 30 cm) which had been previously equilibrated with a solution of 0.1 M sucrose, 10 mM Tris-Cl, pH 7.5, containing 10 mg of soybean lipids/ml, 0.1% sodium cholate, and 6 mM octyl glucoside. The column was eluted at 12 ml/h and 1 ml fractions were collected. After the protein-column profile was determined, neighboring fractions were pooled, diluted 12-fold with 150 mM NaPi, 150 mM KPi, pH 7.5, and centrifuged at 150,000 *g* for 60 min. The pelleted material was resuspended in 0.1 M sucrose, 150 mM NaPi, 150 mM KPi, pH 7.5. The suspension of pellets 12–15 were pooled to make fraction A and those of pellets 20–23 were combined to make fraction B (see Figs. 8 and 9). As a control for stability, an aliquot of the original supernatant, diluted five-fold with 20 mM Tris-Cl solution to make the detergent concentrations equal to those of the equilibration solution, was kept until the column fractions were obtained. Then, it was diluted again simultaneously with the column fractions, centrifuged, and suspended likewise (sample S of Fig. 9). All procedures were carried out close to 0°C. Before reconstitution, aliquots of S, A, and B, were incubated at 35°C for 10 min and immediately placed back in ice (Villegas *et al.*, 1980; Villegas and Villegas, 1980).

Electrophoresis of original membrane, supernatant, and fractions was ordinarily carried out in 9% polyacrilamide slabs in the presence of 0.1% sodium dodecylsulfate with the

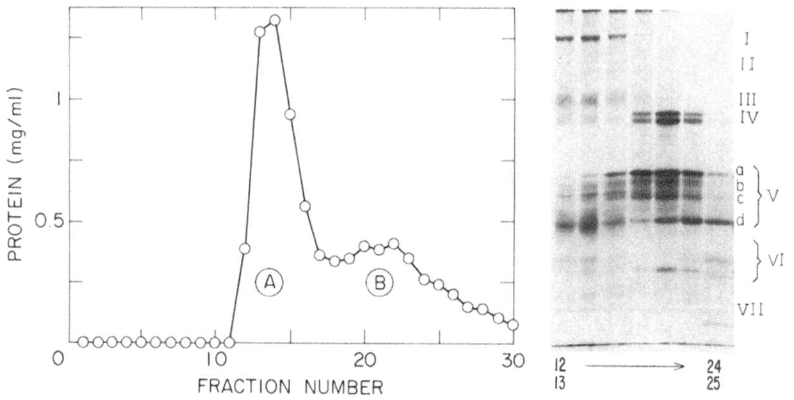

Figure 8. Sepharose 6B column profile of membrane protein (left) and gels performed on pooled aliquots from neighboring fractions as indicated (right). Pairs 12–13 and 14–15 were combined to make fraction A, and pairs 20–21 and 22–23 were mixed to obtain fraction B. The activity and gels of fractions A and B are depicted in Fig. 9.

discontinuous buffer system and gel formulation of Laemmli (1970). The protein samples were dissociated prior to electrophoresis by heating about 40 μl of 400–500 μg/ml of protein at 100°C for 2 min in the presence of sodium dodecylsulfate and 2-mercaptoethanol, and 7.5 or 10 μg of protein of each sample were placed on the gel. The reference proteins used and their subunit molecular weights were: bovine serum albumin (monomer, 67,000; dimer 134,000; trimer, 202,000), chicken ovalbumin (115,000), and chymotrypsinogen (25,000).

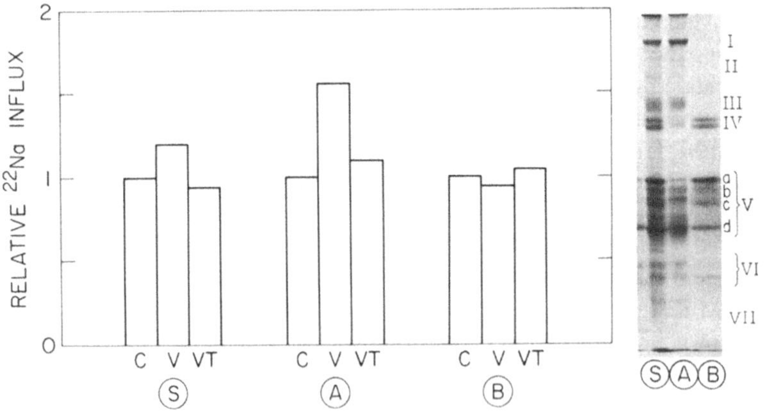

Figure 9. Relative ^{22}Na flux into vesicles reconstituted with soybean lipids (40 mg/ml) and membrane protein (0.25 mg/ml) either of the unfractionated supernatant (S) or of each of the two fractions, A and B. The fractions were obtained from the supernatant by gel exclusion chromatography, as indicated in the text and in Fig. 8. In each case the ^{22}Na influx was measured in a control sample, C, and in samples treated with 0.5 mM veratridine, V, or with 0.5 mM veratridine plus 100 nM tetrodotoxin. On the right side, a photograph of the electrophoretic patterns on a SDS-9% acrylamide gel of the unfractionated supernatant (S) and of fractions A and B.

B. Na-Channel Activity of the Partially Purified Membrane Material

For reconstitution, one volume of the unfractionated supernatant (S) or one volume of the fractions A or B, was added to one volume of sonicated liposomes (80 mg/ml) to obtain concentrations of 0.25–0.75 mg of protein per ml and 40 mg of soybean lipids per ml. Aliquots of 0.4 ml of the protein–lipid mixtures were placed into different test tubes, one without drugs as a control, the others containing either veratridine (0.5 mM) or veratridine plus tetrodotoxin (100 nM). After freeze-thawing for 1 min, the mixtures were sonicated for 15 sec at 4°C, and the Na channel was assayed as described above by measuring the tetrodotoxin-sensitive ^{22}Na influx caused by veratridine.

After the Na-channel activity of the material present in the unfractionated supernatant (S) was demonstrated, the fractions A and B obtained from it were assayed separately. The elution profile from the Sepharose 6B column and the corresponding electrophoretic patterns of the pairs of neighboring fractions are shown in Fig. 8. Pairs 12–13 and 14–15 were combined to make fraction A, and pairs 20–21 and 22–23 were mixed to obtain fraction B. The Na-channel activity of the vesicles reconstituted with S (unfractionated) and with fractions A and B, as well as the electrophoretic patterns of S, A, and B are depicted in Fig. 9. The data of Fig. 9 were obtained with vesicles which were reconstituted with 0.25 mg of protein/ml of S, A, or B, and 40 mg of soybean lipids/ml. As is shown in Fig. 10, increasing the amount of fraction A protein used for reconstitution up to 0.75 mg/ml (the soybean lipids kept at 40 mg/ml) produced a linear increase in the Na-channel activity of the reconstituted vesicles. The drugs produced no effect on vesicles reconstituted in the absence of fraction A. Preliminary results on the blocking by tetrodotoxin of the ^{22}Na flux into the fraction A reconstituted vesicles gave a K_i value of 14 nM tetrodotoxin, similar to the K_i value of 12 nM determined in the vesicles spontaneously formed by the isolated lobster nerve membranes (see above).

Which are the polypeptides responsible for the Na-channel activity exhibited by the vesicles reconstituted with fraction A? The polypeptide composition of fraction A is characterized by the presence and abundance of band I and by the presence of band III, both of which are absent in fraction B. Fraction A also has small quantities of peptides of lower molecular weights, particularly of Vd. A component which does not enter the SDS-9%

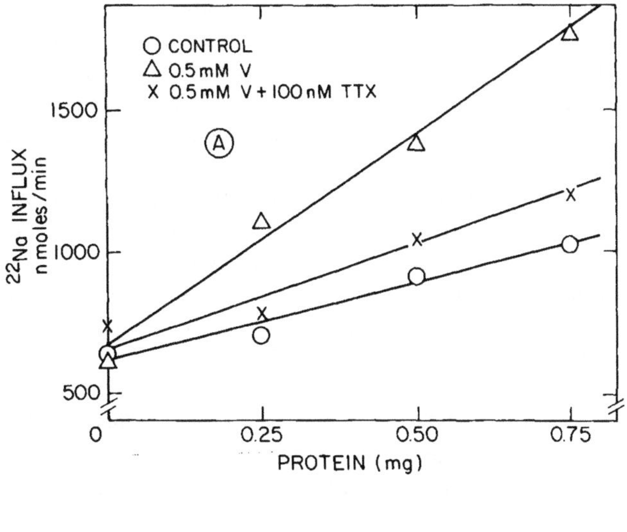

Figure 10. Influx of ^{22}Na as a function of the concentration of fraction A protein used for reconstitution. The concentration of soybean lipids was kept constant at 40 mg/ml.

polyacrylamide is present in S and in fraction A, and absent in fraction B. This latter component could be an aggregate since it diminishes when the polyacrylamide concentration in the running gel is lowered. The apparent molecular weights of the peptides estimated from the electrophoretic mobilities in the SDS-9% polyacrylamide gel were 200,000 daltons for band I, 120,000 for band III, and 48,000 for band Vd. In experiments in which a 3–15% polyacrylamide gradient was utilized in the running gel and additional high molecular weight reference proteins (thyroglobulin, 300,000; ferritin, 220,000 and 18,500) were used, the molecular weight for band I was estimated to be approximately 250,000 daltons.

The apparent molecular weight of the peptide of band I is similar to that estimated for the tetrodotoxin receptor of the electric eel electroplax by the method of irradiation inactivation (Levinson and Ellory, 1973), for the solubilized and partially purified tetrodotoxin receptor of the same electric eel electroplax (Agnew *et al.*, 1980a,b), for the solubilized tetrodotoxin receptor of crab nerve (Lazdunski *et al.*, 1979), and for the solubilized and purified saxitoxin receptor extracted from rat brain (Catterall *et al.*, 1979; Hartshorne and Catterall, 1981). The apparent molecular weight of the peptides of band I is also similar to the scorpion toxin receptor of neuroblastoma cells and rat brain synaptosomes (Beneski and Catterall, 1980). Whether any of the other nerve membrane peptides present in fraction A, in addition to the 250,000 daltons peptide, is also a constituent of the Na channel is unknown at present. However, it should be noticed that peptides of 59,000 and 42,000 daltons were also present in the tetrodotoxin receptor preparations of electric eel (Agnew *et al.*, 1980a,b); a peptide of 38,300 daltons was also present in the purified preparations of rat brain saxitoxin receptor (Hartshorne and Catterall, 1981), and a peptide of 32,000 daltons was also present in the experiments on the scorpion toxin receptor of rat brain (Beneski and Catterall, 1980).

Molecular weight estimates of large peptides in gel electrophoresis are not very precise and similar electrophoretic mobilities cannot be considered as a sufficient argument to propose the same composition and function of peptides. However, it is interesting to point out the similarity in molecular weights of the tetrodotoxin–saxitoxin receptor, the scorpion toxin receptor, the peptide of 250,000 daltons in fraction A, and the active Na-channel fraction of our experiments.

VI. SUMMARY

This chapter describes experiments on membrane material extracted with detergents and fractionated by gel exclusion chromatography carried out in our laboratory. This work consists of (1) the isolation of lobster nerve plasma membranes with Na-channel activity, (2) the incorporation of membrane fragments containing Na channels into soybean liposomes, and (3) the reconstitution of the Na channel by incorporating into soybean liposomes. The active fraction obtained by gel exclusion chromatography is characterized by its content of two high molecular weight peptides that are absent in the inactive fraction: one of 250,000 daltons (band I) and another of 120,000 (band III). The active fraction also has a large component that does not enter the gel and small quantities of peptides of lower molecular weights. Peptides of approximately 250,000 daltons have been proposed as neurotoxin receptors (tetrodotoxin–saxitoxin receptor, scorpion toxin receptor) in other nerve membrane preparations.

ACKNOWLEDGMENTS. We wish to express our thanks to Mr. Andrés Buonanno for his critical reading of the manuscript. The technical assistance of Mr. Henry Parada and the secretarial help of Miss Isabel Otaegui are greatly appreciated. Z.S.-M is a visiting scientist from the

Universidad Simón Bolivar. This work was supported by Grant 31-26.S1-0702 of the Consejo Nacional de Investigaciones Científicas y Tecnológicas (CONICIT) of Venezuela.

REFERENCES

Agnew, W. S., Moore, A. C., Levinson, S. R., and Raftery, M. A., 1980a, Identification of a large molecular weight peptide associated with a tetrodotoxin binding protein from the electroplax of *Electrophorus electricus*, *Biochem. Biophys. Res. Commun.* **92:**860–866.

Agnew, W. S., Moore, A. C., Levinson, S. R., and Raftery, M. A., 1980b, Biochemical characterization of a voltage-sensitive sodium channel protein from the electroplax of *Electrophorus electricus*, in: *Nerve Membrane, Biochemistry and Function of Channel Proteins* (G. Matsumoto and M. Kotani, eds.), pp. 25–44, University of Tokyo Press, Tokyo.

Albuquerque, E. X., 1972, The mode of action of batrachotoxin, *Fed. Proc.* **31:**1133–1138.

Albuquerque, E. X., Daly, J. W., and Witkop, B., 1971, Batrachotoxin: Chemistry and pharmacology, *Science* **172:**995–1002.

Armstrong, C. M., 1975, Ionic pores, gates, and gating currents, *Quart. Rev. Biophys.* **7:**179–210.

Armstrong, C. M., 1981, Sodium channels and gating currents, *Physiol. Rev.* **61:**644–683.

Balerna, M., Fosset, M., Chicheportiche, R., Romey, G., and Lazdunski, M., 1975, Constitution and properties of axonal membranes of crustacean nerves, *Biochemistry* **14:**5500–5511.

Barnola, F. V., and Villegas, R., 1976, Sodium flux through the sodium channels of axon membrane fragments isolated from lobster nerve, *J. Gen. Physiol.* **67:**81–90.

Barnola, F. V., Villegas, R., and Camejo, G., 1972, Tetrodotoxin receptors in plasma membranes isolated from lobster nerve fibers, *Biochim. Biophys. Acta* **298:**84–94.

Beneski, D. A., and Catterall, W. A., 1980, Covalent labeling of the protein components of the Na^+ channel with photoactivable derivative of scorpion toxin, *Proc. Natl. Acad. Sci. USA* **77:**639–643.

Benzer, T. I., and Raftery, M. A., 1972, Partial characterization of a tetrodotoxin-binding component from nerve membrane, *Proc. Natl. Acad. Sci. USA* **69:**3634–3637.

Bergman, C., Dubois, J. M., Rojas, E., and Rathmayer, W., 1976, Decreased rate of sodium conductance inactivation in the node of Ranvier induced by a polypeptide toxin from sea anemone, *Biochim. Biophys. Acta* **455:**173–184.

Camejo, G., Villegas, G. M., Barnola, F. V., and Villegas, R., 1969, Characterization of two different membrane fractions isolated from the first stellar nerves of the squid *Dosidicus gigas*, *Biochim. Biophys. Acta* **193:**247–279.

Catterall, W. A., 1975a, Activation of the action potential sodium ionophore of cultured neuroblastoma cells by veratridine and batrachotoxin, *J. Biol. Chem.* **250:**4053–4059.

Catterall, W. A., 1975b, Cooperative activation of the action potential Na^+ ionophore by neurotoxins, *Proc. Natl. Acad Sci. USA* **72:**1782–1786.

Catterall, W. A., 1977a, Membrane potential dependent binding of scorpion toxin to the action potential sodium ionophore. Studies with a toxin derivative prepared by lactoperoxidase-catalyzed iodination, *J. Biol. Chem.* **252:**8660–8668.

Catterall, W. A., 1977b, Activation of the action potential Na^+ ionophore by neurotoxins. An allosteric model, *J. Biol. Chem.* **252:**8669–8676.

Catterall, W. A., 1979, Neurotoxins as allosteric modifiers of voltage-sensitive sodium channels, in: *Neurotoxins: Tools in Neurobiology* (B. Cecarelli and F. Clementi, eds.), pp. 305–316, Raven Press, New York.

Catterall, W. A., 1980, Neurotoxins that act on voltage-sensitive sodium channels in excitable membranes, *Annu. Rev. Pharmacol. Toxicol.* **20:**15–43.

Catterall, W. A., Ray, R., and Morrow, C. S., 1976, Membrane potential dependent binding of scorpion toxin to the action potential sodium ionophore, *Proc. Natl. Acad. Sci. USA* **73:**2682–2686.

Catterall, W. A., Morrow, C. S., and Hartshorne, R. P., 1979, Neurotoxins binding to receptor sites associated with voltage-sensitive sodium channels in intact, lysed, and detergent solubilized brain membranes, *J. Biol. Chem.* **254:**11379–11387.

Chacko, G. K., Barnola, F. V., Villegas, R., and Goldman, D. E., 1974a, The binding of tetrodotoxin to axonal membrane fraction isolated from garfish olfactory nerve, *Biochim. Biophys. Acta* **373:**308–312.

Chacko, G. K., Goldman, D. E., Malhotra, H. C., and Dewey, M. M., 1974b, Isolation and characterization of plasma membrane fractions from garfish *Lepisosteus osseus* olfactory nerve, *J. Cell Biol.* **62:**831–843.

Chandler, W. K., and Meves, H., 1965, Voltage clamp experiments on internally perfused giant axons, *J. Physiol. (London)* **180:**788–820.

Cole, K. S., 1968, *Membranes, Ions and Impulses*. University of California Press. Berkeley, CA. USA.

Colquhoun, D., Henderson, R., and Ritchie, J. M., 1972, The binding of labeled tetrodotoxin to nonmyelinated nerve fibers, *J. Physiol (London)* **227**:95–126.

Condrescu, M., and Villegas, R., 1982, Ionic selectivity of the nerve membrane sodium channel incorporated into liposomes, *Biochim. Biophys. Acta*, in press.

Conti, F., Hille, B., Neumcke, W., Nonner, W., and Stampfli, R., 1976, Conductance of the sodium channel in myelinated nerve fibers with modified sodium inactivation, *J. Physiol. (London)* **262**:729–742.

Denburgh, J. L., 1972, An axon plasma membrane preparation from the walking legs of the lobster *Homarus americanus, Biochim. Biophys. Acta* **282**:453–458.

Eisenman, G., 1962, Cation selective glass electrodes and their mode of operation, *Biophys. J.* **2**(2):259–323.

Fischer, S., Cellino, M., Zambrano, F., Zampighi, G., Telleznagel, M., Marcus, D., and Canessa-Fischer, M., 1970, The molecular organization of nerve membranes. I. Isolation and characterization of plasma membranes from the restinal axons of the squid: An axolemma-rich preparation, *Arch. Biochem. Biophys.* **138**:1–15.

Gasko, O. D., Knowles, A. F., Shertzer, H. G., Soulinna, E. M., and Racker, E., 1976, The use of ion exchange resins for studying ion transport in biological systems, *Anal. Biochem.* **72**:57–65.

Hartshorne, R. P., and Catterall, W. A., 1981, Purification of the saxitoxin receptor of the sodium channel from rat brain, *Proc. Natl. Acad. Sci. USA* **78**:4620–4624.

Henderson, R., and Strichartz, G., 1974, Ion fluxes through sodium channels of garfish olfactory nerve membranes, *J. Physiol. (London)* **238**:329–342.

Henderson, R., Ritchie, J. M., and Strichartz, G., 1973, The binding of labeled saxitoxin to the sodium channel in nerve membrane, *J. Physiol (London)* **235**:783–804.

Herzog, W. H., Feibel, R. M., and Bryant, S. H., 1964, Effect of aconitine on the giant axon of the squid, *J. Gen. Physiol.* **47**:719–733.

Hille, B., 1966, The common mode of action of the agents that decrease the transient change in sodium permeability in nerves, *Nature* **210**:1220–1222.

Hille, B., 1968, Pharmacological modifications of the sodium channel of the frog nerve, *J. Gen. Physiol.* **51**:199–219.

Hille, B., 1971, The permeability of the sodium channel to organic cations in myelinated nerve, *J. Gen. Physiol.* **58**:599–619.

Hille, B., 1972, The permeability of the sodium channel to metal cations in meylinated nerve, *J. Gen. Physiol.* **59**:637–658.

Hille, B., 1975, Ion selectivity, saturation, and block in sodium channels. A four barrier model, *J. Gen. Physiol.* **66**:535–560.

Hille, B., 1976, Gating in sodium channels of nerve, *Annu. Rev. Physiol.* **38**:139–152.

Hille, B., 1978, Ionic channels in excitable membranes. Current problems and biophysical approaches, *Biophys. J.* **22**:283–294.

Hodgkin, A. L., 1964, *The Conduction of the Nervous Impulse*, Thomas, Springfield, Illinois.

Kao, C. Y., 1966, Tetrodotoxin, saxitoxin, and their significance in the study of the excitation phenomena, *Pharmacol. Rev.* **18**:997–1049.

Kasahara, M., and Hinkle, P. C., 1976, Reconstitution of D-glucose transport catalyzed by a protein fraction from human erythrocytes in sonicated liposomes, *Proc. Natl. Acad. Sci. USA* **73**:396–400.

Khodorov, B. I., 1978, Chemicals as tools to study nerve fiber sodium channels, in: *Membrane Transport Processes*, Vol. 2 (D. C. Tosteson, Yu. A. Ovchinnikov, and R. LaTorre, eds.), pp. 153–174, Raven Press, New York.

Laemmli, W. K., 1970, Cleavage of structural proteins during the assembly of the bacteriophage T4, *Nature* **227**:680–685.

Lazdunski, M., Balerna, M., Chicheportiche, R., Fosset, M., Jacques, Y., Lombet, A., Romey, G., and Schweitz, H., 1979, Interactions of the neurotoxins with the selectivity filter and the gating system of the sodium channel, in: *Neurotoxins: Tools in Neurobiology* (B. Cecarelli and F. Clementi, eds.), pp. 353–361, Raven Press, New York.

Levinson, S. R., and Ellory, J. C., 1973, Molecular size of the tetrodotoxin binding site estimated by irradiation inactivation, *Nature* **245**:122–123.

Miller, C., and Racker, E., 1976, Fusion of phospholipid vesicles reconstituted with cytochrome C oxydase and mitochondrial hydrophobic protein, *J. Membr. Biol.* **26**:319–333.

Moore, J. W., Anderson, N. C., Blaustein, M. P., Takata, M., Lettvin, J. Y., Pickard, W. F., Bernstein, T., and Pooler, J., 1966, Alkali cation selectivity of squid axon membrane, *Ann. N.Y. Acad. Sci.* **137**:818–829.

Mozhayeva, G. N., Naumov, A. P., Negulijaew, Y. A., and Nosyreva, E. D., 1977, The permeability of aconitine-modified sodium channels to univalent cations in myelinated nerves, *Biochim. Biophys. Acta* **466**:461–473.

Mullins, L. J., 1959, The penetration of some cations into muscle, *J. Gen. Physiol.* **42**:817–829.

Mullins, L. J., 1968, Ion fluxes in dialyzed squid axons, *J. Gen. Physiol.* **51**(5, Part 2):146–148.

Mullins, L. J., 1975, Ion selectivity of carriers and channels, *Biophys. J.* **15**:921–931.

Munson, R., Westermark, B., Glaser, L., 1979, Tetrodotoxin-sensitive sodium channels in normal human fibroblasts and normal human glia-like cells. *Proc. Natl. Acad. Sci. USA* **76**:6425–29.

Nakamura, Y., Nakajima, S., and Grundfest, H., 1965, The action of tetrodotoxin on electrogenic components of squid giant axons, *J. Gen. Physiol.* **48**:985–996.

Narahashi, T., 1974, Chemicals as tools in the study of excitable membranes, *Physiol. Rev.* **54**:813–889.

Narahashi, T., 1979, Modulation of nerve membrane sodium channels by neurotoxins, in: *Neurotoxins: Tools in Neurobiology* (B. Cecarelli and F. Clementi, eds.), pp. 293–303, Raven Press, New York.

Narahashi, T., and Seyama, I., 1974, Mechanism of nerve membrane depolarization caused by grayanotoxin I, *J. Physiol. (London)* **242**:471–487.

Narahashi, T., Deguchi, T., Urukawa, N., and Ohkubo, Y., 1960, Stabilization and rectification of muscle fiber membrane by tetrodotoxin, *Am. J. Physiol.* **198**:934–938.

Narahashi, T., Moore, J. W., and Scott, W. R., 1964, Tetrodotoxin blockage of sodium conductance increase in lobster giant axons, *J. Gen. Physiol.* **47**:965–974.

Ohta, M., Narahashi, T., and Keeler, R. F., 1973, Effect of veratrum alkaloids on membrane potential and conductance of squid and crayfish giant axons, *J. Pharmacol. Exp. Ther.* **184**:143–154.

Ray, R., and Catterall, W. A., 1978, Membrane potential dependent binding of scorpion toxin to the action potential sodium ionophore. Studies with 3(4-hydroxy ^{3}(^{125}I)iodophenyl) propionyl derivative, *J. Neurochem.* **31**:397–407.

Ray, R., Morrow, C. S., and Catterall, W. A., 1978, Binding of scorpion toxin to receptor sites associated with voltage-sensitive sodium channels in synaptic nerve ending particles, *J. Biol. Chem.* **253**:7307–7313.

Ritchie, J. M., Rogart, R., and Strichartz, G. R., 1976, A new method for labelling saxitoxin and its binding to nonmyelinated fiber of the rabbit vagus, lobster walking leg and garfish olfactory nerves, *J. Physiol (London)* **261**:477–494.

Romey, G., Chicheportiche, R., Lazdunski, M., Rochat, H., Miranda, F., and Lissitzky, S., 1975, Scorpion neurotoxin—a presynaptic toxin which affects both Na^+ and K^+ channels in axons, *Biochem. Biophys. Res. Commun.* **64**:115–121.

Romey, G., Abita, J. P., Schweitz, H., Wunderer, G., and Lasdunski, M., 1976, Sea anemone toxin: A tool to study molecular mechanisms of nerve conduction and excitation-contraction coupling, *Proc. Natl. Acad. Sci. USA* **73**:4055–4059.

Sevcik, C., 1976, Binding of tetrodotoxin to squid nerve fibers. Two kind of receptors, *J. Gen. Physiol.* **68**:95–103.

Ulbricht, W., 1965, Voltage clamp studies of veratrinized frog nodes, *J. Cell. Comp. Physiol.* **66**(6, Part 2):91–98.

Ulbricht, W., 1972, Rate of veratridine action on the nodal membrane. II. Fast and slow phase determined with periodic impulses in the voltage clamp, *Pflügers Arch.* **336**:201–212.

Villegas, R., and Villegas, G. M., 1980, Incorporation into liposomes of sodium channels present in preparations of nerve membranes and of membrane particles extracted with detergents, in: *Nerve Membrane, Biochemistry and Function of Channel Proteins* (G. Matsumoto and M. Kotani, eds.), pp. 45–55, University of Tokyo Press, Tokyo.

Villegas, R., and Villegas, G. M., 1981, Nerve sodium channel incorporation in vesicles, *Annu. Rev. Biophys. Bioeng.* **10**:387–419.

Villegas, R., Barnola, F. V., Sevcik, C., and Villegas, G. M., 1976, Action of the sterol-binding form of filipin on the lobster axon membrane, *Biochim. Biophys. Acta* **426**:81–87.

Villegas, R., Villegas, G. M., Barnola, F. V., and Racker, E., 1977, Incorporation of the sodium channel of lobster nerve into artificial liposomes, *Biochem. Biophys. Res. Commun.* **79**:210–217.

Villegas, R., Villegas, G. M., Barnola, F. V., and Racker, E., 1979, Studies on the incorporation of the sodium channel of lobster nerve into soybean liposomes, in: *Neurotoxins: Tools in Neurobiology* (B. Cecarelli and F. Clementi, eds.), pp. 373–385, Raven Press, New York.

Villegas, R., Villegas, G. M., Condrescu-Guidi, M., and Suárez-Mata, Z., 1980, Characterization of the nerve membrane sodium channel incorporated into soybean liposomes: A sodium channel active particle, *Ann. N.Y. Acad. Sci.* **358**:183–203.

24

Tyrosinated Tubulin Necessary for Maintenance of Membrane Excitability in Squid Giant Axon

Gen Matsumoto, Hiromu Murofushi, Sachiko Endo, Takaaki Kobayashi, and Hikoichi Sakai

I. THE ROLE OF MICROTUBULES

Microtubules are known to play important roles in many cellular functions including motility, mitosis, transport, and the maintenance of cell shape by forming cytoskeletons. Nerve tissues contain a large amount of tubulin and microtubules, and evidence has been accumulated in support of the view that neuronal microtubules are directly involved in axonal transport (Abe *et al.*, 1973; James *et al.*, 1970).

Cytoskeletal networks lying immediately beneath the axolemma of the squid giant axon are composed of interwoven filaments which consist partly of actin-like filaments (Metuzals and Tasaki, 1978). As observed by electron microscopy, the network is firmly attached to the axolemma. The idea that the cytoskeletal networks may play a significant role in the maintenance of membrane excitability has been proposed by Metuzals and Tasaki (1978). Furthermore, Yoshioka *et al.* (1978) found that a loss of axon excitability during internal perfusion was accompanied by the release of a 56,000-dalton protein from the axon.

We have recently identified numerous microtubules which are distributed under the axolemma in the direction parallel to the longitudinal axis of the axon (Endo *et al.*, 1979). Subsequently, we have described the phenomenon of suppression of action potentials by intracellular application of microtubule inhibitors (Matsumoto and Sakai, 1979a; Baumgold *et al.*, 1978). Next, based on the results of comparison between the media favorable for axoplasmic microtubule assembly *in vitro* and those beneficial for maintaining membrane excitability of the axon, we have proposed that axoplasmic microtubules underlying the axolemma participate in the process of excitation in the axon (Matsumoto and Sakai, 1979a).

Gen Matsumoto • Electrotechnical Laboratory, Tsukuba Science City, Ibaraki 305, Japan. *Hiromu Murofushi, Sachiko Endo, and Hikoichi Sakai* • Department of Biophysics and Biochemistry, Faculty of Science, University of Tokyo, Bunkyo-ku, Tokyo 113, Japan. *Takaaki Kobayashi* • Department of Biochemistry, Jikei University, Minato-ku, Tokyo 105, Japan.

This proposal was supported by the subsequent finding that restoration of excitability could be achieved by intracellularly perfusing the axon with a reaction medium which favorably sustains enzymatic activity of tubulin–tyrosin ligase (Matsumoto and Sakai, 1979b) or by perfusing directly with a solution containing both tubulin and tubulin-tyrosine ligase under the condition that assembly of microtubules from tyrosinated tubulin is favored (Matsumoto *et al.*, 1979).

In this paper, we summarize the results that indicate involvement of microtubules in the excitation process of the axon. Based on data obtained by the voltage-clamp method; we then discuss the role of axoplasmic microtubules in the process of restoration of membrane excitability.

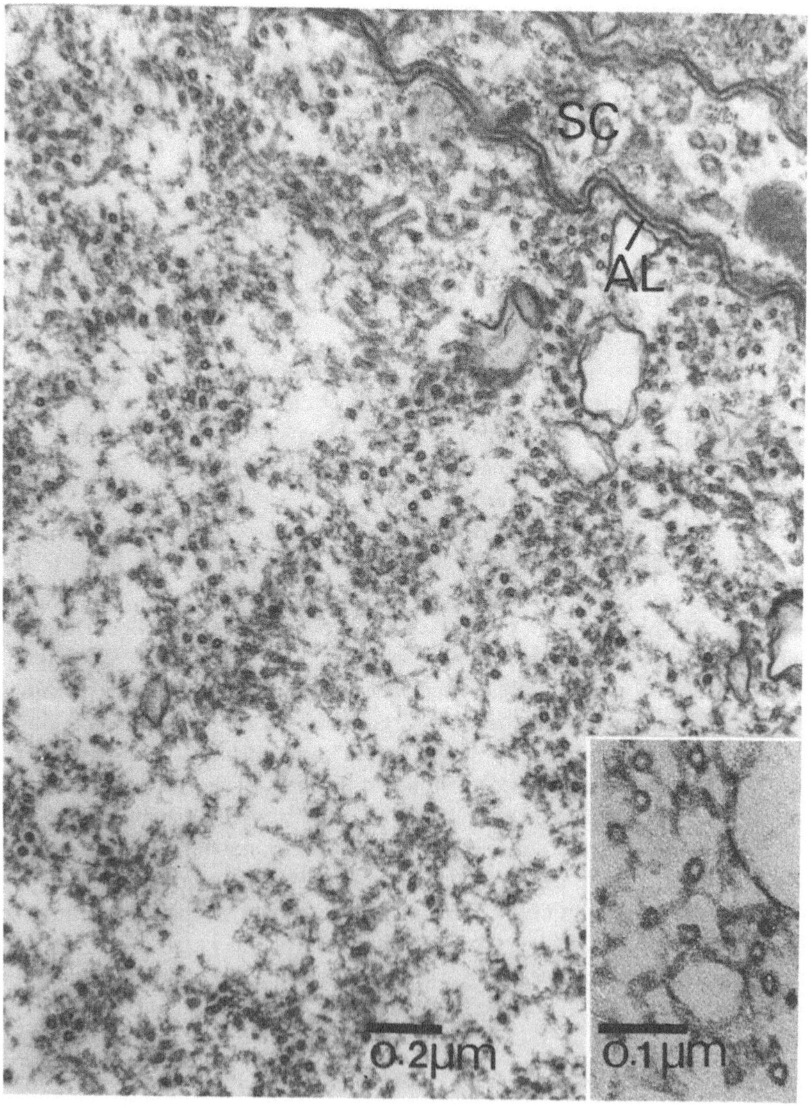

Figure 1. Cross section of a giant axon. The number of microtubules is more numerous in the peripheral axoplasm (about 1 µm thick) than in the interior. Note the network of interwoven thin filaments with microtubules (inset). AL, axolemma; SC, Schwann cell.

II. DISTRIBUTION OF MICROTUBULES IN SQUID GIANT AXON

In earlier studies on microtubule assembly, brains or nerve tissues have been used as starting material for purifying microtubule proteins, because of the high content of tubulin in those tissues (Borisy and Taylor, 1967). Microtubules as well as neurofilaments are abundantly observed by electron microscopy in many kinds of nerve cells (Berthold, 1978; Wuerker and Kirkpatrick, 1972).

In squid giant axon, however, early electron microscopic observations revealed only a few microtubules in the axoplasm (Takenaka *et al.*, 1968; Villegas and Villegas, 1960). This scarcity of microtubules was apparently due to poor preservation of the axoplasmic microtubules during fixation for electron microscopy. We found that addition of dimethylsulfoxide to the fixative (Endo *et al.*, 1979) or of tannic acid, which stabilizes microtubules, greatly improves preservation of microtubules in the squid giant axon.

Figure 1 shows a cross section of an axon in which we see numerous microtubules mostly interconnected with thin filaments forming loose bundles. Direct counting of the number of microtubules in the axoplasm from the periphery toward the interior of the axon provides some information about the spatial distribution of microtubules (Fig. 2). The peripheral region of the axoplasm contains about 90 microtubules per μm^2. This value corresponds to the mean number of microtubules detectable in unmyelinated axons (Berthold, 1978). The density of microtubules decreases gradually toward the center of the axon. It is unlikely that this regional difference is due to nonuniform preservation of microtubules (which depends on the rate of penetration of the fixative).

When the axoplasm is extruded with a roller and the resulting nearly axoplasm-free axon is perfused for 6 min with the standard perfusion medium to which colchicine (1 mM) is added, the membrane excitability is lost; action potentials can not be evoked, and even the resting potential is diminished under these conditions. Cross sections of such a deteriorated axon reveal a significant decrease in the number of microtubules (Fig. 2). Only a few

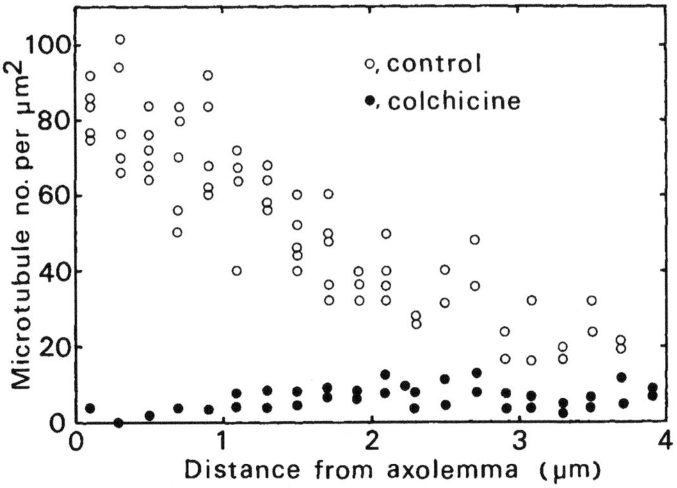

Figure 2. Distribution of microtubules in the peripheral region of the axon and decrease in the number of microtubules after internal perfusion of the axon with colchicine. The axoplasm was extruded by a roller and the axoplasm-free axon was internally perfused first with the standard perfusion medium and then with the standard perfusion medium containing 1 mM colchicine for 6 min, followed by fixation with dimethylsulfoxide-containing fixative (Endo *et al.*, 1979).

microtubules remain throughout the peripheral layer of the axoplasm, and the number of the remaining microtubules per μm^2 is roughly constant regardless of the distance from the axolemma. In contrast, neurofilaments seem to remain intact.

A longitudinal section of a fresh axon is shown in Fig. 3. The characteristic image of the axon at low magnification is that most of the numerous microtubules lie parallel to the longitudinal axis of the axon. The thin filaments correspond to those seen in the cross sections, namely, to the crosslinked microtubules observed in the peripheral region of the axoplasm. In the same region, microtubules and neurofilaments seem to associate with each other with crossbridges made of thin filaments. On the axoplasmic side of the axolemma,

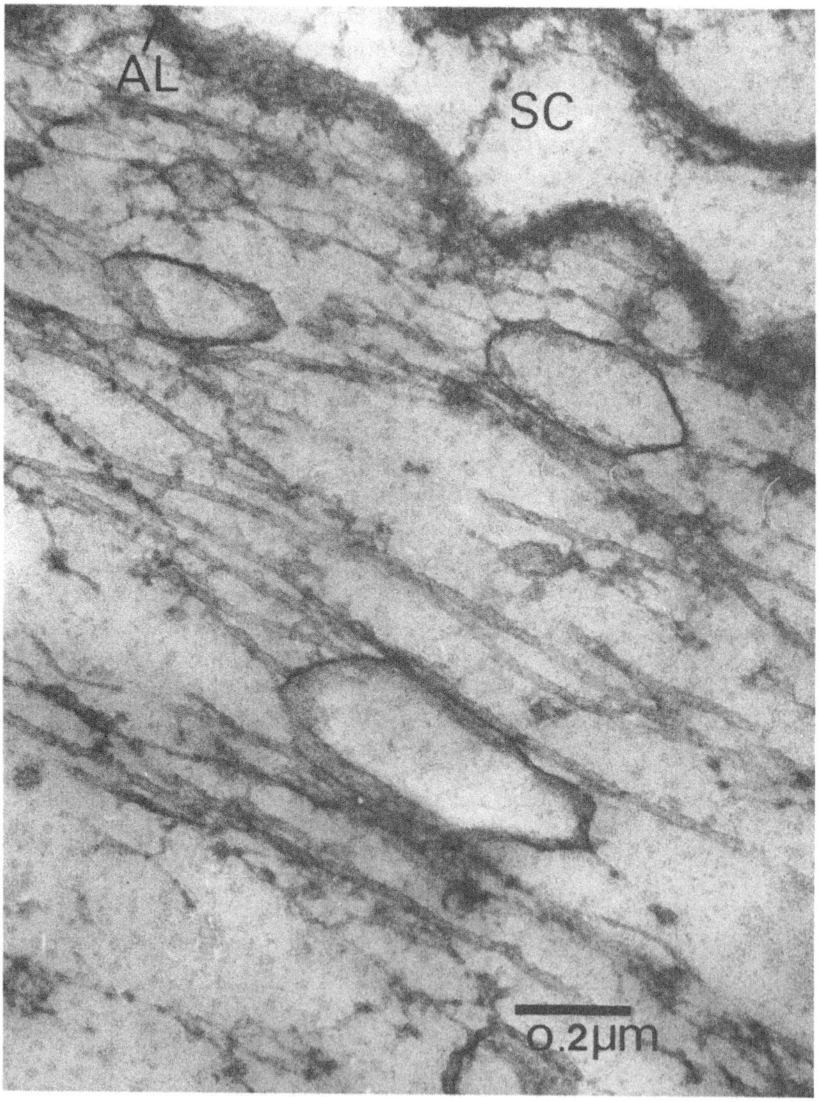

Figure 3. Longitudinal section of an axon. Fibrous structure composed of microtubules and neurofilaments runs parallel to the longitudinal axis of the axon. AL, axolemma; SC, Schwann cell.

some microtubules terminate at the axolemma, but most of them are connected with the axolemma through thin filaments.

III. CORRESPONDENCE OF MEDIA FAVORABLE FOR AXOPLASMIC MICROTUBULE ASSEMBLY IN VITRO TO THOSE FAVORABLE FOR MAINTENANCE OF MEMBRANE EXCITABILITY

Microtubule proteins of giant axons can be purified by a cycle of temperature-dependent assembly and disassembly (Sakai and Matsumoto, 1978; see Fig. 4). Halide ions in the medium enhance the assembly process in the order $F^- > Cl^- > Br^- > I^-$. This order agrees with that for the survival time of squid giant axons with respect to membrane excitability (Tasaki *et al.*, 1965). When one compares the concentrations of KF and K-glutamate favorable for microtubule assembly *in vitro* with those required for maintaining the excitability of the axon, one finds that there is no significant difference between the two, the optimal

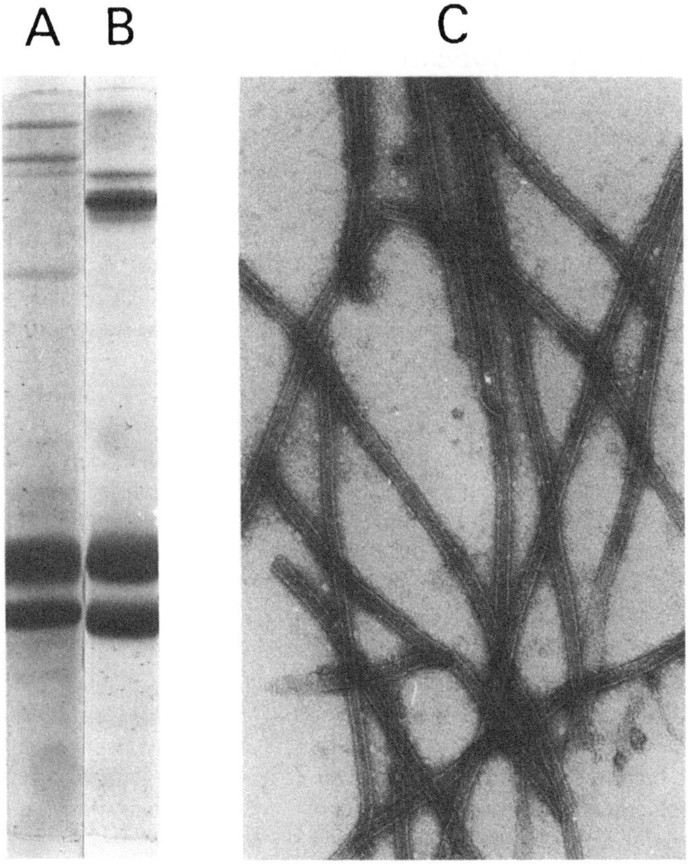

Figure 4. One-cycled microtubule proteins of axoplasm (A), in comparison with two-cycled porcine brain microtubule proteins (B), and assembled microtubules from axoplasmic tubulin (C).

concentrations being about 0.1 M for KF and 0.3 M for K-glutamate (Matsumoto and Sakai, 1979a).

Microtubules are known to be depolymerized when modified by sulfhydryl reagent (Kuriyama and Sakai, 1974; Mellon and Rebhun, 1976). The membrane excitability of the giant axon can be suppressed by internal perfusion with a medium containing PCMB (parachloromercuribenzoate), NEM (N-ethylameimide), diamide, DTNB (dithionitrobenzene), or other sulfhydryl reagents (Baumgold *et al.*, 1978; Oxford *et al.*, 1978; Matsumoto and Sakai, 1979a). Colchicine, podophyllotoxin, and vinblastine also affect the membrane excitability. However, griseofulvin has a very weak effect on excitability of the giant axon because the reagent is not a potent depolymerizer of microtubules. Furthermore, Ca ions, which quickly induce depolymerization of microtubules, bring about a rapid deterioration of the axon excitability. These results suggest that the suppression of action potentials is directly related to the loss of microtubules.

IV. EFFECT OF COLCHICINE ON MEMBRANE EXCITABILITY MEASURED BY VOLTAGE-CLAMP METHOD

Effect of Colchicine on Sodium Current

In light of the results mentioned above, it is important to determine whether or not microtubules or tubulin molecules actually participate in the process of excitation of the axon. In the following experiment, colchicine was used to suppress axon excitability because it is specific for tubulin. Its effects on sodium and potassium currents were studied by voltage-clamping axons under internal perfusion with a medium containing 10 μM colchicine (Matsumoto *et al.*, 1980). Initially, the axoplasm was extruded using a roller, and the resulting nearly axoplasm-free axon was internally perfused with the standard perfusion medium consisting of 370 mM KF, 30 mM K-Hepes and 4% (v/v) glycerol at pH 7.2 with the perfusion flow rate of 20–50 μl/min.

The voltage-dependence of the sodium conductance $g_{Na}(V)$ was determined by the equation, $g_{Na}(V) = I^P_{Na}/(V\text{-}V_{Na})$, where I^P_{Na} represents the peak inward current and V_{Na} the reversal potential (Hodgkin and Huxley, 1952a). Figure 5 shows the relation between the sodium conductance and the membrane potential in an axon perfused with a medium containing 100 μm colchicine. Judging from the amount of tubulin in the axoplasm (> 150 μM), the concentration employed expected to be substoichiometric for the tubulin molecules remaining in the ectoplasm (Sakai and Matsumoto, 1978). Colchicine lowers the apparent maximal sodium conductance down to about two thirds of the original level after 10 min of perfusion. When the observed sodium conductance is normalized, it is revealed that colchicine diminishes the steepness of the $g_{Na}(V)$ profile significantly (data not shown). It should be noted here that even 10 μM colchicine significantly diminishes the sodium conductance.

In contrast, colchicine has no effect on the sodium inactivation, even when its concentration is as high as 1 mM. Figure 6 shows the observed peak sodium current as a function of pre-pulse potentials. We saw that the effect of 1 mM colchicine upon the time course of sodium inactivation is negligibly small (data not shown).

Thus, we find that the effects of colchicine on the sodium current are: (1) a reduction of the maximal sodium conductance, and (2) diminution of the voltage-dependence of the sodium activation. At concentrations higher than 1 mM, colchicine increases the leakage

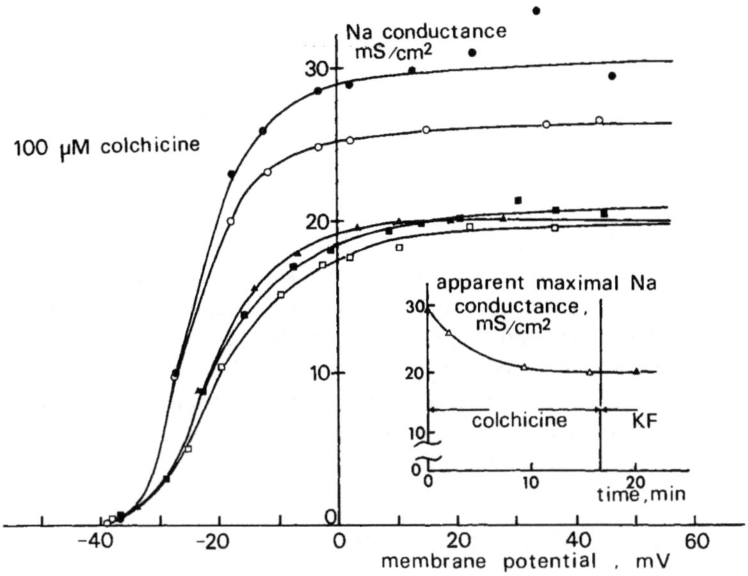

Figure 5. Voltage-dependence of the sodium conductance g_{Na} (V) when the axon was internally perfused with the standard perfusion medium containing 100 μM colchicine. The data were obtained 0 (●), 2 (○), 9.5 (■), and 15.5 (□) min after the onset of perfusion with the colchicine-containing medium. Further data (▲) were obtained 3.5 min after switching the colchicine-containing medium to the standard perfusion medium again. Temperature 20°C. Inset: time course of changes in the apparent maximal sodium conductance.

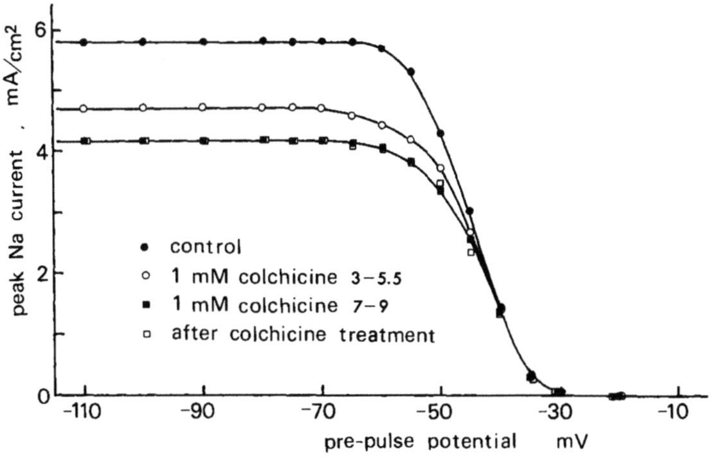

Figure 6. Peak sodium current as a function of a pre-pulse potential when the axon was perfused with the standard perfusion medium containing 1 mM colchicine. The pre-pulse and test pulse durations were fixed at 40 and 2 msec, respectively. The test pulse amplitude was 45 mV. The holding potential was −-50 mV. The data were obtained 9 (●), 3–5.5 (○), and 7–9 (■) min after the onset of perfusion with the 1 mM colchicine-containing medium. Further data (□) were taken after switching from the colchicine solution to the standard perfusion medium.

current, reduces the reversal potential, and eventually brings about a loss of electrical excitability.

V. RESTORATION OF MEMBRANE EXCITABILITY

A. Internal Perfusion of Axon with Purified Microtubule Proteins

The experimental results described above led us to postulate that cytoskeletal networks composed mainly of microtubules are, in some manner, involved in the process of excitation in the giant axon. In order to render further experimental support to this postulate, we made the following observations. Initially, the ability of the axon to develop normal action potentials was suppressed by introducing a microtubule inhibitor into the internal perfusion medium. Then, the axon was perfused with a solution containing purified porcine brain microtubule proteins dissolved in the standard perfusion medium. We found that excitability of the axon can never be steadily restored by this procedure.

B. Requirement of Factors Other than Purified Tubulin

Addition of tyrosine, ATP and Mg^{2+} to the tubulin-containing perfusion medium was found to be very favorable for restoration of the excitability which had previously been suppressed by a microtubule inhibitor (Matsumoto and Sakai, 1979b). Furthermore, addition of partially purified tubulin-tyrosine ligase was found to bring about a pronounced favorable effect on the process of restoration (Matsumoto *et al.*, 1979).

The enzyme had been highly purified from porcine brains by Murofushi (1980) using Sepharose-sebacic acid hydrazide-ATP affinity chromatography and Sepharose-tubulin affinity chromatography. The purified enzyme migrates as a single band on SDS-polyacrylamide gel electrophoresis showing a molecular weight of 46,000 (Fig. 7). The pH optimum for the activity is around 8 and a second activity peak is observed at pH 6.5. The purified enzyme catalyzes tyrosination of α tubulin *in vitro*. The Michaelis constants, K_ms are 8.5 μM for ATP and 30 μM for tyrosine.

We found that restoration is favored by another protein factor which can be extracted from the axon (Sakai and Matsumoto, 1978). Because the excitability is quickly suppressed

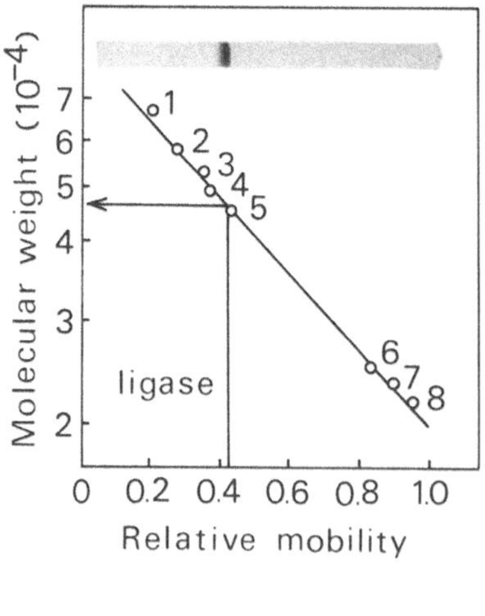

Figure 7. Molecular weight estimation of purified tubulin–tyrosine ligase by SDS-polyacrylamide gel electrophoresis. 1, bovine serum albumin; 2, tubulin β subunit, 3, tubulin α subunit, 4, γ-globulin heavy chain; 5, ovalbumin; 6, γ-globulin light chain; 7, trypsin; 8, soybean trypsin inhibitor.

GIANT AXONS

↓

AXOPLASMS EXTRUDED

↓

SHEATHS

↓

EXTRACTED WITH 10 mM $CaCl_2$-0.55 M NaCl-10 mM HEPES (pH 7.3)

↓

HIGH SPEED LIQUID CHROMATOGRAPHY (G 3000 SW)

↓

CON A-SEPHAROSE AFFINITY CHROMATOGRAPHY

↓

PURIFIED 260 K PROTEIN

Figure 8. Procedure for purifying 260 k protein and SDS-polyacrylamide gel electrophoresis of the purified protein.

by perfusing the axon with a Ca^{2+}-containing medium, the endoplasm-free axons (each cut longitudinally into two) were treated with a Ca^{2+}-containing buffer solution to extract this protein factor. The extract contains a major protein component having a molecular weight of 260,000. We have recently purified this protein by using Con A-Sepharose affinity chromatography (Fig. 8). The adsorbed protein was eluted by α-methylmannoside, which indicates that the 260,000 dalton protein is a glycoprotein; it is PAS(periodic acid Schiff)-positive.

Eventually, we came to the conclusion that the factors necessary for the process of restoration are the following.

1. Porcine brain tubulin which is to be tyrosinated by tubulin–tyrosine ligase in the presence of ATP, Mg^{2+}, and K^+ is required for restoration. This tyrosinated tubulin will be polymerized into microtubules under this experimental condition to about half the optimum level (Sakai and Matsumoto, 1978).

2. The 260,000 protein fraction is necessary for maintaining and stabilizing the restored excitability. If this protein fraction is omitted, the degree of restoration is limited and the restored excitability can not be maintained (Matsumoto *et al.*, 1979). The possibility exists that this protein is located on the axoplasmic surface of the axolemma.

3. Since restoration is further improved by addition of cAMP, phosphorylation of some components in the reaction system seems to be necessary for the restoration process. We note that internal perfusion of fresh axons with a K-glutamate solution to which phosphatases are added suppresses the axon excitability within a short period of time.

When the complete restoration mixture described above is employed, the ability of the axon to generate action potentials is restored satisfactorily after a lag phase (Fig. 9). The threshold falls significantly. It is also noticed that the resting potential recovers to a level even below that of intact axons. The lag phase seems to be due to the necessity of a microtubule to be assembled in the vicinity of the axolemma.

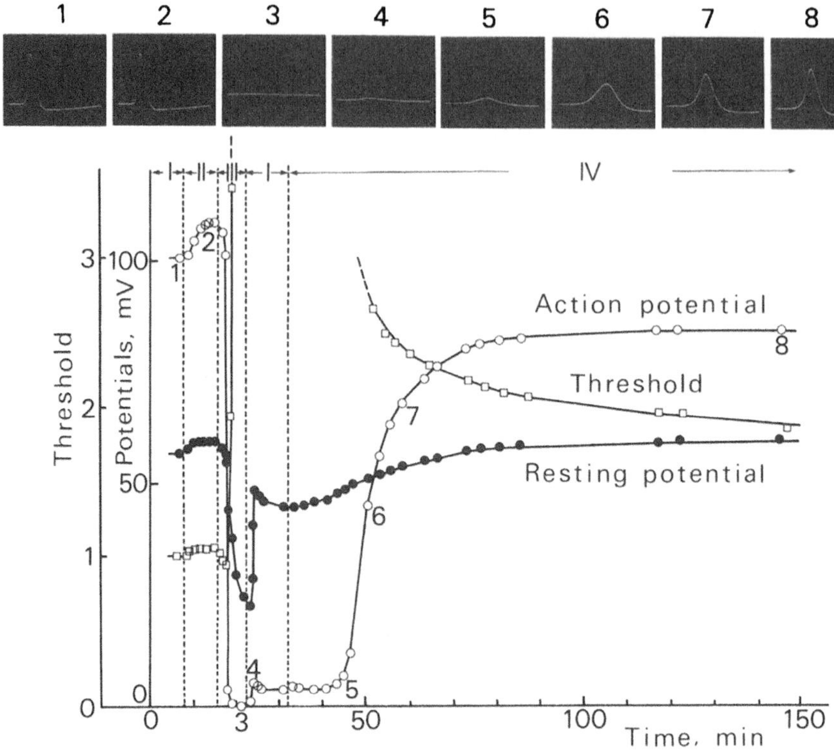

Figure 9. Restoration of action potential, resting potential, and threshold (in arbitrary units) by tubulin–tyrosine ligase, microtubule proteins, axonal 260 k protein fraction, reagents activating the ligase, and cAMP. Solution I: 360 mM KF, 40 mM K-phosphate buffer, 4% (V/V) glycerol, pH 7.2: Solution II: 360 mM K-glutamate, 40 mM K-Hepes, pH 7.2; Solution III: solution II plus 0.2 mM $CaCl_2$ and 1 mM $MgCl_2$; Solution IV: standard perfusion medium containing 1 mg/ml three-cycled microtubule proteins, 0.1 mg/ml axonal 260 k protein fraction, 3 units/ml tubulin–tyrosine ligase, 0.5 mM ATP, 30 μM tyrosine, 0.5 mM $MgCl_2$, 3μM and 0.2 mM GTP. (From Matsumoto *et al.*, 1979).

C. Attempt to Restore Excitability under Voltage Clamp

In the experiments described above, recovery of the action potential amplitude, the resting potential, and the threshold was taken as criteria for restoration. In order to obtain evidence for restoration by taking a different criterion, voltage-clamping of the axon was performed under internal perfusion.

Because the effect of Ca ions is not specific for microtubules, colchicine was chosen to bring about suppression of excitability. When the axon was internally perfused for 38 min with the standard medium to which colchicine (18.5 mM) was added, both inward sodium and outward potassium currents quickly decreased, and, when the perfusion medium was replaced with the standard medium (without colchicine), both currents once recovered but soon began to degradate (Fig. 10). The restoration medium was then introduced into the axon. The restoration medium employed consisted of three-cycled microtubule proteins from squid optic ganglia, 260 k protein purified by Con A-Sepharose affinity chromatography (Fig. 8), GTP, Mg^{2+}, and K^{+}. Figure 10 shows a sign of recovery of the sodium current (see also Fig. 11). In the same axon, the leak conductance and resting potentials were found to degradate with time during and after the colchicine treatment (Fig. 11). Upon switching

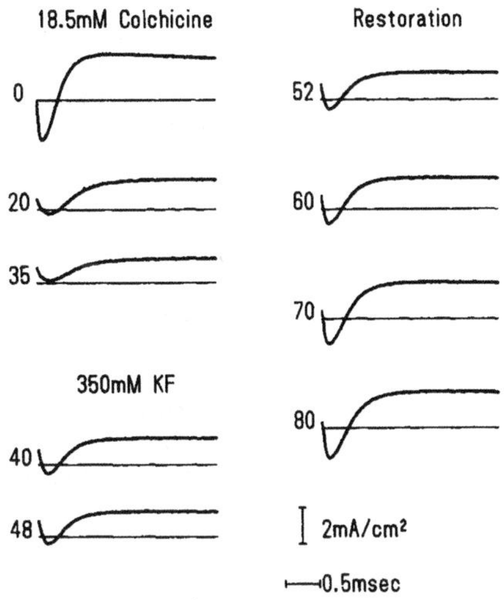

Figure 10. Suppression and restoration of membrane excitability observed under voltage-clamp conditions. The axon was voltage-clamped at membrane potential of 20 mV after preconditioned with a pulse of −100 mV for 30 msec. Suppression of the membrane excitability was induced by internally perfusing the axon with the standard perfusion medium containing 18.5 mM colchicine. On the upper left, three current records were taken, 0, 20 and 35 min after the onset of perfusion with the colchicine-containing medium. At 38 min after the onset of the colchicine perfusion, the internal solution was switched to the standard perfusion medium. The membrane excitability once recovered but soon began to degradate, as seen in the current records taken at 40 and 48 min (lower left). Then at 50 min after the onset of the colchicine perfusion, the internal solution was switched to the restoration medium containing squid brain three-cycled microtubule proteins (1 mg/ml), purified 260 k protein (40 μg]l), $MgCl_2$ (3 mM), and GTP (0.5 mM). Restoration of the initial inward current (right) correlated well with that of the leak current (Fig. 12). Temperature 21°C.

to the restoration medium, the leakage conductance began to decrease simultaneously with a full recovery of the resting potential.

In the experiment described above, tubulin–tyrosine ligase was not used, since we know that more than 94% of α tubulin prepared from squid optic ganglia or axoplasma (Table I) possess carboxyl terminal tyrosine (Kobayashi and Matsumoto, 1982). In tubulin molecules freshly purified from porcine brain, α subunits of about 50% of the tubulin molecules are devoid of tyrosine residues. With such a preparation, tubulin–tyrosine ligase is required for restoration of the action potential. (In this connection, we have noted that carboxypeptidase A brings about a decrease in the maximum sodium and potassium conductances and an increase in the leak current. Carboxypeptidase A has no effect upon the Hodgkin–Huxley gates *m*, *h*, and *n*, while other proteolytic enzymes affect the sodium inactivation (Sevcik and Narahashi, 1975)).

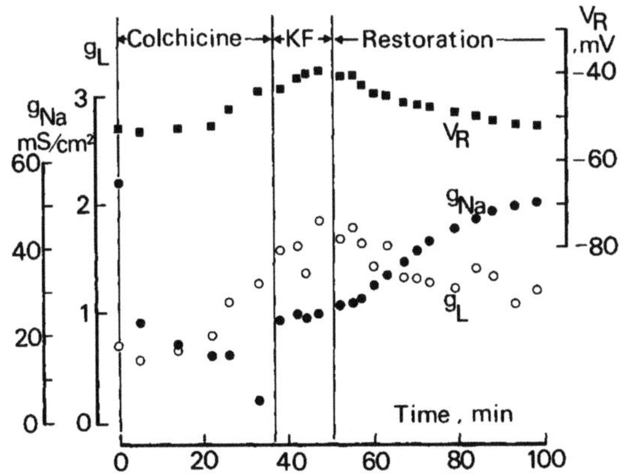

Figure 11. Time course of changes in the sodium (G_{Na}) and leak (G_L) conductances when the axon was voltage-clamped by perfusion first with colchicine, with standard internal solution, and then with the restoration medium, in succession. The same axon as in Fig. 10. Time course of changes of resting potentials (V_R) are also illustrated.

Table I. Acceptor Capacity for Tyrosine of Several Tubulin Species before and after Treating with Carboxypeptidase A

	Acceptor capacity (mole tyrosine/110,000 g protein)	
Tubulin species	Before treatment	After treatment
Porcine brain tubulin[a]	0.31	0.49
Porcine brain tubulin[b]	0.28	0.62
Porcine brain membrane tubulin	0.36	0.44
Squid optic ganglion tubulin[a]	0.03	0.53
Squid optic ganglion membrane tubulin	0.23	0.29
Squid axoplasmic tubulin[a]	0.01	0.48

[a] Purified by cycled assembly–disassembly.
[b] Purified by precipitation with vinblastine.

We have noted that the restoration medium devoid of GTP is ineffective. This finding is quite consistent with our conclusion that the microtubules underlying the axolemma play an important role in maintaining the membrane excitability.

VI. CONCLUSION

We have shown that there is a close correlation between the composition of the media favorable for assembling axoplasmic microtubules and that for maintaining membrane excitability. Furthermore, we have demonstrated that tyrosylated tubulin or microtubules are required for restoration of the resting and action potentials in axons of which the excitability has been almost completely suppressed by disruption of microtubules. We have also shown that the addition of a 260,000 dalton protein fraction and cAMP significantly improve recovery of excitability in axons.

By using the voltage-clamp technique under internal perfusion, we found that a solution containing tyrosylated squid tubulin mixed with GTP and a purified 260,000 dalton protein fraction is capable of maintaining membrane excitability in colchicine-treated axons. These results indicate that tyrosylated microtubules and 260,000 dalton proteins are required for action potential production.

REFERENCES

Abe, T., Haga, T., and Kurokawa, M., 1973, Rapid transport of phosphatidylcholine occurring simultaneously with protein transport in the frog sciatic nerve, *Biochem. J.* **136**:731–740.

Baumgold, J., Matsumoto, G., and Tasaki, I., 19978, Biochemical studies of nerve excitability: The use of protein modifying reagents for characterizing sites involved in nerve excitation, *J. Neurochem.* **30**:91–100.

Berthold, C-H., 1978, Morphology of normal peripheral axons, in: *Physiology and Pathobiology of Axons* (S. G. Waxman, ed.), pp. 3–63, Raven Press, New York.

Borisy, G. G., and Taylor, E. W., 1967, The mechanism of action of colchicine. Binding of colchicine-^{3}H to cellular protein, *J. Cell Biol.* **34**:525–533.

Endo, S., Sakai, H., and Matsumoto, G., 1979, Microtubules in squid giant axon, *Cell Struct. Funct.* **4**:285–293.

Hodgkin, A. L., and Huxley, A. F., 1952a, Current carried by sodium and potassium ions through the membrane of the giant axon of *Loligo, J. Physiol. (London)* **116**:449–472.

Hodgkin, A. L., and Huxley, A. F., 1952b, A quantitative description of membrane current and its application to conduction and excitation in nerve, *J. Physiol. (London)* **117**:500–544.

James, K. A., Bray, J. J., Morgan, I. G., and Austin, L., 1970, The effect of colchicine on the transport of axonal protein in the chicken, *Biochem. J.* **117**:767–771.

Kobayashi, T., and Matsumoto, G., 1982, Tubulin from squid nerve cytoplasm fully retains c-terminus tyrosine, *J. Biochem.* **92**:647–652.

Kuriyama, R., and Sakai, H., 1974, Role of tubulin -SH groups in polymerization to microtubules. Function -SH groups in tubulin for polymerization, *J. Biochem.* **76**:651–654.

Matsumoto, G., and Sakai, H., 1979a, Microtubules inside the plasma membrane of squid giant axons and their possible physiological function, *J. Membr. Biol.* **50**:1–14.

Matsumoto, G., and Sakai, H., 1979b, Restoration of membrane excitability of squid giant axons by reagents activating tyrosine-tubulin ligase, *J. Membr. Biol.* **50**:15–22.

Matsumoto, G., Kobayashi, T., and Sakai, H., 1979, Restoration of the excitability of squid giant axon by tubulin-tyrosine ligase and microtubule proteins, *J. Biochem.* **86**:1155–1158.

Matsumoto, G., Murofushi, H., and Sakai, H., 1980, The effects of reagents affecting microtubules and microfilaments on the excitation of the squid giant axon measured by the voltage-clamp method, *Biomed. Res.* **1**:355–358.

Mellon, M. G., and Rebhun, L. I., 1976, Sulfhydryls and the *in vitro* polymerization of tubulin, *J. Cell Biol.* **70**:226–238.

Metuzals, J., and Tasaki, I., 1978, Subaxolemmal filamentous network in the giant nerve fiber of the squid *(Loligo pealei* L.) and its possible role in excitability, *J. Cell Biol.* **78**:597–621.

Murofushi, H., 1980, Purification and characterization of tubulin-tyrosine ligase from porcine brain, *J. Biochem.* **87**:979–984.

Oxford, G. S., Wu, C. H., and Narahashi, T., 1978, Removal of sodium channel inactivation in squid giant axons by N-bromoacetamide, *J. Gen. Physiol.* **71**:227–247.

Sakai, H., and Matsumoto, G., 1978, Tubulin and other proteins from squid giant axon, *J. Biochem.* **83**:1413–1422.

Sevcik, C., and Narahashi, T., 1975, Effects of proteolytic enzymes on ionic conductances of squid axon membranes, *J. Membr. Biol.* **24**:329–339.

Takenaka, T., Hirakow, R., and Yamagishi, S., 1968, Ultrastructural examination of the squid giant axons perfused intracellularly with protease, *J. Ultrastruct. Res.* **25**:408–416.

Tasaki, I., 1968, *Nerve Excitation: A Macromolecular Approach,* Charles C. Thomas, Springfield, Illinois.

Tasaki, I., Singer, I., and Takenaka, T., 1965, Effects of internal and external ionic environment on excitability of squid giant axon: A macromolecular approach, *J. Gen. Physiol.* **48**:1095–1123.

Villegas, G. M., and Villegas, R., 1960, The ultrastructure of the giant nerve fibre of the squid: Axon-Schwann cell relationship, *J. Ultrastruct. Res.* **3**:362–373.

Wilson, L., Bamburg, J. R., Mizel, S. B., Grisham, L. M., and Creswell, K. M., 1974, Interaction of drugs with microtubule proteins, *Fed. Proc.* **33**:158–166.

Wuerker, R. B., and Kirkpatrick, J. B., 1972, Neuronal microtubules, neurofilaments and microfilaments, in: *International Review of Cytology,* Vol. 33 (G. H. Bourne, J. F. Danielli, and K. W. Jeon, eds.), pp. 45–75, Academic Press, New York.

Yoshioka, T., Pant, H. C., Tasaki, I., Baumgold, J., Matsumoto, G., and Gainer, H., 1978, An approach to the study of intracellular proteins related to the excitability of the squid giant axon, *Biochim. Biophys. Acta* **538**:616–626.

Index

GPSR Compliance

The European Union's (EU) General Product Safety Regulation (GPSR) is a set of rules that requires consumer products to be safe and our obligations to ensure this.

If you have any concerns about our products, you can contact us on ProductSafety@springernature.com

In case Publisher is established outside the EU, the EU authorized representative is:

Springer Nature Customer Service Center GmbH
Europaplatz 3
69115 Heidelberg, Germany

Batch number: 10838062

Printed by Printforce, the Netherlands